A TEXTBOOK OF

WORKSHOP TECHNOLOGY

(Manufacturing Processes)

(A Textbook for the Engineering Degree and Diploma Students)

R.S. KHURMI
J.K. GUPTA

S Chand And Company Limited
(ISO 9001 Certified Company)

S Chand And Company Limited
(ISO 9001 Certified Company)
Head Office: D-92, Sector–2, Noida – 201301, U.P. (India), Ph. 91-120-4682700
Registered Office: A-27, 2nd Floor, Mohan Co-operative Industrial Estate,
New Delhi – 110 044, Phone: 011-49731800
www.**schandpublishing.com**; e-mail: **info@schandpublishing.com**

S. CHAND PUBLISHING

Marketing Offices:

Chennai : Ph: 23632120; chennai@schandpublishing.com
Guwahati : Ph: 2738811, 2735640; guwahati@schandpublishing.com
Hyderabad : Ph: 40186018; hyderabad@schandpublishing.com
Jalandhar : Ph: 4645630; jalandhar@schandpublishing.com
Kolkata : Ph: 23357458, 23353914; kolkata@schandpublishing.com
Lucknow : Ph: 4003633; lucknow@schandpublishing.com
Mumbai : Ph: 25000297; mumbai@schandpublishing.com
Patna : Ph: 4011400; patna@schandpublishing.com

First Edition 1981
Subsequent Editions and Reprints 1983, 85, 87, 89, 91, 93, 97, 99, 2000, 2001, 2002, 2004, 2006, 2007, 2008 (Twice), 2009 (Twice), 2010 (Thrice), 2011, 2012, 2013, 2014 (Twice), 2015, 2017, 2018 (Thrice), 2024, 2025
Reprint 2026

ISBN: 978-81-219-0868-9 **Product Code:** H3WST62MAFT10ENAF0XO

PRINTED IN INDIA
By Shri Shyamlal Printing Press Private Limited, Plot 20/4, Site-IV, Industrial Area Sahibabad, Ghaziabad – 201 010 and Published by S Chand And Company Limited, A-27, 2nd Floor, Mohan Co-operative Industrial Estate, New Delhi – 110 044.

PREFACE TO THE SIXTH EDITION

We feel satisfied in presenting the new edition of this standard treatise. The favourable and warm reception which the previous editions and reprints of this book has enjoyed all over India and abroad, is a matter of great satisfaction for us.

The present edition of this treatise has been thoroughly revised and brought up-to-date. The Objective Type Questions have been added at the end of each chapter. The mistakes which had crept in have been eliminated. The following two new chapters have been added to widen the scope and utility of this treatise :

1. Properties, Testing and Inspection of Metals; and 2. Quality Control.

We wish to express our sincere thanks to numerous professors and students, both at home and abroad, for sending their valuable suggestions and recommending the book to their students and friends. We hope that they will continue to patronise this book in future also.

Any error, omission and suggestions for the improvement of this volume brought to our notice, will be thankfully acknowledged and incorporated in the next edition.

R.S. KHURMI
J.K. GUPTA

PREFACE TO THE FIRST EDITION

We take an opportunity to present this standard treatise entitled 'A Textbook of Workshop Technology' (Manufacturing Processes) to the students of Degree and Diploma of all the Indian and foreign universities. The object of this book is to present the subject matter in a most concise, compact, to the point and lucid manner.

While writing the book, we have constantly kept in mind the various requirements of the students. No effort has been spared to enrich the book with simple language and self-explanatory diagrams. Every care has been taken not to make the book voluminous, as the students have also to face other subjects of equal importance. In short, it is hoped that the book will embrace the requirements of all the engineering students, and will earn appreciation of all the fellow teachers.

Although every care has been taken to check the mistakes and misprints, yet it is difficult to claim perfection. Any errors, omissions and suggestions for the improvement of this volume, will be thankfully acknowledged and incorporated in the next edition.

R.S. KHURMI
J.K. GUPTA

CONTENTS

CONTENTS

1

Introduction

1. Introduction. 2. Engineering Metals. 3. Classification of Manufacturing Processes. 4. Plant Layout. 5. Objectives of Plant Layout. 6. Advantages of a Good Plant Layout. 7. Types of Plant Layout. 8. Fixed Position Layout. 9. Process Layout or Functional Layout. 10. Product Layout or Line Layout. 11. Combination Layout or Group Layout. 12. Shop Layout. 13. Layout of a Carpentry and Pattern Making Shop. 14. Layout of a Foundry Shop. 15. Layout of a Smithy and Forging Shop 16. Layout of a Welding Shop. 17. Layout of a Fitting Shop. 18. Layout of a Sheet Metal and Painting Shop.

1·1 Introduction

The subject of Workshop Technology has become increasingly important to the engineer, supervisor or worker, engaged in the production of various types of machines or tools. It has been established since long, that the ability to work skillfully, with hands, can be developed more readily and accurately, when the work to be performed in the shop is understood both in its practical and theoretical aspects.

The study of Workshop Technology has, therefore, been made compulsory these days for a worker, foreman and engineer so that he can make himself acquainted with basic knowledge of manufacturing processes and materials.

1·2 Engineering Metals

The engineering metals always play an important role in an industry, because the process of all manufacturing starts with the raw material. The materials chiefly used in the shops are metals which may be broadly divided into the following two groups :

1. *Ferrous metals.* The metals which contain iron as their chief constituent are called ferrous metals. The various ferrous metals used in industry are pig iron, cast iron, wrought iron, steel etc.

2. *Non-ferrous metals.* The metals which contain a metal other than iron as their chief constituent are called non-ferrous metals. The various nonferrous metals used in industry are aluminium, copper, zinc, lead, brass, tin etc.

1·3 Classification of Manufacturing Processes

The manufacturing plays an important role in engineering because it changes the form of material into the final product. The development of modern manufacturing is dependent on research in materials and the products require a variety of manufacturing processes for these materials. The various processes used in manufacturing are classified into the following five groups :

1. *Primary shaping processes.* The following primary shaping processes are commonly used in engineering practice :

(*a*) Casting	(*b*) Forging	(*c*) Rolling	(*d*) Bending
(*e*) Drawing	(*f*) Squeezing	(*g*) Extruding	(*h*) Shearing
(*i*) Forming	(*j*) Piercing	(*k*) Spinning	(*l*) Crushing

2. *Machining processes.* The machining operations are performed on castings, rollings and forgings in order to obtain the desired accuracy and shape. The following machining operations are

used for this purpose :

(*a*) Turning	(*b*) Drilling	(*c*) Grinding	(*d*) Threading
(*e*) Boring	(*f*) Slotting	(*g*) Planing	(*h*) Milling
(*i*) Hobbing	(*j*) Sawing	(*k*) Broaching	(*l*) Knurling

3. *Surface finishing processes.* These processes are used effectively to provide a good surface finish to the metal surface of the product. In a surface finishing process, either a negligible amount of metal is removed or a small amount of material is added to the surface of the product. It may be noted that there is no appreciable change in dimensions by the surface finishing processes.

The following processes are commonly used for obtaining the desired surface finish :

(*a*) Sand blasting	(*b*) Buffing	(*c*) Lapping	(*d*) Honing
(*e*) Belt grinding	(*f*) Polishing	(*g*) Electroplating	(*h*) Metal spraying
(*i*) Tumbling	(*j*) Superfinishing	(*k*) Anodising	(*l*) Phosphating
(*m*) Pickling	(*n*) Hot dipping	(*o*) Galvanizing	(*p*) Parkerising
(*q*) Sheradising	(*r*) Painting	(*s*) Inorganic coating	

4. *Joining processes.* These processes are used for joining two or more pieces of metal parts. The following processes are commonly used for joining the metal parts :

(*a*) Welding	(*b*) Soldering	(*c*) Brazing	(*d*) Riveting
(*e*) Sintering	(*f*) Adhesive joining	(*g*) Screwing	(*h*) Pressing

5. *Processes affecting change in properties.* These processes are used to impart certain specific properties to the metal part for specific conditions of use. The following processes are commonly used for such purposes :

(*a*) Heat treatment	(*b*) Hot working	(*c*) Cold working	(*d*) Shot peening

1·4 Plant Layout

The plant layout is defined as the physical arrangement of buildings, machinery, equipment, work places and other facilities of production to accomplish the most efficient utilisation of men, machines and materials. It is one of the most important phases to be undertaken while planning a new factory. The efficient layout planning is the key for the efficient utilisation of resources. The expensive equipment, intricate tools, skilled manpower and a well designed product, may all fail to achieve high productivity due to poorly laidout machinery and facilities.

1·5 Objectives of Plant Layout

The objectives of a plant layout are as follows :

1. The entire space of the plant is economically and effectively used.
2. The investment in equipments is minimum.
3. The handling and transportation of material is minimum.
4. The flow of material is smooth and rapid.
5. The supervision is easy, simple and effective .
6. The utilisation of men and machines is optimum.
7. The maintenance of equipments and control of production is easy.
8. The plant provides adequate safety, healthy and comfortable working condition to all employees.
9. There is a flexibility of rearrangement of production plans.
10. There is a provision for future expansion of the plant.

1·6 Advantages of a Good Plant Layout

Though a good plant layout has many advantages, yet the following are the most important :

1. It provides better health and safety of employees.
2. It improves morale of the workers.
3. It avoids the production delays.
4. It requires less supervision.
5. It reduces indirect cost of labour.
6. It minimises material handling time.
7. It provides easy and quicker adjustment to changed condition.
8. It increases overall efficiency of the plant.

1·7 Types of Plant Layout

The plant layout may be broadly classified into the following four groups :

1. Fixed position layout or static product layout ;
2. Process layout or functional layout ;
3. Product layout or line layout ; and
4. Combination layout or group layout.

These types of plant layout are discussed, in detail, in the following pages.

1·8 Fixed Position Layout

In the fixed position layout, the product (because of its heavy weight) is kept at one place. The men, machine, material and tools are brought to this place and operations are performed. This type of layout is to be used under the following conditions :

(*a*) When the product is very large.

(*b*) When the quantity is relatively small and the operation needs assembly with hand tools.

(*c*) When the cost of moving the product is higher than the cost of moving the components and tools etc.

The examples of this type of layout are : Ship building, aircraft assembly, heavy machine tool making etc.

The advantages and disadvantages of the fixed position layout are as follows :

Advantages

1. The movement of material, men and tooling is reduced to a minimum.
2. It is flexible to changes in the product design and the related processes.
3. Since the highly skilled workers are made to work at one point, therefore, responsibility for quality of work can be fixed.

Disadvantages

1. The movement of machines to the production point is costly and time consuming.
2. The highly skilled workers are required.

1·9 Process Layout or Functional Layout

In the process layout or functional layout, all similar machines or operations are grouped in one area or department. For example, all drilling machines are located in one department and all lathes in the other department. Thus, in a plant laid by process, the products that require holes to be drilled in them are moved to the drilling department and then moved to the subsequent departments for other work. This type of layout is commonly found in workshops involved in job order type of work and in plants where a limited number of unlike products are produced.

The advantages and disadvantages of process layout are as follows :

Advantages

1. The better and efficient utilisation of machines requires low machine investment.

2. The continuity of production is easy in case of

   (*a*) breakdown in machines, (*b*) shortage of raw materials, and (*c*) absenteeism of workers.
3. The changes in variety of product, sequence of operations and volume of production can be easily accomodated.
4. It provides opportunity for specialisation in one type of process or machine.
5. Since the workers attend to the same type of machines and operations repeatedly, therefore, the supervision is easy and the quality of product is better.

Disadvantages

1. The materials handling and transportation costs are high.
2. The manufacturing time increases due to large amount of work in process.
3. The production control is difficult.
4. It requires large storage space because of slow movement of materials.
5. It requires large in-process inventory of materials.

1·10 Product Layout or Line Layout

In product or line layout, the machines and other manufacturing facilities are located in the order of sequence required to manufacture the product. Thus, if the sequence of operations for a given part consists of drilling a hole, milling a slot and forming the part, then the arrangement or layout of machines to manufacture this part will consist of

(*a*) a drilling machine to drill the hole,

(*b*) a milling machine to mill the slot, and lastly

(*c*) a press to form the part.

Therefore, in layout by product, different machines are grouped in an area to process only one product or a few of similar products. This type of layout is useful only in the manufacture of large quantities of a given product where balance in production lines is possible.

This type of layout is found in assembling process, such as, automobile assembly and in forming and treating process such as machining of medium size components.

The advantages and disadvantages of product layout are as follows :

Advantages

1. The material handling is less. Thus there is less damage to materials.
2. The work in-process inventory is less.
3. The production control is better due to easier supervision and less paper work.
4. The manufacturing time is less because there are no work delays.
5. The manufacturing costs are low at high volume of production.
6. The machines can be operated by less skilled workers.

Disadvantages

1. The capital investment in machines is high.
2. The manufacturing cost increases, if the production is limited.
3. The layout is less flexible to changes in product variety and design.
4. The breakdown of a single machine may effect the whole production.

1·11 Combination Layout or Group Layout

The process layout and the product layout are the classic types of layout. These layouts are more often found in combination than in their pure forms. These days, most of the manufacturing industries have adopted a combination layout. In this type of layout, a set of machines or equipment is grouped together in the section so that each set or group of machines or equipment is used to perform similar operations to produce a family of components. Since this type of process layout involves the *group technology* technique, therefore it is also known as *group layout*. The flow of product follows the product layout sequence. Hence it is called combination or group layout.

1·12 Shop Layout

The technique for achieving an economical arrangement of manufacturing processes in conjuction with the plant services is called *shop layout*. Its main objective is to make efficient use of various machines and equipments in a particular shop so as to facilitate the processing of the product in the most efficient and economical manner in the minimum possible time.

The layouts for some shops is discussed in the following pages. It may be noted that the size of the shop is not standardised, but it depends upon the need of the individual.

1·13 Layout of a Carpentry and Pattern Making Shop

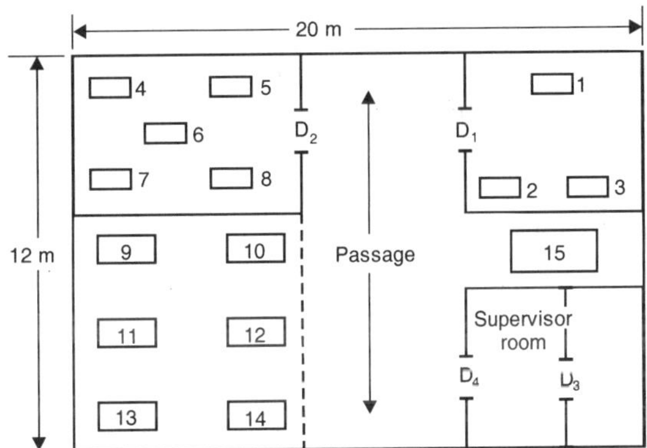

1. Circular saw. 2. Universal wood working machine. 3. Band saw. 4 to 8. Wood working tables with two vices each. 9 to 14. Carpenter's tables with two vices each. 15. Inspection table. D_1, D_2, D_3 and D_4 are Doors.

Fig. 1.1. Layout of a carpentry and pattern making shop.

The various operations performed in carpentry shop are sawing, planing, joining and many other similar works required in the making of doors, windows, furniture etc. In carpentry, the basic material used is timber. In a pattern shop, the patterns, core boxes are made. Fig. 1.1 shows the layout and the equipment required for the carpentry and pattern shop.

1·14 Layout of a Foundry Shop

The various operations performed in a foundry shop are moulding and casting. The moulds are made in dry, green and loam sands. The core-making is an other important work done in this shop. Fig. 1.2 shows the layout of a foundry shop.

1·15 Layout of a Smithy and Forging Shop

The various operations performed in a smithy and forging shop are upsetting, drawing out, forge welding, cutting, bending, swaging, cold and hot riveting etc. Fig. 1.3 shows the layout of a smithy and forging shop.

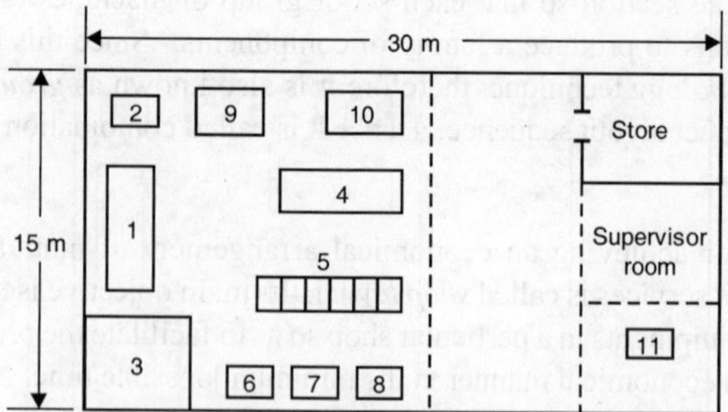

1. Cupola. 2. Blower for cupola. 3. Coke and raw material store. 4. Sand bed. 5. Bench for moulding. 6. Oil-firedtilting furnace. 7, 8. Pit furnaces. 9. Drying oven. 10. Core making table. 11. Inspectiontable.

Fig. 1.2. Layout of a foundry shop.

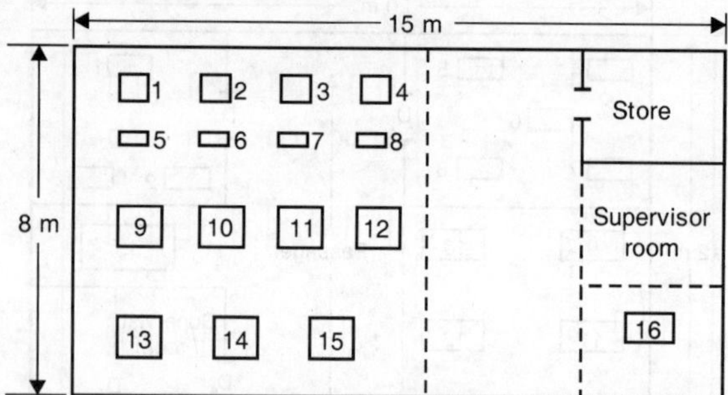

1 to 4. Furnaces. 5 to 8. Anvil. 9. Pedestal grinder. 10 to 12. Swage blocks. 13 to 14. Power hammers. 15. Pneumatic riveting machine. 16. Inspection table.

Fig. 1.3. Layout of a smithy and forging shop.

1·16. Layout of a Welding Shop

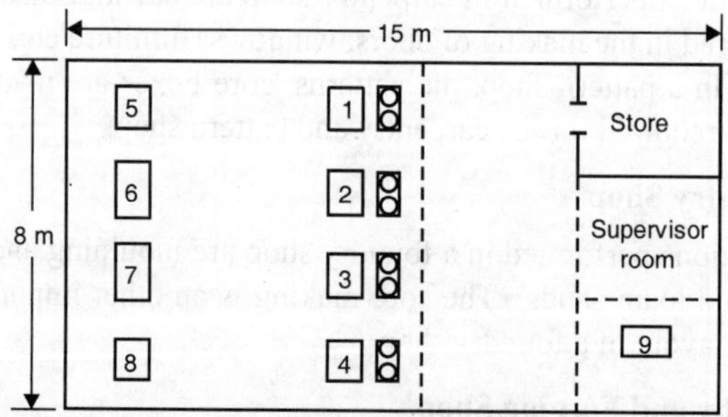

1 to 3. Welding tables with acetylene and oxygen cylinders. 4. Gas cutting with acetylene and oxygen cylinders. 5. Pedestal grinder. 6, 7. Electricwelding set. 8. Spot welding. 9. Inspectiontable.

Fig. 1.4. Layout of a welding shop.

The various operations performed in a welding shop are joining and cutting metal parts by the application of heat. It is a very useful shop in fabrication, erection and maintenance work. Fig. 1.4 shows the layout of a welding shop.

1·17 Layout of Fitting Shop

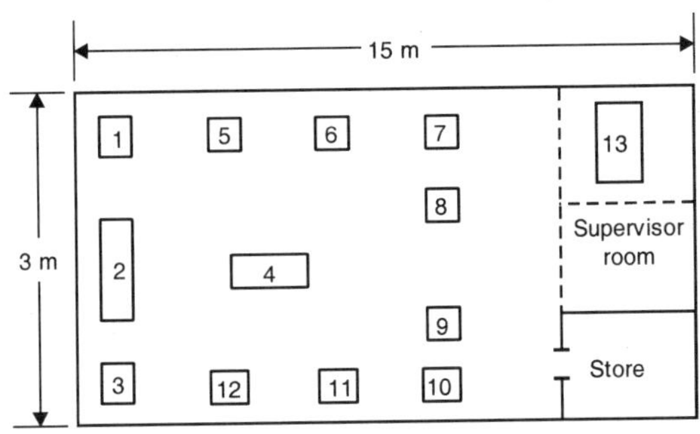

1. Grinder. 2. Power saw. 3. Bench drill. 4. Marking table with surface plate. 5 to 12. Working tables with four bench vices each. 13. Inspection table.

Fig. 1.5. Layout of a fitting shop.

The various operations performed in a fitting shop are filing, scraping, grinding, sawing, marking, drilling, tapping, etc. The fitting shop plays a very important role in all the engineering concerns dealing with manufacturing or repairing work. Though the work can be sufficiently finished by various machining operations, yet the final finishing is done by hand operations in a fitting shop. Fig. 1.5 shows the layout of a fitting shop.

1·18 Layout of a Sheet Metal and Painting Shop

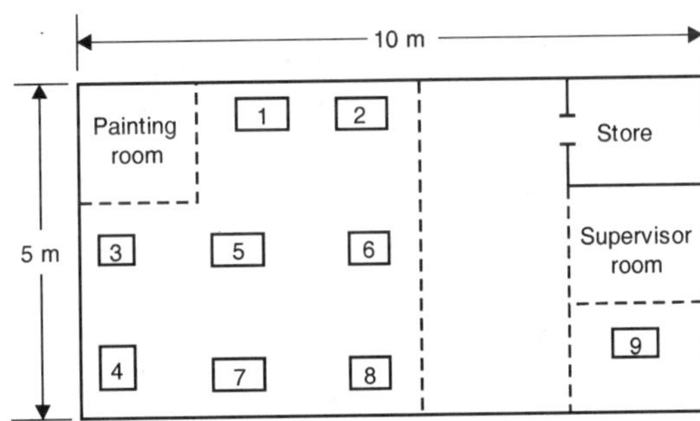

1. Shearing machine. 2. Bending machine. 3. Circle cutting machine. 4. Buffing and polishing machine. 5 to 8. Working tables. 9. Inspection table.

Fig. 1.6. Layout of a sheet metal and painting shop.

The various operations performed in a sheet metal shop are cutting, shearing, bending, etc. The painting on the metal work is done in a painting shop. The layout of a sheet metal and painting shop is shown in Fig. 1.6.

QUESTIONS

1. Discuss the various manufacturing processes.
2. List at least ten processes each under the primary shaping processes and machine processes.

3. What do you understand by plant layout ? What are its objectives ?
4. Explain the use of joining processes, surface finishing processes and processes affecting change in properties.
5. What are the advantages of a good plant layout ?
6. Describe the various types of plant layout. In what situation the each type of plant layout is used ?
7. What do you understand by shop layout ?
8. Prepare a layout for a carpentry shop having the following machines :
   (*i*) Wood working lathes ... ... 4 Nos.
   (*ii*) Band saw ... ... 1 No.
   (*iii*) Circular saw ... ... 1 No.
   (*iv*) Drilling machine ... ... 1 No.
   (*v*) Surfacing machine ... ... 1 No.
9. List out the machines and hand tools required for a pattern shop. Draw also the layout for such a shop.
10. What equipment is required for a smithy shop ? Prepare the layout for this shop.
11. Prepare a layout for a small welding shop having electric arc welding set and a gas welding plant. This shop is used for weld repair of bicycle frames and other small components. List the necessary equipment for this welding shop.
12. Draw the layout of a fitting shop for a polytechnic.
13. A painting and a sheet metal shop is to be set up. Prepare the layout for this shop.
14. List out the equipment required for a foundry shop and draw the layout.

OBJECTIVE TYPE QUESTIONS

1. Casting is a
   (*a*) surface fimihing process (*b*) machining process
   (*c*) primary shaping process (*d*) joining process
2. Threading is a
   (*a*) surface finishing process (*b*) machining process
   (*c*) primary shaping process (*d*) joining process
3. Which of the following is a surface finishing process ?
   (*a*) Rolling (*b*) Bending (*c*) Broaching (*d*) Sheradising
4. Which of the following process affect the change in properties ?
   (*a*) Hot working (*b*) Soldering (*c*) Brazing (*d*) Painting
5. A good plant layout
   (*a*) increases indirect cost of labour (*b*) decreases overall efficiency of the plant
   (*c*) requires less supervision (*d*) increases materials handling time.

ANSWERS

1. (*c*) 2. (*b*) 3. (*d*) 4. (*a*) 5. (*c*)

2

Industrial Safety

1. Introduction. 2. Objectives of Industrial Safety. 3. Accident. 4. Common Causes of Accidents. 5. Good Housekeeping. 6. Safety Guards. 7. General Safety Precautions. 8. Factories Act 1948, Regarding Safety.

2·1 Introduction

In an industry, there are number of machines and other equipments which have large number of moving parts and other dangerous projections which may cause hazardous accidents, fatal injuries if proper measures are not taken for safeguarding against them. It is thus very essential for a worker to know the basic engineering principles and he should be aware of the dangers which may exist in an industry. Every worker should be made to understand and practice the safe working procedures, to avoid accidents, not only for himself but also for other workers.

2·2 Objectives of Industrial Safety

The objectives of industrial safety are as follows :

1. To increase production as a means to a higher standard of living.
2. To reduce cost of production.
3. To conserve and make the best use of labour available.
4. To reduce needless sufferings and human wastage.
5. To have better human relations in the industry.
6. To have better morale of employees.

2·3 Accident

An accident is a mishap which causes some injury to the person, damage to machines, tools and equipment, which results in loss of production. The accident may be due to worker's own fault which may be prevented by his own precautions or it may be due to his employer's fault for not making safe working conditions. The accidents may be minor, serious or fatal.

Note : In the Employees State Insurance Act 1948, the employment injury is defined as "A personal injury or occupation disease caused to a worker by accident arising out of and in the course of his employment for which compensation is recoverable by a workman from his employer under the workmen's compensation Act 1923".

2·4 Common Causes of Accidents

The following are the common causes of accidents :

1. Poor housekeeping.
2. Lack of skill or knowledge *i.e.,* unskilled worker.
3. Loose clothing of the worker.
4. Insufficient working space.
5. Carelessness of the worker.
6. Improper or defective tools.
7. Poor working conditions such as bad lighting and ventilation.
8. Unguarded machinery.

9. Improper maintenance of machines.
10. Overloading of machines.
11. Operating of machines at unsafe speed.
12. Bad layout, planning, processes etc.
13. Violation of safety rules and non-observance of safety precautions.

Note : According to H.W. Heinrich who carried out the analysis of accidents, observed that

(*a*) 88 percent accidents occur due to human causes such as unsafe practices, lack of foresight, wrong attitude etc.

(*b*) 10 percent accidents occur due to mechanical failure such as plant condition, tool or equipment failure.

(*c*) 2 percent accidents occur due to nature such as weather, floods, cyclones etc.

2·5 Good Housekeeping

House-keeping implies an orderly arrangement of operations, tools, equipment, storage facilities and supplies. It is a practical method of getting high production and improvement in employee's morale and personnel relations.

A good housekeeping is very essential in the prevention of accidents. The following points should be observed to have good housekeeping :

1. The working place should be kept clean and free from oil, grease in order to avoid slipping of workers.
2. The work benches and machines should be clean.
3. The tools and measuring instruments should be kept at their proper places.
4. The scrap should be kept in scrap boxes at all times.
5. There should be a sufficient space for the worker to move around.
6. The material handling should be minimum.
7. The material should not be stored in passages.
8. There should be a suitable and sufficient lighting in every part of work place or passage.
9. The adequate ventilation for fresh air shall be provided.
10. The noise level should not distract the worker.

2·6 Safety Guards

The moving parts of the machine must be properly guarded to maintain safety. The type of a guard should be selected with regard to the nature of the machine, purpose and mode of operation. The main types of guards are as follows :

1. *Fixed guards.* The fixed guards are most commonly used in workshops to prevent access to the dangerous parts of a machine from all directions and during all operations. These guards are of robust construction and tightly secured to the machine. Sheet metal, steel sheet, and sometimes cast iron are the suitable material for the fixed guards. When a clean view of the moving parts is essential during the operation, then metal of small mesh is suitable.

2. *Interlocking guards.* When it is impracticable to use a fixed guard, then wherever possible an interlocking guard is fitted. This type of guard forms an integral part of the machine and cannot be removed. Before the machine is operated, these guards are closed to prevent access to the dangerous part and are kept in closed position until the dangerous part is at rest.

3. *Automatic guards.* The automatic guard is operated by the moving part of the machine in which it is fitted and these two are so linked that the part will automatically bring the guard in protective position before the machine starts operating. The automatic guards are suitable on medium and heavy presses.

4. *Trip guards.* A trip guard is provided with tripping device or safety switch inside the guard which enables quick stopping of the machine as soon as the operator approaches within the reach of dangerous parts, thus safeguarding the operator from revolving and dangerous parts.

5. *Distance guards.* The distance guard is fencing made of grills, bars or rails placed at suitable height and distance from the machine so that even if the operator, by chance, extends hands over it, his clothes, fingers or other parts of the body does not reach within the area of dangerous parts.

2·7 General Safety Precautions

The following safety precautions should be strictly observed by a worker while working in an industry.

1. Always wear proper dress and shoes in a workshop. A worker should not wear loose clothes, necktie etc., to guard against getting caught with moving machinery.
2. Before operating a machine, it should be cleaned and oiled, if necessary.
3. The machines should have effective and properly working guards in proper place.
4. The heavy work should not be lifted by hands.
5. Before operating a machine, the work and tool should be clamped securely.
6. The machines should be operated only by a skilled workman.
7. The machines should not be left in running position while going for an urgent call.
8. The operator should not lean towards the machine.
9. When the machine is running, the operator should not indulge in mischief with other operators.
10. Wear safety goggles or face shield to protect eyes from the welding arc.

2·8 Factories Act 1948, Regarding Safety

The salient provisions relating to safety under Chapter IV of the Indian factories Act 1948, which is still in force, are as follows :

1. *Fencing of machinery*

In every factory, the following parts should be fenced :

(*a*) Every moving part of a prime mover.

(*b*) Every flywheel connected to a prime mover.

(*c*) The headrace and tailrace of every water wheel and water turbine.

(*d*) Any part of a stock bar which projects beyond the headstock of a lathe.

The following parts should be securely fenced by safeguards of substantial construction which shall be kept in position while the parts of the machinery they are fencing, are in motion and use :

(*a*) Every part of an electric generator, a motor or rotary converter.

(*b*) Every part of transmission machinery.

(*c*) Every dangerous part of any other machinery.

2. *Work on or near the machinery in motion*

In any factory, when it becomes necessary to examine any part of a machinery, while the machinery is in motion, it should be carried out only by specially trained adult male worker wearing tight-fitting clothes and whose name has been recorded in the register. Such a worker

(*a*) Should not handle a belt at a moving pulley unless the belt is less than 150 mm in width and unless the belt joint is either laced or flush with the belt.

(*b*) Every set screw, bolt and key on any revolving shaft, spindle, wheel or pinion should be securely fastened.

(c) All the revolving parts such as shafts, spindles and all spur, worm and other toothed or friction gearing in motion should be securely fenced to prevent any contact of the worker or his clothes.

3. *Striking gear and devices for cutting off power*

(a) A suitable striking gear or other efficient mechanical appliance should be used to shift driving belts to and from fast and loose pulleys which form part of the transmission machinery. Such gear or appliance should be so constructed, placed and maintained as to prevent the belt from creeping back on to the fast pulley.

(b) The driving belts, when not in use, should not be allowed to rest or ride upon shafts in motion.

(c) In every factory, suitable devices for cutting off power in emergencies from running machinery should be provided and maintained in every work room.

4. *Casing of new machinery*

(a) Every set screw, bolt or key on any revolving shaft, spindle, wheel or pinion should be so sunk or encased or otherwise effectively guarded as to prevent danger.

(b) All spur, worm and other toothed or friction gearing, which does not require frequent adjustment while in motion, should be completely encased unless it is so situated as to be as safe as it would be if they were completely encased.

5. *Hoists and lifts*

In a factory, every hoist and lift should be of good mechanical construction, sound material and adequate strength. They should be properly maintained and thoroughly examined by a competent person at least once in every period of six months.

6. *Lifting machinery, chains, ropes and lifting tackles*

In any factory, the following provision should be complied with in respect of every lifting machine (other than a hoist and lift) and every chain and lifting tackle for the purpose of raising or lowering persons, goods or materials.

All parts including the working gear, whether fixed or movable, of every lifting machine and every chain, rope or lifting tackle should be of good construction, sound material and adequate strength. They should be properly maintained and thoroughly examined by a competent person at least once in a year.

7. *Revolving machinery*

It should always be ensured that safe working peripheral speed is not exceeded for every revolving machinery like a cage, flywheel, pulley, grinding wheel or similar appliance driven by power.

8. *Pressure plant*

The pressure plant should not be operated at a pressure higher than the specified safe working pressure.

9. *Precautions against dangerous fumes*

In any factory, the following provisions should be complied with :

(a) No person should be permitted to enter any chamber, tank, pit, pipe flue or any other confined space in which dangerous fumes are likely to be present to such an extent as to involve risk of the person, unless it is provided with a manhole of adequate size or other effective means of ingress.

(b) No portable electric light of voltage exceeding twenty four volts should be permitted in any factory for use inside any confined space as mentioned above. Only a light or lamp with flame proof construction should be permitted to be used where fumes are likely to be inflammable.

(*c*) No person in any factory should be permitted to enter any confined space as mentioned above, until all practicable measures have been taken to remove any fumes which may be present and to prevent any ingress of fumes.

(*d*) The suitable breathing apparatus, reviving apparatus, belts and ropes should be kept ready in every factory for use in case of emergency. A sufficient number of workers should be trained to use such apparatus and in the method of restoring respiration.

10. *Explosive or inflammable dust, gas etc*

When in a factory, any manufacturing process produces dust, gas, fume or vapour of such a character and to such extent as to likely explode on ignition, all practicable measures should be taken to prevent such explosion by the :

(*a*) Effective enclosure of the plant or machinery used in the process.

(*b*) Removal or prevention of the accumulation of such dust, gas, fume or vapour.

(*c*) Exclusion or effective enclosure of all possible sources of ignition.

11. *Precautions in case of fire*

In a factory, the following precautions and measures against fire, should be observed :

(*a*) In every factory, all the doors and windows, except they are of sliding type, should be so constructed as to open outwards.

(*b*) The doors and windows to be used as exit in case of fire, should be distinctively marked in a language understood by majority of the workers and in red letters of adequate size.

(*c*) There should be effective and clearly audible means of giving warning in case of fire to every person employed in the factory.

(*d*) The maximum number of workers and officers should be trained in fire fighting systems.

12. *Safety of building and machinery*

If an inspector finds that any part of building, machinery or plant is in such a condition that is dangerous to human life and safety, he should then serve a written notice to the manager of the factory specifying the measures which in his opinion should be adopted and requiring them to be carried out before a specified date.

QUESTIONS

1. What do you understand by industrial safety ? What are its objectives ?
2. What is an accident ? Give in brief the causes which lead to an accident in a workshop.
3. What are the basic requirements of a good housekeeping ?
4. Describe the different types of guards used for safeguarding against dangerous parts.
5. Enumerate the safety precautions to be taken in a workshop.
6. What are the salient provisions of Factories Act 1948, regarding safety ?

OBJECTIVE TYPE QUESTIONS

1. The objectives of industrial safety are

   (*a*) to reduce cost of production

   (*b*) to increase production

   (*c*) to have better human relations in the industry

   (*d*) all of the above

2. Which of the following statement is correct ?

(*a*) An accident is a mishap which causes some injury to the person, damage to machines, tools and equipment, which results in loss of production.

(*b*) 10 percent accidents occur due to human causes.

(*c*) 88 percent accidents occur due to mechanical failure.

(*d*) all of the above

3. The interlocking safety guard

(*a*) is provided with a tripping device inside the guard

(*b*) is operated by the moving part of the machine in which it is fitted.

(*c*) forms an integral part of the machine and can not be removed.

(*d*) none of the above

4. According to factories Act 1948, regarding safety against dangerous fumes, no portable electric light of voltage exceeding ......should be permitted in any factory for use inside any confined space.

(*a*) 24 volts (*b*) 40 volts (*c*) 60 volts (*d*) 100 volts

ANSWERS

1. (*d*) 2. (*a*) 3. (*c*) 4. (*a*)

3

Fundamentals of Metals and Alloys

3·1 Introduction

The materials are broadly divided into two groups, *i.e.*, *metals* and *non-metals*, in accordance to their properties. The metals are usually solid at room temperature and are malleable and ductile. They are good electrical and thermal conductors. The other properties of metals include good lustre, high density, high melting point, good strength and hardness, low specific heat etc. Most of the metals are found to be elastic to a certain limit and they deform plastically at higher stresses. A metal can exist in the solid, liquid or gaseous states depending upon the temperature and pressure exerted on the metal. Some of the common metals and alloys are iron, copper, aluminium, nickel, tin, brass, bronze etc.

The non-metals in the solid state are mostly brittle and are poor electrical and thermal conductors. They do not form alloy but combine chemically to form compound. Some of the non-metals are glass, plastics, wood, concrete etc.

3·2 Structure of Solids

All solid substances are either amorphous solids or crystalline solids. In the *amorphous solids,* the atoms* are arranged chaotically, *i.e.*, the atoms are not arranged in a systematic order. The common amorphous solids are wood, plastics, glass, paper, rubber etc. In *crystalline solids*, the atoms making up the crystals arrange themselves in a definite and orderly manner and form. All solid metals such as iron, copper, aluminium etc. are crystalline solids. The arrangement of atoms in the amorphous and crystalline solids is shown in Fig. 3.1

---

* The *atom* is the smallest and chemically invisible particle of matter and the smallest quantity of a substance, which can exist freely by itself, in a chemically recognisable form, known as *molecule.* The molecules are composed of atoms of different elements in a fixed proportion. The molecules may contain one atom (known as monoatomic), two atoms (known as diatomic), three atoms (known as triatomic) or even more atoms (known as polyatomic).

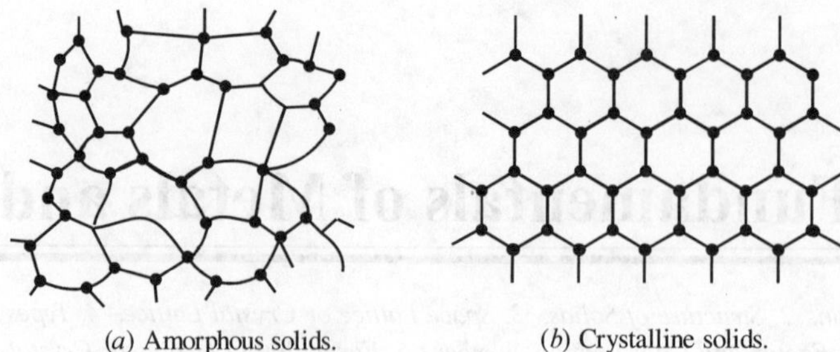

(a) Amorphous solids. (b) Crystalline solids.

Fig. 3.1. Structure of solids.

3·3 Space Lattice or Crystal Lattice

We have already discussed that all solid metals are made up of a number of crystals arranged themselves in a definite and orderly manner and form and each crystal consists of a large number of atoms. This definite and orderly manner and form of atoms producing a geometrical shape in the aggregate is called *space lattice* or *crystal lattice.* The space lattice of a crystal is described by means of a three-dimensional coordinate system, in which the coordinate axes coincides with any three edges of the crystal that intersect at one point and do not lie in a single plane.

A crystal is composed of *unit cells.* A unit cell contains the smallest number of atoms, which when taken together have all the properties of the crystals of the particular metal. The unit cells are arranged like building blocks in a crystal, *i.e.*, they have the same orientation and their similar faces are parallel. A unit cell may also be defined as the smallest parallelopiped which could be transposed in three coordinate directions to build the space lattice. The space lattice of various substances differ in the size and shape of their unit cells.

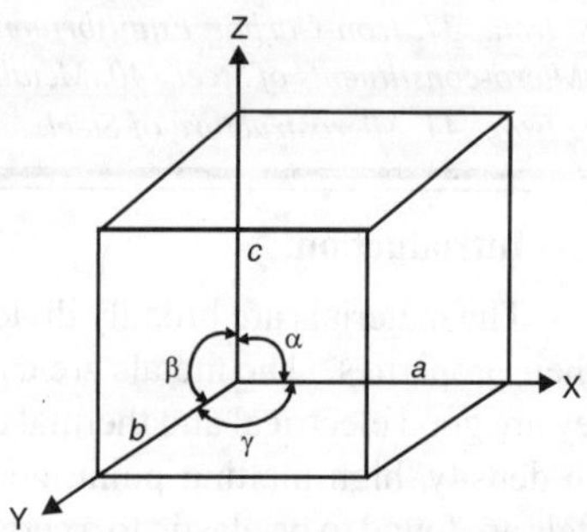

Fig. 3.2. Lattice arrangement of a unit cel

Fig. 3.2 shows a unit cell of a three dimensional crystal lattice. It is formed by *primitives* or intercepts *a*, *b* and *c* along the three axes respectively. The three angles (α, β and γ) are called interfacial angles. Both the intercepts and interfacial angles constitute the lattice parameters of the unit cell. It is thus obvious that if the values of intercepts and interfacial angles are known, we can easily determine the form and actual size of the unit cell.

3·4 Types of Space Lattices

There are 32 classes of crystal systems based on the geometrical considerations (*i.e.*, symmetry and internal structure). But it is a common practice to divide all the crystal systems into seven groups or basic systems. These seven basic crystal systems are distinguished from one another by the angles between the three axes and the intercepts of the faces along them. These seven basic crystal systems are

1. Cubic ; 2. Tetragonal ; 3. Hexagonal ; 4. Orthorhombic ; 5. Rhombohedral ; 6. Monoclinic ; 7. Triclinic.

According to Bravais (a scientist), there are fourteen possible types of space lattices in these seven crystal systems. But the following three types are usually found in most of the metals.

1. *Body centred cubic* (*B.C.C.*) *space lattice.* In a unit cell of body centred cubic space lattice, there are nine atoms. The eight atoms are located at the corners of the cube and one atom at

it centre, as shown in Fig. 3.3 (*a*). This type of lattice is found in *alpha (α)-iron, tungsten, chromium manganese, molybdenum, tantalum, barium, vanadium etc.

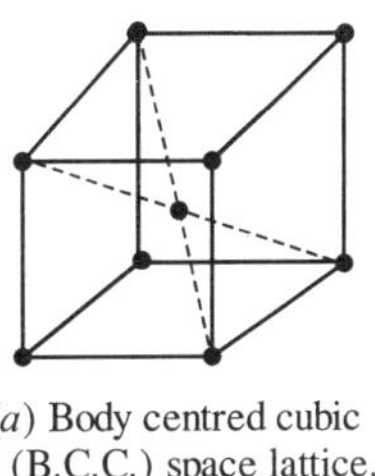

(*a*) Body centred cubic (B.C.C.) space lattice.

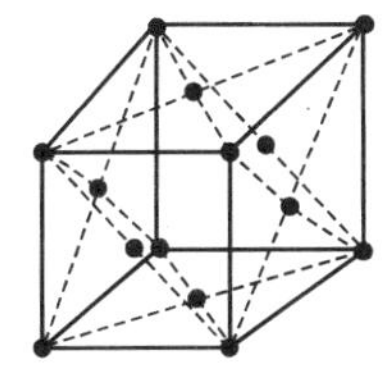

(*b*) Face centred cubic (F.C.C.) space lattice.

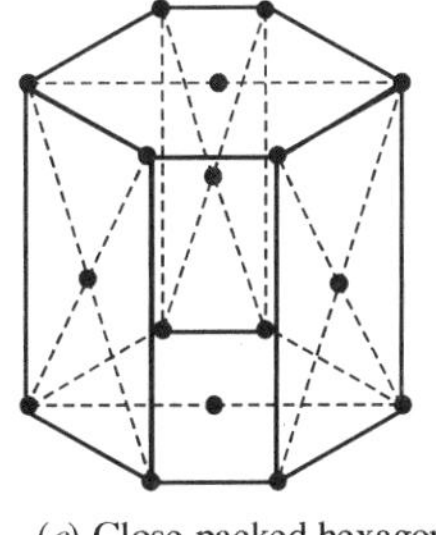

(*c*) Close packed hexagonal (C.P.H.) space lattice.

Fig. 3.3. Types of space lattices.

2. *Face centred cubic (F.C.C.) space lattice.* In a unit cell of face centred cubic space lattice, there are fourteen atoms. The eight atoms are located at the corners of the cube and six atoms at the centres of six faces, as shown in Fig. 3.3 (*b*). This type of lattice is found in gamma (γ) iron, aluminium, copper, lead, silver, nickel, gold, platinum, calcium etc.

3. *Close packed hexagonal (C.P.H.) space lattice.* In a unit cell of close packed hexagonal space lattice, there are seventeen atoms. The twelve atoms are located at the twelve corners of the hexagonal prism, one atom at the centre of each of the two hexagonal faces and three atoms are symmetrically arranged in the body of the cell as shown in Fig. 3.3 (*c*). This type of lattice is found in zinc, magnesium cobalt, cadmium, antimony, bismuth, beryllium, titanium, zirconium etc.

3·5 Atomic Packing Factor and Coordination Number

The *atomic packing factor* is defined as the ratio of the volume occupied by the atoms to the total volume of the unit cell. This is also known as *packing efficiency.*

The *coordination number* is defined as the number of equally spaced nearest neighbouring atoms by which a given atom is surrounded. This number describes the packing of the crystal lattices. It may be noted that higher the coordination number, the more closely packed are the atoms in a crystal lattice.

We know that in a body centred cubic (B.C.C.) crystal lattice, each corner atom is surrounded by eight other body centred unit cells. Moreover, their body centred atoms are more near the corner atoms than those in the other corners. Therefore, the nearest adjacent neighbours are the body centred atoms of the eight surrounding unit cells. Thus the coordination number of B.C.C. crystal lattice is eight. Similarly, in a face centred cubic (F.C.C.) crystal lattice, each corner atom is surrounded by eight other face centred unit cells. It may be noted that twelve face centred atom are more near the corner atom than those in the other corners. Thus the coordination number of F.C.C. crystal lattice is twelve.

3·6 Formation of Grains or Crystals (Solidification of Metals)

We have already discussed that all solid metals are crystalline and the crystals or grains are made up of several atoms. Let us now study how the grains are formed when the liquid metal solidifies. The process of solidification or crystallisation will proceed only if the liquid metal is cooled below the **equilibrium temperature of solidification.

* See Art. 3·36.

** It is the temperature at which the given metal exist simultaneously in a liquid and solid phases.

The solidification commences by the formation of small nuclei scattered at random in the cooling liquid. At these points, a few atom assume an orderly arrangement to give the unit cubic structure and the growth takes place in three dimensions as shown in Fig. 3.4. From the main arms of the crystal, secondary growth occur to give tree-like crystals known as *dendrites* (from a Greek word *dendron* meaning tree) and process of solidification is, therefore, termed as *dendritic solidification.*

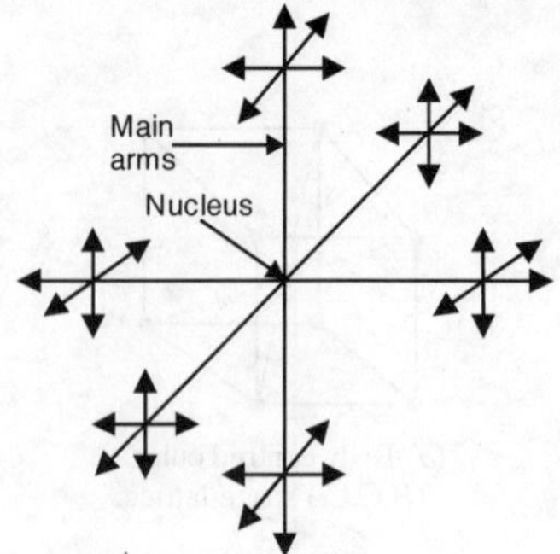

Fig. 3.4. Solidification of metals.

The formation of these grains from the nucleus by dendritic growth is shown in Fig. 3.5. The crystals nuclei form in the cooling liquid is shown in Fig. 3.5(*a*) and then proceed to form axes of solidification and the dentritic crystal grow as shown in Fig. 3.5(*b*). First of all the long branches known as primary axes of the dendrite are formed and then the branches evolve from their edges and grow in a perpendicular direction (*i.e.*, secondary axes are formed). Lastly, ternary axes evolve and grow on the secondary axes. This action called nucleation or crystal growth continues until all the metal between the branches has solidified. The crystals outward growth is stopped by impingement on adjacent crystals also growing by the nucleation. The contact surfaces formed by adjacent crystals (because of different orientation of the grains) are called *grain boundaries* as shown in Fig. 3.5(*c*).

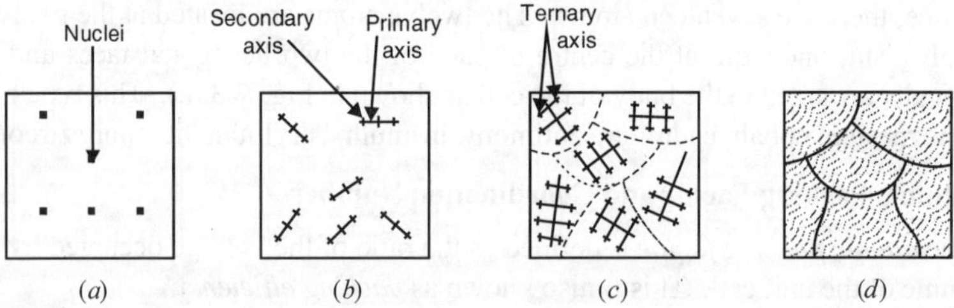

Fig. 3.5. Formation of grains.

Each crystal or the grain is built up of thousands of small unit cells or cubes. The axes of the cubes all point in the same direction, but this direction varies from one crystal to another. This effect is called orientation of the atoms. The orientation of the atomic pattern is constant in a given crystal but varies form grain to grain.

The liquid between the branches of the dendrites solidifies, giving homogeneous grains with no evidence of dendritic grown as shown in Fig. 3.5(*d*), except where shrinkage occurs with the formation of interdendritic porosity. In actual practice, impurities exist in the metal and these are frequently pushed by growing crystal to the grain boundaries and are trapped in between the branches of the dendrites. The distribution of these impurities, therefore, betrays the dendritic growth.

3·7 Effect of Grain Size on Mechanical Properties

The grain size has an important effect on the mechanical properties of a metal such as strength, ductility, hardness etc. The size of the grains depends upon a number of factors, but the principal one being the heat treatment to which the metal has been subjected.

When a low carbon steel is heated, there is no change in grain size upto the *lower critical point. At this temperature birth of new grains take place. At the upper critical point, the average grain

* The temperature point at which the change starts on heating is called *lower critical point* and the temperature point where this change ends in heating is called *upper critical point*.

size is a minimum. Further heating of the steel causes an increase in the size of the grains, which in turn governs the final size of the grains when cooled. All steels do not start growing large crystals immediately upon being heated above the upper critical range. Some steels like medium carbon steel and many alloy steels when heated to a higher temperature, known as *coarsening temperature,* the grain size increases very rapidly. The coarsening temperature is not a fixed temperature and may be changed by prior hot or cold working and heat treatment. The hot working on steel is started at temperatures well above the critical range, with the steel in a plastic state, and has the effect of refining the grain structure and eliminating any coarsening effect due to the high temperature. The hot forging or rolling should not continue below the critical temperature.

The quenching of steel from the upper critical point results in a *fine grained structure,* whereas slow cooling or quenching from a higher temperature yields a *coarse grained structure.*

The coarse-grained steels are less tough and have a greater tendency for distortion than those having a fine grain. However, they have better machinability and greater depth hardening power. The fine-grained steel, in addition to being tougher, are more ductile and have less tendency to distort or crack during heat treatment. The control of grain size is possible through regulation of composition in the initial manufacturing procedure, but after the steel is made, the control is through proper heat treatment.

3·8 Bonds in Solids

It is a well known fact that an atom, by itself rarely exists independently. Most of the elements are found in the form of clusters or aggregates of atoms. Any such cluster, in which atoms are held together and is electrically neutral, is known as a molecule. It has been observed that when the atoms of the elements combine to form molecules, the forces of attractions and repulsion which are basically electrostatic in nature are developed between the atoms, which hold them together. These forces are known as *bonds* and the process of holding is known as *bonding.*

The bonds may be of the following two types :

1. Primary bonds; and 2. Secondary bonds.

The *primary bond* (also known as *attractive bond*) is an interatomic bond, in which electrostatic force holds the atoms together. It may be noted that the primary bond is strong and more stable.

The *secondary bond* is an intermolecular bond, in which the weak forces hold the molecules together. The weak forces are also known as Van der Waals forces. It may be noted that a secondary bond is weak and less stable than the primary bond.

3.9 Types of Primary Bonds

The primary bonds are of the following three types :

1. *Ionic bond or electrostatic bond.* An ionic bond is formed when one or more electrons, from he outermost shall of one atom are transferred to the outermost shell of another atom. We know that the atom, which loses the electron, acquires a positive charge and becomes a positive ion and the atom, which gains the electron, acquires a negative charge and becomes a negative ion. In this way, both the atoms become oppositely charged ions. The electrostatic attraction, between the oppositely charged ions, forms the ionic bond. The following two conditions are necessary for the bond formation:

(*a*) The atom, which loses electron, should have low ionization energy. It means that a small amount of energy is required to remove the electron from its outermost energy level.

(*b*) The atom which gains electron, should have high electron affinity. In means that the atom should be able to readily accept an electron in its outermost energy level.

2. *Covalent bond.* A covalent bond is formed when two or more electrons of an atom, in its outermost energy level, are shared by electrons of other atoms. It happens, when the sharing of electrons takes place in such a way that each atom, in the resulting molecule, sets a stable configuration.

Sometimes, a covalent bond is also formed when two atoms of different non-metals share one or more pairs of electrons in their outermost energy level. It may be noted that when the atoms share a single pair of electrons, the bond formed is known as *single covalent bond.* Similarly, when the two atoms share two pairs of electrons, the bond formed is known as *double bond* and when the atoms share the three pairs of electrons, the bond formed is called *triple bond.*

3. *Metallic bond.* It has been established that in all the metal atoms, the electrons in their outermost energy levels (known as valence electrons) are loosely held by their nuclei. As a result of this, the electrons in the outermost energy level require very small amount of energy to detach themselves from their nuclei. At room temperature, all the metal atoms lose electrons from their outermost energy levels, which form an electron cloud or common pool of electrons. These atoms after leaving their outermost electrons acquire positive charges and become positive ions. Thus a metal may be considered as positive ions surrounded by a large number of free electrons forming electron gas or electron cloud.

The electrostatic force of attraction between the electron cloud and positive ions forms a bond, which is known as metallic bond (as it takes place in metals only).

3·10 Comparison of Properties between Ionic, Covalent and Metallic Solids

The following table gives the comparison of properties between the ionic, covalent and metallic solids :

S.No.	*Ionic solid*	*Covalent solid*	*Metallic solid*
1.	The ionic solids have very low electrical and thermal conductivities.	The covalent solids have low electrical and thermal conductivities.	The metallic solids have high electrical and thermal conductivities.
2.	They have high hardness due to their crystalline structure.	They have low hardness except diamond, silicon, carbide etc.	They have crystalline structure but are soft.
3.	These are not malleable and ductile.	These are also not malleable and ductile.	These are malleable and ductile.
4.	They have high melting and boiling temperatures.	They have lower melting and boiling temperatures than ionic solids.	They have slightly lower melting and boiling temperatures than ionic solids.
5.	The ionic compounds exist in the form of solids only.	The covlent compounds exist in the form of solids, liquids and gases.	The metallic compounds also exist in the form of solids only.
6.	These are soluble in water.	There are not soluble in water but are soluble in organic solvents such as benzene, toluene etc.	These are neither soluble in water nor in organic solvents.

3·11 Types of Secondary Bonds

The secondary bonds are of the following three types :

1. *Dispersion bond.* We know that in a symmetrical molecule, the electrons are uniformly distributed in the space around the nucleus and they are constantly in motion, therefore, in a symmetrical molecule, the centres of positive charges and negative charges coincide with each other. But it has been observed that at certain times, the distribution of electrons in the molecules is not symmetrical around its nucleus. This results in the displacement of the centres of positive and negative charges. The electronic imbalance of the charge is known as polarisation. This polarisation is of the fluctuating nature and is known as dispersion effect. Due to this effect, there exists a weak force of attraction

between the two molecules of the same element and a bond is formed between them which is known as *dispersion bond.* The molecules of noble gases which consist of single atoms are held together by dispersion bonds when they are solidified at very low temperatures.

2. *Dipole bond.* We have already discussed that a covalent bond is formed when the outermost electrons of an atom are shared by two or more atoms. But some times there is an unequal sharing of electrons between two atoms. It has been observed that the unequal sharing of electrons takes place only in those substances in which one of the atoms, in a molecule, has a high affinity to attract electrons than the other. The effect of this unequal sharing of electrons is to create opposite charges on the parent atoms. As a result of this, permanent dipoles are produced. It has been observed that such dipoles attract each other and a bond is formed, which is known as *dipole bond.*

3. *Hydrogen bond.* A hydrogen bond is a particular type of dipole bond, in which one of the atom is hydrogen atom. The other atom has a high affinity to attract electron from the hydrogen atom.

3·12 Defects (or Imperfections) in Crystals

In has been observed, that the crystals are rarely found to be perfect. The atoms do not have their full quota of electrons in the lowest energy level. But the atoms vibrate due to thermal effect, and the electrons also change their positions. There are many other types of defects found in the structure of the crystals. It will be interesting to know that the type and magnitude of any defect, in the structure of the metal crystals, plays an important role at the time of the selection as well as actual use. Though there are many types of defects in the crystals, yet we shall discuss only those defects which are found in the arrangement of atoms. Following three types of defects are important from the subject point of view :

1. Point defects. 2. Line defects. 3. Surface defects.

These defects are discussed, in detail, in the following pages.

3·13 Point Defects

The defects, which take place due to imperfect packing of atoms during crystallisaiton, are known as *point defects*. The point defects also take place due to vibrations of atoms at high temperatures. The presence of this defect in a crystal results in an increase in internal energy of the crystal and decrease in its free energy. Following types of point defects are important from the subject point of view :

1. *Vacancies.* Whenever one or more atoms are missing from a normally occupied position, as shown in Fig. 3.6(*a*), the defect caused is known as vacancy. Such defects can be a result of imperfect packing during the formation of crystals. They also arise from thermal vibration of the atoms at high temperatures. It may be noted that there may be a single vacancy (if one atom is missing), di-vacancies (if two atoms are missing), tri-vacancies (if three atoms are missing) and so on.

2. *Interstitial defects.* Whenever an extra atom occupies interstitial position (*i.e.* voids) in the crystal system, without dislodging the parent atom as shown in Fig. 3.6(*b*), the defect caused is known as interstitial defect. It may be noted that the atom, which occupies the interstitial position is generally smaller than the parent atoms. In close packed structures (such as FCC and HCP), the largest size of the atom that can fit in the voids have radius about 22.5% of the radii of parent atoms. However, if an atom, with larger size than that mentioned above fits in the voids, they produce distortion in atoms.

3. *Frenkel defect.* Whenever a missing atom (responsible for vacancy) occupies interstitial position (responsible for interstitial defect) as shown in Fig. 3.6(*c*), the defect caused is known as frenkel defect. It may be noted that a frenkel defect is a combination of vacancy and interstitial defects. This type of defect is more common in ionic crystals, because the positive ions, being smaller in size, get lodged easily in the interstitial positions.

4. *Substitutional defect.* Whenever a foreign atom (*i.e.*, other than the parent atoms) occupies a position, which was initially meant for a parent atom (or in other words, replaces a parent atom), as

shown in Fig. 3.6(*d*), the defect caused is known as substitutional defect. It may be noted that in this type of defect, the atom which replaces the parent atom may be of the same size or slightly smaller or larger than that of the parent atom.

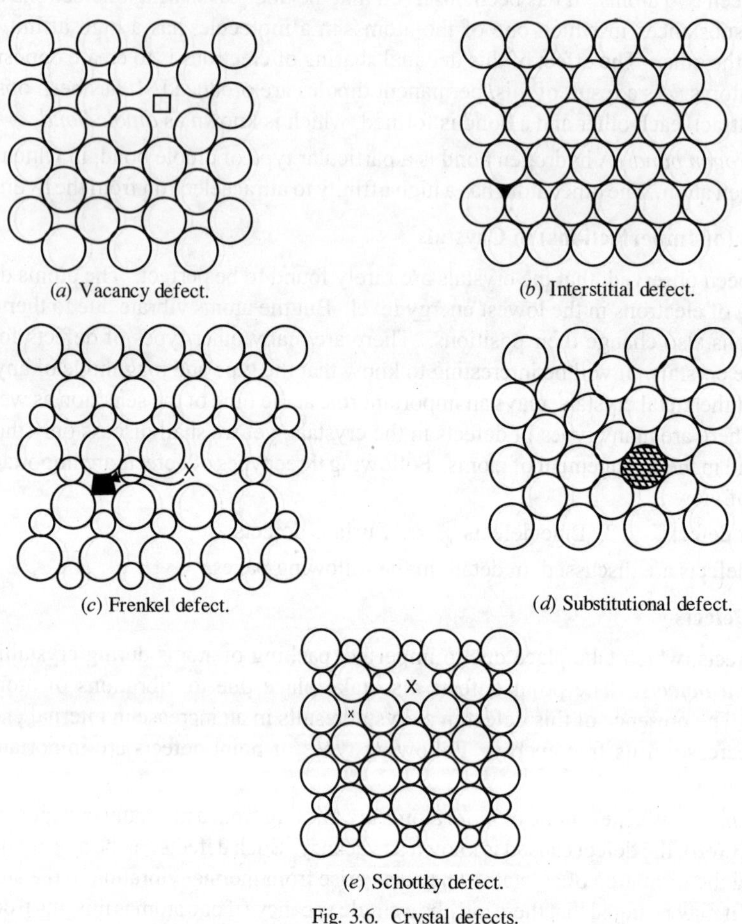

(*a*) Vacancy defect. (*b*) Interstitial defect.

(*c*) Frenkel defect. (*d*) Substitutional defect.

(*e*) Schottky defect.

Fig. 3.6. Crystal defects.

5. *Schottky defect.* Whenever a pair of positive and negative ions are missing from a crystal, as shown in Fig 3.6(*e*), the defect caused is known as schottky defect. It may be noted that in this type of defect, the crystal is electrically neutral.

6. *Phonon.* Whenever a group of atoms is displaced from its ideal location, the defect caused is known as phonon. It may be noted that such a defect is caused by thermal vibrations. It happens as the atoms interact with one another, they tend to vibrate in synchronism in the same way as waves on the ocean surface. It will be interesting to know that this defect effects electrical and magnetic properties to a great extent.

3·14 Line Defects

The defects, which take place due to dislocation or distortion of atoms along a line, in some direction, are called as *line defects*. The line defects also take place when a central portion of a crystal lattice slips without effecting the outer portion. Following two types of line defects are important from the subject point of view :

1. *Edge or Taylor Orwan dislocation.* Whenever a half plane of atom is inserted between the planes of atoms in the perfect crystals, the defect so produced is known as edge dislocation.

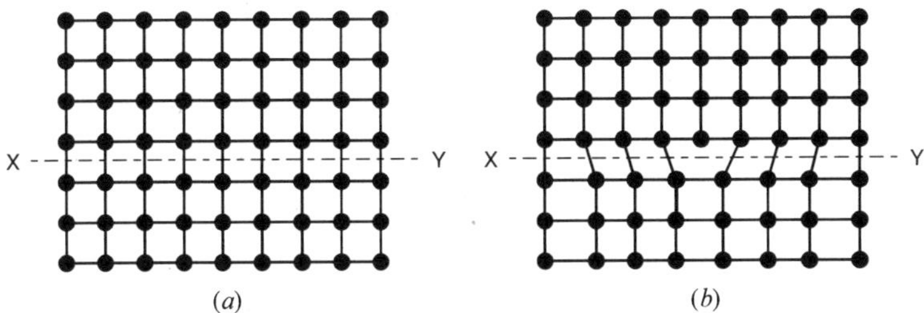

Fig. 3.7. Edge dislocation.

Fig. 3.7 (*a*) shows a cross section of a crystal where dots represent atoms arranged in an orderly manner. Fig. 3.7(*b*) shows the displacement of atoms when an extra half plane is inserted from the top. It may be noted from this figure that top and bottom of the crystal above and below the line *XY* appears perfect. If the extra half plane is inserted from top, the defect so produced, is represented by ⊥ (inverted Tee) as shown in figure. And if the extra half plane is inserted from the bottom, the defect so produced is represented by ⊤(Tee).

2. *Screw or Burger's dislocation.* Whenever the atoms are displaced in two separate planes perpendicular to each other the defect so produced is known as screw dislocation. Fig. 3.8(*a*) represents an isometric view of a perfect crystal. Fig. 3.8(*b*) shows the displacement of atoms in the region *ABC*. In screw dislocation, the arrangement of atoms appear like that of a screw or a helical surface.

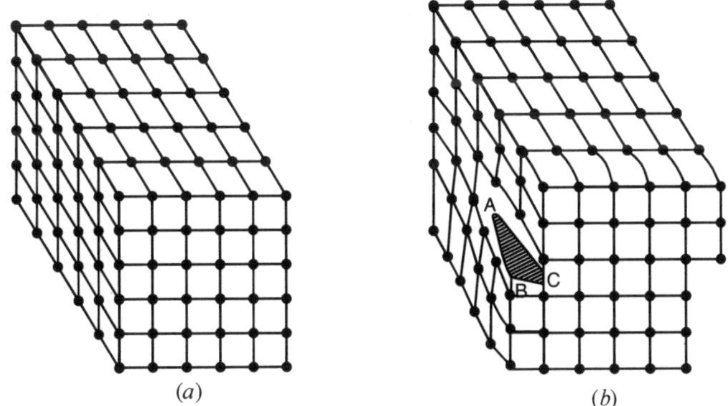

Fig. 3.8. Screw dislocation.

3·15 Surface Defects

The defects, which takes place on the surface of a material, are known as *surface defects* or *plane defects.* It may be noted that surface defects take place either due to imperfect packing of atoms during crystallisation or defective orientation of the surface. Following types of surface defects are important from the subject point of view :

1. *Grain boundary.* Whenever grains of different orientation separate the general pattern of atoms and exhibits a boundary, as shown in Fig. 3.9, the defect caused is known as grain boundary.

This type of defect generally takes place during the solidification of the liquid metal.

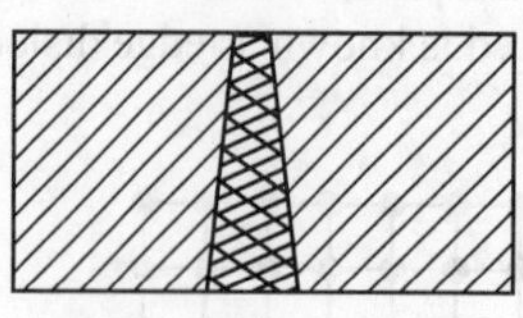

Fig. 3.9. Grain boundary.

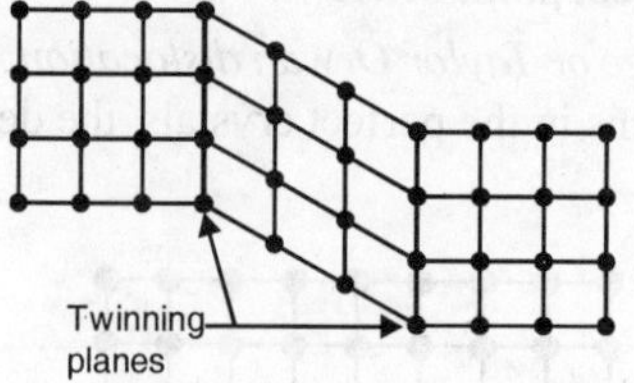

Fig. 3.10. Twin boundary

2. *Twin boundary.* When the boundaries in which the atomic arrangement on one side of the boundary is a somewhat mirror image of the arrangement of atoms of the other side, as shown in Fig. 3.10, the defect caused is known as twin boundary.

3. *Stacking fault.* Whenever the stacking of atoms is not in proper sequence throughout the crystal, the fault caused is known as stacking fault.

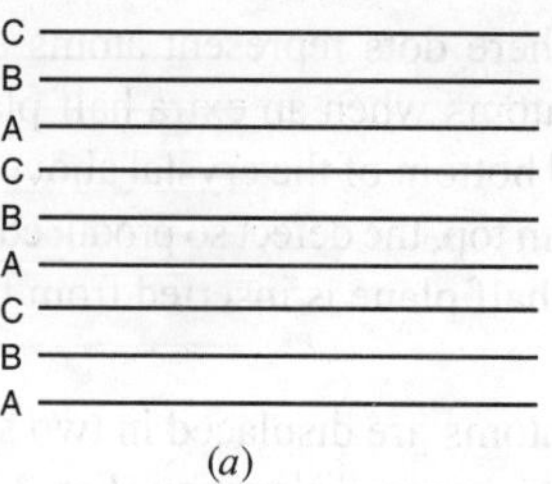

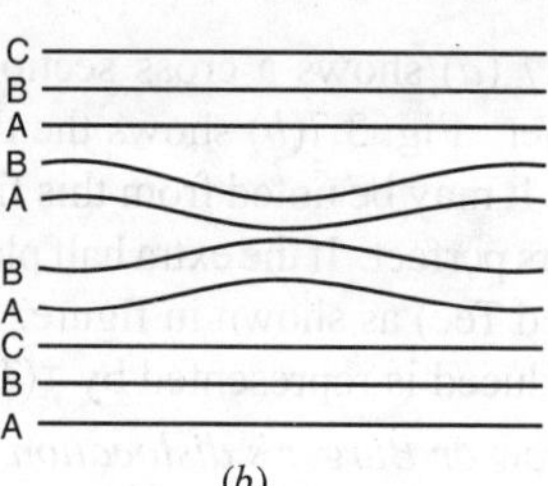

Fig. 3.11. Stacking fault.

Fig. 3.11(*a*) shows the proper sequence of atomic planes, if we read from bottom to top is *A-B-C-A-B-C-A-B-C*. Fig. 3.11(*b*) shows the sequence of atomic planes as *A-B-C-A-B-A-B-A-B-C*. The region in which the stacking fault occurs (*A-B-A-B*) forms a thin region of hexagonal close packing in a F.C.C. crystal.

3.16 Deformation of Metals

The changes produced in the shape of a metal piece under the action of a single force or set of forces, is known as *deformation* or mechanical deformation. The deformation of metals or their alloys is essential to give a desired shape to the finished surfaces. It has been experimentally found that the deformed metals are superior to the cast metals from which they are produced. The various operations used in the deformation of metals are rolling, forging, spinning, drawing etc.

The deformation of metals may be classified into the following two types depending upon the nature of strain produced during deformation :

1. Elastic deformation; and 2. Plastic deformation :

These deformations are discussed, in detail, in the following pages.

3·17 Elastic Deformation

The elastic deformation (also called temporary or recoverable deformation) may be defined as the process of deformation which appears and disappears simultaneously with the application and removal of external force or stress.

It has been observed that whenever a stress of low magnitude is applied to a piece of metal, it causes displacement of atoms from their original positions. But on the removal of stress, the atoms spring back and occupy their original positions. In elastic deformation, the tensile strain is due to a slight elongation of the unit cell in the direction of the tensile load Similarly, the compressive strain is due to a slight contraction of the unit cell in the direction of the compressive load.

Fig. 3.12 shows the affect of tensile and compressive loads respectively on the atoms of the unit cell respectively. Fig. 3.12(*a*) represents the front face of the F.C.C. unit cell before loading. In this figure, the circles represent the atoms of metal specimen. Fig. 3.12(*b*) shows a slight elongation of the unit cell in the direction of tensile load. Similarly Fig. 3.12(*c*) represents the front face of the F.C.C. unit cell before loading. And Fig. 3.12(*d*) shows a slight contraction of the unit cell in the direction of compression load. The unit cell, after removal of the load, returns back to the normal positions as shown in Fig. 3.12(*a*) and (*c*).

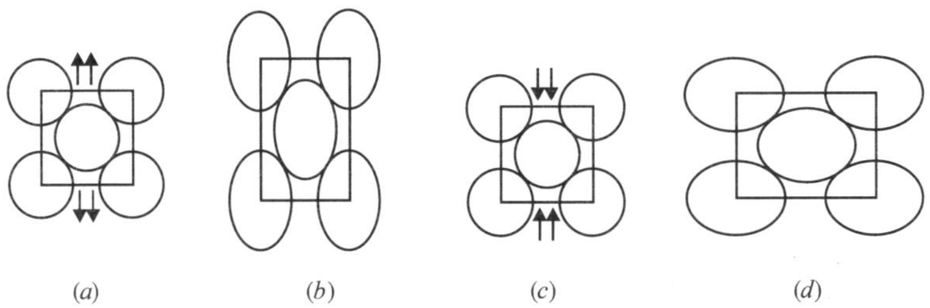

(*a*) (*b*) (*c*) (*d*)

Fig. 3.12. Elastic deformation.

It may be noted, from the above discussion, that any elongation or contraction, in the crystal structure of a metal, in one direction, due to uniaxial load produces an adjustment in dimensions at right angles to the force. This type of change in dimension (*i.e.,* at right angles to the applied load) is due to lateral strain. The ratio of lateral strain to the original strain is known as Poisson's Ratio.

It has been experimentally found that the strain is nearly proportional to the applied stress (or load). The ratio between stress and strain is known as Modulus of Elasticity or Young's Modulus. It is an important characteristic of a metal, which gives an idea about the amount of elasticity in it. It has been observed that the crystal structures of metals are also subjected to shearing loads. This produces a displacement of one plane of atoms, relative to the adjacent plane of atoms. This type of displacement in the crystal structure is known as shearing strain. The ratio between shearing stress and shearing strain is known as Shear Modulus or Modulus of Rigidity.

3·18 Plastic Deformation

The plastic deformation may be defined as the process of permanent deformation, which exists in a metal, even after the removal of the stress. It is due to this property, that the metals may be subjected to various operations like rolling, forging, drawing, spinning etc.

The plastic deformation may occur under the tensile, compressive or torsional stresses. It is a function of stress, temperature and rate of straining. There are two basic modes of plastic deformation, namely slip or gliding and twinning, as discussed below :

1. *Deformation by slip.* The term 'slip' may be defined as a shear deformation, which moves the atoms through many interatomic distances, relative to their initial positions. The deformation by slip is as shown in Fig. 3.13.

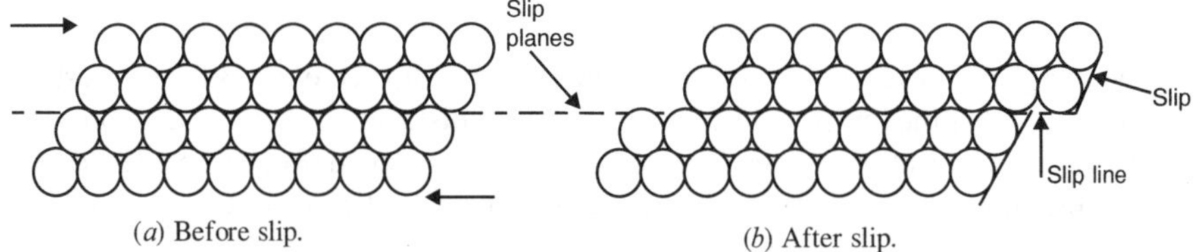

(*a*) Before slip. (*b*) After slip.

Fig. 3.13. Deformation by slip.

Fig. 3.13(*a*) shows the adjacent planes of a hypothetical crystal. A shearing stress, acting as indicated by the arrows, tends to move the atoms of the upper planes to the right. The movement of atoms or slip occurs only when the shear stress exceeds a critical value. The atoms move an integral number of atomic distances along the slip plane and a step is produced as shown in Fig. 3.13(*b*). The phenomenon of slip or glide may be considered as analogous to the distortion produced in a pile of cards when it is pushed from one side.

A careful examination of the surface of a deformed crystal, under the microscope, shows a group of parallel lines, which correspond to a step on the surface. These are called *slip lines*. The slip occurs most readily in specific directions on certain crystallographic planes, Generally the *slip plane* is the plane of greatest atomic density and the *slip direction* is the closest-packed direction within the slip plane. Since the planes of greatest atomic density are also the most widely spaced planes, in the crystal structure, therefore, the resistance to slip is generally less for these planes than for any other set of planes. The slip plane together with the slip direction is called the *slip system*.

It will be interesting to know that number of slip systems in F.C.C. metals are 12, in B.C.C. metals, 48 and in H.C.P. metals only 3. Certain metals show additional slip systems with increased temperature. However, it may be noted that in all cases, the slip direction remains the same, while the slip plane changes with the temperature.

Fig. 3.14. Mechanism of slip.

The simplified mechanism of slip is shown in Fig. 3.14. In this case, the slip is shown to occur by the translation of one plane of atoms over another. Fig. 3.14(*a*) shows the position of atoms in two adjacent planes of a hypothetical crystal before slip. Fig. 3.14.(*b*) shows the movement of atoms of the upper plane towards right on the application of shear stress. Similarly, Fig. 3.14. (*c*) shows the new position of atoms after the slip has taken place.

The value of shear stress required for such a movement of atoms in a perfect lattice (called theoretical shear strength) is approximately equal to the value of shear modulus divided by six (*i.e.*, *G*/6). However, in actual practice, the value of shear stress required to produce a slip is found to be at least 100 times smaller than the theoretical value. It means that a mechanism, responsible for slip, is not due to the bodily shearing of planes of atoms but due to some other cause.

The experimental evidence shows that the mechanism of slip is actually due to the movement of dislocations in the crystal lattice. The shear stress required for producing a slip due to the movement of dislocations is a small fraction of the theoretical value (*i.e., G*/6) and it matches the observed shear strengths of metals. Since the mechanism of slip requires the growth and movement of dislocation line, therefore, energy is required for this purpose. The energy of a dislocation line is given by the relation,

$$E \propto l.G.\, b^2$$

where l = Length of dislocation line,

G = Shear modulus, and

b = Unit slip vector or *Burgers vector.

The energy required will be minimum, when the unit slip vector (b) is the shortest one and the shear modulus (G) has the lowest value. It means that the dislocations having the shortest slip vector (b) are the easiest dislocations to generate and expand for plastic deformation. The directions in a metal, with the shortest slip vector, will be the directions with the greatest linear density of atoms.

---

* See Art. 3.20.

Similarly, the lowest value of shear modulus (G) corresponds to the planes, which are widely spaced and hence have the greatest planar density of atoms.

2. *Deformation by twinning.* The term 'twinning' may be defined as the plastic deformation, which takes place along two planes due to a set of forces applied on given metal piece. The process of twinning is shown in Fig. 3.15.

Fig. 3.15(a) shows the circles, which indicate the arrangement of atoms before twinning. Fig. 3.15(b) shows the arrangement of atoms after twinning along the planes AB and CD. It may be noted that the process of deformation between the two planes AB and CD is similar to that of slip. Whereas the arrangement of atoms on either side of the twinning planes (*i.e.*, towards left of AB and right of CD) remains unaffected.

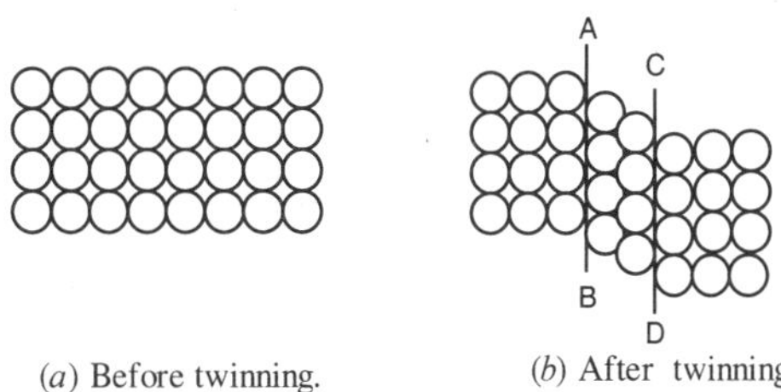

(a) Before twinning. (b) After twinning

Fig. 3.15. Deformation by twinning.

It has been observed that a metal, usually, deforms by twinning only if it is unable to slip. Moreover, the deformation produced by twinning is small. But it places the slip planes in more favourable orientation causing the deformation to take place through slip.

It has been observed that the twinning may be produced either by mechanical deformation or annealing. The twins produced due to mechanical deformation, are known as *mechanical twins*. These are produced in B.C.C. (*i.e.,* body centred cubic) and H.C.P. (*i.e.*, hexagonal close packed) metals unders the conditions of shock loading at room temperature. The twins produced due to annealing of cold work metal are known as *annealing twins*. These are produced in certain F.C.C. (*i.e.,* face centred cubic) metals like copper, silver etc.

Like slipping the twinning also occurs along certain crystallographic planes and directions. These are known as *twin planes* and *twin directions* respectively. In twinning process the movement of atoms is only a fraction of interatomic distance. Fig. 3.16 shows the dark circles, which indicate the arrangement of atoms. The lines AB and CD represent the planes of symmetry, from where the twinning starts and ends respectively. These planes are known as twinning planes.

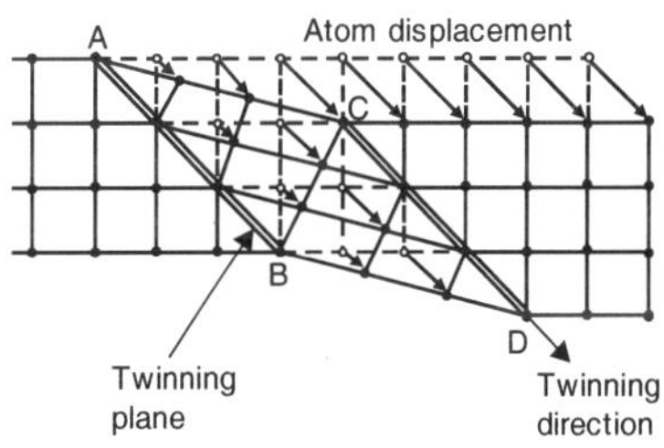

Fig. 3.16. Mechanism of twinning.

It has been observed that the crystals twin about the twinning planes. The atoms in the regions to the left of the twinning plane AB and right of the twinning plane CD remain undisturbed. Whereas in the twinned region, each atom moves by a distance proportional to its distance from the twinning plane AB. The blank circles indicate the new position of the atoms. The twinning occurs due to the growth and movement of dislocations in the crystal lattice.

3·19 Dislocations

The term 'dislocation' may be defined as a linear disturbance of the atomic arrangement in a crystal. It may occur in metallic crystals during their growth from its molten or vapour state. They may also be produced during slipping of the atomic planes over the others. The dislocation can move very easily on the slip planes through the crystals. This movement may be prevented by the presence of boundaries. It has been observed that the dislocations pile up at the grain boundaries, which act as barriers or obstacles for their further movement. The dislocations have a tendency to multiply during the process of deformation. It may be noted that the presence of dislocations, in a metal crystal, reduces its strength even lower than its theoretical value.

The two main kinds of dislocations are edge dislocation and screw dislocation, as discussed below :

1. *Edge dislocation.* It is the plastic movement of atoms, which start within the crystal and their effects can be seen on the edge of the crystal.

Fig. 3.17(*a*) shows the circles, which indicate the arrangement of atoms. In this figure, the dislocation line is shown by the atoms 2-6-7 and slip plane by X-Y. Fig. 3.17 (*b*) and (*c*) show the process of dislocation and Fig. 3.17.(*d*) the arrangement after dislocation.

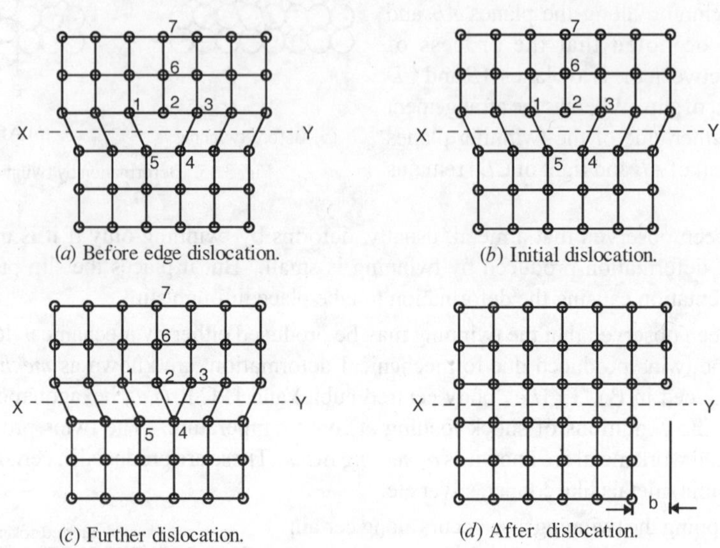

Fig. 3.17. Edge dislocation.

It may be noted that due to shear stress, the atom '2' moves towards the atom '3' and away from '1'. At the same time, the movement of atoms below the slip plane is such that the separation between atoms 1-5 decreases; whereas the distance between atoms 3-4 increases as shown in Fig. 3.17(*b*). On further increasing the shear stress, the atom '2' continues to move towards right, until it bonds with atom '4' as shown in Fig. 3.17(*c*). It may be noted that at this stage, the dislocation has moved to the right side by one atomic distance. This procedure is repeated, until the dislocation has moved to extreme right corner giving a step of height '*b*' as shown in Fig. 3.17(*d*). From the above discussion, we see that in an edge dislocation, the movement of dislocation takes place in a direction perpendicular to the slip plane. It has been observed that an edge dislocation requires much lower yield stress to produce deformation than that required for slipping.

2. *Screw dislocation.* It is the plastic movement of atoms, within the crystal, along two separate planes perpendicular to each other. The effect of screw dislocation can be seen on the sides of the crystal. It has been observed that the effect on the crystal appears like that of a screw or a helical surface.

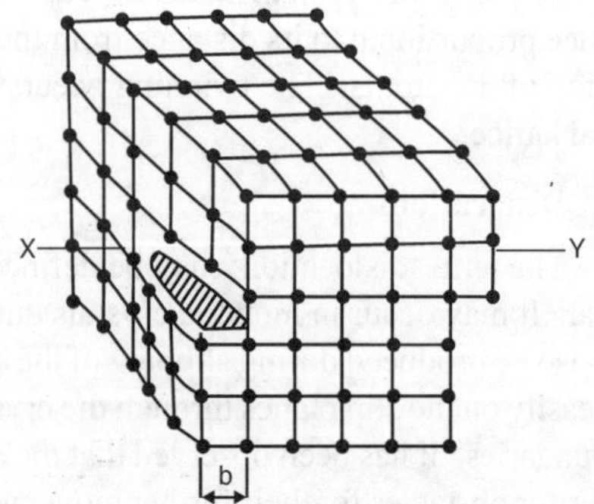

Fig. 3.18. Screw dislocation in a simple cubic structure

Fig. 3.18 shows the screw dislocation in a simple cubic crystal. In this figure, the dark circles indicate the arrangement of atoms in cubic crystal and *X-Y* represents slip plane. It will be interesting to know that in screw dislocation, the movement of dislocation takes places in a direction parallel to the slip plane.

3·20 Burgers Vector

It is also called as slip vector. The Burgers vector (*b*) is the vector, which defines the magnitude and direction of slip. In other words, it indicates how much above the slip plane and in what direction the lattice appears to have been shifted with respect to the lattice below the slip plane. The Burgers vector is an important property of a dislocation, because if the Burgers vector and the orientation of a dislocation line are known, the dislocation is completely described.

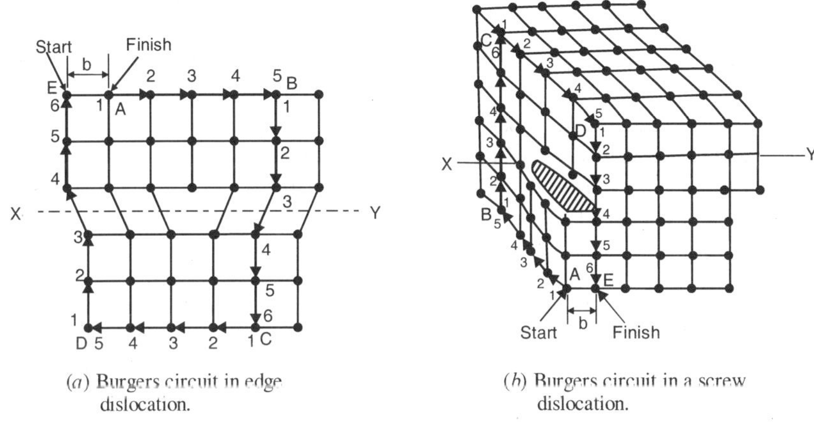

(*a*) Burgers circuit in edge dislocation.

(*b*) Burgers circuit in a screw dislocation.

Fig. 3.19. Burgers vector.

A convenient way of defining the Burgers vector of a dislocation is by means of a Burgers circuit. The Burgers circuit may be obtained by considering atomic arrangement around an edge dislocation or screw dislocation. Starting from a lattice point, trace a path from atom to atom, an equal distance in each direction, always in the direction of one of the vectors of the unit cell. If the region enclosed by the path does not contain a dislocation, the Burgers circuit will close. However, if the path encloses a dislocation, the Burgers circuit will not close. The closure failure of the Burgers circuit is the Burger's vector (*b*).

Fig. 3.19 (*a*) and (*b*) show the Burgers circuit in an edge dislocation and screw dislocation respectively. The Burgers circuit is the path traced by counting equal number of atoms in all the four directions. In Fig. 3.19(*a*), let us start the Burger circuit from one of the corner, say *A*. Let the circuit cover five unit cells and reach B. Similarly, from B, let the circuit cover another six units and reach *C*. Again from *C*, the circuit cover five units (*i.e.,* equal number of units as that of *AB*) and reach *D*. Finally, let the circuit cover six units. (*i.e.,* equal number of units as that of *BC*) and reach *E*. The distance *EA* is known as Burgers vector. It will be interesting to know that if the region enclosed by the circuit, does not contain any dislocation, then the Burgers circuit must close.

It will be interesting to know that an edge dislocation lies perpendicular to its Burgers vector. On the other hand, a screw dislocation lies parallel to its Burgers vector. Similarly, an edge dislocation moves in the direction of Burgers vector, while a screw dislocation moves in a direction perpendicular to the Burgers vector.

3·21 Deformation of Polycrystalline Materials

As a matter of fact, the deformation of a polycrystalline material is different than that of single crystalline material. It is due to the reason that the deformation of a polycrystalline material takes place due to movement of dislocations. It has been observed that these dislocations move under the action of applied load and cause slip. If the movement of dislocations does not take place, the stress required to produce deformation will increase. This phenomenon may be better understood from the following example :

In a crystalline material, the grains have random orientations. On the application of a force, the grains which have their slip planes parallel to the direction of a force, deform more easily. But their deformation is restricted by the adjacent grains having different orientation. As a result of this, the grains develop a complex state of stresses.

The deformation of a polycrystalline material is also affected by the pressure of grain boundaries. These areas act as physical barriers to the movement of dislocations. This results in piling up of dislocations at grain boundaries, which offer resistance to slipping. Thus a high stress is required to cause plastic deformation.

It will be interesting to know that greater the number of grain boundaries in a polycrystalline material, higher will be the resistance offered to the dislocation movement. Thus greater will be the stress required to produce plastic deformation. It is due to this fact, that a fine-grained material possess high tensile strength and better mechanical properties as compared to a coarse-grained material.

3·22 Changes in Properties of Metals due to Deformation

Following are some important changes in properties of polycrystalline metals due to plastic deformation :

1. *Work hardening.* It is also called *strain hardening.* The term 'work hardening' may be defined as a process of deforming a metal at room temperature to improve its hardness, tensile and fatigue strengths etc. We know that whenever a metallic piece is subjected to a load, beyond its elastic limit, some plastic deformation takes place. As a matter of fact, this deformation takes place due to slipping of the atomic planes. It has been observed that the slipping takes place more easily in those planes, which have favourable orientation *i.e.,* the planes which are parallel to the direction of the load. As the deformation proceeds, the availability of slip planes decreases. This decreases the resistance of metal against plastic deformation. And the work piece is known to be work-hardened or strain-hardened.

It has been observed that the work hardening produces internal stresses in metal. These stresses are produced due to :

1. Piling up of dislocations at grain boundaries.
2. Distortion of grains due to loading.

It will be interesting to know that the work hardening improves the hardness, yield point and strength of metals like iron, copper, aluminium, nickel etc. But it decreases the ductility and electrical conductivity of the metals.

2. *Cold working.* The term 'cold working' may be defined as the process of deforming plastically, a piece of metal below its recrystallisation temperature. The recrystallisation of metal, usually, takes place within the temperature range of 0.3 to 0.5 times the melting temperature of metal. Some of the important processes of cold working are drawing, squeezing, shearing, bending, extruding etc.

It has been observed that the process of cold working changes the various properties of metals such as hardness, ductility, tensile and fatigue strength etc. It increases the hardness, tensile and fatigue strength and the electrical resistance of a metal. But it decreases the ductility and creep resistance. The proportionality limit can also be increased with heavy cold working. The metals such as phosphor bronze and low carbon steel possess high proportionality limit, and are used as spring materials.

It has also been observed that during cold working, a certain amount of work done on the metal is stored internally in the form of strain energy. This energy produces internal stresses in a cold worked metal. It is due to this fact, that these metals are liable to crack when subjected to heat treatment. The internal stresses can be removed by particular process of heat treatment called *annealing.

3. *Preferred orientation.* When a polycrystalline specimen of any pure metal or an alloy is deformed, the slip is expected to occur on the same slip system in every crystal present in the specimen. Therefore, independent of their original orientation, every crystal in a metal being extended in a specific direction should undergo similar rotations relative to that direction. If the deformation is extensive, then every crystal present in the specimen should ultimately approach the same orientation relative to the axis of the principal strain. The above phenomenon occurs in actual practice and the orientation, which takes place is known as *preferred orientation* or *texture*. The preferred orientation is very important with regard to the physical and mechanical properties of the section as a whole.

The preferred orientation, resulting from deformation, is strongly dependent on the slip and twinning systems available for deformation. But it is not affected by processing variables such as die angle, roll diameter, roll speed etc. However, the direction of flow is the most important process variable.

It has been observed that preferred orientation in metals or alloys results in an *anisotropy* in mechanical properties. The anisotropy means that the material properties vary in different directions. But when these crystals are combined in a random fashion into a polycrystalline aggregate, the mechanical properties of the aggregate tend to be isotropic. However, the crystal alignment, which accounts for the preferred orientation, again introduces anisotropy in mechanical properties. As a result of this, different mechanical properties in different directions can result in an uneven response of the metal during forming and fabrication operations. The preferred orientation is of great advantage in magnetic applications, where it is used to produce grain-oriented silicon-iron transformer sheets.

3·23 Recovery, Recrystallisation and Grain Growth

We have discussed in the previous articles, the mechanism of plastic deformation through slip and twinning. We have also discussed that during cold working, the various properties of the metal changes and a certain amount of work done on the metal is stored internally in the form of strain energy. This energy produces internal stresses in a cold worked metal which leads to cracking of metals. This renders the cold worked metals thermodynamically unstable. It is, therefore, necessary to bring back the metal approximately to the same state as it was before deformation. In order to achieve this (*i.e.*, to relieve the metal from internal stresses), a particular process of heat treatment called annealing is used. In this process, the deformed metal is heated to a temperature below the melting point. In doing so, the metal loses its stored energy and comes back to almost in the same equilibrium state as it was prior to deformation. It has been observed that metal loses its stored energy in the three stages (*i.e.*, recovery, recrystallisation and grain growth) as discussed below :

1. *Recovery.* The term 'recovery' may be defined as the process of removing internal stresses, in a metal, by heating it to a relatively low temperature. This temperature is, usually, below the melting point.

It has been observed that the recovery process does not affect the grain structure. But it removes the internal stresses only. Moreover, the recovery process does not affect the hardness and strength, but it increases the ductility of the metal. As a result of cold working, the dislocations pile up at grain boundaries. During the process of recovery, these dislocations start reducing and rearrange

---

* For details, please refer Chapter 7 on Heat Treatment of Metals and Alloys.

themselves. They do so, through mechanism known as *polygonization.* In this mechanism, the dislocation climb out of their slip planes and rearrange themselves in a lower energy configuration.

2. *Recrystallisation.* The term 'recrystallisation' may be defined as the process of forming strain-free new grains, in a metal, by heating it to a temperature known as recrystallisation temperature. It may be noted that the recrystallisation temperature is, usually, the temperature at which about 50% of the cold worked metal recrystallises in one hour. It has been observed that the formation of new grains in a recrystallisation takes place through the following three processes :

1. Nucleation. 2. Primary grain growth. 3. Secondary grain growth.

It will be interesting to know, that during nucleation, small strain-free nuclei are formed at the grain boundaries where the deformation is very high. These nuclei grow into strain-free grains during primary grain growth. These grains meet each other and replace the old grains by the new ones. During the secondary grain-growth, these new grains grow at the expense of others. At the end of this stage, the grains are usually, of very small size but of the same shape. The factors, which affect recrystallisation process, are time of heating, temperature prior to deformation, impurity and alloy addition etc.

It has been observed that the increase in the time of heating or the annealing reduces the recrystallisation temperature. A certain amount of deformation is required before the crystallisation may take place. The addition of impurity or alloy increases the recrystallisation temperature. It may be noted that the recrystallisation does not produce new structures. But it produces strain-free new grain and complete elimination of internal stresses. It results in a sharp decrease in hardness and strength as well as increase in ductility.

3. *Grain growth.* The term 'grain growth' may be defined as the process of forming strain-free grains larger in size in metal by heating it to a temperature above that of recrystallisation. It may be noted that the recrystallisation produces strain-free new grains. These grains are of smaller size, but of equal shape. When the temperature is increased above that of recrystallisation, these grains grow in size. The grain growth takes place even during recrystallisation. But the growth rate is slow and becomes rapid with the increase of temperature. The grain growth takes place due to the combination of individual grains, thereby reducing their boundary area. As a result of this, the total energy decreases and the grains become stable. The factors, which affect the growth rate, are time of heating, temperature, degree of cold work and addition of impurities. It has been observed that the grain growth results in the decrease of hardness and strength as well as increase in ductility.

The schematic diagram indicating recovery, recrystallisation and grain growth and the variation in the chief mechanical properties in each region, is shown in Fig. 3.20.

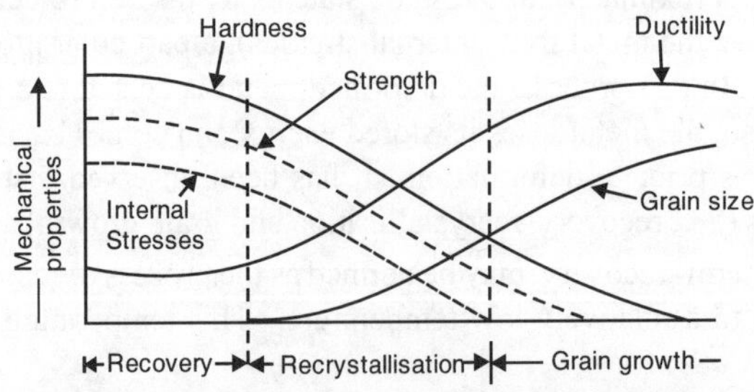

Fig. 3.20. Recovery, recrystallisation and grain growth and variation in chief mechanical properties.

3·24 Fracture of Metals

The fracture of a material may be defined as its fragmentation or separation, under the action of an external force, into two or more parts. It may occur as a sudden breaking up of a material either as the end result of extensive plastic deformation or as a result of fatigue in a part of the material. In actual practice, the fracture of materials is a serious problem, and should always be avoided. It has been observed that any type of fracture consists of two main components, *i.e.*, crack initiation and crack propagation. The fractures may be broadly divided into the following four types :

1. *Brittle fracture.* The brittle fracture may be defined as a fracture which takes place by the rapid propagation of crack with a negligible plastic deformation. It has been observed that in amorphous materials, the fracture is completely brittle. But in crystalline materials, it occurs after a small deformation.

It may be noted that in crystalline materials, the fracture takes place normal to the specific *crystallographic planes* (*i.e.*, planes with lesser interatomic bonds). These planes are called *cleavage planes* and the corresponding fracture is known as *cleavage fracture*. The brittle fracture, generally, occurs in body centred cubic (B.C.C.) and hexagonal close packed (H.C.P.) single crystals at very low temperatures. It may also occur in crystals of graphite, talc and mica. In polycrystalline materials, the fracture takes place along the grain boundaries because of the tendency of brittle fracture increase with the decrease in temperature and higher rate of straining.

2. *Ductile fracture.* The ductile fracture may be defined as the fracture which takes place by a slow propagation of crack with appreciable plastic deformation. It has been observed that the ductile fracture, generally, takes place in metals which do not work-harden much.

It has been seen that the ductile fracture is the end result of extensive plastic deformation of a specimen in a tensile test. If we note the deformation of the specimen and the corresponding force (*i.e.,* load) and plot a graph with strain along the horizontal axis and stress along the vertical axis, we shall obtain a curve as shown in Fig. 3.21.

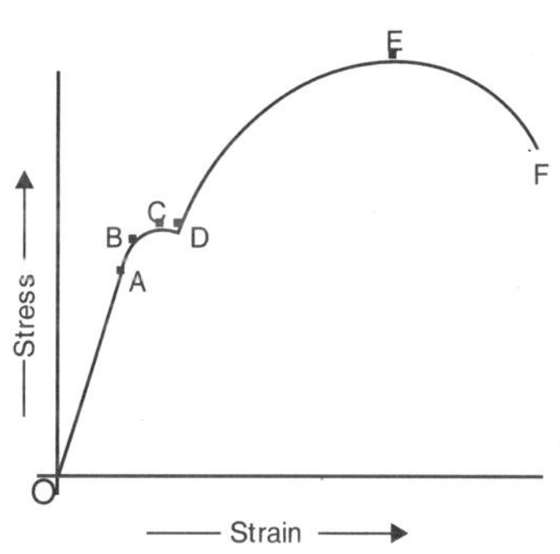

Fig. 3.21. Ductile fracture.

The curve *OA* indicates that the ratio of stress to the strain is constant. The curve *AB* indicates that the ratio of stress to the strain is not constant. The point '*B*' is known as elastic point. The curve *BC* indicates that the strain increases more quickly than the stress (in comparison to *OA* or *OB*). The point '*C*' is known as yield point. The curve *CD* indicates that strain increases without any increase in stress. The curve *DE* indicates that higher stresses are required for higher values of strain. The stress, corresponding to the point '*E*', is known as ultimate tensile stress. As the stress is increased beyond this point, the fracture takes place.

We have already discussed that when the tensile stress, across the specimen, is increased, beyond the elastic limit, there is a uniform reduction in its cross-sectional area. The process continues till the ultimate tensile strength is reached. If the tensile stress is increased beyond this value, a neck is formed somewhere near the middle of the specimen. It has been observed that during the formation of neck, small cracks are formed due to the combination of dislocations which were formed during the manufacture of the specimen as shown in Fig. 3.22(*a*).

It has been observed that the continuation of the plastic deformation produces cavities or bigger cracks in the specimen as shown in Fig. 3.22(*b*). These cavities grow outward and form a central crack as shown in Fig. 3.22(*c*). On further increase of stress, the crack propagates on the surface of the specimen, which results in a cup and cone type of fracture as shown in Fig. 3.22(*d*).

It may be noted that a ductile fracture occurs by a slow propagation of crack (*i.e.*, slow tearing of the metal) with a considerable amount of energy expenditure. The appearance of the fractured surface is rough and dull. Moreover, a considerable amount of slip is apparent. It may be noted that in a ductile fracture, the crack continues to propagate so long as the material is strained. If we stop the straining of material, the development of crack also stops. In actual practice, the ductile fracture is not of much significance.

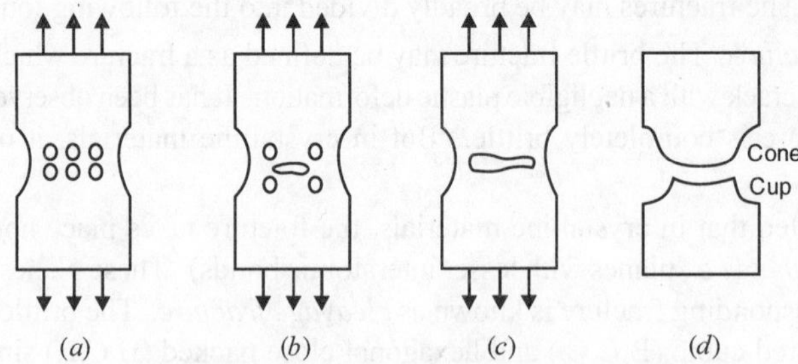

Fig. 3.22. Mechanism of ductile fracture.

3. *Creep fracture.* The creep fracture may be defined as the fracture, which takes place due to excessive creeping of materials, under steady loading. It is always exhibited in metals like iron, nickel, copper and their alloys at higher temperatures. But some metals like zinc, tin, lead and their alloys also creep at room temperature. It has been observed that the tendency of creep fracture increases with the increase in temperature, higher rate of straining etc. It has been observed that the creep resistance may be increased by the addition of certain elements such as cobalt, nickel, manganese, tungsten etc.

The creep fracture may take place due to the following two processes :

1. Sliding of grain boundaries; and
2. Movement of dislocations, from one slip to another, by climbing.

It has been observed that at moderate stresses and temperatures, the creep occurs due to the sliding of grain boundaries only. But at higher stresses and temperatures, it occurs due to the movement of dislocations by climbing. It has also been observed that a creep fracture, which takes place due to continuation of creep, is known as *intercrystalline fracture,* as it takes place along the grain boundaries.

It may be noted that some fractures initiate along the grain boundaries, which act as points of high stress concentration. In such cases, the crack generally initiates along the lines where three or more boundaries meet or intersect. But sometimes, the fractures initiate due to the growth of voids along grain boundaries. With the passage of time, these voids grow and combine with each other to form cracks. This type of fracture, usually, occurs when small stresses are applied for a longer period.

The factors which affect creep fracture are grain size, strain hardening heat treatment, alloying etc. But by controlling all these factors, the tendency to creep fracture may be greatly avoided.

4. *Fatigue fracture.* The fatigue fracture may be defined as the fracture which takes place under repeatedly applied fatigue stresses. It occurs at stresses well below the tensile strength of the materials. The fatigue stress is always exhibited in some parts of the machine parts such as axles, shafts, crank shafts etc.

The fatigue fracture takes place due to the initiation of microcracks at the surface of the materials. It results in to and fro motion of dislocations near the surface. The microcracks act as the points of stress concentration. This excessive stress helps the propagation of crack with every cycle of stress application. The extent, to which a crack may propagate, depends upon the brittleness of the material. In brittle materials, the crack first grows to a critical size and then propagates rapidly through the material. But in ductile materials, the crack grows slowly until the remaining cross-sectional area can not support the applied stress. The fracture then takes place, rapidly, in ductile manner.

It has been observed that the tendency of fatigue fracture increases with the increase in temperature, higher rate of straining etc. The fatigue resistance of a material may be increased by a careful design of the machine parts, so that the points of stress concentraction are avoided.

3·25 Metal Alloys

The pure metals are rarely used in engineering work because of their physical and mechanical properties. In order to make these metals suitable for engineering use, some other materials are added to them to form an alloy. Thus, an alloy is a substance which possesses metallic properties and is composed of two or more elements, of which atleast one is metal. The metal which is present in the alloy in large proportions is known as *base metal,* whereas all other elements present in the form of metal or non-metal are called *alloying elements.* The alloying elements are added intentionally to the base metal in order to obtain certain desired properties which are not found in the base metal. The undesirable elements which are either added accidently in the process of alloying or are carried away by the base metal due to imperfect smelting or refining process are known as *impurities.*

The addition of alloying elements to the base metal either in small proportions or high proportions may result in significant changes in the properties of the base metal. The resulting structure obtained after adding the alloying elements to the base metal determines the ultimate properties of the alloy as a whole. The type and extent of change of properties depend upon whether the alloying elements in an alloy are completely soluble or insoluble in the base metal or whether they form a completely new base with the base metal.

It may be noted that each constituent of an alloy is known as a *component.* A pure metal consists of a single component, whereas an alloy may be binary (or two component), ternary (or three component), etc.

3·26 Solidification of Alloys and Cooling Curves

The graphical representation of variation in temperature with time, when the material is cooled down from its liquid state, is known as *cooling curve.* It helps in studying the important structural changes that takes place during cooling or solidification of metals and alloys. The cooling curves are very useful in drawing the equilibrium or phase diagrams. The three types of typical cooling curves, as shown in Fig. 3.23, are as follows :

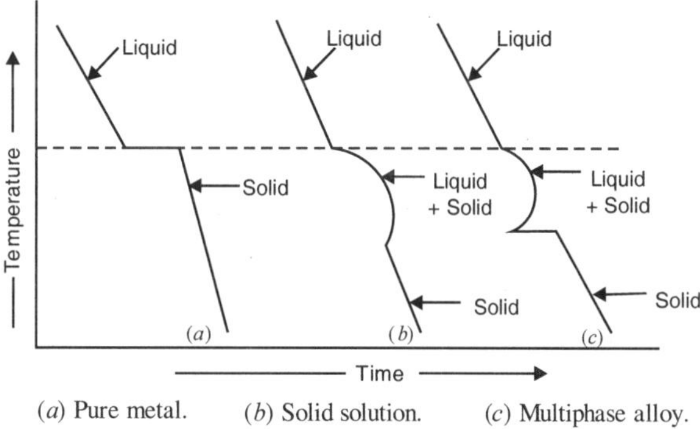

(*a*) Pure metal. (*b*) Solid solution. (*c*) Multiphase alloy.

Fig. 3.23. Cooling curves for pure metals and alloys.

1. The cooling curve for the pure metal is shown in Fig. 3.23(*a*),
2. The cooling curve for the binary solid solution is shown in Fig. 3.23 (*b*), and
3. The cooling curve for the multiphase alloy (also called binary eutectic system) is shown in Fig. 3.23(*c*).

3·27 Types of Alloys

We have seen in the last article of cooling curves that as a result of cooling, the final form of the metal or alloy is that of a solid. A solidified alloy may exist in the following three forms :

1. As a *solid solution* in which the atoms of the alloying elements may be selectively soluble in the atoms of base metal.

2. As an *intermetallic compound* in which the atoms of the alloying elements may form a compound with the atoms of base metal.

3. As a *mechanical mixture* in which the atoms of the alloying elements may be relatively insoluble in the atoms of the base metal.

3·28 Solid Solutions

We have already discussed that when the atoms of the alloying elements and the atoms of the base metal dissolve completely in each other, then the resulting mixture is known as *solid solution.* The metal which is present in a smaller quantity in a solid solution is called *solute*, while the metal which is present in a greater quantity is called *solvent* or the host structure. For example, in a solid solution of brass, 60 percent are copper atoms and the remaining 40 percent are zinc atoms. Thus, the copper is a solvent and zinc is a solute. The solid solutions form more readily when the solvent and solute atoms have smaller sizes and comparable electron structures. The solid solutions are of the following two types :

1. Interstitial solid solution, and 2. Substitutional solid solution.

These two types of solid solutions are discussed, in detail, in the following pages.

3·29 Interstitial Solid Solution

A solid solution in which solute atoms are located in the interstices (or voids) of the solvent lattice is called *interstitial solid solution.* Fig. 3.24 illustrates the atomic arrangement in an interstitial solid solution. The solid solvent of carbon, nitrogen and hydrogen, in iron, are some of the common examples of interstitial solid solutions. Interstitial solid solution can form only when the solute atom is small enough to fit into the space between the solvent atoms. There can be appreciable solid solubility only if the radii of the solvent atoms is about 60% or less of the radii of the solvent atoms. The radii of the atoms of commercial alloys is in the range of 0.12 nm to 0.16 nm. Therefore, the only atoms small enough to form interstitial solid solutions are the hydrogen, oxygen, nitrogen, carbon and boron.

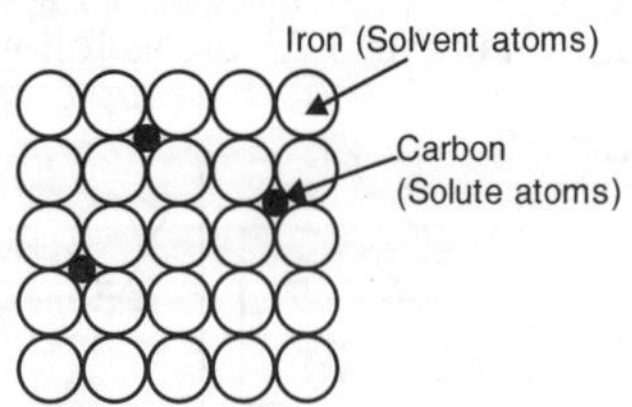

Fig. 3.24. Interstitial solid solution.

3·30 Substitutional Solid Solution

A solid solution in which solute atoms replace some solvent atoms so that they lie at normal atom sites of the crystal structure are called as *substitutional solid solution.* Fig. 3.25 illustrates the atomic arrangement in a substitutional solid solution of brass. The solution is formed when zinc atoms substitute for copper atoms in the crystal structure. This type of solid solution is very common among metal systems. The solution of copper and nickel to form cupronickel is another example. In this solution, any fraction of the atoms in the original copper structure may be replaced by nickel. Thus copper-nickel, solid solutions may range from no nickel and 100% copper to 100 percent nickel and no copper. Thus there is no solubility limit of either nickel-in-copper or copper-in-nickel solid solution. The copper-nickel alloys have FCC structure. On the other hand, there is definite limit to the amount of tin that may replace copper to form bronze and retain the FCC structure of copper. The substitution of solute atoms, in a solvent, may take place in any one of the following two ways :

1. *Random substitution.* When the solute atoms replace the solvent atoms at random, it is called random (or disordered) substitution. The solid solution, so formed, is called *random substitutional solid solution.* Fig. 3.25(*a*) shows the random substitution of zinc atoms for copper in a crystal structure.

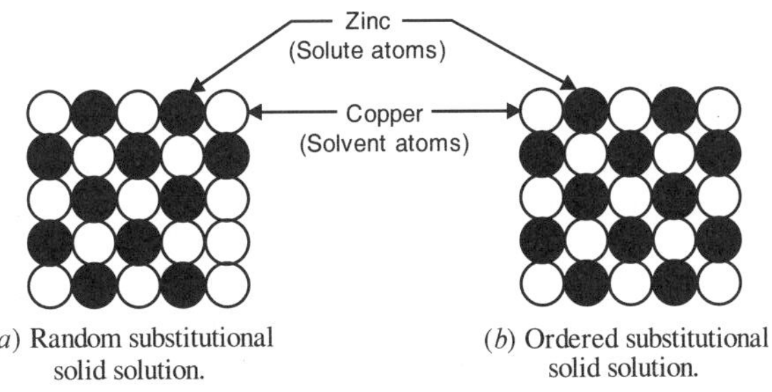

(*a*) Random substitutional solid solution.

(*b*) Ordered substitutional solid solution.

Fig. 3.25. Substitutional solid solution.

2. *Ordered substitution.* When the solute atoms replace the solvent atoms in order, it is called an ordered substitution. The solid solution, so formed is, called an *ordered substitutional solid solution.* Fig 3.25(*b*) shows an ordered substitution of zinc atoms for copper in crystal structure.

It may be noted that random solution structure is more common at high temperatures while ordered solution structure is more common at low temperatures. It is due to the fact that at high temperatures, the greater thermal agitation tends to destroy the orderly arrangement.

For example, the solid solution of copper and zinc mixed in equal proportions (*i.e.*, 50 percent copper and 50 percent zinc) forms a random body centred cubic (BCC) structure above 470°C. However below 470°C, the alloy becomes ordered.

3·31 Intermetallic Compounds

When the difference in the *electronegativity of the elements forming an alloy is very large, there is a greater tendency to form intermetallic compounds (also called intermediate compounds or chemical compounds). These compounds have crystal structures and properties different from either of the pure elements. The term intermetallic is applicable when both the elements are metallic. For example, an alloy of copper and zinc mixed in equal proportion (*i.e.*, 50 percent copper and 50 percent zinc) is an intermetallic compound that forms a system of limited solubility. The resulting structure of the intermetallic compounds is body centred cubic (BCC) which is different from that of either copper which is face centred cubic (FCC) or zinc which is hexagonal close packed (HCP) crystal lattice.

However, many intermetallic compounds of engineering importance contain a non-metallic element, especially carbides and nitrides respectively. A common example of this is iron-carbide (Fe_3C). The intermetallic compounds are of the following three types :

1. *Valency compounds.* These are the compounds formed by mixing of electropositive elements (*e.g.*, Na, K, Al, Mg etc.) with electronegative elements (*e.g.*, Sn, Pb, As, Sb etc.). The valency compounds are usually hard, brittle and poor conductors. They have sharp melting points which may be above the melting point of either pure element.

2. *Electron compounds.* These compounds do not obey the normal valency rules. These compounds are formed in those metals which have similar electrochemical properties and a favourable size factor, but have different numbers of valence electrons. These compounds are essentially metallic in properties and are not of exact fixed composition. The crystal structure of these compounds

* The tendency of an atom to attract electrons to itself during the bond formation with other atoms is called *electronegativity* of that atom.

depends upon the ratio of total valence electrons to the total number of atoms.

3. *Interstitial compounds.* These are the compounds which form between the elements (which can go into interstitial solid solution) and metals in which they have little solubility. If the size factor is favourable, the structure of these compounds is simple. However, in some cases such as iron carbide (Fe_3C), the structure is complex.

3·32 Mechanical Mixtures

In some cases, when the liquid solution of an alloy is cooled, the crystals of its constituents separate simultaneously from one another and form a close *mechanical mixture.* This is also known as *eutectic mixture.* In mechanical mixtures, the metals retain their same mechanical properties and structural identities.

3·33 Phase Diagrams

The phase diagrams are convenient graphic representation of changes in the state of alloys due to variation in temperature and concentration. These diagrams are also called *constitutional* or *equilibrium diagrams* because they enable the phase content of the alloy to be determined at any temperature and composition. They enable the phase transformation to be followed in heating or cooling the alloy under equilibrium conditions, *i.e.*, when all the processes in a given system are reversible. This means that changes occurring in a system as a result of processes proceeding in one direction are fully compensated by changes due to reversal of the process in the system.

The phase diagrams are powerful tools in the hands of scientists and engineers who design materials for specific applications as well as for others who anticipate the stability of specific materials while designing products for service environments. Following are some of the important applications or uses of the phase diagrams :

1. To predict what phases are in equilibrium for selected alloy compositions at desired temperatures.
2. To determine the chemical composition of each phase.
3. To calculate the quantity of each phase that is present.
4. To predict phase transformations which occur if the material is subjected to very slow heating or cooling.
5. To determine the microstructure as a result of phase transformation.

3·34 Classification of Phase Diagrams

The phase diagrams are classified on the basis of number of compounds present in the alloy system. These are of the following three types :

1. *Uniary phase diagrams.* These diagrams are meant for single-component systems, such as pure element or compound *i.e.*, iron, magnesium etc. In uniary phase diagrams, there is no composition variable. The only variables are pressure and temperature. Therefore, these phase diagrams are also known as pressure-temperature diagrams.

2. *Binary phase diagrams.* These diagrams are meant for two-component systems. The binary phase diagrams have all the three variables, *i.e.*, composition, pressure and temperature. Therefore, we need a three dimensional diagram to plot the variation in composition, pressure and temperature. This complexity is avoided by plotting the phase diagrams at constant pressure (usually the pressure is kept constant and equal to one atmosphere). This assumption is valid because most of the materials are used under constant pressure (*i.e.*, isobaric) conditions. Therefore, for a two-component system, we get a two-dimensional diagram (called binary phase diagram) of temperature versus composition. The binary phase diagrams are quite often referred to as *constitutional diagrams* or *equilibrium diagrams*

3. *Ternary phase diagrams.* These diagrams are meant for three-component system. The ternary phase diagrams have three independent variables. Two variables specify the composition and the third is temperature. Therefore, a three dimensional space model is necessary to show the phase present for any composition and temperature.

Note : Out of the above three phase diagrams, the binary phase diagrams are in maximum use because the most commonly used alloys are two-component alloys.

3·35 Important Terms Used in Phase Diagrams

The following important terms should be clearly understood in connection with the phase diagrams :

1. *System.* A system means the whole complex of phases of one or several components at different pressures and compositions. It may be composed of solids, liquids, gases or their combinations. It may have metals or non-metals separately or in any combination. The system is isolated from the surrounding in such a way that it is not affected by its surroundings. However, the system is subjected to some change in overall composition, temperature, pressure, total volume etc.

2. *Components.* The components are the substances (chemical elements or compounds) whose presence is necessary and sufficient to make up a system. For example, pure metals such as Cu, Ni, Pb, Sn etc., are all components. The pure metal constitutes a single-component system. Similarly, a combination of two metals such as Cu-Ni, Cu-Zn, Pb-Sn etc. constitute a two-component (or binary) system.

3. *Phase.* A phase is a physically and chemically homogeneous portion of a system, separated from the other portions by a surface, the interface. For example, a homogeneous liquid solution is a single phase system, a mixture of crystals of two types differing in composition and structure and separated by an interface comprise two phase systems.

4. *Phase rule.* All changes which take place in a system consisting of several phases, in accordance with external conditions (temperature and pressure), conform to the phase rule. The phase rule establishes the relationship between the number of degrees of freedom, the number of components and the number of phases. Mathematically,

$$F = C + N - P$$

where F = Number of degrees of freedom. It is the quantity of independent external or internal variable factors (temperature, pressure and concentration) which may be altered without causing the disappearance of a phase or the formation of a new phase in the system,

C = Number of components in the system,

P = Number of phases in equilibrium, and

N = Number of external factors (temperature and pressure).

In applying the phase rule to metal systems, the effects of pressure may be neglected leaving only one variable external factor *i.e.*, temperature. Since the degrees of freedom cannot be less than zero, therefore, the number of phases in a system cannot exceed the number of components plus one. Therefore, no more than three phases may be in equilibrium in a binary system, no more than four in a ternary system, etc. In cases when the maximum number of phases is in equilibrium, the number of degrees of freedom equals to zero. This is called *non-variant equilibrium.*

A system in non-variant equilibrium may exist only under entirely definite conditions, at a constant temperature and at a definite composition of all phases involved.

3·36 Allotropic Forms of Pure Iron

We know that pure substances may exist in more than one crystalline form. Each such crystalline form is stable over more or less well defined limits of temperature and pressure. This is known as *allotropy* or *polymorphism.* The pure iron exists in three allotropic forms *i.e.*, alpha (α), gamma (γ)

and delta (δ).

Fig. 3.26 shows the heating and cooling curves for pure iron, indicating the temperature ranges over which each of these crystallographic forms are stable at atmospheric pressure. From the heating curve, it is evident that from the room temperature to 910°C, pure iron has a body-centred cubic (BCC) structure and is called α-iron. The α-iron is ferromagnetic at room temperature. But on heating to 768°C (called curie point), the ferromagnetism disappears. However the crystal structure remains BCC. Non-magnetic α-iron is stable upto 910°C. Above 910°C, it is transformed into face-centred cubic (FCC) structure called γ-iron. Upon heating to 1404°C, the γ-iron is transferred back into the BCC structure called δ-iron. It is stable upto the melting point (1539°C) of pure iron. The BCC structure of δ-iron has a longer cube edge (*i.e.,* lattice parameter) than BCC structure of α-iron.

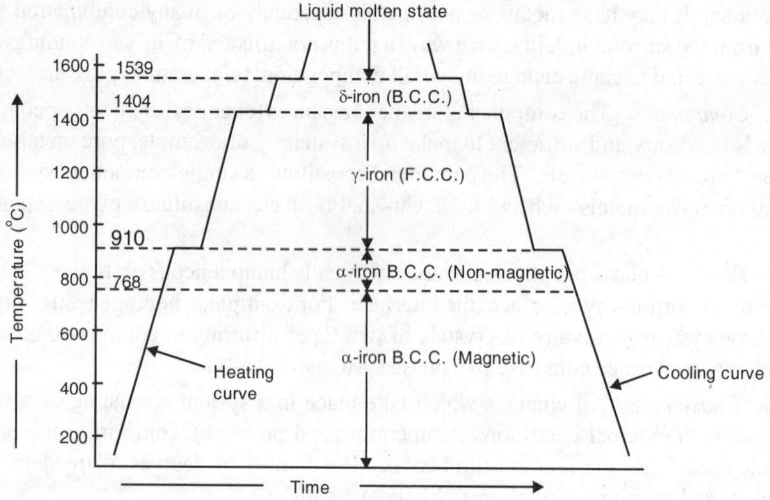

Fig. 3.26. Heating and cooling curves for pure iron.

Note : From the cooling curve, it is noticed that when the iron from the liquid state (above 1539°C) is cooled down to room temperature, the same changes, as discussed above, takes place in the reverse order.

3·37 Iron-Carbon Equilibrium Diagram

The iron-carbon (or iron carbide) equilibrium diagram is shown in Fig. 3.27. It concerns transformations that occur in alloys having compositions from pure iron to cementite (6.67 percent carbon). Any point in the diagram represents a definite alloy at definite temperature. The carbon contents are shown on the horizontal axis and temperature on the vertical axis. Whenever an alloy is heated or cooled so that a line on the diagram is crossed, a phase change occurs.

The point A (1539°C) on the diagram is the melting point of pure iron. The points N (1401°C) and G(910°C) correspond to the allotropic transformations of α-iron into γ-iron. The point E shows the solubility limit of carbon in γ-iron at 1130°C (1.7 percent). The point H represents the maximum solubility of carbon (0.10 percent) in δ-iron at 1490°C. All alloys represented by composition and temperatures in the region above ABCD are completely liquid. When liquid alloys containing upto 4.3 percent carbon are cooled to temperature on the curve ABC, solid crystals begin to form. ABC is called *liquidus* for these alloys.

At temperatures on the line AB, the crystals of the solid solution of carbon in α-iron (δ-solution) begin to precipitate from the liquid alloy. A chemical reaction proceeds along the line HJB at constant temperature of 1490°C between the crystals of δ-iron saturated with carbon (point H) and the liquid containing 0.5 percent carbon (point B). The result of this reaction is the formation of solid solution of carbon in γ-iron *i.e.,* austenite. The content of the austenite formed along the line HJB is 0.18 percent (point J). Since the maximum solubility of carbon in austenite is 1.7% at 1130°C (point E), therefore, the crystals of austenite with carbon content ranging from 0.18 to 1.7% separate the melt along the line BC. The complete solidification of these alloys proceeds along the line HJEC called *solidus.* The alloys containing 0.18 to 1.7% carbon become solid along the part of the HJE and those with carbon content from 1.7 to 4.3% are completely solid on the line EC (1130°C).

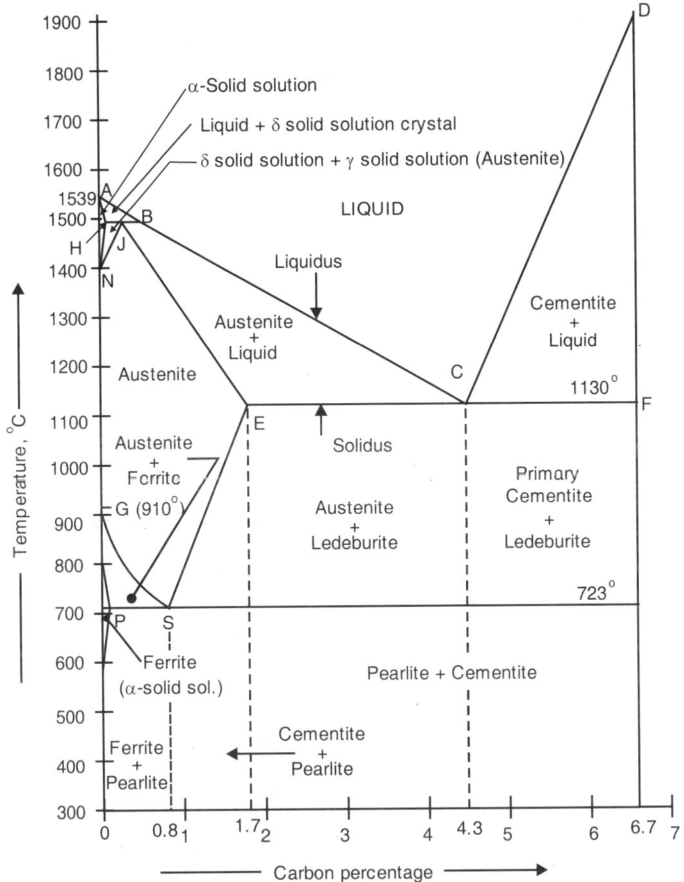

Fig. 3.27. Iron-carbon equilibrium diagram.

When liquid alloys containing more than 4.3 percent carbon are cooled to temperatures on the line CD, solid crystals or cementite (iron carbide, Fe_3C) begin to form. Hence the line CD is called liquidus of these alloys. When any alloy containing more than 4.3 percent carbon is cooled to temperature on the line CF (1130°C), it becomes completely solid. The alloy containing 4.3 percent carbon is called *eutectic alloy* (also called *ledeburite*) and it will therefore, solidify completely at C (1130°C) with the simultaneous formation of austenite and cementite (the solid eutectic mixture).

The region ABJN of the diagram (Fig. 3.27) in which δ-iron may exist is not very important in practice because temperatures between the melting point of iron and 1401°C are not employed in various mechanical working or heat treatment processes. Therefore, we shall discuss all the important practical aspects of the primary solidification (transformation from the liquid to the solid state) and by a modified iron-carbon diagram as shown in Fig. 3.28. The iron carbon alloys containing up to 1.7 percent carbon are called *steels* and those containing over 1.7 percent carbon are called *cast irons.*

Transformation from the liquid to the solid state

Let us consider the course of events when liquid alloys of various carbon contents are cooled to a temperature just below the eutectic temperature, 1130°C, represented by the line EF in Fig. 3.28. If an alloy containing 0.8% carbon is cooled from a point (say m) above the liquidus lines AC, it will remain liquid until cooled to temperature t_1 on the liquidus AC. At this temperature, crystals of austenite of composition m (given by drawing a horizontal line to the solidus), begins to precipitate from the liquid alloy. The amount of austenite continuously increases as the temperature falls and its composition varies along the solidus AE while that of the liquid phase varies along the liquidus AC.

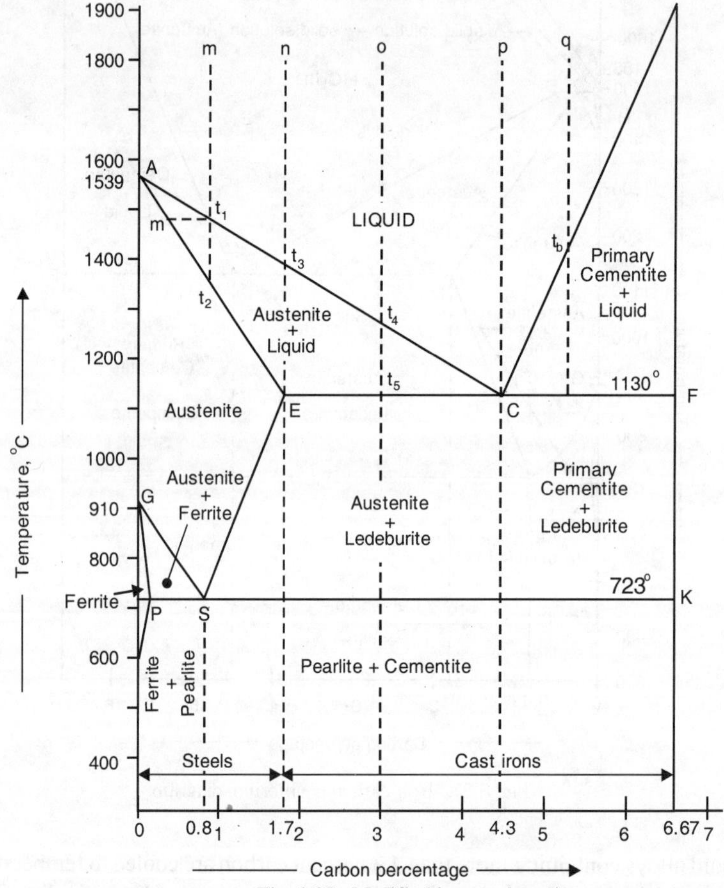

Fig. 3.28. Modified iron-carbon diagram.

When the temperature decrease to a value (t_2) at which the percentage of carbon on the solidus curve AE is the same as for the original liquid alloy (0.8%C), the alloy will become completely solid and will consist entirely of crystals of austenite. Steel of any composition will solidify exactly in the same manner and will consist of only austenite, at the moment when the process is completed. When a liquid alloy containing 1.7% of carbon is cooled from point n, a process similar to the above will take place. The austenite crystals will begin to form at temperature t_3 lying on the liquidus AC. As the crystals separate, theliquid becomes richer in carbon and the last drop of the liquid containing 4.3% carbon will solidify when the eutectic temperature (1130°C) is reached. At this temperature, the completely solidified alloy will consist entirely of austenite crystals containing 1.7% carbon dissolved in solid solution in γ-iron.

When a liquid alloy containing 3 percent carbon (point o in the figure) is cooled, it begins to precipitate austenite crystals, from the liquid alloy, at temperature t_4 lying on the liquidus AC. In this case also, the austenite crystals continuously increases as the temperature falls. When the temperature reaches to a value t_5 (1130°C), the alloy will become completely solid. At this temperature, the liquid phase enriched in carbon upto 4.3 percent is finally solidified. Form this, saturated austenite crystals (1.7 percent carbon) precipitate simultaneously with cementite and form the eutectic mixture, ledeburite. Therefore, the austenite formed at 1130°C (eutectic temperature) is called *primary austenite.* The cast iron of any composition between 1.7 to 4.3 per cent carbon will solidify in this manner.

If an alloy of exactly the eutectic composition (4.3 precent carbon) is cooled from a temperature corresponding to p in the Fig. 3.28, it will remain liquid until the eutectic temperature (1130°C) is reached. At this temperature, the alloy will solidify completely forming ledeburite (eutectic of saturated austenite and cementite). At the eutectic, the liquid phase is in equilibrium with two solid phases (cementite and saturated austenite). With these three phases in equilibrium in the two component system, the system will have zero degrees of freedom (*i.e.* nonvariant). Therefore, the liquid must change to solid at constant temperature when heat is lost by the system.

Finally, consider a liquid alloy containing 5 percent of carbon at a temperature indicated by the point q in Fig. 3.28. On cooling this alloy remains liquid until the temperature t_6 on liquidus CD is reached, where the primary cementite (Fe_3C) begins to precipitate from the liquid alloy. Since cementite is phase richer in carbon (6.67% carbon), its precipitation will decrease the carbon content of the liquid alloy with the drop in temperature along the liquidus CD. When the eutectic temperature (1130°C) is reached, the liquid alloy will have the eutectic composition (4.3% carbon) and will solidify at this constant temperature forming the solid eutectic of saturated austenite and cementite (lederburite). The completely solid alloy will consist of primary cementite and ledeburite.

Note : Iron-carbon alloys containing 4.3 percent carbon are called *eutectic cast irons,* above 4.3 percent carbon are termed as *hyper-eutectic cast iron* and those in the range of 1.7 to 4.3 percent carbon are known as *hypo-eutectic cast irons.*

3·38 Transformations in Solid Iron-Carbon Alloys (Steel)

We have discussed in Art. 3.37, that the temperatures at which the structural changes in steel on heating and cooling occur are called *critical points* or *arrest points*. These are designated by the symbols Ac (for heating) and Ar (for cooling) where A stands for arrest, c stands for *chauffage* (a french word used for heating) and r stands for *refroidissment* (a word used for cooling).

Let us now consider the transformations (known as *secondary crystallisation*) which occur in the solid state. Consider a steel with 0.3 percent carbon being cooled from a temperature above the line GS (where the steel is entirely austenitic) as shown in Fig. 3.29. A little consideration will show that nothing will happen to this until a temperature of 800°C on the line GS is reached. At this temperature, the gamma-iron in austenite will begin to transform into alpha-iron, crystal of ferrite (α-iron containing only a minute amount of dissolved iron) will form and the austenite will increase in carbon content with further cooling along the line GS. The critical points along the line GS are designated as Ac_3 in heating and as Ar_3 in cooling. When a temperature 723°C (point S) is reached, the remaining austenite containing 0.8 percent carbon decomposes with simultaneous precipitation of ferrite and cementite which form the *eutectoid mixture* known as *pearlite*. The point S is called the *eutectoid point.* Since the formation of pearlite proceeds at a constant temperature (723°C) in the presence of three phases (austenite, ferrite and cementite) which are in equilibrium in a two component system, therefore, the number of degrees of freedom of the system will be zero. Hence to change the temperature, at least one phase must disappear.

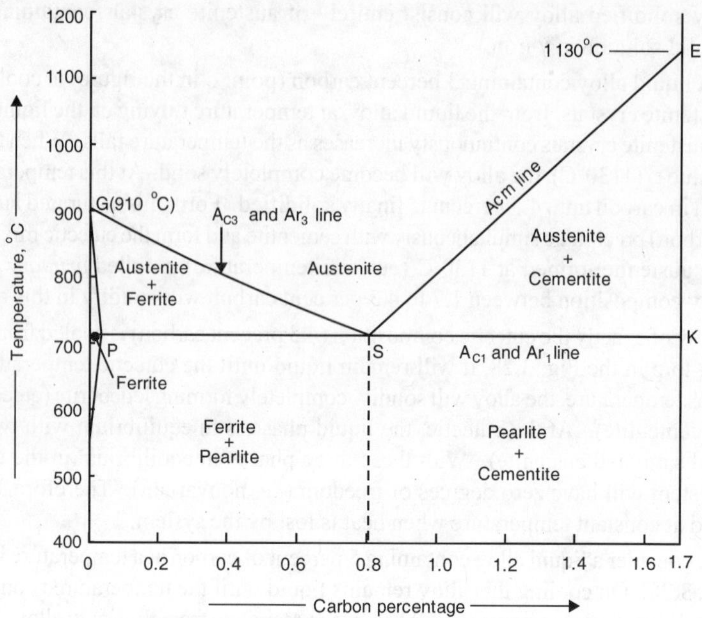

Fig. 3.29. Transformation in solid iron-carbon alloys.

In structure, pearlite consists of thin alternating plates or lamellae of cementite and ferrite called lamellar pearlite. The ferrite that separates from the austenite before the eutectoid point is reached, is called *proeutectoid ferrite* which is a part of pearlite and is known as eutectoid ferrite.

The decomposition of austenite with the formation of pearlite corresponds to the line PSK (723°C) for all iron-carbon alloys. The temperature (critical point) at which pearlite is formed in cooling is designated as Ar_1 and the transformation of pearlite into austenite in heating occurs at a temperature designated as Ac_1.

When a steel in the austenitic state containing exactly 0.8 percent carbon is cooled, no change occurs until the eutectoid point S is reached. The further loss of heat will cause complete transformation of the eutectoid austenite into pearlite.

Now consider austenite containing 1.3 percent carbon. When this alloy is cooled to 960°C (on the curve SE, called Acm line), the austenite begins to decompose with the precipitation of excess carbon as cementite. Since cementite contains 6.67 percent carbon, its separation will cause a progressive decrease in carbon content of the austenite with the drop in temperature along the line SE. When the eutectic temperature (723°C) is reached, the remaining austenite will have eutectic composition (0.8 percent carbon) and will transform into pearlite at this constant temperature. The cementite that separates from the austenite, before the eutectoid temperature is reached, is called *proeutectoid cementite.*

The point P in Fig. 3.29 (shown at larger scale in Fig. 3.30), indicates the maximum solubility of carbon in α-iron (Ferrite) at 723°C (0.025 percent). When this alloy is called, the solubility of carbon in α-iron decreases. Therefore, alloys having a carbon content between points P and Q will consist of ferrite and excess cementite which usually precipitates along the boundaries of ferrite grains.

The composition of the ferrite is indicated by the line PQ. After being completely cooled, alloys of composition to the left of point Q will consist of only solid solution of carbon in α-iron *i.e.,* ferrite grains.

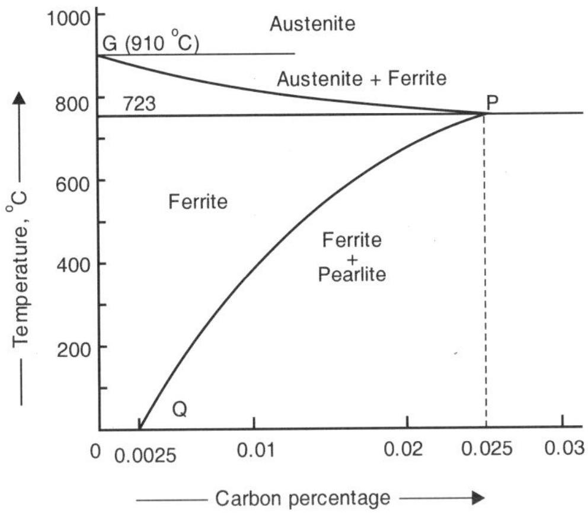

Fig. 3.30. Enlarged view of point P.

From the above discussion, we conclude that

1. The steels which contain less than 0.8% carbon (known as *hypo-eutectoid steels*) are composed of proeutectoid ferrite and pearlite after transformation.

2. The steels which contain 0.8% carbon (known as *eutectoid steels*) transform entirely into pearlite, and

3. The steels which contain above 0.8% carbon (known as *hyper-eutectoid steels*) transform into proeutectoid cementite and pearlite.

Note : When steels are very slowly heated from temperatures below 723°C, the reverse of the changes discussed above, will occur. An eutectoid steel, composed entirely of pearlite, will change completely into austenite when heated above the eutectoid temperature. The steel with 1.3 percent carbon will convert entirely into austenite when heated to slightly above 960°C, and heating of steel 0.3 percent carbon to a little above 800°C makes it completely austenite.

3·39 Micro-constituents of Steel

The constituents of steel as seen under a microscope may be classified as follows :

1. *Ferrite.* It is the name given to the grains or crystals of solid solution of carbon in alpha iron. The solubility of carbon in alpha iron is 0.025 percent at 723°C. When low carbon steels (less than 0.8% carbon) are cooled slowly from temperatures above or within the critical range, they consist chiefly of ferrite. It is soft, ductile and highly magnetic. It has a low tensile strength of approximately 280 MPa.

2. *Cementite.* When carbon goes into a definite combination with iron, it forms iron carbide (Fe_3C) which is called cementite. It contains 6.67 percent carbon and 93.33 percent iron. It is found in steels containing over 0.8 percent carbon when they are cooled slowly from within or just above the critical range. The amount of cementite increases with the increase in percentage of carbon in iron. It is identified as round particles in the structure. It is extremely hard and brittle. It is magnetic below 210°C.

3. *Pearlite.* It is a mechanical mixture of about 87 percent ferrite and 13 percent cementite. It is found in all steels when they are cooled slowly below 723°C. A steel with 0.8 percent carbon is

wholly pearlite. The soft steels containing less than 0.8 percent carbon contains ferrite and pearlite, whereas hard steels containing more than 0.8% carbon are mixtures of pearlite and cementite. The structure of pearlite consists of thin alternating plates or lamellae of cementite and ferrite. It is a strong metal and may be cut reasonably well with cutting tools. It has a tensile strength of 875 MPa.

4. *Austenite.* It is a solid solution of carbon or iron carbide (Fe_3C) in gamma-iron. When low carbon steel is heated, its constituents remain practically unchanged until the temperature of 723°C (*i.e.,* lower critical point) is reached. At this temperature, the pearlite contents of the metal get completely changed into a new substance called austenite. On further heating the metal to upper critical temperature, the remaining ferrite or cementite is absorbed by austenite and the metal is wholly composed of austenite. The maximum solubility of carbon in austenite is 1.7 percent at 1130°C. It is generally soft, ductile and non magnetic. It is denser than ferrite.

5. *Ledeburite.* It is an eutectic mixture of austenite and cementite. It is formed at about 1130°C and contains 4.3 percent carbon.

6. *Martensite.* It is the product obtained on account of the decomposition of austenite when it is cooled rapidly. The cooling of austenite is considered to have started at 325°C and is completed at about 200°C. It is strong, extremely hard and brittle. It has a needle like structure and is magnetic. It is a desirable constituent of tool steel, which requires great hardness, but little ductility such as files and razor blades etc.

7. *Troostite.* It is a product obtained by further decomposition of austenite by cooling the metal slowly until the transformation begins and then cooling rapidly to prevent its completion. It occurs in irregular grains. It has dark appearance on etching. It is somewhat softer, weaker and more ductile than martensite.

8. *Sorbite.* It is produced when steel is heated at a fairly rapid rate from the temperature solid solution to normal room temperature. Its properties are intermediate between those of pearlite and troostite.

3·40 Metallography

The study of the internal structure of a metal or alloy, in relation to its physical and mechanical properties, under a microscope is called *metallography.*

When the structure of a metal is seen with the naked eye or by low power magnification (magnification of × 10 or less) then it is said to be *macrography* and the observed structure is called *macrostructure.* On the other hand, when the structure of a metal is seen at high magnification (from × 50 to × 2000) then it is said to be *micrography* and the observed structure is called *microstructure.*

3·41 Macrography

The macrography reveals the size, form and arrangement of crystallites (dendrites) in cast metals, fibres in deformed metal, shrinkage, porosity and gas cavities, cracks appearing during certain fabrication processes etc.

The macrostructure may be studied either directly on the surface of the work (for example a casting) or on a fracture or on specimens cut out of a large billets (ingots, forgings, etc.). The surface of the specimen is ground and then etched by special reagents. The action of etching reagents is based on their ability to colour and dissolve various constituents in different manner and to widen microcavities, cracks and similar defects.

In order to reveal the dendritic structure of cast steel, a reagent consisting of 10 or 20 percent aqueous solution of ammonium persulphate [$(NH_4)_2\ S_2O_8$] is employed. A Keshian reagent (25 parts H_2O + 10 parts H_2SO_4 + 65 parts HCl) is used to reveal fibres in deformed metal, as well as cracks, cavities, porosity and chemical non homogeneity.

The degree of non homogeneity in the distribution of sulphur (or phosphorus) in steel is revealed by sulphur printing. The method consists of applying ordinary silver bromide photographic paper, soaked in a 5% solution of sulphuric acid (H_2SO_4), to the polished surface of the sample. At the sulphur rich areas, the paper will darken due to the formation of silver sulphide.

3·42 Micrography

The micrography is the most important branch of metallography and yields valuable metallurgical information concerning the metal. The absolute necessity for micro-examination arises from the fact that many microscopically observed structural characteristics of a metal, such as size and shape of the crystal, nonmetallic inclusions microcracks and other hetrogeneous conditions, exert a powerful influence on the mechanical properties of the metal. If the effect of such structural characteristics on the properties and the extent to which they are present is known, then it is possible to predict as to how the metal will behave under a specific stress. The microstructure throws considerable light on the mechanical and thermal treatment undergone by the metal.

The microstructure examination is conducted on specimen called *microsection.* The specimen of a size which can easily be handled (say 1 cm diameter and 1.5 cm high or 1cm cube) is first obtained by gentle cutting or machining. The surface to be examined is formed by filing or grinding and then on various grades of special polishing emery paper, arranged on a thick glass plate or secured to a rapidly rotating wheel. The specimen is ground first by the coarsest emery paper and successively by finer and finer grades until the finest grade is reached. The direction of grinding is changed through 90° while passing from one grade to another so that the scratches given by one grade are completely removed by the next grade. Finally, the specimen is polished on a revolving disc covered with sylet broad cloth or felt holding in suspension in water a fine grade of polishing powder like alumina, chromium oxide etc.

The finished microsection is first examined before etching to reveal nonmetallic inclusions, such as graphitic inclusions in cast iron.

The microstructure is revealed by etching the polished surface by etching reagents which attack the micro-constituents differentially corroding one leaving the other stand in relief. A few of the etching reagents and their applications are shown in table 3.1. The specimen after etching is carefully washed in water and alcohol in a blast of hot air.

Table 3.1. Etching reagents and their applications.

S. No.	*Etching reagents*	*Composition*		*Applications*
1.	Nitric acid	Nitric acid (concentrated) Methyl alcohol	5 cc 95 cc	General structure of iron and steel
2.	Picric acid	Picric acid Ethyl alcohol	4 gm 95 cc	–do–
3.	Marbles reagent	Copper sulphate Hydrochloric acid Water	4 gm 20 cc 20 cc	Stainless steel
4.	Ferric chloride	Ferric chloride Hydrochloric acid Water	15 cc 30 cc 200 cc	Copper products such as sheets, rods etc.
5.	Ferric chloride for brasses and bronzes	Ferric chloride Hydrochloride Water	10 gm 30 cc 120 cc	Brasses, bronzes, german silver, copper aluminium alloys
6.	Sodium hydroxide	Sodium hydroxide	10 gm	Aluminium alloys.

Now the study of the structure is carried out with the aid of metallurgical microscope. Since the metals are opaque, therefore their structural constituents are studied under reflected light.

Note : A photograph of microstructure may be obtained by placing a camera at the microscope eyepiece. It provides a permanent record of the structure and facilitates study. A photograph of the microstructure is called *photomicrograph.*

3·43 Microstructure of Pure Iron

The microstructure of a pure iron, which is a soft metal having crystalline structure known as ferrite, is shown in Fig. 3.31. The diagram has been drawn to represent a photomicrograph. The microscope reveals the structure to an enlarged scale, and the magnification is usually stated below the photograph (*e.g.,* × 200) as shown in Fig. 3.31.

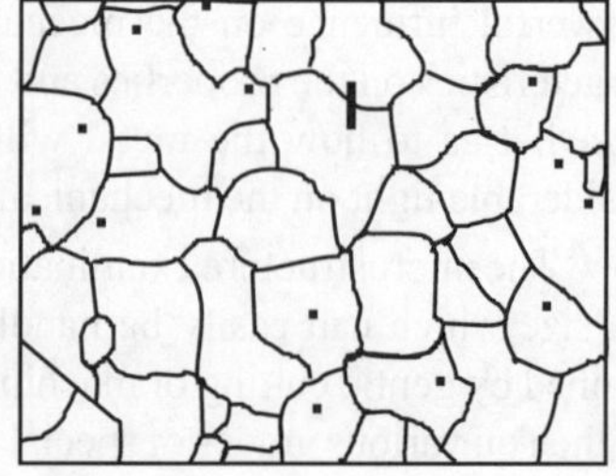

Fig. 3.31. Microstructure of pure iron × 20(

A little consideration will show that the structure of pure iron looks like a map with fields separated by hedges. The shapes looking like the fields are the crystals, and the black lines like hedges are where they join together. If the average size of the crystals (obtained by measuring on Fig. 3.31) is divided by the magnification factor, then the crystal size of the specimen is obtained.

3·44 Microstructure of Steel

The pure iron (or ferrite) is very soft and ductile. As soon as carbon is added to this iron, a great change occurs in its structure and properties. Fig. 3.32(*a*) shows the microstructure of a mild steel containing 0.25% to 0.3% carbon. The white constituent is the ferrite while the dark patches represents that part of the structure which contains the carbon. It must be remembered that these dark patches are not actual carbon but contain it in a chemically combined form. This substance is called *pearlite.*

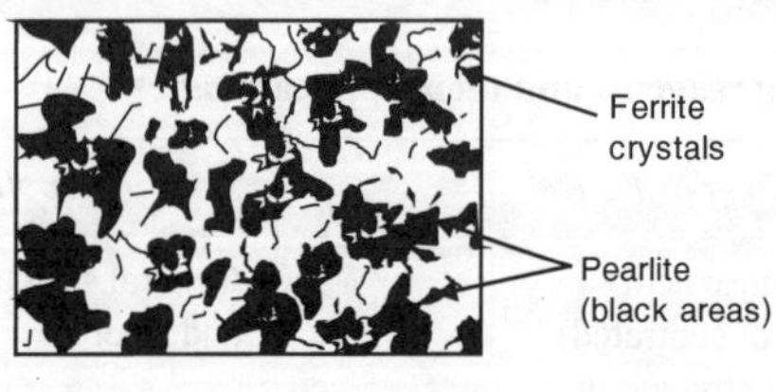

(*a*) Mild steel (0.25.0.3%) × 100. (*b*) Structure of pearlite × 10(

Fig. 3.32. Microstructure of mild steel.

If we further magnify one of the dark patches, it will be seen that the pearlite is a substance built up of alternate light and dark plates, as shown in Fig. 3.32(*b*). These layers are alternately *ferrite* and *cementite.* Thus pearlite is a combination of ferrite and cementite in the proportion of 87% ferrite and 13% cementite. In other words, 100 parts of pearlite contains 13 parts of cementite. Since *cementite consists of 1 part of carbon to 14 parts of ferrite (*i.e.,* 15 parts of cementite has 1 part of carbon and 14 parts of ferrite), therefore, 13 parts of cementite in 100 parts of pearlite will contain 13/15 of carbon *i.e.,* 0.87% carbon. But the basic form of pearlite being an equilibrium product mostly derived at slow rates of cooling has 0.8 percent carbon. Therefore, a steel with 0.25% carbon

* Cementite contains 6.67% carbon and 93.33% ferrite (iron) by weight.

contains 0.25/0.8 = 30% of pearlite and 70% of ferrite. The pearlite is a strong metal and may be cut reasonably well with cutting tools.

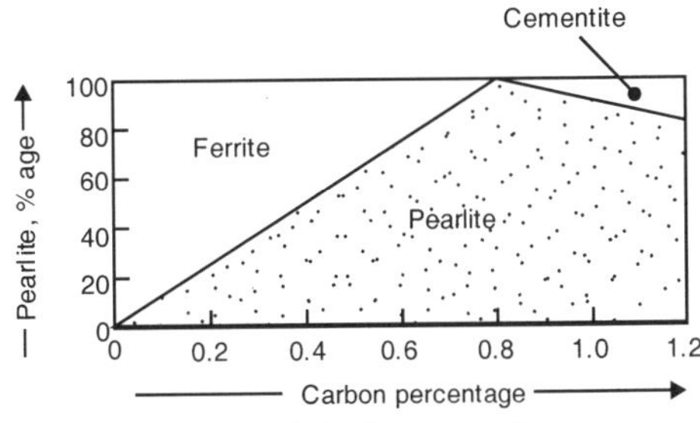

Fig. 3.33. Relation between carbon percentage and pearlite percentage.

When the carbon content of steel increases, the proportion of pearlite also increases, until when the steel contains 0.8% of carbon. At this stage, the structure consists entirely of pearlite. If the carbon content is still further increased, there will be some cementite left over and the percentage of pearlite decreases as shown in Fig. 3.33.

QUESTIONS

1. What are the various forms in which the solid substances exist ?
2. Define the term 'crystal lattice'. What is a unit cell ?
3. Discuss, with the help of neat sketches, the arrangement of atoms in metals.
4. Explain the formation of grains when the liquid metal solidifies. How grain size effects mechanical properties of a metal ?
5. What do you understand by the term 'bonds' as applicable to solids ? Explain the various types of primary and secondary bonds.
6. What is meant by crystal imperfection ? Explain the different types of imperfections.
7. Define 'elastic deformation' and 'plastic deformation' of metals.
8. Explain the process of plastic deformation of metals by twinning.
9. Distinguish between edge dislocation and screw dislocation. What is a Burgers vector ?
10. Write a short note on the deformation of polycrystalline materials.
11. What changes in mechanical properties of polycrystalline metals take place due to plastic deformation ?
12. Differentiate between recovery, recrystallisation and grain growth.
13. What is fracture ? Explain the various types of fractures that occur in metals.
14. How an alloy differ from a pure metal ? Explain the process of solidification of alloys with the help of suitable cooling curves.
15. What is a solid solution ? Explain interstitial and substitutional solid solutions.
16. Write short notes on 'intermetallic compounds' and 'mechanical mixtures'.
17. What do you understand by a phase diagram ? Write their important applications. How are these classified ?
18. What is a phase rule ? What information does it provide ?
19. Draw the iron- carbon equilibrium diagram and explain the transformations that occur in alloys having compositions from pure iron to cementite.

20. What do you understand by eutectoid steel, hypo-eutectoid steel and hyper-eutectoid steel ?

21. What are the various constituents of steel ?

22. Differentiate between 'metallography' and 'macrography'.

23. How a microstructure examination of a specimen is conducted ?

24. Draw the microstructure of pure iron and mild steel.

OBJECTIVE TYPE QUESTIONS

1. A unit cell is

(*a*) the smallest group of atoms, which when regularly repeated, forms the crystal
(*b*) a group of atoms which form a cubic arrangement
(*c*) a unit cube containing the smallest number of atoms
(*d*) none of the above

2. In a unit cell of face centred cubic space lattice, there are

(*a*) six atoms (*b*) nine atoms
(*c*) fourteen atoms (*d*) seventeen atoms

3. The type of space lattice found in alpha (α) iron is

(*a*) face centred cubic space lattice
(*b*) body centred cubic space lattice
(*c*) close packed hexagonal space lattice
(*d*) none of the above

4. The coordination number of face centred cubic space lattice is

(*a*) six (*b*) twelve (*c*) eighteen (*d*) twenty

5. The bond formed by transferring electrons from one atom to another is called

(*a*) ionic bond (*b*) covalent bond (*c*) metallic bond

6. Which of the following solids are malleable and ductile ?

(*a*) ionic solids (*b*) covalent solids (*c*) metallic solids

7. The defect which takes place due to imperfect packing of atoms during crystallisation, is known as

(*a*) line defect (*b*) surface defect (*c*) point defect

8. A screw dislocation has

(*a*) an extra plane from the top (*b*) an extra plane form the bottom
(*c*) two extra planes (*d*) no extra planes

9. The number of slip systems in F.C.C. metals are

(*a*) 12 (*b*) 24 (*c*) 36 (*d*) 48

10. The work hardening process of deforming a metal .... the ductility and electrical conductivity of metals

(*a*) increases (*b*) decreases

11. The process of deforming, plastically, a piece of metal below its recrystallisation temperature, is known as

(*a*) cold working (*b*) hot working

12. The creep fracture takes place due to

(*a*) initiation of microcracks in the surface (*b*) necking near the middle of the specimen
(*c*) sliding of grain boundaries (*d*) none of the above

13. A solid solution in which solute atoms replace some solvent atoms so that they lie at normal atom sites of the crystal structure is called

(*a*) Interstitial solid solution (*b*) substitutional solid solution

(*c*) intermetallic compounds (*d*) mechanical mixtures

14. Which of the following iron exists at 910°C ?

(*a*) α-iron (*b*) β-iron (*c*) γ-iron (*d*) δ-iron

15. Iron-carbon alloys containing 1.7 to 4.3 percent carbon are known as

(*a*) eutectic cast irons (*b*) hypo-eutectic cast irons

(*c*) hyper-eutectic cast irons (*d*) none of these

16. An eutectoid steel consists of

(*a*) wholly pearlite (*b*) wholly austenite

(*c*) pearlite and ferrite (*d*) pearlite and cementite

17. Cementite consists of

(*a*) 13 percent carbon and 87 percent ferrite

(*b*) 13 percent cementite and 87 percent ferrite

(*c*) 13 percent ferrite and 87 percent cementite

(*d*) 6.67 percent carbon and 93.33 percent iron

18. A steel with 0.8 percent carbon and 100 percent pearlite is called

(*a*) eutectoid steel (*b*) hypo-eutectoid steel

(*c*) hyper-eutectoid steel (*d*) none of these

19. The essential constituent of a hardened steel is

(*a*) pearlite (*b*) austenite (*c*) martensite (*d*) troostite

(*e*) sorbite

20. The crystal structure of a material is, generally, examined by

(*a*) naked eye (*b*) optical microscope

(*c*) metallurgical microscope (*d*) X-ray techniques

ANSWERS

1. (*a*)	2. (*c*)	3. (*b*)	4. (*b*)	5. (*a*)
6. (*c*)	7. (*c*)	8. (*d*)	9. (*a*)	10. (*b*)
11. (*a*)	12. (*c*)	13. (*b*)	14. (*a*)	15. (*b*)
16. (*a*)	17. (*d*)	18. (*a*)	19. (*c*)	20. (*d*)

4

Properties, Testing and Inspection of Metals

1. Introduction. 2. Mechanical Properties of Metals. 3. Stress and Strain. 4. Hooke's Law and Modulus of Material. 5. Linear and Lateral Strain. 6. Bulk Modulus. 7. Types of Mechanical Properties of Metals. 8. Strength. 9. Elasticity. 10. Stiffness. 11. Plasticity. 12. Ductility. 13. Malleability. 14. Brittleness 15. Toughness. 16. Hardness. 17. Resilience. 18. Creep. 19. Fatigue. 20. Technological Properties of Metals. 21. Machinability. 22. Weldability. 23. Formability or Workability. 24. Castability. 25. Testing of Metals. 26. Tensile Test. 27. Important Terms used in Tensile Test. 28. Compression Test. 29. Hardness Test. 30. Brinell Hardness Test. 31. Rockwell Hardness Test. 32. Vicker's Hardness Test. 33. Microhardness Test or Knoop's Hardness Test. 34. Shore or Rebound Hardness Test. 35. Scratch Hardness Test. 36. Impact Test. 37. Charpy Impact Test. 38. Izod Impact Test. 39. Fatigue Test. 40. Creep Test. 41. Non-destructive Tests. 42. Visual Inspection. 43. Radiographic Test. 44. Ultrasonic Test. 45. Liquid Penetrant Test. 46. Magnetic Particle Test.

4.1 Introduction

The term 'property', in a broader sense, may be defined as the quality which defines the specific characteristics of a metal. The different materials possess different properties in varying degrees and behave in different ways under given conditions. The detailed study of all the properties of a metal provides a sound basis for predicting its behaviour in manufacturing shop and also in actual use. As a matter of fact, the following properties of a metal are important to know for an engineer or designer, to enable him in selecting suitable metals for his various jobs :

1. Physical properties; 2. Mechanical properties; 3. Technological properties; 4. Electrical properties; 5. Thermal properties; 6. Chemical properties and 7. Magnetic properties.

In this chapter, we shall discuss only the mechanical and technological properties of metals.

4·2 Mechanical Properties of Metals

The mechanical properties of a metal are those properties, which completely define its behaviour under the action of external loads or forces. In other words, the mechanical properties are those properties which are associated with its ability to resist failure as well as behaviour under the action of the external forces. A sound knowledge of these properties is very essential for an engineer to enable him in selecting a suitable metal for his various structures or various components of a machine. Most of the mechanical properties of the metals are generally expressed in terms of stress, strain or both.

4·3 Stress and Strain

The *stress* is defined as the internal resistance set up by the molecules of a material to resist deformation, due to the application of external forces. Mathematically, stress is expressed as the force or load per unit area of cross-section, *i.e.*,

$$\text{Stress, } \sigma = P/A$$

where P = Force or load applied, and

A = Cross-sectional area.

The *strain* is defined as the deformation or change in length per unit length, under the action of the external forces. It is also called axial strain, linear strain or longitudinal strain. Mathematically,

$$\text{Strain,} \qquad \varepsilon = \frac{L - L_0}{L_0} = \frac{\delta L}{L_0}$$

where
L_0 = Original length,
L = Final length, and
δL = Change in length = $L - L_0$

According to the nature of load applied, there are three types of stresses and the corresponding three types of strains, as discussed below :

1. *Tensile stress and strain.* When a body is subjected to two equal and opposite axial pulls P(also called tensile load) as shown in Fig. 4.1(*a*), then the stress induced at any section of the body is known as *tensile stress.* A little consideration will show that due to the tensile load, there will be a decrease in cross-sectional area and increase in length of the body. The ratio of the increase in length to the original length is known as *tensile strain.*

2. *Compressive stress and strain.* When a body is subjected to two equal and opposite axial pushes P (also called compressive load) as shown in Fig. 4.1(*b*), then the stress induced at any section of the body is known as *compressive stress.* A little consideration will show that due to the compressive load, there will be increase in cross-sectional area and a decrease in length of the body. The ratio of the decrease in length to the original length is known as *compressive strain.*

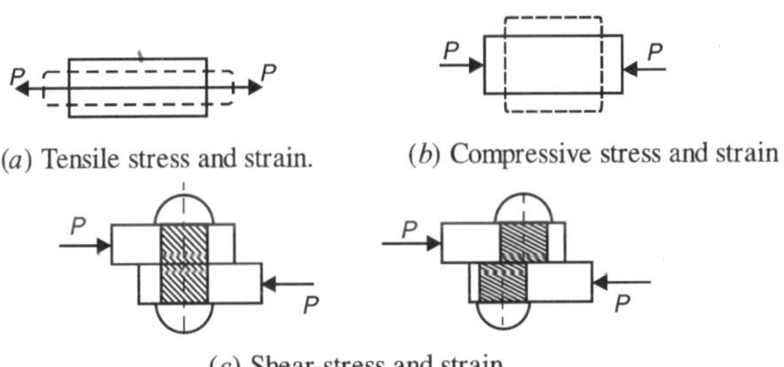

(*a*) Tensile stress and strain. (*b*) Compressive stress and strain

(*c*) Shear stress and strain.

Fig. 4.1. Types of stress and strain.

3. *Shear stress and strain.* When a body is subjected to two equal and opposite forces, acting tangentially across the resisting section, as a result of which the body tends to shear off the section as shown in Fig. 4.1(*c*), then the stress induced is called *shear stress.* The corresponding strain is known as shear strain and it is measured by the angular deformation accompanying the shear stress.

4·4 Hooke's Law and Modulus of Material

The Hooke's law states when a material is loaded within its elastic limit, the stress is proportional to strain. Mathematically,

$$\frac{\text{Stress}}{\text{Strain}} = E = \text{a constant}$$

It may be noted that Hooke's law holds good for tention as well as compression.

The proportionality constant (E) is called modulus of material. It is different for different materials and different types of stresses, *e.g.*, for tensile and compressive stresses, it is called *modulus of elasticity* or *Young's modulus* (E). For shear stress, it is known as *modulus of rigidity* (G) and for volumetric distortion, it is called *bulk modulus* (K). It is expressed in N/mm^2 or MPa (Mega Pascal).

4·5 Linear and Lateral Strain

We have already discussed that when a body is subjected to the tensile force, its length increases and the diameter decreases. Similarly, when the body is subjected to the compressive force, its length decreases and the diameter increases. It is thus obvious, that every direct stress is accompanied by a strain in its own direction which is known as *linear* or *longitudinal strain* and an opposite kind of strain in every direction, at right angles to it, is known as *lateral strain.* The ratio of lateral strain to the linear strain is called *Poissons ratio*. Its value is constant for a particular material but varies for different materials. For most of the materials commonly used in engineering practice, the value of the Poisson's ratio varies from 0.3 to 0.6.

4·6 Bulk Modulus

We have already discussed that when a body is subjected to a system of forces, it undergoes some changes in its dimensions. In other words, the volume of the body is changed. The ratio of the change in volume to the original volume is known as *volumetric strain* and the ratio of the direct stress to the volumetric strain is called *Bulk modulus*.

4·7 Types of Mechanical Properties of Metals

Though there are many mechanical properties of a metal which an engineer should know, yet the following are important from the subject point of view :

1. Strength; 2. Elasticity; 3. Stiffness; 4. Plasticity; 5. Ductility; 6. Malleability; 7. Brittleness; 8. Toughness; 9. Hardness; 10. Resilience; 11. Creep; 12. Fatigue.

These properties are discussed, in detail, in the following pages.

4·8 Strength

It is defined as the property of a material by virtue of which it can withstand or support an external force or load without rupture. The strength of a metal plays an important role in selecting a material for designing various structures and components.

A metal has innumerable types of strengths which may be broadly grouped into the following two types :

(*a*) Depending upon the value of stress, the strength of a metal may be elastic or plastic.

(*b*) Depending upon the nature of stress, the strength of a metal may be tensile, compressive, shear, bending and torsional.

4·9 Elasticity

It is defined as the property of a material by virtue of which it is able to retain its original shape and size after the removal of the load. In nature, no material is perfectly elastic, over the entire range of stress, up to rupture. The elasticity is a tensile property of a material. Steel and some other metals have a wide range over which they appear to be elastic. The elasticity is always desirable in metals used in machine tools and other structural members.

4·10 Stiffness

It is defined as the property of a material by virtue of which it resists deformation or deflection under stress. It is also known as rigidity. The stiffness of a material is measured by the Young's modulus (E) when it follows the Hooke's law. The higher the value of Young's modulus, the stiffer is the material. In case of tensile and compressive stresses, it is indicated by modulus of elasticity, in case of shear stresses, it is indicated by modulus of rigidity and in case of volumetric deformation, it is indicated by bulk modulus.

4·11 Plasticity

It is defined as the property of a material by virtue of which a permanent deformation (without fracture) takes place, under the action of external forces. The plasticity of a material depends upon its nature and the environmental conditions, *i.e.*, whether the metal is shaped red hot or in cold.

Most of the materials have been found to possess a good plasticity. Lead has good plasticity even at room temperature, while cast iron does not possess any appreciable plasticity even when red hot. In general, plasticity increases with the increase in temperature. This property finds its use in forming, shaping and extruding operations of metals.

4·12 Ductility

It is defined as the property of a material by virtue of which it can be drawn into wires or elongated with the application of a tensile force, before rupture takes place. It depends upon the grain size of the metal crystals. A ductile material must be strong and plastic. The ductility of a material is measured by its percentage elongation and percentage reduction in the cross-sectional area before rupture. The term percentage elongation is the maximum increase in the length expressed as percentage of original length. Mathematically, percentage elongation

$$= \frac{\text{Increase in length}}{\text{Original length}} \times 100$$

Similarly, the term percentage reduction of cross-sectional area is the maximum decrease in cross-sectional area, expressed as the percentage of the original cross-sectional area. Mathematically, percentage reduction in cross-sectional area

$$= \frac{\text{Decrease in cross - sectional area}}{\text{Original cross - sectional area}} \times 100$$

It may be noted that a material with a good percentage of elongation or reduction in cross-sectional area explains its high ductility. The materials with more than 15% elongation are considered as ductile and the materials with 5 to 15% elongation are considered of intermediate ductility. But the materials with less than 5% elongation are considered as brittle ones.

The ductile materials commonly used in engineering practice (in order of diminishing ductility) are mild steel, copper, aluminium, nickel, zinc, tin and lead.

4·13 Malleability

It is defined as the property of a material by virtue of which it can be rolled or hammered into thin sheets. It depends upon the crystal structure of the material. All the materials having small grain size are used for thin sheets, whereas the materials with larger grain size are used for thick sheets. A malleable material should be plastic but it is not essential to be so strong.

The malleable materials commonly used in engineering practice (in order of diminishing malleability) are lead, soft steel, wrought iron, copper and aluminium.

4·14 Brittleness

It is defined as the property of a material by virtue of which it will fracture or break without any appreciable deformation. This property is opposite to ductility of a material. It may be noted that brittle materials when subjected to tensile loads, snap off without giving any sensible elongation. Cast iron, glass and concrete are the examples of brittle materials. This property of a material finds its importance for the design of machine tools, which are subjected to sudden loads. We have already discussed that the materials with less than 5 percent elongation are known to be brittle.

4·15 Toughness

It is defined as the property of a material by virtue of which it can absorb maximum energy before fracture takes place. The energy absorbed by the material is usually expressed as *modulus of toughness.* In order to understand clearly this property, consider a piece of mild steel (tough material) and the glass (brittle material). When a sudden load is applied to these materials, we see a mild steel piece absorbs a substantial amount of energy before it fails, while the glass piece breaks suddenly. Thus it is said that the mild steel is much tougher than glass. It has been found that the value of toughness decreases when the material is heated.

The importance of toughness is in the selection of a material for structural and machine parts which have to withstand shock and vibrations such as power press punch and pneumatic hammer etc. It is, therefore, commonly associated with shock or impact loading, *i.e.*, *impact strength.*

4·16 Hardness

It is defined as the property of material by virtue of which it is able to resist wear, scratching, indentation (or penetration), deformation and machinability. It also means the ability of a material to cut another material. There are several tests used to determine the hardness of material, but the following tools are important from the subject point of view :

(*a*) Brinell hardness test; (*b*) Rockwell hardness test; (*c*) Vicker's hardness test; and (*d*) Knoop's hardness test.

These hardness tests are discussed, in detail, later in this chapter.

Note : The term used to indicate the degree of hardness that the metal (particularly steel) can acquire by the process of hardening (*i.e.,* heating and quenching) is known as *hardenability.* A metal which is capable of being hardened throughout its structure is said to have high hardenability.

4·17 Resilience

It is defined as the property of a material by virtue of which it stores energy and resists shock and impact loads. It is measured by the amount of *strain energy stored per unit volume within the elastic limit. This property is essential for spring materials.

Note : The maximum amount of energy which can be stored in the material (body) upto the elastic limit is called *proof resilience.* The amount of proof resilience per unit volume of the material is known as *modulus of resilience.* It is an important property of a material which gives capacity of the material to bear impact or shock loads.

4·18 Creep

It is defined as the property of a metal by virtue of which it undergoes a slow and permanent deformation at constant stress. Generally, the creep occurs in a material when it is subjected to high temperature for a long period of time. This property is always considered while designing internal combustion engines, boilers, turbines etc. The creep is of great importance in the following cases :

1. Soft metals used at about room temperature such as lead pipes and white metal bearings etc.
2. Steam and chemical plants operating at 450 to 550°C.
3. Gas turbines working at higher temperatures.
4. Rockets, missiles, supersonic jets.
5. Nuclear reactor field.

The creep in metals occur in three stages. In the first stage, the material elongates rapidly but at a decreasing rate. In the second stage, the elongation is constant and in the third stage, the elongation increases rapidly until the material fails. The stress for a definite rate of strain at a constant temperature is known as *creep strength.*

4·19 Fatigue

It is defined as the property of a metal by virtue of which it fails at a stress below the yield point stress, when the material is subjected to repeated stress. The failure is caused by means of progressive crack formation which is usually fine and of microscopic size. The failure may occur even without

---

* When a body or material is loaded within elastic limit, it changes its dimensions and on the removal of the load, it regains its original dimensions. So long as it remains loaded, it has stored energy in itself. On removing the load, the energy stored is given off as in the case of a spring. This energy, which is absorbed in a body, when strained within elastic limit is known as *strain energy*. The strain energy is always capable of doing some work.

prior indication. This property of a metal is of great importance in the design of shafts, connecting rods, springs, gears etc.

Notes :1. The stress at which the material fails due to fatigue is known as *fatigue strength.*

2. The maximum value of the stress that can be applied for indefinite number of times without causing the failure, is known as *fatigue* or *endurance limit.* It has been found that endurance limit for ordinary steel is approximately one-half the tensile strength.

4·20 Technological Properties of Metals

The technological properties of a metal are those properties, which completely define its behaviour in shaping, forming and fabrication operation during the manufacturing processes. A sound knowledge of these properties is also essential for an engineer to enable him in selecting a suitable metal for his various structures or various components of a machine. Though there are many technological properties of a metal, an engineer should know, yet the following are important from the subject point of view :

1. Machinability; 2. Weldability; 3. Formability or Workability; and 4. Castability.

These properties are discussed, is detail, in the following pages.

4·21 Machinability

It is defined as the property of a metal, which indicates the ease with which it can be cut or removed by cutting tools in various machining operations such as turning, drilling, boring, milling etc. The machinability of a metal depends upon mechanical and physical properties of a metal, chemical composition of the metal, microstructure of the metal and the cutting conditions.

The machinability may be improved by adding certain elements in the parent metal, *e.g.*, various steels, copper and its alloys. Lead is usually added to copper, brass, bronze, nickel etc. to improve their machinability. Sometimes selenium and tellurinium are also added for improving the machinability of the parent metal. Sulphur is the most suitable element for steels. Manganese (as manganese sulphide) or oxysulphide is effective for free machining steel. Sometimes, phosphorus is also added upto 0.1% to improve machinability of steel. It will be interesting to know that the addition of these elements helps in lowering the mechanical properties, *e.g.*, low hardness, low ductility and low tensile strength. They also act as lubricants. It has been observed that the grey cast iron possesses very good machinability. But machinability of white cast iron is very poor. Mild steel has better machinability than medium carbon steel, high carbon steel, high speed and most of the alloy steels.

4·22 Weldability

It is defined as the property of a metal, which indicates the ease with which two similar or dissimilar metals are joined by fusion (with or without the application of pressure) and with or without the use of filler metal.

Strictly speaking, a metal has good weldability, if it can be easily welded in a fabricated structure. The various factors affecting weldability of a metal are :

1. Composition of the metal.
2. Brittleness and strength of the metal at elevated temperatures.
3. Thermal properties of the metal.
4. Welding techniques, fluxing material and filler material.
5. Proper heat treatment before and after the deposition of the metal.

The common metals having weldability in the descending order are iron, steel, cast iron, low alloy steels and stainless steel.

4·23 Formability or Workability

It is defined as the property of a metal which indicates the ease with which it can be formed (*i.e.,* pressed by forging) into different shapes and sizes. The various factors affecting the formability of a metal are :

1. Crystal structure of the metal.
2. Grain size of the metal.
3. Hot and cold working.
4. Alloying element present in the parent metal.

It will be interesting to know that the metals with small grain size are suitable for shallow forming, while metals with relatively large grain size are suitable for heavy forming. The distortion of grains is an important factor which affect the formability of a metal, as the distortion of grains takes place due to hot working of a metal. The cold working of a metal results in less ductility and more distortion than the hot working.

It has been observed that the addition of some alloying elements reduces the ductility and hence formability of the parent metal. Low carbon steels possess good formability, which does not interfere with the slip planes. Medium carbon steels, on the other hand, are not suitable for cold forming.

4·24 Castability

It is defined as the property of a metal, which indicates the ease with which it is casted into different shapes and sizes from its liquid state. The various factors affecting the castability of a metal are :

1. Solidification rate (fluidity of metal).
2. Gas porosity.
3. Segregation.
4. Shrinkage.

4·25 Testing of Metals

In the previous articles, we have discussed the important mechanical properties of metals. Almost all the mechanical properties are established by conducting tests on various testing machines and all the tests have been standardised. In India, these tests are standardised by the Bureau of Indian Standards (previously known Indian Standard Institution). All the mechanical tests are divided into the following two groups :

1. *Destructive tests.* In destructive tests, the specimen of the material is tested until its complete structural failure, in order to derive its mechanical properties and characteristics. In these types of tests, the material may be rendered useless after the test is over or may be transformed to possess mechanical properties very different from the previous one. The destructive tests include tensile test, compression test, impact test, hardness test, fatigue test, creep test etc.

2. *Non-destructive tests.* In non-destructive test, the material under test possesses the same mechanical properties before and after the tests and the material is not rendered unserviceable even after it has been subjected to loading conditions. The non-destructive tests include visual examination, radiographic test, ultrasonic test, liquid penetrant test, magnetic particle test.

4·26 Tensile Test

The tensile test of a metal is generally performed to determine :

1. Proportional and elastic limit.
2. Yield point.
3. Ultimate tensile strength.
4. Percentage elongation and reduction of area.

The results obtained by the tensile test are widely used in the design of materials for structures and other purposes. In this test, the test piece is pulled out at a constant rate by gradually increasing the axial pull, till the rupture takes place.

The tensile test for a ductile material is, generally, carried out with the help of a *universal testing machine on the specimen made from the material to be tested. The schematic working arrangement of a universal testing machine is shown in Fig. 4.2. This machine has two units, one is the testing unit and the other is control unit.

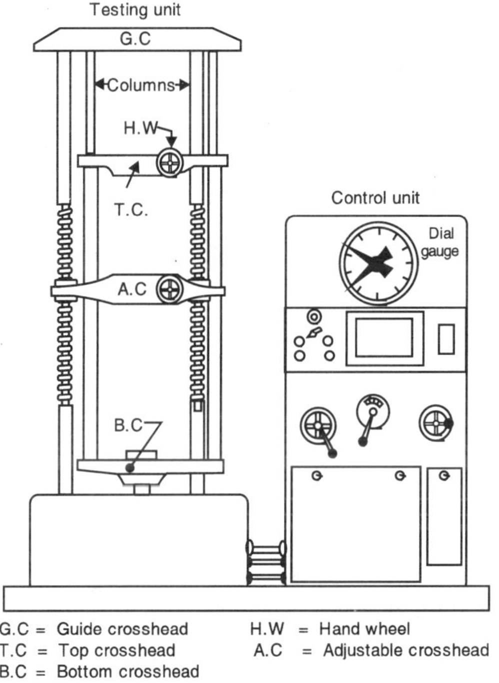

Fig. 4.2. Universal testing machine.

The dimensions and form of the specimen varies according to the size and shape of the material to be tested and the main objective in view. The test is carried out on a specimen having uniform cross-section throughout the gauge length as shown in Fig. 4.3.

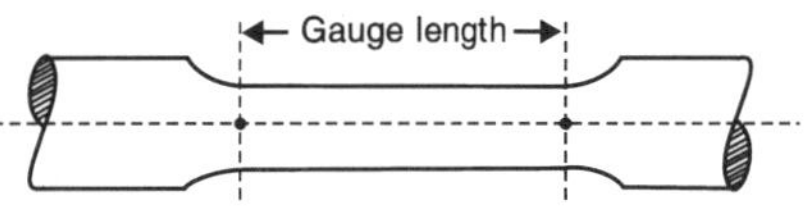

Fig. 4.3. Specimen for tensile test (before test).

The Bureau of Indian Standards (previously Indian Standards Institution) has recommended

---

* The testing machines are classified into the following two groups :

1. Specific purpose testing machine; and 2. Universal testing machine.

A machine on which a specific test or a single test can be performed, is known as *specific purpose testing machine.* For example, Brinell hardness testing machine is a specific purpose testing machine which is only used to determine the Brinell hardness number of a metal.

A machine on which more than one test can be performed, is known as *Universal testing machine.* This machine may be used, apart from tensile test, for compression test, bending test, shear test etc.

These days, electronic universal testing machines with microprocessers are also available in this country.

several sizes of the specimen, but the standard practice is to use a specimen whose gauge length in mm,

$$L_0 = 5.65\sqrt{A_0} = 5.0\, d_0$$

where A_0 = Area of cross-section in mm^2, and

d_0 = Diameter of the specimen in mm.

It means for a gauge length of 50 mm, the specimen diameter should be 10 mm.

The tensile test is performed as discussed below :

1. First of all, one end of the specimen is gripped in the jaws provided in the adjustable crosshead and then after lifting this crosshead to the appropriate height (depending upon the length of the specimen), the other end of the specimen is fixed in jaws in the top crosshead.

2. The tensile load is now applied hydraulically to the specimen by turning the handwheel (towards right) provided in the control unit. The load measuring gauge incorporated in the control unit shows the magnitude of the applied load.

3. The load is gradually increased until the specimen breaks and the corresponding extensions are recorded. A strain gauge or an extensometer is attached to the specimen to show the extensions.

A little consideration will show that with the increase in load, there is a corresponding increase in length between the two extremities of the gauge length. In other words, we can say that the increase in length (or elongation) of the specimen depends upon the load applied. Now if we plot a graph with loads or stresses (*i.e.*, load divided by the original cross-sectional area of the specimen) along the vertical axis and corresponding elongations or strains (*i.e.*, elongation or change in length divided by the gauge length or original length) along the horizontal axis, we shall obtain a curve as shown in Fig. 4.4(*a*). Such a graph is known as *Stress-strain diagram* for a given ductile material (say mild steel). It gives us information about the following important points :

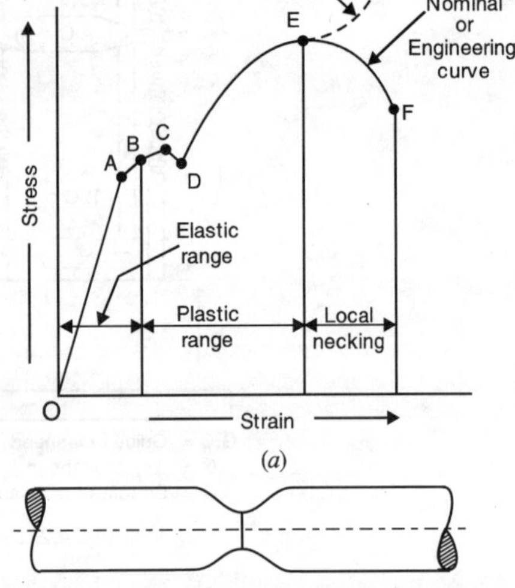

(*b*) Shape of specimen after elongation.

Fig. 4.4. Stress-strain diagram for a mild steel.

(*i*) *Proportional limit or limit of proportionality.* We see from the diagram that from point *O* to *A* is a straight line which represents that the stress is proportional to strain. Beyond point *A*, the curve slightly deviates from the straight line. It is thus obvious that Hooke's law hold good upto point *A* and it is known as proportional limit or limit of proportionality. It is defined as that stress at which the stress-strain curve begins to deviate from the straight line.

(*ii*) *Elastic limit.* It may be noted that even if the load is increased beyond point *A* up to the point *B*, the material will regain its shape and size when the load is removed. This means that the material has elastic properties up to the point *B*. This point is known as *elastic limit.* It is defined as the stress developed in the material without any permanent set.

Notes : (*a*) In some materials, the proportional limit and the elastic limit are almost same, but in most of the materials, the elastic limit is higher than the proportional limit.

(*b*) At elastic limit, the elongation of different materials is different. But in case of mild steel specimen, the elongation is about 2%.

(*iii*) *Yield point.* If the material is stressed beyond point *B*, the plastic stage will reach *i.e.,* on the removal of the load, the material will not be able to recover its original size and shape. A little consideration will show that beyond point *B*, the strain increases at a faster rate with any increase in the stress until the point *C* is reached. At this point, the material yields before the load and there is an appreciable strain without any increase in stress. In case of mild steel, it will be seen that a small load drops to *D*, immediately after yielding commences. Hence for mild steel there are two yield points *C* and *D*. The point *C* and *D* are called the *upper* and *lower yield points* respectively. The stress corresponding to yield point is known as *yield point stress.*

(*iv*) *Proof stress.* It has been observed that most of the materials do not have a clear cut yield point and the stress-strain curve is as shown in Fig. 4.5. For such materials, a term known as *proof stress* is calculated. It is defined as the stress at which the material undergoes a specified value of strain which is usually 0.2 percent for steel and 0.5 percent for cast iron. In order to determine the value of proof stress, an off-set method is employed. In this method, a point such as *K* is marked on the strain axis which represents a specified value of strain. From this point *K*, straight line is drawn parallel to straight line portion (*OA*) of the stress-strain curve meeting the curve at point *M*. The stress corresponding to point *M* is known as *proof stress* or *yield strength.*

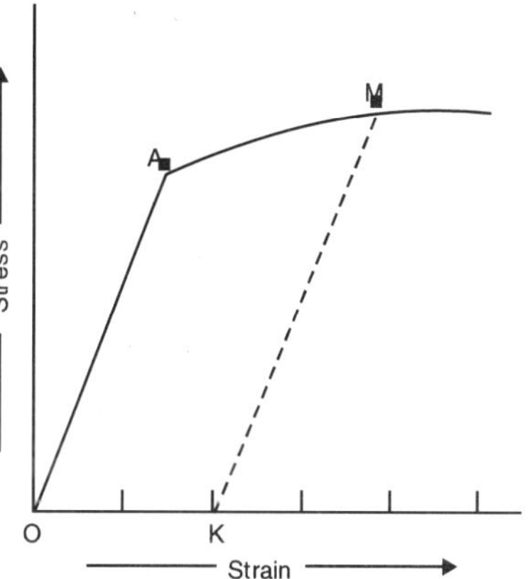

Fig. 4.5. Stress-strain showing proof stress

(*v*) *Ultimate stress.* At *D*, the specimen regains some strength and higher values of stresses are required for higher strains, than those between *A* and *D*. The stress (or load) goes on increasing till the point *E* is reached. The gradual increase in the strain (or length) of the specimen is followed with the uniform reduction of its cross-sectional area. The work done, during stretching the specimen, is transformed largely into heat and the specimen becomes hot. At *E*, the stress, which attains its maximum value is known as *ultimate tensile stress* or *tenacity* which is a measure of tensile strength of a material. It is defined as the largest stress obtained by dividing the largest value of the load reached in a test to the original cross- sectional area of the test piece.

(*vi*) *Breaking stress.* After the specimen has reached the ultimate stress, a neck is formed, which decreases the cross-sectional area of the specimen, as shown in Fig. 4.4(*b*). A little consideration will show that the stress (or load) necessary to break away the specimen, is less than the maximum stress. The stress is, therefore, reduced until the specimen breaks away at point *F*. The stress corresponding to point *F* is known as *breaking stress*.

Note : The breaking stress (*i.e.*, stress at *F* which is less than at *E*) appears to be somewhat misleading. As the formation of a neck takes place at *E* which reduces the cross-sectional area, it causes the specimen suddenly to fail at *F.* If for each value of the strain between *E* and *F*, the tensile load is divided by the reduced cross-sectional area at the narrowest part of the neck, then the curve will follow the dotted line *EG*. This curve *EG* is called the true curve while the curve *EF* is known as nominal or engineering curve. It is an established practice, to calculate strains on the basis of original cross-sectional area of the specimen.

4·27 Important Terms used in Tensile Test

The stress-strain curve as discussed in the previous article, gives a valuable information about mechanical properties of a metal. The following terms are important in the tensile test of a specimen :

1. *Elastic stress.* The elastic load (corresponding to point *B* in Fig. 4.4) divided by the original cross-sectional area of the specimen is known as elastic stress. Mathematically,

$$\text{Elastic stress, } \sigma_0 = \frac{\text{Elastic load}}{\text{Original cross-sectional area}}$$

The elastic stress is expressed in Mega Pascal (MPa).

2. *Working stress.* It is the stress, which is used by the engineers for designing their structures or machine components. The working stress is kept much below the elastic stress, and is also known as *safe stress*.

3. *Yield stress.* The yield load (corresponding to point *C* in Fig. 4.4) divided by the original cross-sectional area of the specimen is known as yield stress. Mathematically,

$$\text{Yield stress, } \sigma_y = \frac{\text{Yield load}}{\text{Original cross-sectional area}}$$

The yield stress is also expressed in MPa.

Notes : **(a)** The yield stress value depends upon whether the stress is calculated on the basis of upper or lower yield point load as the case may be.

(b) The yield stress for low carbon steels is usually equal to 0.6 to 0.7 times the ultimate stress.

4. *Ultimate stress.* The ultimate load (corresponding to point *E* in Fig. 4.4) divided by the original cross-sectional area of the specimen is known as ultimate stress. Mathematically,

$$\text{Ultimate stress, } \sigma_u = \frac{\text{Ultimate load}}{\text{Original cross-sectional area}}$$

The ultimate stress is also called ultimate tensile stress, the maximum stress or tenacity. It is expressed in MPa.

5. *Factor of safety.* The ratio of ultimate stress to the working stress is known as factor of safety. Mathematically,

$$\text{Factor of safety} = \frac{\text{Ultimate stress}}{\text{Working stress}}$$

These days, the value of factor of safety for structural steel is taken as 2 or 2.5. But in the case of cast iron, concrete, wood etc., the value of factor of safety is taken as 4 to 5.

6. *True stress.* The load at any instant (called instantaneous load) divided by the cross-sectional area of the specimen at that instant (called instantaneous cross-sectional area of the specimen) is known as true stress. Mathematically,

$$\text{True stress, } \sigma_T = \frac{\text{Instantaneous load}}{\text{Instantaneous cross-sectional area}}$$

The true stress is expressed in MPa.

7. *Nominal stress.* The load at any instant (called instantaneous load) divided by the original cross-sectional area of the specimen is known as nominal stress. Mathematically,

$$\text{Nominal stress, } \sigma_N = \frac{\text{Instantaneous load}}{\text{Original cross-sectional area}}$$

The nominal stress is expressed in MPa. The value of nominal stress, in tension, is always smaller than the true stress. It is because of the fact that the instantaneous (or actual) cross-sectional area of the specimen becomes less than the original cross-sectional area on the application of load. It may be noted that it is the nominal stress, which is used in most of the engineering calculations for design purposes. Therefore, nominal stress is also called *engineering stress*.

8. *True breaking stress.* The breaking load (corresponding to point *F* in Fig. 4.4) divided by the cross-sectional area at the neck of the specimen is known as true breaking stress. Mathematically,

$$\text{True breaking stress} = \frac{\text{Breaking load}}{\text{Cross-sectional area at the neck}}$$

The true breaking stress is expressed in MPa.

9. *Nominal breaking stress.* The breaking load (corresponding to point F in Fig. 4.4) divided by the original cross-sectional area of the specimen is known as nominal breaking stress. Mathematically,

$$\text{Nominal breaking stress} = \frac{\text{Breaking load}}{\text{Original cross-sectional area}}$$

10. *Percentage reduction of area.* The reduction in cross-sectional area divided by the original cross-sectional area of the specimen, expressed as percent, is known as percentage reduction of area. Mathematically, the percentage reduction of area

$$= \frac{\text{Reduction in cross-sectional area}}{\text{Original cross-sectional area}} \times 100$$

Let A = Original cross-sectional area, and

a = Cross-sectional area at the neck (*i.e.*, where the fracture of the specimen takes place).

∴ Reduction in cross-sectional area

$$= A - a$$

and percentage reduction in area $= \dfrac{A-a}{A} \times 100$

11. *Percentage elongation.* The elongation or increase in the standard gauge length divided by the original length of the specimen, expressed as percent, is known as percentage elongation. Mathematically, the percentage elongation

$$= \frac{\text{Increase in length of the specimen}}{\text{Original length of the specimen}} \times 100$$

Let l = Gauge length or original length of the specimen, and

L = Length of the specimen after fracture or final length. This length is obtained by measuring the fractured specimen after bringing the broken parts together.

∴ Elongation $= L - l$

and percentage elongation $= \dfrac{L-l}{l} \times 100$

12. *Modulus of resilience.* The maximum amount of energy which can be stored in a body upto the elastic limit is known as *proof resilience*. The proof resilience per unit volume of a material is known as *modulus of resilience* or *elastic toughness*. It is determined by the area under the stress-strain curve upto the elastic limit.

13. *Modulus of toughness.* The expenditure or dissipation of energy per unit volume of a material necessary to cause a complete failure, is known as *modulus of toughness* or *fracture toughness*. It is determined by the total area under the stress-strain curve. The approximate value of the modulus of toughness may be obtained from the following relation :

Modulus of toughness or fracture toughness

$$= \frac{\text{Tensile strength} \times \text{Percentage elongation at failure}}{100}$$

Note : For extremely ductile materials, the ordinary stress-strain diagram will not give a correct measure of fracture energy. For such material, it is necessary to use true stress-strain diagram.

4·28 Compression Test

The compression is regarded as the reverse of tensile test. It is generally performed for testing *brittle materials such as cast iron, concrete, stones, bricks, mortars, ceramics, etc. The specimens

* This is because the compressive strength of brittle materials is much more than the tensile strength.

used in this test are, usually made of cubical or cylindrical shape. It has been observed that some errors always creep in the compressive test due to the following practical difficulties :

1. Since the top and bottom faces of the given specimen are seldom absolutely parallel to each other, therefore it is very difficult to ensure axial loading on the specimen.
2. Since the length of the given specimen is kept short enough (not more than twice of its diameter) to avoid its buckling, therefore, within the elastic limit, a small compression takes place which is difficult to measure accurately.
3. The friction between the ends of the given specimen and clutches of the machine prevents the dimensions of the specimen ends from increasing. This results in the lateral expansion to take place more in the centre instead of uniform increase in diameter throughout the whole length. Such an effect, which is called barrel effect, is not a case of an ideal compression.

The compression test may be performed on a Universal testing machine. We have already discussed that in a tensile test, the specimen is held between the jaws provided in the adjustable crosshead and the top crosshead and the load tends to pull the specimen apart. But in case of compression test, the specimen is placed between the compression plates provided in the adjustable cross head and the bottom crosshead. After the specimen is properly placed and gripped firmly between the compression plates, a compressometer or strainometer (a specially designed strain gauge) is attached to measure the compressive strain. The load is gradually applied on the specimen and the corresponding strain is recorded at regular intervals. Now with the recorded data, a graph is plotted with stresses along the vertical axis and the corresponding strains along the horizontal axis. If we draw a curve passing through the vicinity of all such points, we shall obtain a graph as shown in Fig. 4.6.

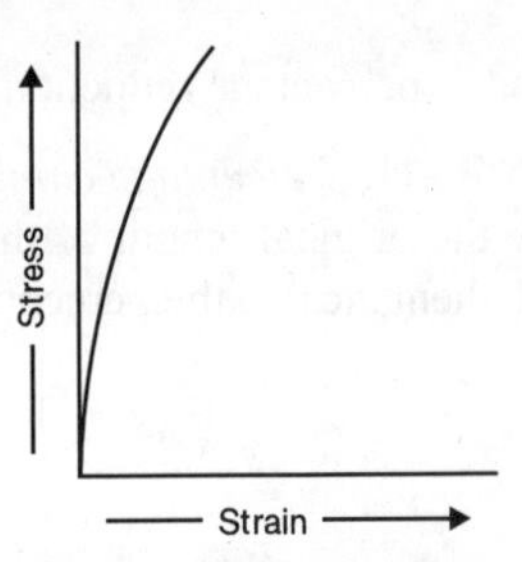

Fig. 4.6. Stress-strain diagran for brittle materials.

From the graph, we see that there is a little strain as compared to the stress. There is always a point, where the specimen will fail due to shear along a diagonal plane. The value of breaking stress for different materials is different. But the general pattern for the stress-strain diagram is approximately the same as shown in Fig. 4.4.

4·29 Hardness Test

The hardness, in a general sense, represents an aggregate of properties of a material more or less related to each other. The resistance to abrasion in case of stones etc., resistance to scratching related to metals and minerals, ability of a metal to cut other metals is a measure of hardness. The hardness values of engineering materials is mostly obtained by the penetration method.

In this method, the material is subjected to a known magnitude of load mechanically applied through a very hard and specifically chosen indenting tool called as a *penetrator.* This penetrator, under the influence of impressed load, penetrates into the surface of the material thereby leaving a permanent impression clearly visible to the naked eye or magnified through some optical instruments. The dimensional measurements of such an impression are put to valuable consideration for deciding the hardness value of a material.

This method of determining the hardness value has been widely used in engineering practice for all metals hard or soft and also on nonmetallic material such as timber, plastic etc. The types of tests included in the category of penetration hardness are :

(*a*) Brinell hardness test; (*b*) Rockwell hardness test; (*c*) Vicker's hardness test (also known as diamond pyramid hardness test); and (*d*) Microhardness or Knoop's hardness test.

The other hardness tests are Shore's or rebound hardness test and scratch hardness test.

All these tests are discussed, in detail, in the following pages.

4·30 Brinell Hardness Test

The Brinell's hardness test is the oldest method for finding hardness of metals by penetration. The test is related to applying a test load normally to the surface of specimen of metals through a steel ball or known diameter for a given time, as shown in Fig. 4.7. The steel ball being of hardness greater than that of the specimen, shall give an impression on the specimen which forms part of the solid spherical ball.

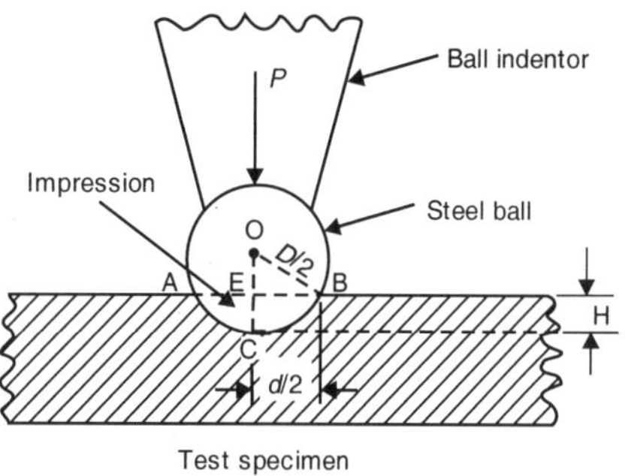

Fig. 4.7. Brinell hardness test.

The Brinell hardness number (B.H.N.) is defined as the ratio of the applied load in newtons to the surface area of impression expressed in sq. mm. Thus, B.H.N. is also a measure of indentation stress which can be expressed in units of N/mm^2. Mathematically, Brinell hardness number,

$$\text{B.H.N.} = \frac{\text{Load in newtons}}{\text{Surface area of impression in mm}^2} = \frac{P}{\pi D H}$$

Now $$H = EC = OC - OE = \frac{D}{2} - \sqrt{\left(\frac{D}{2}\right)^2 - \left(\frac{d}{2}\right)^2}$$

$\therefore$ $$\text{B.H.N.} = \frac{P}{\pi D\left[\frac{D}{2} - \sqrt{\left(\frac{D}{2}\right)^2 - \left(\frac{d}{2}\right)^2}\right]} = \frac{P}{\frac{\pi D}{2}\left[D - \sqrt{D^2 - d^2}\right]}$$

where P = Applied load in newtons,

D = Diameter of steel ball in mm,

d = Diameter of impression in mm.

From the above, we see that following are the main requirements of the Brinell hardness test :

1. *Steel ball.* The diameter of the steel ball is generally 10 mm ± 0.01 mm. For testing harder steels, tungsten carbide balls are widely used.

2. *Test specimen.* The thickness of the test specimen should always be more than 8 times the depth of indentation. The test surface should be free from any oxide film.

3. *Load.* For ferrous metals, the load should be applied for a period of 15 seconds. But for soft metals, the load should be applied for a minimum period of 30 seconds. Moreover, the load should be applied gradually and smoothly.

For the sake of uniformity in results with this test, the relationship between the load and ball diameter has to be fixed. This relationship is given by

$$\frac{P}{D^2} = 300$$

For example, for 10 mm ball diameter,

$$\frac{P}{(10)^2} = 300 \text{ or } P = 30\,000 \text{ N},$$ *i.e.*, load applied should be 30 000 N.

The values of P/D^2 ratio for representative materials are given in the following table :

Table 4.1. Values of P/D^2 ratio for representative materials.

Materials	*Approximate B.H.N.*	*P/D^2 ratio*	*Time lag in seconds*
Steel and Cast iron	Above 100	300	10
Brass, Copper alloys and Aluminium alloys	30–200	100	30
Aluminium	15–100	50	30
Lead, Tin, and their alloys	3–20	10	30

The Brinell hardness test is mostly used for determining the hardness of metallic materials. This test is not recommended for materials having B.H.N. above 630 and also for very thin specimens. Morever, it is a time consuming test and requires an expensive equipment. As the test leaves a large impression on the specimen, therefore, it cannot be adopted in the industry.

Note : The ultimate tensile strength (U.T.S.) may approximately be estimated from the following relation :

$$\text{U.T.S.} = \text{B.H.N.} \times 3.5 \text{ N/mm}^2 \text{ or MPa}$$

4·31 Rockwell Hardness Test

The Rockwell's hardness test is, generally, performed when quick and direct reading is desirable. This test is also performed when the materials have hardness, beyond the range of Brinell's hardness test. It differs from the Brinell's test that in this test the loads for making indent are smaller, and thus make smaller and shallower indents. It is because of these reasons that the Rockwell's hardness test is widely used in the industry.

In this test, either a standard ball indentor of hardened tool steel or tungsten carbide of 1.588 mm diameter or a diamond cone indentor of 120° angle with spherical tip is employed. The testing machine has a dial indicator on the front which has mostly three scales of hardness (*i.e.*, *A*, *B* and *C*), depending upon the material. During the test, the specimen is placed on the anvil and it is raised by the hand wheel till it comes in contact with the indentor. A minor load of 100 N is applied on the specimen and the small pointer indicates 'set'. Now the main pointer is also brought to the 'set' position. The major load is now applied and allowed to continue for three to five seconds. The depth of indentation (in mm) is read from the small pointer. Now the Rockwell number is obtained, mathematically, from the following relation :

Rockwell hardness number on *B* scale,

$$\text{HRB} = 130 - \frac{\text{Depth of indentation in mm}}{0.002}$$

and Rockwell hardness number on *C* scale,

$$\text{HRC} = 100 - \frac{\text{Depth of indentation in mm}}{0.002}$$

The following precautions should be taken in the Rockwell's hardness test :

1. The indentor and anvil should be clean and well placed.
2. The surface of the specimen should be flat, clean, dry, smooth and should be placed perpendicular to the indentors.
3. The thickness of the specimen should be more than 10 times the depth of indentation. If the specimen is curved in shape, then the indent should be made on the concave side.
4. The test should be carried out preferably at a temperature of 27° ± 2°C.

Notes : (*a*) The minor load for each scale of hardness is 100 N.

(*b*) The major load for *A* scale is 500 N and diamond cone indentor is used. This scale is used for testing thin heat treated and hardened steel plates.

(*c*) The major load for *B* scale is 900 N and a ball indentor is used. This scale is used for testing brass, bronze, unhardened steel as well as cast metals.

(*d*) The major load for *C* scale is 1400 N and a diamond cone indentor is used. This scale is used for testing heat treated and hardened steel.

(*e*) The relation between Brinell hardness number (BHN) and Rockwell hardness number on *B* scale (HRB) is given as

$$\frac{1}{\text{BHN}} = \frac{130 - \text{HRB}}{7300}$$

4·32 Vicker's Hardness Test

The Vicker's hardness test is the most accurate test, which has a fairly continuous scale of hardness (Vicker's hardness number of 5 to 1500). The test makes the use of a diamond square based pyramid indentor with 136° angle between the opposite faces. The load range is variable from 50 to 1200 N, in steps of 5 N. A piston and a dashpot of oil is used for controlling the rate and duration of the loading.

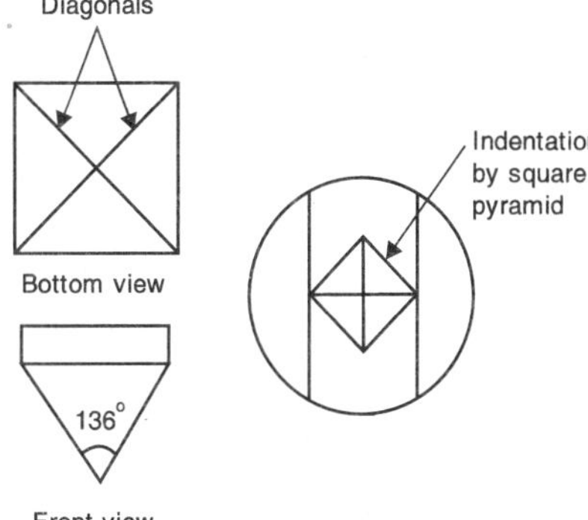

Fig. 4.8. Vickers hardness test.

The test is performed by placing the specimen on an anvil and raised till it is close to the indentor point. The load is then gradually applied to the indentor and then removed. The magnitude of the load applied depends upon the thickness and hardness of the material. The impression made by the indentor on the surface of the specimen is as shown in Fig. 4.8. The diagonal of the square indentation is measured to 0.001 mm length. Now the Vicker's hardness number (VHN) or Diamond pyramid hardness number (DPN) is obtained, mathematically, from the relation :

$$\text{VHN or DPN} = \frac{P}{A} = \frac{2P\sin(\theta/2)}{L^2}$$

$$= \frac{1.854\,P}{L^2}$$

where P = Load applied in newton,

A = Surface area of indentation in square mm,

L = Length of the diagonal in mm, and

θ = Angle between opposite faces of diamond pyramid (equal to 136°).

This test is very suitable for testing polished and hardened material or nitrided surface due to small impression made on the test specimen. This test is accurate and is suitable for metals as thin as 0.15 mm. In spite of these advantages, the Vicker's hardness test is not widely adopted because of its being slow.

All the precautions necessary to perform the test are similar to those in a Rockwell's hardness test.

4·33 Microhardness Test or Knoop's Hardness Test

The microhardness test or Knoop's hardness test is suitable for extremely thin metal plates, exceptionally hard and brittle, very shallow carburised or nitrided surface. The testing machine commonly used for this purpose is called Tukon testing machine. The load applied in this test is kept below 35 N.

In this test, a diamond pyramidal indentor (known as knoop indentor) with short depth diagonals in the ratio of 7 : 1 is used to indent the specimen. The length of the diagonal is read under microscope. Now the Knoop's hardness number (KHN) is obtained, mathematically, from the relation :

$$\text{KHN} = \frac{\text{Load in newton}}{\text{Unrecovered projected area of the impression in mm}^2}$$

4·34 Shore or Rebound Hardness Test

The testing equipment used in this test is known as Shore's scleroscope after its inventor's name A.F. Shore. In this test, an indentor having a small diamond tip weighing 40 grains is allowed to fall freely on the specimen, in a vertically aligned glass tube, from a height of 250 mm. This height of fall is divided into 140 equal scale divisions. The hammer rebounds from the surface of the specimen and the height of rebound is noted in terms of number of scale divisions on the glass tube. This height of rebound represents the relative hardness of the material.

The Shore's scleroscope is widely employed for hardness test as it is easy to operate, easy to read hardness and above all it is a portable instrument. In practice it has been found experimentally that Shore's hardness number is about one-sixth to one-seventh of the Brinell hardness number in case of steels only.

4·35 Scratch Hardness Test

This method is used for stones, minerals and brittle materials such as glass, bakelite etc. In this method, the hardness is measured on Mohs' scale (devised by a German scientist Friedrich Mohs). The Mohs scale carries 10 numbers from 1 to 10. Number one is given to Talc stone which is considered to be the softest material, while number ten is given to Diamond which is the hardest stone. Sometimes, the scratch hardness of a test specimen is obtained in terms of the load in newtons under which a conical diamond will cause a scratch 0.01 mm wide. This method is not popularly used in engineering practice because it can not measure the hardness accurately of the material.

4·36 Impact Test

Many machine parts are subjected to suddenly applied loads called impact blows. It has been observed that a metal may be hard, strong or of high tensile strength. But it may be unsuitable for uses where it is subjected to sharp blows. The capacity of a metal to withstand such blows without fracture is known as an *impact resistance* or *impact strength.* It is an indicative of the toughness of the metal *i.e.*, the amount of energy absorbed by the metal during plastic deformation. The S.I. unit for expressing impact strength is meganewton per square metre, *i.e.*, MN/m^2.

The following two types of impact tests are performed on all metals and sometimes for non-metals :

1. Charpy test; and 2. Izod test.

These tests are performed on a pendulum type impact testing machine, as shown in Fig. 4.9. It consists of the frame, the scale, the pendulum with a striking hammer, the pendulum bearings, the pendulum brake and the clamping parts. The scale

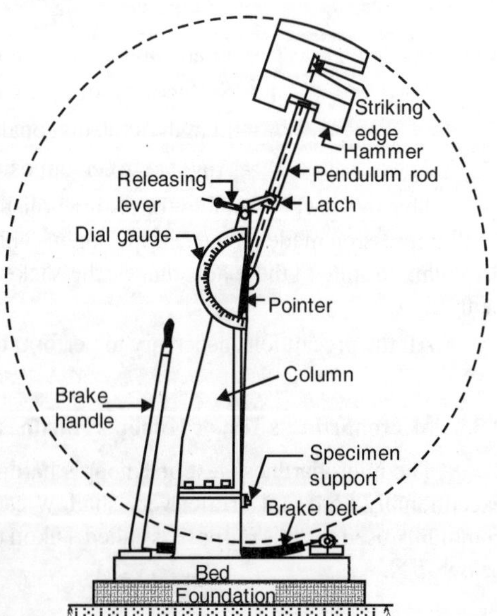

Fig. 4.9. Pendulum type impact testing machine.

is fastened to the front of the frame. It is subdivided from 0 to 150 N-m and 0 to 300 N-m for indication of used impact energy. The scale is also divided from 0 to 160 degrees for the indication of angle of swing.

4·37 Charpy Impact Test

The Charpy test is carried out on a specimen, which is 55 mm × 10 mm × 10 mm in size and has a 2 mm deep notch at its centre making an angle of 45° as shown in Fig. 4.10(*a*).

The specimen is placed horizontally as a simply supported beam between two anvils 40 mm apart in such a way that the striking hammer strikes the specimen on the face which is opposite to the notch as shown in Fig. 4.10 (*b*).

The pendulum is released from the right side from a *known angle. The pendulum, after breaking the specimen, continues to swing on the opposite side through some angle as shown in Fig. 4.10(*c*). The energy absorbed by the specimen, during breaking, is given by the difference between the angle through which the pendulum was released and the angle through which the pendulum has reached after breaking the specimen.

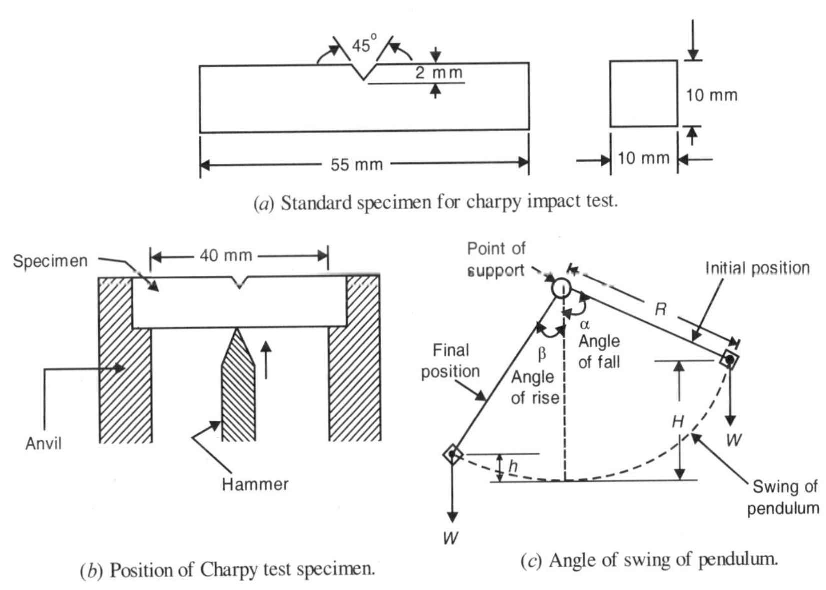

(*a*) Standard specimen for charpy impact test.

(*b*) Position of Charpy test specimen.

(*c*) Angle of swing of pendulum.

Fig. 4.10. Charpy impact test.

Neglecting losses, the energy used in breaking away the specimen is determined as given below :

Let W = Weight of the pendulum in newtons,

α = Angle through which the pendulum falls,

β = Angle through which the pendulum rises,

R = Distance between the centre of gravity of the pendulum and the axis of rotation in metres.

From the geometry of the Fig. 4.10(*c*), we know that initial energy of the pendulum

= Potential energy at height H

$= WH = WR\,(1 - \cos\alpha)$

* The maximum angle, from which the pendulum is released, is 160°.

and energy after breaking the specimen

$$= Wh = WR\,(1 - \cos\beta)$$

∴ Energy required to break away the specimen

$$= WH - Wh = WR\,(1 - \cos\alpha) - WR\,(1 - \cos\beta)$$

$$= WR\,(\cos\beta - \cos\alpha)$$

4·38 Izod Impact Test

The izod test is carried out on a specimen, which is 75 mm × 10 mm × 10 mm in size and has a 2 mm deep notch making an angle of 45° as shown in Fig. 4.11(*a*).

The specimen is held vertically as a cantilever between two jaws, in such a way that the striking hammer strikes the specimen on the same face as that of notch as shown in Fig. 4.11(*b*).

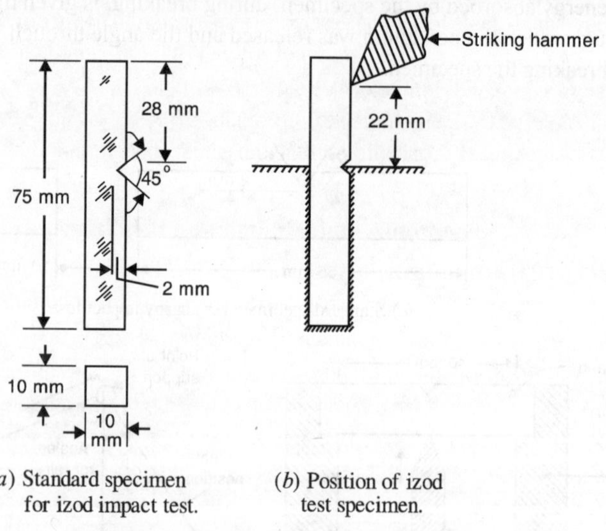

(*a*) Standard specimen for izod impact test.

(*b*) Position of izod test specimen.

Fig. 4.11. Izod impact test.

The remaining procedure for obtaining the energy required to break away the specimen is the same as discussed in the Charpy test.

4·39 Fatigue Test

The failure of a material, under repeatedly applied stress, is called fatigue. It has been observed that some of the machine parts such as axles, shafts, crankshafts, connecting rods, springs, pinion teeth etc., are subjected to varying stresses. It includes the variation in the intensity of the same type of stress as well as different types of stresses (*i.e.*, change of stress from tensile to compressive and vice versa). The varying stresses may be broadly classified into the following four types :

1. The stress varying between two limits of equal value, but of opposite sign.
2. The stress varying between two limits of unequal values, but of opposite sign.
3. The stress varying between zero and a definite value.
4. The stress varying between two limits of unequal values, but of same sign.

Though there are a numerous ways of fatigue test in a laboratory, yet the basic principle is the same. The most common test is conducted on a rotating beam fatigue testing machine as shown in Fig. 4.12.

The test piece (or specimen) is loaded in pure bending. Now the test piece is rotated with the help of a motor. A little consideration will show, that with each rotation, all points on the circumference of the specimen will be alternately in tension and compression. Thus each revolution of the specimen will constitute a complete cycle of stress reversal. The speed of the motor will indicate the frequency of the stress reversal.

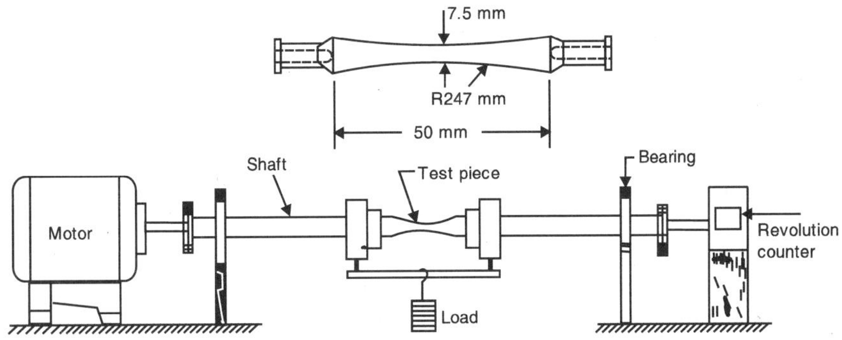

Fig. 4.12. Rotating beam fatigue testing machine.

In this test, a number of identical test pieces or specimens (say 8 to 10) are made from the same material. The first test piece is fixed in the machine, and is loaded as shown in the figure. Generally, this load is not less than that which can produce a stress equal to 3/4 of the tensile strength of the material upto its elastic limit. The speed of the motor shaft is kept constant. It has been found that after a sufficient number of stress reversals (or rotations) a crack is formed, on the outer surface, in the form of a ring. This crack goes on extending towards the centre of the test piece, till it breaks away. It has also been observed that the speed of the motor (or frequency of stress reversal) has no effect on the result. But it is the load or intensity of the stress, which controls the result.

After the first test piece breaks away, the second piece is tested with a decreased load. Similarly third, fourth... etc., test pieces are then tested with still decreased loads. It will be interesting to know that the number of stress reversals will go on increasing each time with the decrease in load. After some tests, a limit is reached, when the stress is not sufficient to break the test piece even after 10×10^6 stress reversals (in some countries this limit is 20×10^6). This safe stress, which after reversing for 10×10^6 times does not cause the test piece to break, is called *endurance limit* or *fatigue limit*.

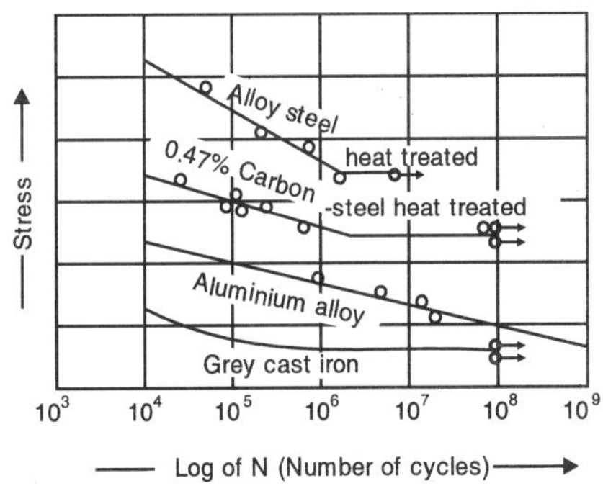

Fig. 4.13. Fatigue test curve.

Now the fatigue test results are plotted on diagrams known as *S-N* diagrams. In this diagram, the stress is plotted against vertical axis and log of *N*(*i.e.*, number of cycles) is plotted against horizontal axis as shown in Fig. 4.13.

The given figure shows the test results for steel, grey cast iron and aluminium alloy.

4·40 Creep Test

The continuous deformation of a metal, under a steady load, is known as creep. This test is very essential to predict the working life of some members or machine components which are subjected to creep. It is always exhibited in metals like iron, nickel, copper and their alloys at increased temperatures. But some metals like zinc, tin, lead and their alloys also creep at room temperatures. It has been observed that some organic metals such as plastics and rubber are very sensitive to creep.

Strictly speaking, the creep in metals is caused by slip occurring along crystallographic directions in the individual crystals together with some deformation of grain boundary. As a matter of fact, most of the deformation in case of metals, is non-recoverable. However, a small fraction of this plastic deformation is recovered with the time after the load is removed.

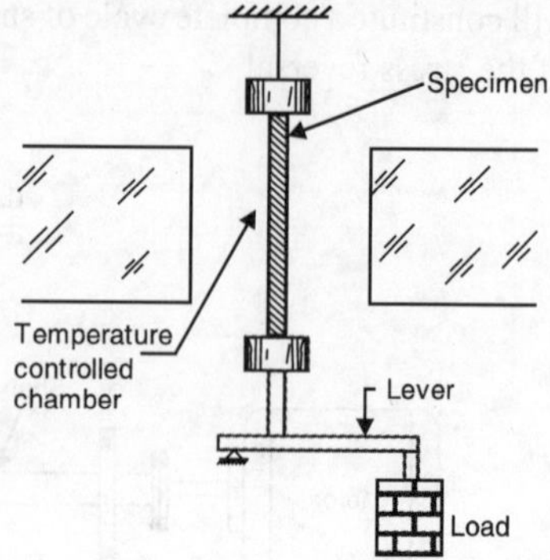

Fig. 4.14. Creep test.

The creep test is generally performed by applying a static load to one end of the lever system. The other end is attached to the specimen, under test in the furnace, and held at constant temperature as shown in Fig. 4.14.

The axial deformation is read, periodically, throughout the test, a curve is plotted between extension (*i.e.*, strain) along the vertical axis and time along the horizontal axis. This procedure is repeated for different loads at the same temperatures. The maximum permissible strain and working life can be estimated from these curves.

Fig. 4.15 shows typical creep curves representing the extension or strain along the vertical axis and the time along the horizontal axis under the application of load. The creep curve indicates the following three different stages of creep :

1. *Primary stage.* This is the initial part of the curve. In this part, the curve indicates rapid extension of the specimen. This part of the curve helps us in providing proper clearance in the various components of machines.

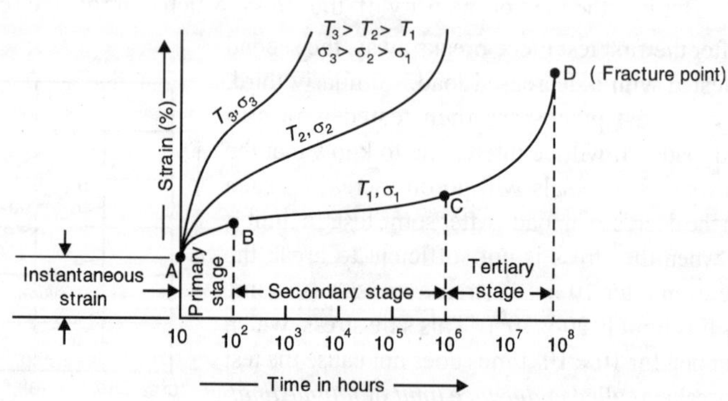

Fig. 4.15. Creep curves.

2. *Secondary stage.* This the most important stage, which indicates that the creep occurs at more or less constant rate, known as minimum creep rate. It will be interesting to know that it is the secondary stage of the creep, which has maximum practical utility.

3. *Tertiary stage.* This is the part of the curve, which indicates rapid extension of the specimen, which finally leads to rupture.

It may be noted from the creep curve that the portions *AB* and *CD* (*i.e.*, primary stage and tertiary stage) are short time periods as compared to portion *BC* which represents the entire life period of the specimen. A little consideration will show, that this period is of great importance, because the life time of a machine component depends upon the rate of this extension. If any attempt is made to decrease the tensile strength, the slope *BC* will decrease accordingly. It is thus obvious, that in such cases, the design must be based on the assumption of a definite period of service and definite amount of permissible distortion. For example, in the design of moving parts in steam turbines and automobile industry, the creep should not exceed 1 percent in 1000 hours.

It may also be noted that as the temperature (T) or the applied stress (σ) increases, the creep curve shifts upwards and the time duration of the various stages reduces significantly as shown in the figure. This indicates that the value of creep, increases with the increase in temperature as well as stress. Therefore, the materials to be used at a high temperature must have high melting point.

4·41 Non-Destructive Tests

We have already discussed that in non-destructive tests, the material or component is tested for its surface defects and internal soundness, without destroying them. The commonly used non-destructive tests are :

1. Visual inspection; 2. Radiographic test; 3. Ultrasonic test; 4. Liquid penetrant test; and 5. Magnetic particle test

These tests are discussed, in detail, in the following pages.

4·42 Visual Inspection

This is one of the first tests to be carried out on the manufactured component. It is a simple, cheap and easy method to inspect any component. For visual inspection, the surface of the component is observed with the naked eye or by optical aids like hand lens or a microscope, in order to detect the surface defects. The inspection should be carried out carefully and systematically on every component. The quality of result solely depends upon the knowledge and skill of the observer.

4·43 Radiographic Test

In this test, high frequency (or short wavelength) radiations of constant intensity are made to fall on the component to be tested. Some of the radiations, while passing through the component, are absorbed by the defects and the remaining radiations emerges out from the other side of the metal component. The intensity of the transmitted radiations are usually recorded on a sensitive photographic film. It may be noted that the darker region on the photographic film represents a defective portion and the less darker region represents the perfect portion. It is due to the fact that more radiations pass through the defective area than the perfect one. The developed image of the photographic film is called *radiography*, and the various techniques to determine the flaws or cracks in the material are known as *radiographic techniques*. The radiographic techniques may be broadly classified into the following three categories, depending upon the nature of high frequency radiation sources :

1. X-radiography; 2. Gamma radiography; and 3. Neutron radiography.

In *X-radiography*, the X-rays are used as a high frequency radiations. This technique is used to detect internal defects in castings, forgings, and components produced by mechanical working processes. But this technique has a drawback that it cannot be used to detect smaller defects. This is due to the fact that due to scattering of X-rays, the small defects are not detected.

In *gamma radiography*, the radioactive elements like radium, cobalt etc. are used as a source of producing radiations of very high frequency (even higher than X-rays). Since the gamma rays have more penetrating power than X-rays, therefore, they are more effective for testing of thicker sections. The gamma radiography is quicker and less expensive than X-radiography. But it requires a longer period to obtain radiography as compared to X-radiography.

In *neutron radiography*, a neutron beam is used as a source of high frequency radiations. This technique is very useful for inspection of very light material such as plastics, explosives, rubber components etc. where X-radiography and gamma radiography can not be used. It may also be used for the detection of voids and detonators as well as for inspection of rubber and paper products etc.

4·44 Ultrasonic Test

The term ultrasonic means a sound wave with a frequency above the upper limit of human hearing. The ultrasonic waves are usually produced by the piezoelectric effect within the crystal probe which is placed on the top surface of the component. In this test, ultrasonic or high frequency (0.5 to 25 MHz) vibrations are transmitted to surface of the component to be tested. These vibrations are reflected partially or fully after striking the defective or other surface of the component. The reflected vibrations are transferred to oscilloscope, which are finally interpreted to get the results.

There are two methods of transmitting and receiving the ultrasound vibrations. In the first method, separate transducers are used for transmitting and receiving vibrations. In the second method, the same transducer is used for transmitting and receiving vibrations. The ultrasonic testing is very useful for detecting voids, cracks and other defects near or far below the surface of any material. This method has the advantage of detecting air gaps even of the order of 0.003 mm.

4·45 Liquid Penetrant Test

The liquid penetrant test is used to detect the cracks which extends upto the surface of the material. This method is specially used for non-ferrous metals and non-metallic substances. In this simple test, some coloured liquid (known as *penetrant*) is sprayed on the surface of the cleaned and dried component to be tested. Some time is allowed for the penetrant to flow into the cracks of the component through capillary action. The excess penetrant is now wiped off from the surface and then coated with a thin layer of white chalk. The penetrant trapped in the cracks seeps out slowly and discolour the chalk at the site of the cracks. Thus the presence of defect is revealed.

In an alternative method, a fluorescent material (instead of a coloured liquid) is most widely used. In this method, a component is dipped into a liquid containing a fluorescent compound such as anthracence. The solution penetrates into the cracks. The excess liquid at the surface is wiped off and dried. The cracks are revealed under the ultraviolet radiation of the light due to the presence of flourescent material.

4·46 Magnetic Particle Test

The magnetic particle test is normally used for detecting invisible cracks only which are either on the surface or extend upto the surface of the components made of ferrous alloys or magnetic materials such as iron, steel, nickel alloys etc. This test is based on the principle that if a magnetic field is passed through a component having a crack or flaw, the magnetic lines of flux will get distorted near the crack. It may be noted that the magnetic lines of flux will be uniform for perfectly sound component.

Another method of detecting the cracks is by magnetising the component and then sprinkling the fine magnetic particles (iron powder) over the whole surface. The magnetic poles formed at the crack attract the iron powder towards it. Thus, there will be a concentration of the iron particles at the location of the cracks.

In order to increase the visibility of a crack, fluorescent magnetic particles or ultra violet light may be used. This test is carried out by a special equipment known as *magnaflux*.

QUESTIONS

1. Name the important properties of a material. Why it is necessary for an engineer to know the mechanical properties of a metal.
2. Define the terms stress and strain. Explain their different types.
3. What do you understand by Modulus of elasticity and bulk modulus.

4. Explain the term Poisson's ratio. What is its value for commonly used engineering materials ?
5. Name the various types of mechanical properties. Describe them briefly.
6. What is the difference between (*a*) ductility and malleability; (*b*) brittleness and hardness; (*c*) stiffness and toughness.
7. Define resilience. Why this property is important for a material ?
8. Explain the terms creep and fatigue. What do you understand by the terms creep strength and fatigue strength ?
9. Describe the various technological properties of metals.
10. What is destructive testing ? List the different types of destructive tests.
11. Draw a stress-strain diagram for a mild steel specimen under tensile testing. What different tensile properties are determined from this test ?
12. What is meant by true stress and true strain ? How it differs from engineering stress and strain?
13. Name the different types of hardness tests. Describe the method of performing Brinell hardness test.
14. Explain the process of Rockwell hardness testing and Vickers hardness testing. What is the relation between the Brinell hardness number and the Rockwell hardness number on B scale ?
15. Describe briefly the microhardness test, rebound hardness test and scratch hardness test.
16. What is an impact test ? How the charpy and izod impact tests are performed with the pendulum type impact testing machine ?
17. Explain the purpose and method of fatigue test. What is S-N curve ?
18. Discuss the method of conducting a creep test. Draw a typical creep curve showing the various stages of creep. What important information it gives ?
19. What is non-destructive testing ? Name the various types of non-destructive tests.
20. Describe briefly the following non- destructive tests :

    (*a*) Ultrasonic test; (*b*) Liquid penetrant test; (*c*) Magnetic particle test.

OBJECTIVE TYPE QUESTIONS

1. The property of a material by virtue of which it can withstand or support an external force or load without rupture, is known as

    (*a*) stiffness (*b*) strength (*c*) elasticity (*d*) ductility
2. The property of a material by virtue of which it can be rolled or hammered into thin sheets, is called

    (*a*) malleability (*b*) plasticity (*c*) toughness (*d*) brittleness
3. The plasticity is that property of a material by virtue of which

    (*a*) it can absorb maximum energy before fracture takes place

    (*b*) it stores energy and resists shock and impact loads

    (*c*) it will fracture or break without any appreciable deformation

    (*d*) it deforms permanently (without fracture) under the action of external forces
4. The creep is that property of a material by virtue of which

    (*a*) it fails at a stress below the yield point stress, when the material is subjected to repeated stress.

    (*b*) it undergoes a slow and permanent deformation at constant stress.

    (*c*) it will fracture or break without any appreciable deformation

    (*d*) it stores energy and resists shock and impact loads

5. The ultimate stress of a material is .... than the breaking stress.

(*a*) higher (*b*) lower

6. The maximum amount of energy which can be stored in a body upto the elastic limit is known as

(*a*) modulus of resilience (*b*) modulus of elasticity

(*c*) modulus of toughness (*d*) proof resilience

7. Which of the following statement is correct ?

(*a*) The tensile test is generally performed on brittle materials

(*b*) The compression test is generally performed on ductile materials

(*c*) The compressive strength of brittle materials is much more than the tensile strength

(*d*) all of the above

8. The relation between the ultimate tensile strength (U.T.S.) in MPa and Brinell hardness number (B.H.N.) is given as

(*a*) U.T.S. = B.H.N. × 0.35 (*b*) U.T.S. = B.H.N. × 3.5

(*c*) U.T.S. = B.H.N. × 0.65 (*d*) U.T.S. = B.H.N. × 6.5

9. In order to find the Brinell hardness number of a steel with a indentor of 10 mm diameter, the load applied is

(*a*) 30 kN (*b*) 40 kN (*c*) 50 kN (*d*) 60 kN

10. In a Vicker's hardness test, a diamond square based pyramid indentor is used. The angle between the opposite faces of diamond pyramid is

(*a*) 36° (*b*) 72° (*c*) 108° (*d*) 136°

11. In a Charpy impact test, the specimen is placed .... as a simply supported beam between two anvils.

(*a*) vertically (*b*) horizontally

12. The size of a specimen for izod impact test is

(*a*) 45 mm × 4 mm × 4 mm (*b*) 55 mm × 6 mm × 6 mm

(*c*) 65 mm × 8 mm × 8 mm (*d*) 75 mm × 10 mm × 10 mm

13. The resistance to fatigue of a material is measured by

(*a*) endurance limit (*b*) elastic limit

(*c*) Young's modulus (*d*) ultimate tensile strength

14. Which of the following is a non-destructive test ?

(*a*) Tensile test (*b*) Ultrasonic test

(*c*) Compression test (*d*) Creep test

15. In a radiographic test for the detection of internal defects,

(*a*) ultrasonic vibrations are transmitted to the surface of the component

(*b*) some coloured liquid is sprayed on the surface of the component

(*c*) X-rays or gamma rays are passed through the component

(*d*) a special equipment known as magnaflux is used

ANSWERS

1. (*b*)	2. (*a*)	3. (*d*)	4. (*b*)	5. (*a*)
6. (*d*)	7. (*c*)	8. (*b*)	9. (*a*)	10. (*d*)
11. (*b*)	12. (*d*)	13. (*a*)	14. (*b*)	15. (*c*)

5

Ferrous Metals and Alloys

1. Introduction. 2. Pig Iron. 3. Manufacture of Pig Iron. 4. Description of Blast Furnace. 5. Melting of Iron Ore in Blast Furnace. 6. Modren Methods of Producing Pig Iron. 7. Composition of Pig Iron. 8. Classification of Pig Iron. 9. Cast Iron. 10. Manufacture of Cast Iron (Cupola). 11. Effect of Impurities on Cast Iron. 12. Types of Cast Iron. 13. Effect of Alloying Elements on Cast Iron. 14. Wrought Iron. 15. Manufacture of Wrought Iron. 16. Steel 17. Classification of Steel Depending upon the Carbon Content. 18. Classification of Steel Depending upon the Degree of Deoxidation. 19. Indian Standard Designation of Steel. 20. Steels Designated on the Basis of Mechanical Properties. 21. Steels Designated on the Basis of Chemical Composition. 22. Manufacture of Steel. 23. Cementation Process. 24. Crucible Process. 25. Bessemer Process. 26. Open Hearth Process. 27. Duplex Process. 28. L-D Process (Linz-Donawitz Process). 29. Electric Process. 30. Heroult Direct Arc Type Electric Furnace. 31. Coreless Induction Type Electric Furnace. 32. Effect of Impurities on Steel. 33. Bright Steel. 34. Free Cutting Steels. 35. Alloy Steel. 36. Effect of Alloying Elements on Steel 37. Indian Standard Designation of Low and Medium Alloy Steels, 38. Special Steels. 39. Stainless Steel. 40. Heat Resisting Steels. 41. Indian Standard Designation of High Alloy Steels. (Stainless Steel and Heat Resisting Steel). 42. High Speed Tools Steels. 43. Indian Standard Designation of High Speed Tool Steel. 44. Spring Steels. 45. Magnet Steel. 46. Maraging Steel. 47. Cutting Alloys.

5·1 Introduction

The metals play an important role in the industries of a nation. Thus it is essential for an engineer to know about the sources and techniques of its production. The metals used in the industry are of the following two types :

1. Ferrous metals ; and 2. Non-ferrous metals.

The **ferrous metals* are defined as those metals which contain iron as their main constituent such as pig iron, cast iron, wrought iron, steel and their alloys. The principal raw material for all ferrous metals is pig iron.

The *non-ferrous* metals are defined as those metals which contain a metal other than iron as their main constituent such as copper, aluminium, brass, tin, zinc etc.

The *non- ferrous metals* are discussed, in detail, in the next chapter.

5·2 Pig Iron

It is the crude form of iron and is used as a raw material for the production of various other ferrous metals, such as cast iron, wrought iron and steel. The pig iron is obtained by smelting iron ores in a blast furnace. The three principal raw materials required in the production of pig iron are iron ore, fuel and flux. These raw materials are discussed, in detail, as follows :

1. *Iron ore.* Any mineral body from which metal can be economically recovered is called as *ore*.

The iron ores are found in various forms as shown below :

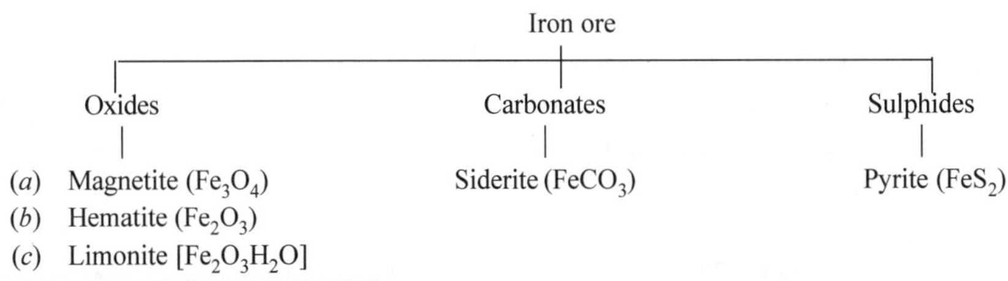

* The word 'ferrous' is derived from a latin word 'ferrum' which means iron.

The iron ores in our country, are mostly found in Bihar, Orissa, Madhya Pradesh, Andhra Pradesh, Karnataka and Tamilnadu. We shall now discuss these iron ores as follows :

(*a*) *Magnetite.* It is also called *load stone* or *magnetic iron ore* and it contains about 62 to 72 per cent of iron when pure. This is a black coloured ore and is available in the form of crystals. Its specific gravity is about 5.2. Since the production of pig iron from magnetite ore is difficult, therefore, it is not recommended for use.

(*b*) *Hematite.* It is also called *red hematite ore* because of its red colour. It contains about 60 to 70 percent of iron when pure. Its specific gravity is about 5. This ore is widely used for the production of pig iron.

(*c*) *Limonite.* It is also called *brown hematite ore* and contains about 40 to 60 percent iron when pure.

(*d*) *Siderite.* It is also known as *spathic iron.* It contains about 35 to 48 percent of iron. This is a yellowish white or grey coloured ore and is available is crystalline structure. Its specific gravity is about 3.7.

(*e*) *Pyrite.* It contains about 30 to 40 percent of iron. Its specific gravity is about 4.8. Since the iron obtained from this type of ore is of very low quality due to high sulphur content, therefore, it is not used for manufacturing pig iron.

Note : The iron ores are never found in pure state but are always associated with earthy matter called the 'Gangue' of the ore. The chief gangue-minerals associated with iron ores are in the form of oxides such as those of silica, alumina, lime, soda, potash, magnesia, manganese. Sulphur is present in the form of sulphides and sulphates, and phosphorus as phosphates.

2. *Fuel.* The most commonly used fuel in a blast furnace is *hard coke*. It serves the following two purposes :

(*a*) It provides the required heat for melting the iron ore, and

(*b*) It combines with the oxygen of the ore in order to minimise the same.

The coke used should have the following properties :

(*i*) It should be hard.

(*ii*) It should have a high calorific value and low ash contents.

(*iii*) It should have high density and good porosity.

(*iv*) It should be free from phosphorous and sulphur.

(*v*) It should be of uniform size.

3. *Flux.* The most commonly used flux in a blast furnace is *limestone,* but dolomite is also sometimes used. The flux is used for removing impurities such as ash, sulphur and the residues of the burnt fuel. The flux melts and drives away the impurities in the form of slag leaving pure iron. It may be noted that for each tonne of pig iron produced, about 500 kg to 750 kg of limestone is added.

5·3 Manufacture of Pig Iron

The pig iron is obtained from the iron ores in the following steps :

1. *Concentration.* First of all, the iron ore as obtained from mines is broken into pieces of sizes 25 mm to 50 mm. After this, other adhering impurities like clay, sand etc. are removed by washing with water and the ore is thus concentrated.

2. *Calcination or roasting.* The concentrated ore is then heated in shallow kilns in the presence of excess air. During this process, most of the moisture, carbon dioxide, sulphur and arsenic are expelled out and ferrous oxide is changed to ferric oxide. It is a porous, less fusible and the desired form of the iron oxide.

3. *Smelting.* The calcined ore is now submitted to the process of smelting *i.e.,* reducing the ore with carbon in the presence of a flux. The smelting is carried out in a large tower called *blast furnace.*

Note : When the calcined ore is in powder form, then it must be put to a process, known as *sintering,* before it is fed for smelting. In a sintering process, the ore is fed on grate in a shallow box and a blast of air is supplied from the bottom. This causes an agglomeration of the powdery ore and reduces the sulphur content.

5·4 Description of Blast Furnace

A blast furnace, as shown in Fig. 5.1, is a chimney-like structure made of heavy-steel plates, lined inside with fire bricks to a thickness of 1.2 to 1.5 metres. It is about 30 metres high with a maximum internal diameter of 9 metres at its widest cross-section. The portion of the furnace above its widest cross-section is called *stack* and is supported on cast iron pillars. The topmost portion of the stack is called *throat* through which the charge is fed into the furnace. The portion of the furnace, below its widest cross-section, is known as *bosh* or the burning zone (or zone of fusion) which rests on a short cylindrical section called the *hearth.* The bosh is provided with holes for a number of water-jacketed iron blowing pipes known as *tuyers.* The heated air under pressure is forced through the water cooled tuyers for the purpose of burning the fuel in the furnace. The tuyers are 12 to 15 in number and are connected to bustle pipe surrounding the furnace. The bustle pipe is, in turn, connected to special heaters (known as blast stoves) through a blast pipe.

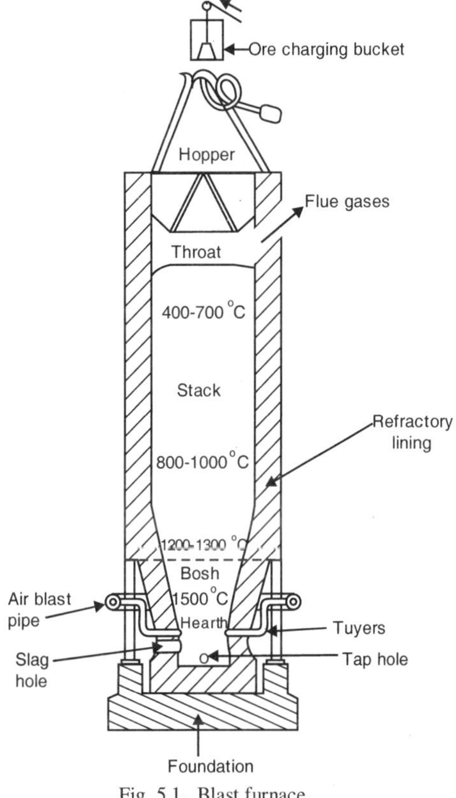

Fig. 5.1. Blast furnace.

The stoves are vertical cylinders lined inside with refractory material and are used for heating the air blast. Generally, three stoves are used for one blast furnace. Each stove is as high as blast furnace and about 6 metres in diameter. The air blast is pumped through the stoves into a 90 cm diameter blast pipe from where it is carried to the tuyers. During functioning of the blast furnace, the stoves are heated by the waste gas from the blast furnace. The top of the furnace is provided with a double cup and cone arrangement which prevents the escape of gases while charging the furnace.

5·5 Melting of Iron Ore in Blast Furnace

The charge of the blast furnace consists of the following :

1. Calcined ore (8 parts),
2. Coke (4 parts), and
3. Lime stone (1 part).

The mixed charge with the above composition is carried to the top of the furnace by means of an inclined hoist or skip. It is first allowed to fall on the upper cone (or hopper) which is then lowered to admit the charge into the charging bell which serves as seal. The upper cone is now raised to close the furnace top and the lower cone is lowered to admit the charge into the furnace. In this way, the escape of gases during the charging is prevented. It may be noted that the charge is fed in successive layers and upto a level just below the throat. The blast of *hot air at about 850°C is forced into the furnace, under a pressure of 0.14 N/mm^2 through the number of tuyers provided at the bottom of the furnace. The following reactions take place in the blast furnace.

1. In the lower part of the furnace (called *zone of fusion*), the hot air blast promotes the combustion of coke and helps to carry the gases upwards. The oxygen in the hot air blast combines with the carbon in the coke to form carbon dioxide producing lot of heat.

$$C + O_2 \longrightarrow CO_2 + 400\ \text{kJ}$$

Here the temperature is 1500°C. A little above this, the temperature prevailing is 1200 to 1300°C. In this region, the porous iron sliding down from above melts. Any oxide which might have escaped oxidation is also reduced by coke.

2. In the middle part of the furnace (called *zone of absorption*), the temperature is 800°C to 1000°C. In this region, the resulting carbon dioxide rising up is reduced to carbon monoxide accompanied by heat absorption. The limestone sliding down decomposes to give lime and carbon dioxide. Lime thus obtained acts as a flux and combines with silica (present as impurity *i.e.,* gangue) to form calcium silicate. The chemical reactions are as follows :

$$CO_2 + C \longrightarrow 2\,CO + 160\ \text{kJ}$$
$$CaCO_3 \longrightarrow CaO + CO_2$$
$$CaO + SiO_2 \longrightarrow CaSiO_3\ \text{(slag)}$$

3. In the upper part of the furnace (called *zone of reduction*), the temperature is 400°C to 700°C. In this region, the iron oxide is reduced to spongy iron by carbon monoxide rising up.

$$Fe_2O_3 + 3CO \longrightarrow 2Fe + 3CO_2$$

The reduction of ferric oxide (Fe_2O_3) by carbon monoxide takes place in the following stages :

(*a*) $$3\,Fe_2O_3 + CO \longrightarrow 2Fe_3O_4 + CO_2$$
(*b*) $$Fe_3O_4 + CO \longrightarrow 3FeO + CO_2$$
(*c*) $$FeO + CO \longrightarrow Fe + CO_2$$

At the bottom of the furnace the molten iron sinks down while above this floats the fusible slag which protects the molten iron from oxidation. The slag is tapped off at intervals of 2 to 3 hours through a slag hole provided just below the tuyers, while the molten iron is drawn off at intervals of 4 to 6 hours through a separate hole provided near the bottom of the hearth. The molten iron thus produced in called *pig iron.* It is either taken in the ladles to be immediately converted into steel or poured into sand moulds of suitable size prepared on the ground near the blast furnace. The iron blocks, so obtained, are known as *pigs.* Each pig is usually one metre long with a D-shaped cross-section measuring about 100 mm each way and the mass in about 36 kg.

Note : The slag from the blast furnace consists of calcium, aluminium and ferrous silicates. It is used as ballast for rail roads, mixed with tar for road making and in the cement manufacture for which purpose it is burned with lime and then finely powdered.

5·6 Modren Methods of Producing Pig Iron

The following are few modren methods of producing pig iron :

1. The electric furnaces are widely used, in places, where the metallurgical coke is not available in sufficient quantities and the electricity is cheap. In this furnace, preheating of the iron ore is done and only its reduction is done by coke. The advantage of using the electric furnace is that the coke consumption is reduced by 50 percent.

2. Now-a-days, low shaft design of blast furnaces are developed so that low grade coke and

---

* The air is heated in a series of blast stoves and passes through a steel pipe (about 90 cm in diameter and lined with fire bricks) to a number of tuyers.

iron ore may be used. The use of oxygen is also increased to help the reactions.

3. Another new process known as *sponge iron process* is being developed. This process enables direct treatment and reduction of low grade iron ores without using blast furnace. It is of great advantage to a country like ours where large quantities of low grade iron ore remain unused.

5·7 Composition of Pig Iron

The pig iron from the blast furnace contains 90 to 92 percent of iron. The various other elements present in pig iron are carbon, silicon, manganese, sulphur and phosphorus. The average composition of pig iron is as follows :

Iron-ferrite (Fe) = 93 to 94 percent
Carbon (C) = 1 to 5 percent
Silicon (SiO_2) = 1 to 2 percent
Manganese (Mn) = 1 to 2 percent
Sulphur (S) and Phosphorus (P)
= 1 to 2 percent

Note : Carbon plays a very important role in iron. It exists in iron in two forms :

1. in a free form *i.e.,* as *graphite,* and
2. in a combined form *i.e.,* as *cementite* and *pearlite.*

The presence of free carbon in iron imparts softness and a coarse crystalline structure to the metal, while the combined carbon makes the metal hard and gives a fine grained crystalline structure. The presence of carbon in iron is influenced by silicon. In molten condition, silicon throws excess of free carbon out of solution and precipitates it until the percentage of carbon is almost the same as its own value. It has also a very powerful influence in decomposing the carbon and iron.

5·8 Classification of Pig Iron

The pig iron is classified according to the contents of free and combined carbon as follows. These classifications are termed as grades.

1. *Grey pig iron* (*Grades 1, 2 and 3*). These grades of pig iron contain almost all of free carbon or graphite. The percentage of graphite carbon is more than 3 percent and combined carbon less than 1 percent. The percentage of silicon is almost the same as that of free carbon. This is a soft variety and has bigger size crystals.

2. *White pig iron* (*Grades 5, 6 and 7*). These grades of pig iron contain almost all of the carbon in the combined state. The percentage of free carbon is less than 1 percent; combined carbon more than 3 percent and silicon less than 1 percent. The crystals are close grained. The metal is hard and strong and can be easily melted. The white variety of pig iron is used for making wrought iron.

3. *Mottled pig iron* (Grade 4). It is in between the grey and white variety. It contains equal proportions of free and combined carbon. It has a mottled appearance and an average hardness.

5·9 Cast Iron

The cast iron is obtained by remelting pig iron with coke and limestone in a furnace known as cupola. It is primarily an alloy of iron and carbon. The carbon contents in cast iron varies from 1.7 percent to 4.5 percent. It also contains small amounts of silicon, manganese, phosphorus and sulphur. The carbon in a cast iron is present either of the following two forms :

1. Free carbon or graphite, and 2. Combined carbon or cementite.

Since the cast iron is a brittle material, therefore, it cannot be used in those parts which are subjected to shocks. The properties of cast iron which makes it a valuable material for engineering purposes are its low cost, good casting characteristics, high compressive strength, wear resistance, and excellent machinability. The compressive strength of cast iron is much greater than tensile strength. Following are the values of ultimate strengths of cast iron :

Tensile	—	100 to 200 MPa
Compressive	—	400 to 1000 MPa
Shear	—	120 MPa

5·10 Manufacture of Cast Iron (Cupola)

The pig iron obtained from blast furnace is most impure and needs refining before using it for making suitable castings. The process involved in refining is remelting of pig iron with coke, limestone (or shell lime) and scrap castings in a cupola, as shown in Fig. 5.2.

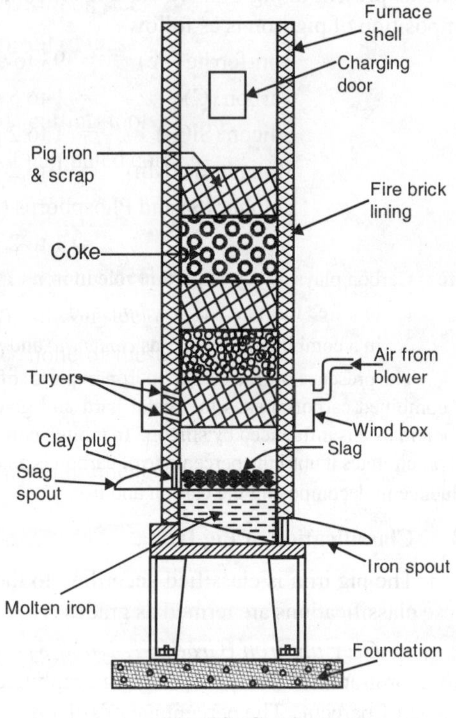

Fig. 5.2. Cupola.

It is very similar to a blast furnace of small size and is much simpler in construction. It consists of a vertical steel cylindrical shell of uniform diameter of about 1 metre and 4 metres high. It is lined with fire bricks and rests on a square bed plate which is supported on four cast iron pillars, above the ground. The square bed plate has a circular opening of the size of the inside dimensions of the furnace which is closed by two semicircular cast iron hinged doors. Generally cupolas are not worked continuously as are blast furnaces but are run for such period as required. Therefore, before using the furnace, it is cleaned of all the unburnt coke, slag etc., which settles at the floor of the furnace, by dropping down the semicircular hinged doors. The tuyers are placed regularly round the circumference of the furnace near the bottom for blowing air into the furnace. The furnace is open at the top and is provided with a side door near the top for charging. The different projecting spouts are provided at the bottom in opposite directions and at different levels for tapping off the slag and the molten iron.

The charge consisting of alternate layers of coke, pig iron mixed with scrap castings and limestone is fed to the furnace through the charging door. The scrap iron is used in order to improve quality of the cast iron and also to effect economy in the cost of its production. The* air for combustion is forced under pressure through the tuyers. The limestone is added as a flux, which combines with the impurities left in the pig iron and removes them in the form of slag which floats above the molten iron. The molten iron thus obtained is called cast iron and is cast into moulds.

5·11 Effect of Impurities on Cast Iron

We have discussed that the cast iron contains small percentages of silicon, sulphur, manganese and phosphorus. The effect of these impurities on cast iron are as follows :

1. *Silicon.* It may be present in cast iron upto 4%. It provides the formation of free graphite

* The air is not preheated.

which makes the iron soft and easily machinable. It also produces sound castings free from blow holes because of its high affinity for oxygen.

2. *Sulphur.* It makes the cast iron hard and brittle. Since too much sulphur gives unsound casting, therefore, it should be kept well below 0.1% for most foundry purposes.

3. *Manganese.* It makes the cast iron white and hard. It is often kept below 0.75%. It helps to exert a controlling influence over the harmful effect of sulphur.

4. *Phosphorus.* It aids fusibility and fluidity in cast iron, but induces brittleness. It is rarely allowed to exceed 1%. Phospheric irons are useful for casting of intricate design and for many light engineering castings when cheapness is essential.

5·12 Types of Cast Iron

Though there are many types of cast iron, yet the following are important from the subject point of view :

1. *Grey cast iron.* It is an ordinary commercial iron having the following compositions :

Carbon = 3 to 3.5%; Silicon = 1 to 2.75%; Manganese = 0.40 to 1%; Phosphorus = 0.15 to 1%; Sulphur = 0.02 to 0.15%; and the remainder is iron.

The grey colour is due to the fact that the carbon is present in the form of *free graphite. It has a low tensile strength, high compressive strength and no ductility. It can be easily machined. A very good property of grey cast iron is that the free graphite in its structure acts as a lubricant. Due to this reason, it is very suitable for those parts where sliding action is desired. The grey iron castings are widely used for machine tool bodies, automotive cylinder blocks, housings, flywheels, pipes and pipe fittings and agricultural implements.

According to Indian standard specifications (IS : 210- 1993), the grey cast iron is designated by the alphabets 'FG' followed by a figure indicating the minimum tensile strength in MPa or N/mm^2. For example, 'FG 150' means grey cast iron with 150 MPa or N/mm^2 as minimum tensile strength. The seven recommended grades of grey cast iron with their tensile strength and Brinell hardness number (B.H.N.) are given in the following table :

Table 5.1. Grey iron casting as per IS : 210-1993.

Indian standard desiguation	*Tensile strength (MPa or N/mm^2)*	*Brinell hardness number (B.H.N.)*
FG 150	150	130 to 180
FG 200	200	160 to 220
FG 220	220	180 to 220
FG 260	260	180 to 230
FG 300	300	180 to 230
FG 350	350	207 to 241
FG 400	400	207 to 270

2. *White cast iron.* It is a particular variety of cast iron which shows a white fracture. It has the following approximate compositions :

Carbon = 1.75 to 2.3%; Silicon = 0.85 to 1.2%; Manganese = 0.10 to 0.40%; Phosphorus = 0.05 to 0.20%; Sulphur = 0.12 to 0.35%; and the remaining is iron.

The white colour is due to the fact that the carbon is in the form of carbide (known as cementite) which is the hardest constituent of iron. This cementite is caused by quick cooling of molten iron.

---

* When filing or machining cast iron makes our hands black, then its shows that free graphite is present in cast iron.

The white cast iron has a high tensile strength and a low compressive strength. Since it is hard (BHN from 400 to 600), therefore, it cannot be machined with ordinary cutting tools but requires grinding as shaping process. The white cast iron may be produced by casting against metal chills or by regulating analysis. It is used for inferior castings and in places where a hard, wear resisting surface is desired as in the outer surface of car wheels, rolls for crushing grains and jaw crusher plates. The white cast iron is also used as a raw material in the production of malleable cast iron and wrought iron.

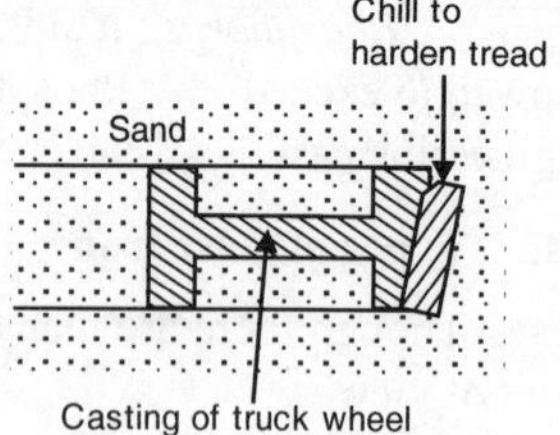

Fig. 5.3. Use of chills.

3. *Chilled cast iron.* It is a white cast iron produced by quick cooling of molten iron. The quick cooling is generally called *chilling* and the iron so produced is known as *chilled cast iron*. All castings are chilled at their outer skin by contact of the molten iron with the cool sand in the mould. But on most castings, this hardness penetrates to a very small depth (less than 1mm). Sometimes a casting is chilled intentionally and sometimes chilled becomes accidently to a considerable depth. The intentional chilling is carried out by putting inserts of iron or steel (chills) into the mould as shown in Fig. 5.3. When the molten metal comes into contact with the chill, its heat is rapidly conducted away and the hard surface is formed. Chills are used on any faces of a casting which are required to be hard to withstand wear and friction.

The process of chilling is used in the casting of rolls for crushing grains and jaw crusher plates. The running surface of rail-carriage wheels is also chilled.

4. *Mottled cast iron.* It is a product in between grey and white cast iron in composition, colour and general properties. It is obtained in castings where certain wearing surfaces have been chilled.

5. *Malleable cast iron.* The malleable iron is a cast iron-carbon alloy which solidifies in the as-cast condition in a graphite free structure, *i.e.,* the total carbon content is present in its combined form as cementite (Fe_3C).

The malleable cast iron is ductile and may be bent without breaking or fracturing the section. Its tensile strength is usually higher than that of grey cast iron and has excellent machining qualities. It is used for making machine parts for which the steel forgings would be too expensive and in which the metal should have a fair degree of accuracy *e.g.*, hubs of wagon wheels, small fittings for railway rolling stock, brake supports, parts of agricultural machinery, pipe fittings, door hinges, locks etc.

The malleable cast iron is obtained from white cast iron by a suitable heat treatment process (*i.e.,* annealing). The annealing process separates the combined carbon of the white cast iron into nodules of free graphite. The following two methods are used for this purpose :

1. Whiteheart process; and 2. Blackheart process.

In a *whiteheart process,* the white iron castings are packed in iron or steel boxes surrounded by a mixture of new and used hematite ore. The boxes are slowly heated to a temperature of 900–950°C and maintained at that temperature for several days. During this period, some of the carbon is oxidised out of the castings and the remaining carbon is dispersed in small specks throughout the structure. The heating process is followed by the cooling process which takes several more days. The result of this heat treatment is a casting which is tough and will stand heat treatment without fracture.

In a *blackheart process,* the castings used contain less carbon and sulphur. They are packed in a netural substance like sand and reduction of sulphur helps to accelerate the process. The castings are heated to a temperature of 850 to 900°C and maintained at that temperature for 3 to 4 days. The carbon in this process transforms into globules, unlike whiteheart process. The castings produced by this process are more malleable.

Note : ***(a)*** According to Indian standard specifications (*IS : 14329 – 1995), the malleable cast iron may be either white heart, blackheart or pearlitic, according to the chemical composition, temperature and time cycle of annealing process.

(b) The *whiteheart malleable cast iron* obtained after annealing in a decarburizing atmosphere have a silvery-grey fracture with a heart dark grey to black. The microstructure developed in a section depends upon the size of the section. In castings of small sections, it is mainly ferritic with certain amount of pearlite. In large sections, microstructure varies from the surface to the core as follows :

Core and intermediate zone : Pearlite + ferrite + temper carbon

Surface zone : Ferrite.

The microstructure shall not contain flake graphite.

(c) The *blackheart malleable cast iron* obtained after annealing in an inert atmosphere have a black fracture. The microstructure developed in the castings has a matrix essentially of ferrite with temper carbon and shall not contain flake graphite.

(d) The *pearlitic malleable cast iron* obtained after heat-treatment have a homogeneous matrix essentially of pearlite or other transformation products of austenite. The graphite is present in the form of temper carbon nodules. The microstructure shall not contain flake graphite.

(e) According to IS : 14329 – 1995, the whiteheart, blackheart and pearlitic malleable cast irons are designated by the alphabets WM, BM and PM respectively. These designations are followed by a figure indicating the minimum tensile strength in MPa or N/mm^2. For example 'WM 350' denotes whiteheart malleable cast iron with 350 MPa as minimum tensile strength. The following are the different grades of malleable cast iron :

Whiteheart malleable cast iron — WM 350 and WM 400

Blackheart malleable cast iron — BM 300 ; BM 320 and BM 350

Pearlitic malleable cast iron — PM 450 ; PM 500; PM 550; PM 600 and PM 700

6. *Nodular or spheroidal graphite cast iron.* The nodular or spheroidal graphite cast iron is also called *ductile cast iron* or *high strength cast iron.* This type of cast iron is obtained by adding small amounts of magnesium (0.1 to 0.8%) to the molten grey iron just after tapping. The addition of magnesium causes the **graphite to take form of small nodules or spheroids instead of the normal angular flakes. It has high fluidity, castability, tensile strength, toughness, wear resistance, pressure tightness, weldability and machinability. It is generally used for castings requiring shock and impact resistance along with good machinability, such as hydraulic cylinders, valves, pipes and pipe fittings, cylinder heads for compressors and diesel engines, rolls for rolling mill and centrifugally cast products.

According to Indian standard specifications (IS : 1865 - 1991), the nodular or spheroidal graphite cast iron is designated by the alphabets 'SG' followed by the figures indicating the minimum tensile strength in MPa or N/mm^2 and the percentage elongation. For example, SG 400/15 means spheroidal graphite cast iron with 400 MPa as minimum tensile strength and 15 percent elongation. The Indian standard (IS : 1865 - 1991 recommends nine grades of spheroidal graphite cast iron based on mechanical properties measured on separately-cast test samples and six grades based on mechanical properties measured on cast-on sample as given in Table 5.2.

The letter A after the designation of the grade indicates that the properties are obtained on cast-on test samples to distinguish them from those obtained on separately-cast samples.

---

* This standard (IS : 14329 - 1995) supersedes the previous three standards, *i.e.*,

(*a*) IS : 2107-1977 for whiteheart malleable iron casting,

(*b*) IS : 2108-1977 for blackheart malleable iron casting, and

(*c*) IS : 2640-1977 for pearlitic malleable iron casting.

** The graphite flakes in cast iron act as discontinuities in the matrix and thus lower its mechanical properties. The sharp corners of the flakes also act as stress raisers. The weakening effect of the graphite can be reduced by changing its form from a flake to a spheroidal form.

Table 5.2. Recommended grades of spheroidal graphite cast iron as per IS : 1865-1991.

Grade	*Minimum tensile strength (MPa)*	*Minimum percentage elongation*	*Brinell hardness number (BHN)*	*Predominant constituent of matrix*
SG 900/2	900	2	280 – 360	Bainite or tempered martensite
SG 800/2	800	2	245 – 335	Pearlite or tempered structure
SG 700/2	700	2	225 – 305	Pearlite
SG 600/3	600	3	190 – 270	Ferrite + Pearlite
SG 500/7	500	7	160 – 240	Ferrite + Pearlite
SG 450/10	450	10	160 – 210	Ferrite
SG 400/15	400	15	130 – 180	Ferrite
SG 400/18	400	18	130 – 180	Ferrite
SG 350/22	350	22	≤ 150	Ferrite
SG 700/2A	700	2	220 – 320	Pearlite
SG 600/3A	600	2	180 – 270	Pearlite + Ferrite
SG 500/7A	450	7	170 – 240	Pearlite + Ferrite
SG 400/15A	390	15	130 – 180	Ferrite
SG 400/18A	390	15	130 – 180	Ferrite
SG 350/22A	330	18	≤ 150	Ferrite

7. *Alloy cast iron.* The cast irons as discussed above contain small percentages of other constituents like silicon, manganese, sulphur and phosphorus. These cast irons may be called as *plain cast irons.* The alloy cast iron is produced by adding alloying elements like nickel, chromium, molybdenum, copper and vanadium in sufficient quantities. These alloying elements give more strength and result in improvement of properties. The alloy cast iron has special properties like increased strength, high wear resistance, corrosion resistance or heat resistance. The alloy cast irons are extensively used for gears, automobile parts like cylinders, pistons, piston rings, crank cases, crankshafts, camshafts, sprockets, wheels, pulleys, brake drums and shoes, parts of crushing and grinding machinery etc.

8. *Meehanite cast iron.* The cast iron treated with calcium silicide as graphitizer, is known by the trade name of meehanite cast iron. The grey iron used for graphitisation should have low silicon and carbon in the range of 2.5 to 3%. If the grey iron having this range of carbon is cast through the normal process of casting, then it results in the production of white cast iron. But when calcium silicide is used as a graphitizer, it produces a cast iron of fine graphite structure which possesses excellent mechanical properties. It may be noted that it is not entirely due to the use of calcium silicide alone, but it is also due to the effective control of all other factors involved in the melting of iron and in moulding of casting.

The meehanite cast iron has high strength, toughness, ductility and machinability. It responds well to heat treatment and has similar electrical and magnetic properties as that of normal cast iron. It is suitable for machine tool castings.

There are large number of meehanite cast iron available under the following five categories :

1. General engineering; 2. Heat resisting; 3. Corrosion resisting; 4. Wear resisting; and 5. Nodular 'S' type.

5·13 Effect of Alloying Elements on Cast Iron

We have already discussed that alloy cast iron is produced by adding alloying elements like nickel, chromium, molybdenum, copper and vanadium. The effect of these alloying elements on the cast iron are as follows :

1. *Nickel.* It may be present in cast iron from 0.25 to 2%. It acts as a graphitizer in cast iron. It increases tensile strength, hardness, wear resistance and machinability. It also refines the grain size and increases the fluidity of the molten metal to make it suitable for castings of intricate design.

2. *Chromium.* It acts as a carbide stabilizer in cast iron. It increases the tensile strength, wear resistance, hardness and aids chilling. The addition of chromium ranges from 0.15 to 0.90% with or without the other alloying elements. When chromium is used with nickel in the ratio of 1 : 3, the improvement in wearing resistance is achieved.

3. *Molybdenum.* It may be present in cast iron between 0.25 to 1.5%. It is added to improve toughness, tensile strength, fatigue strength and machinability of cast iron. It also increases hardness and shock resistance of castings.

4. *Copper.* It promotes graphite formation and improves strength. It also increases toughness of cast iron. It is usually present in cast iron from 0.25 to 2.5%.

5. *Vanadium.* It is present in cast iron from 0.1 to 0.5%. It increases heat resistance, hardness, strength and machinability of cast iron.

Note : The cast irons alloyed with 10% nickel and 6% manganese become non-magnetic. This special type of non-magnetic cast iron is commercially known as *Nomag*.

5·14 Wrought Iron

It is the purest iron which contains at least 99.5% iron but may contain upto 99.9% iron. The typical composition of a wrought iron is

Carbon = 0.02%; Silicon = 0.12; Sulphur = 0.018%; Phosphorus = 0.02%; Slag = 0.07%; and the remaining is iron.

The wrought iron is a tough, malleable and ductile material. It cannot stand sudden and excessive shocks. Its ultimate tensile strength is 250 MPa to 500 MPa and the ultimate compressive strength is 300 MPa. It can be easily forged or welded. The melting point of wrought iron is about 1530°C. It is used for bolts and nuts, chains, crane hooks, railway couplings, water and steam pipes. It is also used as a raw material for crucible steel or in cementation process.

Note : The presence of phosphorus and sulphur in wrought iron produces *cold shortness* and *red shortness* respectively. The cold shortness renders the metal brittle and weak, thus making it unsuitable for cold working. The red shortness renders the metal brittle, thus making it unworkable at red heat. Thus these two constituent must be properly controlled.

5·15 Manufacture of Wrought Iron

The chief point in the manufacture of wrought iron is the oxidation of nearly all the carbon and other elements from pig iron. The following two processes are commonly used in the manufacture of wrought iron :

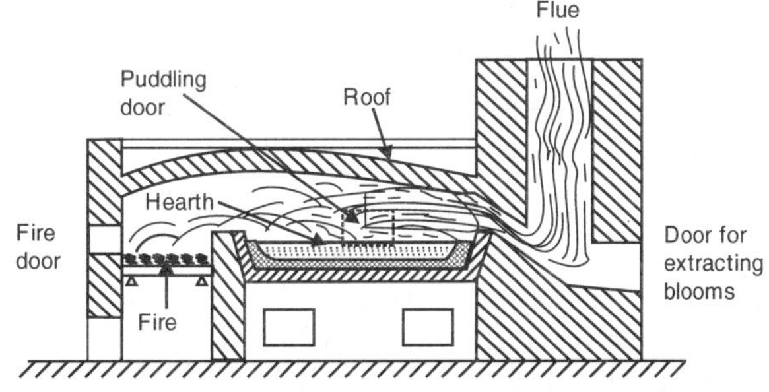

Fig. 5.4. Puddling furnace.

1. *Puddling process.* The wrought iron is produced from pig iron by remelting it in the puddling furnace. This is a coal fired, reverberatory furnace as shown in Fig. 5.4. The term reverberatory is applied to furnaces in which the charge is not in actual contact with the fire, but receives its heat by reflection from the roof of the shaped furnace.

In this process, the charge consisting of pig iron and scrap iron is fed on the furnace hearth which is lined with iron oxide. When the charge melts, most of the elements like silicon, manganese and carbon are removed by oxidation due to their contact with the lining of iron oxide. As the melting proceeds, the puddler continuously stirs the molten metal with the help of a long rake, through the puddling door. When the melting is complete, the impurities form a slag with some of the oxides present and this slag floats on the top of the hearth. The slag is not drawn off, but more mill scale is thrown into the furnace and stirred with the molten metal from time to time. The addition of this oxide to the slag, already rich in oxide causes the carbon of the pig iron to react with the iron oxide and the formation of carbon monoxide takes place. This causes the molten mass to boil up. At this stage, some of the slag rises up and flows out from the slag hole. As the carbon becomes oxidised it is washed away, the bubbling subsides and the iron becomes stiffer and pasty in form. The temperature of the furnace is now raised to the highest possible point. During this period the charge is continuously puddled until it becomes a mass of pasty iron intermixed with slag. It is removed from the furnace in the form of balls (called blooms) having a mass of about 45 kg. These balls, in white hot condition, are then hammered to squeeze out most of the slag in the iron. This operation is called *shingling* or *squeezing.* The squeezed balls still in white hot condition are then passed through rolling mills to give the shape of bars known as *muck bars* or *puddle bars* which have a cross-section of approximately from 15 mm to 100 mm.

A number of these muck bars are piled in cross-wise layers and further reheated to white, welding heat in the furnace and re-rolled. This process eliminates more slag and single refined iron is obtained. The bars produced by this iron are known as *merchant bars.* The repeated operation of piling, reheating and re-rolling eliminates more of the slag and produces double refined iron. The bars produced by this iron are known as *best bars.* It is an expensive method of manufacturing wrought iron and is rarely used now-a-days.

2. *Aston or Byers process.* In *Aston or Byers process, the pig iron is first melted in a cupola and then refined in a bessemer converter. This refined metal is then poured into a ladle (known as shotting) containing a required amount of slag previously prepared in a small open hearth furnace. The temperature of the slag is maintained at very low temperature than the freezing point of the refined metal. Therefore, the refined metal solidifies rapidly, liberating the dissolved gases with sufficient force, producing small pieces of the metal. These pieces settle to the bottom of the ladle and weld together as spongy mass. Each ball of iron collected in the ladle of 3 to 4 tonnes and the rate of production is about one ball after every five minutes. The balls of spongy mass are pressed to remove some of the slag. The pressed block is known as *bloom.* The hot bloom is then passed through rolling mills to produce products of wrought iron of different shapes and sizes. The composition of wrought iron produced by this process is as follows :

Carbon – 0.02%; Silicon – 0.15%; Manganese – 0.03%; Sulphur – 0.02%; Phosphorus – 0.12%; Slag – 3% (by mass) and remaining is iron.

5·16 Steel

It is an alloy of iron and carbon, with carbon content up to a maximum of 1.5%. The carbon occurs in the form of iron carbide, because of its ability to increase the hardness and strength of the steel. Other elements *e.g.*, silicon, sulphur, phosphorus and manganese are also present to greater or lesser amount to impart certain desired properties to it. Most of the steel produced now-a-days is *plain carbon steel* or simply *carbon steel.* A carbon steel is defined as a steel which has its properties mainly due to its carbon content and does not contain more than 0.5% of silicon and 1.5% of manganese.

---

* The large scale production of wrought iron resulted from the work of Dr. James Aston and his associates at the A.M. Byers Co.

5·17 Classification of Steel Depending upon the Carbon Content

According to Indian standard (IS : 7598-1974), steel is classified as :

1. Unalloyed steel or plain carbon steel; and 2. *Alloy steel.

The unalloyed or plain carbon steels varying from 0.06% carbon to 1.5% carbon are divided into the following types depending upon the carbon content :

1. Dead mild steel – up to 0.15% carbon
2. Low carbon or mild steel – 0.15% to 0.45% carbon
3. Medium carbon steel – 0.45% to 0.8% carbon
4. High carbon steel – 0.8% to 1.5% carbon

5·18 Classification of Steel Depending upon the Degree of Deoxidation

The steels, depending upon the degree of deoxidation occurring during its production are classified as follows :

1. *Killed steel.* Since this steel is strongly deoxidised, therefore, it evolves no gas during solidification and the solidified metal is from usual casting defects like blow holes, porosity, segregation etc. Most of the steels containing more than 0.25 percent carbon are killed. The killed steel is denoted by the symbol K.

2. *Rimmed steel.* This steel is partially deoxidised. The purpose of rimming the steel is to produce a clean surface carrying low carbon content. The rimmed steel is preferred for those manufacturing processes which produce the products of good surface finish and the formability is easy. This steel is, therefore, ideal for rolling, deep drawing, spinning etc. The rimmed steel is denoted by the symbol R.

3. *Semi-killed or balanced steel.* This steel falls in the intermediate category of killed steel (which is completely deoxidised) and rimmed steel (which is partially deoxidised). Almost 90 percent of the total production of steel is semi-killed steel. Structural steels containing 0.15 to 0.25 percent carbon are generally semi-killed. This type of steel has a sound surface of considerable thickness and free from blow holes. No symbol is used to denote the semi-killed steel.

5·19 Indian Standard Designation of Steel

According to Indian standard **[IS : 1762 (Part-I)-1974], a new system of designating the steel is recommended. According to this standard, steels are designated on the following two bases :

(*a*) On the basis of mechanical properties, and

(*b*) On the basis of chemical composition.

We shall now discuss, in detail, the designation of steel on the above two basis, in the following pages.

5·20 Steels Designated on the Basis of Mechanical Properties

These steels are carbon and low alloy steels where the main criterion in the selection and inspection of steel is the tensile strength or yield stress. According to Indian standard ***IS : 1570 (Part-I)-1978 (reaffirmed 1993), these steels are designated by a symbol 'Fe' or 'Fe E' depending on whether the steel has been specified on the basis of minimum tensile strength or yield strength, followed by the figure indicating the minimum tensile strength or yield stress in N/mm^2. For example

---

* Alloy steel is discussed in Art. 5.35.

** This standard was reaffirmed in 1993 and covers the code designation of wrought steel on letter symbols.

*** The Indian standard IS : 1570-1978 (Reaffirmed 1993) on wrought steels for general engineering purposes has been revised on the basis of experience gained in the production and use of steels. This standard is now available in seven parts.

'Fe 290' means a steel having minimum tensile strength of 290 N/mm^2 and 'Fe E 220' means a steel having yield strength of 220 N/mm^2.

Table 5.3 shows the tensile and yield properties of standard steels with their uses according to IS: 1570 (Part-I)-1978 (Reaffirmed 1993).

Table 5.3. Indian standard designation of steel according to IS: 1570 (Part-I)-1978 (Reaffirmed 1993).

Indian standard designation	*Tensile strength (Minimum) N/mm^2*	*Yield stress (Minimum) N/mm^2*	*Minimum percentage elongation*	*Uses as per IS: 1871 (Part I)-1987 (Reaffirmed 1993)*
Fe 290	290	170	27	It is used for plain drawn or enamelled parts, tubes for oil well casing, steam, water and air passage, cycle, motor cycle and automobile tubes, rivet bars and wire.
Fe E 220	290	220	27	
Fe 310	310	180	26	These steels are used for locomotive carriages and car structures, screw stock and other general engineering purposes.
Fe E 230	310	230	26	
Fe 330	330	200	26	
Fe E 250	330	250	26	
Fe 360	360	220	25	It is used for chemical pressure vessels and other general engineering purposes.
Fe E 270	360	270	25	
Fe 410	410	250	23	It is used for bridges and building construction, railway rolling stock, screw spikes, oil well casing, tube piles, and other general engineering purposes.
Fe E 310	410	310	23	
Fe 490	490	290	21	It is used for mines, forgings for marine engines, sheet piling and machine parts.
Fe E 370	490	370	21	
Fe 540	540	320	20	It is used for locomotive, carriage, wagon and tramway axles, arches for mines, bolts and seamless and welded tubes.
Fe E 400	540	400	20	
Fe 620	620	380	15	It is used for tramway axles and seamless tubes.
Fe E 460	620	460	15	
Fe 690	690	410	12	It is used for locomotive, carriage and wagon wheels and tyres, arches for mines, seamless oil well casing and drill tubes, and machine parts for heavy loading.
Fe E 520	690	520	12	
Fe 770	770	460	10	It is used for locomotive, carriage and wagon wheels and tyres, and machine parts for heavy loading.
Fe E 580	770	580	10	
Fe 870	870	520	8	It is used for locomotive, carriage and wagon wheels and tyres.
Fe E 650	870	650	8	

Notes : 1. The steels from grades Fe 290 to Fe 490 are general structural steels and are available in the form of bars, sections, tubes, plates, sheets and strips.

2. The steels of grades Fe 540 and Fe 620 are medium tensile structural steels.

3. The steels of grades Fe 690, Fe 770 and Fe 870 are high tensile steels.

5·21 Steels Designated on the Basis of Chemical Composition

According to Indian standard, IS : 1570 (Part II/Sec I)-1979 (Reaffirmed 1991), the carbon steels are designated in the following order :

(*a*) Figure indicating 100 times the average percentage of carbon content,

(*b*) Letter 'C', and

(*c*) Figure indicating 10 times the average percentage of manganese content. The figure after multiplying shall be rounded off to the nearest integer.

For example 20 C8 means a carbon steel containing 0.15 to 0.25 percent (0.2 percent on average) carbon and 0.60 to 0.90 percent (0.75 percent rounded off to 0.8 percent on an average) manganese.

Table 5.4 shows the Indian standard designation of carbon steel with composition and their uses.

Table 5.4. Indian standard designation of carbon steel according to IS : 1570 (Part II/Sec I)-1979 (Reaffirmed 1991).

Indian standard designation	*Composition*		*Uses as per IS: 1871 (Part II)-1987 (Reaffirmed 1993)*
	Carbon (C)	*Manganese (Mn)*	
4C2	0.08 Max	0.40 Max.	It is a dead soft steel generally used in electrical industry.
5C4	0.10 Max.	0.50 Max.	These steels are used where cold form ability is the primary requirement. In the rimming quality, they are used as sheet, strip, rod and wire especially where excellent surface finish or good drawing qualities are required, such as automobile body, and fender stock, hoods, lamps, oil pans and a multiple of deep drawn and formed products. They are also used for cold heading wire and rivets and low carbon wire products. The killed steel is used for forging and heat treating applications.
7C4	0.12 Max.	0.50 Max.	
10C4	0.15 Max.	0.30 – 0.60	
10C4	0.15 Max.	0.30 – 0.60	The case hardening steels are used for making camshafts, cams, light duty gears, worms, gudgeon pins, spindles, pawls, ratchets, chain wheels, tappets, etc.
14C6	0.10 – 0.18	0.40 – 0.70	
15C4	0.20 Max.	0.30 – 0.60	It is used for lightly stressed parts. The material, although easily machinable, is not designed specifically for rapid cutting, but is suitable where cold web, such as bending and riveting may be necessary.
15C8	0.10 – 0.20	0.60 – 0.90	These are general purposes steels used for low stressed components.
20C8	0.15 – 0.25	0.60 – 0.90	
25C4	0.20 – 0.30	0.30 – 0.60	
25C8	0.20 – 0.30	0.60 – 0.90	
30C8	0.25 – 0.35	0.60 – 0.90	It is used for making cold formed parts such as shift and brake levers. After suitable case hardening or hardening and tempering, this steel is used for making sprockets, tie rods, shaft fork and rear hub, 2 and 3 wheeler lambretta parts such as sprocket, lever, hubs for forks, cams, rocket arms and bushes. Tubes for aircraft, automobile, bicycle and furniture are also made of this steel.

Indian standard designation	Composition		Uses as per IS: 1871 (Part II)-1987 (Reaffirmed 1993)
	Carbon (C)	Manganese (Mn)	
35C4	0.30 – 0.40	0.30 – 0.60	It is used for low stressed parts, automobile tubes and fasteners.
35C8	0.30 – 0.40	0.60 – 0.90	It is used for low stressed parts in machine structures, cycle and motor cycle tubes, fish plates for rails and fasteners.
40C8	0.35 – 0.45	0.60 – 0.90	It is used for crankshafts, shafts, spindles, push rods, automobile axle beams, connecting rods, studs, bolts, lightly stressed gears, chain parts, umbrella ribs, washers etc.
45C8	0.40 – 0.50	0.60 – 0.90	It is used for spindles of machine tools, bigger gears, bolts, lead screws, feed rods, shafts and rocks.
50C4	0.45 – 0.55	0.30 – 0.60	It is used for keys, crankshafts, cylinders and machine parts requiring moderate wear resistance. In surface hardened condition it is also suitable for large pitch worms and gears.
50C12	0.45 – 0.55	1.1 – 1.50	It is a rail steel. It is also used for making spike bolts, gear shafts, rocking levers and cylinder liners.
55C4 55C8	0.50 – 0.60 0.50 – 0.60	0.30 – 0.60 0.60 – 0.90	These steels are used for making gears, coil springs, cylinders, cams, keys, crankshafts, sprockets and machine parts requiring moderate wear resistance for which toughness is not of primary importance. These are also used for cycle and industrial chains, spring, can opener, umbrella ribs, parts of camera and typewriter.
60C4	0.55 – 0.65	0.30 – 0.60	It is used for making clutch springs, hardened screws and nuts, machine tool spindles, couplings, crankshafts, axles and pinions.
65C9	0.60 – 0.70	0.50 – 0.80	It is a high tensile structural steel used for making locomotive carriage and wagon tyres. It is also used for engine valve springs, small washers and thin stamped parts.
70C6	0.65 – 0.75	0.50 – 0.80	It is used for making baffle springs, shock absorbers, springs for seat cushions for road vehicles. It is also used for making rail tyres, unhardened gears and worms, washers, wood working saw, textile and jute machinery parts and clutch plates etc.
75C6	0.70 – 0.80	0.50 – 0.80	It is used for making light flat springs formed from annealed stock. Because of good wear properties when properly heat treated, it is used for making shear blades, rack teeth, scrappers and cultivators' shovels.
80C6 85C6	0.75 – 0.85 0.80 – 0.90	0.50 – 0.80 0.50 – 0.80	These steels are used for making flat and coil springs for automobile and railway vehicles. It is also used for girder rails. The valve spring wire and music wire are special applications of steel 85 C6. After suitable heat treatment, these steels are also used for making scraper blades, discs and spring tooth harrows. It is also used for clutch parts, wood working saw, band saw and textile and jute machinery parts.

Indian standard designation	*Composition*		*Uses as per IS: 1871 (Part II)-1987 (Reaffirmed 1993)*
	Carbon (C)	*Manganese (Mn)*	
98C6 113C6	0.90 – 1.05 1.05 – 1.20	0.50 – 0.80 0.50 – 0.80	These steels in the oil hardened and tempered condition are used for coil or spiral springs. It is also used for pen nibs, volute spring, spring cutlery, knitting needle and hacksaw blades.

5·22 Manufacture of Steel

The principal methods of manufacturing steel are :

1. Cementation process; 2. Crucible process; 3. Bessemer process; 4. Open hearth process; 5. Duplex process; 6. L - D process; and 7. Electric process.

In all of these processes of manufacturing steel, either * carbon is added to wrought iron or by first decarburising pig iron and then added the proper amount of carbon. The cementation and crucible processes are the oldest methods of manufacturing steels from wrought iron on a small scale. Other methods are now-a-days used for manufacturing all grades of steel from pig iron on a large scale. We shall now discuss all the above mentioned processes in the following pages.

5·23 Cementation Process

In this process, bars of wrought iron about 2.5 to 3 metres long and 12 to 18 mm thick are embedded in powdered charcoal in fire brick boxes placed in a furnace. It is subjected to a high temperature (about 1000°C) for about 5 to 10 days depending upon the quality of steel desired. A period of 5 days produces steel for saws, springs etc., 7 to 8 days for shear steel and 10 days for very hard steel suitable for cold chisels.

The steel made by this process is called *cement steel* because ferrite in the wrought iron is converted into cementite (*i.e.*, iron carbide). Since carbon combines with wrought iron, and has its surface covered with blisters, therefore, the steel produced by this process is known as *blister steel.*

5·24 Crucible Process

The raw material used for this process is wrought iron, steel scrap and alloys (like tungsten, chromium or ferro-manganese). These materials with measured quantity of charcoal are placed in fireclay or graphite crucibles having a capacity of about 45 kg and melted in a regenerative furnace. It may be noted that in a crucible process, carbon is added to the iron as the carbon content of wrought iron is low. The necessary carbon is taken up by the metal from the charcoal during melting. After the materials are melted, the crucibles are then taken from the furnace, and the slag is removed from the top of the molten metal. The molten steel is poured into a small cast iron moulds.

The steel produced by this method is very homogeneous, free from slag and dirt and much superior to cement steel. The steel so produced is known as *crucible steel.* It is very hard, tenacious and resistant to corrosion. It is used for making high grade surgical instruments, tools, springs, files, blades etc.

5·25 Bessemer Process

In this process, as shown in Fig. 5.5, cold air under pressure is blown through molten pig iron in a refractory lined egg-shaped steel vessel, called converter, until elements like silicon, manganese and carbon are removed by oxidation. The converter is mounted on trunnions attached to the side in such a manner that it may be tilted or in some cases turned entirely upside down. One of the trunnions

* Wrought iron is too soft and ductile to serve many of the purposes for which it is desirable to use as ferrous material. Therefore, to obtain the desired properties (strength, hardness etc.) it is necessary to add the proper amounts of alloying elements usually carbon.

is hollow and serves to connect the blast main to a pipe leading down the side of the converter to the wind box which is attached to the bottom of the converter. About 20 fire brick cylinders, called *tuyers,* lead from the wind box through the bottom lining to the interior of the converter. The capacity of the converters vary from 10 to 25 tonnes. The following are the three distinctive stages used to convert molten pig iron into steel.

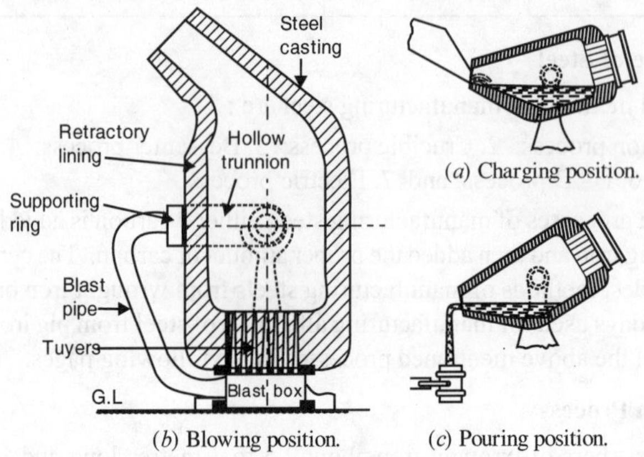

(*a*) Charging position.

(*b*) Blowing position. (*c*) Pouring position.

Fig. 5.5. Bessemer process.

1. The first stage (known as *charging position*) starts by pouring molten pig iron into the converter until the level is just below the blast holes, when a blast of air at about 0.14 N/mm^2 pressure is forced to bustle through the molten iron.

2. In the second stage (known as *blowing position*), the converter is tilted to the vertical position and the air blast turned on. Under the heat of the molten pig iron, the in-rushing air burns out most of the impurities, increasing still further the heat of the molten metal. First of all silicon and manganese burns out which is indicated by the brown smoke rising up through the mouth of the converter. After this, the carbon is next to oxidise which is indicated by a white flame.

3. In the third stage (known as *pouring position*), the white flame of the burning carbon drops and the contents of the converter are poured in a ladle. Now a small quantity of some alloy rich in carbon and manganese (*i.e.*, spiegeleisen or ferro-manganese) is added. The manganese takes up some of the oxygen of the burnt iron and the final product is a steel of quite good strength and ductility.

Note : The Bessemer process may be *acidic* or *basic* depending upon the lining of the furnace. In the *acidic bessemer process,* the furnace is lined with silica bricks. The slag produced in this process contains a large amount of silica. Since phosphorus in a pig iron cannot be removed by this process, therefore, acidic bessemer process is unsuitable for producing steel from pig iron containing large quantities of phosphorus.

In *basic bessemer process,* also known as *Thomas process,* the furnace is lined with a mixture of tar and burned dolomite. This process is applicable for making steel from pig iron which contain more than 1.5% phosphorus. After charging the molten pig iron, lime is also added to combine with acidic materials (like silica, phosphorus and sulphur) and to make a basic slag.

5·26 Open Hearth Process

The open hearth process for steel making is sometimes called *'Siemens-Martin Process'. The furnace is of reverberatory type, but instead of having a fire place, it has gas and air chambers at either

* This is named after Frederick Siemens who devised the furnace and the French metallurgists the Martin brothers, who devised the pig and scrap process in 1865.

end as shown in Fig. 5.6. The chambers are known as regenerators and are partly filled with built-up-rows of a cheeker work of silicon bricks in order to permit easy and free flow of gases. The charge consisting of either iron ore and pig iron or scrap and pig iron is fed into the hearth through the charging door. The preheated gaseous fuel (*i.e.*, coal gas and air) is conveyed to the hearth through the chambers, for maintaining suitable temperatures (1600°C to 1650°C).

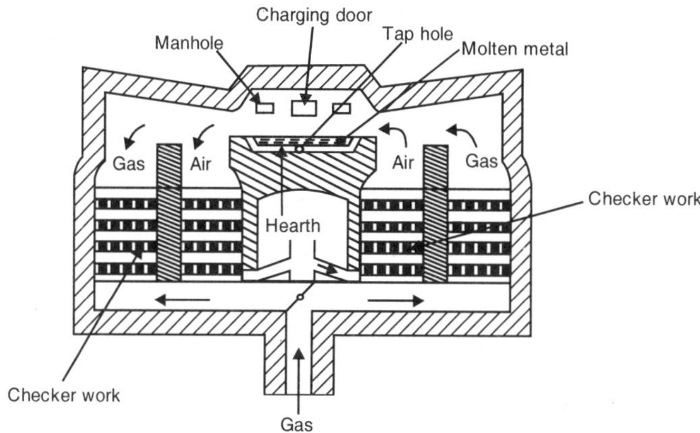

Fig. 5.6. Reverberatory furnace.

During the first half cycle, the waste hot gases of combustion before escaping to the chimney are made to pass through the two chambers on the left hand side and heat up the checkered brick work. The gaseous fuel is conveyed to the furnace by the other two chambers on the right hand side. During the next half cycle, the direction of flow is reversed and the gaseous fuel is made to pass through the chambers on the left hand side. In this way, the gaseous fuel gets preheated by taking up the heat from the checkered brick work. When they reach the furnace, their heat is sufficient to melt the charge on the hearth.

The refining of the metal is effected by the oxidation of the excess quantities of the impurities like silicon, manganese, phosphorus and passes off as a slag. The carbon escapes in the form of gaseous oxide and when all the carbon is removed, leaving practically pure iron, a measured quantity of ferro-manganese is added to the molten metal to obtain steel of desired specifications. The molten metal is tapped off into ladles and then poured into ingot moulds.

Notes : 1. The open hearth process is more suitable than the Bessemer process when a large quantity of mild steel, with definite quality and chemical composition, is required. This is due to the fact that in open hearth process, the samples of the metal can be taken out from time to time and tested.

2. As in the case of Bessemer process, the open hearth process may also be acidic or basic according to the presence or absence of phoshprus in the pig iron as discussed earlier.

5.27 Duplex Process

The Indian pig iron contains on an average 0.3% phosphorus and is unsuitable for steel conversion by acidic and basic bessemer process and acidic open hearth process. In order to produce steel low in phosphorus, only the open hearth process has to be adopted which is a slow process. Therefore, to make up the production, a combination of acidic bessemer and basic open hearth known as *duplex process* is often employed. This process is in operation at Tata Iron and Steel works, Jamshedpur (Bihar).

In this process, the pig iron is first treated in an acid lined bessemer converter until the silicon, manganese and a part of the carbon have been oxidised. The molten metal is then transferred to a large basic open hearth furnace where the sulphur, phosphorus and the remaining carbon are removed.

The calculated, amount of ferro-manganese is now added in the molten metal to produce steel of desired specifications. Finally, the liquid steel is poured into large rectangular moulds.

The duplex process reduces the time required for refining the metal by about half the time taken by open hearth process. It also permits the use of low grade high phosphorus type of pig iron to obtain steel of desired specifications.

5·28 L-D Process (Linz-Donawitz Process)

It is the latest development in steel making processes and is now adopted at Rourkela steel plant where three converters of 40 tonnes capacity are working. Actually, the L-D process was orignated in 1953 at the Iron and steel works of Linz and Donawitz in Austria.

In this process a jet of very pure oxygen (99.5%) is introduced at a pressure of about 0.7 N/mm^2 to 1.05 N/mm^2 through a specially designed water cooled oxygen nozzle (called lance). It is suspended vertically along the central line of the converter with its end about 1 metre above the bath. The converter has a basic lining such as dolomite ($CaCO_3$. $MgCO_3$) or magnesite ($MgCO_3$) and is similar in shape to a bessemer converter but is much simpler in design. The oxygen steam leaves the nozzle in the form of a cone at supersonic speed (faster than sound) and on striking the bath produces an intense heat having a temperature of about 2500°C to 3000°C. The high temperature developed, burns away impurities like carbon, iron, silicon, manganese, phosphorus and sulphur. The following is the sequence of operations of the L-D process :

1. The titled converter, as shown in *Fig. 5.5 of 40 tonnes capacity while hot from the previous blow, is charged with scrap (about 15%) followed by molten pig iron.

2. The converter is brought to the vertical position and oxygen at a pressure of 1.05 N/mm^2 is blown through a water-cooled nozzle followed by addition of lime through a chute.

3. The blow is continued for about 18 minutes. In the beginning a short flame is obtained, which becomes longer and luminous after 4-5 minutes. The dropping of the flame indicates the end of the refining period.

4. The supply of oxygen is stopped and the converter is tilted and samples of slag and steel are drawn. The liquid slag is now stiffened with burnt lime or crushed cold slag and the steel is carefully poured out into a ladle retaining the major portion of the slag in the converter.

Notes : **1.** The L-D process is the only process where sulphur can be effectively reduced under oxidising conditions.

2. The steel produced by L-D process is superior to that produced by bessemer or open hearth process.

5·29 Electric Process

In 1878, Siemens was the first to apply electric current for melting iron. Until 1920, the electric process of steel making could not become popular, due to high cost of electrical energy. Later, with the development of cheap electric power, this process is mainly used for the preparation of high quality and special alloy steels of high melting point. The various advantages of the electric process over other processes are :

1. It can be easily operated with cold charges or hot metal.
2. The generation of heat in the furnace is perfectly under control.
3. It is the cleanest process, as the fuel (*i.e.*, electricity) does not contaminate the charge.
4. It is possible to have control over the addition of oxidising, reducing and alloying elements. Therefore, steel of desired quality can be obtained without any loss of alloying elements and the fuel.

Due to these advantages, most high grade alloy steels, stainless and heat-resisting steels are at

* Please refer to Bessemer converter.

present manufactured in the electric furnace. The process, by proper handling, can produce steel equal in every respect to the best crucible steel.

The electric process may be acidic or basic, as in other steel making processes, but basic electric process is mostly used because it permits extensive elimination of impurities. The acid furnace is essentially a melting furnace and is chiefly used is steel foundries. The electric furnaces are of the following two types :

1. Heroult direct-arc type, and
2. Coreless (*i.e.*, high frequency) induction type.

These electric furnaces are discussed, in detail, in the following pages.

5·30 Heroult Direct Arc Type Electric Furnace

The Heroult direct arc type electric furnace is shown in Fig. 5.7. It consists of a cylindrical steel shell with a bowl-shaped bottom. The top consists of a dome-shaped fire brick cover. The furnace may be provided with either a basic or an acid lining. The electrodes of graphite or amorphous carbon (about 2m long) are suspended through the roof, which in many furnaces may be removed for charging the furnace. The furnaces with capacities up to 60 tonnes have three electrodes, but larger furnaces have six electrodes. The electrodes can be raised mechanically or automatically. This arrangement permits to attain desired temperature in the furnace by varying the distance between the metal bath and the electrodes. The furnace is mounted on roller or rocker so that it may be tilted for discharging the molten metal and the slag.

The charge is usually a steel scrap, but it may consist of molten metal from open hearth furnace also. If scrap forms the whole charge, heavy scrap is charged at the bottom and lighter scrap above it. This makes a better contact with the electrodes, and reduces the high current surges which are likely in initial period of melting. The charging is done either from the top (where roof is removable) or through the working door (in small furnaces).

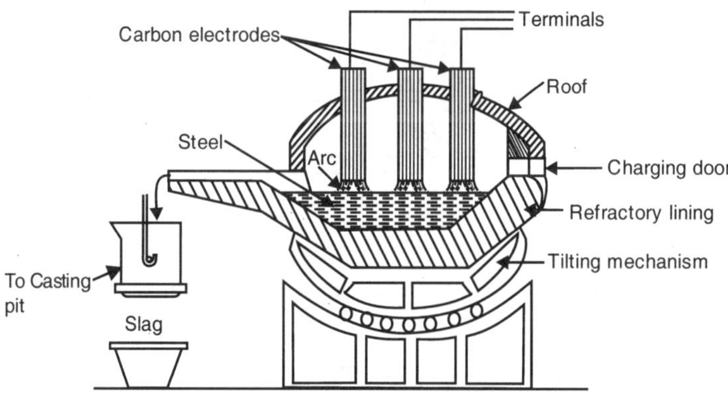

Fig. 5.7. Heroult direct arc type electric furnace.

When the charging is over, the current in the furnace is turned on and the electrodes are lowered and adjusted till an arc is struck. The intense heat of the electric furnace causes the charge to melt (if not already in molten state) and separate out the impurities in the form of slag which floats above the molten metal. When the metal is rendered practically free from carbon and other impurities, ferro-manganese is added in small quantity to supply the desired carbon percentage to the molten steel. If alloyed steel is required, then alloying element is added in required proportions to the molten steel. In order to ensure thorough mixing of the various elements, the metal is tapped about half an hour after the addition of the alloying element. The refined or the alloyed steel is drawn now by tapping or by tilting the furnace.

Note : The basic-lined furnace of the Heroult type is especially adopted to the production of best quality carbon and alloy steels of the following varieties :

1. Stainless steel, containing 11 to 14% chromium.
2. Corrosion and heat resisting steels, containing 17 to 26% chromium and 8 to 13% nickel.
3. Alloy and carbon tool steels.
4. High speed tool steels.
5. Manganese steel, containing 12 to 15% manganese.
6. Magnet steels.
7. Alloy steels for special machinery parts.

5·31 Coreless Induction Type Electric Furnace

This furnace, often called *high frequency induction* furnace, consists of a crucible lined with an acid or basic refractory as shown in Fig. 5.8. The crucible is surrounded by a water cooled coil of flattened copper tubing which serves as the primary for the high frequency current. The frequency for the primary current may range from 500 to 60 000 cycles, but the most common frequency is about 1000 cycles. A special refractory is placed inside the copper coil and the space between the coil and the crucible is packed with granular refractory. The bottom rests on fire clay brick that is supported by the outer shell of the furnace. The entire furnace may be tilted on trunnions for pouring the molten metal.

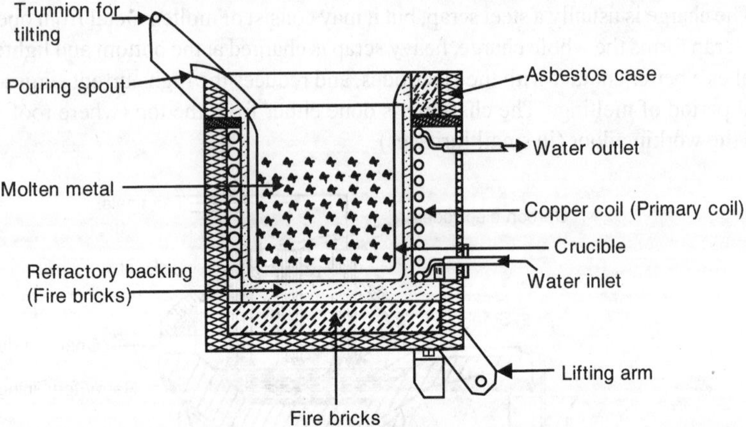

Fig. 5.8. Coreless induction type electric furnace.

When a metallic charge is placed in the crucible, the high frequency primary current is supplied. The magnetic field set up around the primary coil induces eddy currents in the secondary circuit which comprises the crucible containing the metal charge. The eddy currents rapidly melts the charged metal and brings it to a pouring temperature.

The induction furnace is used principally for melting a charge of carefully selected materials without much change in composition. The following materials are successfully made in the induction furnace :

1. High quality, complex alloy tool steels.
2. Extremely low carbon alloys *e.g.*, magnetic alloys and stainless irons.
3. High grade die steel, heat-resisting steel etc.
4. Hard carbides, *e.g.*, the carbides of tungsten, chromium and cobalt.

5·32 Effect of Impurities on Steel

The following are the effects of impurities like silicon, sulphur, manganese and phosphorus, on steel.

1. *Silicon.* The amount of silicon in the finished steel usually ranges from 0.05 to 0.30%. Silicon is added in low carbon steels to prevent them from becoming porous. It removes the gases and oxides, prevent blow holes and thereby makes the steel tougher and harder.

2. *Sulphur.* It occurs in steel either as iron sulphide or manganese sulphide. Iron sulphide because of its low melting point produces red shortness, whereas manganese sulphide does not effect so much. Therefore, manganese sulphide is less objectionable in steel than iron sulphide.

3. *Manganese.* It serves as a valuable deoxidising and purifying agent, in steel. Manganese also combines with sulphur and thereby decreases the harmful effect of this element remaining in the steel. When used in ordinary low carbon steels, manganese makes the metal ductile and of good bending qualities. In high speed steels, it is used to toughen the metal and to increase its critical temperature.

4. *Phosphorus.* It makes the steel brittle. It also produces cold shortness in steel. In low carbon steels, it raises the yield point and improves the resistance to atmospheric corrosion. The sum of carbon and phosphorus usually does not exceed 0.25%.

5·33 Bright Steel

The steel produced by various cold working processes is known as bright steel. It has a very clean and smooth surface and possesses close dimensional accuracy. This steel is available in very wide range of sizes such as round, hexagonal and square bars, flats and special sections. The advantage of bright steel is that the parts made out of these sections may either be directly used in assemblies or machined to required size and shape easily, if required.

5·34 Free Cutting Steels

The free cutting steels (sometimes known as free machining steels) contain sulphur and phosphorus. These steels have higher sulphur content than other carbon steels. In general, the carbon content of such steels vary from 0.1 to 0.45 percent and sulphur from 0.08 to 0.3 percent. These steels are used where rapid machining and high quality surface finish after machining is the prime requirement. It may be noted that the presence of sulphur and phosphorus causes long chips in machining to be easily broken and thus prevent clogging of machines. Now-a-days, lead is used from 0.05 to 0.2 percent instead of sulphur, because lead also greatly improves the machinability of steel without the loss of toughness. It may be noted that free cutting steels have low dynamic strength and are more liable to corrosion. These steels are frequently supplied in the cold drawn form and have high tensile strength and hardness but less ductile when compared to ordinary carbon steels.

According to Indian standard, IS : 1570 (Part III)-1979 (Reaffirmed 1993), carbon and carbon-manganese free cutting steels are designated in the following order :

1. Figure indicating 100 times the average percentage of carbon,
2. Letter 'C',
3. Figure indicating 10 times the average percentage of manganese, and
4. Symbol 'S' followed by the figure indicating 100 times the average content of sulphur. If instead of sulphur, lead (Pb) is added to make the steel free cutting, then symbol 'Pb' may be used.

Table 5.5 shows the composition and uses of carbon and carbon-manganese free cutting steels, as per IS : 1570 (Part III)-1979 (Reaffirmed 1993).

Table 5.5. Indian standard designation of carbon and carbon-manganese free cutting steels according to IS : 1570 (Part III)-1979 (Reaffirmed 1993).

Indian standard designation	*Composition in percentages*					*Uses as per IS : 1871 (Part III)-1987 (Reaffirmed 1993)*
	Carbon (C)	*Silicon (Si)*	*Manganese (Mn)*	*Sulphur (S)*	*Phosphorus (P) Max.*	
10C8S10	0.15 Max.	0.05 – 0.30	0.60 – 0.90	0.08 – 0.13	0.06	It is used small parts to be cyanided or carbonitrided.
14C14S14	0.10 – 0.18	0.05 – 0.30	1.20 – 1.50	0.1 – 0.18	0.06	It is used for parts where good machinability and finish are important.
25C12S14	0.20 – 0.30	0.25 Max.	1.00 – 1.50	0.10 – 0.18	0.06	It is used for bolts, studs and other heat treated parts of small section. It is suitable in either cold drawn, normalised or heat treated condition for moderately stressed parts requiring more strength than mild steel.
40C10S18	0.35 – 0.45	0.25 Max.	0.80 – 1.20	0.14 – 0.22	0.06	It is used for heat treated bolts, engine shafts, connecting rods, miscellaneous gun carriage, and small arms parts not subjected to high stresses and severe wear.
11C10S25	0.08 – 0.15	0.10 Max.	0.80 – 1.20	0.20 – 0.30	0.06	It is used for lightly stressed components not subjected to shock (nuts, studs etc.) and suitable for production on automatic lathes. It is not recommended for general case hardening work but should be used when ease of machining is the deciding factor.
40C15S12	0.35 – 0.45	0.25 Max.	1.30 – 1.70	0.08 – 0.15	0.06	It is used for heat treated axles, shafts, small crankshafts and other vehicle parts. It is not recommended for forgings in which transverse properties are important.

5·35 Alloy Steel

An alloy steel may be defined as a steel to which elements other than carbon are added in sufficient amount to produce an improvement in properties. The alloying is done for specific purposes to increase wearing resistance, corrosion resistance and to improve electrical and magnetic properties, which cannot be obtained in plain carbon steels. The chief alloying elements used in steel are nickel, chromium, molybdenum, cobalt, vanadium, manganese, silicon and tungsten. Each of these elements confer certain qualities upon the steel to which it is added. These elements may be used separately or in combination to produce the desired characteristic in steel.

5·36 Effect of Alloying Elements on Steel

The effect of various alloying elements as discussed in the previous article, are as follows :

1. *Nickel.* It is one of the most important alloying elements. These steels contain 2 to 5% nickel and 0.1 to 0.5% carbon. In this range, nickel improves tensile strength, raises elastic limit, imparts hardness, toughness and reduces rust formation. It is largely used for boiler plates, automobile engine parts, large forgings, crankshafts, connecting rods etc. When nickel is added to steel in appreciable proportions (about 25%) it results in higher strength steels with improved shock and fatigue resistance. It makes the steel resistant to corrosion and heat. It is used in the manufacture of boiler tubes, valves for use with superheated steam, valves for internal combustion engines, pump barrels, sparking plugs for petrol engines, liners and pump parts etc. A nickel steel alloy containing about 36% nickel and 0.5% carbon is known as *invar.* It can be rolled, forged, turned and drawn. It has nearly zero coefficient of expansion. So it is widely used for making pendulums of clocks, precision measuring instruments etc.

2. *Chromium.* The addition of chromium to steel increases strength, hardness and corrosion resistance. A chrome steel containing 0.5 to 2% chromium and 0.1 to 1.5% carbon is used for balls, rollers and races for bearings, dies, rolls for rolling mills, permanent magnets etc.

A steel containing 3.25% nickel, 1.5% chromium and 0.25% carbon is known as *nickel-chrome steel.* The combination of toughening effect of nickel and the hardening effect of chromium produces a steel of high tensile strength with great resistance to shock. It is extensively used for armour plates, motor car crankshafts, axles and gears requiring great strength and hardness.

3. *Vanadium.* It is added in low and medium carbon steels in order to increase their yield and tensile strength properties. A very useful effect of vanadium is the improvement it imparts to the hardenability of steel. In constructional steels, it is added to an extent of 0.25% while for tool steels and other special steels, increased percentage of vanadium is used. In combination with chromium (*i.e.,* chrome-vanadium steel containing about 0.5 to 1.5% chromium, 0.15 to 0.3% vanadium and 0.13 to 1.1% carbon), it produces a marked effect on the properties of steel and makes the steel extremely tough and strong. These steels are largely used for making spring steels, high speed tool steels, crankshafts, locomotives and wagon axles, chassis and other parts of automobiles.

4. *Tungsten.* The addition of tungsten raises the critical temperature of steel and hence it is used in increasing the strength of the alloyed steels at high temperature. It imparts cutting hardness and abrasion resistance properties to steel. When added to the extent of 5 to 6%, it gives the steel good magnetic properties and as such it is commonly used for magnets, in electrical instruments etc.

It is usually used in conjuction with other elements. Steel containing 18% tungsten, 4% chromium, 1% vanadium and 0.7% carbon is called as *tool steel* or *high speed steel.* Since the tools made with this steel have the ability to maintain its sharp cutting edge even at elevated temperature, therefore, it is used for making high speed cutting tools such as cutters, drills, dies, broaches, reamers etc.

5. *Manganese.* It is added to steel in order to reduce the formation of iron sulphide by combining with sulphur. It is usually added in the form of ferro-manganese or silico-mangenese. It

makes the steel hard, tough and wear resisting. The manganese alloy steels containing over 1.5% manganese with a carbon range of 0.40 to 0.55% are widely used for gears, axles, shafts and other parts where high strength combined with fair ductility is required.

Steel containing manganese varying from 10 to 14% and carbon from 1 to 1.3% form an alloy steel which is extensively hard and tough and a high resistance to abrasion. It is largely used for mining, rock crushing and railway equipment.

6. *Silicon.* It increases the strength and hardness of steel without lowering its ductility. Silicon steels containing from 1 to 2% silicon and 0.1 to 0.4% carbon have good magnetic permeability and high electrical resistance. It can withstand impact and fatigue even at elevated temperature. These steels are principally used for generators and transformers in the form of laminated cores.

7. *Cobalt.* It is added to high speed steel from 1 to 12%, to give red hardness by retention of hard carbides at high temperatures. It tends to decarburise steel during heat treatment. It increases hardness and strength but too much of it decreases impact resistance of steel. It also increases residual magnetism and coercive magnetic force in steel for magnets.

8. *Molybdenum.* A very small quantity (0.15 to 0.30%) of molybdenum is generally used with chromium and manganese (0.5 to 0.8%) to make molybdenum steel. These steels possess extra tensile strength and is used for air plane fuselage and automobile parts. It can replace tungsten in high speed steels.

5·37 Indian Standard Designation of Low and Medium Alloy Steels

According to Indian standard, IS : 1762 (Part I)-1974 (Reaffirmed 1993), low and medium alloy steels shall be designated in the following order :

1. Figure indicating 100 times the average percentage carbon.
2. Chemical symbol for alloying elements each followed by the figure for its average percentage content multiplied by a factor as given below :

Element	*Multiplying factor*
Cr, Co, Ni, Mn, Si and W	4
Al, Be, V, Pb, Cu, Nb, Ti, Ta, Zr and Mo	10
P, S and N	100

For example 40 Cr 4 Mo 2 means alloy steel having average 0.4% carbon, 1% chromium and 0.25% molybdenum.

Notes : 1. The figure after multiplying shall be rounded off to the nearest integer.

2. Symbol 'Mn' for manganese shall be included in case manganese content is equal to or greater than 1 percent.

3. The chemical symbols and their figures shall be listed in the designation in the order of decreasing content.

Table 5.6 (Pages 103 and 104) shows the composition and uses of some low and medium alloy steels according to Indian standard IS : 1570-1961 (Reaffirmed 1993).

5·38 Special Steels

In order to meet specific requirement in respect of specific properties and applications, special steels are used. The most important of these steels are stainless steel, heat resisting steels, high speed tool steels, spring steels, magnet steels and maraging steels. These steels are discussed, in detail, in the following pages.

Table 5.6. Composition and uses of alloy steels according to IS : 1570-1961 (Reaffirmed 1993)

Indian standard designation	*Composition in percentages*						*Uses as per IS : 1871–1965*
	Carbon (C)	*Silicon (Si)*	*Manganese (Mn)*	*Nickel (Ni)*	*Chromium (Cr)*	*Molybdenum (Mo)*	
11Mn2	0.16 Max.	0.10 – 0.35	1.30 – 1.70	–	–	–	It is a notch ductile steel for general purposes. It is also used in making filler rods, colliery cage suspension gear tub, mine car draw gear, couplings and rope sockets.
20Mn2 27Mn2	0.16 – 0.24 0.22 – 0.32	0.10 – 0.35 0.10 – 0.35	1.30 – 1.70 1.30 – 1.70	– –	– –	– –	These are used for welded structures, crankshafts, steering levers, shafting spindles etc.
37Mn2	0.32 – 0.42	0.10 – 0.35	1.30 – 1.70	–	–	–	It is used for making axles, shafts, crankshafts, connecting rods etc.
47Mn2	0.42 – 0.52	0.10 – 0.35	1.30 – 1.70	–	–	–	It is used for tram rails and similar other structural purposes.
40Cr1	0.35 – 0.45	0.10 – 0.35	0.60 – 0.90	–	0.90 – 1.20	–	It is used for making gears, connecting rods, stub axles, steering arms, wear resistant plates for earth moving and concrete handling equipment etc.
50Cr1	0.45 – 0.55	0.10 – 0.35	0.60 – 0.90	–	0.90 – 1.20	–	It is a spring steel. It is used in a helical automobile front suspension springs.
35Mn2Mo28 35Mn2Mo45	0.30 – 0.40 0.30 – 0.40	0.10 – 0.35 0.10 – 0.35	1.30 – 1.80 1.30 – 1.80	– –	– –	0.20 – 0.35 0.35 - 0.55	These are used for making general engineering components such as crankshafts, bolts, wheel studs, axle shafts, levers and connecting rods.
40Cr1Mo28	0.35 – 0.45	0.10 – 0.35	0.50 – 0.80	–	0.90 – 1.20	0.20 – 0.35	It is used for making axle shafts, crankshafts, connecting rods, gears, high tensile bolts and studs, propeller shaft joints etc.

Indian standard designation	Composition in percentages						Uses as per *IS : 1871–1965*
	Carbon (C)	*Silicon (Si)*	*Manganese (Mn)*	*Nickel (Ni)*	*Chromium (Cr)*	*Molybdenum (Mo)*	
15Cr3Mo55 25Cr3Mo55	0.10 – 0.20 0.20 – 0.30	0.10 – 0.35 0.10 – 0.35	0.40 – 0.70 0.40 – 0.70	0.30 Max. 0.30 Max.	2.90 – 3.40 2.90 – 3.40	0.45 – 0.65 0.45 – 0.65	These are used for components requiring medium to high tensile properties. In the nitrided condition, it is used for crankshafts, cylinder liners for aero and automobile engines, gears, and machine parts requiring high surface hardness and wear resistance.
40Ni3	0.35 – 0.45	0.10 – 0.35	0.50 – 0.80	3.20 – 3.60	0.30 Max.	–	It is used for parts requiring excessively high toughness. In particular, it is used for components working at low temperatures (in refrigerators, compressors, locomotives and aircraft) and for heavy forging, turbine blades, severely stressed screws, bolts and nuts.
30Ni4Cr1	0.26 – 0.34	0.10 – 0.35	0.40 – 0.70	3.90 – 4.30	1.10 – 1.40	–	It is used for highly stressed gears and other components requiring high tensile strength of the order of 16 N/mm^2 and where minimum distrotion in heat treatment is essential.
35Ni1Cr60	0.30 – 0.40	0.10 – 0.35	0.60 – 0.90	1.00 – 1.50	0.45 – 0.75	–	It is used in the construction of aircraft and heavy vehicles for crankshafts, connecting rods, gear shafts, chain parts, clutches, flexible shafts for plenary gears, camshafts etc.
40Ni2Cr1Mo28	0.35 – 0.45	0.10 – 0.35	0.40 – 0.70	1.25 – 1.75	0.90 – 1.30	0.20 – 0.35	It is used for high strength machine parts, collects, spindles, screws, high tensile bolts and studs, gears, pinions, axle shafts, tappets, crankshafts, connecting rods, boring bars, arbours etc.

5·39 Stainless Steel

It is defined as that steel which when correctly heat treated and finished, resists oxidation and corrosive attack from most corrosive media. The different types of stainless steels are discussed below :

1. *Martensitic stainless steel.* The chromium steels containing 12 to 14 percent chromium and 0.12 to 0.35 percent carbon are the first stainless steels developed. Since these steels possess martensitic structure, therefore, they are called *martensitic stainless steels.* These steels are magnetic and may be hardened by suitable heat treatment and the hardness obtainable depends upon the carbon content. These steels can be easily welded and machined. When formability, softness etc. are required in fabrication, steel having 0.12 percent maximum carbon is often used in soft condition. With increasing carbon, it is possible by hardening and tempering to obtain tensile strength in the range of 600 to 900 N/mm^2, combined with reasonable toughness and ductility. In this condition, these steels find many useful general applications where mild corrosion resistance is required. Also, with the higher carbon range in the hardened and lightly tempered condition, tensile strength of about 1600 N/mm^2 may be developed with lowered ductility.

These steels may be used where the corrosion conditions are not too severe, such as for hydraulic, steam and oil pumps, valves and other engineering components. However, these steels are not suitable for shafts and parts working in contact with non-ferrous metals (*i.e.*, brass, bronze or gun metal bearings) and with graphite packings because electrolytic corrosion is likely to occur. After hardening and light tempering, these steels develop good cutting properties. Therefore, they are used for cutlery, springs, surgical and dental instruments.

Note : The presence of chromium provides good resistance to scaling upto a temperature of about 750°C, but it is not suitable where mechanical strength in the temperature range of 600 to 750°C is required. In fact, creep resistance of these steels at this temperature is not superior to that of mild steel. But at temperature below 600°C, the strength of these steels is better than that of carbon steels and up to 480°C is even better than that of austenitic steels.

2. *Ferritic stainless steel.* The steels containing greater amount of chromium (from 16 to 18 percent) and about 0.12 percent carbon are called *ferritic stainless steels.* These steels have better corrosion resistant property than martensitic stainless steels. But, such steels have little capacity for hardening by heat treatment. However, in the softened condition, they possess good ductility and are mainly used as sheet or strip for cold forming and pressing operations for purposes where moderate corrosion resistance is required. They may be cold worked or hot worked. They are ferro-magnetic, usually undergo excessive grain growth during prolonged exposure to elevated temperatures and may develop brittleness after electric arc resistance or gas welding. These steels have lower strength at elevated temperatures than martensitic steels. However, resistance to scaling and corrosion at elevated temperatures are usually better. The machinability is good and they show no tendency to intercrystalline corrosion.

Note : When nickel from 1.5 to 2.5 percent is added to 16 to 18 percent chromium steel, it not only makes more resistant to corrosion than martensitic steel but also makes it hardenable by heat treatment. Such a steel has good resistance to electrolytic corrosion when in contact with non-ferrous metals and graphite packings. Thus it is widely used for pump shafts, spindles and valves as well as for many other fittings where a good combination of mechanical and corrosion properties are required.

3. *Austenitic stainless steel.* The steel containing high content of both chromium and nickel are called *austenitic stainless steels.* There are many variations in chemical composition of these steels but the most widely used steel contains 18 percent chromium and 8 percent nickel with carbon content as low as possible. Such a steel is commonly known as *18/8 steel.* These steels can not be hardened by quenching, infact they are softened by rapid cooling from about 1000°C. They are non-magnetic and possess greatest resistance to corrosion and good mechanical properties at elevated temperature.

These steels are very tough and can be forged and rolled but offer great difficulty in machining. They can be easily welded, but after welding, it is susceptible to corrosive attack in an area adjacent to the weld. This susceptibility to corrosion (called intercrystalline corrosion or weld decay) may be removed by softening after welding by heating to about 1100°C and cooling rapidly. These steels are used in the manufacture of pump shafts, rail road car frames and sheathing, screws, nuts and bolts and small springs. Since 18/8 steel provide excellent resistance to attack by many chemicals, therefore, it is extensively used in chemical, food, paper making and dyeing industries.

Note : When increased corrosion resistance properties are required, for some purposes, then molybdenum from 2 to 3 per cent may be added.

5·40 Heat Resisting Steels

The steels which can resist creep and oxidation at high temperatures and retain sufficient strength are called *heat resisting steels.* A number of heat resisting steels have been developed as discussed below :

1. *Low alloy steels.* These steels contain 0.5 percent molybdenum. The main application of these steels are for superheater tubes and pipes in steam plants, where service temperatures are in the range of 400°C to 500°C.

2. *Valve steels.* The chromium-silicon steels such as *silchrome* (0.4% C, 8% Cr, 3.5% Si) and *Volmax* (0.5% C, 8% Cr, 3.5% Si, 0.5% Mo) are used for automobile valves. They possess good resistance to scaling at dull red heat, although their strength at elevated temperature is relatively low. For aeroplane engines and marine diesel engine valves 13/13/3 nickel-chromium tungsten valve steel is usually used.

3. *Plain chromium steel.* The plain chromium steel consists of

(*a*) Martensitic chromium steel with 12-13% Cr, and

(*b*) Ferritic chromium steels with 18-30% Cr.

These steels are very good for oxidation resistance at high temperatures as compared to their strength which is not high at such conditions. The maximum operating temperature for martensitic steels is about 750°C, whereas for ferritic steels, it is about 1000 – 1150°C.

4. *Austenitic chromium-nickel steels.* These steels have good mechanical properties at high temperatures with good scaling resistance. These alloys contain a minimum of 18 percent chromium and 8 percent nickel stabilised with titanium or niobium. Other carbide forming elements such as molybdenum or tungsten may also be added in order to improve creep strength. Such alloys are suitable for use upto 1100°C and is used for turbine discs and blades.

5·41 Indian Standard Designation of High Alloy Steels (Stainless Steel and Heat Resisting Steel)

According to Indian standard, IS : 1762 (Part I)-1974 (Reaffirmed 1993), the high alloy steels (*i.e.*, stainless steel and heat resisting steel) are designated in the following order :

1. Letter 'X'.
2. Figure indicating 100 times the percentage of carbon content.
3. Chemical symbol for alloying elements each followed by a figure for its average percentage content rounded off to the nearest integer.
4. Chemical symbol to indicate specially added element to allow the desired properties.

For example, X 10 Cr 18 Ni 9 means alloy steel with average carbon 0.10 percent, chromium 18 percent and nickel 9 percent.

Table 5.7 shows the composition and uses of some types of the stainless steels and heat resisting steels according to Indian standard IS : 1570 (Part V)-1985 (Reaffirmed 1991).

Table 5.7. Indian standard designation of high alloy steels (stainless steel and heat resisting steels) according to IS : 1570 (Part V)-1985 (Reaffirmed 1991).

Indian standard designation	*Composition in percentages*						*Uses as per IS : 1871-1965*
	Carbon (C)	*Silicon (Si)*	*Manganese (Mn)*	*Nickel (Ni)*	*Chromium (Cr)*	*Molybdenum (Mo)*	
30Cr13	0.26 –0.35	1.0 Max.	1.0 Max.	1.0 Max.	12.0 – 14.0	–	It is used for structural parts with high strength and kitchen utensils.
15Cr16Ni2	0.10 – 0.20	1.0 Max.	1.0 Max.	1.5 – 3.0	15.0 – 18.0	–	It is used for aircraft fittings, wind shield wiper arms, bolting materials, paper machinery etc.
07Cr18Ni9	0.12 Max.	1.0 Max.	2.0 Max.	8.0 – 10.0	17.0 – 19.0	–	It is used for aircraft fire walls and cawlings, radar and microwaves antennae, jewellery, household novelties, automotive wheel covers, refrigerator trays, kitchen utensils, railway passenger car bodies, ice making equipment, tubular furniture, screen door and storm window frames, electric switch parts, flexible couplings etc.
04Cr17Nil2 Mo2	0.08 Max.	1.0 Max.	2.0 Max.	10.5 – 14.0	16.0 – 18.5	2.0 – 3.0	It is used for high temperature chemical handling equipment for rayon, rubber and marine industries, photographic developing equipment, pulp handling equipment, steam jacketed kettles, coke plant equipment, food processing equipment, edible oil storage tanks.
45Cr9Si9	0.40 –0.50	3.25 – 3.75	0.30 – 0.60	0.50 Max.	7.50 – 0.50	–	It is used for heat resisting outlet valves in oil engines, lorries and cars.
80Cr20Si2Ni1	0.75 – 0.85	1.75 – 2.25	0.20 – 0.60	1.20 – 1.70	19.0 – 21.0	–	It is used for highly stressed outlet valves in high speed carburators and heavy oil engines.

Table 5.8. Indian standard designation of high speed tool steel according to IS : 7291-1981 (Reaffirmed 1993).

Indian standard designation	Chemical composition in percentages								Brinell hardness in annealed condition (HB) Max.
	Carbon (C)	*Silicon (Si)*	*Manganese (Mn)*	*Chromium (Cr)*	*Molybdenum (Mo)*	*Vanadium (V)*	*Tungsten (W)*	*Cobalt (Co)*	
XT 72 W 18 Cr 4 V 1	0.65 – 0.80	0.15 – 0.40	0.20 – 0.40	3.75 – 4.50	–	1.00 – 1.25	17.50 – 19.0	–	255
XT 75 W 18 Co 5 Cr 4 Mo V 1	0.70 – 0.80	0.15 – 0.40	0.20 – 0.40	3.75 – 4.50	0.40 – 1.00	1.00 – 1.25	17.50 – 19.0	4.50 – 5.50	269
XT 80 W 20 Co 12 Cr 4 V 2 Mo 1	0.75 – 0.85	0.15 – 0.40	0.20 – 0.40	4.00 – 4.75	0.40 – 1.00	1.25 – 1.75	19.50 – 21.0	11.00 – 12.50	302
XT 125 W Co 10 Cr Mo 4 V 3	1.20 – 1.30	0.15 – 0.40	0.20 – 0.40	3.75 – 4.75	3.00 – 4.00	2.80 – 3.50	8.80 – 10.70	8.80 – 10.70	269
XT 87 W 6 Mo 5 Cr 4 V 2	0.82 – 0.92	0.15 – 0.40	0.15 – 0.40	3.75 – 4.75	4.75 – 5.50	1.75 – 2.05	5.75 – 6.75	–	248
XT 90 W 6 Co Mo 5 Cr 4 V 2	0.85 – 0.95	0.15 – 0.40	0.20 – 0.40	3.75 – 4.75	4.75 – 5.50	1.70 – 2.20	5.75 – 6.75	4.75 – 5.25	269
XT 110 Mo 10 Co 8 Cr 4 W 2	1.05 – 1.15	0.15 – 0.40	0.15 – 0.40	3.50 – 4.50	9.0 – 10.0	0.95 – 1.35	1.15 – 1.85	7.75 – 8.75	269

Notes : 1. For all steels, sulphur (*S*) and phosphorus (*P*) is 0.030 percent Max.

2. If sulphur is added to give free machining properties, then it shall be between 0.09 and 0.15 percent.

5·42 High Speed Tool Steels

These steels are used for cutting metals at a much higher cutting speed (two to three times higher) than ordinary carbon tool steels. The carbon steel cutting tools do not retain their sharp cutting edges under heavier loads and higher speeds. This is due to the fact that at high speeds, sufficient heat may be developed during the cutting operation and causes the temperature of the cutting edge of the tool to reach a red heat. This temperature would soften the carbon tool steel and thus the tool will not work efficiently for a longer period. The high speed steels have the valuable property of retaining their hardness even when heated to red heat. Most of the high speed steels contain tungsten as the chief alloying element, but other elements like cobalt, chromium, vanadium etc. may be present in some proportion. Following are the different types of high speed steels :

1. *18-4-1 High speed steel.* This steel, on an average, contains 18 percent tungsten, 4 percent chromium and 1 per cent vanadium. It is considered to be one of the best of all purpose tool steels. It is widely used for drills, lathe, planer and shaper tools, milling cutters, reamers, broaches, threading dies, punches etc.

2. *Molybdenum high speed steel.* This steel, on an average, contains 6 percent tungsten, 6 percent molybdenum, 4 percent chromium and 2 percent vanadium. It has excellent toughness and cutting ability. The molybdenum high speed steels are better and cheaper than other types of steels. It is particularly for drilling and tapping operations.

3. *Supper high speed steel.* This steel is also called *cobalt high speed steel* because cobalt is added from 2 to 15 percent in order to increase the cutting efficiency especially at high temperatures. This steel, on an average, contains 20 percent tungsten, 4 percent chromium, 2 percent vanadium and 12 percent cobalt. Since the cost of this steel is more, therefore, it is principally used for heavy cutting operations which impose high pressure and temperatures on the tool.

4. *Vanadium high speed steel.* This steel, on an average, contains more than 1 per cent vanadium. This steel has better abrasive resistance than 18-4-1 steel. It is preferred for machining difficult-to-machine materials.

5·43 Indian Standard Designation of High Speed Tool Steel

According to Indian standard, IS : 1762 (Part I)-1974 (Reaffirmed 1993), the high speed tool steels are designated in the following order :

1. Letter 'XT'.
2. Figure indicating 100 times percentage of carbon content.
3. Chemical symbol for alloying elements each followed by the figure for its average percentage content rounded off to be nearest integer, and
4. Chemical symbol to indicate specially added element to attain the desired properties.

For example, XT 75 W 18 Cr 4 V 1 means a tool steel with average carbon content 0.75 percent, tungsten 18 percent, chromium 4 percent and vanadium 1 percent.

Table 5.8 (Page 108) shows the composition of high tool steels as per Indian standard, IS : 7291-1981 (Reaffirmed 1993).

5·44 Spring Steels

The most suitable material for springs are those which can store up the maximum amount of work or energy in a given weight or volume of spring material, without permanent deformation. These steels should have a high elastic limit as well as high deflection value. The spring steel, for aircraft and automobile purposes should possess maximum strength against fatigue effects and shocks. The steels most commonly used for making springs are as follows :

1. *High carbon steels.* These steels contain 0.6 to 1.1 percent carbon, 0.2 to 0.5 percent silicon and 0.6 to 1 percent manganese. These steels are heated to 780 – 850°C according to the

composition and quenched in oil or water. It is then tempered at 200 – 500°C to suit the particular application. These steels are used for laminated springs for locomotives, carriages, wagons, and for heavy road vehicles. The higher carbon content oil hardening steels are used for volute, spiral and conical springs and for certain types of petrol engine inlet valve springs.

2. *Chrome-vanadium steels.* These are a high quality spring steels and contain 0.45 to 0.55 percent carbon, 0.9 to 1.2 percent chromium, 0.15 to 0.20 percent vanadium, 0.3 to 0.5 percent silicon and 0.5 to 0.8 percent manganese. These steels have high elastic limit, resistance to fatigue and impact stresses. Moreover, these steels can be machined without difficulty and can be given a smooth surface free from tool marks. These are hardened by oil quenching at 850 – 870°C and tempered at 470 – 510°C for vehicle and other spring purposes. These steels are used for motor car laminated and coil springs for suspension purposes, automobile and aircraft engine valve springs.

3. *Silicon manganese steels.* These steels contain 1.8 to 2.0 percent silicon, 0.5 to 0.6 percent carbon and 0.8 to 1 percent manganese. These steels have high fatigue strength, resistance and toughness. These are hardened by quenching in oil at 850 – 900°C and tempered at 475 – 525°C. These are the usual standard quality modern spring materials and are much used for many engineering purposes.

5·45 Magnet Steel

A typical magnet steel contains 15 to 40 percent cobalt, 0.4 to 10 percent tungsten, 1.5 to 9 percent chromium and upto 1 percent carbon. This steel is mainly used for making permanent magnets for electrical engineering instruments, loud speakers, magnetos etc.

5·46 Maraging Steel

These steels are iron based alloys containing 16 to 20 percent nickel, 6 to 10 percent cobalt, 4 to 6 percent molybdenum with small amount of other elements such as aluminium and titanium and less than 0.03 percent carbon. The structure of this alloy changes to martensite when air cooled from a temperature of 815°C. Its yield strength and elongation properties may be improved by age-hardening at 480°C. These steels are used for those components which require extremely high strength and good toughness such as air frame and engine components.

5·47 Cutting Alloys

The cutting alloys are, infact, nonferrous alloys which are developed specially to operate at very high cutting speeds and to retain their strength at extremely high temperatures (1000°C to 1100°C). Since these alloys have high brittleness and cannot be machined by ordinary methods, therefore, solid tools of these alloys are not made. They are used as small bits which are either brazed or held mechanically to the tool shank or body.

According to the method of production, the cutting alloys are classified into the following two groups :

1. Cast cutting alloys (stellites); and 2. Cemented carbides.

These alloys are discussed, in detail, as follows :

1. *Cast cutting alloys (Stellites).* These are excellent materials for cutting tools and contain principally cobalt, tungsten and chromium with small percentages of one or more elements like boron, nickel, niobium, molybdenum and tantalum. These alloys are cast to shape and finished by grinding. They retain their hardness and maintain good cutting edges on tool at much higher temperatures than high speed steel. The tools made of this material can be safely used at twice the cutting speed than high speed steel tools and still maintain the same feed. These alloys are more brittle and do not respond to heat treatment. They are costlier than high speed steel and are used for machining hard materials like cast iron, alloy steels, plain carbon steels and nonferrous alloys.

2. *Cemented carbides.* These alloys are made from the powders of tungsten carbides, titanium

carbide, boron carbide, and tantalum carbide with cobalt as binder. The powders of these carbides are first compacted with cobalt binder (3 to 15 percent) and then *sintered. These alloys are also known as *sintered carbides.* They are extremely hard alloys and retain their hardness even at temperatures upto 1200°C. Since these alloys are very brittle, therefore, they are used only as bits. These bits may be of different sizes and shapes like triangular, square, round etc. These bits are either brazed or held mechanically on the tool shanks. The carbide tools can be operated at cutting speeds two to three times that of stellite tools, but at such speeds, a much smaller feed must be used. The high wear resistance of these alloys make them suitable for wire drawing dies and other similar applications.

QUESTIONS

1. Give the classification of metals.

2. What are the principal iron ores used for the extraction of iron ? Describe in brief the process of extraction of iron from its ores.

3. Describe with the help of a neat sketch, the process of melting of iron in a blast furnace.

4. What is cast iron ? How it is manufactured ? Mention its important characteristics and uses.

5. Write short notes on :

(*a*) Grey cast iron; (*b*) White cast iron; (*c*) Chilled cast iron;
(*d*) Malleable cast iron; and (*e*) Meehanite cast iron.

6. What do you understand by the term 'alloy cast iron' ? Name the alloying elements and discuss their effects.

7. What are the constituents of wrought iron ? Describe briefly, with the aid of sketches, the manufacturing processes of wrought iron. Mention the important uses of the metal.

8. Define mild steel. How it differs from cast iron and wrought iron ?

9. What is the difference between the killed steel and rimmed steel ?

10. What is the effect of carbon, sulphur and phosphorus on the properties of steel ?

11. What do you understand by the term acid or basic furnace ? Describe in detail the manufacture of steel by bessemer process.

12. Describe with the help of a neat sketch the open hearth process of making steel.

13. What is the L-D process of making steel ? State the advantages of this process.

14. List the advantages of the electric process of making steel over other processes.

15. Explain with the help of a neat sketch, the principle of operation of a coreless induction type electric furnace.

16. Write short notes on

(*a*) Bright steel; and (*b*) Free cutting steel.

17. Define alloy steel. Why the alloying elements are added to steel.

18. Explain the effect of nickel, chromium, cobalt, manganese on the properties of steel.

19. (*a*) State the typical composition of the following :

(*i*) Stainless steel; and (*ii*) Spring steel

(*b*) Write the uses for which they are commonly used in engineering practice.

20. Write short notes on

(*a*) Heat resisting steel; (*b*) Magnetic steel; and (*c*) Maraging steel.

---

* The compacting and sintering are well known powder metallurgy processes. For complete details, please refer chapter on 'Powder Metallurgy'.

21. (*a*) What are high speed tool steels ? What are the principal alloying elements in them ?

(*b*) What are cutting alloys ? Describe cast cutting alloys and cemented carbides.

22. Name the ferrous metals and alloys used for the following engineering parts :

(*a*) Crankshaft; (*b*) Connecting rod; (*c*) Pistons; (*d*) Piston rings; (*e*) Lathe spindle; (*f*) Lathe tool; (*g*) Drills; (*h*) Hacksaw blades; (*i*) Chisels; and (*j*) Leaf springs.

OBJECTIVE TYPE QUESTIONS

1. The iron ore mostly used for the production of pig iron is

(*a*) magnetite (*b*) hematite (*c*) limonite
(*d*) siderite (*e*) pyrite (*f*) all of these

2. The portion of the blast furnace below its widest cross-section is called

(*a*) hearth (*b*) stock (*c*) bosh
(*d*) any one of the above (*e*) none of the above

3. The cupola is used to manufacture

(*a*) pig iron (*b*) cast iron (*c*) wrought iron
(*d*) steel (*e*) all of these

4. The coke in the charge of blast furnace

(*a*) controls the grade of pig iron
(*b*) supplies heat to reduce ore and melt the iron
(*c*) acts as an iron-bearing mineral
(*d*) forms a slag by combining with impurities
(*e*) none of the above

5. Combined carbon in iron makes the metal

(*a*) soft and gives a coarse grained crystalline structure
(*b*) soft and gives a fine grained crystalline structure
(*c*) hard and gives a coarse grained crystalline structure
(*d*) hard and gives a fine grained crystalline structure

6. Chilled cast iron is produced

(*a*) by adding magnesium to the molten cast iron
(*b*) by quick cooling of molten cast iron
(*c*) from white cast iron by annealing process
(*d*) none of the above

7. Malleable cast iron is produced

(*a*) by adding magnesium to the molten cast iron
(*b*) by quick cooling of molten cast iron
(*c*) from white cast iron by annealing process
(*d*) none of the above

8. Nodular cast iron is produced by adding .... to the molten cast iron.

(*a*) nickel (*b*) chromium (*c*) copper
(*d*) molybdenum (*e*) magnesium

9. Grey cast iron is .... than white cast iron.

(*a*) softer (*b*) harder

10. According to Indian standard specifications, a plain carbon steel designated by 40C8 means that

(*a*) carbon content is 0.04 percent and manganese is 0.08 percent

(*b*) carbon content is 0.4 percent and manganese is 0.8 percent

(*c*) carbon content is 0.35 to 0.45 percent and manganese is 0.60 to 0.90 percent

(*d*) carbon content is 0.60 to 0.80 percent and manganese is 0.8 to 1.2 percent

11. In basic bessemer process, the furnace is lined with

(*a*) silica bricks | (*b*) a mixture of tar and burnt dolomite bricks

(*c*) either (*a*) or (*b*) | (*d*) none of the above

12. Which of the following process of steel making is in operation at Tata Iron and Steel Works, Jamshedpur ?

(*a*) Bessemer process | (*b*) Open hearth process

(*c*) Duplex process | (*d*) Electric process

(*e*) L-D process | (*f*) none of these

13. Duplex process of steel making is a combination of

(*a*) basic bessemer and acid open hearth processes

(*b*) acid bessemer and basic open hearth processes

(*c*) acid bessemer and acid open hearth processes

(*d*) basic bessemer and basic open hearth processes

14. The electric process of steel making is especially adopted to

(*a*) alloy and carbon tool steel | (*b*) magnet steel

(*c*) high speed tool steel | (*d*) stainless steel

(*e*) all of the above | (*f*) none of the above

15. The steel produced by bessemer or open hearth process is .... to that produced by L-D process.

(*a*) superior | (*b*) inferior

16. Silicon is added in low carbon steels to

(*a*) make the steel tougher and harder | (*b*) make the steel of good bending qualities

(*c*) raise the yield point | (*d*) all of the above

17. Which of the following steel making process is being adopted at Rourkela Steel Plant ?

(*a*) Bessemer process | (*b*) Open-hearth process

(*c*) Electric process | (*d*) L-D process

(*e*) Duplex process

18. A steel alloy containing 36% nickel is called

(*a*) stainless steel | (*b*) high speed steel

(*c*) Invar | (*d*) heat resisting steel

19. Vanadium when added to steel

(*a*) increases tensile strength | (*b*) decreases tensile strength

(*c*) raises critical temperature | (*d*) lowers critical temperature

20. The machinability of steel is improved by adding

(*a*) nickel | (*b*) chromium

(*c*) nickel and chromium | (*d*) sulphur, lead and phosphorous

(*e*) cobalt and molybdenum

21. Shock resistance of steel is increased by adding

(*a*) nickel (*b*) chromium

(*c*) nickel and chromium (*d*) sulphur, lead and phosphorous

(*e*) cobalt and molybdenum

22. The steel widely used for motor car crankshaft is

(*a*) nickel steel (*b*) chrome steel

(*c*) nickel-chrome steel (*d*) silicon steel

(*e*) high speed steel (*f*) all of these

23. The silicon steel is widely used for

(*a*) connecting rods (*b*) cutting tools

(*c*) generators and transformers in the form of laminated cores

(*d*) motor car crankshafts

24. The cutting tools are made from

(*a*) nickel steel (*b*) chrome steel

(*c*) nickel-chrome steel (*d*) silicon steel

(*e*) high speed steel (*f*) none of these

25. Free cutting steels

(*a*) contain carbon in free form

(*b*) is used where rapid machining is the prime requirement

(*c*) require minimum cutting force

(*d*) can be cut freely

(*e*) all of the above

26. Ball bearings are, usually, made from

(*a*) low carbon steel (*b*) high carbon steel

(*c*) medium carbon steel (*d*) high speed steel

(*e*) chrome steel (*f*) all of these

27. The alloying element which increases residual magnetism and coercive magnetic force in steel for magnets is

(*a*) chromium (*b*) nickel

(*c*) vanadium (*d*) magnanese

(*e*) cobalt (*f*) none of these

28. 18-4-1 high speed steel contains

(*a*) vanadium 4% chromium 18% and tungsten 1%

(*b*) vanadium 1%, chromium 4% and tungsten 18%

(*c*) vanadium 18%, chromium 1% and tungsten 4%

(*d*) none of the above

29. A steel containing 15 to 20 percent nickel and about 0.1 percent carbon is called

(*a*) ferritic stainless steel (*b*) austenitic stainless steel

(*c*) martensitic stainless steel (*d*) nickel steel

30. 18/8 steel contains

(*a*) 18 percent nickel and 8 percent chromium

(*b*) 18 percent chromium and 8 percent nickel

(*c*) 18 percent nickel and 18 percent chromium

(*d*) 8 percent nickel and 8 percent chromium

ANSWERS

1. (*b*)	2. (*c*)	3. (*b*)	4. (*b*)	5. (*d*)
6. (*b*)	7. (*c*)	8. (*e*)	9. (*a*)	10. (*c*)
11. (*b*)	12. (*c*)	13. (*b*)	14. (*e*)	15. (*b*)
16. (*a*)	17. (*d*)	18. (*c*)	19. (*b*)	20. (*d*)
21. (*c*)	22. (*b*)	23. (*c*)	24. (*e*)	25. (*b*)
26. (*e*)	27. (*e*)	28. (*b*)	29. (*a*)	30. (*b*)

6

Non-ferrous Metals and Alloys

1. Introduction. 2. Aluminium. 3. Copper. 4. Lead. 5. Tin. 6. Zinc. 7. Nickel. 8. Magnesium. 9. Cadmium. 10. Vanadium. 11. Antimony. 12. Aluminium Alloys. 13. Copper Alloys. 14. Brass. 15. Bronze. 16. Gun Metal. 17. Lead Base Alloys. 18. Tin Base Alloys (Babbit Metal). 19. Bearing Metals. 20. Zinc Base Alloys. 21. Nickel Base Alloys. 22. High Temperature Alloys. 23. Materials for Nuclear Energy.

6·1 Introduction

We have already discussed that the non-ferrous metals are those which contain a metal other than iron as their chief constituent.

The non-ferrous metals are usually employed in industry due to the following characteristics :

1. Easy to fabricate *i.e.*, casting, rolling, forging, welding and machining,
2. High resistance to corrosion,
3. Very good electrical and thermal conductivity,
4. Low weight, and
5. Attractive appearance.

The various non-ferrous metals used in engineering practice are aluminium, copper, lead, tin, zinc, nickel etc., and their alloys. We shall now discuss these non-ferrous metals and their alloys, in detail, in the following pages.

6·2 Aluminium

The chief source of aluminium is a clayey mineral called *bauxite which is a hydrated aluminium oxide ($Al_2O_3.2H_2O$). In India, it is found in some districts of Bihar, Karnataka, Madhaya Pradesh, Maharashtra and Tamil Nadu.

Manufacture. The bauxite is first purified and then dissolved in fused cryolite (double fluoride of aluminium and sodium, ($AlF_3.3NaF$). The aluminium is separated, from this solution, by electrolysis at about 900°C.

Properties and uses. It is a silvery white, light metal having specific gravity 2.7 and melting point 658°C. Its tensile strength varies from 90 MPa to 150 MPa.

In its pure state, the metal would be weak and soft for most purposes, but when mixed with small amounts of other alloys, it becomes hard and rigid. So, it may be blanked, formed, drawn, turned, cast, forged and die cast. Its good electrical conductivity is an important property and is widely used for overhead cables. The high resistance to corrosion and its non-toxicity make it a useful metal for cooking utensils under ordinary conditions. It is extensively used in aircraft and automobile components where saving of weight is an advantage.

6·3 Copper

It is one of the most widely used non-ferrous metal in industry. It is not found in pure state from under the earth. It occurs in some minerals such as copper glance (Cu_2S), copper pyrites ($CuFeS_2$), malachite ($CuCO_3.CuO_2H_2$) and azurite ($3CuCO_3.CuO_2H_2$). In India, copper ores are found at Ghatsila

* It takes its name from the village of Baus, near Arles, in France, where it was first found by Berthier.

in Bihar and Khetri in Rajasthan.

Manufacture. The extraction of copper is a laborious process. The ores (usually copper pyrites) are first calcined, then smelted in a blast furnace with silica (sand) and coke. It is finally oxidised in a bessemer converter to obtain *blister copper* which contains iron and other impurities. Blister copper is refined in a current of air in a reverberatory furnace, treated with coal and stirred with a green pole to obtain crude copper. The pure copper (99.9 percent) is obtained from crude copper by electrolysis.

Properties and uses. It is a soft, malleable and ductile metal with a reddish-brown appearance. Its specific gravity is 8.9 and melting point is 1083°C. It is a good conductor of electricity. It is largely used in making electric cables and wires, for electric machinery and appliances, in electrotyping and electroplating, in making coins and household utensils.

It may be cast, forged, rolled and drawn into wires. It is non-corrosive under ordinary conditions and resists weather very effectively. Copper in the form of tubes is used widely in mechanical engineering. It is also used for making munitions. It is used for making useful alloys with tin, zinc, nickel and aluminium.

6·4 Lead

The most important ore of lead is lead sulphide (PbS) known as *galena* or *galenite*.

Manufacture. The ore is first concentrated and then smelted in a reverbertory furnace at moderate temperature and controlled supply of air, so that only a part of the ore may be roasted and converted to lead oxide and sulphate. The supply of air is then cut off and temperature is raised to bright red heat (about 1000°C) and little quantity of lime and lead sulphide is added. The lead oxide and sulphate formed react with lead sulphide giving metallic lead.

Properties and uses. It is a bluish grey metal having specific gravity 11.36 and melting point 326°C. It is so soft that it can be cut with a knife. It is resistant to corrosion and many acids have no chemical reaction on it. It is extensively used for making solders, as a lining for acid tanks, cisterns, waterpipes, and as coating for electrical cables.

6·5 Tin

The only ore of tin is stannic oxide (SnO_2) called *tin stone* or *cassiterite*. Small quantities of tin stone are found in Hazaribagh (Bihar). Large deposits of tin stone occur in Tavoy (Burma).

Manufacture. The ore is crushed, roasted to expel sulphur and arsenic, mixed with flux and smelted in a reverberatory furnace. The liquid metal is run into moulds. The moulded ingots are then refined. In order to obtain high purity tin, electro-deposition method is used.

Properties and uses. It is brightly shining white metal. It is soft, malleable and ductile. It has good resistance to acid corrosion. It can be rolled into very thin sheets. It is used for making important alloys, fine solder, as a protective coating for iron and steel sheets and for making tin foil used as moisture proof packing.

6·6 Zinc

The chief ore of zinc is *zinc blende* (also called zinc sulphide, sphalerite and blackjack, ZnS) which contains 67% of zinc. It is often associated with galena and iron and copper pyrites. The other zinc ore of less importance is *Calamine* ($ZnCO_3$) which contains 52% of zinc.

Manufacture. The ore is first roasted in a reverberatory furnace to convert the zinc sulphide (ZnS) into zinc oxide (ZnO) and in case of calamine, to drive off carbonic acid and water. The roasted ore is then mixed with some form of carbon and put into long horizontal fire-clay retorts. The retorts are now strongly heated to about 1100°C in a special type of furnace. The heat distils off the zinc in the form of greeny-white vapour and this is condensed to molten zinc, outside the furnace. The metal so produced is about 98% pure and is known as *spelter*.

Note : Another process for the production of zinc is electrolytic process. In this process, the roasted ore is dissolved in sulphuric acid and after purification of the solution, zinc is deposited electrolytically. The zinc so produced is 99.9% pure.

Properties and uses. The zinc is a bluish white metal which in pure state has bright smooth crystals at its fracture. Its specfic gravity is 7.1 and its melting point is 420ºC. It boils at 940ºC and can easily be distilled. It is not very malleable and ductile at ordinary temperatures, but it can be readily worked and rolled into thin sheets or drawn into wires by heating it to 100º-150º C. At about 200ºC, it becomes so brittle that it may be powdered. Its tensile strength is 19 to 25 MPa. It offers high resistance to atmospheric corrosion.

Due to its high resistance to atmospheric corrosion, it is used for covering steel sheets to form galvanised iron. The covering is done by dipping the sheets into the molten metal after previous fluxing. The galvanised wires, nails etc., are also made by this process. When rolled into sheets, zinc is used for roof covering and for providing a damp-proof non-corrosive lining to containers etc. The other important uses of zinc are in the manufacture of brasses and in the production of zinc base die casting alloys. The oxide of zinc is used as pigment in paints.

6·7 Nickel

The most important ore for nickel is iron sulphides which contain about 3% of nickel. About 90% of the total production of nickel is obtained by this source.

Manufacture. The process of extracting nickel from sulphide ores is very similar as for sulphur bearing copper ores (Art. 6.3). The ores are first roasted to reduce the sulphur content and then smelted to separate the gangue. The *matte delivered by the smelting furnace is then put in a bessemer converter to oxidise the iron. The iron oxide combines with a siliceous flux (limestone and quartz) added for this purpose and form the slag. The bessemer matte consists chiefly of copper and nickel sulphides.

Properties and uses. It is a silvery white metal capable of taking a high polish. Its specific gravity is 8.85 and melting point 1452ºC. It is almost as hard as soft steel. When it contains a small amount of carbon, it is quite malleable. It can be satisfactorily rolled with as little as 0.005 percent of carbon. It is somewhat less ductile than soft steel, but small amounts of magnesium improve the ductility considerably. It is resistant to attack of most acids but dissolves readily in nitric acid.

It is used as an alloying metal in some types of steel and cast iron. It is extensively used as a coating for other metals such as steel, copper, brass etc. for both decorative and corrosion protection purposes.

6·8 Magnesium

The principal ores of magnesium are dolomite ($CaCO_3.MgCO_3$), magnesite ($MgCO_3$), carnallite ($KCl.MgCl_2.6H_2O$), kieserite ($MgSO_4.H_2O$) and brucite ($MgO.H_2O$).

Manufacture. The magnesium is extracted by a chloride process in which a melted mixture of the chlorides of sodium, potassium and magnesium is electrolysed. The electrolysis is carried out in an air-tight iron vessel, the walls of which act as the cathode, with a centrally located graphite anode, which is protected by a procelain hood from which the chlorine escapes. As the magnesium is liberated, it rises to the surface and tapped in laddles. The metal is refined by remelting it in iron pots and fluxing it with sodium magnesium chloride to remove oxide inclusions. If especially pure magnesium is desired, it is prepared by distillation.

Properties and uses. It is the lightest metal used as an engineering material. The tensile strength of the cast metal is 91 MPa which is the same as that of ordinary cast aluminium. It is preferable when reduction of weight is important, because of its low density of 1.74. The tensile

* The impure product of the smelting of sulphide ores especially those of copper or nickel, is known as *matte*.

strength of rolled and annealed magnesium is 175 MPa which is equal to that of a good quality of grey cast iron. It is harder than aluminium. It can be readily machined and under the buffing wheel it takes a high polish.

It readily form alloys with most metals (not with iron and chromium). The metals with which it is mostly alloyed are aluminium, zinc, cadmium and copper. It is employed in the form of sheets, wires, rods, tubes etc. The magnesium ribbon is now chiefly used for the degasification of radio tubes. In the powdered form, mixed with an oxidising agent such as barium nitrate or potassium chlorate, it is employed in the manufacture of flash light powder and in military pyrotechnics for the production of rockets, signals and flares.

6·9 Cadmium

It is probably the first metal that was not discovered in an ore. One cadmium bearing mineral, *greenockite* (C_dS) has been recognised but there are no workable deposit of this.

Manufacture. It is obtained commercially as a by product in the metallurgy of zinc and to some extent of lead. It is recovered from the fume produced in roasting zinc blende, from the blue powder formed during zinc distillation, especially in the earlier stages of the process. It may also be recovered by electrolytic methods or by precipitation with metallic zinc.

Properties and uses. It is a white metal with bluish tinge, capable of taking a high polish. Its specific gravity is 8.65 and melting point 321°C. It is slightly harder than tin but is softer than zinc. It is malleable and ductile and can be readily rolled and hammered into foil and drawn into wire. At 80°C, it becomes so brittle that it can be pulverised under the hammer.

It is chiefly used in antifriction alloys for bearings. It is also used as a rust-proof coating for iron and steel. Bolts, nuts and other small parts employed in automobile manufacture, refrigerator, trimmings, locks and wire products are plated with it.

6·10 Vanadium

The important ore of vanadium is *patronite* (V_2S_9). It occurs is conjunction with iron pyrite, free sulphur and carbonaceous matter.

Manufacture. The preparation of pure vanadium is difficult and very little is produced commercially. For obtaining the metal, acceptable methods are the reduction of the dichloride with hydrogen, and the electrolysis of vanadium trioxide in fused calcium vanadate.

Properties and uses. It is silvery white in colour. Its specific gravity is 5.68 and melting point 1710°C. It is harder than quartz, but is sufficiently malleable and tough so that when heated to a suitable temperature, it may be rolled and hammered into rods and drawn into wires. The chief use of vanadium is in manufacture of alloy steels. Non-ferrous alloys of vanadium are those of copper and aluminium, from which excellent castings may be made.

6·11 Antimony

It is occasionally found in free nature, but the chief ore is *stibnite* (the black trisulphide, Sb_2S_3).

Manufacture. There are several methods for extracting antimony from its ores. The usual process consists of roasting the ore to the oxide and then reducing the oxide with carbon. In another process, the trisulphide is heated with scrap iron which separates metallic antimony. To a small extent, antimony is obtained as a by-product in the refining of other metals such as lead, copper, zinc and silver.

Properties and uses. It is silvery white, hard, highly crystalline and so brittle that it may be readily powdered under the hammer. Its specific gravity is 6.62 and melting point 630°C. It is mostly used as an alloying element with most of the heavy metals. The metals with which antimony is most commonly alloyed are lead, tin and copper.

6·12 Aluminium Alloys

The aluminium may be alloyed with one or more other elements like copper, magnesium, manganese, silicon and nickel. The addition of small quantities of alloying elements converts the soft and weak metal into hard and strong metal, while still retaining its light weight. The main aluminium alloys are discussed below :

1. *Duralumin.* It is an important and interesting wrought alloy. Its composition is as follows:

Copper = 3.5-4.5 % ; Manganese = 0.4-0.7% ; Magnesium = 0.4-0.7% ; and the remaining is aluminium.

This alloy possesses maximum strength (about 400 MPa) after heat treatment and age hardening. After working, if the metal is allowed to age for 3 or 4 days, it will be hardened. This phenomenon is known as *age hardening*.

It is widely used in wrought conditions for forging, stamping, bars, sheets, tubes and rivets. It can be worked in hot condition at a temperature of 500°C. However, after forging and annealing, it can also be cold worked. Due to its high strength and light weight, this alloy may be used in automobile and aircraft components.

2. *Y-alloy.* It is also called *copper-aluminium alloy*. The addition of copper to pure aluminium increases its strength and machinability. The composition of this alloy is as follows :

Copper = 3.5-4.5% ; Manganese = 1.2-1.7% ; Nickel = 1.8-2.3% ; Silicon, Magnesium, iron = 0.6% each ; and the remainder is aluminium.

This alloy is heat treated and age hardened like duralumin. The ageing process is carried out at room temperature for about five days.

It is mainly used for cast purposes, but it can also be used for forged components like duralumin. Since Y-alloy, has better strength (than duralumin) at high temperature, therefore, it is much used in aircraft engines for cylinder heads and pistons.

3. *Magnalium.* It is made by melting the aluminium with 2 to 10% magnesium in a vacuum and then cooling it in a vacuum or under a pressure of 100 to 200 atmospheres. It also contains about 1.75% copper. Due to its light weight and good mechanical properties, it is mainly used for aircraft and automobile components.

4. *Hindalium.* It is an alloy of aluminium and magnesium with a small quantity of chromium. It is the trade name of aluminium alloy produced by Hindustan Aluminium Corporation Ltd., Renukoot (U.P.). It is produced as a rolled product in 16 gauge, mainly for anodized utensil manufacture.

6·13 Copper Alloys

The copper alloys are broadly classified into the following two groups :

1. Copper-zinc alloys (Brasses), in which zinc is the principal alloying metal, and
2. Copper-tin alloys (Bronzes), in which tin is the principal alloying metal.

These alloys are discussed, in detail, in the following articles.

6·14 Copper-Zinc Alloys (Brass)

The most widely used copper zinc alloy is *brass*. There are various types of brasses, depending upon the proportion of copper and zinc. This is fundamentally a binary alloy of copper with zinc each 50%. By adding small quantities of other elements, the properties of brass may be greatly changed. For example, the addition of lead (1 to 2%) improves the machining quality of brass. It has a greater strength than that of copper, but have a lower thermal and electrical conductivity. Brasses are very resistant to atmospheric corrosion and can be easily soldered. They can be easily fabricated by processes like spinning and can also be electroplated with metals like nickel and chromium. The following table shows the composition of various types of brasses according to Indian standards.

Table 6.1. Composition of various types of brasses according to Indian standards.

Indian standard designation	*Composition in percentages*	*Uses*
Cartridge brass	Copper = 70 Zinc = 30	It is a cold working brass used for cold rolled sheets, wire drawing, deep drawing, pressing and tube manufacture.
Yellow brass (Muntz metal)	Copper = 60 Zinc = 40	It is suitable for hot working by rolling, extrusion and stamping.
Leaded brass	Copper = 62.5 Zinc = 36 Lead = 1.5	These are used for plates, tubes etc.
Admiralty brass	Copper = 70 Zinc = 29 Tin = 1	
Naval brass	Copper = 59 Zinc = 40 Tin = 1	It is used for marine castings.
*Nickel brass (German silver)	Copper = 60-45 Zinc = 35-20 Nickel = 5-35	It is used for valves, plumbing fittings, automobile fittings, type writer parts and musical instruments.

6·15 Bronze

The alloys of copper and tin are usually termed as *bronzes*. The useful range of composition is 75 to 95% copper and 5 to 25% tin. The metal is comparatively hard, resists surface wear and can be shaped or rolled into wires, rods and sheets very easily. In corrosion resistant properties, bronzes are superior to brasses. Some of the common types of bronzes are as follows :

1. *Phosphor bronze.* When bronze contains phosphorus, it is called *phosphor bronze.* Phosphorus increases the strength, ductility and soundness of castings. The tensile strength of this alloy when cast varies from 215 MPa to 280 MPa but increases upto 2300 MPa when rolled or drawn. The alloy possesses good wearing qualities and high elasticity. The metal is resistant to salt water corrosion. The composition of the metal varies according to whether it is to be forged, wrought or made into castings. A common type of phosphor bronze has the followig composition according to Indian standards.

Copper = 87-90% ; Tin = 9-10% ; and Phosphorus = 0.1-0.3%

It is used for bearings, worm wheels, gears, nuts for machine lead screws, pump parts, linings and for many other purposes. It is also suitable for making springs.

2. *Silicon bronze.* It contains 96% copper, 3% silicon and 1% manganese or zinc. It has good general corrosion resistance of copper combined with higher strength. It can be cast, rolled, stamped forged and pressed either hot or cold and it can be welded by all the usual methods.

It is widely used for boilers, tanks, stoves or where high strength and good corrosion resistance is required.

3. *Beryllium bronze.* It is a copper base alloy containing about 97.75% copper and 2.25% beryllium. It has high yield point, high fatigue limit and excellent cold and hot corrosion resistance. It is particularly suitable material for springs, heavy duty electrical switches, cams and bushings.

* It is also called nickel silver.

Since the wear resistance of beryllium copper is five times that of phosphor bronze, therefore, it may be used as a bearing metal in place of phosphor bronze. It has a film forming and a soft lubricating property, which makes it more suitable as a bearing metal.

4. *Manganese bronze.* It is an alloy of copper, zinc and little percentage of manganese. The usual composition of this bronze is as follows :

Copper = 60% ; Zinc = 35% ; and Manganese = 5%.

This metal is highly resistant to corrosion. It is harder and stronger than phosphor bronze. It is generally used for bushes, plungers, feed pumps, rods etc. Worm gears are frequently made from this bronze.

5. *Aluminium bronze.* It is an alloy of copper and aluminium. The aluminium bronze with 6-8% aluminium has valuable cold working properties. The maximum tensile strength of this alloy is 450 MPa with 11% of aluminium. They are most suitable for making components exposed to severe corrosion conditions. When iron is added to these bronzes, the mechanical properties are improved by refining the grain size and improving the ductility.

Aluminium bronzes are widely used for making gears, propellers, condenser bolts, pump components, tubes, air pumps, slide valves and bushings etc. Cams and rollers are also made from this alloy. The 6% aluminium alloy has a fine gold colour which is used for imitation jewellery and decorative purposes.

6·16 Gun Metal

It is an alloy of copper, tin and zinc. It usually contains 88% copper, 10% tin and 2% zinc. This metal is also known as *Admirality gun metal*. The zinc is added to clean the metal and to increase its fluidity.

It is not suitable for being worked in the cold state but may be forged when at about 600ºC. The metal is very strong and resistant to corrosion by water and atmosphere. Originally, it was made for casting guns. It is extensively used for casting boiler fittings, bushes, bearings, glands etc.

6·17 Lead Base Alloys

The lead base alloys are employed where a cheap and corrosion resistant material is required. An alloy containing 83% lead, 15% antimony, 1.5% tin and 0.5% copper is used for large bearings subjected to light service.

6·18 Tin Base Alloys (Babbit Metal)

A tin base alloy containing 88% tin, 8% antimony and 4% copper is called *babbit metal*. It is a soft material with a low coefficient of friction and has little strength. It is the most common bearing metal used with cast iron boxes where the bearings are subjected to high pressure and load.

Note : Those alloys in which lead and tin are predominating are designated as *white metal bearing alloys.*

6·19 Bearing Metals

The following are the widely used bearing metals :

1. Copper-base alloys ; 2. Lead-base alloys ; 3. Tin-base alloys, and 4. Cadmium-base alloys.

The copper-base alloys are the most important bearing alloys. These alloys are harder and stronger than the white metals (lead base and tin base alloys) and are used for bearings subjected to heavy pressures. These include brasses and bronzes which are discussed in Art. 6.14 and 6.15. The lead base and tin base alloys are discussed in Art. 6.17 and 6.18. The cadmium base alloys contain 95% cadmium and 5% silver. It is used for medium loaded bearings subjected to high temperature.

The selection of particular type of bearing metal depends upon the conditions under which it is to be used. It involves factors relating to bearing pressures, rubbing speeds, temperatures, lubrication etc. A bearing material should have the following properties :

1. It should have low coefficient of friction.
2. It should have good wearing qualities.
3. It should have ability to withstand bearing pressures.
4. It should have ability to operate satisfactorily with suitable lubrication means at the maximum rubbing speeds.
5. It should have a sufficient melting point.
6. It should have high thermal conductivity.
7. It should have good casting qualities.
8. It should have minimum shrinkage after casting.
9. It should have non-corrosive properties.
10. It should be economical in cost.

6·20 Zinc-Base Alloys

The most of the die castings are produced from zinc base alloys. These alloys can be casted easily with a good finish at fairly low temperatures. They have also considerable strength and low in cost. The ususal alloying elements for zinc are aluminium, copper and magnesium and they are all held in close limits.

The composition of two standard die casting zinc alloys are as follows :

1. Aluminium 4.1%, copper 0.1%, magnesium 0.04% and the remainder is zinc.
2. Aluminium 4.1%, copper 1%, magnesium 0.04% and the remainder is zinc.

Aluminium improves the mechanical properties and also reduces the tendency of zinc to dissolve iron. Copper increases the tensile strength, hardness and ductility. Magnesium has the beneficial effect of making the castings permanently stable. These alloys are widely used in the automotive industry and for other high production markets such as washing machines, oil burners, refrigerators, radios, phonographs, television, business machines etc.

6·21 Nickel-Base Alloys

The nickel-base alloys are widely used in engineering and industry on account of their high mechanical strength properties, corrosion resistance etc. The most important nickel base alloys are discussed below :

1. *Monel metal.* It is an important alloy of nickel and copper. It contains 68% nickel, 29% copper and 3% other constituents like iron, manganese, silicon and carbon. Its specific gravity is 8.87 and melting point 1360°C. It resembles nickel in appearance and is strong, ductile and tough. It is superior to brass and bronze in corrosion resisting properties. It can be cold and hot worked, cast, forged and welded. It is used for making propellers, pump fittings, condenser tubes, steam turbine blades, sea water exposed parts, tanks and chemical and food handling plants.

2. *K-alloy.* It consists of 3% aluminium and 0.5% titanium, in addition to the composition of monel metal. It has better mechanical properties than monel metal. It is used in parts requiring high strength and good corrosion resistance.

3. *Inconel.* It consists of 80% nickel, 14% chromium and remainder is iron. Its specific gravity is 8.55 and melting point 1395°C. This alloy has excellent mechanical properties at ordinary and elevated temperatures. It can be cast, rolled and cold drawn. It is used for making springs which have to withstand high temperatures and are exposed to corrosive action. It is also used for exhaust manifolds of aircraft engines.

4. *Nichrome.* It consists of 65% nickel, 15% chromium and 20% iron. It has high heat and oxidation resistance. It is used in making electrical resistance wire for electric furnaces and heating elements.

5. *Nimonic.* It consists of 80% nickel and 20% chromium. It has high strength and ability to operate under intermittent heating and cooling conditions. It is widely used in gas turbine engines.

6·22 High Temperature Alloys

The high temperature alloys are those alloys which can withstand temperature in excess of 1100ºC. The alloys are used in components of nuclear plants, jet and rocket engines. Some of the high temperature alloys are as follows :

1. *Incoloy.* It is a nickel base alloy. It consists of 42% nickel; 13% chromium; 6% molybdenum; 2.4% titanium; 0.4% carbon and the remaining is iron.

2. *Hastelloy.* It is also a nickel base alloy. It consists of 45% nickel, 22% chromium, 9% molybdenum ; 1.5% cobalt ; 0.5% tungsten ; 0.15% carbon and the remaining is iron.

3. *Vitallium.* The main constituent of this alloy is cobalt. It consists of 62% cobalt ; 28% chromium ; 5.5% molybdenum ; 2.5% nickel ; 1.7% iron and 0.28% carbon.

Note : The Inconel and Nimonic, as discussed in the previous article, are also high temperature alloys.

6·23 Metals for Nuclear Energy

With the development of nuclear power, the demand for metals to be used with nuclear energy is widely created. The various metals for producing nuclear energy are used as raw materials for *moderators, reflectors, fuel elements, fuel canning materials, control elements and pressure vessel materials. The important metals used for nuclear energy are uranium, thorium, plutonium, zirconium, beryllium and niobium. These metals are discussed, in details, as follows :

1. *Uranium.* It is the most important metal used for nuclear energy. It is found in nature as a natural metal. It is used as a nuclear fuel. It is found to be very radioactive and can be easily oxidized. It has poor resistance to corrosion. In its pure form, it is weak. Therefore, it is protected by a thin film of aluminium or zirconium through a rolling process (known as cladding). It is made highly suitable for nuclear power when mixed with some alloying elements like chromium, molybdenum, plutonium, zirconium etc. The uranium compounds have been found to give better service. The uranium oxide is a highly refractory and does not exhibit any change in phase in an inert atmosphere. It possesses a good strength, high corrosion resistance, high coefficient of expansion, low thermal shock resistance and poor thermal conductivity.

2. *Thorium.* It is a less important fuel and is free from phase changes below 1480ºC. It is also a radioactive metal like uranium and carries F.C.C. structure. It is soft and weak in its pure form, but its mechanical properties are greatly affected by adding small amounts of impurities. For example, the addition of 0.2 percent carbon increases its tensile strength from 140 MPa to 380 MPa. It may be noted that the addition of titanium, zirconium and niobium decreases the strength and hardness of the metal.

3. *Plutonium.* It is a concentrated type of fuel which is used in a fast reactor without a moderator. It is highly toxic and chemically very reactive. It can be easily oxidised. Its melting point is 640ºC.

It has a poor resistance to corrosion. It is naturally fissiable and continues to emit high energy alpha, gamma and similar other radiations. It is chiefly used in the production of atomic weapons and in fast breeding atomic reactors. Since some plutonium compounds and alloys emit neutrons, therefore perfect health protection is essential while using these materials.

4. *Zirconium.* In its natural form, it contains 0.5 to 2% hafnium, which is a strong absorber of neutrons. This metal is very hard and its melting point is 1850ºC. It has a poor resistance to CO_2 at elevated temperature but this is improved by adding 0.5% copper and 0.5% molybdenum. The addition of these elements increases the tensile strength upto 510 MPa and the simultaneous increase in creep

* A moderator, in nuclear power plants, is a medium introduced into the fuel mass in order to slow down the speed of fast moving neutrons.

resistance at 450°C. This is specially used in gas cooled reactors.

5. *Beryllium.* It is a coarse grained light and brittle metal. It has a highly developed texture. Its melting point is 1280°C. During its melting, it is necessary to have vacuum or an inert gas atmosphere because it is very reactive and can readily form compounds with the atmosphere inside the furnace or the refractories present inside.

In cast conditions, it is coarse grained and brittle. Thus powder metallurgy techniques are used to fabricate components from beryllium. In its pure form, it oxidises fairly rapidly at about 780°C. It is corroded due to the presence of some traces of water in CO_2. It is used as a moderator, reflector and neutron source.

6. *Niobium.* It has a good strength, ductility and corrosion resistance. It has a high melting point of 2450°C. Its oxidation resistance above 400°C is indifferent, but is greatly improved by alloying. Its workability is badly affected by impurities like oxygen, nitrogen, hydrogen and carbon.

QUESTIONS

1. Describe the manufacturing methods, properties and uses of the following non-ferrous metals :
(*a*) Aluminium ; (*b*) Copper ; (*c*) Lead ;
(*d*) Tin ; (*e*) Zinc ; and (*f*) Magnesium

2. Discuss the various types of aluminium alloys giving their composistion, properties and uses.

3. What is brass ? Describe the composition, properties and uses of important types of brasses.

4. What is bronze ? Describe the composition, properties and uses of different types of bronzes.

5. Name the important properties which a bearing metal should possess. Describe the important bearing metals.

6. Write the composition, properties and uses of the following nickel-base alloys.
(*a*) Monel metal ; (*b*) Inconel ; (*c*) Nichrome ; and (*d*) Nimonic.

7. Write the composition of the following metals :
(*a*) Incoloy ; (*b*) Hastelloy ; and (*c*) Vitallium.

8. What are the main metals used in nuclear energy ? Describe them in brief.

OBJECTIVE TYPE QUESTIONS

1. The aluminium alloy, mainly used, for anodized utensil manufacture, is
(*a*) duralumin (*b*) Y-alloy (*c*) magnalium (*d*) hindalium.

2. Brass is an alloy of
(*a*) copper and zinc (*b*) copper and tin
(*c*) copper, tin and zinc (*d*) none of these

3. An alloy of copper, tin and zinc is known as
(*a*) brass (*b*) bronze (*c*) gun metal (*d*) muntz metal

4. Muntz metal (Yellow brass) contains
(*a*) 70% copper and 30% zinc (*b*) 60% copper and 40% zinc
(*c*) 59% copper, 40% zinc and 1% tin (*d*) 60.45% copper, 35.2% zinc and 5.35% nickel
(*e*) 62.5% copper, 36% zinc and 1.5% lead

5. German silver contains
(*a*) 70% copper and 30% zinc (*b*) 60% copper and 40% zinc
(*c*) 59% copper, 40% zinc and 1% tin (*d*) 60.45% copper, 35.2% zinc and 5.35% nickel
(*e*) 62.5% copper, 36% zinc and 1.5% lead

6. Babbit metal is a

(*a*) lead-base alloy (*b*) copper-base alloy
(*c*) tin-base alloy (*d*) cadmium-base alloy

7. German silver contains

(*a*) 1% silver (*b*) 2% silver (*c*) 5% silver (*d*) 8% silver
(*e*) no silver

8. Monel metal is an alloy of

(*a*) nickel and copper (*b*) nickel and chromium
(*c*) nickel, chromium and iron (*d*) copper and chromium

9. Monel metal contains

(*a*) 65% nickel, 15% chromium and 20% iron
(*b*) 68% nickel, 29% copper and 3% other constituents
(*c*) 80% nickel and 20 % chromium
(*d*) 80% nickel, 14% chromium and 6% iron

10. Inconel contains

(*a*) 65% nickel, 15% chromium and 20% iron
(*b*) 68% nickel, 29% copper and 3% other constituents
(*c*) 80% nickel and 20% chromium
(*d*) 80% nickel, 14% chromium and 6% iron

11. The metal suitable for bearing subjected to light loads, is

(*a*) silicon bronze (*b*) white metal (*c*) monel metal (*d*) phosphor bronze

12. The metal suitable for bearings subjected to heavy loads, is

(*a*) silicon bronze (*b*) white metal (*c*) monel metal (*d*) phosphor bronze

13. Phosphor bronze has

(*a*) good wearing qualities and high elasticity
(*b*) high resitance to corrosion
(*c*) valuable cold working property
(*d*) all of the above
(*e*) none of the above

14. Beryllium bronze has

(*a*) low yield point (*b*) low fatigue limit (*c*) high yield point (*d*) high fatigue limit
(*e*) both (*a*) and (*b*) (*f*) both (*c*) and (*d*)

15. Which of the following metal is used for nuclear energy ?

(*a*) uranium (*b*) thorium (*c*) plutonium (*d*) zirconium
(*e*) all of the above (*f*) none of the above

ANSWERS

1. (*d*)	2. (*a*)	3. (*b*)	4. (*b*)	5. (*d*)
6. (*c*)	7. (*e*)	8. (*a*)	9. (*b*)	10. (*d*)
11. (*d*)	12. (*b*)	13. (*a*)	14. (*f*)	15. (*e*)

7

Heat Treatment of Metals and Alloys

1. Introduction. 2. Behaviour of Steel When Heated and Cooled. 3. Isothermal Transformation Diagram or Time Temperature Transformation (T T T) Diagram. 4. Process of Heat Treatment. 5. Types of Heat Treatment Processes. 6. Annealing. 7. Normalising. 8. Hardening. 9. Hardenability. 10. Testing of Hardenability. 11. Austempering. 12. Martempering. 13. Tempering. 14. Types of Tempering. 15. Tempering Baths. 16. Surface Hardening or Case Hardening. 17. Carburising. 18. Cyaniding. 19. Nitriding. 20. Induction Hardening. 21. Flame Hardening. 22. Precipitation Hardening or Age Hardening of Aluminium Alloy. 23. Heat Treatment Furnaces. 24. Types of Furnaces. 25. Hearth Furnaces. 26. Stationary Hearth Furnaces. 27. Movable Hearth Furnaces. 28. Bath Furnaces. 29. Measurement of Temperature. 30. Thermo-electric Pyrometer. 31. Optical Pyrometer.

7·1 Introduction

The heat treatment is defined as an operation or a combination of operations, involving the heating and cooling of a metal or an alloy in the solid state for the purpose of obtaining certain desirable conditions or properties. The aim of heat treatment is to achieve one or more of the following objects :

1. To increase the hardness of metals.
2. To relieve the internal stresses set up in the material after hot or cold working
3. To improve machinability.
4. To soften the metal.
5. To modify the internal structure of the material to improve its electrical and magnetic properties.
6. To change grain size.
7. To increase the qualities of a metal to provide better resistance to heat, corrosion and wear.
8. To improve mechanical properties like tensile strength, ductility and shock resistance etc.

Before we deal with the mechanism of heat treatment, it is essential to know the behaviour of steel when it is heated and cooled.

7·2 Behaviour of Steel When Heated and Cooled

When steel is heated, a wide range of changes takes place in its internal structure depending upon its carbon content and heating temperature. Consider a piece of steel with 0.3% carbon placed in a furnace is heated at a uniform rate in such a way that the temperature of the steel can be observed at equal intervals of time. If a curve between time and temperature change is drawn, as shown in Fig. 7.1, it will be found that the temperature rises uniformly upto 723°C (a dull red heat). It will, for a short time, remain stationary at this temperature and then rise at a somewhat slower rate until it reaches 800°C (a good red heat). After this, if the heating can be maintained, the temperature will continue to rise approximately at the initial rate upto 910°C. The point at which the temperature received its first check is called *point of decalescence*.

Similarly, when the metal is allowed to cool slowly, the reverse change takes place. The temperature of steel will fall uniformly till it reaches about the temperature where it received its first check when it was being heated, and here, if it is observed carefully, it will be seen that the cooling down has stopped. Not only a check in the cooling down will be observed at this point, but the steel will probably be seen to take an extra glow as if heat had been imparted to it. This point is called *point of recalescence.* After this the rate of cooling will proceed normally until the metal is cold. From above, it is definitely established that a structural change takes place between 723°C and 800°C, (for steel containing 0.3% carbon) otherwise there would have been steady fall in the temperature.

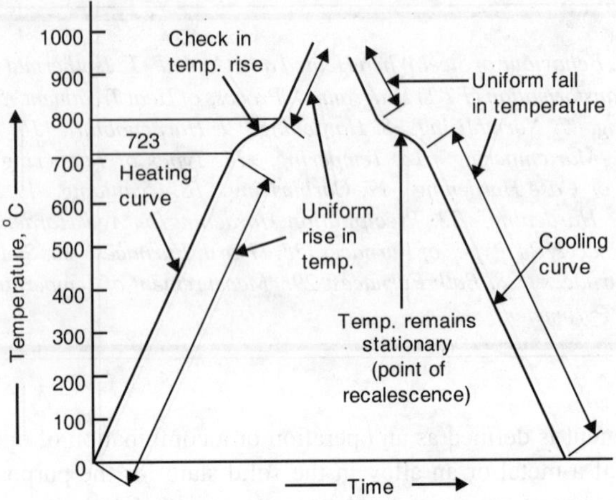

Fig. 7.1. Heating and cooling curves for steel.

We have discussed in the previous chapter that the structure of steel, less than 0.8% carbon, is made up of areas of ferrite (iron) surrounding areas of pearlite which is a substance made up of layers of ferrite and cementite. When steel is heated, this structure remains stable until a temperature of about 723°C is reached. Then, with the addition of further heat, carbon in the pearlite commences to dissolve in iron in the solid form. This change of state continues, until the whole structure of metal consists of a solid solution of carbon in iron which is called *austenite.* When austenite is completely formed, the rise in temperature continues as at first. On cooling, the reverse takes place and at 723°C the austenite changes back to pearlite again.

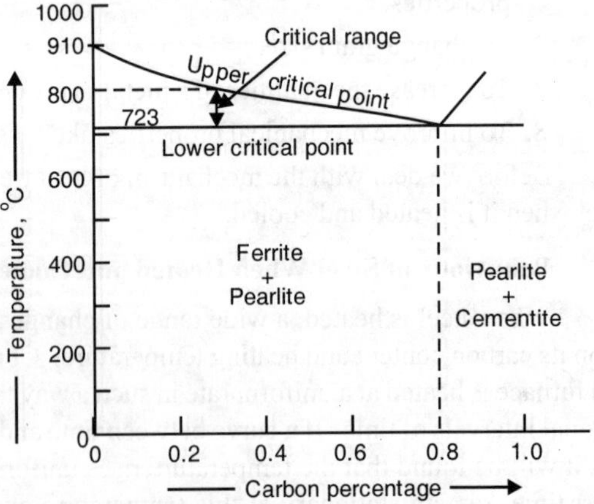

Fig. 7.2. Changes in steel.

The temperature point at which the change starts on heating is called *lower critical point* and temperature point where this change ends in heating is called *upper critical point.* The range between these two critical points is known as *critical range.* The temperature at which the change starts (*i.e.*, lower critical point) is same for all steels, but the ending point of transformation (*i.e.*, upper critical point) varies according to the carbon content in steel. This is shown in Fig. 7.2. It will be seen that for a steel containing 0.8% carbon (wholly pearlite) there is only one critical point, the whole transformation takes place at that temperature. Fig. 7.3 shows the change which occurs to the micro- structure of the steel when the carbon goes into solution. Fig. 7.3(*a*) shows the normal structure of ferrite and pearlite and Fig. 7.3(*b*) shows the austenite at the end of the change (*i.e.*, when the carbon goes into solution). The austenite is a solid solution of carbon in iron. All the ordinary steels above the critical range are in this condition.

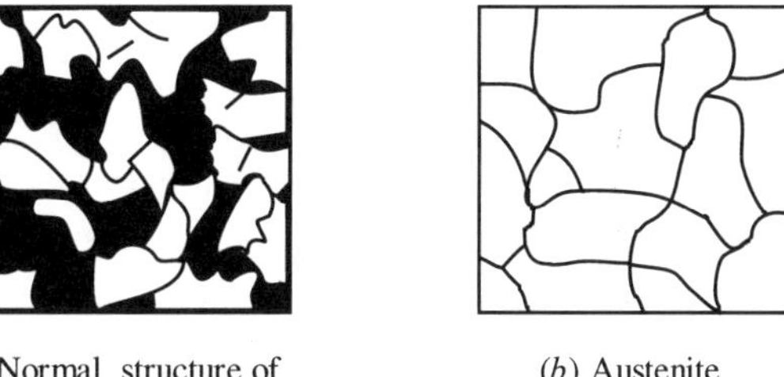

(*a*) Normal structure of ferrite and pearlite. (*b*) Austenite

Fig. 7.3. Change in structure of steel when heated to critical rang

Note : A steel with 0.8% carbon is termed as *eutectoid steel*, under 0.8% carbon as *hypo-eutectoid steel* and above 0.8% carbon as *hyper-eutectoid steel.*

7·3 Isothermal Transformation Diagram or Time Temperature Transformation Diagram (TTT Diagram)

We have already discussed the iron-carbon equilibrium diagram in chapter 3 (Art. 3.37), which shows the changes in the micro-structure of steel under equilibrium conditions. But most of the heat treatment processes are carried out under non-equilibrium conditions, where the variation in the rate of cooling produces corresponding variation in the micro-structure and mechanical properties. For example, when austenite steel (above 723°C) is cooled at different rates, the resulting structure will be as shown in Fig. 7.4.

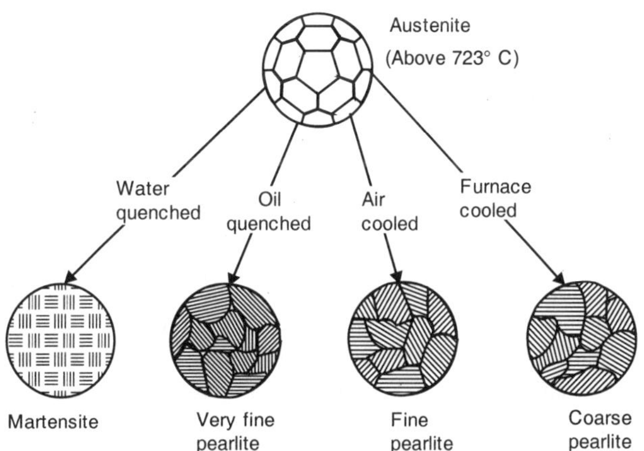

Fig. 7.4. Effect of cooling rate on micro-structure.

From the above, we see that by changing only the rate of cooling, different phases can be achieved. The information on the change of phase with the cooling rate may be conveniently displayed with the help of a *isothermal transformation diagram* which is commonly known as *time-temperature-transformation diagram* (briefly written as TTT diagram). This diagram is also called S-curve or C-curve due to its shape. In such a diagram, the temperature is plotted along the vertical axis using a linear scale, whereas the abscissa represents the time on a logarithmic scale. The TTT diagram for the eutectoid steel is shown in Fig. 7.5. It relates to the transformation of austenite to the time and temperature conditions to which it is subjected. The curve 1 indicates the beginning of transformation of austenite and the curve 2 indicates the end of transformation.

The austenite is stable above the lower critical point (723°C), but at temperature below this, the austenite is unstable and gives rise to the formation of a mixture composed of the ferrite and cementite. But depending upon the transformation temperature, the structures and properties of this ferrite-cementite mixture are substantially different. The various changes in the structure due to the transformation of austenite at different temperatures are shown in Fig. 7.5.

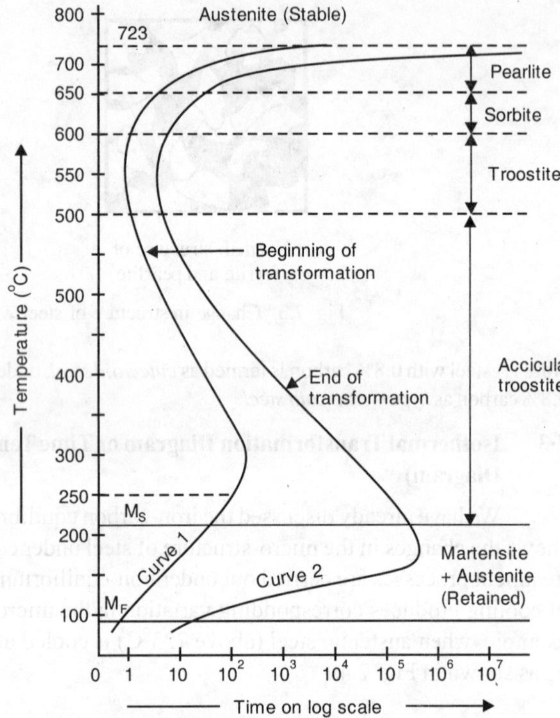

Fig. 7.5. T T T diagram for eutectoid steel.

It has been observed that ferrite-cementite mixture, formed at different temperatures, due to the transformation of austenite, differs primarily in the degree of dispersion (or refinement) of both phases. At lower transformation temperatures, both the phases are more dispersed. But at high transformation temperatures (at about 700°C), the ferrite-cementite mixture becomes sufficiently distinct and is known as *pearlite.*

It has also been observed that at somewhat lower transformation temperatures (at about 650°C), the ferrite-cementite mixture is more dispersed and is known as *sorbite.* At still lower transformation temperatures (at about 600°C), the ferrite-cementite mixture becomes so dispersed that its structure can only be determined by an electronic microscope. Its structure is similar to pearlite and sorbite and is known as **troostite.* At still more lower transformation temperature (from about 550°C to 200°C), the ferrite-cementite mixture is more dispersed and is known as *accicular troostite.* It is a needle like structure. At more lower transformation temperatures (below 200°C), the austenite does not change into ferrite-cementite mixture, but it changes into martensite. The formation of martensite does not take place instantaneously. It depends upon the temperature and occurs over a wide range. The upper and lower limits of the transformation range are known as M_S and M_F respectively. The symbol M_S stands for the temperature at which the formation of martensite begins and symbol M_F stands for the temperature at which the formation of martensite ends.

7·4 Process of Heat Treatment

We have already discussed that the process of heat treatment is carried out first by heating the metal and then cooling it. In fact, the process consists of

1. Heating the metal to a specified temperature;
2. Holding the metal at this temperature for a specified period; and
3. Cooling the metal (*i.e.*, quenching) according to the specified process.

* It is also known as bainite. The sorbite is called upper bainite and the accicular troostite as lower bainite.

It may be noted that the true heat treatment does not involve any chemical change. But it essentially involves time-temperature cycle which means the range and rate of heating and cooling.

The temperature to which a metal or an alloy is heated depends upon its grade, grain size as well as type and shape. In general, the metal is never heated much beyond its upper critical temperature. As a result of this, some plastic deformation of the grains of the metal takes place. This deformation depends upon the chemical composition of the metal or alloy and the temperature to which it is heated.

The metal or an alloy is now held at the required temperature to ensure uniformity of temperature throughout the mass. The period of heating depends upon the size and shape of the component. In general, the time required for heating the component is approximately 1 minute to 3 minutes for 1 mm thickness of the largest section.

The main transformation in the properties of a component takes place in the cooling process of the metal. It depends mostly on the rate at which the cooling takes place. It also depends upon the medium in which the cooling takes place. Now-a-days, only five quenching medias adopted are caustic soda solution, brine, water, oil and air.

7·5 Types of Heat Treatment Processes

Following are the various heat treatment processes commonly employed in engineering practice :

1. Annealing ;
2. Normalising ;
3. Hardening ;
4. Tempering ;
5. Case hardening ;
6. Surface hardening.

We shall now discuss, in detail, all the above mentioned processes in the following pages.

7·6 Annealing

The annealing is one of most important process of heat treatment of steel. It is employed for the following purposes :

1. To soften the steel so that it may be more easily machined or cold worked.
2. To refine the grain size and structure to improve mechanical properties like strength and ductility.
3. To relieve internal stresses which may have been caused by hot or cold working or by unequal contraction in casting.
4. To alter electrical, magnetic or other physical properties.
5. To remove gases trapped in the metal during initial casting.

The following are five types of annealing :

1. *Full annealing.* The purpose of full annealing is to soften the metal, to refine the grain structure, to relieve the stresses and to remove trapped gases in the metal. The process consists of

(*a*) heating the steel 30°-50°C above the upper critical temperature for hypo-eutectoid steel and by the same temperature above the lower critical temperature for hyper-eutectoid steels (see Fig. 7.6).

(*b*) holding it at this temperature for sometime to enable the internal changes to take place. The time allowed is approximately 3 to 4 minutes for each millimeter of thickness of the largest section, and

(*c*) cooling slowly in the furnace.

The rate of cooling varies from 30°C to 200°C per hour depending upon the composition of steel. The cooling is usually carried out in the furnace. The objects may be taken out of the furnace and cooled in ashes so as to prolong the cooling time.

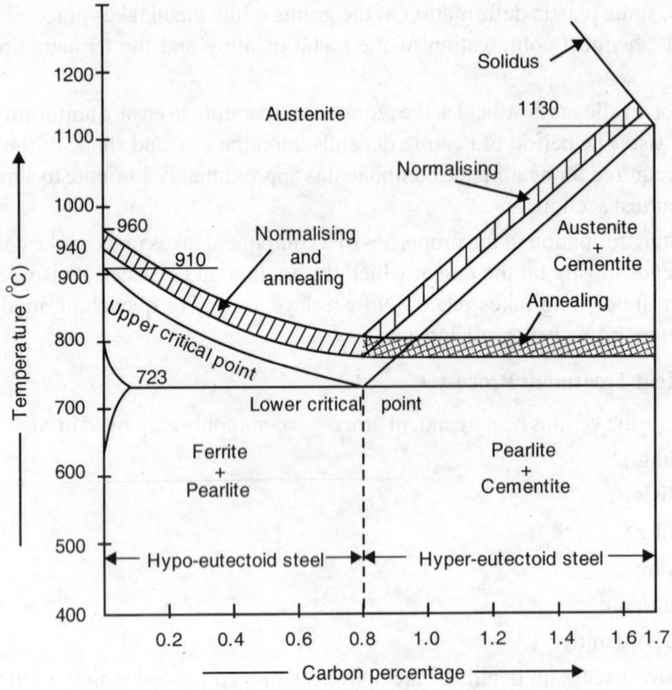

Fig. 7.6. Temperature ranges for various heat treatment processe

In order to avoid decarburisation of the steel during annealing, the steel is packed in a cast iron box containing a mixture of cast iron borings, charcoal, lime, sand or ground mica. The box along with its contents is allowed to cool slowly in the furnace after proper heating has been completed.

When slow cooling is done, the austenite is transformed into a structure consisting of pearlite and ferrite, in case of hypo-eutectoid steel. In eutectoid steel, it is transformed to pearlite and in case of hyper-eutectoid steel, it decomposes to give pearlite and cementite structure.

The following table shows the approximate temperatures for annealing depending upon the carbon contents in steel.

Table 7.1. Annealing temperatures.

S. No.	*Carbon content, percent*	*Annealing temperature, °C*
1.	Less than 0.12 (Dead mild steel)	875 – 925
2.	0.12 to 0.45 (Mild steel)	840 – 970
3.	0.45 to 0.50 (Medium carbon steel)	815 – 840
4.	0.50 to 0.80 (Medium carbon steel)	780 – 810
5.	0.80 to 1.50 (High carbon or tool steels)	760 – 780

2. *Process annealing.* It is also known as low temperature annealing or sub-critical annealing. This process is used for relieving the internal stresses previously set up in the metal and for increasing the machinability of the steel. In this process, steel is heated to a temperature below or close to the

lower critical temperature (generally 550°C – 650°C), held at this temperature for sometime and then cooled slowly. This causes complete recrystallisation in steels which have been severely cold worked and a new grain structure is formed. This process increases ductility and plasticity, improves shock resistance, reduces hardness and tensile strength, improves machinability and percentage elongation and reduction in area. The process annealing is commonly used in the sheet and wire industries.

3. *Spheroidise annealing* (*Spheroidising*). It is another form of annealing in which cementite in the form of small globules or spheroids is produced in the structure of steel. This is usually applied to high carbon tool steels which are difficult to machine. The operation consists of heating the steel to a temperature slightly above the lower critical temperature (730°C to 770°C). It is held at this temperature for some time and then cooled slowly to a temperature of 600°C. The rate of cooling is from 25°C to 30°C per hour.

The spheroidising improves the machinability of steels, but lowers the hardness and tensile strength. These steels have better elongation properties than normally annealed steel.

4. *Diffusion annealing (Homogenization).* This type of annealing produces coarse grain structure. But by subsequent plastic working, fine grain structure can be produced. This process is mainly used for ingots and large castings. After diffusion annealing, the castings undergo full annealing to improve their properties or to refine grain structure.

The process consists of heating the steel to a high temperature (1100°C – 1200°C) at which diffusion occurs quite easily and to some extent, equalises the composition of steels. It is held at this temperature for 8 to 20 hours and then cooled to 800°C – 850°C inside the furnace for a period of about 6 to 8 hours. It is further cooled in the air to room temperature. The total heating, holding, and cooling time ranges from 50 to 120 hours.

5. *Isothermal annealing.* This type of annealing produces a more homogeneous structure and the machinability is improved. The main advantage of this process is that the annealing time is considerably reduced.

The process consists of heating the steel to the austenite state as in case of full annealing and then cooled rapidly in air or by a blast in a furnace to a temperature of 600°C to 700°C. The steel is held at this temperature isothermally (*i.e.*, at constant temperature) for some time. It is then cooled rapidly in air to room temperature. During the time of isothermal holding, complete decomposition to pearlite takes place and hence the name is *isothermal annealing*.

7·7 Normalising

The normalising is done for the following purposes :

1. To refine the grain structure of the steel to improve machinability, tensile strength and structure of weld.
2. To remove strains caused by cold working processes like hammering, rolling, bending etc., which makes the metal brittle and unreliable.
3. To remove dislocations caused in the internal structure of the steel due to hot working.
4. To improve certain mechanical and electrical properties.

The process of normalising consists of heating the steel 30°C – 50°C above its upper critical temperature (for hypo-eutectoid steels) or Acm line (for hyper-eutectoid steels). It is held at this temperature for about fifteen minutes and then allowed to cool down in still air.

This process provides a homogeneous structure consisting of ferrite and pearlite for hypo-eutectoid steels, and pearlite and cementite for hyper-eutectoid steels. The homogeneous structure provides a higher yield point, ultimate tensile strength and impact strength with lower ductility to steels. The process of normalising is frequently applied to castings and forgings etc. The alloy steels may also be normalised but they should be held for two hours at a specified temperature and then cooling in the furnace.

7·8 Hardening

The main objects of hardening are :

1. To increase the hardness of the metal so that it can resist wear.
2. To enable it to cut other metals *i.e.*, to make it suitable for cutting tools.

The process of hardening consists of :

(*a*) heating the metal to a temperature of 30°C to 50°C above the upper critical point for hypo-eutectoid steels and by the same temperature above the lower critical point for hyper-eutectoid steels, as shown in Fig. 7.7.

(*b*) keeping the metal, at this temperature for a considerable time, depending upon its thickness.

(*c*) quenching (cooling suddenly) in a suitable cooling medium.

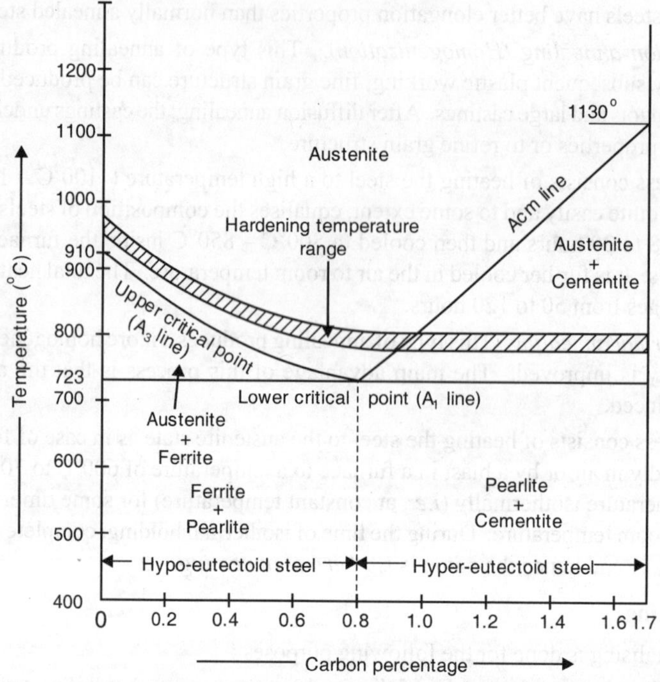

Fig. 7.7. Hardening process.

The heating operation is required for the purpose of transforming the ferrite and pearlite of hypo-eutectoid steels and cementite and pearlite of hyper-eutectoid steel into austenite. However, a large portion of cementite remains as such in case of hyper-eutectoid steels. If the carbon content of the steel is known, then the proper temperature, to which the steel should be heated may be obtained from iron-carbon diagram, as shown in Fig. 7.7. In any heat treating operation, the rate of heating is very important, because heat flows from the exterior to the interior of steel at a definite maximum rate. If the steel is heated too fast, the outside surface becomes hotter than the interior surface and uniform structure can not be obtained. If a piece is irregular in shape, a slow rate is more essential to eliminate warping, and cracking. The heavier the section, the longer must be the heating time to achieve uniform results. Even after the correct temperature has been reached, the piece should be held at that temperature for a sufficient period of time to permit its thickest section to attain a uniform temperature.

The hardness obtained from a given treatment depends upon the rate of cooling, the carbon content and the work size. A very rapid cooling is necessary to harden low and medium plain carbon steels. The quenching in a water or brine solution is a method of rapid cooling which is commonly used. For high carbon and alloy steels, mineral oil is generally used as the quenching medium, because its action is not so severe as that of water. Certain alloy steels can be hardened by air cooling, but for ordinary steels, such a cooling rate is too slow to give an appreciable hardening effect. Large parts are usually quenched in an oil bath. The temperature of the quenching medium must be kept uniform to achieve uniform results. Any quenching bath used in production work should be provided with means for cooling.

A *rapid cooling from the hardening temperature causes the austenite to be transformed into another constituent called *martensite* which is very hard and brittle. The hardening of steel depends entirely upon the formation of martensite, because austenite is comparatively soft and ductile.

It may be noted that the low carbon steels cannot be hardened appreciably, because of the presence of ferrite which is soft and is not changed by the treatment. As the carbon content goes on increasing, the possible obtainable hardness also increases.

The following table shows the effect of cooling rate in different cooling media applied to eutectoid steels.

Table 7.2. Effect of cooling rate on the microstructure and mechanical properties.

Cooling medium	*Structure*	*Ultimate tensile stress (MPa)*	*Yield stress (MPa)*	*Percentage elongation*	*Percentage reduction in area*	*Rockwell hardness (HRC)*
Water quenching	Martensite	1750	–	Low	Low	65
Oil quencing	Very fine pearlite	1120	560	5	22	35
Air cooling	Fine pearlite	880	280	8	27	25
Furnace cooling	Coarse pearlite	520	140	12	30	15

7·9 Hardenability

The **hardenability is that property of steel which determines the depth and distribution of hardness obtained by quenching from the austenising temperature. The ratio of formation of austenite to martensite is the deciding factor for hardenability. With the increase of this formation, the ratio of hardenability also increases. Steel with a high hardenability forms thick layers of martensite. The hardenability of steel depends upon the following factors :

1. The method of manufacture of steel.
2. The amount of carbon content in steel.
3. The homogeneity of the austenite.
4. The grain size of the austenite.
5. The heating rate and duration.
6. The medium of quenching.

---

* If a hypo-eutectoid steel is cooled slowly, the austenite is transformed into ferrite and pearlite which makes the steel soft and ductile.

** The hardness and hardenability are two different terms. The hardness is defined as the resistance to plastic deformation or penetration, whereas hardenability is defined as the ease with which hardness may be obtained by quenching.

7. The rate of quenching.
8. The size of component.
9. The surface conditions.

The cooling rate at the core (centre) of the component is slower which results in the formation of pearlite, which, in turn reduces the hardenability. The alloying elements like magnesium, chromium, molybdenum and nickel increases the hardenability. The grain size of austenite also affects the hardenability. The larger grain size increases the hardenability or depth of hardened layer can even be doubled or tripled by increasing the grain size.

7·10 Testing of Hardenability

The hardenability of steel is generally determined by a standard test known as *Jominy end quench test* as shown in Fig. 7.8 (*a*). In this test, a cylindrical bar (known as Jominy test bar) of 25 mm diameter and 100 mm effective length is heated to above the critical temperature to form austenite. It is then held in a jig and quenched from its one end by throwing a jet of water from the water pipe as shown in Fig. 7.8(*a*). The flow rate, pressure and temperature of water are specified.

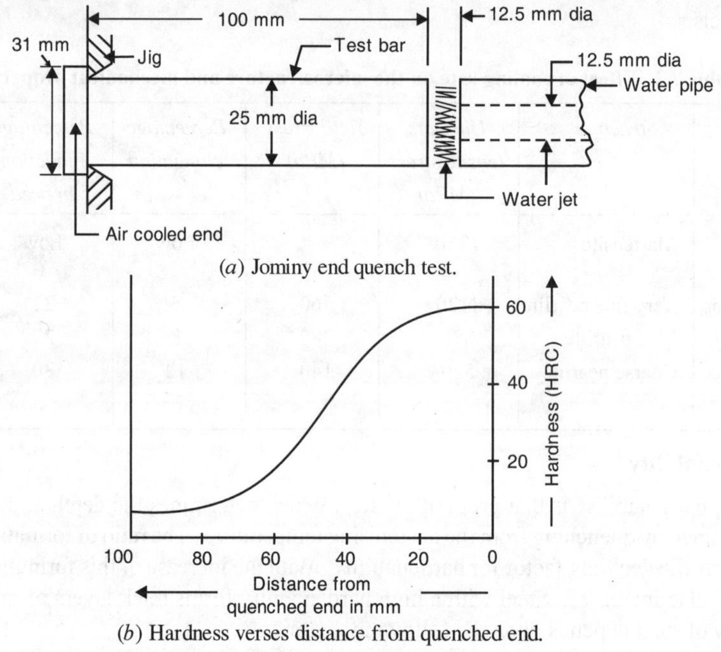

(*a*) Jominy end quench test.

(*b*) Hardness verses distance from quenched end.

Fig. 7.8. Testing of hardenability.

Since the end is cooled very rapidly, therefore it has maximum hardness. The hardness goes on decreasing as the distance from the quenched end increases, as shown in Fig. 7.8(*b*), because the cooling rate goes on decreasing, when the distance from the quenched end increases.

7·11 Austempering

The austempering is misnomer because it is not a tempering process, but a hardening process. It is also known as *isothermal quenching.* In this process, the steel is heated, above the upper critical temperature, at about 875°C where the structure consists entirely of austenite. It is then suddenly cooled by quenching it in a salt bath or lead bath maintained at a temperature of about 250°C to 525°C so as to facilitate the transformation of austenite into *accicular troostite* or *bainite.* After complete transformation, steel is cooled in air. A good impact strength and ductility is obtained in this process and the chances of cracking are also reduced.

7·12 Martempering

This process is also known as *stepped quenching* or *interrupted quenching.* It consists of heating steel above the upper critical point and then quenching it in a salt bath kept at a suitable temperature so that it is in the upper martensite range (usually 150°C to 300°C). After the temperature becomes uniform throughout the steel structure, without the formation of bainite, it is further cooled in air or oil, to room temperature. The steel is then tempered. The martempered steel is free from internal stresses. It avoids cracks and warping etc., usually caused by ordinary hardening. The martempering is mostly used in case of alloy steels.

7·13 Tempering

The steel hardened by rapid quenching is very hard and brittle. It also contains internal stresses which are severe and unequally distributed to cause cracks or even rupture of hardened steel. The tempering (also known as *drawing*) is, therefore, done for the following reasons :

1. To reduce brittleness of the hardened steel and thus to increase ductility.
2. To remove the internal stresses caused by rapid cooling of steel.
3. To make steel tough to resist shock and fatigue.

The tempering process consists of reheating the hardened steel to some temperature below the lower critical temperature, followed by any desired rate of cooling. The exact tempering temperature depends upon the purpose for which the article or tool is to be used.

The tempering temperatures may be judged by the colour formed on the surface of the steel being tempered. The colours are caused by surface oxidation of the steel with the formation of thin films of iron oxide. The various tempering temperatures and colours, for different types of tools made from carbon steels, are given in the following table.

Table 7.3. Tempering temperatures and colours for carbon steel tools.

S. No.	*Type of work*	*Tempering temperature, °C*	*Tempering colour*
1.	Scrapers and lathe tools for brass.	220	Light straw or pale yellow
2.	Hacksaws, steel-engraving tools, circular saws for steel, light turning and parting tools.	225	—
3.	Hammer faces, planers for steel, screwing dies for brass.	230	Straw or dark straw
4.	Wood-engraving tools, paper-cutters, planers for iron.	235	—
5.	Shear blades, milling cutters and hollow mills, bone-cutting tools.	240	Golden yellow
6.	Rock drills, screw cutting dies, boring cutters and reamers.	245	—
7.	Taps and chasers, knurls, pen-knives, mill chisels and picks.	250	—
8.	Moulding and planing cutters for hard wood, punches and dies.	255	Yellowish brown or brown
9.	Plane irons, gouges and brace bits, stone-cutting tools, planing and moulding cutters.	260	—

S. No.	Type of work	Tempering temperature, °C	Tempering colour
10.	Twist drills for wood.	265	Reddish brown dappled with purple
11.	Dental and surgical instruments, augers and pressing cutters.	270	—
12.	Gimlets, axes and hot sets.	275	—
13.	Cold chisels and sets for steel, cold chisels for cast iron, chisels for wood.	280	Purple
14.	Planing cutters for soft wood.	285	—
15.	Cold chisels for wrought iron.	290	Violet
16.	Screw drivers.	295	Blue
17.	Springs.	300	Dark blue

7·14 Types of Tempering

The tempering procedures are classified as follows according to tempering temperatures because tempering temperatures has much effect on the properties of steel.

1. *Low temperature tempering.* In this case, the steel is heated to a low tempering temperature of 150°C to 250°C and the holding time is 1 to 3 hours. This process removes the internal stress, and increases ductility without changing the structure of steel from martensite or reducing its hardness. It also increases the strength and toughness and provides high wear resistance. This method is applied in the manufacture of measuring tools and cutting tools. This is also employed to components which are surface hardened by carburising, cynading and carbonitriding.

2. *Medium temperature tempering.* This type of tempering is performed at 350°C to 450°C. This process alters the structure of steel. The martensite is transformed into *secondary troosite* which results in the increase of ductility, elongation and toughness but the hardness and strength of the metal reduces. It is mainly used for those articles which are to be subjected to impact loads such as chisels, hammers, springs etc.

3. *High temperature tempering.* This type of tempering is performed at 500°C to 650°C. On heating to 500°C, martensite begins changing to fine pearlite or sorbite and the transformation of sorbite is completed on reaching a temperature of 650°C. The internal stresses are completely eliminated in this process. The sorbite steel is employed for making highly stressed parts, because it has a remarkable mechanical properties like ductility, strength, and shock resistance.

7·15 Tempering Baths

The baths using tempering oils may be employed for temperatures upto approximately 230°C. The tempering oils are usually mineral oils having flash points of the order of 300°C. An adequate quantity of oil should be employed and the baths should be provided with wire baskets which when loaded with work may be lowered into the tempering bath. For temperatures above about 230°C, liquid salt baths are preferred. These salt baths usually consist of mixtures of nitrates or nitrates and nitrites. The chlorides and fluorides are usually employed for higher temperatures.

7·16 Surface Hardening or Case Hardening

In many engineering applications, it is desirable that a steel being used should have a hardened surface to resist wear and tear. At the same time it should have soft and tough interior or core so that it is able to absorb any shocks etc. This is achieved by hardening the surface layers of the article while the rest of it is left as such. This type of treatment is applied to gears, ball bearings, railway wheels, etc.

Following are the various surface or case hardening processes by means of which the surface layer is hardened :

1. Carburising,
2. Cyaniding,
3. Nitriding,
4. Induction hardening, and
5. Flame hardening.

We shall now discuss, in detail, these processes in the following articles.

7·17 Carburising

We have discussed in Art. 7·8, that a low carbon steel (containing carbon content upto 0.25 percent) cannot be hardened appreciably by the hardening process. Such steels are enriched in carbon on their surface before hardening by quenching. The process of inducing carbon to low carbon steels in order to give it a hard surface is called *carburising*. The surface is made hard only upto a certain depth. Following two methods are commonly used for carburising :

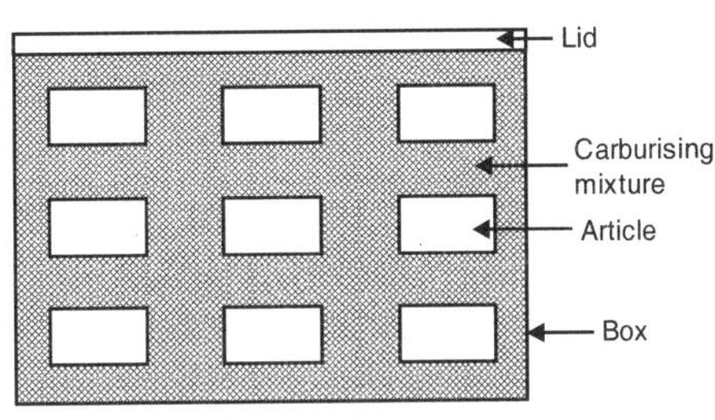

Fig. 7.9. Pack or solid carburising.

1. Pack or solid carburising, and
2. Gas carburising.

1. *Pack or solid carburising.* In this process, the article to be carburised is placed in a carburising box of proper design made of special heat resisting alloy, cast steel or steel sheet. The space between the article and the box is filled with a solid carburising compound, as shown in Fig. 7.9. There are numerous varieties of carburising compounds, but the one having following composition may be used :

Hard charcoal – 53 to 55 per cent; Coke – 30 to 32 per cent;
Sodium carbonate – 2 to 3 per cent; Barium carbonate – 10 to 12 per cent; and
Calcium carbonate – 3 to 4 per cent.

The layer of compound between the article and the box should be as uniform as possible to give uniform transfer of heat from the outer surface of the box to the article. The box is covered with a lid and sealed with clay to eliminate the entry of air. The box is now gradually heated in a furnace to the selected carburising temperature (generally 900°C to 950°C). During heating, carbon monoxide gas is formed which react with the article to form carbon and carbon dioxide gas. The surface of the articles absorbs the carbon and get rich in its carbon content. The carbon thus absorbed impregnates into the body of the article by the diffusion process. The temperature, time and carburising compound composition used depends upon the depth of case (carbon rich surface) desired. For example, with a compound composition (as given above), a case depth of 1 to 1.25 mm can be obtained at 925°C in an overall carburising time of nine hours. Out of these nine hours, five hours are required for heating up the work and four hours represent the time 'at temperature'. After heating, the article may be quenched in oil directly or it may be

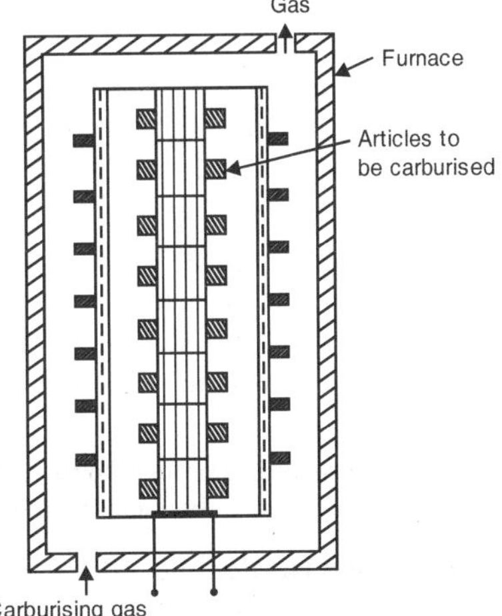

Fig. 7.10. Gas Carburising.

slowly cooled and then subjected to proper heat treatment.

Note : If certain parts of the article are not to be carburised, then those parts may be protected by copper plating or by covering with a non carburising material such as sand and fire clay.

2. *Gas carburising.* This process has gained popularity now-a-days, especially for the production of hard surfaces on light and small articles of low carbon steel. In this process, as shown in Fig. 7.10, the articles to be case hardened are heated to the proper temperature in the furnace having an atmosphere of gases rich in hydrocarbons. The gases used for this purpose may be methane, ethane, propane, butane, coal gas etc., mixed with carbon monoxide or carbon dioxide. The operation is performed in the furnace at 900°C to 950°C. The carbon in the gas combines with the surface of the article and makes it hard. It is found that a case depth of 0.5 mm to 0.75 mm can be obtained in the four hours of heating in the gas carburising temperature of about 925°C. This is a quicker process as compared to the pack carburising and is useful only when mass production of the article is desired.

7·18 Cyaniding

The cyaniding (also called *liquid carburising*) is a case or surface hardening process in which carbon and nitrogen both are absorbed by the metal surface to get it hardened. In this process, the piece of low carbon steel is immersed in a bath of cyanide salt (consisting of sodium cyanide, sodium chloride and sodium carbonate in equal proportions) maintained at 850°C to 950°C. The immersed steel piece is left in the molten cyanide salt bath, at the above temperature, for about 15 to 20 minutes. It is then taken out of the bath and quenched in water or oil. The cyanide yields carbon monoxide and nitrogen which behave as active carburising agents in hardening the surface of steel. Under average conditions, a case depth of 0.125 mm will be produced in about 15 minutes at 850°C. The process can give surface hardness upto a depth of 0.8 mm.

This process is mainly applied to the low carbon steel parts of automobiles (oil pump gears, drive worm screws, change over switch shafts, steps, sleeves, brake cam, speed box gears), some parts of motor cycles (gears, shafts, pins) and agricultural machinery (self propelled gears).

7·19 Nitriding

This is a process of case or surface hardening in which nitrogen gas is employed in order to obtain hard surface of the steel. This process is commonly used for those steels which are alloyed with chromium, molybdenum, aluminium, manganese etc. The steel articles, usually well machined and finished are placed in an air-tight container made of high nickel-chromium steel and provided with inlet and outlet tubes through which the ammonia gas is circulated. The nitriding process is generally carried out in the electric furnace where the temperature in the range of 450°C to 550°C is maintained. The container with the articles is placed in the furnace and ammonia gas is passed through it. The ammonia gas when comes in contact with steel articles, it gets dissociated in the form of nascent nitrogen which react with the surface of the articles and form nitrides which is very hard. This process can give surface hardness upto a depth of 0.8 mm.

The nitriding process is used in the production of machine parts which require high wear resistance at elevated temperatures such as automobile and air plane valves and valve parts, piston pins, crankshafts and cylinder liners. It also finds some application in the production of ball and roller bearing parts, parts to withstand high pressure steam services, die-casting dies, wire-drawing dies etc.

Note : The core should be brought to its original toughness before nitriding, by quenching in oil from about 900°C and tempering from about 600°C to 650°C.

7·20 Induction Hardening

It is a process of surface hardening in which the surface to be hardened is surrounded by an inductor block which acts as a primary coil of a transformer. The inductor block should not touch the surface to be hardened. A high frequency current is passed through this block. The heating effect is due to induced eddy currents and hysteresis losses in the surface material. The inductor block surrounding the heated surface has water connections and numerous small holes on its inside surface. As soon as the surface reaches to the proper temperature (750°C to 760°C for 0.5 percent carbon steel and 790 to 800°C for alloy steel), it is automatically spray-quenched under pressure.

An important feature of this method of hardening is its rapidity of action, because it requires only a few seconds to heat steel to a depth of 3 mm. The actual time depends upon the frequency used, power input and depth of hardening required.

The induction hardening is widely used for hardening surfaces of crankshafts, camshafts, gears, automobile components, spline shafts, spindles, brake drums etc.

7·21 Flame Hardening

Sometimes only those portions of an article are to be hardened which are liable to be subjected to wear, abrasion or shocks. This local hardening is done by a process known as *flame hardening.* In this process, the portions to be hardened are heated with a flame of oxy-acetylene torch above its critical temperature. The heated portion is immediately quenched by means of spray of water directed on the surface. Since the heating is localised, stresses are not developed, therefore, chances of distortion and cracking are reduced.

This process is mostly used for articles like gears, wheels which can not be quenched as a whole but require only local hardening of its portions. The advantage of this method is that the time taken for heating is comparatively less than when the metal is heated in a furnace.

7·22 Precipitation Hardening or Age Hardening of Aluminium Alloy

The precipitation hardening (often called age hardening) is the process of hardening of an alloy by the precipitation of a constituent from a supersaturated solid solution by heating it to some elevated temperature. The precipitation can be achieved only with those alloys in which there is a decrease in solid solubility of one material in another as the temperature is reduced, resulting in a supersaturated solution. The phase diagram, of one such alloy (aluminium copper alloy *i.e.*, duralumin) is shown in Fig. 7.11. Assuming that the composition is 4 percent copper and 96 percent aluminium. In order to harden the alloy, it must be heated to temperature *A* above the solubility line *CD* to allow the constituents to enter into a solid solution. The alloy must be held there for sufficient time until a homogeneous alloy is obtained. It is then cooled rapidly to room temperature leaving the alloy in a supersaturated, unstable state. In this state, the alloy is soft but being unstable, the excess constituents will precipitate out of the solution with the lapse of time. This process, if accomplished at room temperature is known as normal aging. The particles that are precipitated from the solid solution, form at the grain boundaries and slip planes, producing a keying action that reduces slippage between the crystals. The hardness is influenced by the size of the precipitated particles. It is increased as the size of the particle is increased, but after the critical size is reached, further growth results in brittleness and loss of strength.

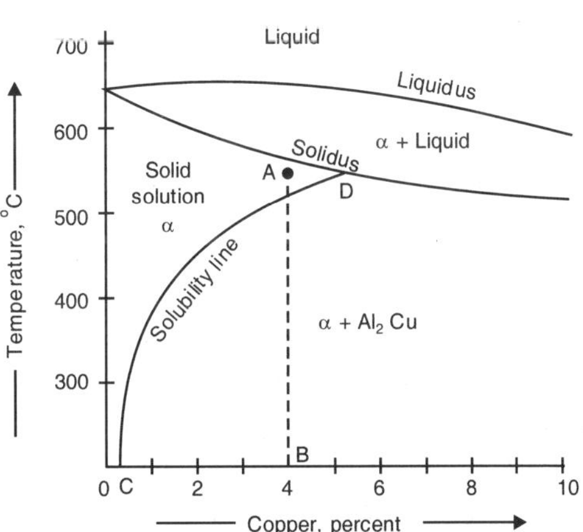

Fig. 7.11. Phase diagram of aluminium-copper alloy.

The *artificial ageing* differs slightly in that an additional heat treatment is used. In this case, the alloy is heated to some elevated temperature that accelerates the precipitation of some constituent from the supersaturated solution. As the temperature is increased, the more rapid is the precipitation and increase in hardness. An equilibrium condition is finally reached that decreases the strength as a result of the large size of the precipitate crystals. This condition which should be avoided, is known as *overageing*.

7·23 Heat Treatment Furnaces

The furnace is the most important equipment used in the heat treatment of steel at elevated temperatures. The various types of furnaces are available, but the selection of the furnace depends upon the following factors :

1. The size and shape of the product,
2. The volume of the production, and
3. The type of heat treatment.

All the furnaces should be capable to meet the following basic requirements :

1. The form of heating should be such that it can raise the temperature of loaded furnace to the required temperature in a reasonably short time.
2. The temperature control should be flexible and accurate.
3. The heating chamber should gain or loose heat uniformly at all surfaces.
4. The controlled atmosphere should be maintained to prevent the oxidation of steel surfaces, the formation of scale, decarburisation etc.
5. It should occupy less space as far as possible.

7·24 Types of Furnaces

The following two types of furnaces are commonly used for heating the metal to a particular temperature.

1. Hearth furnaces, and 2. Bath surfaces.

These furnaces are discussed, in detail, in the following pages.

7·25 Hearth Furnaces

In these furnaces, heating is done by gas (town gas, blast furnace gas or natural gas), fuel oil or electric current. The gas or oil furnaces are used for large work. Now-a-days, electric furnaces are much used due to the following reasons :

1. It is simple and economical,
2. It has high thermal efficiency,
3. It has precision temperature control,
4. It covers minimum space, and
5. Since no combustion occurs, therefore, no waste gases are produced.

According to the arrangement of their heating chamber or combustion space, the hearth furnaces are built in two types :

1. Stationary hearth furnaces, and 2. Movable hearth furnaces.

These furnaces are discussed, in detail, in the following pages.

7·26 Stationary Hearth Furnaces

The stationary hearth furnaces (also known as box type furnaces) consist of a rectangular box of steel lined with fire bricks, as shown in Fig. 7.12(*a*).

Following types of stationary hearth furnaces are important from the subject point of view :

1. *Direct fuel fired furnace.* In this furnace, the metal is heated by the direct contact with the products of combustion from the fuel. These furnaces are mostly used for annealing steel castings and for heating steel to be forged.

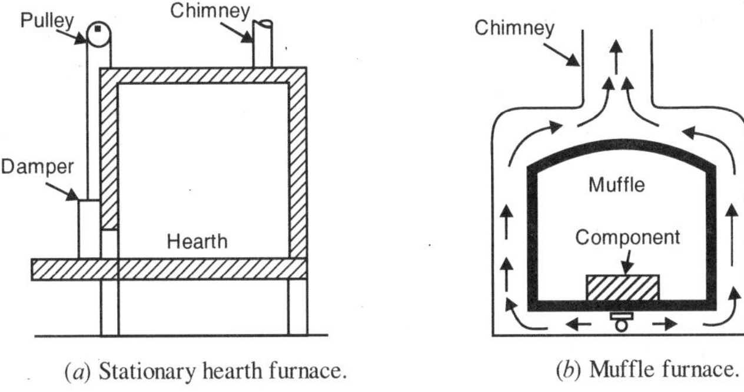

(*a*) Stationary hearth furnace. (*b*) Muffle furnace.

Fig. 7.12. Stationary hearth furnaces.

2. *Indirect fuel fired furnace.* In this furnace, the heating and combustion chambers are made separately. These furnaces are extensively used for heat-treating of the metals upto 1100°C and have the advantage of reduced scaling and contamination from the fuels.

3. *Muffle furnace.* In this furnace, the muffle (*i.e.*, heating chamber) and the combustion chamber are made separately as shown in Fig. 7.12(*b*). The gases of combustion circulate around the muffle and does not come in contact with the steel being heated. In electric muffle furnace, the resistors are usually placed around the outside of the muffle. The muffle is made of a special heat resisting alloy or special refractory material. These furnaces protect the metal from oxidation during heat treating and are commonly employed for heat treating work in tool rooms.

4. *Recirculation furnace.* In this furnace, the gas or oil is burnt in a chamber and the fuel gases are recirculated through the work space by suitably arranged centrifugal fan. It is used for stress relieving, tempering and hardening for temperature ranges below 700°C.

7·27 Movable Hearth Furnaces

These are special types of furnaces and are extensively used for mass production work. These are of the following two types :

1. *Car bottom type.* The car bottom type movable hearth furnace is similar to a fixed hearth furnace except that its bottom has a carriage or car with hearth mounted on it. It can be moved from the furnace for loading and unloading or charging and discharging in case of large heat treating furnaces.

2. *Rotary type.* The rotary type movable hearth furnace has refractory lined shell structure enclosing a rotating hearth. This type of furnace is used for hardening, tempering and carburising processes.

7·28 Bath Furnaces

These furnaces consist of a cylindrical steel casing lined with refractory material and faced with insulating bricks. The heating chamber containing salt, lead or oil (heating media) is suspended in a steel pot containing the articles to be heated. According to the heating media, the following bath furnaces are important from the subject point of view :

1. *Salt bath furnaces.* In these furnaces, a bath of molten salt is used as a heating media. The articles are heated by immersing them in liquid salts along with the wires or wire baskets which hold them, as shown in Fig. 7.13. The salt bath should be such that it does not attack the steel by producing

oxidation, decarburisation, corrosion or erosion. The sodium and potasium chlorides, nitrites and cyanides are the common salts used in salt baths. They are mixed in various proportions to obtain different melting point ranging from 180°C to 1350°C.

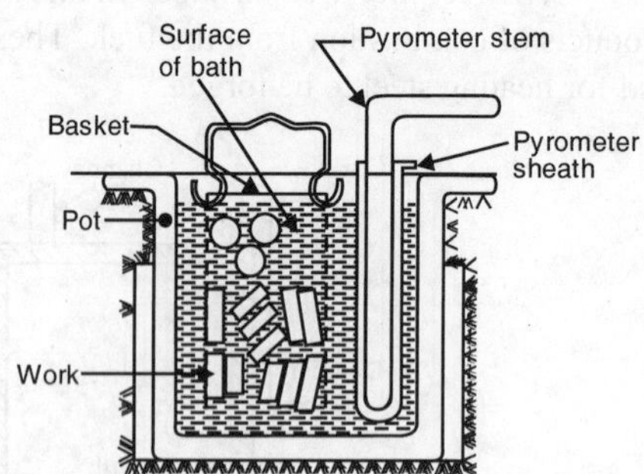

Fig. 7.13. Salt bath furnace.

The sodium and potassium saltpetre are used for tempering springs and similar articles; sodium chloride, calcined soda and barium chloride are used for heating carbon and alloy steels; sodium chloride is only used for high speed steel hardening.

2. *Lead bath furnaces.* In lead bath furnaces, a lead bath tank made of cast steel or some other heat-resisting alloy contains ordinary commercial lead instead of salt. A charcoal covering is placed on the top of the bath in order to decrease oxidation. The lead bath is mainly used for hardening of files, reamers, drills, broaches, axes etc. It may be noted that the articles should be perfectly dry before they are immersed in bath, otherwise formation of steam will take place which is liable to produce dangerous explosions.

3. *Oil bath furnaces.* In an oil bath furnace, mineral oil heated in a metallic container is used. This is suitable only for quenching and tempering steel parts up to 260°C.

4. *Electrically heated bath furnaces.* In these furnaces, electric current is used for heating the salt bath instead of gas or oil. These furnaces are of the following two types :

(*a*) *Externally heated bath furnaces*

In these type of furnaces, as shown in Fig. 7.14, electrical resistance elements are located in the wall of the furnace. The wall is lined with a refractory material in order to prevent rapid loss of heat and to withstand high temperatures. The outside wall of the metallic tank or bath is heated by radiation. These furnaces are used only for quenching and tempering upto a temperature of 700°C.

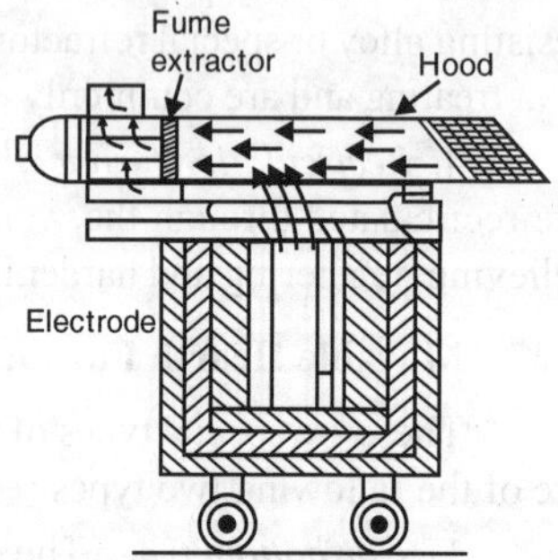

Fig. 7.14. Externally heated bath furnace.

(*b*) *Immersed electrode bath furnaces*

In these type of furnaces, two electrodes of pure iron are immersed in the salt and fed at low voltage from a transformer. The salt is heated by passing alternating current between the electrodes. The molten salt possesses sufficient resistance for providing heat energy to operate the furnace while conducting the current between the electrodes. A carbon rod provided between the electrodes is supplied with current and glows red. This melts the salt, thus increasing its conductivity and making it liquid. These furnaces are used for preheating carburising and hardening of high speed steels. The furnace may be operated over a temperature range between 600°C to 1400°C.

7·29 Measurement of Furnace Temperature

The temperature measurement of the furnace is very important because the successful heat treatment depends upon the correct temperature. There are many methods used for this, but a more scientific and reliable method of measuring furnace temperature is by an instrument called a *pyrometer.* Though there are various types of pyrometers, yet the following two are important from the subject of point view :

1. Thermo-electric pyrometer, and 2. Optical pyrometer.

These pyrometers are discussed, in detail, in the following pages.

7·30 Thermo-electric Pyrometer

The thermo-electric pyrometer, as shown in Fig. 7.15, works on the principle that when two dissimilar wires are joined to form a complete electric circuit and two junctions are maintained at different temperatures, then an electric current flows in the circuit. The magnitude of the current depends upon the metals used and the temperature difference of the junctions. The *hot junction,* which is placed in the furnace, is often made up of wires of platinum and an alloy of platinum and rhodium welded together. It is called a *thermocouple.* The leads from these wires are carried to a sensitive galvanometer which constitutes the cold junction and indicates the current flowing in the circuit due to the difference of temperatures between the two junctions. The galvanometer is so calibrated that instead of indicating electrical units, it reads in degrees of temperature.

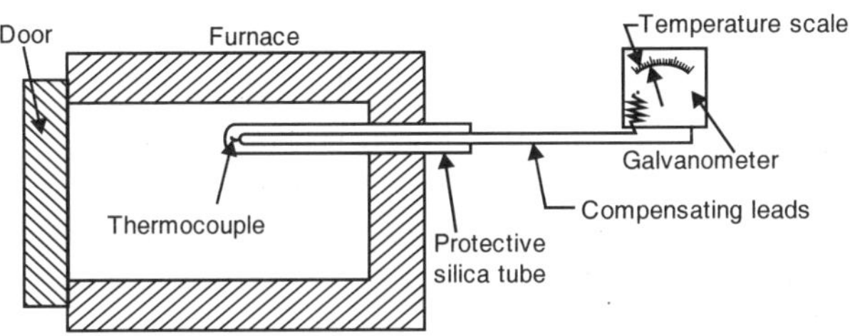

Fig. 7.15. Thermo-electric pyrometer.

7·31 Optical Pyrometer

The optical pyrometer compares the intensity of light being emitted from the furnace with that from some standard source. In the disappearing filament type optical pyrometer, the glow of a standard filament lamp is varied until it matches the light from the furnace and disappears when viewed through the telescope. The instrument, as shown in Fig. 7.16(*a*), is set up in front of the furnace and the light from the furnace is viewed through the eye-piece. The current through the lamp is varied by a resistance and when a colour match is obtained, the lamp filament disappears from sight. The lamp current

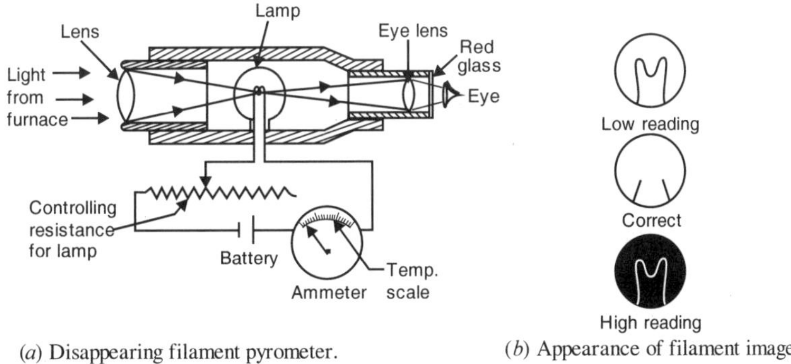

(*a*) Disappearing filament pyrometer. (*b*) Appearance of filament image

Fig. 7.16. Optical pyrometer.

required to cause this is indicated on an ammeter. This instrument may be calibrated in degrees of temperature, instead of in units of electric current. The appearance of the filament, as seen through the telescope, is shown in Fig. 7.16(*b*).

QUESTIONS

1. What is the object of heat treating the metals and alloys ?
2. Explain the structural changes occurring in a plain carbon steel with help of the heating and cooling curves.
3. Define the following terms :
   (*a*) Point of decalescence; (*b*) Point of recalescence;
   (*c*) Upper critical point; and (*d*) Lower critical point.
4. Explain the isothermal transformation occurring in steel, with the help of suitable examples.
5. Why the time-temperature-transformation diagrams (TTT diagram) are constructed ?
6. What is the process of heat treatment ? Name the various heat treatment processes.
7. Why annealing is done to steels ? Discuss the methods for full annealing, process annealing, spheroidise annealing, diffusion annealing and isothermal annealing.
8. Explain, with the help of a diagram, the following heat treatment processes :
   (*a*) Normalising ; and (*b*) Hardening.
9. What is the difference between hardness and hardenability ? State the factors which affect hardenability. How it is measured ?
10. Explain austempering and martempering. Are these hardening or tempering process ? Explain your answer with reasons.
11. What is tempering ? Why it is done ? Describe in brief, the various types of temperings you know.
12. What do you understand by case hardening ? Name the various case hardening processes.
13. Explain the following case hardening processes :
    (*a*) Carburising ; (*b*) Cyaniding ;
    (*c*) Nitriding ; and (*d*) Induction hardening.
14. Describe precipitation hardening with the help of a suitable example.
15. Discuss the various types of furnaces used in heat treatment.
16. How the temperature of the furnace is measured ? Explain themo-electric pyrometer and optical pyrometer.

OBJECTIVE TYPE QUESTIONS

1. The purpose of heat treatment is to
   (*a*) relieve the stresses set up in the material after hot or cold working
   (*b*) modify the structure of the material
   (*c*) change grain size (*d*) soften the metal
   (*e*) all of the above (*f*) none of the above
2. The temperature point at which the change starts on heating the steel is called
   (*a*) lower critical point (*b*) upper critical point
   (*c*) point of recalescence (*d*) point of decalescence
3. When austenite steel (above 723°C) is air cooled, then the resulting structure will be
   (*a*) martensite (*b*) very fine pearlite
   (*c*) fine pearlite (*d*) coarse pearlite

4. In full annealing, the hypo-eutectoid steel is heated from 30°C to 50°C above the upper critical temperature and then cooled

(*a*) in still air (*b*) slowly in the furnace

(*c*) suddenly in a suitable cooling medium (*d*) any one of the above

5. In spheroidising process, the steel is

(*a*) heated below the lower critical temperature and then cooled slowly

(*b*) heated upto the lower critical temperature and then cooled in still air

(*c*) heated slightly above the lower critical temperature and then cooled slowly to a temperature of 600°C

(*d*) none of the above

6. The process which improves the machinability of steels, but lowers the hardness and tensile strength, is

(*a*) normalising (*b*) full annealing

(*c*) process annealing (*d*) spheroidising

(*e*) all of the above (*f*) none of the above

7. The process used for relieving the internal stresses previously set up in the metal and for increasing the machinability of steel, is

(*a*) normalising (*b*) full annealing

(*c*) process annealing (*d*) spheroidising

8. Normalising of steel is done to

(*a*) refine the grain structure

(*b*) remove strains caused by cold working

(*c*) remove dislocations caused in the ınternal structure due to hot working

(*d*) all of the above

(*e*) none of the above

9. A steel is heated at about 875°C where the structure consists of entirely austenite. It is then cooled suddenly at a temperature of about 250°C to 525°C. This process of heat treatment is known as

(*a*) normalising (*b*) annealing

(*c*) austempering (*d*) martempering

(*e*) hardening (*f*) none of these

10. In the austempering process of heat treatment, austenite changes into

(*a*) martensite (*b*) troostite

(*c*) sorbite (*d*) bainite

11. The heat treatment process used for castings is

(*a*) carburising (*b*) normalising

(*c*) annealing (*d*) tempering

12. The heat treatment process used for softening hardened steel is

(*a*) carburising (*b*) normalising

(*c*) annealing (*d*) tempering

13. Which of the following is a case hardening process ?

(*a*) Carburising (*b*) Cyaniding

(*c*) Nitriding (*d*) Flame hardening

(*e*) all of the above

14. The process in which carbon and nitrogen both are absorbed by the metal surface to get it hardened, is known as

(*a*) carburising (*b*) cyaniding

(*c*) flame hardening (*d*) induction hardening

15. Age hardening is related to

(*a*) duralumin (*b*) brass

(*c*) copper (*d*) silver

ANSWERS

1. (*e*)	2. (*a*)	3. (*c*)	4. (*b*)	5. (*c*)
6. (*d*)	7. (*c*)	8. (*d*)	9. (*c*)	10. (*d*)
11. (*b*)	12. (*d*)	13. (*e*)	14. (*b*)	15. (*a*)

8

Mechanical Working of Metals

1. Introduction. 2. Hot Working. 3. Hot Working Processes. 4. Hot Rolling. 5. Types of Rolling Mills. 6. Hot Forging. 7. Hot Spinning. 8. Hot Extrusion. 9. Hot Drawing or Cupping. 10. Hot Piercing. 11. Cold Working. 12. Cold Working Processes. 13. Cold Rolling. 14. Cold Forging. 15. Cold Spinning. 16. Cold Extrusion. 17. Cold Drawing. 18. Cold Bending. 19. Shot Peening.

8·1 Introduction

The mechanical working of metal is defined as an intentional deformation of metals plastically under the action of externally applied forces.

The mechanical working of metals is described as hot working and cold working depending upon whether the metal is worked above or below the recrystallisation temperature. The temperature at which the new grains are formed in the metal is known as *recrystallisation temperature*.

The metal is subjected to mechanical working for the following purposes :

1. To reduce the original block or ingot into desired shapes,
2. To refine grain size, and
3. To control the direction of flow lines.

8·2 Hot Working

The working of metals above the recrystallisation temperature is called *hot working*. This temperature should not be too high to reach the solidus temperature, otherwise the metal will burn and become unsuitable for use. The hot working of metals has the following advantages and disadvantages.

Advantages

1. The porosity of the metal is largely eliminated, thus producing strong and uniform structure.
2. The grain structure of the metal is refined *i.e.*, the coarse grains are converted into fine grains which change the properties of the metal.
3. The impurities like slag are squeezed into fibres and are uniformly distributed throughout the metal.
4. The mechanical properties such as toughness, ductility, percentage elongation, precentage reduction in area, and resistance to shock and vibration are improved due to the refinement of grains,
5. The deformation of the metal is easy, with a small pressure applied on it.

Disadvantages

1. It requires expensive tools.
2. It produces poor surface finish, due to the rapid oxidation and scale formation on the metal surface.
3. Due to the poor surface finish, close tolerances cannot be maintained.
4. The correct temperature range for working is difficult to maintain.

8·3 Hot Working Processes

The following are the various hot working processes :

(*i*) Hot rolling, (*ii*) Hot forging,

(*iii*) Hot spinning, (*iv*) Hot extrusion,

(*v*) Hot drawing or cupping, and (*vi*) Hot piercing.

We shall now discuss the above processes, in detail, in the following pages.

8·4 Hot Rolling

The hot rolling process is the most rapid method of converting large sections into desired shapes. The forming of bars, plates, sheets, rails, angles, I-beams and other structural sections are made by hot rolling.

The operation consists of passing the hot ingot through at least two rolls rotating in opposite directions at the same speed, as shown in Fig. 8.1. The space between the rolls is adjusted to conform to the desired thickness of the rolled section. The rolls, thus, squeeze the passing ingot to reduce its cross-section and increase its length.

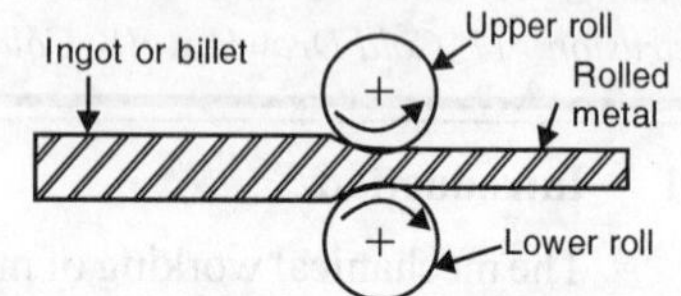

Fig. 8.1. Operation of hot rolling.

The *ingots* are casted in moulds of suitable form and are used in rolling mills as raw material for preparing desired sections. The first operation to the ingot is carried out at the blooming mill where it is rolled to *blooms*. The bloom has a square cross-section with a minimum size of 150 mm × 150 mm. The blooms are cut up in lengths convenient for the subsequent reducing process into *billets*. The billet is smaller than a bloom and has a minimum size of 50 mm × 50 mm.

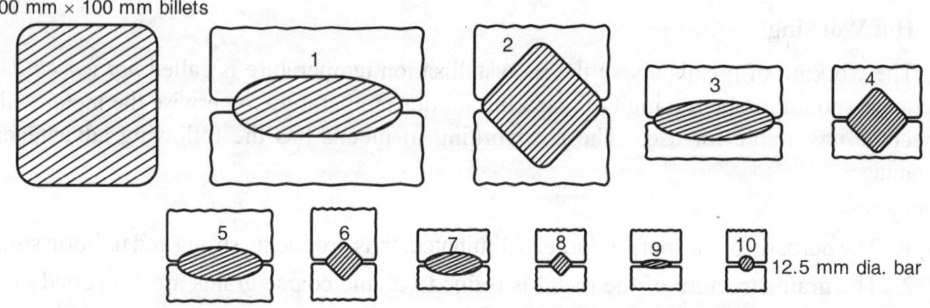

Fig. 8.2. Sequence of operations in reducing of cross-section of 100 mm× 100 mm billet to 12.5 mm diameter bar.

The successive stages in the reduction of a billet to a round bar is shown in Fig. 8.2. The billet is rotated at 90° after each pass.

When the final product of the rolling mill is to be sheets, plates or strips, the ingots or blooms are rolled to slabs. The slabs have rectangular cross-section with a minimum width of 250 mm and thickness of 50 mm.

The materials commonly hot rolled are aluminium, copper, magnesium, their alloys and many grades of steel.

8·5 Types of Rolling Mills

The following types of rolling mills are commonly used for rolling :

1. *Two-high rolling mill*. The two-high rolling mill, as shown in Fig.8.3, consists of two heavy rolls placed exactly one over the other. One of the rolls is fixed while the other is adjustable.

Sometimes both the rolls are adjustable. In this operation, the metal piece is passed between the two rolls rotating at the same speed but in opposite direction. After each pass, the direction is reversed. The metal is turned through 90 degrees at frequent intervals, to keep the section uniform and to refine the metal throughout. It may be noted that about thirty passes are required to reduce a large ingot into a bloom.

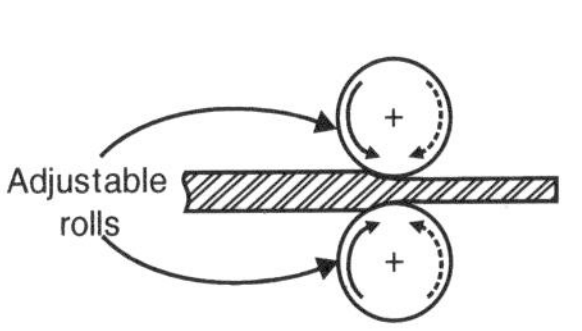

Fig. 8.3. Principle of two high rolling mill.

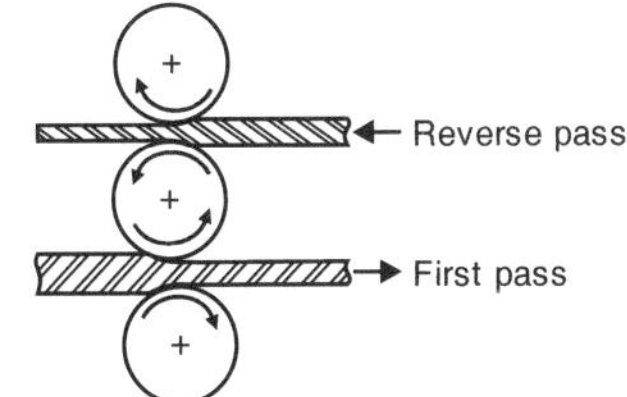

Fig. 8.4. Principle of three-high rolling mill.

2. *Three-high rolling mill.* The three-high rolling mill consists of three rolls, as shown in Fig. 8.4. The upper and the bottom rolls rotate in the same direction whereas the middle roll rotates in the opposite direction. Though there is some difficulty due to lack of correct speed for all passes, yet the three-high mill is less expensive to make and has a higher output than the two-high mill.

3. *Four-high rolling mill.* The four-high rolling mill consists of four rolls, two of which are working rolls and the other two are backing up rolls, as shown in Fig. 8.5.

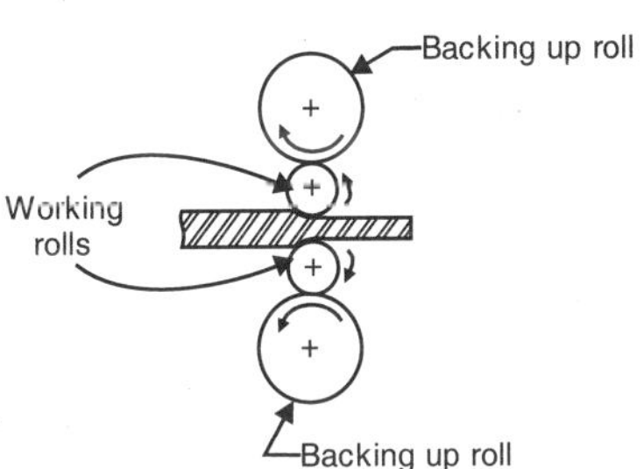

Fig. 8.5. Principle of four-high rolling mill.

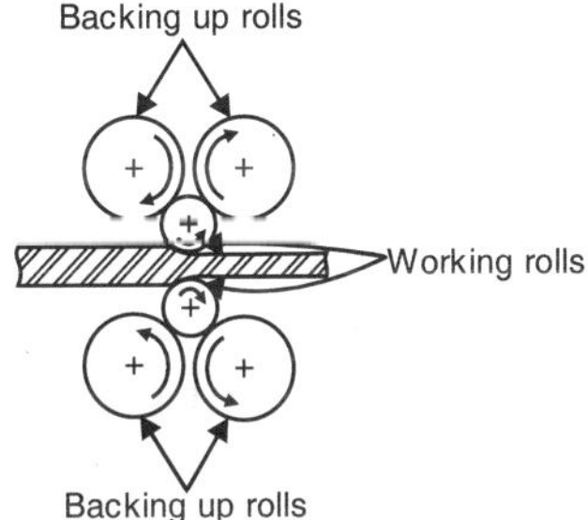

Fig. 8.6. Principle of cluster mill.

The backing up rolls have larger diameter and are used to prevent the deflection of small working rolls. These mills are used for rolling of slabs.

4. *Cluster mill.* The cluster mill consists of two working rolls and four or more backing up rolls. The number of back up rolls depends upon the amount of support required for the working rolls during the operation, as shown in Fig. 8.6. This type of mill is generally used for cold rolling.

8·6 Hot Forging

It consists of heating the metal to plastic state and then the pressure is applied to form it into desired shapes and sizes. The pressure applied in this is not continuous as for hot rolling, but intermittent. The pressure may be applied by hand hammers, power hammers or by forging machines. The following processes are commonly used for hot forging.

1. *Hammer or smith forging.* It is also known as *open die forging* because it involves the compression of the billet between flat dies with no constraint on the spread of the metal. In this process, the metal to be forged, is heated to the appropriate forging temperature in a gas, oil or electric furnace and is placed on the anvil. Then it is hammered to the desired shape. The hammering may be done either by a hand hammer or a mechanical hammer operated by gravity, air, steam or electricity.

2. *Drop or die forging.* The hammer or smith forging is a useful and indispensable process. But it is not suitable for the production of large number of identical forgings. When a large quantity of a certain component with sound and high quality structure is required, a method known as drop or die forging is used. The drop forging is also called closed die forging. The drop forged parts are used in automobiles, aeroplanes, rail road equipments, engines, agricultural implements and household appliances.

The drop forging process consists of hammering heated bars or billets of steel or other metals into closed impression dies. The closed impression dies compress the metal causing it to flow and completely fill the die impression. A complete closed impression die consists of two alloy-steel blocks. One of these blocks is called the top or upper die which is fastened to the ram of a steam drop hammer, board drop hammer or other type of power hammer. The other block is the bottom or lower die and it is fastened to the anvil. The impressions for producing the desired shape to be forged are machined in the working surfaces of each half die. When these two die halves come together in alignment, a completely closed die cavity or die impression is formed. Since machined die impressions are employed, much larger variety of shapes can be forged with greater accuracy and better finish than is possible with smith forging.

3. *Press forging.* The hammer forging dissipates a large amount of energy near the surface of the metal and in the foundation of the machinery. As a result of this, the hammer forging is not used for the forging of large sections. When a large section of high quality is required, a method known as press forging is used. The press forging process consists of squeezing the metal by a press. The press operates by hydraulic, air or steam action or by mechanical means such as crank or screw. Hydraulic presses have slower action than hammers and the pressure application may be closely controlled. This helps in maintaining sustained pressures in hydraulic presses. Mechanical action presses may approach the speed of hammer blows. But they cannot maintain sustained pressures. It press forging, usually the forging dies have only one impression and the forging is completed with one pressing of the hot metal kept within the two die halves.

4. *Upset forging.* This process involves increasing the diameter of the end or central portion of a bar of metal by compressing its length. Originally, this process was developed to form heads on bolts, rivets, valves etc. But these days it is used to produce a variety of useful forgings.

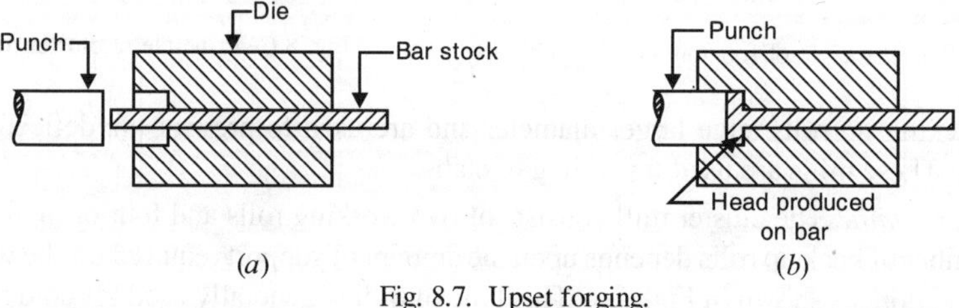

Fig. 8.7. Upset forging.

In upset forging the heated bar of stock is placed between two halves of a die and against a gauge which permits the correct amount of metal to stick out beyond the gripping parts of the dies as shown in Fig. 8.7 (*a*). Then the bar is squeezed into the impression chamber by the header ram (or punch) as shown in Fig.8.7 (*b*). Some of the upset forgings can be completed in a die with one single impression. But in most cases, more than one forging step is necessary. In those cases, the die has two or more forging positions or impressions.

5. *Roll forging.* This process is used to make products such as axles, leaf springs, tapered levers etc. In this process, forging mills are often used as a preliminary shaping process to be followed by press or hammer forging. The forging mills are semicylindrical in shape and contain several grooves through which the stock may be progressively passed.

In roll forging, the heated bar is placed between the rolls while they are in the open position. As the rolls turn one-half revolution, the bar is progressively squeezed and rolled out towards the operator.

6. *Swaging*. This process is a type of forging operation in which rods or tubes are reduced in size by the repeated blows of opposed hammers. The repeated blows cause the metal to flow inward and take the shape of a die.

8·7 Hot Spinning

It consists of heating the metal to forging temperature and then forming it into the desired shape on a spinning lathe as shown in Fig. 8.8. The shaping is done with a blunt pressure tool which contacts the surface of the rotating part and causes the metal to flow to some desired form. The parts of circular cross-section which are symmetrical about the axis of rotation, are made by this process.

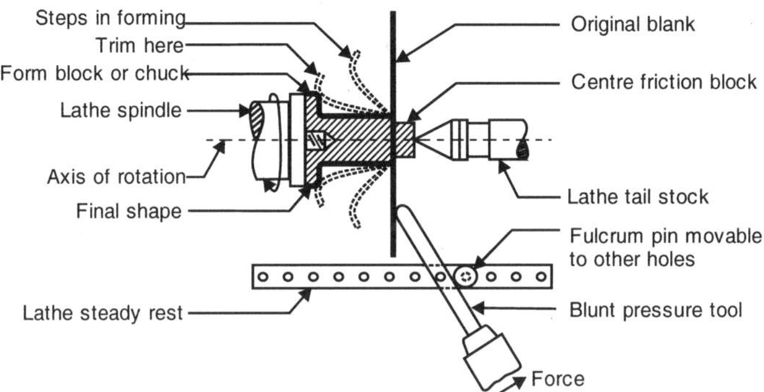

Fig. 8.8. Important features of spinning operation.

This method is generally used for thicker plates and sheets. When working with a relatively thin metal, a handwood bar is commonly used as a back-up support opposite to the tool in order to avoid wrinkling at the outer edge.

8·8 Hot Extrusion

It consists of pressing a metal inside a chamber to force it out by high pressure through an orifice which is shaped to provide the desired form of the finished part. Most commercial metals and their alloys such as steel, copper, aluminium and nickel are directly extruded at elevated temperatures. The rods, tubes, structural shapes, flooring strips and lead covered cables etc., are the typical products of extrusion. The various methods of extrusion are as follows :

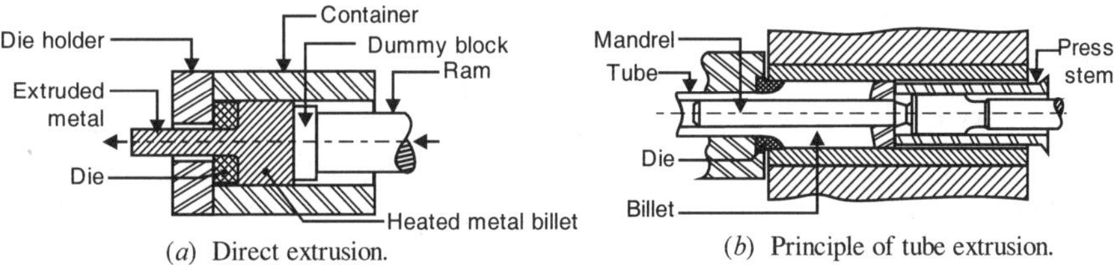

(*a*) Direct extrusion. (*b*) Principle of tube extrusion.

Fig. 8.9. Type of hot extrusion.

1. *Direct extrusion*. It is also called *forward extrusion*. In this method, the heated round billet is placed into the die chamber and the dummy block and ram is placed into position as shown in Fig. 8.9 (*a*). The metal is extruded through the die opening by applying the pressure on the ram. The pressure is applied only a small amount of metal remains unextruded. It is then cut to the required length and removed.

The tube extrusion is a form of direct extrusion but uses a mandrel to shape the inside of the tube. After the heated billet is placed inside, the die containing the mandrel is pushed through the ingot as shown in Fig. 8.9 (*b*). The press stem then advances and extrudes the metal through the die and around the mandrel.

2. *Indirect extrusion.* It is also called *backward extrusion*. It is similar to direct extrusion except that the extruded part is forced through the ram stem as shown in Fig. 8.10. Since, there is no frictional force between the billet and the container wall, therefore, less force is required by this method. The use of this process is limited for the following reasons :

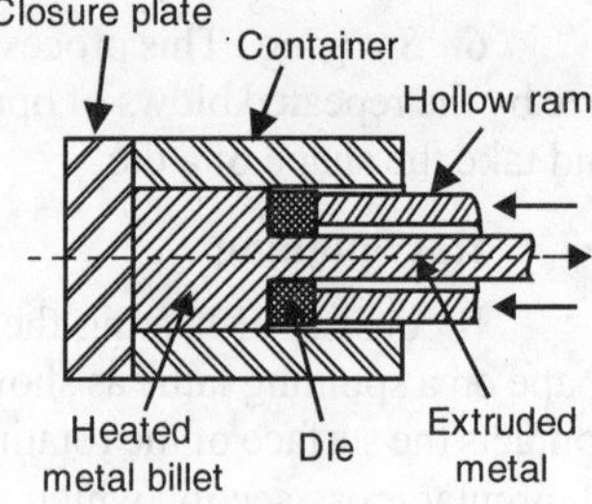

Fig. 8.10. Indirect extrusion.

(*a*) The ram becomes weak when it is made hollow, and

(*b*) The difficulty is experienced in providing support for the extruded part.

8·9 Hot Drawing or Cupping

It is mostly used for the production of thick-walled seamless tubes and cylinders. It is usually performed in two stages. The first stage consists of drawing a cup out of a hot circular plate with the help of a die and punch as shown in Fig. 8.11 (*a*). The second stage consists of reheating the drawn cup and drawing it further to the desired length having the required wall thickness as shown in Fig. 8.11 (*b*). The second drawing operation is performed through a number of dies, which are arranged in a descending order of their diameters, so that the reduction of wall thickness is gradual in various stages.

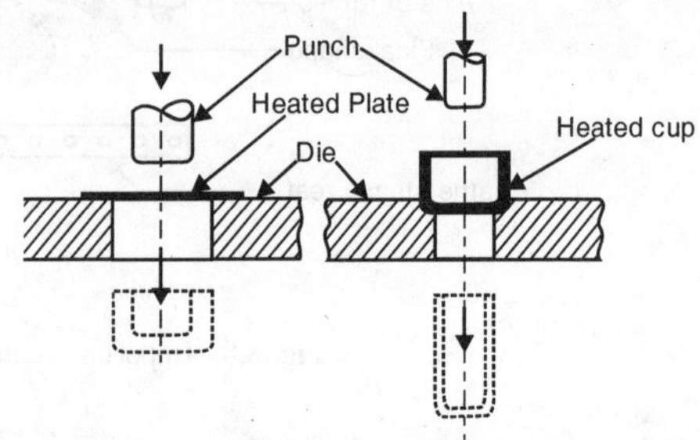

(*a*) First Stage

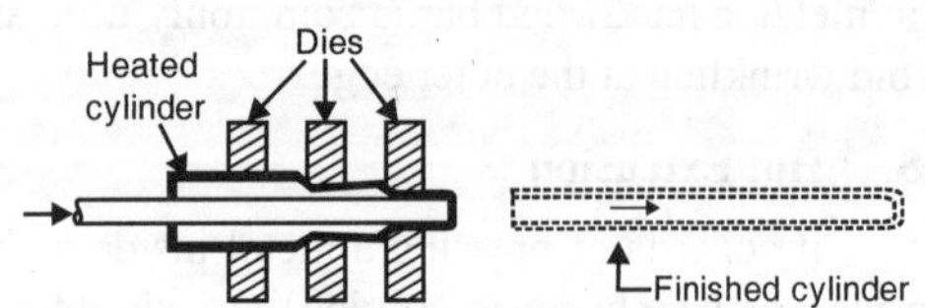

(*b*) Second Stage

Fig. 8.11. Operations of drawings thick-walled cylinders fromheatedplates.

Notes : 1. For long thin-walled cylinders or tubes, repeated heating and drawing is necessary.

2. If the final product is to be a tube, the closed end is cut off and the balance is sent through finishing and sizing rolls as used in piercing process.

3. In order to produce closed-end cylinders (similar to those used for storing oxygen) the open end is swaged to form a neck or reduced by hot spinning.

8·10 Hot Piercing

This process is used for the manufacture of seamless tubes. In its operation, the heated cylindrical billets of steel are passed between two conical shaped rolls operating in the same direction. A mandrel is provided between these rolls which assists in the piercing and controls the size of the hole as the billet is forced over it.

The whole process is carried out in the following steps. This process is generally applied for seamless tubes upto 150 mm in diameter.

1. The solid billet is first centre punched and then brought to forging heat in a furnace before being pierced.

2. It is then pushed into the two piercing rolls, as shown in Fig. 8.12 (*a*), which impart both rotation and axial advance. The alternate squeezing and bulging of the billet open up a seam in its centre, the size and shape of which is controlled by a mandrel. This process produces a thick walled tube.

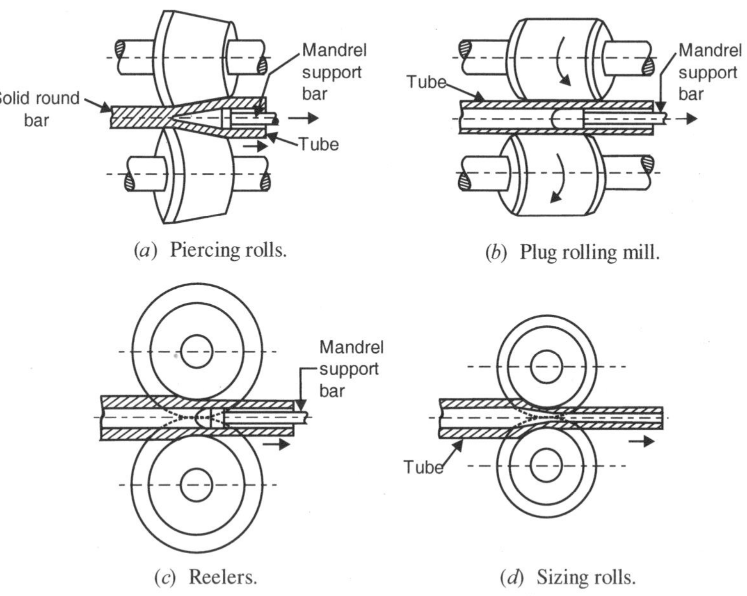

(*a*) Piercing rolls. (*b*) Plug rolling mill.

(*c*) Reelers. (*d*) Sizing rolls.

Fig. 8.12. Steps in the manufacture of seamless tubes.

3. The thick walled tube is now passed between the grooved rolls over a plug held by a mandrel, as shown in Fig. 8.12 (*b*). This operation produces a longer tube with specified wall thickness.
4. The tube is passed through a reeling machine, as shown in Fig. 8.12 (*c*), which further straightens the tube and gives smooth surface to the walls of the tube.
5. The final sizing and finishing is done by passing through the sizing rolls, as shown in Fig. 8.12 (*d*). In this case no mandrel is employed.

It may be noted that

1. In order to produce longer tubes, upto 350 mm in diameter, they are given a second operation on piercing rolls.
2. To produce tubes upto 600 mm in diameter, the double pierced tubes are processed on a rotary rolling mill as shown in Fig. 8.13. They are finally completed by reelers and sizing rolls as discussed above.

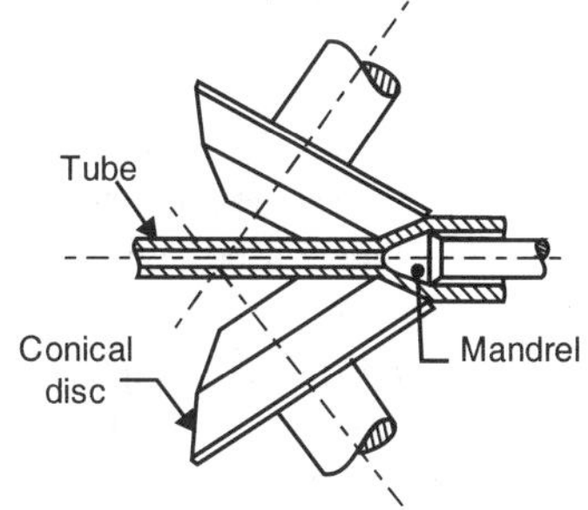

Fig. 8.13. Rotary rolling mill.

8·11 Cold Working

The working of metals below their recrystallisation temperature is known as *cold working*. Most of the cold working processes are performed at room temperature. The cold working distorts the grain structure and does not provide an appreciable reduction in size. It requires much higher pressures than hot working. The extent to which a metal can be cold worked depends upon its ductility. The higher the ductility of the metal, the more it can be cold worked. During cold working, severe stresses known as residual stresses are set up. Since the presence of these stresses is undesirable, therefore, a suitable heat treatment may be employed to neutralise the effect of these stresses. The cold working is usually used as finishing operation, following the shaping of the metal by hot working. It also increases tensile strength, yield strength and hardness of steel but lowers its ductility. The increase in hardness due to cold working is called *work-hardening*.

In general, cold working produces the following effects :

1. The stresses are set up in the metal which remain in the metal, unless they are removed by subsequent heat treatment.
2. A distortion of the grain structure is created.
3. The strength and hardness of the metal are increased with a corresponding loss in ductility.
4. The recrystalline temperature for steel is increased.
5. The surface finish is improved.
6. The close dimensional tolerance can be maintained.

8·12 Cold Working Processes

Though there are various cold working processes, yet the following are important from the subject point of view :

1. Cold rolling,
2. Cold forging,
3. Cold spinning,
4. Cold extrusion,
5. Cold drawing,
6. Cold bending, and
7. Cold peening.

All these cold working processes are discussed, in detail, in the following pages.

8·13 Cold Rolling

It is generally employed for bars of all shapes, rods, sheets and strips, in order to provide a smooth and bright surface finish. It is also used to finish the hot rolled components to close tolerences and improve their toughness and hardness. The hot rolled articles are first immersed in a weak solution of sulphuric acid to remove the scale and washed in water, and then dried. This process of cleaning the articles is known as *pickling*. These cleaned articles are then passed through rolling mills. The rolling mills are similar to that used in hot rolling.

8·14 Cold Forging

The cold forging is also called *swaging*. During this method of cold working, the metal is allowed to flow in some pre-determined shape according to design of dies, by a compressive force or impact. It is widely used in forming ductile metals. Following are the three, commonly used, cold forging processes :

1. *Sizing*. It is the simplest form of cold forging. It is the operation of slightly compressing a forging, casting or steel assembly to obtain close tolerance and a flat surface. The metal is confined only in a vertical direction.

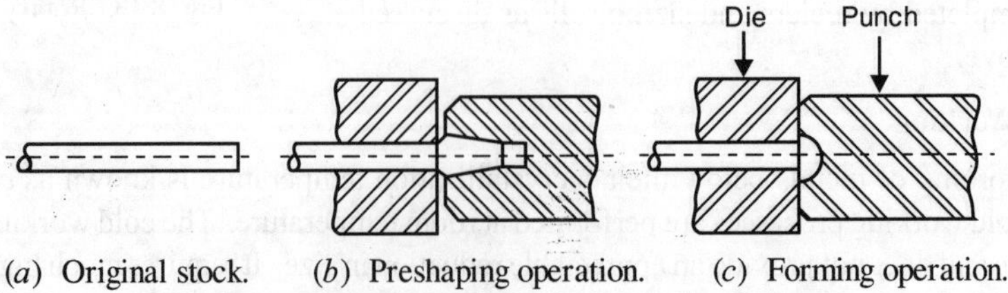

(*a*) Original stock. (*b*) Preshaping operation. (*c*) Forming operation.

Fig. 8.14. Operations performed in a cold header.

2. *Cold heading*. This process is extensively used for making bolts, rivets and other similar headed parts. This is usually done on a cold header machine. Since the cold header is made from unheated material, therefore, the equipment must be able to withstand the high pressures that develope. The rod is fed to the machine where it is cut off and moved into the header die, as shown in Fig. 8.14. The operation may be either single or double and upon completion the part is ejected from the dies.

After making the bolt head, the threads are produced on a *thread rolling machine*. This is also a cold working process. The process consists of pressing the blank between two rotating rolls which have the thread form cut on their surface.

3. *Rotary swaging*. This method is used for reducing the diameters of round bars and tubes by rotating dies which open and close rapidly on the work. The end of rod is tapered or reduced in size by a combination of pressure and impact as shown in Fig. 8.15.

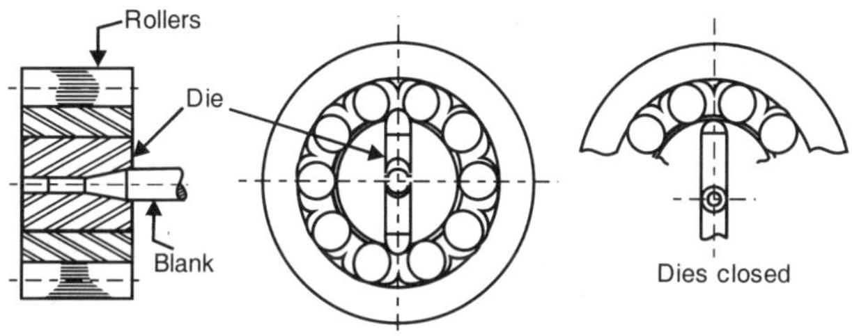

Fig. 8.15. Operation of rotary swaging.

8·15 Cold Spinning

The process of cold spinning is similar to hot spinning except that the metal is worked at room temperature. The process of cold spinning is best suited for aluminium and other soft metals. The commonly used spun articles out of aluminium and its alloys are processing kettles, cooking utensils, liquid containers, and light reflectors etc.

8·16 Cold Extrusion

The principal of cold extrusion is exactly similar to hot extrusion. The most common cold extrusion process is *impact extrusion*. The operation of cold extrusion is performed with the help of a punch and die as shown in Fig. 8.16 (*a*). The work material is placed in position into a die and struck from top by a punch operating at high pressure and speed. The metal flows up along the surface of the punch forming a cup-shaped component as shown in Fig. 8.16 (*b*). When the punch moves up, compressed air is used to separate the component from the punch. The thickness of the side walls is determined by the amount of clearance between the punch and die. The process of impact extrusion is limited to soft and ductile materials such as lead, tin, aluminium, zinc and some of their alloys. The various items of daily use such as tubes for shaving creams and tooth pastes and such other thin walled products are made by impact extrusion.

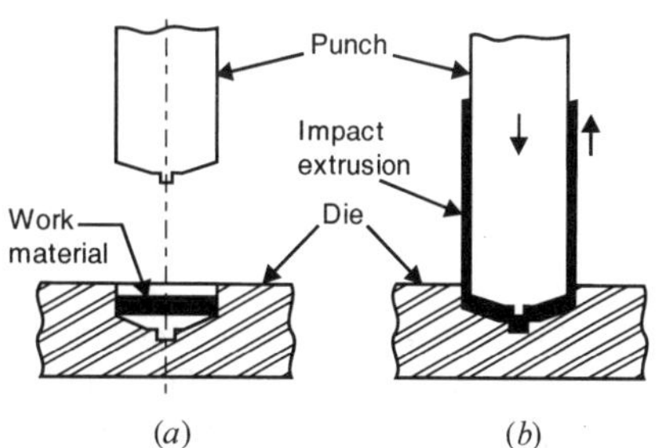

Fig. 8.16. Principles of impact extrusion.

8·17 Cold Drawing

It is generally employed for bars, rods, wires, tubes etc. The important cold drawing processes are as follows :

1. *Bar drawing.* In bar drawing, the rolled bars from the mills are first pickled, washed and coated to prevent oxidation. A draw bench, as shown in Fig. 8.17, is employed for cold drawing. One end of the bar is reduced in diameter by the swaging operation to permit it to enter a drawing die. This end of bar is inserted through the die and gripped by the jaws of the carriage fastened to the chain of the draw bench. The length of the bars which can be drawn is limited by the maximum travel of the carriage, which may be from 15 meters to 30 meters.

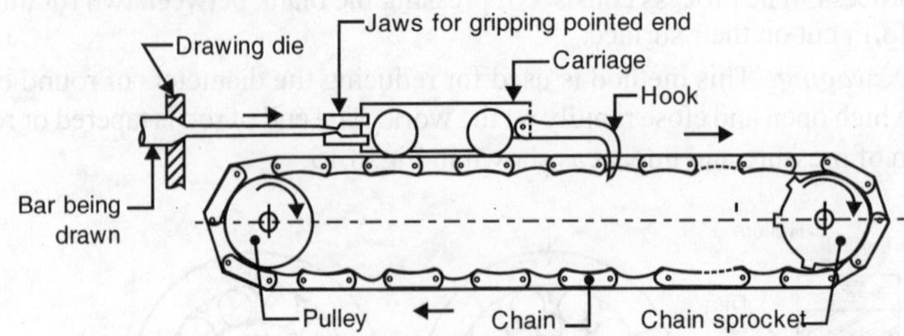

Fig. 8.17. Important features of a draw bench drawing a bar.

2. *Wire drawing.* In wire drawing, the rolled bars from the mills are first picked, washed and coated to prevent oxidation. They are then passed through several dies of decreasing diameter to provide the desired reduction in size. The dies are usually made of carbide materials. The following two processes may be used for wire drawing.

(*a*) *Single draft process.* In this process, a coil is placed on a reel or frame and the end of the rod is pointed so that it enters a die as shown in Fig. 8.18. The end is grasped by jaws on a draw bench and pulled to such length as may be wound around a drawing block or reel. The rotation of the draw block pulls the wire through the die and forms it into a coil. The operation is repeated with smaller dies and blocks until the wire is drawn to final size.

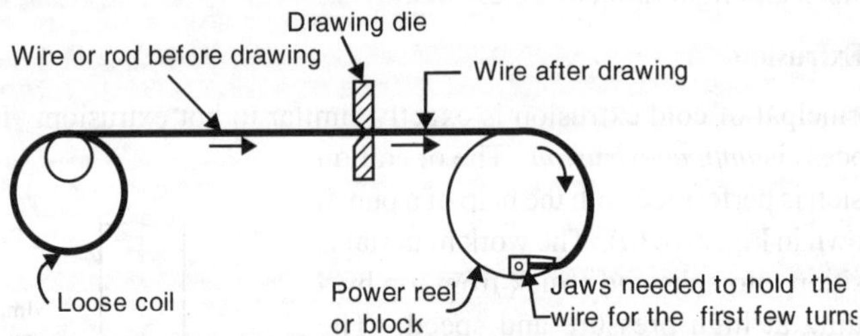

Fig. 8.18. Important features of wire drawing.

(*b*) *Continuous drawing process.* In this process, the wire is fed through several dies and draw blocks in series. This permits of drawing the maximum amount in one operation before annealing is necessary. The number of dies in the series depends upon the kind of metal being processed and may vary from four to twelve successive drafts.

3. *Tube drawing.* The tube drawing is similar to bar drawing and in most cases it is accomplished with the use of a draw bench. The most commonly used method of tube drawing is shown in Fig. 8.19 (*a*). In this method, one end of the tube is reduced in diameter by a swaging operation to permit it to enter a die smaller than the outside diameter of the tube. This end of the tube is gripped by the jaws of the carriage fastened to the chain of the draw bench. The inside surface and diameter is controlled by a fixed mandrel over which the tube is drawn.

The method as shown in Fig. 8.19 (*b*), uses a floating mandrel which adjusts itself to the correct position because of its stepped contour. The method, as shown in Fig. 8.19 (*c*), is usually used for small sized tubing. It may be noted that when several reductions are desired the material must be annealed after each pass.

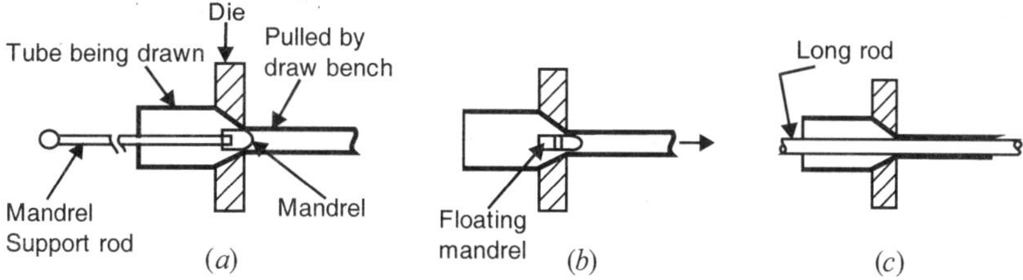

Fig. 8.19. Methods for drawing tubes.

8·18 Cold Bending

The bars, rods, wires, tubes, structural shapes and sheet metal may be bent to many shapes in cold conditions through dies. A little consideration will show that when the metal is bend beyond the elastic limit, the inside of the bend will be under compression while the outside will be under tension. The stretching of the metal on the outside makes the stock thinner. Usually, a flat strip of metal is bend by *roll forming*. The materials commonly used for roll forming are carbon steel, stainless steel, bronze, copper, brass, zinc and aluminium. Some of its products are metal windows, screen frame parts, bicycle wheel rims, trolley rails etc. Most of the tubing is now-a-days roll-formed in cold conditions and then welded by resistance welding.

The roll forming may be subdivided into the following two types :

1. *Cold-roll forming*. In cold roll forming, as shown in Fig. 8.20 (*a*), a flat metal strip is fed

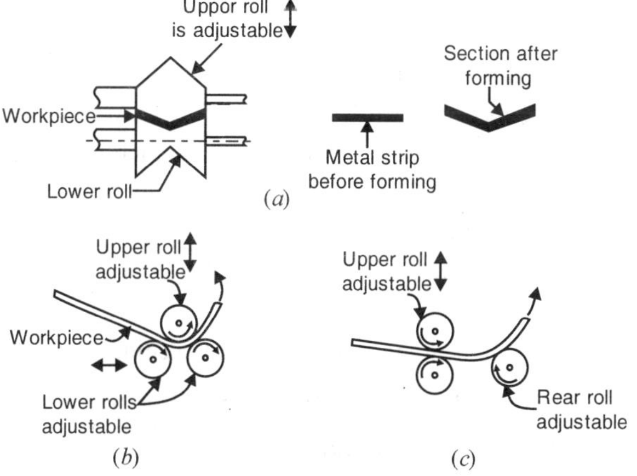

Fig. 8.20. Types of roll forming.

lengthwise between one or more pairs of mating forming rolls. These pairs of rolls are mounted on stands in a straight line like a continuous rolling mill. The pairs of forming rolls change the flat strip into a desired straight section. The section thickness of the metal strip is not purposely changed, and only straight sections are formed by cold-roll forming. The metal strip is usually supplied in large coils. The cold-roll forming is an economical method for forming straight sections in long lengths and large quantities. The rolls, usually, are made for the job except for simple bends. The number of rolls required depends upon the intricacy of the shape, the size of the section, the thickness and type of the material. For a simple channel, four pairs may be used but for more complicated forms several times of this number may be required in addition to the mating horizontal rolls. The cold roll forming machines are equipped with guide rolls mounted vertically to assist in the forming operation and straightening rolls to 'true up' the product as it emerges from the last forming pass.

2. *Roll bending*. Another method of bending metal plates and strips into cylindrical shapes is by means of a *roll bending*. It uses three large rolls, as shown in Fig. 8.20 (*b*) and (*c*), which may be quite long to curve metal sheets and plates. As a metal plate enters and goes through the rolls, its final diameter is determined by the position of the adjustable roll. The closer it is moved to the other rolls, smaller will be the diameter.

8·19 Shot Peening

This process is used to improve the fatigue resistance of the metal by setting up compressive stresses in its surface. This is done by blasting or hurling a rain of small shot at high velocity against the surface to be peened. The shot peening is done by air blast or by some mechanical means. As the shot strikes, small indentations are produced, causing a slight plastic flow of the surface metal to a depth of a few hundreds of a centimetre. This stretching of the outer fibres is resisted by those underneath, which tend to return them to their original length, thus producing an outer layer having a compressive stress while those below are in tension. In addition, the surface is slightly hardened and strengthened by the cold working operation.

QUESTIONS

1. What do you understand by mechanical working of metals ?

2. Define hot working of metals. What are its advantages and disadvantages ?

3. Explain the principle of hot rolling.

4. Describe the different types of rolling mills used for rolling.

5. Write short notes on the following :

(*a*) Hot forging, (*b*) Hot spinning, and (*c*) Hot drawing.

6. Discuss the various methods of hot extrusion.

7. Explain, with the help of neat sketches, the process of hot piercing.

8. What is cold working of metals ? How it differs from hot working ?

9. Discuss the following cold working processes :

(*a*) cold rolling ; (*b*) cold forging, (*c*) cold drawing, and (*d*) cold extrusion.

10. Write short notes on :

(*a*) Roll forming, (*b*) Roll bending, and (*c*) Shot peening.

OBJECTIVE TYPE QUESTIONS

1. The metal is subjected to mechanical working for

(*a*) refining grain size (*b*) reducing original block into desired shape
(*c*) controlling the direction of flow lines (*d*) all of the above
(*e*) none of the above

2. The hot working of metals is carried out

(*a*) at the recrystallisation temperature (*b*) below the recrystallisation temperature
(*c*) above the recrystallisation temperature (*d*) at any temperature

3. During hot working of metals

(*a*) porosity of the metal is largely eliminated
(*b*) grain structure of the metal is refined
(*c*) mechanical properties are improved due to refinement of grains
(*d*) all of the above (*e*) none of the above

4. During hot working of metals

(*a*) poor surface finish is produced (*b*) scale is formed on the metal surface
(*c*) close tolerances can be maintained (*d*) expensive tools are required
(*e*) all of the above

5. Structural sections such as rails, angles, I-beams are made by

(*a*) hot rolling (*b*) hot drawing (*c*) hot piercing (*d*) hot extrusion
(*e*) hot spinning (*f*) any one of the above

6. A two-high rolling mill consists of two rolls which rotate

(*a*) at the same speed and in the same direction
(*b*) at the same speed but in opposite direction
(*c*) at different speeds and in the same direction
(*d*) at different speeds and in the opposite direction

7. In a four-high rolling mill, there are four rolls out of which

(*a*) one is working roll and three are backing up rolls
(*b*) two are working rolls and two are backing up rolls
(*c*) three are working rolls and one is backing up roll
(*d*) all of the four are working rolls
(*e*) all of the four are backing up rolls

8. The parts of circular cross-section which are symmetrical about the axis of rotation are made by

(*a*) hot forging (*b*) hot spinning (*c*) hot extrusion (*d*) hot drawing
(*e*) hot piercing (*f*) any one of the above

9. The cold working of metals is carried out

(*a*) at the recrystallisation temperature
(*b*) below the recrystallisation temperature
(*c*) above the recrystallisation temperature
(*d*) at any temperature

10. Cold working of metal increases

(*a*) tensile strength (*b*) yield strength (*c*) hardness (*d*) all of these

11. During cold working process

(*a*) grain structure is distorted (*b*) strength and hardness of metal increases
(*c*) surface finish is improved (*d*) close dimensional tolerance can be maintained
(*e*) all of the above

12. The process used to improve fatigue resitance of the metal by setting up compressive stresses in its surface, is known as

(*a*) hot piercing (*b*) extrusion (*c*) shot peening (*d*) cold heading

13. The process extensively used for making bolts and nuts is

(*a*) hot piercing (*b*) extrusion (*c*) shot peening (*d*) cold heading

14. The process of increasing the cross-section of a bar at the expense of its length is called

(*a*) drawing down (*b*) upsetting (*c*) spinning (*d*) peening

15. The process of decreasing the cross-section of a bar and increasing its length is called

(*a*) drawing down (*b*) upsetting (*c*) spinning (*d*) peening

ANSWERS

1. (*d*)	2. (*c*)	3. (*d*)	4. (*e*)	5. (*a*)
6. (*b*)	7. (*b*)	8. (*b*)	9. (*b*)	10. (*d*)
11. (*e*)	12. (*c*)	13. (*a*)	14. (*b*)	15. (*a*)

9

Carpentry and Joinery

1. Introduction. 2. Advantages of Timber. 3. Classification of Trees. 4. Structure of a Timber Tree. 5. Types of Timber (Soft and Hard Woods). 6. Sawing of Timber. 7. Seasoning of Timber. 8. Defects in Timber. 9. Preservation of Timber. 10. Methods of Application of Preservatives. 11. Characteristics of Good Timber. 12. Types of Indian Timbers. 13. Market forms of Timber. 14. Adhesives used in Carpentry. 15. Auxiliary Material used in Carpentry. 16. Carpentry Tools. 17. Marking, Measuring and Testing Tools. 18. Cutting Tools. 19. Saws. 20. Chisels. 21. Axe. 22. Planing Tools. 23. Boring Tools. 24. Striking Tools. 25. Holding Tools. 26. Miscellaneous Tools. 27. Setting and Sharpening of Tools. 28. Carpentry Processes. 29. Carpentry Joints. 30. Wood Working Machines. 31. Jointer. 32. Wood Planer. 33. Wood Shaper. 34. Mortiser. 35. Sanding Machines.

9·1 Introduction

The carpentry deals with the constructional work such as making roofs, floors, partitions etc. of a building by means of wood with the help of carpentry tools. The term joinery is used for connecting the wooden parts with the different joints such as making of doors, windows, stairs etc.

The timber is the material used for carpentry and joinery work. It is the wood obtained from *exogenous trees by cutting these trees after their full growth. The following technical terms relating to timber must be clearly understood :

1. *Standing or stationary timber*. It is the timber available from a living tree.
2. *Rough timber*. It is the timber obtained after felling a tree.
3. *Converted timber*. It is the timber which has been sawn into various market sizes such as beams, battens, planks etc.
4. *Dressed timber*. It is the timber which has been sawn, placed and worked to the required condition.
5. *Structural timber*. It is the timber used in framing and load bearing structures.

9·2 Advantages of Timber

The timber has a number of advantages over other materials used in constructional work, the important of which are as follows :

1. It is easily available and can be quickly transported by simple means.
2. It is very easy to be worked on with tools.
3. It is lighter and stronger than most of the materials used in construction work.
4. It has less cost of construction.
5. It has a high salvage value.
6. It is quite suitable for sound proof construction.
7. It is a non-conductor of heat.
8. It responds very well to painting and polishing etc.

* Refer Art. 9·3.

9·3 Classification of Trees

The trees are classified into the following two main classes depending upon their manner of growth :

1. *Exogenous trees.* The exogenous trees are those which grow outward from the centre adding almost concentric layers of fresh wood every year, known as annular rings. The timber obtained from exogenous trees is extensively used in engineering work.

The exogenous trees may be further divided into the following two types :

(*a*) Conifers or evergreen trees ; and (*b*) Deciduous or broad leaf trees.

The *conifer* or *evergreen trees* are those trees which have needle shaped leaves and are quite symmetrical, with branches radiating from the central trunk. These trees yield soft wood which is generally light coloured, resinous, light in weight and weak. They show distinct annual rings. The trees like deodar, pine, chir, kail belong to this category.

The *deciduous or broad leaf trees* are those trees which shed their leaves in autumn and put on new leaves in the spring season. These trees yield hard wood which is usually close grained, strong, heavy, dark coloured, durable and non-resinous. They do not show distinct annual rings. The timber used for engineering purposes is mostly derived from deciduous trees. The trees like teak, sal, shisham, oak, mahogany, mango, neem, babool etc. belong to this category.

2. *Endogenous trees.* The endogenous trees are those which grow inward or endwards. The stems of these trees are light and tough but they are too flexible to be used for engineering purposes. The trees like bamboos, palms, canes etc. belong to this category.

9·4 Structure of a Timber Tree

The cross-section of an exogenous tree is shown in Fig. 9.1. The important parts are described as below :

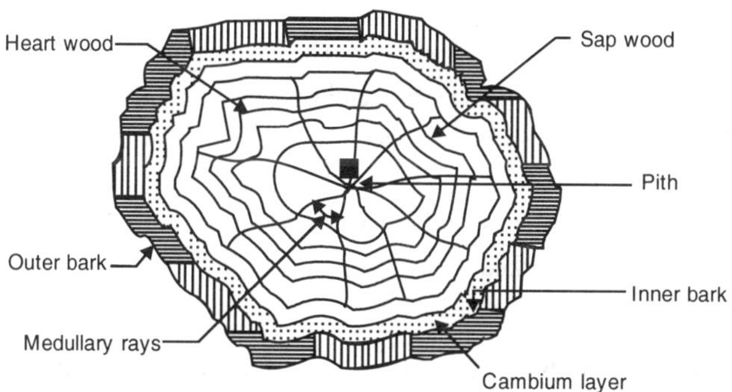

Fig. 9.1. Cross-section of an exogenous tree.

1. *Pith or medulla.* It is the innermost central portion or core of the tree. It consists of only cellular tissues. It contains a lot of fluid which nourishes the plant. The annular rings are added around this every year. When the plant becomes old, the medulla dries up and decays.

2. *Annular rings.* These are concentric circles or rings of woody fibre around the pith. They are formed every year. The number of annular rings indicates the age of a tree.

3. *Heartwood.* It consists of the inner annular rings around the pith. It is compact, strong, durable and darker in colour. The heartwood increases in size as the tree ages. This part provides the best timber for engineering works.

4. *Sapwood.* It consists of the outer annular rings between the heartwood and cambium layer. It is also called *albumum.* It is weaker, softer and lighter in colour as compared to heartwood.

This wood is unsuitable for engineering purposes as it is liable to decay or attack by insects.

5. *Cambium layer*. It is a thin layer just below the bark and not converted to sapwood yet. This is mainly responsible for the circumferential growth of the tree. If the bark is removed, the cambium layer get exposed and the cells cease to be active which results in the death of a tree.

6. *Medullary rays*. These are thin radial fibres extending from the pith to cambium layer. These rays carry sap from the outside to the inner parts of the tree to nourish it. They hold the annular rings together tightly to provide a solid structure to the tree.

7. *Inner bark or bast*. It is the inner skin or layer covering the cambium layer. It protects the cambium layer from injury.

8. *Outer bark or cortex*. It is the outermost cover or skin of stem. It prevents the tree from external injuries and weathering action.

9·5 Types of Timber (Soft and Hard Woods)

The timber is mainly of two types, namely soft wood and hard wood. These two woods are described here.

1. *Soft wood*. The soft wood is obtained from trees having needle shaped leaves or conifers. The various soft woods are Kail, Pine, Deodar, Chir, Walnut, Semal, and Sprue. It is widely used for building construction. The soft wood has the following characteristics :

(*a*) It is light in colour and weight.

(*b*) It has straight fibres and fine texture.

(*c*) It has a good tensile resistance but is weak across the fibres.

(*d*) It readily catches fire because it is a resinous wood.

(*e*) It is easy to be worked.

2. *Hard wood*. The hard wood is obtained from trees having broad leaves or deciduous. The various hardwoods are Sal, Teak, Mahogany, Shisham, Oak, Beach, Ash, Babool, Neem, Mango etc. It is widely used for doors, furniture, joinery etc. The hard wood has the following characteristics :

(*a*) It is dark in colour and heavier in weight.

(*b*) Its fibres are quite close and compact.

(*c*) It is non-resinous wood.

(*d*) It has both good tensile as well as shear resistance.

(*e*) It is difficult to be worked.

9·6 Sawing of Timber

The timber may generally be sawn in one of the following four ways :

1. *Ordinary sawing*. It is also called flat or slab sawing. In this type of sawing, the cuts are made into the log of wood according to the required thickness. The saw cuts are tangential to the annular rings and right through the cross-section of timber piece as shown in Fig. 9.2. The log is moved backward and forward on a saw which cuts parallel slices or planks. The different thickness can be obtained by varying the width of cutting.

It is very easy and quick method of sawing. It is almost an economical method and the wastage of timber is minimum.

2. *Tangential sawing*. It is sometimes known as plain sawing and gives flat grained timber. In this method, the boards or planks are sawn out of the log of wood, tangential to the annular rings as shown in Fig. 9.3. The planks obtained by this method, warp too much. This method is adopted only when the annular rings are very distinct and the medullary rays are not clearly defined.

3. *Radial Sawing*. In this method, the timber is cut parallel to medullary rays and perpendicular

to annular rings, as shown in Fig. 9.4. Since the cut section shrinks at a uniform rate and the warping is less, therefore it can be used for high quality timber works. But there is a considerable quantity of wastage of wood. The cost of sawing is also high because more time and labour are required in turning the wood several times.

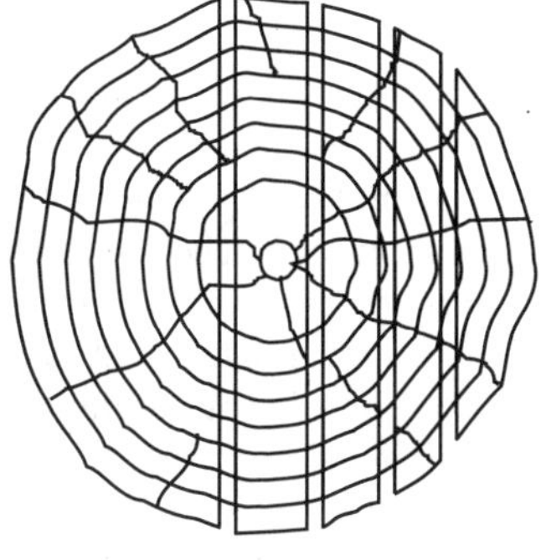

Fig. 9.2. Ordinary sawing.

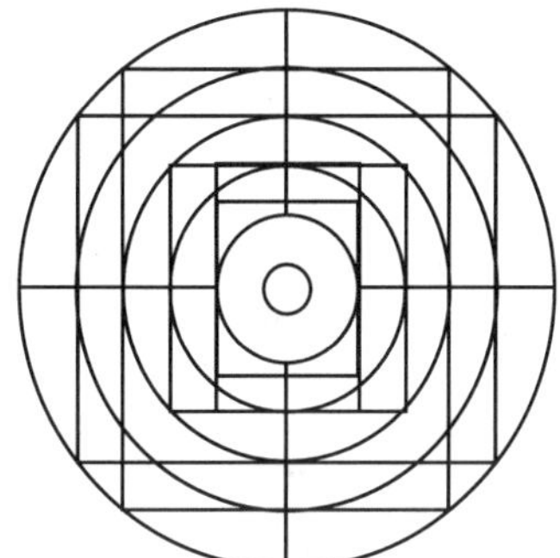

Fig. 9.3. Tangential sawing.

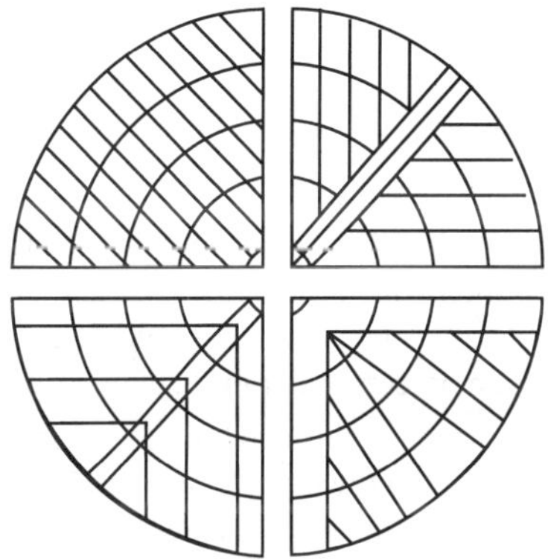

Fig. 9.4. Radial sawing.

Fig. 9.5. Quarter sawing

4. *Quarter sawing*. In this method, the log is cut into four quarters, as shown in Fig. 9.5. These quarters have their faces radial. The boards are cut alternately from each of these faces, thus cutting the annual rings at angles not less than 45°. It may produce very fine figure wood when adopted in case of timber having no distinct medullary rays. In quarter sawing, there is a tendency for the timber to bend in a transverse direction.

9·7 Seasoning of Timber

The seasoning of timber is the process of drying timber or removing moisture or sap, present in a freshly felled tree, under controlled conditions. The following methods are commonly used for seasoning of timber.

1. *Natural seasoning or Air seasoning*. In this method of seasoning, the tree, after felling, is converted into logs, planks or battens. These are stacked in a dry place about 300 mm above the floor level with longitudinal and cross pieces arranged one upon another. While stacking, it should be ensured that there is a space between them for free circulation of air all around each piece. In order to prevent the effect of moisture on the wood from the bottom, a layer of cinder ash or sand is spread

on the levelled plateform before stacking the wood. The stacked wood is turned upside down periodically in order to accelerate the rate of drying. The wood gets dried due to the circulation of free air, which dries up the moisture.

This method of seasoning the wood is simple and cheap, but it is very slow and extends over years depending upon the type of wood and its cross-section. The softwood and their sections dry up easily whereas hardwood and thick sections take more time.

2. *Artificial seasoning or Kiln seasoning*. This method of seasoning is the quickest of all the commonly used wood seasoning processes. It keeps the moisture contents under control. The process is carried out in a chamber under controlled temperature and humidity conditions with proper air circulation and ventilation system. Usually, steam is used for this purpose. The seasoning is started at a comparatively lower temperature and high humidity. The conditions are changed as the timber dries. At the end of seasoning, the air is fairly hot and humidity is low. The required humidity level is maintained to avoid warping and cracking of wood. The drying of wood at a uniform rate is well maintained by the circulating air. The ventilation is provided to avoid over heating and excessive humidity. Before removing the wood, the kiln is allowed to cool till the temperature inside the kiln is within 15°C to 20°C of the outside temperature. The seasoning by this method takes four to five days under normal conditions. It is best suited for mass production work.

9·8 Defects in Timber

The defects in timber are of the following two types :

1. Defects developed during the growth of a tree, and
2. Defects developed after felling of a tree.

These defects are discussed, in detail, as follows :

1. *Defects developed during the growth of a tree*. In all kinds of timber, several natural defects occur. They are caused by the nature of the soil upon which the tree grows and also by the changes of nature to which it is subjected while growing. These defects should be avoided or removed during conversion for use. Following are the most common defects under this category :

(*a*) *Heart shake*. These are the wide splits or cracks running through the heart wood of a tree, as shown in Fig. 9.6 (*a*). These splits radiate from the pith running towards the sapwood. This defect usually occurs in over matured trees and is caused due to the shrinkage of the heartwood.

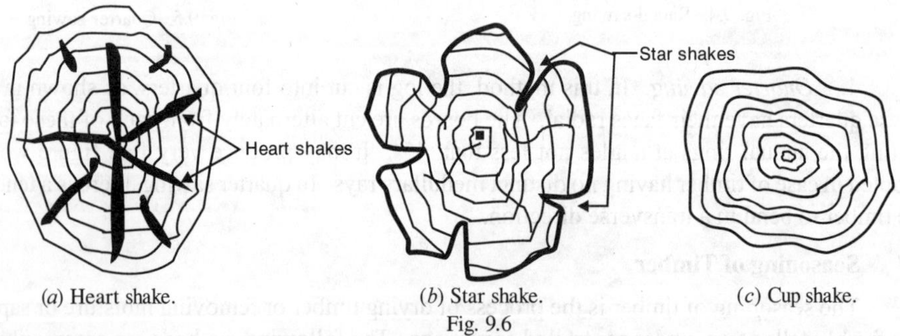

(*a*) Heart shake. (*b*) Star shake. (*c*) Cup shake.

Fig. 9.6

(*b*) *Star shake*. These are radial splits or cracks which are wider at the circumference of the tree and becomes narrower towards the centre as shown in Fig. 9.6(*b*). The star shakes are mostly caused by severe frost or by severe heat of the sun. They are mostly confined to sapwood.

(*c*) *Cup shake or Ring shake*. These are the cavities which appear in between the annual rings, as shown in Fig. 9.6(*c*). These cavities may be completely circular or partial. These are caused by strong winds which sway the tree or due to excessive frost which affects the moisture present in the tree when it is still young.

(*d*) *Rind galls*. These are the peculiar curved swellings formed on the body of a tree. These are caused generally by the growth of layers over the wounds left after branches have been improperly cut-off or removed.

(*e*) *Twisted fibres*. This defect is developed in a tree owing to its peculiar position where the fibres of the wood are twisted by strong winds turning the tree constantly in one direction. The trees on hills are the ones most affected by this defect.

(*f*) *Knots*. These are the roots of small branches of the tree which are embedded in the stem with the formation of annular rings at right angles to those of the stem. When the knot is firmly intact with the surrounding wood and can not be separated, it is then called a *live knot*. On the other hand, a knot which is not held firmly in place and can be separated from the wood, it is then called a *dead knot*.

2. *Defects developed after felling of a tree*. The following are the main defects which occur after felling of a tree.

(*a*) *Dry rot*. This defect is caused by the fungus. It changes the timber into dry powder. The unseasoned timber is easily attacked by fungus. Thus, a well seasoned timber should be used in order to prevent dry rot.

(*b*) *Wet rot*. This defect is caused by alternating drying and wetting of the timber. The timber used in uncovered situation is very much likely to be subjected to wet rot. It is commonly found in a living tree. The wet rot can be avoided by using thoroughly seasoned wood for exterior and under ground work. The seasoned wood should be protected from moisture by tarring or painting. The best remedy for treating wet rot is by using a suitable preservative.

(*c*) *Honey combing*. This defect is caused during seasoning of wood. The incorrect kiln seasoning causes the interior wood to dry quicker than the external surface. This gives rise to internal stresses, causing circular and radial cracks in the wood crossing each other.

(*d*) *Case hardening*. This defect is caused by uneven drying of timber. The outer portion of the wood dries earlier and quicker than the inner portion, causing shrinkage in the outer tissues leading to drying stresses. Due to this, the outer surface of the wood gets hardened.

9·9 Preservation of Timber

In order to protect the timber from internal decay and attack of insects like white ants, some chemical preservations are used to increase the life of timber and to make the timber structures durables. A good preservative should have the following requirements :

1. It should be cheap and easily available.
2. It should not leave any mark on the wood.
3. It should be able to penetrate the wood.
4. It should be non-inflammable.
5. It should be free from obnoxious smell.
6. It should not corrode the metals with which it comes in contact.
7. It should be poisonous for insects and fungi.
8. It should not be washed away by water or deteriorated by heat, light etc.

The following preservatives are commonly used for the preservation of wood :

1. *Tar oils*. It is a very common type of preservative used for rough type of works such as timber fences, ends of door and window frames etc. The *creosote* is the most widely used oil preservative. The three common types of creosote are coal tar, wood tar and water gas-tar creosotes. Out of these, coal tar is the most effective.

2. *Water soluble chemical salts*. The Forest Research Institute, Dehradun has developed a new preservative called ASCU. It is in powder form. It consists of hydrated arsenic pentaoxide, blue

vitriol and potassium dichromate. For use, six parts (by weight) of the powder are dissolved in 100 parts (by weight) of water. The solution is sprayed on the timber. The wood treated by this method can be painted, varnished, polished or waxed.

3. *Organic solvent chemicals.* These chemicals are costilier than other types of preservatives. They include some toxic chemical salts, nepthol, phenol etc. These are dissolved in spirit or alcohol and are applied on wood. The spirit or alcohol evaporates leaving the salts on the surface of the wood. The salts are covered with paint or polish afterwards, to preserve the surface of the wood. These preservatives give very good results and the timber treated by them does not warp or shrink after drying.

9·10 Method of Application of Preservatives

Following are the three most important method of applying preservatives to the timber.

1. *Brush and spray method.* In this method, the timber to be treated is brushed, sprayed or dipped in the preservative solution for a suitable period. Atleast, two coats are applied, the second coat is applied only after the first coat has dried. This method is suitable only when the wood is likely to be subjected to damp conditions. The preservatives may be sprayed by machine in order to have uniform application and better results than hand brushing.

2. *Dipping or soaking treatment.* This method is also known as *steeping process.* In this type of treatment, the timber pieces are submerged in an open tank containing the preservative solution, for sufficiently long time until the required absorption is obtained. The water soluble type preservative is used to avoid excessive evaporation loss. For a seasoned timber, a soaking period of 7 to 10 days is desirable and a penetration of 3 mm to 6 mm is expected. This method gives better penetration and better results as compared to brushing and spraying.

3. *Pressure treatment.* It is the most effective method of treating timber with preservative. In this method, the preservative is injected into the timber under pressure, in airtight steel containers called *autoclaves.* This method proves to be essential for treating non-durable timbers which are to be used at places where there is a danger of attack by fungi and insects.

9·11 Characteristics of Good Timber

The timber used for carpentry should have the following characteristics :

1. It should be from the heart of tree and free from sap.
2. It should be free from defects like knots, shakes, resin pockets, worm holes, twisted fibers etc.
3. It should have uniform dark colour with close fibres.
4. It should not clog the teeth of the saw when it is cut.
5. It should smell sweet on cutting.
6. It should give clean sound when struck with hammer.
7. It should be properly seasoned. It should not warp or twist after seasoning.
8. It should be easily workable.
9. It should be strong and heavy.
10. It should have high resistance to fire.
11. It should not split when nails are driven into it.
12. On planning, it should present a firm bright appearance with a silky lustre.

9·12 Types of Indian Timbers

The trees can be readily identified by their physical appearance *i.e.*, their shape, the shape of their leaves, the appearance and colour of bark and the kind of flower, or fruit that they bear. The most common Indian timbers used in carpentry and joinery are given below :

1. *Babul.* It is obtained from a thorny tree, which bears small leaves and yellow flowers. The wood have close-grained, tough and pale red colour. It grows abundantly all over India and is used for agriculture equipments, tools, handles, bodies of carts and wheels, tent pegs etc.

2. *Deodar.* It is found in Himalayas. It is light-coloured and has close grains. It is light and can be easily worked. It is used for the manufacture of table, chair and furniture etc.

3. *Mahogany.* It has rich golden brown colour, which deepens with age. It contains resinous oil which prevents attack of insects. It is very tough, strong, and highly stable in fashioned form. This wood is extensively used in the manufacture of furnitures, boats, and in pattern making.

4. *Mango.* It is largely found all over India. It has coarse and open grained fibres. It has deep grey colour. It is cheap and readily available in large quantity and is extensively used in manufacture of doors, windows and furniture but not of good quality.

5. *Sal.* The wood is of dark brown colour. It is hard, close-grained, heavy and durable. It is considered to be the most useful and valuable wood for almost all engineering works. This tree is found all over India and is used in the manufacture of beams, posts, roofing, railway sleepers etc.

6. *Teak.* It is native to India and Burma. It is known for its extreme durability and long lasting quality. The wood is moderately hard, strong and straight grained. It can be easily worked. It is oily and fragrant and takes a beautiful polish. It possesses a lovely golden-yellow colour, which darkens with age. The teak is most important ship-building timber, and out door work like gates, external doors, windows etc. are made from teak as well. It is also used for staircases, flooring, panelling and furniture.

7. *Kail.* It is a close-grained, heavy wood. It has brown colour. It is moderately hard and durable wood and does not permit easy workability. It is used for carpentry and joinery, furniture making and for railway sleepers.

8. *Chir or cheel.* It has coarse grained fibres. It is soft and brown in colour. It is mostly found in Himalayan region. It is used for interior wood work in low cost house building.

9·13 Market Forms of Timber

The various market forms in which the timber is cut and sold are given below :

1. *Log.* It is trunk of a felled tree with branches cut off.

2. *Baulk.* It is a roughy squared sawn log.

3. *Batten.* It is a timber whose cross-sectional dimensions do not exceed 175 mm width-wise and 50 mm in thickness.

4. *Plank.* It is a timber piece more than 275 to 450 mm width and 50 to 150 mm thickness. The length varies from 2.5 m to 6.5 m.

5. *Deal.* It is a timber whose thickness is up to 100 mm and width is above 225 mm.

6. *Board.* It is the form of timber whose thickness is less than 35 mm and width is above 125 mm.

7. *Scantlings.* These are pieces of miscellaneous sizes sawn out of a log *i.e.*, they have no standard sizes.

9·14 Adhesives used in Carpentry

The adhesives are used in wood work for joining the wood pieces together. For this purpose, glue is widely used. The glued joint, if properly made, becomes the strongest one. The glue may be applied either in cold or hot condition. There are several types of glues available in the market, but the following are the most prominent and commonly used :

1. *Animal glue.* It is made from hides, bones, hoofs and other parts of animals. These materials

are refined and are available in sheets, flakes or powdered form. This glue is applied when hot (about 70°C) and it sets rapidly. Before applying, it is soaked in cold water overnight and then heated. Commercially, the animal glue is available in liquid form and can be applied in cold condition. Since the animal glue is not water proof, therefore it is not suitable for out door furniture, boats, etc. It is tough and strong. It gives strength even in poorly fitted joints.

2. *Resin glue.* It is made from formaldehyde uric acid and other chemicals. It is available in powder form. It gives high strength and stability at high temperatures and resists moisture.

3. *Casein glue.* It is made by adding an alkali to the curd of skimmed milk. It is available commercially in powder form and can be converted into paste by adding water. It takes about 15 to 20 minutes to set. It is a strong water resistant glue and commonly used for furniture, boat making etc.

4. *Vegetable glue.* It is also known as *starch glue,* as it is made by mixing the starch of roots, grains and corn of trees with hot water. The acid or alkali modifiers are used to make the starch paste thick and tacky. Since the viscosity of this paste is very high, therefore it is applied only mechanically in cold form. It gives strong, water proof wooden joint. It is best suited for plywood work.

5. *Blood albumen glue.* It is made by adding an alkali to beef blood. It is available commercially in flake form. The flakes are dissolved in water one hour before its use. It gives a strong and water proof joint.

6. *Gum arabic.* It is obtained from acacia tree. It contains mixed mineral salts of arabic acids. It forms the most useful natural resin adhesive. It is soluble in water. It is used for joints exposed to tropical conditions.

7. *Contact cement.* It is available commercially under different trade names. The instructions provided by their manufacturers must be strictly followed. This cement is stainless, moisture resistant, heat resistant and provides a good bond.

The contact cements are used when bonding veneers and resin laminates to wood. After applying a uniform coating, the cement is allowed to dry for about 30 minutes and the pieces are placed together to join. Once the parts are in contact, they can not be moved. Care must be taken to position the parts properly before applying the cement.

9·15 Auxiliary Material used in Carpentry

Though timber is the basic material used in carpentry work, yet the other main materials such as nails, screws, bolts and nuts, dowels are also frequently used. These materials are discussed as below :

1. *Nails.* There are many types of nails available, each designed for a different use. The most common types of nails used in wood working are wire nails, cut or clasp nails, oval brads and panel pins. The cut or clasp nails have better holding capacity than wire nails. The wire nails have a round section while the oval brads are of oval section. The panel pins also have a round section, with a round tip, and are used for fine and delicate work. The nails are mainly used for reinforcing the glued joints. These are sold by weight in the market.

2. *Screws.* The wooden screws are available in different lengths, and are sold in the market by numbers. They are used mainly for fixing the metallic fittings like hinges, tower bolts, hasps etc. A hole is drilled before putting the screw which is driven into the wood by a screw driver. The diameter of the hole drilled should be less than the diameter of the screw.

3. *Bolts and nuts.* These are available of different diameters, length of the threads and shape of the heads. These are used to join very heavy components such as wooden roof trusses and folding type furniture etc.

4. *Dowels.* These are wooden nails made from bamboo or some other wood. These are made by the carpenter himself. These are used to join or align the wooden parts. A hole is drilled between

the two parts to be joined and the dowel is driven through them.

9·16 Carpentry Tools

The carpentry tools may be classified in the following main groups :

1. Marking, measuring and testing tools,
2. Cutting tools,
3. Planing tools,
4. Boring and drilling tools,
5. Fixing or striking tools, and
6. Miscellaneous tools.

We shall now discuss these tools, in detail, in the following articles.

9·17 Marking, Measuring and Testing tools

The exact measuring and marking are indispensable for the processing of wood. The use of inaccurate measuring tools and carelessness of marking will result in faulty work pieces.

The following marking, measuring, and testing tools are important from the subject point of view :

1. *Pencil.* It is a carpenter's marking equipment. The pencil gradings of H, HB and F are all popular for marking dimensions, joints and face marks.

2. *Pencil compass.* It is used for marking circles, curves or small circles from 25 mm to 300 mm in diameter. One leg of the compass has a steel point pin and the other leg has a socket for holding the pencil. The dividers are also used for the same purpose.

3. *Trammel points.* These are used for laying out larger arcs and circles. They are also used for laying out various geometrical shapes such as an ellipse or a multi-sided figure.

It consists of a straight edge and trammel points, as shown in Fig. 9.7.

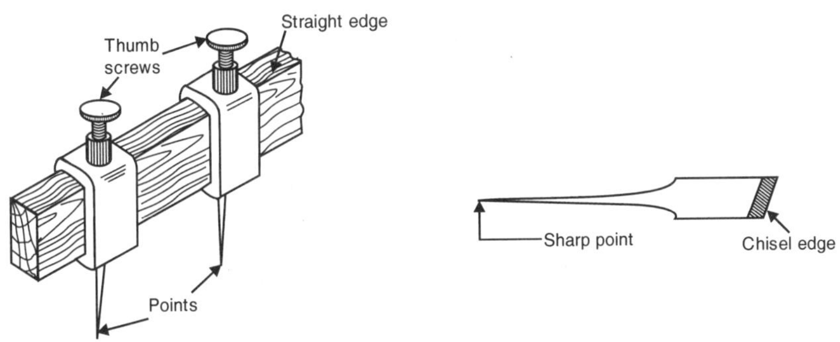

Fig. 9.7. Trammel points.

Fig. 9.8. Marking knife.

4. *Marking knife or scriber.* It may have a sharp point at one end and a slanting chisel edge at the other end as shown in Fig. 9.8. It is mainly used for marking where accuracy is necessary.

5. *Marking gauge.* It is used to draw lines parallel to the edge. It is very important tool for marking and is made of wood. It consists of a stem having rectangular or square cross-section. The stem has a sharp steel marking pin near one end, which projects about 2 mm below the lower surface. The stem carries the head as shown in Fig. 9.9. It can be fixed to any desired position by means of a thumb screw. After the stem has been set as required and the screw tightened, the face of the head is placed against the edge of the job. The marking pin is pressed down firmly which scores a line on the job surface as the head transverses the edge.

6. *Mortise gauge.* It is an improved form of marking gauge which carries two scribing pins instead of one, as shown in Fig. 9.10. One of the pins is movable while the other is fixed.

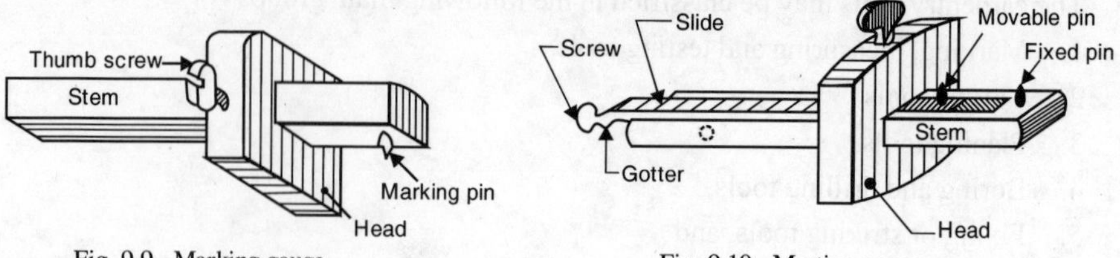

Fig. 9.9. Marking gauge.

Fig. 9.10. Mortise gauge.

The movable pin can be regulated at any required distance from the fixed pin by means of a thumb screw provided at the end of the stem. It is used to mark two parallel lines in a single operation on a face. This is specially used to mark mortises and tenons and other similar joints requiring such parallel lines.

7. *Cutting gauge.* It is similar to a marking gauge except that it has a steel cutter fitted in place of the marking pin. It is mainly used for cutting parallel strips from thin wooden sheets upto 3 mm and for marking deep lines in thicker sections.

8. *Four-fold rule.* It is a wooden scale and consists of four pieces each 15 cm long, hinged together by means of pins as shown in Fig. 9.11.

It is used for measuring and setting down dimensions. It can be easily carried in the pocket.

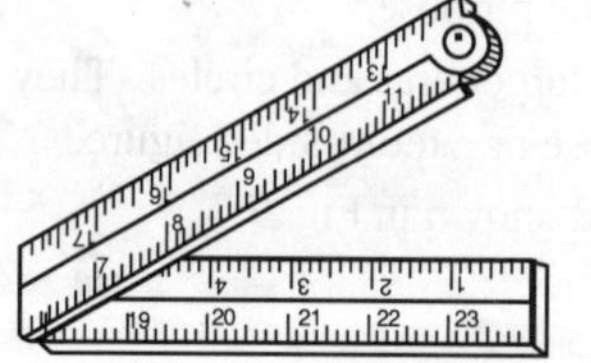

Fig. 9.11. Four-fold rule.

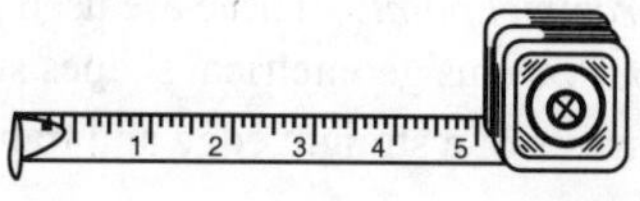

Fig. 9.12. Steel tape.

9. *Steel tape.* A steel tape is shown in Fig. 9.12. It is used for measuring longer dimensions. It is usually available in 2 metre length.

10. *Straight edge.* It is a machined flat piece made of wood or metal having straight edges as shown in Fig. 9.13. It is used for testing trueness of large surfaces and edges.

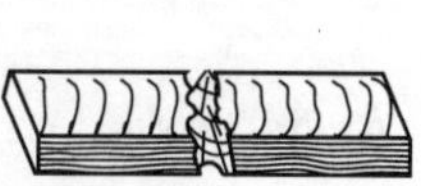

Fig. 9.13. Straight edge

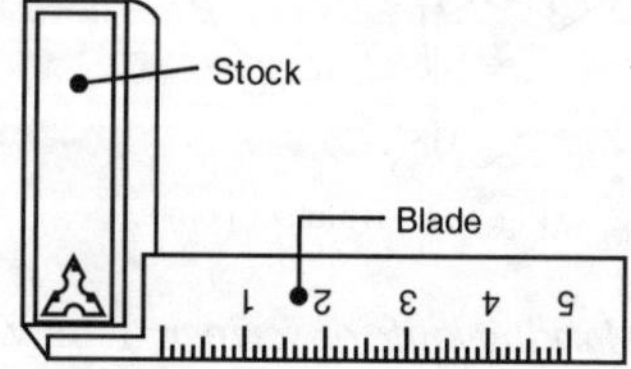

Fig. 9.14. Try square.

11. *Try-square.* It is used for drawing and testing right angles. It consists of a steel blade fixed at right angles to the edge of a machined stock either of metal or wood, as shown in Fig. 9.14. The stock, when in use, acts as a fence and the blade as a guide for marking. It is available in various sizes like 100 mm, 150 mm and 300 mm.

12. *Bevel and mitre square.* The bevel square, as shown in Fig. 9.15(a), is also called *sliding bevel.* It is an adjustable try square for marking angles between 0° and 180°.

It consists of a stock and a sliding blade with a screw to tighten the blade in position after it is set. The various sizes in common use are 100 mm, 150 mm, 200 mm and 250 mm.

The *mitre square,* as shown in Fig. 9.15(*b*), is used to mark angles of 45°.

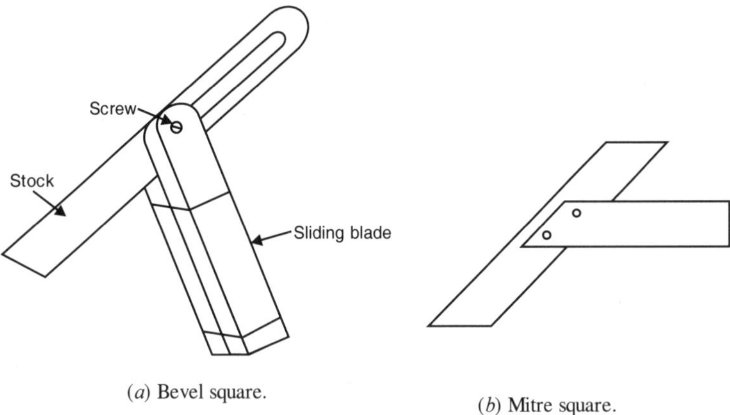

(*a*) Bevel square. (*b*) Mitre square.

Fig. 9.15. Bevel and mitre square.

13. *Combination set.* Some carpenters prefer to use a *combination set which is similar to the one used in bench work. It consists of a rule on which a square head, a bevel protractor and a centre head slide.

14. *Calipers.* There are two types of calipers, one is the *inside caliper* and the other is *outside caliper.* As the name indicates, the inside caliper is used to measure the inside diameters while the outside caliper is used to measure the outside diameters.

15. *Divider.* It is similar to a caliper but its legs have sharp points. It is used for drawing circles and arcs.

16. *Wing compass.* It consists of two finely pointed steel legs and a quadrant, having angles marked on it. The legs after setting to a required angle or distance can be locked by a set screw. It is used to set off a number of equal spaces, making arcs or circles and scribing parallel lines on a curved or straight surface.

17. *Spirit level.* It is used for checking level *i.e.,* to check whether the surface is a true plane or inclined plane. It consists of a wooden body having all parallel sides, with slightly curved longitudinal and transverse level glass tube. It is filled with spirit, ether of alcohol. The air bubble position is calibrated on a levelled surface and marked at the centre.

18. *Plumb bob.* It is used to check the vertical surfaces. It consists of a conical metallic bob tied to a thread. The thread is held against the vertical surface and its position is checked with respect to the thread and the tip of the conical bob.

When spirit level and plumb bob are used in combination, then they can be used to check the right angle of large surfaces where a try square can not be used.

9·18 Cutting Tools

The cutting tools may be broadly classified into the following three groups :

1. Saws, 2. Chisels, and 3. Axe.

These cutting tools are discussed, in detail, in the following pages.

---

* See Chapter 16 on Bench Work and Fitting.

9·19 Saws

A saw consists of a thin spring steel blade which has a series of sharp teeth on one edge as shown in Fig. 9.16. Each tooth cuts or tears away a small portion of wood. The teeth are bent alternately to the right and left of the blade to enable it to pass through the cut being formed in the timber with the minimum of friction. The bending of teeth in this manner is called *setting of saw.* The blade is fixed to the wooden handle by riveting. The front portion of the saw is called *toe* and the back portion as *heel.* The saws vary greatly in shape and size. The size of the saw is generally specified by the length of the blade and pitch of teeth in millimetre.

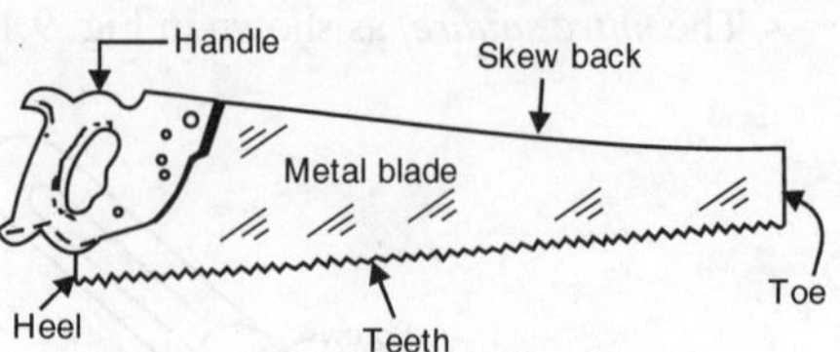

Fig. 9.16. Parts of a saw.

There are many types of saws used for different types of work. But, in general, the saws are divided into the following two groups :

(*a*) *Pull saws or draw saws.* When the cutting of wood takes place during the return stroke of the saw (*i.e.*, when the saw is being pulled by the operator towards his side), it is then known as pull saw or draw saw. It may be noted that when thin wooden sheets are cut by the pull saw, the pressure is not only exerted on the wooden sheet but also on the hand by which the saw is held. Thus there is no possibility of bending and breaking of thin wooden sheet when cut by pull saws. Thus these saws are used to cut thin wooden sheets.

(*b*) *Push saws.* When the cutting of wood takes place during the forward stroke of the saw (*i.e.*, when the saw is being pushed by the operator), it is known as push saw. These saws are very useful for general cutting of wood. The cutting action is fast in this case. The only draw back with these saws is that the thin wooden sections cannot be cut satisfactorily because due to pressure during forward stroke, there is a possibility of bending and breaking of thin wooden sections. The push saws, being very useful, are more in common use than pull saws.

Following are the different types of saws commonly used by the carpenter :

1. *Rip saw.* It is used for cutting along the grains of wood. Its blade is about 700 mm long and has 3 to 5 points or teeth per 25 mm length. The teeth of the rip saw have a series of chisel edge as shown in Fig. 9.17, which tear out shavings each equal to width of a tooth.

Fig. 9.17. Rip saw.

The width of blade is about 120 mm to 200 mm near the handle and 60 mm to 100 mm at the toe. These saws should be kept at an angle of 35° to 40° for cutting of wood.

2. *Cross-cut or hand saw.* It is used for cutting across the grains of wood, but it is also used for general sawing. The blade length of this saw is 500 mm to 700 mm with 8 to 10 points or teeth per

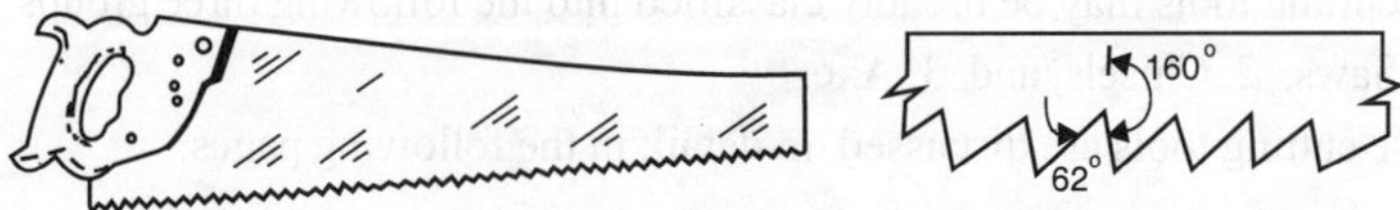

Fig. 9.18. Cross-cut or hand saw.

25 mm length. The cross-cut saw resembles the rip saw, but the shape of its teeth is quite different. The common shape of the cross-cut saw and its teeth are shown in Fig. 9.18.

The teeth are like a series of small triangular knives which severe the fibres and force out the waste wood in the form of saw dust. A blade with finer pitch is considered best for hard woods and a blade with coarse pitch is used for soft wood.

3. *Panel saw.* It is similar to the cross-cut saw but its blade is thinner. The blade is about 500 mm long with 10 to 12 teeth per 25 mm length. It is used for accurate work. It is mostly used for cutting panels for the door shutters. It is sometimes used for ripping as well as cross-cutting.

4. *Tenon saw or back saw.* It has a parallel blade of width 60 mm to 100 mm, length 250 mm to 400 mm and 12 to 20 points or teeth per 25 mm length. The teeth are shaped like a cross-cut saw, in form of an equilateral triangle as shown in Fig. 9.19.

Fig. 9.19. Tenon or back saw.

The top of the blade is stiffened with a solid strip which acts as a reinforcement so that the blade does not bend during the operation. It is generally used for cutting shoulders to tenons and when a clean cut is necessary. It is suitable to cut wood across the grains.

5. *Dovetail saw.* It is just a small tenon saw having 50 mm to 70 mm width of blade and an open wooden handle as shown in Fig. 9.20. The length of blade is about 200 mm to 300 mm and has about 14 teeth per 25 mm length. The blade is thinner and narrower than that of tenon saw and has no reinforcement at the top.

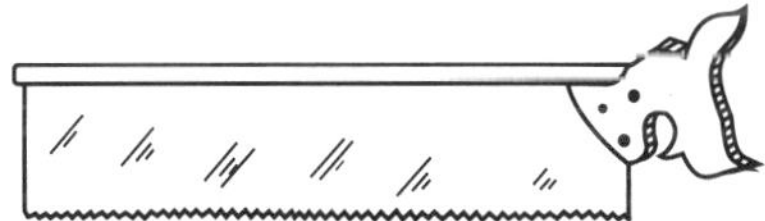

Fig. 9.20. Dovetail saw.

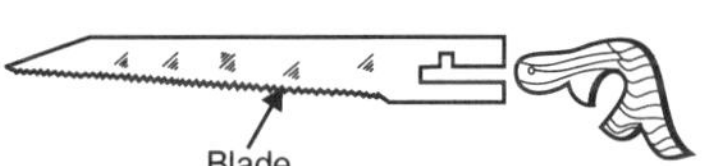

Fig. 9.21. Compass or turning saw.

It is used for very fine work as for forming dovetail joints in drawers and cutting shoulders on narrow rails.

6. *Compass or turning saw.* It consists of narrow tapering blade about 250 mm to 400 mm long which is fitted and screwed to the open handle as shown in Fig. 9.21. The width of blade is about 40 mm to 50 mm near the handle and 20 mm to 30 mm at the toe. There are two types of compass saws. In one type, the blade is fixed while the other type is provided with three blades of different sizes which are interchangeable. Since the blade is quite flexible, therefore it is widely used for cutting curves on outside or inside of the wood. It is mostly used in pattern making.

7. *Bow saw.* It consists of a narrow blade about 3 mm to 12.5 mm and 200 mm to 350 mm long with two wooden handles as shown in Fig. 9.22. The blade has about 10 teeth per 25 mm length. A string about 3 metres long is wound round the upper ends of the sides and a short wooden lever is passed between it.

The blade is tightened to the required tension by twisting the string by means of the lever. The bar acts as a fulcrum and restrains the lever to prevent the winding of the string. It is used for cutting quick curves.

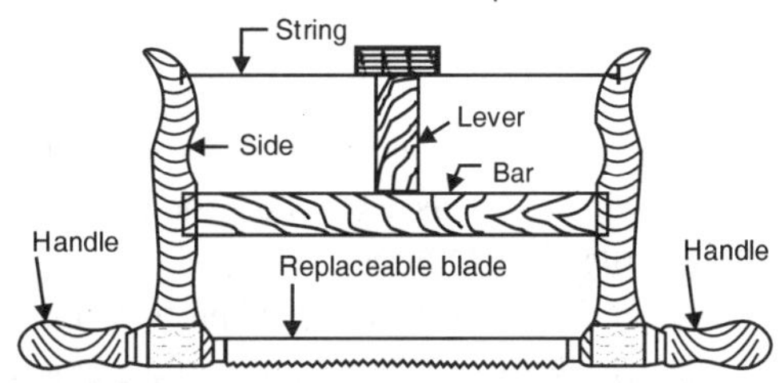

Fig. 9.22. Bow saw.

8. *Pad or key hole saw.* It is smallest saw having a blade tapering in width and 200 mm to 300 mm long, as shown in Fig. 9.23. The width of the blade is about 6 mm near the handle and 3 mm at the tip and has 8 to 15 teeth per 25 mm length. The blade is fastened to the slotted wooden or metallic handle (also known as pad) by means of two screws, so that the length of the blade can be adjusted according to the requirement. This saw is very useful in forming key holes and similar curved work.

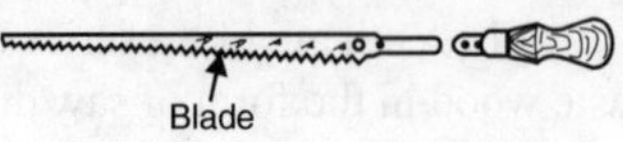

Fig. 9.23. Pad or key hole saw.

9. *Coping saw.* It consists of a thin narrow blade, held in a U-shaped metal frame, as shown in Fig. 9.24. The blade is fixed in the frame with teeth pointing towards the handle and the blade is tensioned by screwing the handle. The length of the blade is about 150 mm to 200 mm. It is used for cutting small radius curves.

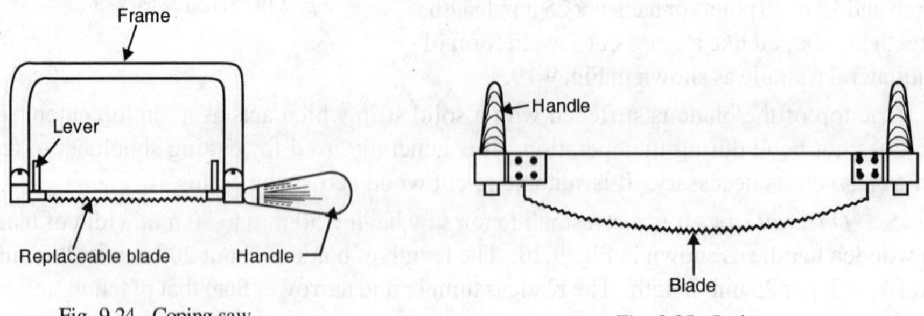

Fig. 9.24. Coping saw.

Fig. 9.25. Jack saw.

10. *Jack saw.* It is a two-man saw with thick untensioned saw blade and two wooden handles as shown in Fig. 9.25. The length of the blade is about 900 mm to 1350 mm and width is more than 75 mm. It is used for rough cuts with round or square timbers, across and inclined to the grains.

9·20 Chisels

The chisels are cutting tools and are used for fitting, shaping and surface decoration. It is a flat thick piece of tool steel, whose one edge is ground to an accurate level to form a cutting edge and the other end is provided with a wooden handle.

The various parts of the chisel, as shown in Fig. 9.26, are as follows :

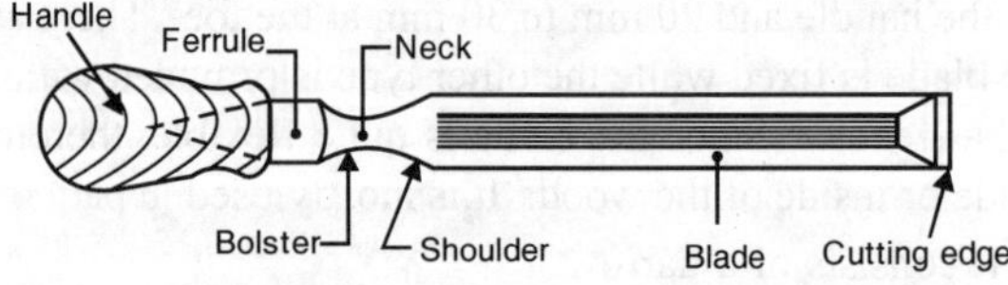

Fig. 9.26. Different parts of a chisel.

(*a*) *Handle.* It is made of hard wood.

(*b*) *Tang.* It is made in the shape of a square pyramid to fix into the handle.

(*c*) *Ferrule.* It is a brass or iron ring fitted at the bottom of the handle to prevent it from splitting during the operation.

(*d*) *Bolster.* It is an enlarged portion of the neck of the chisel blade, just below the tang.

(*e*) *Blade.* It is the main part of the chisel made of tool steel and has a cutting edge at the bottom.

(*f*) *Cutting edge.* It is main working part of the chisel used for cutting the wood.

The size of the chisel is specified by the width of the blade, type and the purpose for which it is used. The different types of chisels are as follows :

1. *Firmer chisel.* The firmer chisel, as shown in Fig. 9.27(*a*), is a medium duty tool used for general work. It may be used either by hand pressure or by blows from a mallet for cutting square recesses *i.e.*, in mortising. The width of the chisel blade varies from 5 mm to 50 mm and length is about 125 mm. The both sides of the body are planed and the end is bevelled to form the cutting edge.

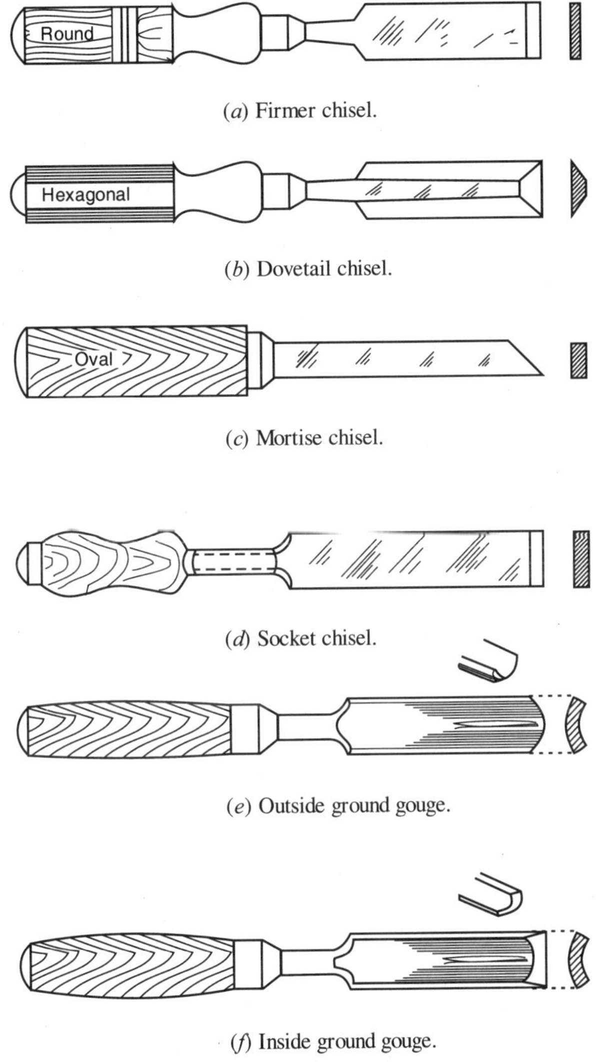

(*a*) Firmer chisel.

(*b*) Dovetail chisel.

(*c*) Mortise chisel.

(*d*) Socket chisel.

(*e*) Outside ground gouge.

(*f*) Inside ground gouge.

Fig. 9.27. Types of chisels.

2. *Dovetail chisel or bevelled edge firmer chisel.* The dovetail chisel or bevelled edge firmer chisel, as shown in Fig. 9.27(*b*), is similar to firmer chisel but with a bevelled edge to reduce the blade thickness on the sides. It is generally used to finish narrow corners where the ordinary firmer chisel is unsatisfactory due to its thick sides.

3. *Paring chisel.* The paring chisels are similar to firmer chisels, but they have a long thin blade. The length varies from 225 mm to 500 mm and width from 5 mm to 50 mm. It is used for deep work, such as housings, where the shorter firmer chisel cannot reach. The edge of the blade may or may not be bevelled. It is manipulated by steady pressure by the hand. It should not be driven by blows.

4. *Mortise chisel.* The mortise chisel, as shown in Fig. 9.27(*c*), is a heavy duty tool used to withstand severe strain as in framing where deep cuts are necessary. An iron ridge is fitted to the end of the handle to withstand the force of heavy hammer used in driving the tool into the wood. The one side of the mortise chisel is plane and the other is gradually tapered towards the cutting edge. The blade width of the chisel varies from 3 mm to 15 mm and thickness from 6 mm to 15 mm. The squared recesses (mortise) through or blind are usually cut by these chisels and hence are called mortise chisels.

5. *Socket chisel.* The socket chisel, as shown in Fig. 9.27(*d*), is similar to firmer chisel but at its end the blade is provided with a socket type construction in place of tang in firmer chisels. In this type of chisel, the wooden handle is inserted into a socket on the end of the blade. This prevents the handle from breaking or splitting as the chisel is to be worked under heavy blows. It functions as a mortising chisel and may be used for all general cutting work. The blade width of the socket chisels vary from 20 mm to 50 mm. The both sides of the chisel are plane and the end bevelled to obtain a cutting edge.

6. *Gouges.* The gouges are chisels with a hollow shaped blade for scooping or cutting round holes. It is mostly used for wood turning. There are two kinds of gouges, *i.e.*, outside ground gouge as shown in Fig. 9.27(*e*) and inside ground gouge or scribing gouge as shown in Fig. 9.27(*f*). The gouges vary in sizes from 5 mm to 25 mm.

9·21 Axe

The *axe*, as shown in Fig. 9.28(*a*), is a cutting tool made of carbon steel. The both sides of the axe are bevelled to form a cutting edge. It is used for splitting wood along the grains for rough work.

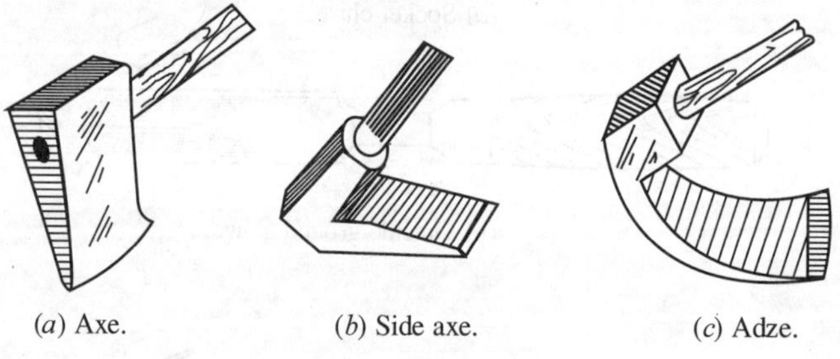

(*a*) Axe. (*b*) Side axe. (*c*) Adze.

Fig. 9.28. Types of axes.

The *side axe,* as shown in Fig. 9.28(*b*), has one side plane and the other bevelled to form a cutting edge. It is used for making the surface roughly plane.

An *adze*, as shown in Fig. 9.28(*c*), has outer face convex, inner face concave and edge is bevelled to form a cutting edge. It is used to chop inside curves and to produce concave surfaces.

9·22 Planing Tools

The planing tools or planes are usually cutting tools used for shaving or smoothing plane surfaces. The various types of planes commonly used in carpentry are as follows :

1. *Wooden jack planes.* A wooden jack plane, as shown in Fig. 9.29(*a*), consists of a body or stock made of beach wood. The bottom face of the stock is called *sole*. The handle is glued into a slot and a hole is formed for the double iron and wedge. The double iron consists of a cutting iron or

blade [Fig. 9.29(*b*)] and a back or cap iron [Fig. 9.29(*c*)] which are made of crucible cast steel. The jack planes are available from 350 mm to 425 mm in length and with blades from 50 mm to 75 mm in width. The back iron helps to support and strengthen the cutting blade and to turn the shavings upwards. It may be noted that without the back iron the wood would be torn. A brass nut is attached to the iron and receives the screw which connects both the plates together. The irons are fixed in the plane at an angle of 45° to the sole, by means of the wedge as shown Fig. 9.29(*d*). The distance that the edge of the cutting iron projects beyond that of the back iron is called the *set* of the iron and it depends upon the character of wood to be planed and the thickness of the desired shaving. The set is approximately 3 mm for soft wood and 1.5 mm for hard wood. These planes are used for rough planing.

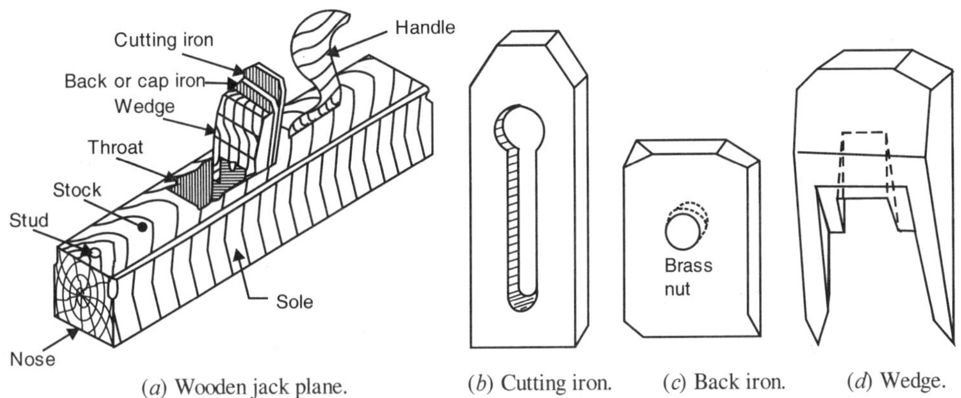

(*a*) Wooden jack plane. (*b*) Cutting iron. (*c*) Back iron. (*d*) Wedge.

Fig. 9.29. Wooden jack plane and its parts.

2. *Metal jack plane.* The metal jack planes, as shown in Fig. 9.30, are used for the same purpose as the wooden jack planes. The metal planes are more durable and produces high degree of finish. The body is made of cast steel, gun metal or malleable iron. It is provided with a wooden handle at the back and a wooden knob at the front for holding it by both hands. The other parts of the metallic jack plane are cutting iron, cap iron for reinforcement, the lever cap and the adjusting screw. These planes are 125 mm to 250 mm in length with blade width from 30 mm to 60 mm.

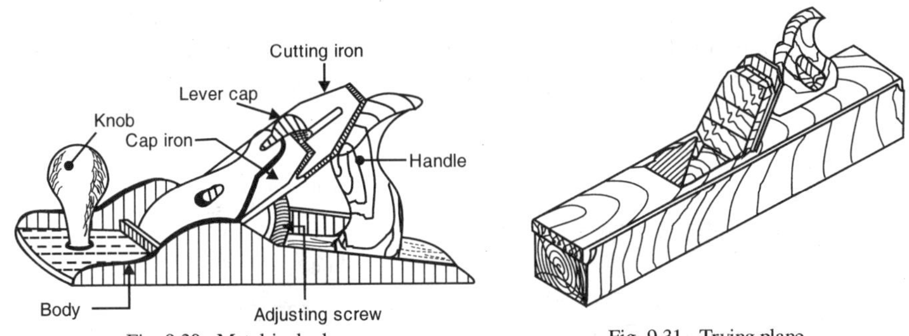

Fig. 9.30. Metal jack plane.

Fig. 9.31. Trying plane.

3. *Trying plane.* The trying plane, as shown in Fig. 9.31, is nothing but a jack plane having the length from 550 mm to 650 mm and blade width of 60 mm. It is used for finishing large surfaces.

4. *Smoothing plane.* A smoothing plane, as shown in Fig. 9.32, is used for smoothing the wooden surfaces after the jack and trying planes have been applied. Its length varies from 200 mm to 250 mm and the blade width is 70 mm. It has no handle. In its operation, the stock or body itself is

held in both hands. The front and back of the body is slightly curved so that it can be easily held in hands.

5. *Rebate plane.* A rebate plane, as shown in Fig. 9.33, is used to cut the edge of a board into the form of a step so that it may fit into another board similarly cut. It has only a single iron and is fixed by a wedge as shown in Fig. 9.33.

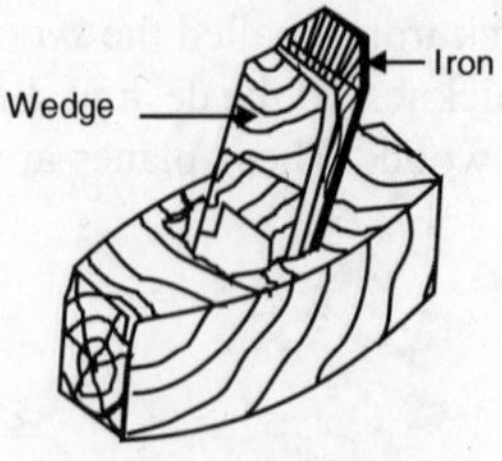

Fig. 9.32. Smoothing plane.

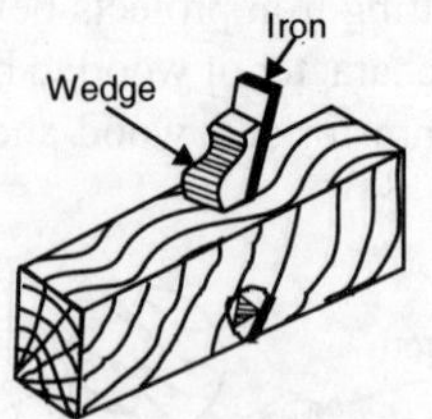

Fig. 9.33. Rebate plane.

6. *Hollow and round plane.* The hollow plane, as shown in Fig. 9.34(*a*), is used for producing convex surfaces on the wood and concave surfaces are formed by the round plane as shown in Fig. 9.34(*b*).

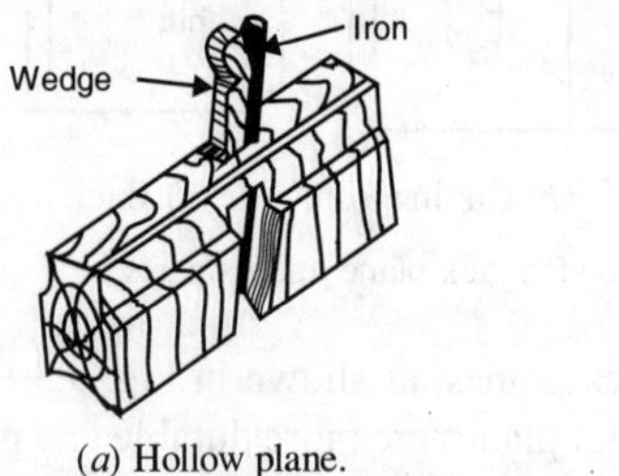

(*a*) Hollow plane. (*b*) Round plane.

Fig. 9.34. Hollow and round plane.

7. *Plough plane.* A plough plane, as shown in Fig. 9.35, is used to cut channels of various shapes to fix the panels in a shutter frame and to produce grooves for a tongue and groove joint. The plane works like a plough that excavates wood. The depth of the groove is controlled by a depth gauge which is fixed on the body of the frame and operated by a thumb screw. These planes are provided with interchangeable blades varying in width from 3 mm to 15 mm.

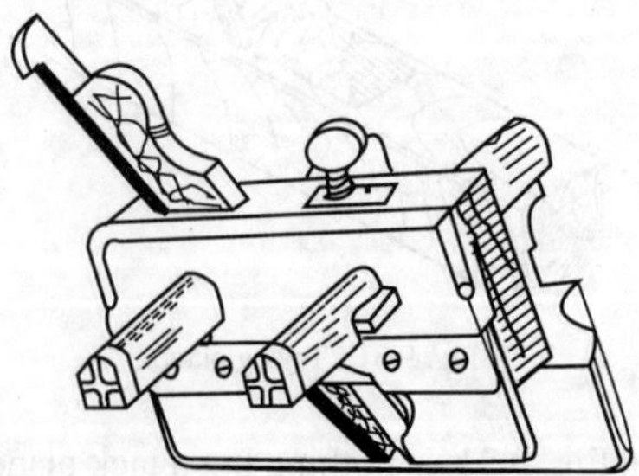
Fig. 9.35. Plough plane.

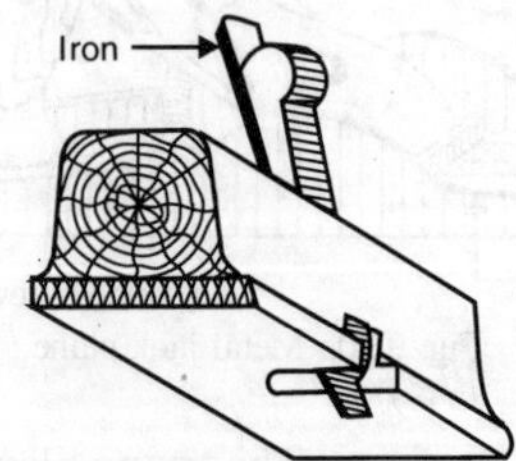

Fig. 9.36. Router plane.

8. *Router plane.* A router plane, as shown in Fig. 9.36, is used for increasing uniform depth grooves formed previously by another tool. The width of the blade varies from 3 mm to 15 mm.

9. *Block plane.* The block plane, as shown in Fig. 9.37, is used for planing against the grain. It can be easily handled because of its small size. It has only a single iron or cutter which is inclined at 12° to 20°. It is especially used for small work which is not readily accessible and for preparing mitres of hard wood mouldings.

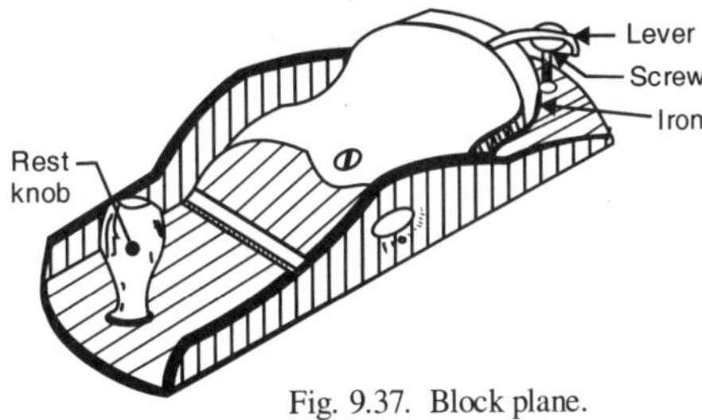

Fig. 9.37. Block plane.

10. *Spoke shave.* The spoke shave, as shown in Fig. 9.38(*a*) and (*b*), is used for planing circular work having quick curves. The tool is held in both hands and is moved along the work in the same direction as the grain.

(*a*) Metal spoke shave.

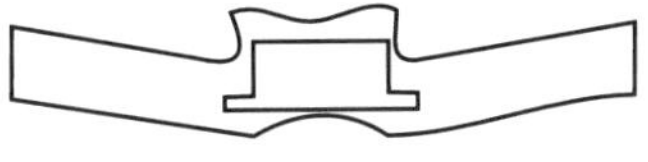
(*b*) Wooden spoke shave.

Fig. 9.38. Spoke shaves.

11. *Draw knife.* The draw knife, as shown in Fig. 9.39, is similar to spoke shave in operation and use. It has a fixed blade whereas the blade of spoke shave is adjustable. The blade is about 200 mm to 300 mm long and has two handles on both sides of the blade.

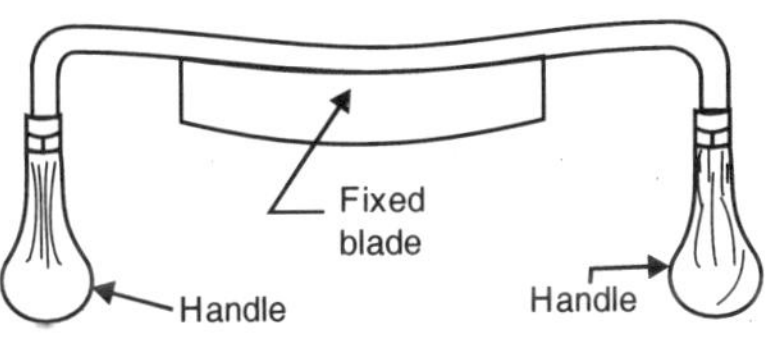

Fig. 9.39. Draw knife.

9·23 Boring Tools

The boring tools are used for producing holes in wood. Following are the various boring tools used by a carpenter.

1. *Bradawl.* A bradawl is used for boring small holes, upto 5 mm diameter and 6 mm deep, in soft wood. It is a steel bar whose bottom end may have a point or flat and a wooden handle is fixed

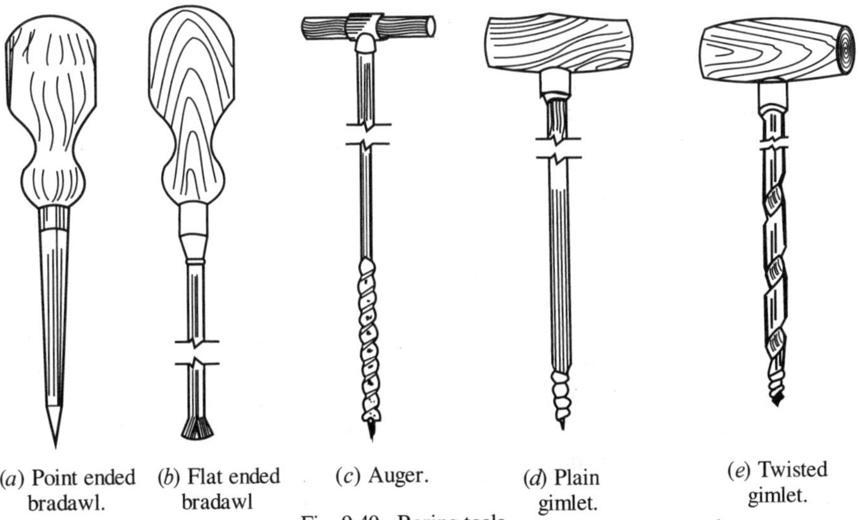
(*a*) Point ended bradawl. (*b*) Flat ended bradawl (*c*) Auger. (*d*) Plain gimlet. (*e*) Twisted gimlet.

Fig. 9.40. Boring tools.

at the top end, as shown in Fig. 9.40(*a*) and (*b*) respectively. In its operation, the point is pressed on the wood and the handle is rotated by both hands, simultaneously pressing downwards. The flat ended bradawl is used to enlarge the holes already made by point ended bradawl.

2. *Auger.* An auger, as shown in Fig. 9.40(*c*), is used to bore holes upto 25 mm diameter in sections upto 150 mm. It is a steel bar having a fluted body for about half of its length from the bottom. A screw point is provided at the bottom of the tool which helps in centring the tool. The top end of the tool is provided with an eye to fix the handle. In its operation, the handle of the tool is held in both hands, and rotated clockwise, simultaneously pressing downward.

3. *Gimlet.* A gimlet is used to bore small holes in wood. There are two types of gimlet, *i.e.,* plain and twisted, as shown in Fig. 9.40(*d*) and (*e*) respectively. A plain gimlet consists of a steel bar carrying a fluted body upto a small length from the bottom where as the twisted gimlet carries a spiral fluted body over the whole length. A screw point is provided at the bottom of the tool which helps in its centring. In its operation, it is also held in both hands and rotated clockwise, simultaneously pressing downwards.

4. *Brace and bits.* The brace is a holding and turning tool for a bit used for boring holes. These are two types of braces commonly used in carpentry *i.e.*, ratchet brace and wheel brace.

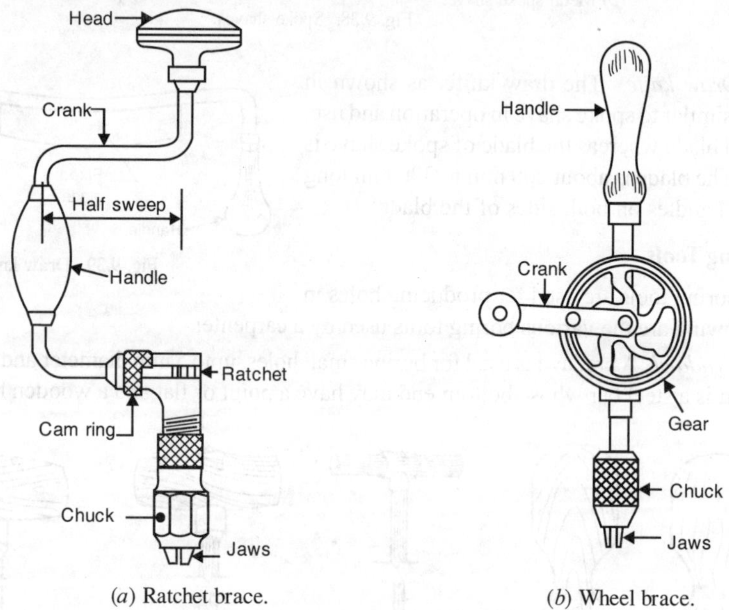

(*a*) Ratchet brace. (*b*) Wheel brace.

Fig. 9.41. Types of braces.

The *ratchet brace,* as shown in Fig. 9.41(*a*) consists of a steel crank having a wooden hemispherical head at the top, a wooden handle in the middle and a chuck for the bit at the bottom end. The ratchet arrangement is provided just above the chuck, so that when the crank is rotated in the reverse direction, the chuck and the bit does not rotate in the reverse direction. It is very useful for drilling holes in corners or difficult positions where the space does not permit complete revolution of the crank. In its operation, the proper size of the bit is gripped in the jaws of the chuck and kept over the place where hole is desired. The head of the brace is pressed downwards by one hand and the handle is rotated by the other hand.

A *wheel brace* (commonly known as hand drill), as shown in Fig. 9.41(*b*), is used for drilling small holes upto 6 mm diameter. It consists of two bevel gear wheels. The crank is fitted at the centre of the larger wheel. A wooden handle is provided at the top of the wheel brace. In its operation, the handle is pressed downwards with one hand and the crank handle is rotated by the other hand. This rotates the larger gear wheel and in turn rotates the smaller gear wheel. Thus the bit gripped in the jaws of the chuck also rotates and a hole is drilled.

Following are the various types of bits used with braces.

(*a*) *Shell bit.* The shell bit, as shown in Fig. 9.42(*a*), is used for boring holes upto 12 mm diameter and which do not require a high degree finish.

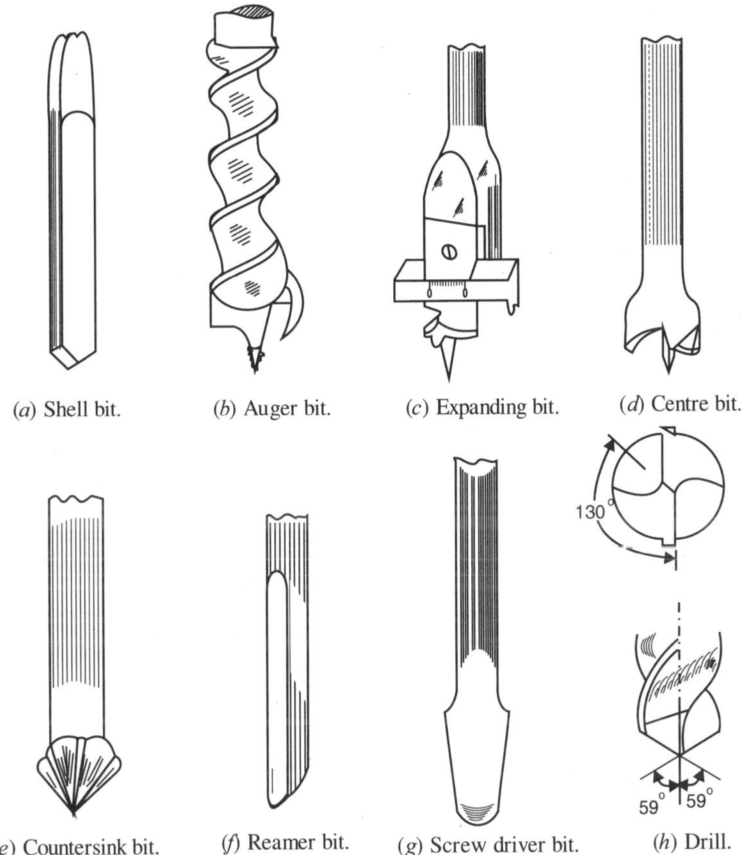

(*a*) Shell bit. (*b*) Auger bit. (*c*) Expanding bit. (*d*) Centre bit.

(*e*) Countersink bit. (*f*) Reamer bit. (*g*) Screw driver bit. (*h*) Drill.

Fig. 9.42. Types of bits.

(*b*) *Auger or twist bit.* The auger bit, as shown in Fig. 9.42(*b*), has a screw point at its bottom and a fluted body with two cutting lips. It is obtainable in sizes from 6 mm to 35 mm diameter. It produces a long, clean and accurate hole either with or across the grain. The shorter auger bits are called *dowel bits* and is used for producing accurate holes for dowel pins.

(*c*) *Expanding bit.* The expanding bit, as shown in Fig. 9.42(*c*), is used for boring holes of any size between certain limits. These bits are available in different sizes varying from 12 mm to 35 mm diameter.

(*d*) *Centre bit.* The centre bit, as shown in Fig. 9.42(*d*), is used for drilling large diameter (but not very deep) holes across the grains. Its diameter varies from 3 mm to 35 mm. The bottom end of this bit resembles that of twist bit except that its one end contains the cutting edge and the other end a spur. The spur is kept larger than the cutting edge which helps in producing a clear and smooth hole.

(*e*) *Countersink or rose bit.* The countersink or rose bit, shown in Fig. 9.42(*e*), is used for shaping a hole to fit the head of a countersunk headed screw.

(*f*) *Reamer bit.* The reamer bit, as shown in Fig. 9.42(*f*), is used for enlarging holes. These are tapered bits shaped like shell bits.

(*g*) *Screw driver bit.* The screw driver bit, as shown in Fig. 9.42(*g*), is employed to force in screw when the pressure on the ordinary screw driver is inadequate. It is a time saver when there are many screws to be installed.

(*h*) *Drill.* The morse drill, as shown in Fig. 9.42(*h*), is used for making screw holes. These drills are used with a wheel brace.

9·24 Striking Tools

The striking tools include hammers and mallets. These are discussed as follows :

1. *Claw hammer.* The claw hammer, as shown in Fig. 9.43, is frequently used by a carpenter. It is made of cast steel and has a head at one end and a claw at the other end. The head is used to drive nails and the claw end is used for pulling out nails from the wood. The size of the hammer is designated by its weight and is available from 250 gm to 750 gm.

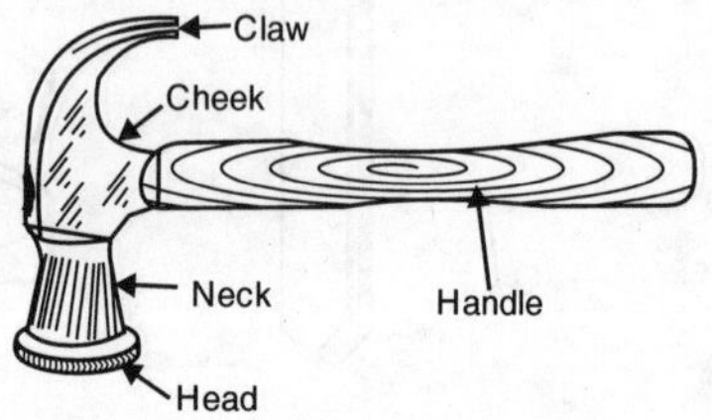

Fig. 9.43. Claw hammer.

2. *Mallet.* It is made of hard wood and may be round or rectangular in shape, as shown in Fig. 9.44. These are used to strike the cutting tools which have wooden handle such as chisels and gouges. The head has an eye (through hole) for fixing the handle to hold it comfortably.

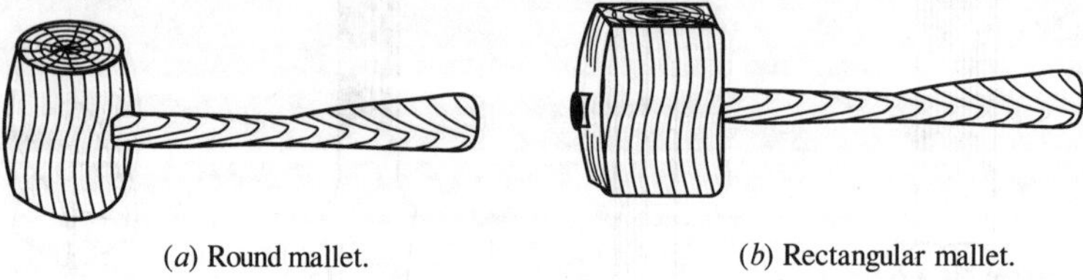

(*a*) Round mallet. (*b*) Rectangular mallet.

Fig. 9.44. Types of mallets.

9·25 Holding Tools

The holding tools are used to hold the work so that it can be cut accurately. Following are the different types of holding tools and devices :

1. *Bench vice.* The bench vice, as shown in Fig. 9.45, is fixed to the carpenter's bench on which work is done. It has two jaws, one of which is kept fixed to the side of the table while the other is kept movable by means of a square threaded screw and a handle. The jaws are lined with hardwood. The size of the vice is indicated by the width of the jaw and the distance it can open.

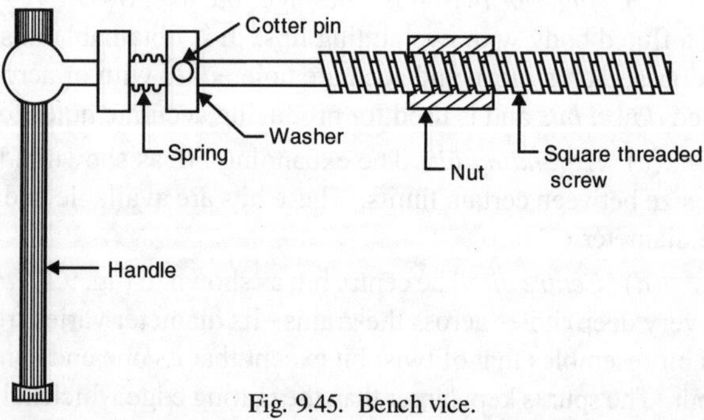

Fig. 9.45. Bench vice.

2. *Bench stop.* The bench stop is a mild steel plate and has teeth at its one end. It is fixed on one corner of the work table. It is used to prevent the work from moving forward when being planed.

3. *Bench hook.* The bench hook, as shown in Fig. 9.46, is a wooden plank of size about 250 mm × 125 mm × 20 mm. Two wooden pieces of size 25 mm × 25 mm × 100 mm are fixed on this plank by means of screws.

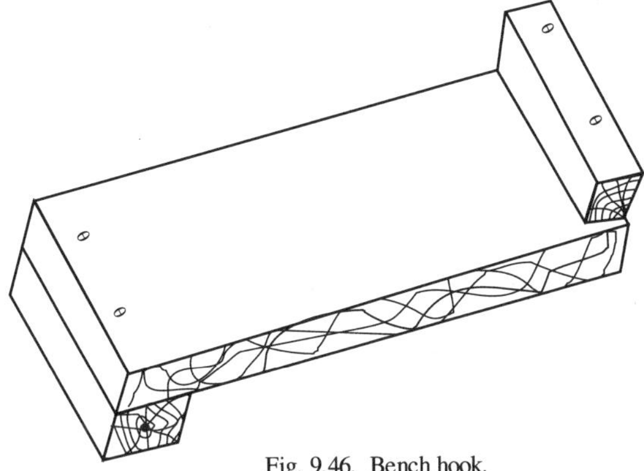

Fig. 9.46. Bench hook.

It is also used to prevent the work from moving forward when being planed.

4. *T-cramp.* The T-cramp, as shown in Fig. 9.47, is also called *sash cramp* or *bar cramp*. It is a straight machined T-section piece of steel with a series of holes along its length at regular intervals. One end of this is forged to form a head through which a long screw passes. The outer end of the screw is provided with a handle and the inner end carries a fixed jaw. Another jaw known as sliding saw is also mounted on the iron which can be adjusted anywhere on a hole according to the size of work and secured at that position with the help of a metal pin. It is specifically used for holding glued pieces tightly. The two pieces, after gluing, are kept between the jaws and the fixed jaw is moved by the handle till the two pieces are held tightly. It is kept in this position till the glue is dry. This cramp is also used for holding unglued pieces firmly for fixing dowels.

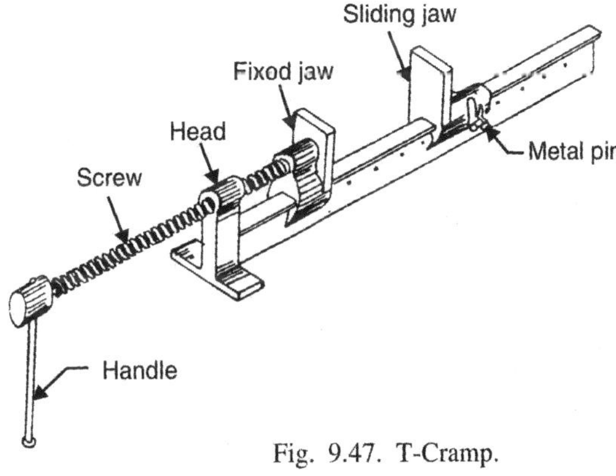

Fig. 9.47. T-Cramp.

5. *G-cramp.* The G-cramp, as shown in Fig. 9.48, is used for small work. One of the jaw is cast with the body while the other jaw is adjusted by means of square threaded screws and a handle. These are available in different sizes varying from 125 mm to 300 mm.

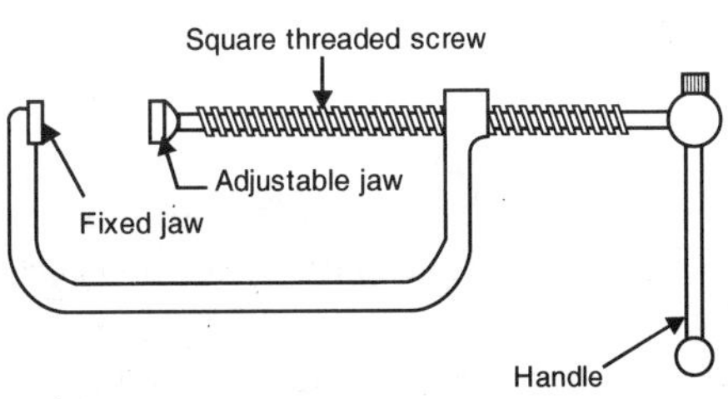

Fig. 9.48. G-cramp.

6. *Hand screw cramp.* The hand screw cramp, as shown in Fig. 9.49, is used for applying pressure over a large area. These cramps have two wooden jaws with the two steel screws, one having right hand threads and the other left hand threads. The common sizes have opening from 150 mm to 350 mm.

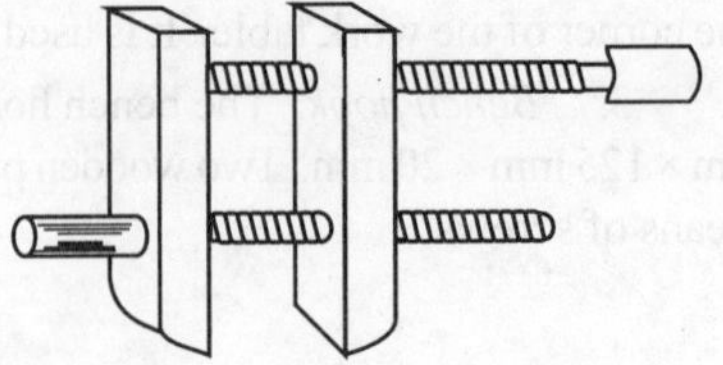

Fig. 9.49. Hand screw cramp.

9·26 Miscellaneous Tools

Following are the various tools which are also used by a carpenter, in addition to the previously discussed tools.

1. *Screw driver.* The screw driver (simple and ratchet type) as shown in Fig. 9.50, are used for driving the wooden screws into the wood or for unscrewing them. These are available in various sizes and shapes.

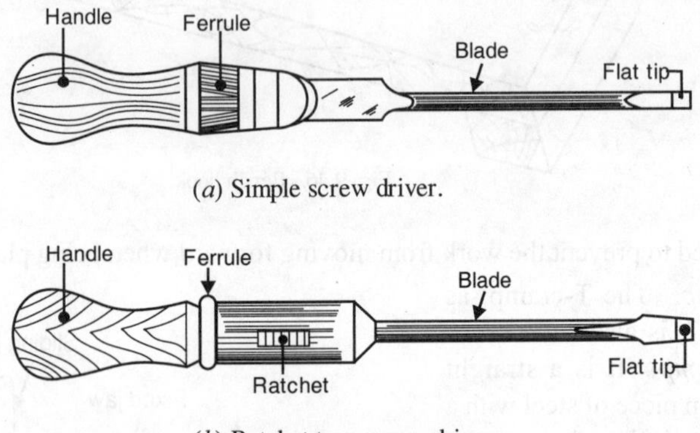

(*a*) Simple screw driver.

(*b*) Ratchet type screw driver.

Fig. 9.50. Screw drivers.

2. *Pincer.* The pincer, as shown in Fig. 9.51, is used for pulling out nails from wood. It has two arms of forged steel hinged crosswise. The two jaws have their outer faces plane and inner faces are bevelled to obtain an edge.

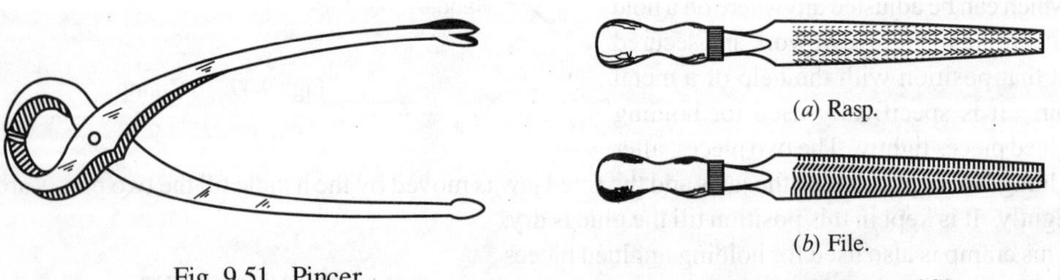

(*a*) Rasp.

(*b*) File.

Fig. 9.51. Pincer.

Fig. 9.52. Rasp and file.

3. *Rasp and file.* A rasp, as shown in Fig. 9.52(*a*), is a file used for finishing. The rasp has sharp cutting teeth on its surface for this purpose. The file, as shown in Fig. 9.52(*b*), is used for removing rasp marks and finally the scratches left by the file are removed with the scraper and glass paper.

4. *Scraper.* The scraper is a thin piece of steel hardened and tempered. It is usually used to remove torn grain and plane marks left after hardwoods have been planed. The pressure of the thumbs on the scraper cutting edge removes fine shavings.

5. *Glass paper.* The glass paper is available in paper sheets to which small particles of glass are struck. The sharp edges of these particles cut the wood. For flat surfaces, the glass paper is folded around a wooden block. It should not be used for loose joints. It is available in various grades like 0,00,1 etc.

6. *Oil stone.* The oil stone is an essential tool used for providing keen edges on cutting tools. The oil stones may be artificial or natural stones. The carborundum and India are the best artificial stones where as Washita and Arkansas are the natural stones. These stones are available in coarse, medium or fine grades.

9·27 Setting and Sharpening of Tools

The setting and sharpening of saws, chisels and planes is done as discussed below :

1. *Setting and sharpening of saws.* The saws get blunt after using and their teeth have to be resharpened and reset properly. The sequence of operations for setting and sharpening a saw are as follows :

(*a*) The saw is fixed firmly in the clamp as shown in Fig. 9.53(*a*).

(*b*) The points of the teeth are put in alignment with a flat file as shown in Fig. 9.53(*a*). This process of levelling the teeth is called *topping*.

(*c*) The topped teeth are shaped with a triangular file held at right angles to the plane of the saw blade and moved in a straight line forwards and backwards.

(*d*) After shaping, the teeth have to be set. The setting of teeth consists of bending the alternate teeth of the saw in opposite direction. This is done in order to make the cut wider than the thickness of the blade.

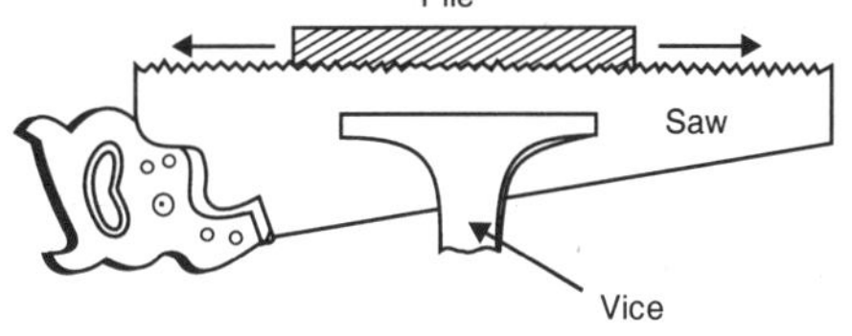

(*a*) Levelling of teeth.

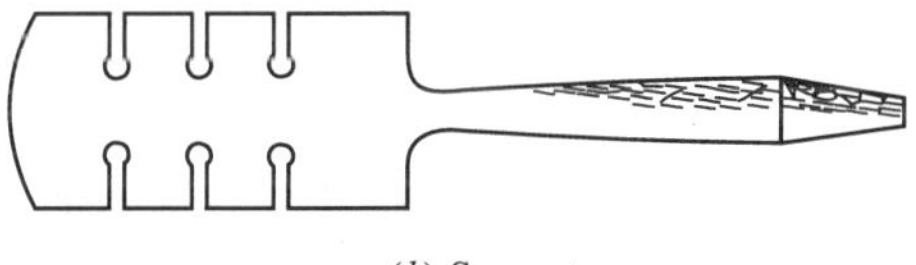
(*b*) Saw set.

Fig. 9.53. Sharpening methods of saws.

There are many ways of setting the teeth of a saw, but the most common method is by using a gate or lever type hand set, as shown in Fig. 9.53(*b*). It consists of a steel plate and a handle. The plate has a number of gaps or gates along its edge, according to the thickness of the blade of the saw. The saw set is used by slipping over a suitable gap over the blade tooth and then bent with the help of a handle. The amount of set depends upon the type of work to be done.

(*e*) The sides of the teeth are dressed with an oil stone to remove only burrs etc.

(*f*) Finally, the sharpening is done with a triangular file, but at this time the file is held pointing upwards at an angle of about 10° to the horizontal and pointing crosswise towards the saw handle at about 80° to the line of blade.

2. *Setting and sharpening of chisels.* It is a fact that no good job can be done with a blunt or badly sharpened chisel. Hence it is necessary that a carpenter should know its sharpening technique for producing good work. While sharpening on a grind stone or oil stone, the following two angles are maintained.

(*a*) The ground angle (*X*) which varies from 20° to 25°, is made on a grinder or grind stone as shown in Fig. 9.54(*a*) and (*b*).

(*b*) The sharpening angle (*Y*) which varies from 30° to 35°, is made with oil stone as shown in Fig. 9.54(*c*) and (*d*).

When the chisel blade is ground on a pedestal grinder, it is held in both hands with its bevelled edge facing the grinding wheel. The blade is moved sideways on a rotating grinding wheel at an angle of 20° to 25°. During this operation, the blade should be lightly pressed against the wheel. The blade is frequently dipped in water during grinding.

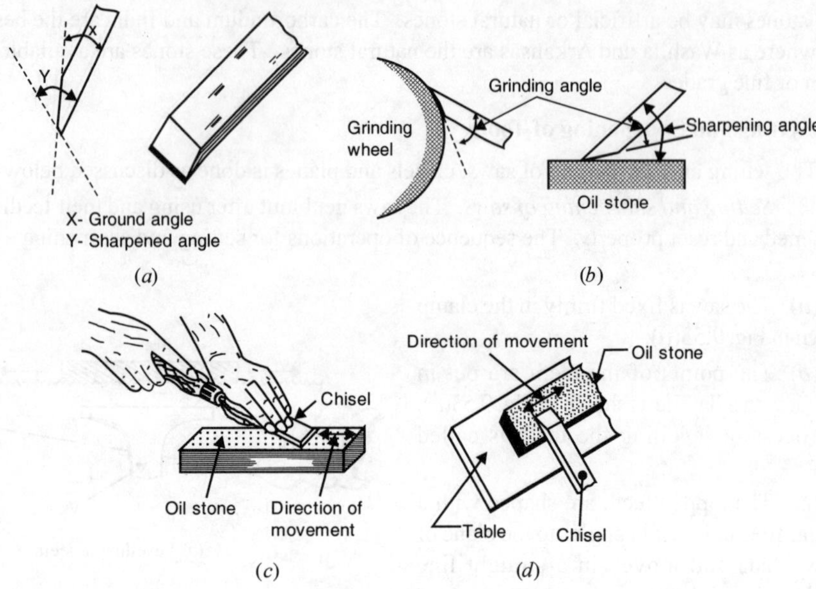

Fig. 9.54. Sharpening methods of chisels.

After grinding, the blade is held at an angle of 30° to 35° on an oil stone and moved to and fro to sharpen the edge.

The gouges are also ground on grind stone and then sharpened on a flat oil stone having curved shape.

3. *Setting and sharpening of planes.* The setting or cutting adjustment of planes is most important. The efficient working of planes depends upon the following factors :

(*a*) The cutting angle of the plane. In case of planes with back iron, the cutting angles are 47.5° and 42.5° respectively.

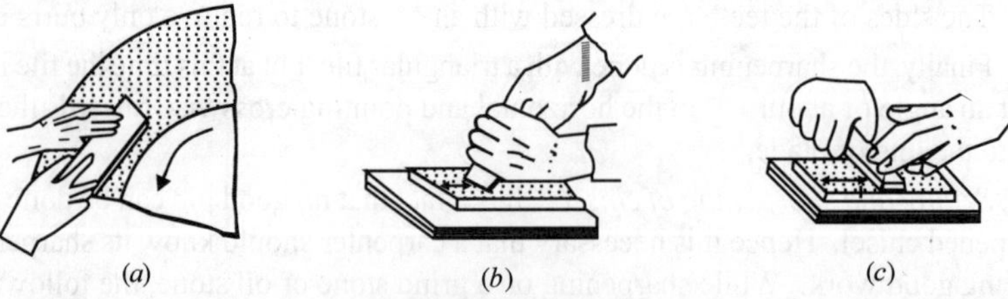

Fig. 9.55. Sharpening methods of planes.

(*b*) The space between the cutting edge and body of the plane. For jack plane, it should be atleast 3 mm so that thick shavings can easily pass through the opening and 0.4 mm for very thin shaving.

(*c*) The back iron is set 3 mm to 0.4 mm from the cutting edge depending upon the coarseness or fineness of the work.

The blade of planes is sharpened in the similar way as the chisel blade. First of all, the cutting angle is maintained by grinding the iron blade on the grinding wheel. During grinding, keep the stone wet to avoid spoiling the temper of the tool by heat. The grinding operation is shown in Fig. 9.55(*a*).

After grinding, the edge is then sharpened on an oil stone, as shown in Fig. 9.55(*b*), by pushing backwards and forwards in an inclined position. The burr formed on the back side of the blade is removed by moving the back sideways on the stone as shown in Fig. 9.55(*c*).

9·28 Carpentry Processes

The following principal processes are used in the carpentry and joinery work.

1. *Marking and measuring.* It is the process of setting out dimensions on a piece of wood for required shape. It is done to guide the operator for further operations. These dimensions, are taken from a model or working drawings. The marking is done with the help of marking tools like pencil, marking gauge, marking knife etc. while measuring is done with the help of measuring tools like tape, try square etc. Before marking, one of the edge and face is planed. The angles are marked with bevel square. While marking, care should be taken to avoid defects such as large knots and shakes.

2. *Sawing.* It is the process of cutting wood of required sizes and shapes such as straight, inclined or curved. The wood may be sawn along the grains or across the grains. In order to cut the short wood or board along the grains, it is first fixed in a vice to prevent vibration and the wood is moved up in the vice as the sawing proceeds. This operation is shown in Fig. 9.56. It may be noted that the waste wood is kept on the right side so that the sawing line can be plainly seen. The wood is always sawn on the waste side. The left hand thumb is placed on the blade to guide the saw. When sawing long wood or board, it is supported on trestlers or boxes and the cut is first started at a low angle and then straighten upto 45°. The cut is tested occasionally with a try square to see if it is straight. As the sawing proceeds, move the supports to saw between them. A wedge is inserted between the saw cut for ease in cutting. The sawing operation across the grains is shown in Fig. 9.57.

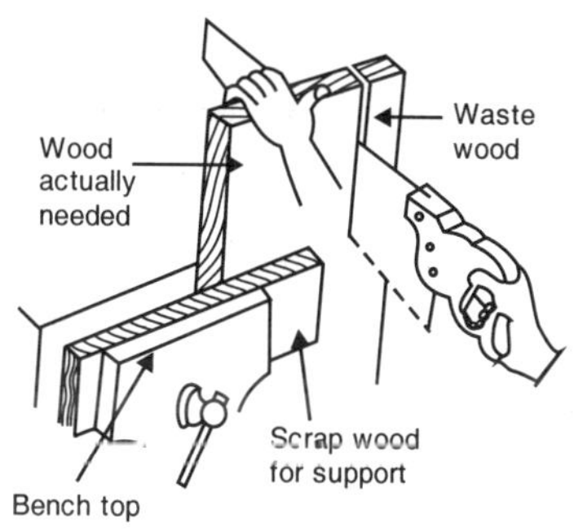

Fig. 9.56. Operation for sawing along the grains

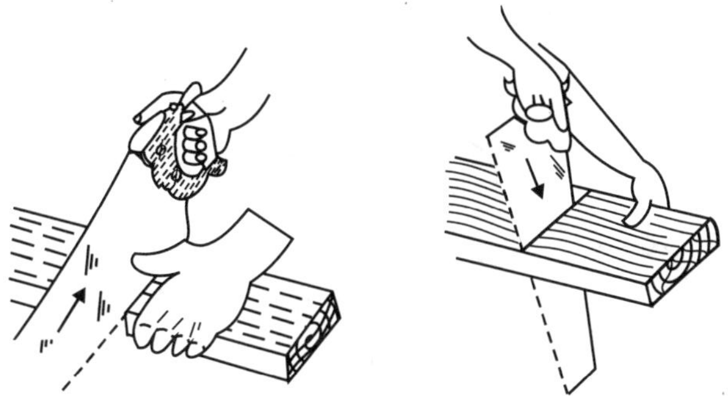
Fig. 9.57. Operation for sawing across the grains.

3. *Planing.* It is the operation used for obtaining a true surface of wood by using a planer. The planing is done along the grains otherwise a smooth surface cannot be achieved.

The planing involves the following sequence of operations :

(*a*) Select that surface of wood which appears to be best.

(*b*) Plane it flat using jack plane in as few strokes as possible. It may be remembered that every shaving off makes the wood thinner. In planing, the pressure is applied by left hand to keep the plane level with the wood and the right hand pushes forward, as shown in Fig. 9.58(*a*). The pressure is relieved during return stroke as shown in Fig. 9.58(*b*).

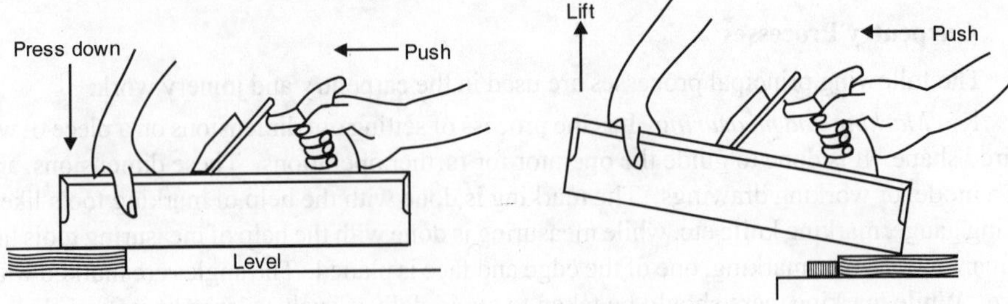

(*a*) Pressure applied during forward stroke. (*b*) Pressure relieved during return stroke.

Fig. 9.58. Planing processes.

(*c*) Check the planed surface with a steel ruler or try square on length, across diagonals and across width.

(*d*) After the best side has been planned true, it should be marked with a face side mark. The trued face side is now the working side.

(*e*) Now plane the edge straight and square. It should be tested at several places on the edge with a try square. When a square and straight edge has been obtained, it is marked with a cross to represent face edge mark.

(*f*) After planing the face side and face edge, the second edge and second side is planed.

4. *Chiselling.* It is the operation used for cutting small stock of wood to produce the desired shape. The chiselling consists of the following operations :

(*a*) First of all, the marking is done on the side where chiselling is to be performed.

(*b*) Saw cuts are made on the side of the lines upto the required depth which prevent the groove being too wide.

(*c*) Hold the chisel flat side up, as shown in Fig. 9.59(*a*), and slice off the wood in thin shavings. A mallet may be used to drive the chisel in hard woods.

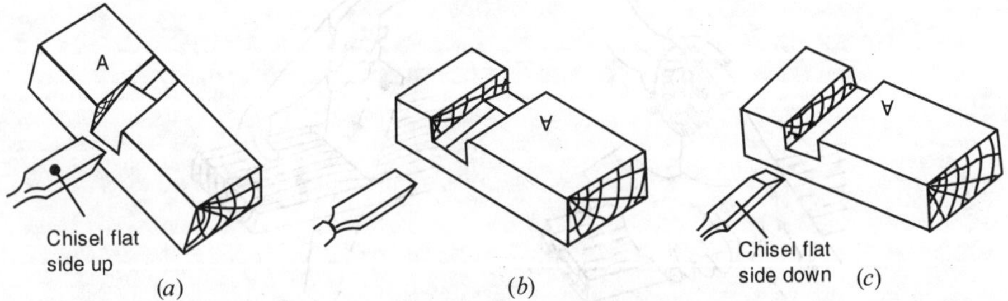

Fig. 9.59. Operations of chiselling.

(*d*) Now turn the piece around and repeat the chiselling, as shown in Fig. 9.59(*b*), from the other edge.

(*e*) There will remain a middle hump. In order to remove this, hold the chisel flat side down and apply pressure by hand or mallet, as shown in Fig. 9.59(*c*). The few cuts may be done by holding the chisel near the blade end and slicing side ways.

5. *Mortising and tenoning.* The mortising is the operation of producing a mortise *i.e.*, a cavity or recess in a wooden piece as shown in Fig. 9.60(*a*). It is done by firmer or mortise chisels and a mallet. The tenoning is the operation of producing tenon. A tenon is a projected piece of wood which fits into the corresponding mortise, as shown in Fig. 9.60(*b*). A tenon saw is used for cutting the excess wood on the sides to produce the tenon. It may be noted that when a chiselled hole or mortise has to be made, cutting must be done from both sides of the wooden piece in order to avoid splintering.

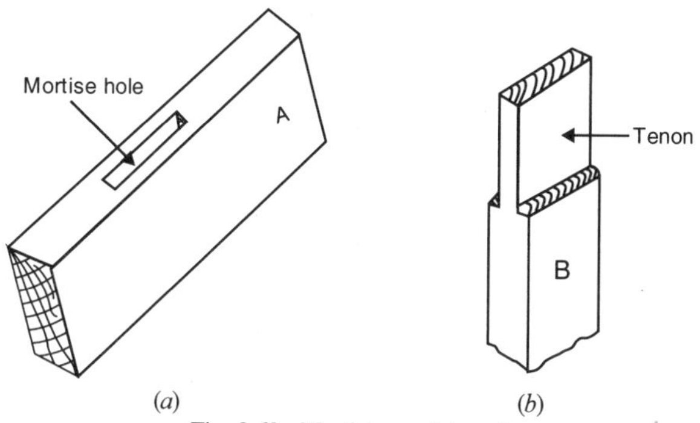

(*a*) (*b*)

Fig. 9.60. Mortising and tenoning.

6. *Grooving and tongueing.* The grooving is the process of producing channels in a wooden piece and tongueing is the process of producing corresponding projections of wood to fit into channels, as shown in Fig. 9.61. The grooves are cut with a plough plane and the tongues are made with a moulding plane.

The grooves and tongues are usually cut on the edges of planks and boards to join them together for increasing the width.

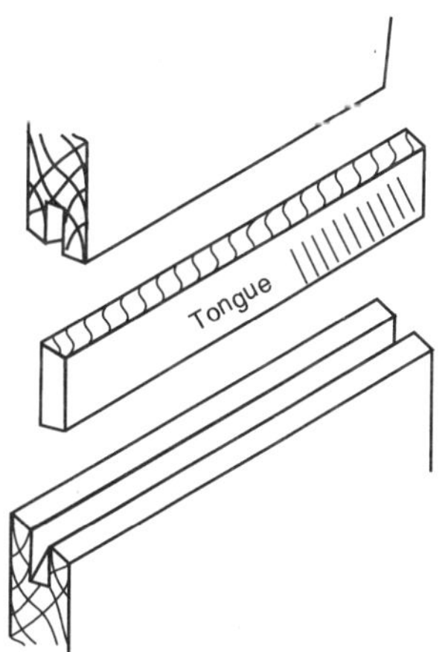

Fig. 9.61. Grooving and tongueing process.

7. *Recessing.* It is similar to mortising operation but the cavity produced is blind and not through as for mortising.

8. *Boring.* It is process of producing round holes, through or blind, in wood. The small holes are made with a bradawl and gimlet where for larger holes the braces and bits or drills are required. The boring operation consists of:

(*a*) The centre of the hole is marked on the work and it is firmly secured in a holding tool.

(*b*) The boring tool is turned constantly in one direction and after a short time it is withdrawn to remove the waste wood from the hole. It is again put in operation till the hole of required depth is made.

9. *Rebating.* It is the process of producing a rebate or recess along the edge of a wooden piece. The rebating is done by a rebating plane and may be either longitudinal, cross or both.

10. *Moulding.* It is the process of cutting surface along the length of a wooden piece. It is done by a moulding plane. It is used in producing decorative work.

9·29 Carpentry Joints

The following joints are important from the subject point of view:

1. *Lap joint.* The lap joint is formed by cutting identical steps over a considerable length and held in position by glue. These joints are used to secure the corners and intersections of the framing and at the same time keep all the face flush, *i.e.*, in the same plane. The lap joints are usually glued and may be reinforced by means of concealed screws. It may be noted that dovetail lap joints are stronger than other lap joints and offer greater resistance to tension. The various lap joints as shown in Fig. 9.62, are T-lap joint, dovetail lap joint, corner lap joint and cross lap joint.

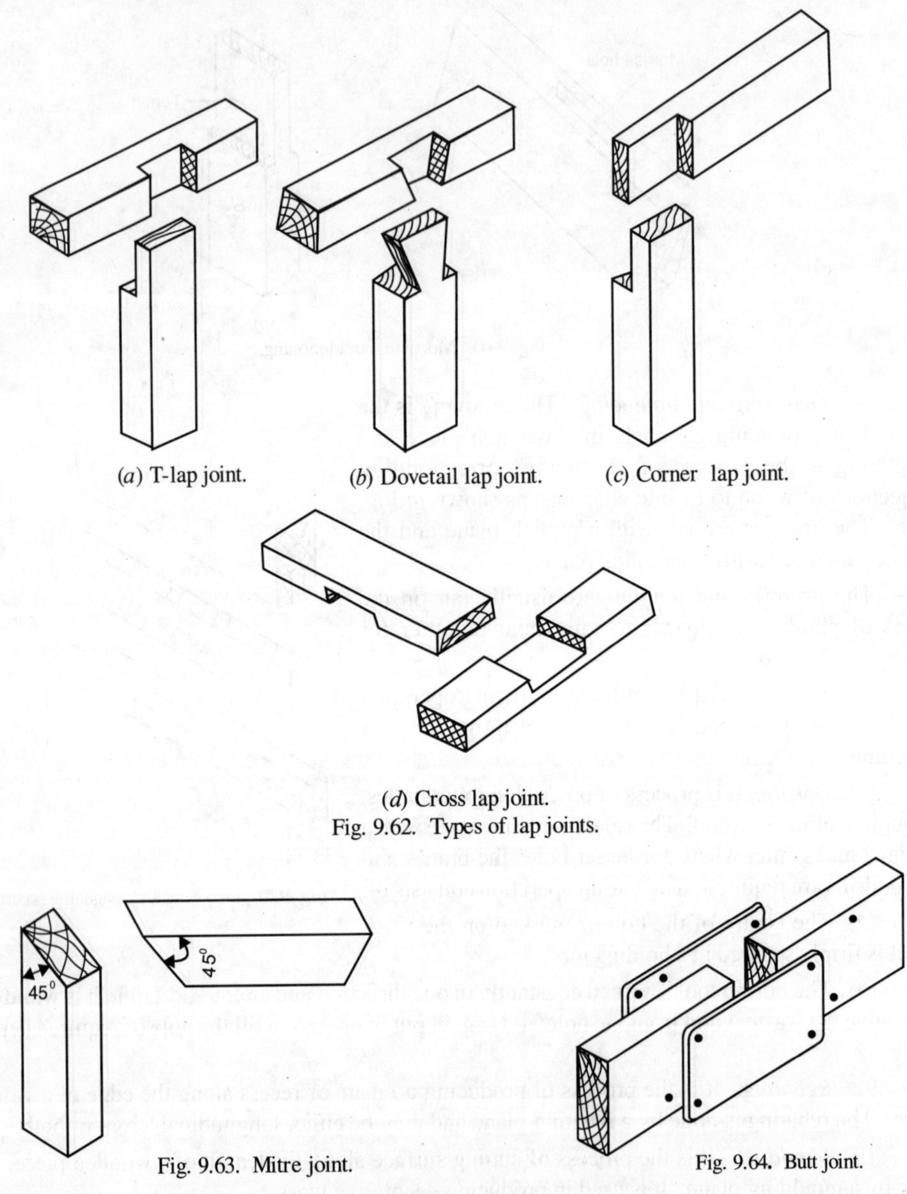

(*a*) T-lap joint. (*b*) Dovetail lap joint. (*c*) Corner lap joint.

(*d*) Cross lap joint.

Fig. 9.62. Types of lap joints.

Fig. 9.63. Mitre joint. Fig. 9.64. Butt joint.

2. *Mitre joint.* This joint is formed by cutting the ends of wooden members at an angle and then joining them by dowels, nails or screws. This type of joint is mostly used in light work such as

in photo-frames. The mitre joint is shown in Fig. 9.63.

3. *Butt joint.* The butt joint is formed by joining the wooden members edge to edge by means of glue or by screwing fish plates, (*i.e.*, steel plates) on two opposite sides, as shown in Fig. 9.64.

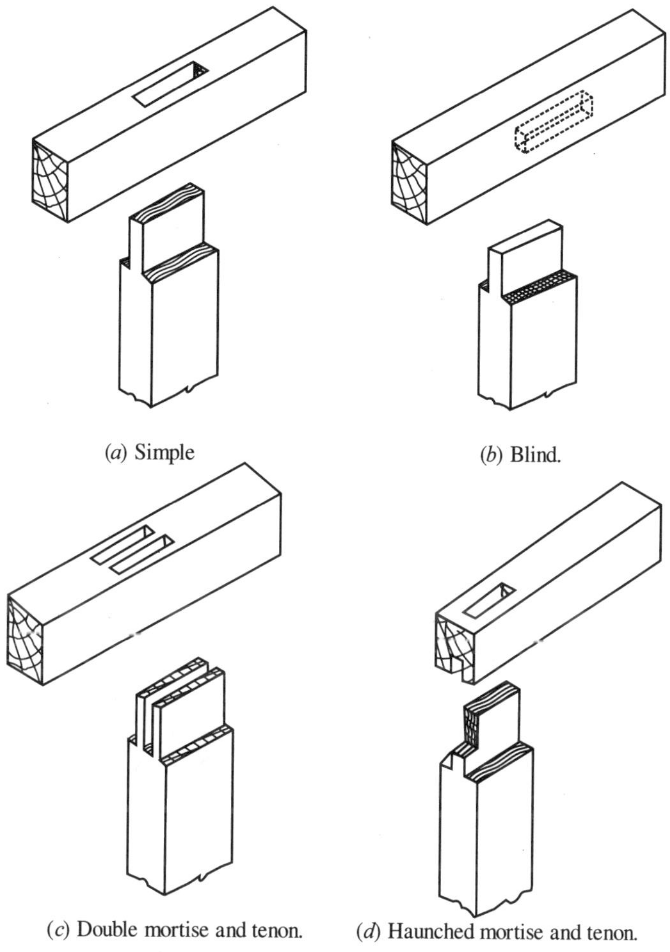

(*a*) Simple (*b*) Blind.

(*c*) Double mortise and tenon. (*d*) Haunched mortise and tenon.

Fig. 9.65. Types of mortise and tenon joints.

4. *Mortise and tenon joint.* This is the most common type of the joint used in many constructions. It consists of a rectangular peg (called tenon) which fits into a corresponding rectangular hole (called mortise). This joint proves to be one of the strongest joint, if it is provided with due reinforcement by means of glueing and dowels etc. The various mortise and tenon joints, shown in Fig. 9.65, are simple, blind, double mortise and tenon, and haunched mortise and tenon joints.

5. *Groove and tongue joint.* This joint is often used in edge to edge joining of wood members, such as drawing boards, door panels, floor boards etc. This is easy joint to make and assemble. The width of the tongue should not be more than one half of the thickness of the board. A groove and a tongue joint is shown in Fig. 9.66.

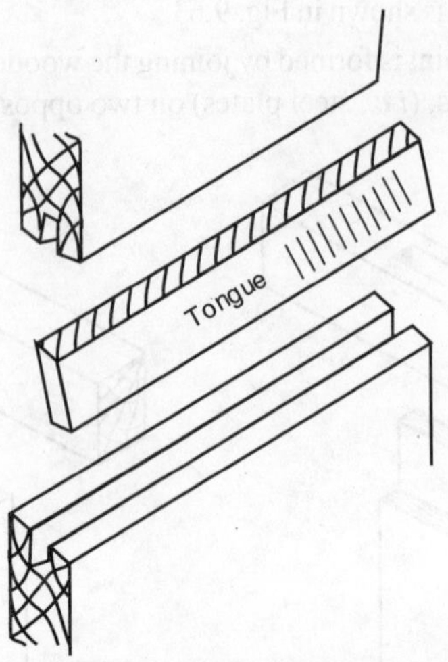

Fig. 9.66. Groove and tongue joint.

6. *Dovetail joint.* The dovetail joint is used because of their high strength and durability. It is the most common type of joint used in furniture and box making.

These joints are made in number of varieties and styles, such as common or through dovetail, lapped dovetail, lapped secret dovetail and secret dovetail etc. The various dovetail joints are shown in Fig. 9.67.

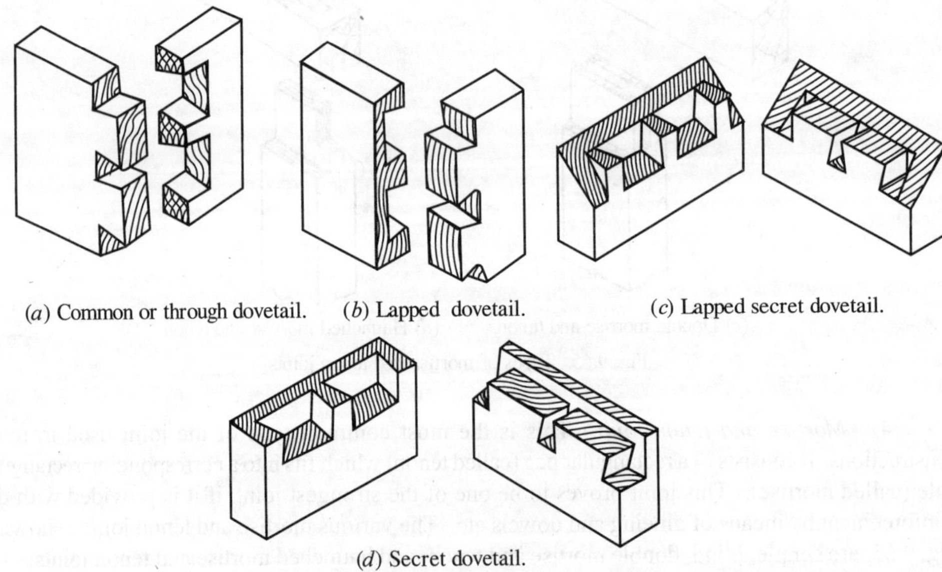

Fig. 9.67. Types of dovetail joint.

7. *Briddle joint.* The briddle joint is reverse of mortise and tenon joint. It is also called an open mortise and tenon joint. This joint is used to connect a rafter to a tie beam. A tenon is provided in the tie beam and a recess in the rafter end. The two types of briddle joints, shown in Fig. 9.68, are corner briddle joint and T-briddle joint.

8. *Dowel joint.* The dowels are thin, small and round hard wood sticks.

This joint is often used as a substitute for mortise and tenon joints, for securing loose fittings and for strengthening joints in circular, square and irregular work. The typical dowel joint is shown in Fig. 9.69.

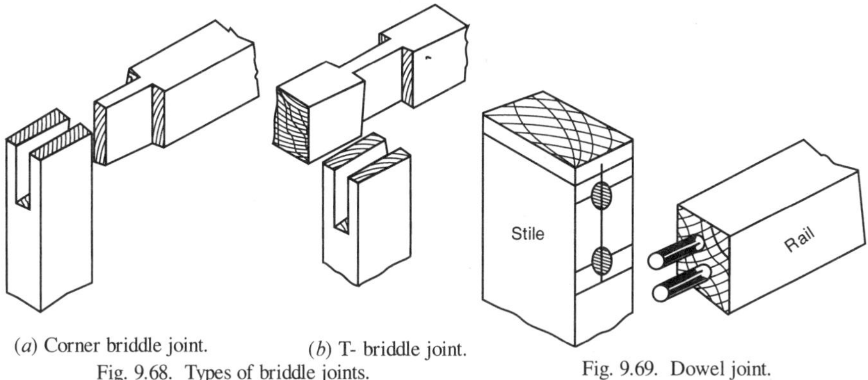

(*a*) Corner briddle joint. (*b*) T- briddle joint.
Fig. 9.68. Types of briddle joints.

Fig. 9.69. Dowel joint.

9·30 Wood Working Machines

In addition to the hand tools as discussed in the previous articles, a large number of power driven machines are used for large scale production and accuracy of work. The size and capacity of the machines depends upon the quality of work required, the quantity of work to be handled and the type of wood to be worked upon. The machines must be well guarded to protect the worker from the hazards of operation. Following are some of the commonly used machines :

1. *Wood working lathe.* It is one of the most important machines used in a carpentry shop. It is employed primarily for turning wooden parts to give round and cylindrical shapes. The irregular shapes can also be easily turned with the help of special form tools.

A wood working lathe, as shown in Fig. 9.70, resembles an engine lathe which is most frequently used in a machine shop. It consists of a cast iron bed, head stock, tail stock, live and dead centres, tool rest, cone pulley, motor and a speed control device. The workpiece is rotated between the two centres (*i.e.*, live centre on the headstock and dead centre in the tail stock) and the turning is done by a tool held in the tool post. The size of the wooden lathe is specified by the maximum length of the job which can be turned between centres and the swing *i.e.*, the maximum diameter of the work that can be turned.

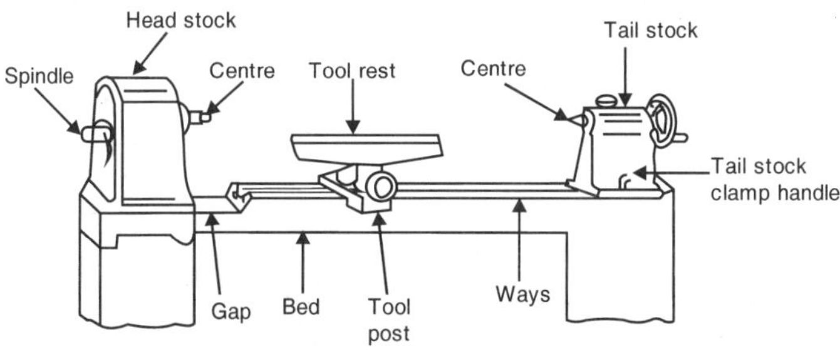

Fig. 9.70. Wood working lathe.

2. *Circular saw.* It is one of the basic machines used in carpentry shop. It is also called a *table* or *bench saw*. It is used for large number of operations like ripping, cross-cutting, levelling,

mitering, grooving and rebating. A circular saw consists of a cast iron table, a circular cutting blade, cut-off guides, main motor, saw guide, elevating hand wheel and tilting hand wheel as shown in Fig. 9.71. The ripping fence can be adjusted to any desired distance to act as a gauge for cutting boards of required width. The work is held on the table and moved against the circular saw to perform different operations. The saw hand wheel is used to adjust the height of the saw above the surface of the table.

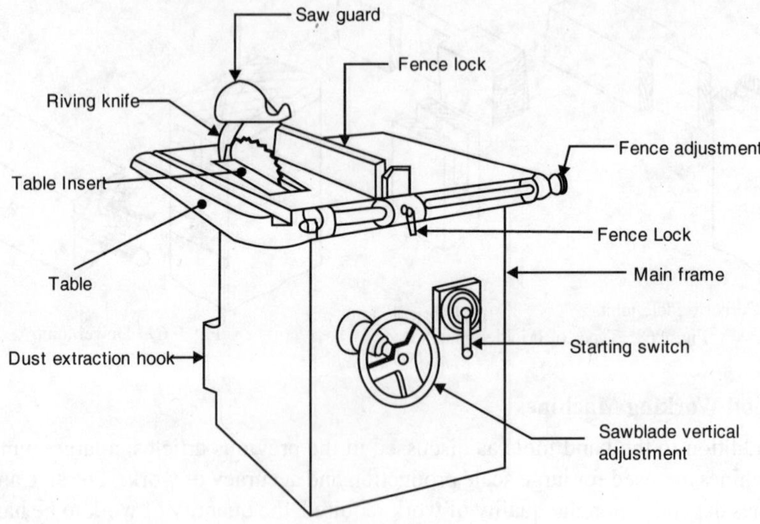

Fig. 9.71. Circular saw.

Many saws usually have provisions for tilting the table upto 45° to enable the machine to cut at different angles required during mitering, levelling etc. The size of the circular saw is determined by the size of the saw blade.

3. *Band saw.* The band saw is shown in Fig. 9.72. It is one of the most versatile machines. It consists of two wheels of equal diameter, a frame table, saw guider, saw tension arrangement, wheel guard and an endless steel band or blade having teeth over it. This blade moves over the two wheels just in the same way as a flat belt moves over the two pulleys. The width of the blade varies from 6 mm to 25 mm and its thickness varies from 1 to 2 mm.

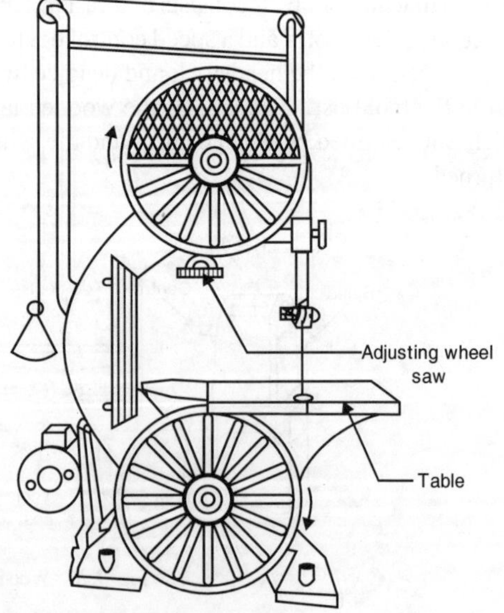

Fig. 9.72. Band Saw.

The band saw may be either vertical or horizontal. In a vertical band saw, the two wheels are arranged one above the other and the moving band cuts the stock in the down ward direction. In a horizontal band saw, the two wheels are placed side by side on the table and the moving band cuts the stock in the horizontal direction. The table can be tilted to any angle for angular cuts.

4. *Jig saw or Scroll saw.* It is used to cut fine internal and external curves on thin wooden

pieces. It is actually a band saw but of much smaller size. It consists of a base, frame, table, upper and lower chucks, guide assembly and blade. The shape of the blade is like a hand hacksaw blade. The blade is held between the chucks with its teeth pointing downwards. The rotary motion of the motor spindle is converted into vertical up and down motion of the blade, cutting the wood only in the downward stroke. The size of the jig saw is specified by the distance between its arm and the blade.

5. *Radial saw.* In the radial saw, the motor and blade are suspended above the work. The blade can be raised, lowered, turned or tilted to make different types of cut. This type of saw facilitates a clear vision of the cutting action of the saw blade. The radial saw requires three adjustments. The depth of cut is determined by lowering or raising the arm at the shoulder. The direction of the cut is set at the elbow by swiveling the motor to any angle upto 360° to cross cut, rip and miter. The motor and blade can be tilted to any angle for the bevel cut. The size and capacity of the radial saw is designated by the size of the saw blade and rating of the motor.

9·31 Jointer

A jointer, as shown in Fig. 9.73, is a power driven machine which does the work of a hand plane. It is used for planing straight edges and surfaces of boards. It does a very accurate work and is best suited for production work. It consists of a heavy cast iron base, front and rear tables, cutter

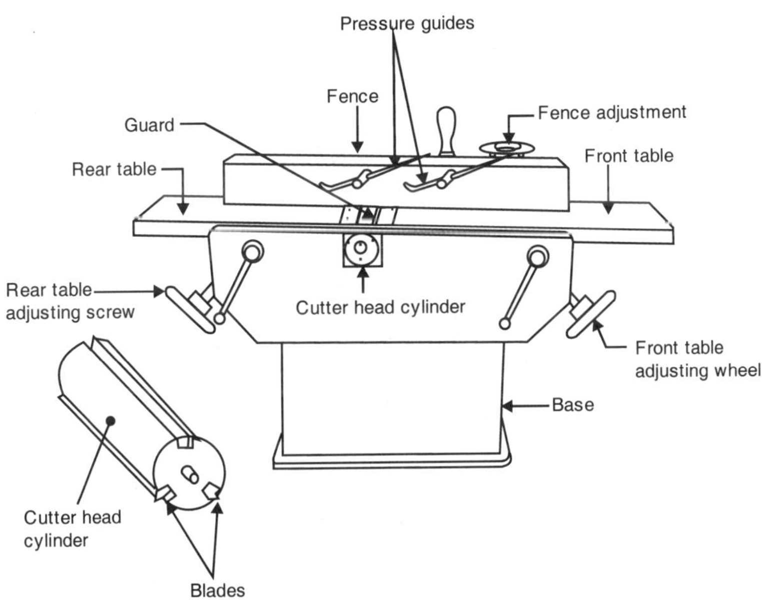

Fig. 9.73. Jointer.

head, fence and a safety guard. The long revolving cutter head is fitted with two or three cutting knives. The speed of the cutter head ranges from 3600 r.p.m. to 6000 r.p.m. A cutter head with three knives, if turned at a speed of 5000 r.p.m. will give 15000 cuts per minute. In order to obtain a smooth surface, the work should be fed slowly. The size of the jointer is specified by the length of knife. The common size of the knife varies from 100 mm to 900 mm.

9·32 Wood Planer

The wood planer is a machine in which wooden planks and boards are planed to a desired thickness at a comparatively faster rate. The main parts of a wood planer are frame, table, in-feed and out-feed rollers, friction rollers, cutter head, pressure bar etc. A board or plank to be planed is fed by means of rollers into the throat of the planer against a revolving cutter head. The cutter head is mounted on an overhead shaft which is adjustable for regulating the depth of cut. When the cut is complete, the board or plank is released on the opposite side of the table. It is necessary to hold the board as it leaves the cutter. In order to accomodate large and heavy stock, the table of the planer is made much wider and longer than that of a jointer. The size of the planer is specified by the maximum thickness and width of stock that it can plane.

9·33 Wood Shaper

It is also called moulding or wood shaping machine. It is a vertical spindle machine used for cutting, moulding and shaping grooves of various shapes and sizes. A wood shaper may be of single spindle or double spindle type. Fig. 9.74 shows a small bench type single spindle shaper. The single spindle shaper is used for light work whereas a double spindle type is used for large work and mass production work. The principal parts of the shaper are frame, vertical spindle, cutter head, table, spindle height adjusting wheel, fence and guard. The cutter knives are fitted in the cutter head, mounted on the spindle, which revolves at a very high speed ranging from 8000 r.p.m. to 10000 r.p.m., making smooth and clean cuts.

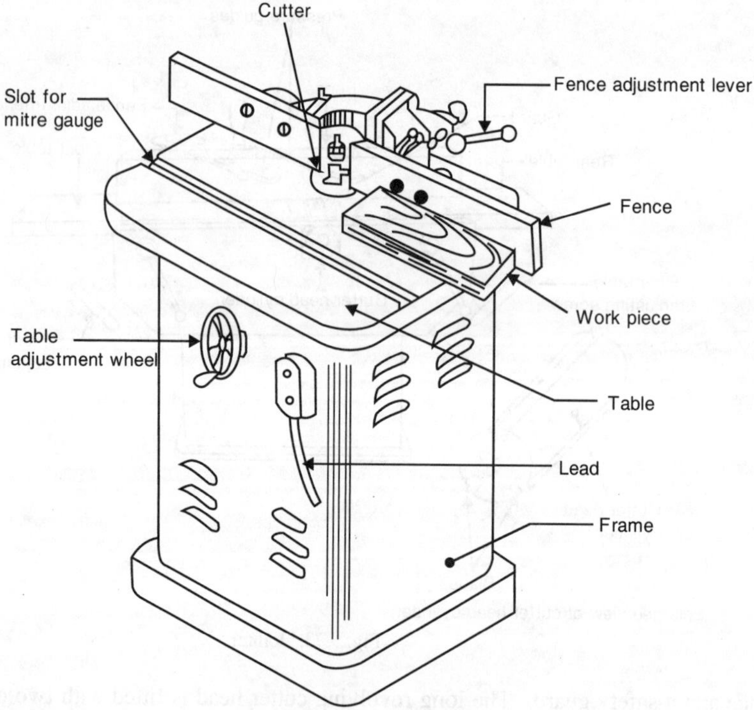

Fig. 9.74. Wood shaper.

9·34 Mortiser

The mortiser or mortising machine is used to make mortises in a wooden piece, for the mortise and tenon joints. There are three types of mortisers, as discussed below.

1. *Chain mortiser.* It is used for making mortises in doors and windows. The mortiser consists of an endless chain having saw type teeth on its outer surface. The chain revolves around a guide bar and cuts the stock. The process is very fast and the mortise produced by this method has a round bottom.

2. *Hollow chisel mortiser.* It is the most commonly used machine. It consists of a heavy cast iron column, table, mortising head with hollow chisel, driving motor, adjustable fence, work holding device and various controls for longitudinal table feed, cross feed and vertical table transverse. The mortising head carries two chucks. The upper chuck grips the auger bit whereas the lower chuck holds the hollow chisel. The auger bit revolves at a very high speed inside the hollow chisel. The chisel is brought down and pressed into the wood by a foot lever. The auger bit and chisel work together to make a square hole. The rotating bit cuts a round hole and square chisel cuts out the corners as it is forced through the hole, thus making all the sides and bottom square. The table can be moved longitudinally, crosswise and vertically up and down. These movements facilitate the proper centering of the job and also helps in accomodating stocks of different sizes.

3. *Oscillating bit mortiser.* It carries an oscillating router bit which produces comparatively small mortises suitable for small cabinets and chair work.

9·35 Sanding Machines

The sanding machines are used to produce smooth surface on wooden articles, by using a sand paper. There are many types of the sanding machines, but the commonly used machines are of the following four types :

1. *Belt sander.* It consists of an endless abrasive belt moving over a pair of drums. It is used for sanding and shaping flat surfaces. The moving belt is pressed against the work by applying the pressure on the back side of the belt with the help of a pressure plate held in the hand.

2. *Disc sander.* It consists of a motor driven flat disc to which abrasive paper is cemented. It is used for sanding curved surfaces. The work is laid on the table and moved backward and forward across the revolving disc. In a disc sander, the table is tilted for sanding inclined surfaces.

3. *Drum sander.* It is largest in size of all the sanding machines. It consists of a horizontally revolving cylinder about 300 mm in diameter over which the abrasive material is fixed. It is used for finishing both inside and outside curves.

4. *Spindle sander.* It is similar to a drum sander but of smaller size. It consists of a vertical spindle projecting through a horizontal table. The spindle revolves and oscillates vertically up and down. A sand paper is wrapped and fastened around the spindle. The table can be tilted for sanding inclined surfaces.

QUESTIONS

1. Why timber is used in constructional work ? Write its advantages.
2. Define exogenous trees and endogenous trees.
3. Draw the cross-section of an exogenous tree. Explain the various parts of this tree.
4. Differentiate between soft and hard wood.
5. Describe the different method of seasoning timber. Which method is best and why ?
6. Explain with the help of suitable sketches, the defects in timber.
7. Why preservation of timber is necessary ? Describe in brief of few methods of preservation.
8. What are the characteristics of a good timber ? Discuss the various types of Indian timber.
9. Write short notes on adhesives used in carpentry.
10. What are the common auxiliary materials used in carpentry ?
11. List out the marking and measuring tools. Describe, with neat sketches, the use of any four of them.
12. How does a rip saw differ from cross cut saw ? Explain with neat sketches.
13. Describe, with neat sketches, the construction and uses of tenon saw, bow saw and key hole saw.

14. What do you understand by setting of saw teeth ? Why it is done ?

15. Classify the chisels used in carpentry. Describe them briefly with neat sketches.

16. Name the different types of planers used in carpentry. Describe, with sketches the following :

(*a*) Wooden jack plane, and

(*b*) Iron jack plane,

(*c*) Smoothing plane.

17. Name the different types of boring tools used in carpentry. Explain them with neat sketches.

18. Describe, with the help of a neat sketch, the construction and working of a ratchet brace.

19. Write short notes on the following :

(*a*) Claw hammer, (*b*) Mallet,

(*c*) Carpenter's vice, (*d*) G-Cramp,

(*e*) Bench stop, and (*f*) T-Cramp.

20. Sketch and describe the following :

(*a*) Pincer; (*b*) Rasp and file, and (*c*) Scraper.

21. Describe the process of sharpening of saws, chisels and planes.

22. Explain the various operations done in carpentry.

23. Explain with sketches any two types of joints used in carpentry.

24. Sketch and describe the following joints made in a carpentry shop.

(*a*) Lap dovetail joint;

(*b*) Mortise and tenon joint,

(*c*) Briddle joint,

(*d*) Mitre joint, and

(*e*) Groove and tongue joint.

25. Describe with the help of a neat sketch, the working of a wood working lathe.

26. Sketch and describe the construction and working of the following wood working saws :

(*a*) Jig saw; (*b*) Circular saw; (*c*) Radial saw; and (*d*) Band saw.

27. What are the common types of mortising machines used in wood working ?

28. Why sanding machines are used ? Explain briefly the different types of sanding machines.

OBJECTIVE TYPE QUESTIONS

1. Which of the following is an exogenous tree ?

(*a*) Bomboo tree (*b*) Teak tree (*c*) Palm tree (*d*) Cane tree

2. The thin radial fibres extending from the pith to the cambium layer, is called

(*a*) bast (*b*) cortex (*c*) heart wood (*d*) medullary rays

3. The purpose of seasoning of wood is to

(*a*) reduce the voids (*b*) remove the curves

(*c*) reduce the moisture content (*d*) change the direction of grains

4. The tensile strength of wood is

(*a*) maximum in the direction parallel to the grains

(*b*) minimum in the direction parallel to the grains

(*c*) maximum in the direction across the grains

(*d*) same in all the directions

5. The age of a tree can be estimated by the
(*a*) diameter of the bark (*b*) height of the tree
(*c*) number of annular rings (*d*) none of these

6. The defect caused by imperfect seasoning is called
(*a*) wet rot (*b*) dry rot
(*c*) honeycombing (*d*) case hardening

7. A saw which cuts wood during the return stroke of the saw is known as
(*a*) push saw (*b*) pull saw or draw saw

8. The mortise gauge is a
(*a*) striking tool (*b*) planing tool (*c*) boring tool (*d*) marking tool

9. A rip saw
(*a*) is a two man saw
(*b*) is used for cutting along the grains of wood
(*c*) has a narrow blade with two wooden handles
(*d*) all of the above

10. A gimlet is a .... tool.
(*a*) boring (*b*) marking

11. A hollow plane is used for producing .... surfaces on the wood.
(*a*) concave (*b*) convex

12. Which of the following statement is wrong ?
(*a*) The mortise gauge has two scribing pins.
(*b*) The gouges are chisels with a hollow shaped blade.
(*c*) An auger is a planing tool.
(*d*) all of the above

13. The sharpening angle of chisels is
(*a*) 20° to 25° (*b*) 25° to 30° (*c*) 30° to 35° (*d*) 35° to 40°

14. The dowels are
(*a*) wooden nails (*b*) box nails (*c*) wire nails (*d*) none of these

15. The joint used for making drawing boards, is
(*a*) briddle joint (*b*) dowel joint
(*c*) mortise and tenon joint (*d*) tongue and groove joint.

ANSWERS

1. (*b*)	2. (*d*)	3. (*c*)	4. (*a*)	5. (*c*)
6. (*c*)	7. (*b*)	8. (*d*)	9. (*b*)	10. (*a*)
11. (*b*)	12. (*c*)	13. (*c*)	14. (*a*)	15. (*d*)

10

Pattern Making

1. Introduction. 2. Pattern Materials. 3. Pattern Making Tools. 4. Pattern Allowances. 5. Types of Patterns. 6. Solid or Single Piece Pattern. 7. Split Pattern. 8. Match Plate Pattern. 9. Cope and Drag Pattern. 10. Loose Piece Pattern. 11. Gated Pattern. 12. Sweep Pattern. 13. Skeleton Pattern. 14. Shell Pattern. 15. Segmental Pattern. 16. Follow Board Pattern. 17. Lagged-up Pattern. 18. Left and Right Hand Pattern. 19. Core Boxes. 20. Colour Coding for Pattern and Core Boxes.

10·1 Introduction

The pattern making deals chiefly with the construction of patterns. A *pattern* may be defined as a model or replica of desired casting which when moulded in sand forms an impression called *mould*. The mould when filled with molten metal forms casting after solidification of the poured metal. The quality and accuracy of casting depends upon the pattern making.

10·2 Pattern Materials

There are several materials commonly used in the construction of patterns. The modern pattern making employs the use of materials that can be easily shaped and are durable. The type of pattern material chosen depends upon the following factors :

(*a*) The design of casting,

(*b*) The number of castings to be produced,

(*c*) The type of casting and moulding process used in foundry, and

(*d*) The degree of accuracy and surface finish required.

Following are the commonly used materials for pattern making

1. *Wood*. It is widely used material for patterns. It is used when small number of castings are to be produced. The wood as a pattern material has the following advantages and disadvantages.

Advantages

(*a*) It is cheap and light.

(*b*) It can be easily worked and shaped as desired.

(*c*) It can be cut and fabricated into numerous forms by gluing, bending and carving.

(*d*) It is easily planed and sanded to smooth surface and can be preserved fairly for a long time with shellac.

Disadvantages

(*a*) It has the tendency to wear out by the constant contact with damp sand.

(*b*) Since it can not withstand the continuous abrasive action of sand, therefore it is unsuitable for repetition work.

The pine, deodar, teak, kail, shisham and mahogany are the commonly used woods for patterns, but mahogany is more durable for heavy duties of repetition moulding. The wood selected for pattern making should be free from knots and it should be properly seasoned before use.

2. *Metal*. When large number of castings are required, the pattern is made of a metal. The metal patterns are more durable, have longer life and produces moulds to a close dimensional accuracy.

Following materials are commonly used for pattern making.

(*a*) *Cast iron.* It is cheap, easy to file and fit. It is strong, gives a good smooth mould surface with sharp edges and is resistant to abrasive action of the sand. But the cast iron patterns are heavy and are easily broken. Though the cast iron can be cast into fairly thin sections, but there is a danger that the edges of thin sections, which cool very quickly, may be too hard to machine. The patterns for small and light work are not ordinarily made of cast iron because of their tendency to rust.

(*b*) *Brass.* It is strong, tough, rust proof and takes a better surfaces finish than cast iron. It has the ability to withstand wear of the moulding sand. It can be used successfully for casting very thin sections without danger of encountering unmachinable edges. The small patterns can be easily rectified, built up or fitted by soldering. Since the brass patterns are heavier and costlier than cast iron, therefore it is used only for patterns or sections of patterns and not for the plates on which the patterns are mounted.

(*c*) *Aluminium and its alloy.* The aluminium and its alloy having 80 percent aluminium, 14 percent zinc, 3 percent copper, 1.5 percent iron and 1.5 percent silicon, are widely used for production moulding due to its light weight, strength, corrosion resistance and machining characteristics. Since the aluminium and its alloys are softer than brass or cast iron, therefore the patterns are liable to be damaged by accidental contact with the steel ramming tools. Though they are fairly resistant to sand abrasion, yet it is advisable to protect critical locations with steel or iron insertions.

(*d*) *White metal.* It is not used to any great extent for patterns, but is the best material used for lining stripping plates, because it can be cast into the narrow cavities between the master pattern and frame of the stripping plate. It is an alloy of lead, copper and antimony and has a low melting point about 260° C. The white metal patterns are soft and easily worn away by the moulding sand. Since it is commonly used for die casting, therefore, it may be called as die casting alloy.

3. *Plaster.* The plaster of paris or gypsum cement is successfully used as a pattern material because it can be easily casted into intricate shapes and can be easily worked. It has a high compressive strength (upto 30 MPa) and controlled expansion. It is used for making small patterns and core boxes of intricate shapes.

4. *Plastics.* The thermo-setting resins, usually phenolic resin plastics, are now gaining popularity as a plastic material for patterns. The plastics are light in weight, have high strength, high wear resistance, high corrosion resistance, low solid shrinkage and have very smooth surface finish. In order to make plastic patterns, first of all a master pattern from wood is made and then a plaster of paris mould is prepared from this master pattern. The plastic resin is now poured in the mould and heated to some specific temperature. After solidification, it produces a plastic pattern. In large scale production, numerous plastic patterns of similar shape and size can be mounted on match plates so as to reduce the unit cost of casting.

5. *Wax.* It is used for *investment casting process. It helps in imparting a high degree of surface finish and dimensional accuracy to castings. The wax pattern is prepared by pouring heated wax into the split moulds or a pair of dies while the latter are kept water cooled. The dies after having been cooled down are parted off. The wax pattern is now taken out and used for moulding.

Note : Sometimes old castings are used as a pattern in case of emergency. The casting to be used as patterns must be well cleaned and made smooth so that they can be drawn out of the mould easily. In order to increase the thickness and other dimensions so as to provide for finish, machining and shrinkage etc., the surfaces of old castings are lagged with thin strips of wood or leather.

10·3 Pattern Making Tools

A pattern maker is basically a carpenter. Therefore, a pattern maker uses the same tools, as discussed in the previous chapter. In addition to these tools a special scale known as *shrink scale* or

* For details, refer Chapter 13 on Special Casting Processes.

contraction scale is used by a pattern maker.

Since the molten metal on cooling shrinks, therefore the final size of the casting produced will be smaller than the cavity of the mould. Thus it is necessary for a pattern maker to make a pattern of the larger size than the size of the final casting, by an amount equal to the shrinkage that takes place on solidification of molten metal. In order to compensate for the known shrinkage of the metal, the pattern maker uses the appropriate shrink scale (which is longer than the ordinary rule of the same length) in laying out the dimensions of a pattern. The shrink scales are graduated to shrinkage allowance directly and are available in 300 mm to 500 mm in length.

10·4 Pattern Allowances

We have already discussed in the previous article that a pattern is always made larger than the required size of casting considering the various allowances. The following allowances are usually provided in a pattern.

1. *Shrinkage or contraction allowance.* The various metals used for casting contract after solidification in the mould. Since the contraction is different for different metals, therefore their corresponding allowances also differ and there is a shrink or contraction scale for each type of metal used in a casting. The following table shows the contraction allowances for castings of different materials in sand moulds.

Table 10.1. Shrinkage or contraction allowances.

Type of material	*Contraction allowance (mm/ metre)*	*Type of material*	*Contraction allowance (mm/metre)*
Grey cast iron	7 to 10.5	Steel	20
White cast iron	21	Zinc and lead	24
Malleable iron	15	Bronze	10.5 to 21
Copper, brass and aluminium	16	Magnesium	18

Notes : (*a*) The total contraction is volumetric, but the correction for it is expressed linearly.

(*b*) A separate scale is available for each allowance. Thus the dimensions are set out directly during laying out a pattern.

(*c*) If a pattern is first made of wood and then from some other metal then double allowances are provided on the wooden pattern. For example, if an aluminium pattern made from a wooden master pattern is to be used for grey iron castings, then a total shrinkage allowance of 26 mm/metre (16 mm/metre for aluminium and 10 mm/metre for grey iron), may be allowed on the wooden pattern.

2. *Draft allowance.* It is a taper which is given to all the vertical walls of the pattern for easy and clean withdrawal of the pattern from the sand without damaging the mould cavity. It may be expressed in millimeters per metre on a side or in degrees. The amount of taper varies with the type of patterns. The wooden patterns require more taper than metal patterns because of the greater frictional resistance of the wooden surfaces. The taper on the inner surfaces must be greater than on the outside surfaces. The amount of taper varies from ½ to 1½ degrees. It may be reduced to less than ½ degree for larger castings.

3. *Finish or machining allowance.* This allowance is provided on the pattern if the casting is to be machined. This allowance is given in addition to shrinkage allowance. The amount of this allowance varies from 1.6 to 12.5 mm which depends upon the type of casting metal, size and shape of casting, method of casting used, method of machining to be employed and the degree of finish required. The ferrous metals require more machining allowance than nonferrous metals. Fig. 10.1 shows the shrinkage, draft and machining allowances provided in a pattern for vertical casting.

4. *Distortion or camber allowance*. This allowance is provided on patterns used for castings of such design in which the contraction is not uniform throughout. For example, consider a long I-section beam having the top flange much thicker than the bottom one, as shown in Fig. 10.2 (*a*).

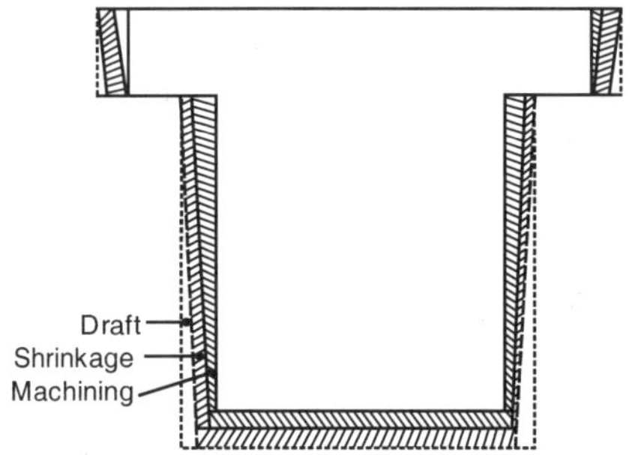

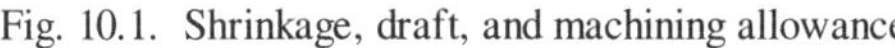

Fig. 10.1. Shrinkage, draft, and machining allowance

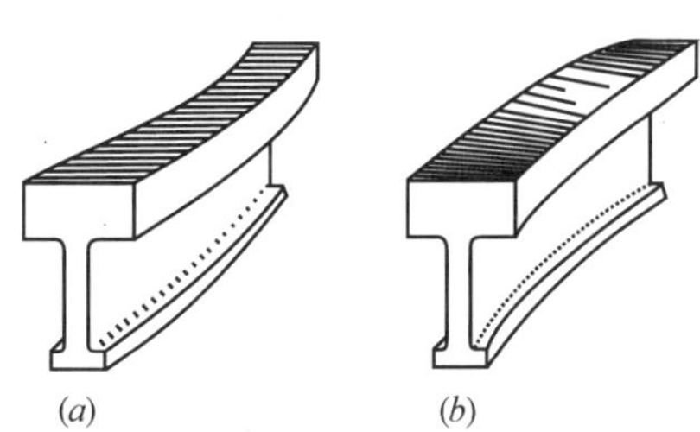

Fig. 10.2. Distortion or camber allowance.

The thinner flange will freeze and contract first and when the top flange freezes it contracts more than the thin flange because the rate of cooling of this thin flange is retarded by the transmitted heat from the thick section. Due to this unequal contraction, the I-section will be distorted or cambered. The thicker flange will be concave and the thinner flange convex. In order to compensate this, the distortion allowance is provided in the opposite direction as shown in Fig. 10.2 (*b*)

5. *Rapping or shaking allowance*. This allowance is provided in the pattern to compensate for the rapping of mould because the pattern is to be rapped before removing it from the mould. In small and medium-sized casting, this allowance can be neglected. But in larger casting this allowance is considered by making the pattern slightly smaller than the casting.

10·5 Types of Patterns

The type of pattern selected for a particular casting depends upon the following conditions.

1. The shape and size of casting,
2. The number of castings required, and
3. The method of moulding employed.

The common types of patterns are discussed in detail in the following pages :

10·6 Solid or Single Piece Pattern

A solid or single piece pattern, as shown in Fig. 10.3, has a compact form, It has no joints, partings or loose pieces in its construction. It has usually one broad surface that serves as a parting surface in the mould. This type of pattern is used for a limited number of castings because most of the moulding operations like parting surface formation, cutting of gating system, providing runners and risers, withdrawal of pattern etc. is done by hand.

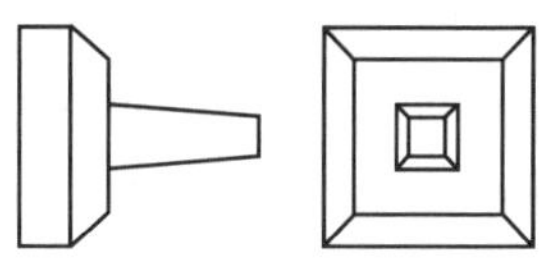

Fig. 10.3. Solid pattern.

10·7 Split Pattern

The most of the patterns are not made in a single piece because of the difficulties encountered in moulding them. In order to eliminate this difficulty, some patterns are made into two or more pieces. A pattern consisting of two pieces is called a *two piece split pattern, as* shown in Fig. 10.4. One-half of the pattern rests in the lower part of the moulding box known as *drag* and the other half in the upper part of the moulding box known as a *cope.*

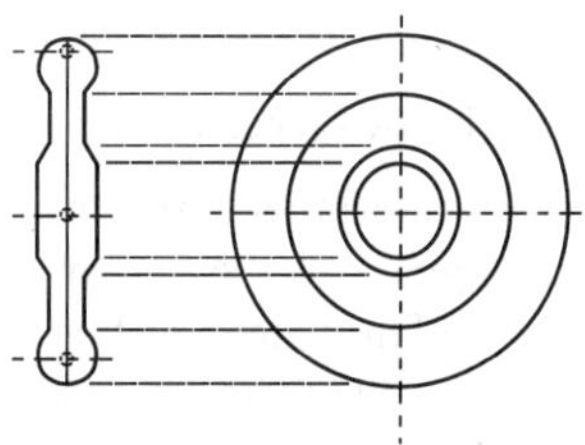

Fig. 10.4. Split pattern.

The line of separation of the parts is called *parting line* or *parting surface*.

Sometimes a pattern for complex castings is made in three parts as shown in Fig. 10.5. Such a pattern is called a *multi-piece pattern*. A three piece pattern requires a moulding box with three parts, the middle one being called as *cheek*.

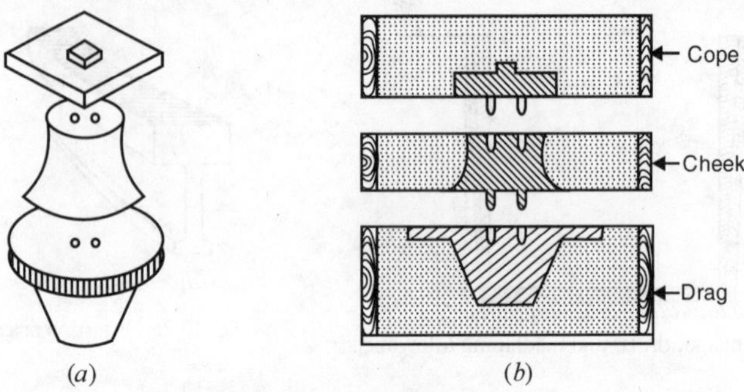

Fig. 10.5. Multipiece pattern.

The split patterns are commonly used for casting of spindles, cylinders, steam valve bodies, water stop-cocks and taps, bearings, small pulleys and wheels.

10·8 Match Plate Pattern

The match plate patterns are used on machine moulding for quantity production of castings. A single pattern or a number of patterns may be mounted on a match plate. When the cope and drag portions of the split pattern are mounted on opposite sides of the wooden or metal plate (usually aluminium plate), the pattern is called a *match plate pattern*. The gates and runners are permanently fastened to the drag side of the plate in their correct positions in order to form a complete match plate. When the match plate is lifted off the mould, all patterns are withdrawn and the gates and risers are completed in one operation.

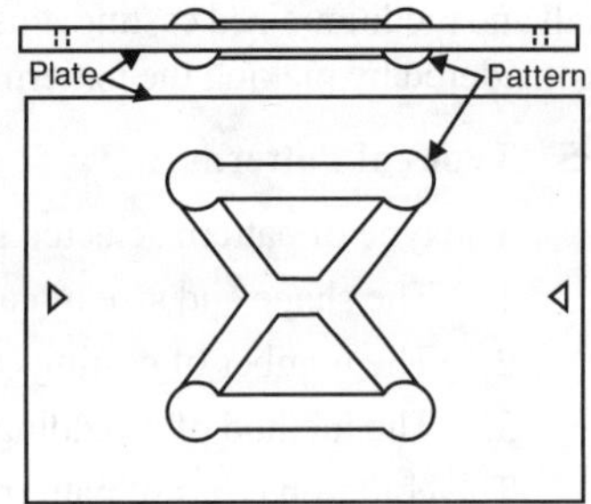

Fig. 10.6. Match plate pattern.

Fig. 10.6 shows a match plate pattern upon which the patterns of two small dumb bells are mounted. Though the cost of construction of the match plate patterns is quite high, yet it is easily compensated by the high rate of production, greater dimensional accuracy and minimum machining requirement of casting.

10·9 Cope and Drag Pattern

When very large castings are to be made, the complete mould becomes too heavy to be handled by a single operator. In order to ease this problem, the cope and drag pattern is used. It is nothing but a two-piece pattern, split on a convenient joint line. One part is moulded in a cope and the other part in a drag of the moulding box.

10·10 Loose Piece Pattern

Sometimes a pattern has to be made with projections or overhanging parts, as shown in Fig. 10.7. These projections make the removal of the pattern difficult. Therefore, such projections are made in loose pieces and are fastened loosely to the main pattern by means of wooden or wire dowel pins. These pins are taken out during the moulding operation.

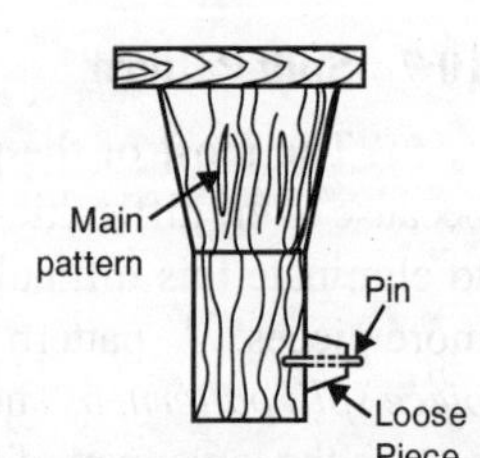

Fig. 10.7. Loose piece pattern

After moulding, the main pattern is withdrawn first and then the loose piece is removed with the help of a lifter.

10·11 Gated Pattern

The gated pattern, as shown in Fig. 10.8 is used for mass production of small castings. When a number of small patterns are placed in a single mould, then each pattern may be provided with a gate pattern with it. It consists of pieces of wood or metal fixed to the patterns to form the runner and rising channels in the mould. In this way, full supply of the molten metal flows into every part of the mould.

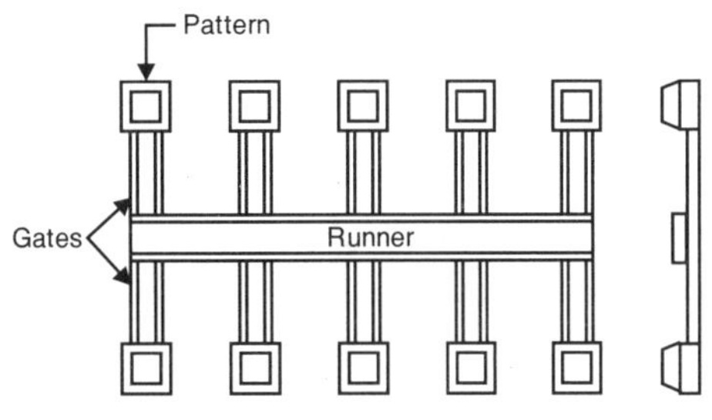

Fig. 10.8. Gated pattern.

The gated pattern eliminates hand cutting of gates and thus makes the moulding easy, If a group of patterns is to be placed in one mould, the gate pattern has a further function of holding the patterns in the proper position with respect to each other.

10·12 Sweep Pattern

The moulds of large size and symmetrical in shape particularly of circular section can be easily prepared by using a sweep instead of a full pattern. It is not considered as a true pattern when compared with others. A sweep is a template of wood or other material which has the shape corresponding to the shape and size of casting. It is arranged to rotate about a central axis by mounting it on a spindle. The sand is rammed in place with a cavity of approximately similar shape and size to that of the required mould. Now by revolving the sweep, the desired shape of the mould is generated. The sweep and the spindle are then removed and the hole in the centre is filled up.

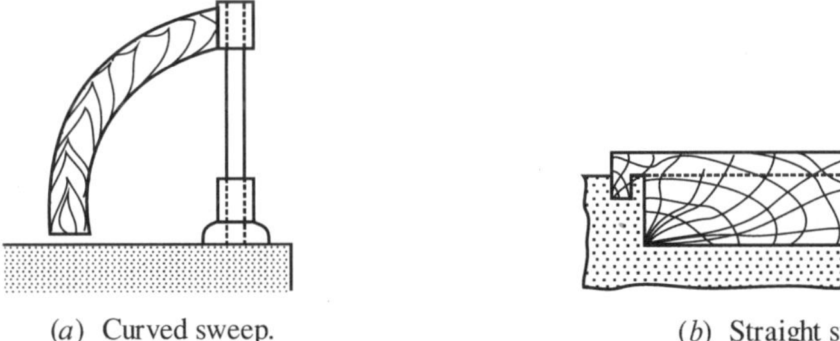

(*a*) Curved sweep. (*b*) Straight sweep.

Fig. 10.9. Sweep pattern.

Fig. 10.9 (*a*) shows a curved sweep which may be used to form part of the mould for a large cast-iron kettle. Fig. 10.9 (*b*) shows a straight sweep for any type of groove or ridge. The principal advantage of this pattern is that it eliminates expensive pattern construction.

10·13 Skeleton Pattern

When a few and large-sized castings are required, it is not advisable to use a large solid pattern of that size, as it will require a lot of wood and time to make a full pattern. In such cases, a skeleton pattern in the hollow form, consisting of a wooden frame and strips, as shown in Fig. 10.10, is used.

The frame work is filled and rammed with loam sand and a strickle board, as shown in Fig. 10.10, is used to scrap the excess sand out of the spaces between the ribs so that the surface is even with the outside of the pattern. The skeleton pattern is made in two halves, one for the cope and the other for the drag. After taking the impression, the cope and drag are assembled together with the

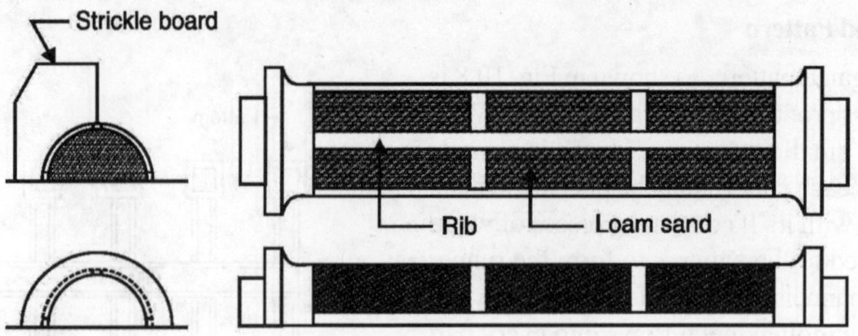

Fig 10.10. Skeleton pattern.

core in position to form the complete mould. The skeleton patterns are used for castings of hollow cast iron pipe (as shown in Fig. 10.10), bends, valve bodies and boxes.

10·14 Shell Pattern

The shell pattern, as shown in Fig. 10.11, is used largely for drainage fittings and pipe work. This type of pattern is usally made of metal mounted on a plate and parted along the centre line, the two sections being accurately doweled together. The short bends are usually moulded and cast in pairs. The shell pattern is a hollow construction like shell. The outside shape is used as a pattern to make the mould while the inside is used as a core box for making cores.

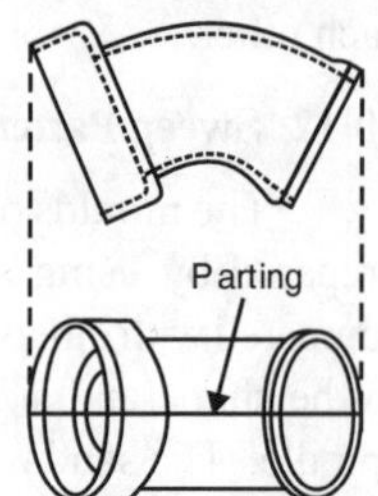

Fig. 10.11. Shell pattern.

10·15 Segmental Pattern

The segmental patterns, as shown in Fig. 10.12, are also known as *part patterns*. These patterns are sections of a pattern so arranged as to form a complete mould by being moved to form each section of the mould. These patterns are usually applied to circular work, such as rings, wheel rims, gears etc.

When a mould is to be made using a segmental pattern, a vertical spindle is firmly fixed in the centre of a drag flask. The mould bottom is rammed, and swept level. Now the segmental pattern is fastened to the spindle. The moulding sand is rammed between the outside of the pattern and the flask and on the inside except at the ends of the pattern. After ramming one segment, the next segment is rammed, and so on until entire perimeter of the mould is completed.

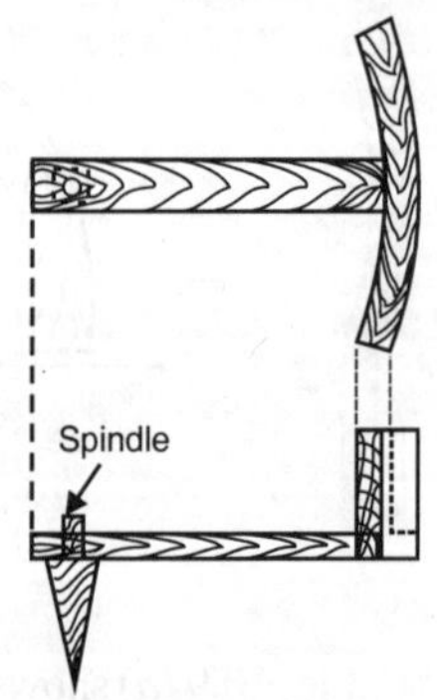

Fig. 10.12. Segmental pattern.

10·16 Follow Board Pattern

A follow board pattern as shown in Fig. 10.13, is used for solid patterns having an irregular parting line. It may be used with either single or multiple gated patterns. The patterns requiring follow boards are usually somewhat difficult to make as a split pattern. The board is routed out so that the patterns rests in it up to the parting line and this board then acts as a moulding board for the first moulding operation.

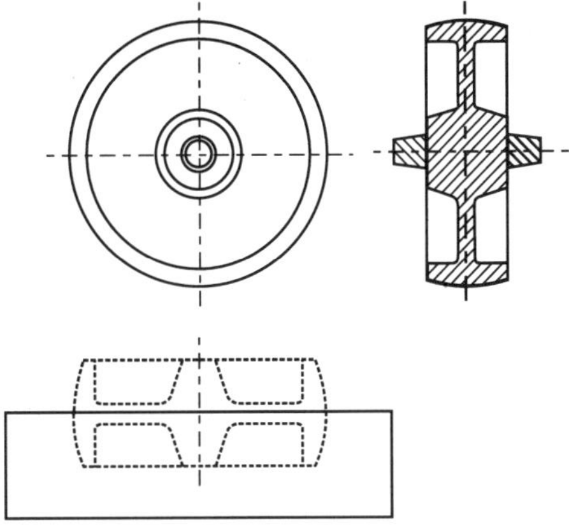

Fig. 10.13. Follow board pattern.

10·17 Lagged-up Pattern

When a pattern or core box is so large or of such a form that it cannot be made economically from a solid piece or when such a method would result in a pattern of little strength or excessive weight, it is necessary to use a lagged or staved pattern.

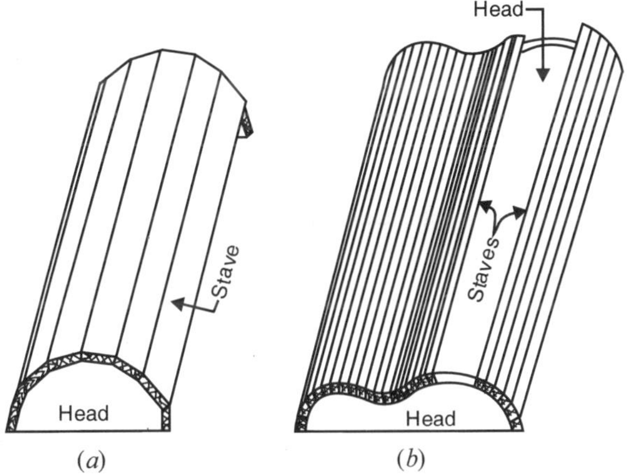

Fig. 10.14. Lagged up pattern.

The cylindrical patterns *e.g.*, barrels, pipes or columns are built up with lag or stave construction to ensure proper shape. The lags are longitudinal strips of wood bevelled on each side for making the joint tight outside. These lags are glued and nailed or screwed to the wooden end pieces called heads, as shown in Fig. 10.14 (*a*). Such a construction gives the maximum amount of strength and permits building close to the finished outline of the pattern so that there is comparatively little excess stock to be removed to bring it to the required form. In building pieces that are not cylindrical, the heads are frequently cut to follow a line parallel to the finished outline of the pattern, so that the staves, when fastened in place, will closely approximate this outline, as shown in Fig. 10.14 (*b*).

10·18 Left and Right Hand Pattern

Some patterns are required to be in pairs, and when their form is such that they cannot be reversed and have the centres of hubs, bosses etc. opposite and in line, then they must be made right and left hand separately. A few examples where a pair of left and right hand patterns is required are legs for wood turning lathe, J-hangers for overhead shafting, legs for garden bench, legs for paddle type sewing machine, brackets for luggage racks in the railway carriages etc.

A bracket, as shown in Fig. 10.15, is an example of left and right hand pattern. The hub and the foot flanges are fastened with screws and are moved from side to side to make the pattern right and left hand, as shown by dotted lines in the figure. The ribs are also loose. The rib *C* is reversed, but right and left hand ribs *D* are required because of the angle at the lower edge.

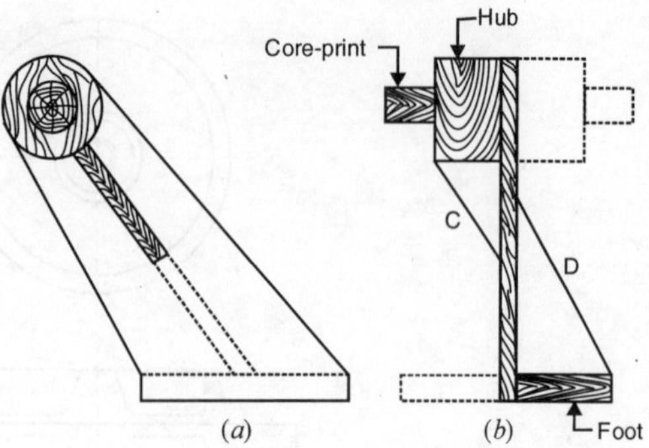

Fig. 10.15. Left and right hand pattern.

10·19 Core Boxes

The core boxes, like patterns, may be made either of wood or metal. The core boxes are used for a casting requiring cores. Wood is generally used for making core boxes, but metal core boxes are preferred where core are to be prepared in large numbers on mass production basis. Following are the various types of core boxes commonly used :

1. *Half core box.* A half core box, as shown in Fig. 10.16, is the most common type of core box. It is used for preparing the two identical halves of a symmetrical core. The two haves are then pasted together after baking to form a complete core.

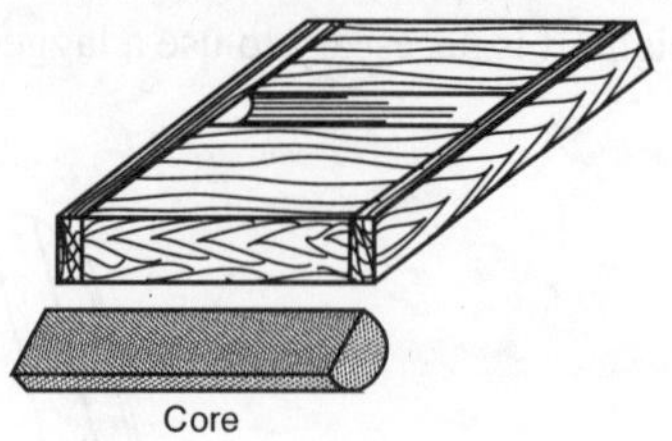

Fig. 10.16. Half core box.

2. *Dump core box*. A dump core box is similar to half core box, but the core produced by this core box do not require pasting and they are complete by themselves. If the core produced is in the shape of slab, then it is called as a *slab box* or a *rectangular box*.

3. *Spilt core box*. A split core box, as shown in Fig. 10.17, is in two parts and a complete core is produced in a single ramming. The two parts are held in position by means of clamps and alignment is maintained by means of dowel pins. Sometimes, the parting surface of the two halves is made along an irregular line to eliminate the use of dowels. For preparing the core, the core sand is rammed from one side. After ramming and striking off the excess sand, the core box is unclamped and rapped. Thus, a complete core is produced in a single ramming.

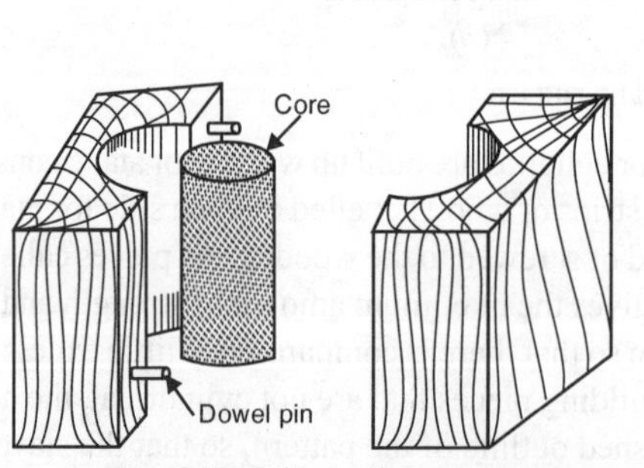

Fig. 10.17. Split core box.

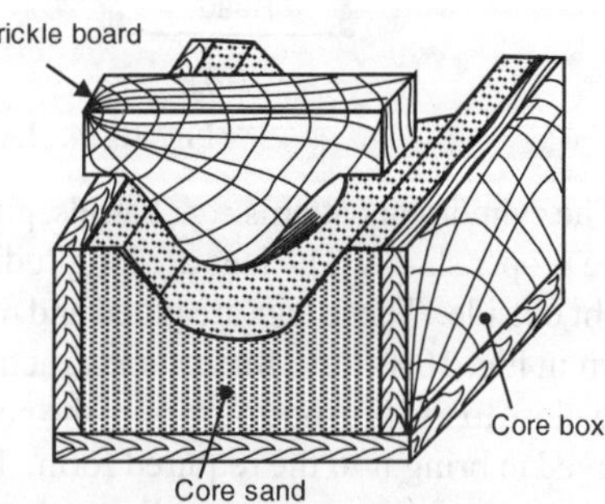

Fig. 10.18. Strickle core box.

4. *Strickle core box.* A strickle core box, as shown in Fig. 10.18, is used when the core is required to have an irregular shape which cannot be easily rammed by other methods. The desired irregular shape is achieved by striking off the core and from the top of the core box with a piece of wood called a *strickle board.* The strickle is cut to correspond exactly to the contour of the required core.

5. *Right and left hand core box.* The right and left hand core boxes as shown in Fig. 10.19, are used when the cores are not symmetrical about the centre line and the two halves of the core made in the same core box are not identical, they cannot be pasted together to form an entire core.

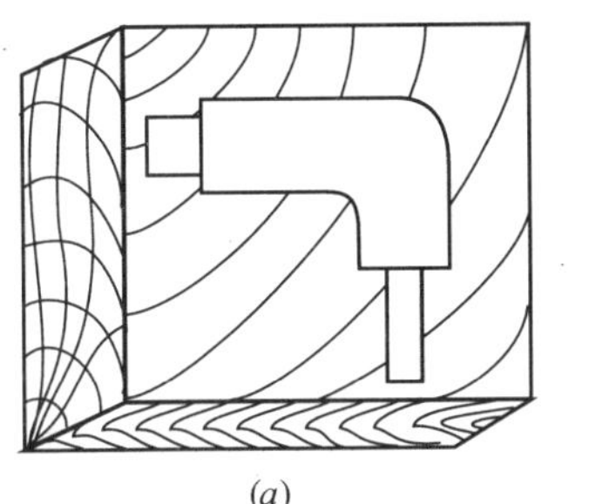

(*a*)

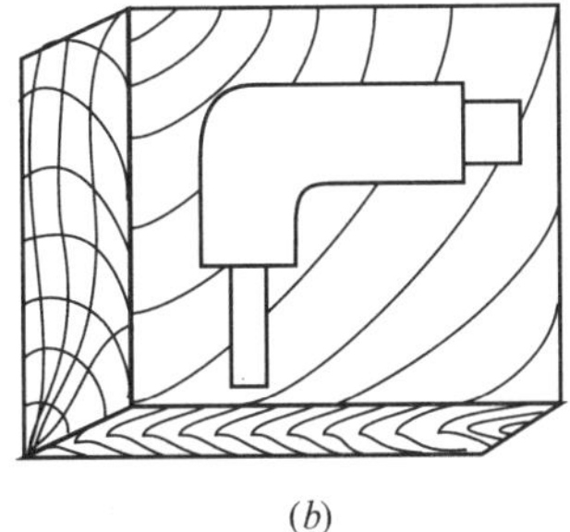

(*b*)

Fig. 10.19. Right and left hand core box.

6. *Loose piece core box.* A loose piece core box is used for making cores when provision for bosses, hubs etc. is required. In some cases, both halves of the right and left core can be prepared from a single core box with the help of loose pieces as shown in Fig. 10.20. One half of the core is prepared with the loose piece placed in the left hand recess and the other half is prepared by shifting the loose piece to the right hand recess.

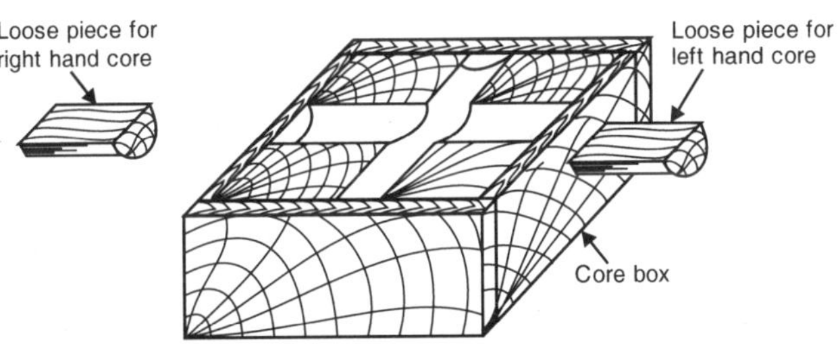

Fig. 10.20. Loose piece core box.

7. *Gang core box.* A gang core box, as shown in Fig. 10.21, is used when large number of small sized cores are to be prepared. In this type of core box, several core cavities are rammed in a single operation.

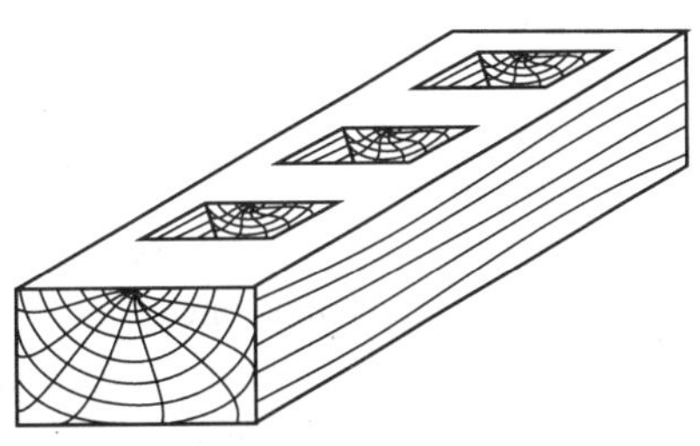

Fig. 10.21. Gang core box.

10·20 Colour Coding for Patterns and Core Boxes

The important parts of a pattern and core box are painted with different colours for the identification of their different parts.

Though there is no universally accepted standard for colouring the surfaces, yet the following colour code is widely used :

1. The surface to be left unmachined are painted with *black* colour.
2. The surface to be machined are painted with *red* colour.
3. The core prints and seats for loose core prints are painted with *yellow* colour.
4. The seats for loose pieces are marked with red strips on yellow base.
5. The stop offs are indicated by diagonal black strips on yellow base.
6. The parting surface is indicated by no colour.

QUESTIONS

1. How a pattern is defined ?

2. What are the common materials, used for pattern making ? Discuss their relative merits and demerits.

3. Describe, with sketches, the important tools used in pattern making.

4. Explain the common allowances provided on patterns.

5. What are the various types of patterns ? Discuss the use of a solid pattern.

6. Discuss the advantages of split and multi-piece patterns.

7. What is the specific advantage of match plate pattern ? Describe how they are used for making the moulds.

8. Sketch and describe the use and advantages of a gated pattern.

9. How are skeleton patterns made ? What is the advantage of using such patterns ?

10. Describe, with the help of a neat sketch, the working of a sweep pattern stating its advantages.

11. Write short notes on the following giving suitable sketches wherever necessary :
(*a*) Cope and drag patterns,
(*b*) Follow board patterns,
(*c*) Segmental patterns, and
(*d*) Core boxes.

12. Write the colour coding for patterns and core boxes.

OBJECTIVE TYPE QUESTIONS

1. When a pattern is made in three parts, the top part is known as a
(*a*) drag (*b*) cheek (*c*) cope (*d*) any one of the above.

2. A taper provided on the pattern for its easy and clean withdrawal from the mould is known as
(*a*) machining allowance (*b*) draft allowance
(*c*) shrinkage allowance (*d*) distortion allowance.

3. If an aluminium pattern made from a wooden master pattern is to be used for grey iron castings, then the shrinkage allowance allowed on the wooden pattern should be
(*a*) 10 mm/m (*b*) 16 mm/m (*c*) 20 mm/m (*d*) 26 mm/m

4. The surface to be left unmachined is marked on the pattern by
(*a*) red colour (*b*) yellow colour (*c*) black colour (*d*) blue colour
(*e*) red strips on yellow base

5. The surface to be machined is marked on the pattern by
(*a*) red colour (*b*) yellow colour (*c*) black colour (*d*) blue colour
(*e*) red strips on yellow base.

ANSWERS

1. (*c*) 2. (*b*) 3. (*d*) 4. (*c*) 5. (*a*)

11

Foundry Tools and Equipments

1. Introduction. 2. Foundry Tools and Equipments. 3. Foundry Hand Tools. 4. Moulding Boxes (Flasks). 5. Moulding Machines 6. Melting Equipment. 7. Pouring Equipment.

11·1 Introduction

A foundry is a place where castings are produced. The *casting* is a process of pouring molten metal into a mould and allowing it to solidify. By this process, intricate parts can be given strength and rigidity which is not frequently obtainable by any other method. The mould or cavity into which the material is poured is made of some heat-resisting material. Sand is widely used, as it can be readily packed to shape and resists high temperatures.

Though all the metals can be cast, but cast iron is widely used for casting. This is due to the fact that in casting with this metal, it is possible to have easy control on its properties which include fluidity, rate of shrinkage, strength and rigidity. The shape of the object is determined by the shape of the mould cavity.

11·2 Foundry Tools and Equipments

The foundry tools and equipments are divided into the following five groups :

1. Hand tools,
2. Moulding boxes (Flasks),
3. Moulding machines,
4. Melting equipment, and
5. Pouring equipment.

We shall now discuss these tools and equipments, in detail, in the following pages.

11·3 Foundry Hand Tools

The hand tools commonly employed in preparing the mould in foundry are as follows :

1. *Shovel.* A shovel, as shown in Fig. 11.1, consists of a square pan fitted with a wooden handle. It is used for mixing and for moving the sand from one place to another in the foundry.

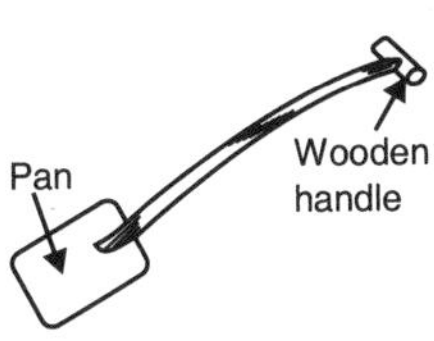

Fig. 11.1. Shovel.

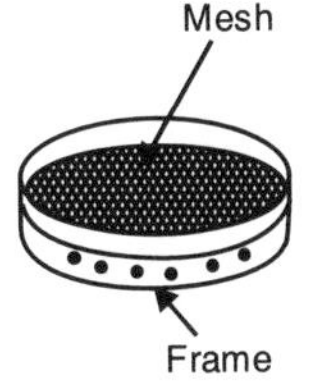

Fig. 11.2. Riddle.

2. *Riddle.* A riddle, as shown in Fig. 11.2, has a standard wire mesh fixed into a circular or square wooden frame. It is used for cleaning the moulding sand. The riddles are specified by the diameter of the frame and the mesh number. When large volumes of sand are to be cleaned, then power operated riddles are used.

3. *Rammer.* A *hand rammer*, as shown in Fig. 11.3(*a*), is made of wood or metal. It has a wedge shaped construction at one end (called peen) and the cylindrical shape at the other end (called butt). It is a short rammer and is used for packing and ramming the sand for bench moulding. The *floor rammer,* as shown in Fig. 11.3(*b*), is similar in construction, but have long handles. It is used for floor moulding or for ramming large moulds. The *peen rammer,* as shown in Fig. 11.3(*c*), has a wedge shaped end and is used for setting into corners and pockets of the mould. The *pneumatic rammers,* as shown in Fig. 11.3(*d*) are automatic rammers operated by compressed air. They have long or short shafts according to the requirements with peen or flat ends. It is used in large moulds saving considerable labour and time.

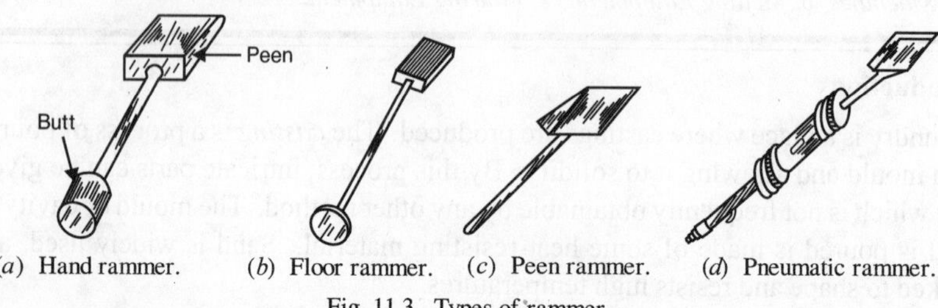

(*a*) Hand rammer. (*b*) Floor rammer. (*c*) Peen rammer. (*d*) Pneumatic rammer.

Fig. 11.3. Types of rammer.

4. *Vent wire.* A vent wire, as shown in Fig. 11.4, is similar to a knitting needle. It has a pointed edge at one end and a handle at the other end. It is used to pierce holes in the rammed sand to provide artificial vents which permit the easy escape of steam and gases generated by the hot metal in contact with the sand.

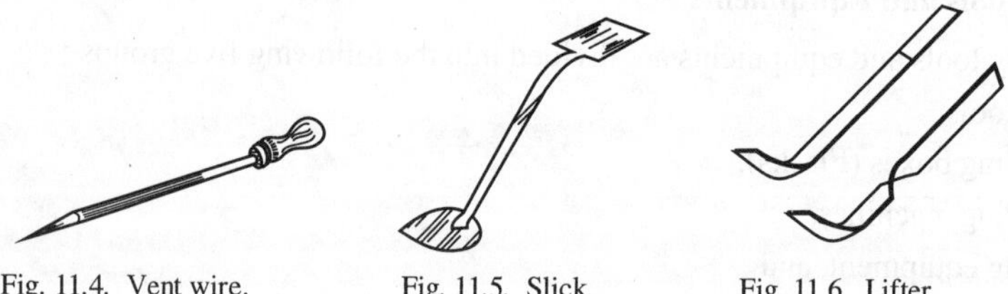

Fig. 11.4. Vent wire. Fig. 11.5. Slick. Fig. 11.6. Lifter.

5. *Slick.* A slick, as shown in Fig. 11.5, is a small double ended tool having a flat on one end and a spoon on the other. This tool is also made in a variety of other shapes and is used for repairing and finishing the mould surfaces after the pattern is withdrawn.

6. *Lifter.* A lifter, as shown in Fig. 11.6, is made of thin sections of steel of various width and lengths with one end bent at right angles. It is used for smoothing and cleaning out depressions in the mould.

7. *Swab.* A *simple swab*, as shown in Fig. 11.7(*a*), is a small brush having long hemp fibres.

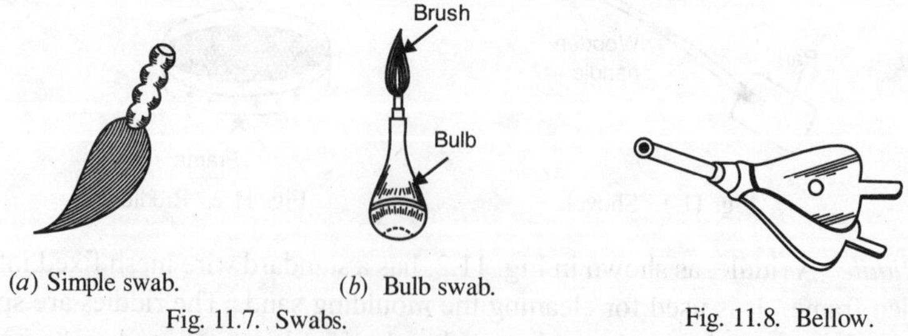

(*a*) Simple swab. (*b*) Bulb swab.

Fig. 11.7. Swabs. Fig. 11.8. Bellow.

A *bulb swab,* as shown in Fig. 11.7(*b*), has a rubber bulb to hold the water and a soft hair brush at the open end. It is used for moistening the sand around the edge before the pattern is removed.

8. *Bellow.* The hand operated bellow, as shown in Fig. 11.8, is used to blow loose particles of sand from the cavities and surface of the mould.

9. *Trowel.* The trowels, as shown in Fig. 11.9, consist of a metal blade with a wooden handle. The small trowels of various shapes are used for finishing and repairing mould cavities as well as for smoothing over the parting surface of the mould. The usual trowel is rectangular in shape and has either a round or a square end.

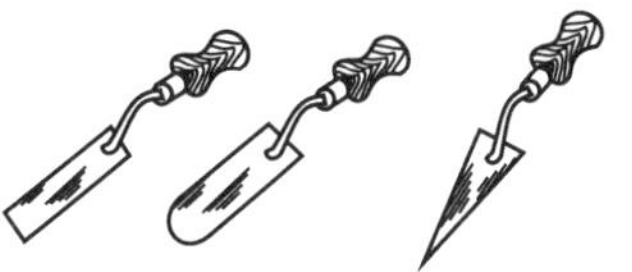
Fig. 11.9. Trowels.

10. *Gate cutter.* A gate cutter, as shown in Fig. 11.10, is a U-shaped piece of thin sheet. It is used for cutting a shallow trough in the mould to act as a passage for the hot metal.

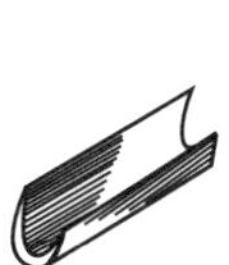
Fig. 11.10. Gate cutter.

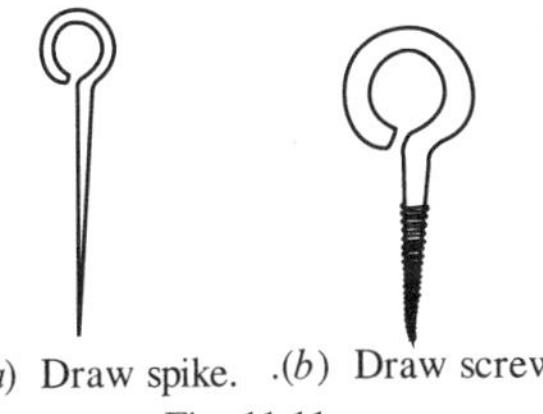
(*a*) Draw spike. (*b*) Draw screw
Fig. 11.11.

11. *Draw spike and screw.* A *draw spike*, as shown in Fig. 11.11(*a*), is a pointed steel rod, with a loop at one end. It is driven into a wooden pattern to hold it when the sand is withdrawn. The *draw screw*, as shown in Fig. 11.11(*b*), is similar in shape but threaded on the end to engage metal patterns.

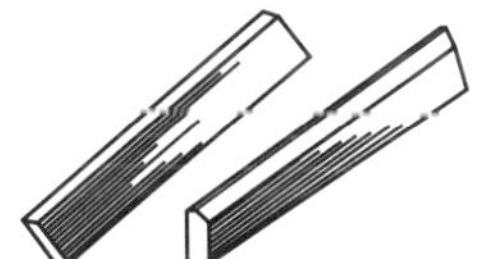
Fig. 11.12. Strike-off-bar.

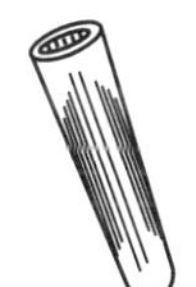
Fig. 11.13. Sprue cutter.

Fig. 11.14. Mallet.

12. *Strike-off-bar.* A strike-off-bar, as shown in Fig. 11.12, is a straight bar of wood or steel usually of rectangular cross-section. It is used to strickle or strike off excess sand to provide a level and smooth surface.

13. *Sprue cutter.* A sprue cutter (also called a runner peg), as shown in Fig. 11.13, is a tapered wooden peg. It is forced into the top part of the mould (known as cope) at the correct position. When the peg is withdrawn, it leaves a cavity (known as down gate) through which the molten metal is poured.

14. *Mallet.* A raw hide mallet, as shown in Fig. 11.14, is used to loosen the pattern in the mould so that it can be withdrawn easily.

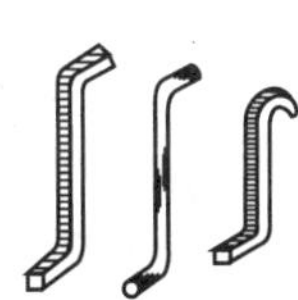
Fig. 11.15. Gaggers.

Fig. 11.16. Clamps.

15. *Gaggers.* The gaggers (also called lifters), as shown in Fig. 11.15, are iron rods bent at one end or both ends. It is used for reinforcement of sand in the top part of a moulding box and to support

hanging bodies of sand. The length of the gaggers vary from 125 mm to 600 mm and they are coated with clay wash to cause the sand to adhere to them.

16. *Clamps.* The clamps, as shown in Fig. 11.16, are used for holding the cope and drag of the mould so that the cope should not rise when the molten metal is poured into the mould.

17. *Pouring weight.* It is simply a cast iron plate with a cross-shaped opening casting in it to give considerable liberty in placing the runner in the mould. It is used on the top of the mould for giving a weight to prevent the pressure of the molten metal forcing the mould apart, causing run out.

18. *Rapping plate.* A rapping plate (also called lifting plate) is made of sheet and are firmly fixed to the top of the pattern by screws or bolts. It is used for lifting and rapping large patterns. These are available in different shapes, as shown in Fig. 11.17.

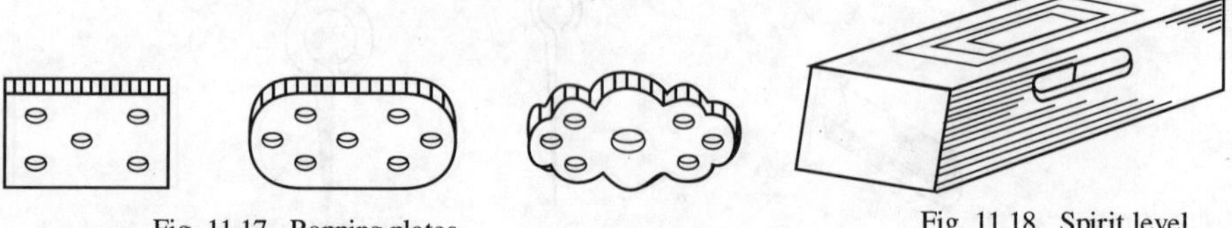

Fig. 11.17. Rapping plates.

Fig. 11.18. Spirit level.

19. *Spirit level.* A spirit level, as shown in Fig. 11.18, consists of an air bubble in a curved glass tube protected by a wooden or metal flame. It is used by the moulder to ensure that his bed of sand, moulding box or moulding machine table is horizontal.

11·4 Moulding Boxes (Flasks)

The sand moulds are prepared in a specially constructed boxes called *flasks,* which are open at the top and bottom. They are made in two parts, held in alignment by dowel pins. The top part is called the *cope* and the lower part as *drag.* If the flask is made in three parts, the intermediate part is called a *cheek.*

These flasks can be made of either wood or metals. Though the wood is the cheapest material and flasks can be made quickly, but they have the disadvantage of wearing out rapidly and are destroyed by the contact with hot metal. The metal flasks of steel, cast iron, magnesium or aluminium alloys are widely used in production work because of their rigidity. Following two types of flasks are widely used in foundry :

1. *Box type flask.* The box type flask (also known as tight or permanent flask), as shown in Fig. 11.19, is used for small and medium-sized castings.

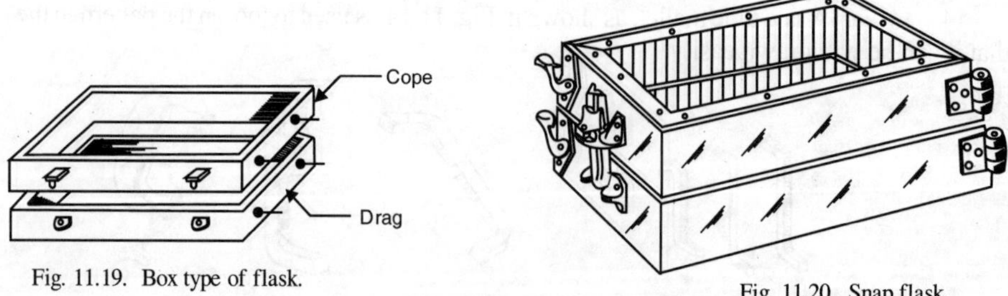

Fig. 11.19. Box type of flask.

Fig. 11.20. Snap flask.

These flasks should not be removed till the pouring of molten metal is completed. Hence a number of tight flasks are required for any one moulding sets up, one for each mould being processed. These flasks are generally made of steel.

2. *Snap flask.* The snap flask, as shown in Fig. 11.20, is often used in the production of small casting and in machine moulding. These flasks are fitted with a hinge at one corner and a fastener at the corner diagonally opposite. After the mould has been made, it is carried to the position where it is to be cast and the snap flask is removed to be used for the following moulds. Since one flask is employed for many moulds, therefore, there is a considerable economy.

A typical wooden moulding box is shown in Fig. 11.21. The side timbers are continued beyond the ends of the box to form two handles at each end. The sides are held together by strong cross timbers, reinforced with bolts. The crossbars prevent the sand from falling out when the top part of the box (*i.e.*, cope) has been rammed up and is lifted away from the bottom part of the box (*i.e.*, drag).

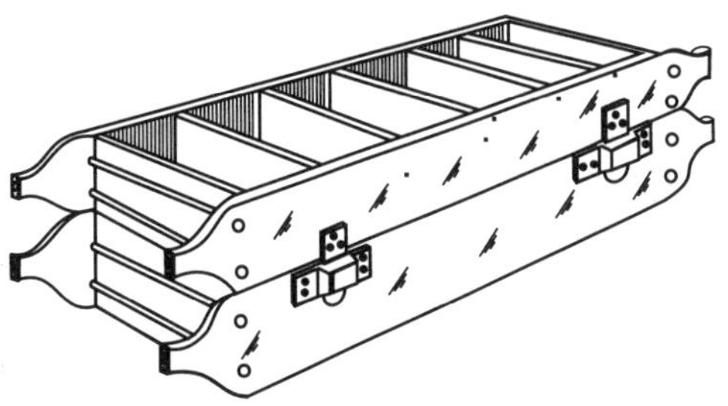

Fig. 11.21. Typical wooden moulding box.

A *moulding board* and a *bottom board* complete the flask. The moulding board is a smooth board on which the flask and patterns are placed when the mould is started. It should be perfectly flat and well reinforced with cleats on the bottom. When the mould is turned over, the function of this board is ended and the mould is placed on a similar board called bottom board which acts as a support until it is poured.

11·5 Moulding Machines

The moulding machines perform the following basic operations :

(*i*) Ramming of sand in the mould.

(*ii*) Lifting or drawing of pattern from the mould, and

(*iii*) Rolling over the mould section.

The two main classes of moulding machines are

(*a*) Hand moulding machines, and

(*b*) Power moulding machines.

In *hand moulding machines,* though the actual ramming is manual operation but the operation of drawing the pattern from the mould is mechanical and it is more accurate than by hand. The *power moulding machines* are used to eliminate the manual labour of ramming and at the same time produce better moulds.

The power moulding machines are classified into the following two groups :

(*a*) According to the method of ramming the sand; and

(*b*) According to the method of removing the pattern from the mould.

These classification of moulding machines are discussed, in detail, in the following pages.

(*a*) *Classification of moulding machines according to the method of ramming the sand.* The moulding machines for ramming the sand fall into five classes as discussed below :

1. *Jolt machine.* This machine is used for ramming of sand in the mould. It consists of an air operated cylinder and piston, at the top of which a machine platen or table is attached. The flask and pattern with sand over it, is placed on the table. In the operation of the machine, as shown in Fig. 11.22(*a*), the compressed air is admitted from the bottom of the cylinder which lifts the piston and the table upwards. After lifting the piston to a certain height (about 30 to 80 mm), the air is released through the exhaust pipe and the table suddenly drops down. This action causes the sand to be packed evenly about the pattern. These machines ram the sand harder at the pattern face with decreasing

hardness towards the back of the mould. These machines can handle only one part of the mould at a time and are especially adopted to large work.

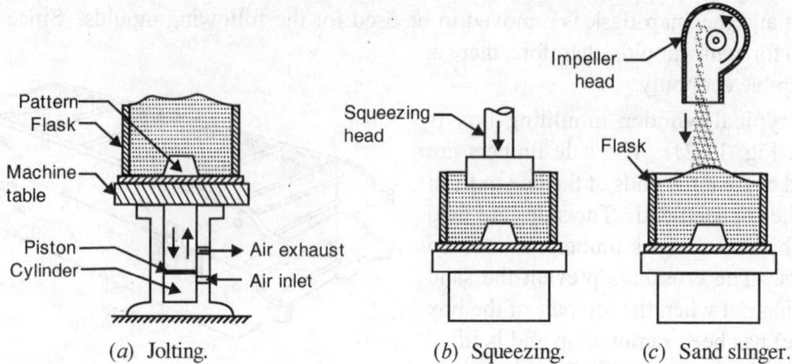

(*a*) Jolting. (*b*) Squeezing. (*c*) Sand slinger.

Fig. 11.22. Power-ramming moulding machines.

2. *Squeezing machine.* The ramming of the sand may also be done by a squeezing machine. In the operation of this machine, as shown in Fig. 11.22(*b*), the table is lifted up by the compressed air against the stationary squeezing head. The sand is, thus, compressed between the machine table and the squeezing head.

These machines ram the sand harder at the back of the mould and softer on the pattern face. These machines are very useful for shallow patterns.

3. *Jolt-squeeze machine.* We have discussed above that the jolt machines ram the sand harder at the pattern face and softer at the back of the mould while the squeezing machines ram the sand harder at the back of the mould and softer on the pattern face. In order to overcome these drawbacks of both the machines and to produce more uniform hardness throughout the mould, the jolt-squeeze machine is used. In the operation of this machine, the flask is assembled with the match plate between the cope and drag. The assembly is placed upside down (*i.e.,* drag on the top and cope at the bottom) on the machine table. The sand is filled into the drag and levelled off and a bottom board is placed on the top. The sand is rammed into the drag by jolting action. Now the assembly is turned over and the cope is filled with sand and levelled off. The pressure board is placed on the top of the flask. The sand is now rammed by squeezing between the pressure board and the machine table. The cope is lifted from the match plate and swung on the side. The match plate is vibrated and removed from the drag by lowering the drag. The cope is now swung to its place, the drag is raised and finally the mould is closed for pouring.

4. *Sand slinger.* The uniform packing of sand in the moulds is an important operation in the production of castings. In order to accomplish this operation in a satisfactory manner, particularly for large moulds, a mechanical device known as *sand slinger* is used. In this operation, as shown in Fig. 11.22(*c*), the moulding sand is delivered to the impeller through the opening by means of conveyor buckets. This sand is thrown in small handfuls by means of a high speed rotary impeller plate, into the flask over the pattern. In this way, high rates of ramming (0.25 m^3 per minute or 500 kg per minute) are achieved. The ramming of sand is controlled by the speed of the impeller which normally varies from 1200 to 1800 r.p.m.

5. *Diaphragm moulding machine.* It is a recent development in moulding machines. It utilises a pure gum rubber diaphragm for ramming sand over the pattern, as shown in Fig. 11.23. In this, the match plate and the flask are mounted on a trolley which can be moved along the machine bed ways by means of an air operated piston. In the operation, first the flask is filled with sand and then the trolley and the flask is moved to the right side so that the flask comes under the diaphragm

pressure head. The air pressure is now used to force the rubber diaphragm over the entire surface of the pattern, regardless of the pattern contour. This results in uniform ramming and hardness of the sand in the flask.

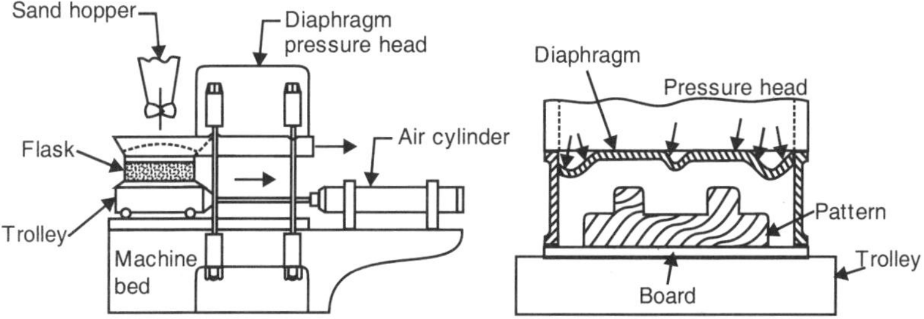

Fig. 11.23. Diaphragm moulding machine.

(*b*) *Classification of moulding machines according to the method of removing the pattern from the mould.* The moulding machines according to the method of removing the pattern from the mould are of three types as discussed below :

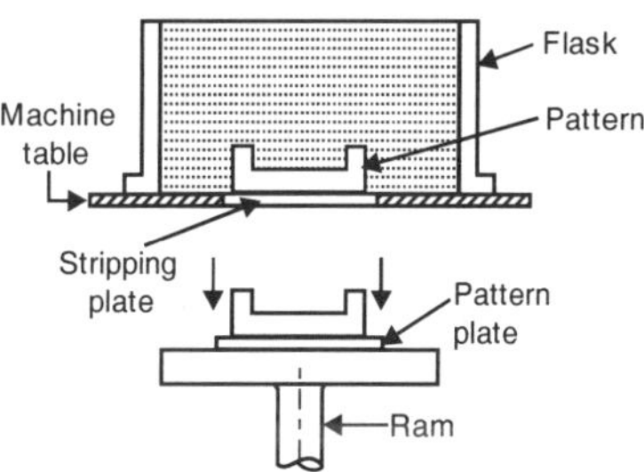

Fig. 11.24. Stripping-plate machine.

1. *Stripping plate machine.* It is used for drawing the pattern from the mould. It consists of a stripping plate having an opening of size equal to that of pattern plate, as shown in Fig. 11.24. The pattern plate with the half pattern screwed on it is raised from below and brought in line with the machine table. The flask is kept on the table and the sand is rammed into it. When the mould is ready, the pattern is withdrawn downwards through the stripping plate which supports the mould when the pattern is drawn.

2. *Pin-lift or push-off machine.* It is also called a *straight draw moulding machine.* This machine, as shown in Fig. 11.25, draws the mould away from the pattern in the upward direction.

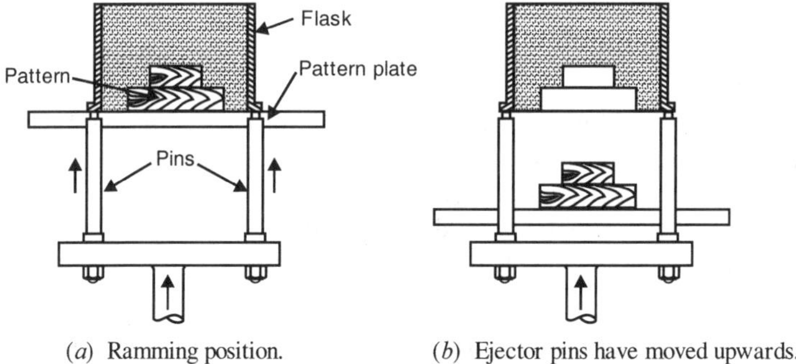

(*a*) Ramming position. (*b*) Ejector pins have moved upwards.

Fig. 11.25. Pin-lift or push-off machine.

After the mould is ready, the flask or moulding box is pushed upwards from the pattern by four pins or posts which pass through the pattern plate and lift the four corners of the flask. This may be a manual operation for small machines, but for a greater weight a long lever is necessary. If the moulding boxes are too heavy to lift even with the lever, then the ejector pins may be operated by a ram working in a pneumatic cylinder. Fig. 11.25(*b*) shows that the ejector pins have moved upwards to lift the mould away from the pattern.

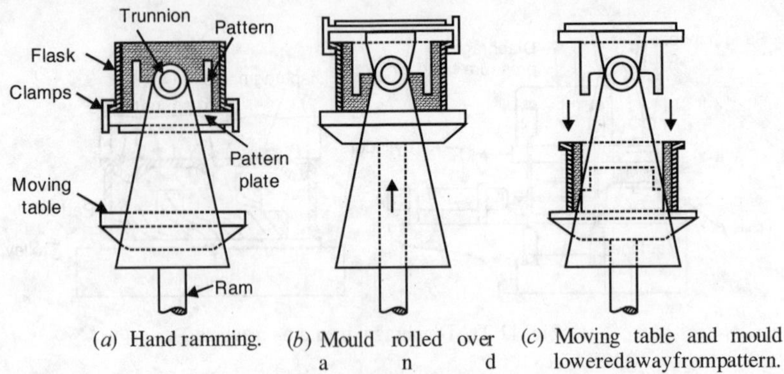

(*a*) Hand ramming. (*b*) Mould rolled over and (*c*) Moving table and mould lowered away from pattern.

Fig. 11.26. Roll over machine.

3. *Roll over machine.* It is a convenient machine for making drags because the mould is rolled over ready for placing on the casting floor. In one type of the roll over mechanism, as shown in Fig. 11.26(*a*), the pattern is placed on the pattern plate or machine plate. The flask or moulding box is clamped to the pattern plate and swivels on two trunnions fixed at the centre of gravity. The sand is filled in it and rammed. After ramming, the flask and the pattern plate is rotated by hand through 180°, as shown in Fig. 11.26(*b*), and locked in a truly horizontal position by screws in the trunnion brackets. The moving table is raised to butt against the flask. The flask is unclamped from the pattern plate and the moving table with the mould is now lowered from the pattern, as shown in Fig. 11.26(*c*). The upward and downward motion of the moving table may be manually operated by means of a system of levers with a counter balance weight or may be actuated by a pneumatic ram.

11·6 Melting Equipment

The metal is required to be melted at a correct temperature before pouring into the mould. The following furnaces are commonly used for melting the metal.

1. Cupola furnaces,
2. Open hearth furnaces,
3. Electric furnaces, and
4. Crucible furnaces.

The first three types of furnaces are mostly used for melting ferrous metals and have been discussed in chapter on ferrous metals. The crucible furnaces are used for melting non-ferrous metals and alloys. The crucible furnace has a special crucible pot made of a refractory material (such as graphite and clay or silicon carbide) or mild steel when melting aluminium and magnesium. The following are the three types of crucible furnaces :

(*a*) Pit type,

(*b*) Stationary type (or bale-out furnaces), and

(*c*) Tilting type.

In a *pit type crucible furnace,* as shown in Fig. 11.27(*a*), the crucible is placed in a pit in the floor. It is usually fired with coke, sufficient coke being packed round and above the crucible pots to melt and superheat the charge without re-coking. The natural drought provided by a tall chimney is controlled by means of loose brick or damper at the foot of the stack.

The *stationary of bale-out furnaces,* as shown in Fig. 11.27(*b*), are used for melting alloys, aluminium, magnesium, zinc, lead and other alloys which have low melting point. The crucible is usually made of mild steel. These furnaces are often fired by oil or gas. They are made cylindrical in shape so that the flame may sweep around and uniformly heat the crucible. The forced draught is provided by a small fan and the waste gases are taken away by the flue outlet.

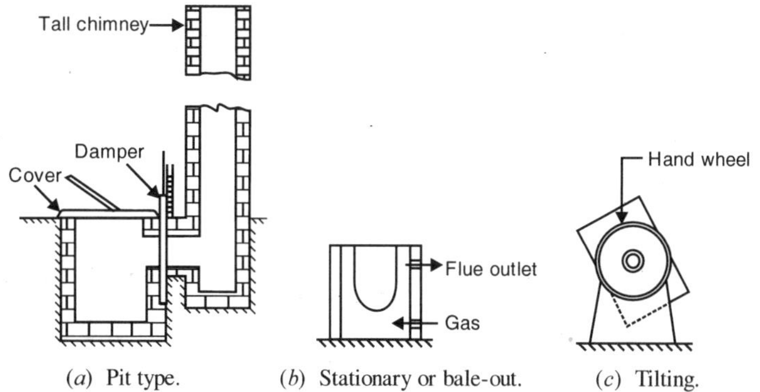

(*a*) Pit type. (*b*) Stationary or bale-out. (*c*) Tilting.

Fig. 11.27. Crucible furnaces.

The *tilting furnaces,* as shown in Fig. 11.27(*c*), are raised above the floor level, mounted on two pedestals and rotated by means of a geared hand wheel or a hydraulic ram. The furnace is fired with coke, oil or gas and the forced draught is employed as for bale out furnaces. The coke-fired tilting furnaces are provided with an enclosed ash pan. The disadvantage of using coke for this type of furnace is that ample coke space is essential. The burners on oil or gas-fired furnaces are arranged so that the heating can be continued even when the furnace is tilted.

11·7 Pouring Equipment

The commonly used pouring equipment, for pouring the molten metal from the furnace, are ladles. The common practice is to store the molten metal temporarily into large *holding ladles,* from which it is tapped off as needed. These holding ladles are constructed of steel plate lined with a suitable refractory such as fire brick. The other ladles used are follows :

1. Hand or shank ladles
2. Bull ladles,
3. Crane or monorail ladles,
4. Lip-pouring ladles,
5. Tea-pot ladles, and
6. Bottom stoppered ladles.

The *hand* or *shank ladles* as shown in Fig. 11.28(*a*), are used for casting very small moulds. These ladles can hold about 15 kg of metal and has only one handle or shank. The larger sizes of shank ladle capable of holding 30 kg to 60 kg of molten metal are called *bull ladles.* These ladles have a single handle attached to one end of the shank ladle and a double handle to the other end, as shown in Fig. 11.28(*b*). These ladles require two men to carry. The man operating the double end rotates the shank to pour the metal, while the man at the single end holds his end steady.

Still larger ladles holding 250 kg or more metal are handled by overhead crane or monorail as shown in Fig. 11.28(*c*). These ladles are fed from the furnaces.

In ordinary *lip pouring ladles*, it is almost difficult to pour clean metal free from the molten slag floating on top. Thus the metal at the pouring lip is skimmed clean, by means of metal skimming bars, before pouring.

In *tea-pot ladles,* as shown in Fig. 11.28(*d*), the pouring spout is usually inside the body. In pouring, these ladles are tilted by a hand-wheel connected to a pinion, worm and gear wheel.

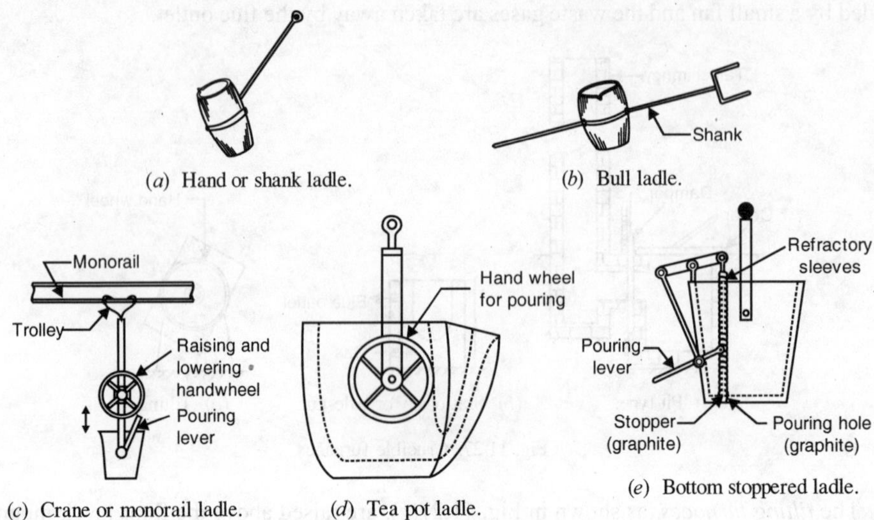

(*a*) Hand or shank ladle. (*b*) Bull ladle. (*c*) Crane or monorail ladle. (*d*) Tea pot ladle. (*e*) Bottom stoppered ladle.

Fig. 11.28. Types of ladles.

In *bottom-stoppered ladles,* as shown in Fig. 11.28(*e*), there is a hole in the bottom of the ladle. The pouring is done by raising a vertical rod and plug covering the hole.

Since in tea-pot and bottom-stoppered ladles, the metal is poured from the bottom and does not disturb the slag which floats on the top, therefore these ladles may be called *self-skimmed* ladles. The lip-pouring, tea-pot and bottom-stoppered ladles are operated by cranes or monorails.

QUESTIONS

1. What are the various types of tools and equipments used in a foundry ?
2. List out the foundry hand tools and describe them with neat sketches.
3. Describe, with neat sketches, the different types of moulding boxes.
4. What is a moulding machine ? What main function does it perform ?
5. Explain the working principles and uses of different moulding machines.
6. How does a sand slinger differ from other moulding machines ?
7. Discuss the working principle of different pattern draw machines.
8. Explain the principle of working of a roll over machine.
9. Describe the various types of furnaces used for melting the metal in a foundry.
10. What are ladles ? Describe the different types of ladles used for pouring the metal.

OBJECTIVE TYPE QUESTIONS

1. Riddle is used for
   (*a*) Smoothing and cleaning out depressions in the mould
   (*b*) cleaning the moulding sand
   (*c*) moistening the sand around the edge before removing pattern
   (*d*) reinforcement of sand in the top part of moulding box
   (*e*) repairing and finishing the mould

2. Swab is used for

(*a*) smoothing and cleaning out depressions in the mould

(*b*) cleaning the moulding sand

(*c*) moistening the sand around the edge before removing pattern

(*d*) reinforcement of sand in the top part of moulding box

(*e*) repairing and finishing the mould

3. In order to produce uniform packing of sand in the mould,

(*a*) a sand slinger is used
(*b*) a squeezing machine is used
(*c*) a jolt machine is used
(*d*) a stripper- plate machine is used.

4. In order to deliver molten metal from pouring basin to gate,

(*a*) a riser is used (*b*) a sprue is used (*c*) a core is used (*d*) a gagger is used

5. Which of the following statement is wrong ?

(*a*) A slick is used for repairing and finishing the mould.

(*b*) A gagger is used for cleaning the moulding sand.

(*c*) A stripper-plate machine is used to draw the pattern from the mould.

(*d*) None of the above

ANSWERS

1. (*b*) 2. (*c*) 3. (*a*) 4. (*b*) 5. (*b*)

12

Moulding and Core Making

12·1 Introduction

The moulding is a process of making a cavity (or mould) out of sand by means of a pattern. The molten metal is poured into the moulds to produce castings. Sometimes, a casting is to be made hollow or with cavities in it. Such type of castings require the use of cores. A core is defined as a sand shape which is exactly similar to the cavities or holes to be provided in the castings. The cores are generally made separately *i.e.*, they are not moulded with the pattern. The process of making cores is called *core making*.

12·2 Moulding Materials

Following are the moulding materials commonly used in foundry practice :

1. Moulding sand,
2. Sand binders, and
3. Additive materials.

We shall now discuss, in detail, these materials in the following pages.

12·3 Moulding Sand

The principal material used in making a mould is sand. The sand is defined as the granular particles resulting from the breakdown of rocks. Quartz and other silica rocks are the source of silica sand which is commonly used for moulding. The silica sand is found in nature on the bottoms and banks of rivers, lakes and larger bodies of water. The silica sand comprises from 50 to 95 percent of the total material in any moulding sand. A good moulding sand contains the following ingredients :

Silica sand (SiO_2)	—	80.8 %
Alumina (Al_2O_3)	—	14.9 %
Iron oxide (Fe_2O_3)	—	1.3 %
Combined water	—	2.5 %
Other inert materials	—	1.5 %

In India, the main sources of foundry sands are as follows :

(*a*) Allahabad and Hardwar in Uttar Pradesh

(*b*) Avadi and Veeriyambakam in Madras

(*c*) Barkar and Damodar areas in Bihar

(*d*) Batala in Panjab

(*e*) Bhavnagar and Satara in Maharashtra

(*f*) Kumardhubi and Hoogli in Bengal
(*g*) Rourkela in Orissa
(*h*) Secundrabad in Andhra Pradesh.

12·4 Sand Binders

The pure silica sand is not used because it lacks in binding quality. Any material added to the sand or provided by nature in the sand that imparts cohesiveness to it, is called a *binder*. The binder holds the sand grains together, imparts strength, resistance to erosion and breakage and degree of collapsibility. The binders are classified as :

1. Clay-type binders,
2. Organic-type binders, and
3. Inorganic-type binders.

The clay is a natural earthy material which becomes plastic when moistened. The fire clay is a *clay-type binder* and widely used with silica sand.

The *organic-type binders* are mostly produced from vegetables or wood. For example, molasses (from sugarcane), starches (from maize corn and potatoes), dextrines (intermediate product in the degradation of starch), linseed oil (from flax seed), resins (from trees), pitch (from coal). They are mainly used in making *oil sands* for cores. The oil sands have become very popular because of the following reasons :

(*a*) They are easy to use for core-making than a clay bonded sand.
(*b*) An oil sand core is more collapsible than a clay-bonded core.
(*c*) The green and dried strengths of the oil sand mixture can be controlled by means of simple variations in the proportions of dextrine and oil respectively.
(*d*) The baked cores are very hard and not easily damaged in handling.

The *inorganic binders* include portland cement, bentonites (a form of clay), fire clays (kaolinite), sodium silicate and iron oxide. These binders have considerable strength at high temperatures. They may be used to obtain green strength, baked strength, hot strength or smooth finish.

12·5 Sand Additives

A material which do not promote binding action, when added to the sand by facing the mould or by mixing with heap sand for improving some special features, is called a *sand additive*. There are many materials in the market that may be used as moulding sand additives. Some commonly used additives are sea coal, wood flour and silica flour.

The *sea coal* is a finely pulverised bituminous coal. It improves the surface finish, aids in cleaning of castings and prevents burning on sand.

The *wood flour* is a finely pulverised or ground soft wood. It improves finish, prevents burning in, aids cleaning and improves collapsibility.

The *silica flour* is a finely ground silica sand of less than 54 microns. It fills interstices (voids) and thus reduces metal penetration.

12·6 Properties of Moulding Sand

The moulding sand must posses the following properties :

1. *Porosity or permeability*. It is that property of sand which permits the steam and other gases to pass through the sand mould. When hot molten metal is poured into the sand mould, it evolves a great amount of steam and other gases while coming in contact with the moist sand. If these gases does not escape completely through the mould, the casting will contain gas holes and pores. Thus the sand from which the mould is made must be sufficiently porous or permeable. The porosity

of sand depends upon its grain size, grain shape, moisture and clay contents in the moulding sand. The extent of ramming of sand directly affects the porosity of the mould. If the sand is too fine, its porosity will be low.

2. *Plasticity*. It is that property of sand due to which it flows to all portions of the moulding box or flask and acquires a predetermined shape under ramming pressure and retain this shape when the pressure is removed. The sand must have sufficient plasticity to produce a good mould. The plasticity is increased by adding water and clay to sand.

3. *Adhesiveness*. It is that property of the sand due to which it adhere or cling to the sides of the moulding box. A good sand must have sufficient adhesiveness so that heavy sand masses can be successfully held in the moulding box or flask without any danger of its falling out when the flask is removed.

4. *Cohesiveness*. It is that property of that sand due to which the sand grains stick together during ramming. It may be defined as the *strength* of the moulding sand. It is of the following three types :

(*a*) *Green strength*. The green sand, after water has mixed to it, must have adequate strength and plasticity for making and handling of the mould. The green strength depends upon the grain shape and size, amount and type of clay and the moisture content.

(*b*) *Dry strength*. When the molten metal is poured, the sand adjacent to the hot metal quickly loses its water as steam. The dry sand must have strength to resist erosion and also the metallostatic pressure of the molten metal, otherwise the mould may enlarge.

(*c*) *Hot strength*. After the moisture has evaporated, the sand may be required to possess strength at some elevated temperature, above 100°C. If the sand does not possesses hot strength, the metallostatic pressure of the liquid metal bearing against the mould walls may cause mould enlargement, or if the metal is still flowing, it may cause erosion, cracks or breakage.

5. *Refractoriness*. It is that property of the sand which enables it to resist high temperature of the molten metal without breaking down or fusing. The higher pouring temperature, such as those for ferrous alloys, require great refractoriness of the sand.

The degree of refractoriness depends upon the quartz content, and the shape and grain size of the particles. The higher the quartz content and the rougher the granulometric composition, the higher is the refractoriness of the sand. The refractoriness is measured by the sinter point of the sand rather than its melting point. The sinter point of the sand must be higher than the temperature at which the molten metal is poured into the sand mould.

6. *Flowability*. It is that property of sand due to which it behaves like a fluid so that, when rammed, it flows to all portions of a mould and distribute the ramming pressure evenly. Generally, sand particles resist moving around corners or projections. In general, flowability increases with decrease in green strength and decrease in grain size. It also varies with the moisture content.

7. *Collapsibility*. It is that property of the sand due to which the sand mould collapses automatically after the solidification of the casting in order to allow free contraction of the metal. This property of sand is dependent upon the amount and type of binder, the temperature to which it is heated in contact with the metal and the time of contact. For example, oil sand core is more collapsible than clay-bonded core, because after the metal has been in contact with it for a short time, the oil burns out quickly, leaving loose sand in the castings which is easily cleaned out.

12·7 Classification of Moulding Sand

The moulding sand may be classified generally into the following three different types :

1. Natural sand ; 2. Synthetic sand ; and 3. Special sand.

The *natural sand* is one which contains sufficient clay as mined from the sand pit so that it can be used directly. It needs only to be tempered and conditioned. Since the natural sands have the advantages of simplicity in their preparation, handling and use, therefore these are used for most of the ferrous and non-ferrous castings.

The *synthetic sand* is one which is artificially compounded by mixing sand grains and selected type of clay. These sands have the advantages of lower cost in larger volume, wide spread availability and the possibility of sand reclamation and reuse.

The *special sand* contain mixtures of inorganic compounds. Though these sands cost more, yet they offer better high-temperature stability than ordinary silica sand. These sands produce better cast surfaces. The special sands used most often are zircon, olivine, chamotte, chromite and chrome magnesite.

The moulding sands, according to their use, are further classified as below :

1. *Green sand.* The sand in its natural or moist state is called *green sand.* It is also called tempered sand. It is a mixture of silica sand with 20 to 30 percent clay, having total amount of water from 6 to 10 percent. The moulds prepared with this sand are called *green sand moulds*. The molten metal is poured in the green sand moulds without any prior baking. The green sand moulds are used for small size castings of ferrous and non-ferrous metals.

2. *Dry sand.* The green sand moulds when baked or dried before pouring the molten metal are called *dry sand moulds*. The sand in this condition is called dry sand. The dry sand moulds have greater strength, rigidity and thermal stability. These moulds are used for large and heavy castings.

3. *Loam sand.* A mixture of 50 percent sand grains and 50 percent clay is called *loam sand.* It is used for loam moulding of large grey-iron castings.

4. *Facing sand.* A sand used for facing of the mould is called *facing sand.* It is a specially prepared sand from silica sand and clay, without the addition of used sand. The thickness of layer of facing sand in a mould ranges from 20 to 30 mm and is used directly next to the surface of the pattern. Since it comes in contact with the molten metal when poured, therefore it must possess high strength and refractoriness.

5. *Backing or floor sand.* A sand used to back up the facing sand and not used next to the pattern, is called *backing sand.* The sand which have been repeatedly used, may be employed for this purpose. It is sometimes called *black sand* because of its black colour.

6. *System sand.* A sand employed in mechanical sand preparation and handling system is called *system sand.* This sand has high strength, permeability and refractoriness.

7. *Parting sand.* A sand employed on the faces of the patterns before moulding is called *parting sand.* The function of the parting sand is that when the pattern is withdrawn, the face of the mould impression is damaged by adhesion of grains of sand or sections of the mould. The parting sand consists of dried silica sand, sea sand or burnt sand.

8. *Core sand.* A sand used for the preparation of cores is called *core sand.* It is sometimes called *oil sand**. It is the silica sand mixed with linseed oil or any other oil as binder.

9. *Molasses sand.* A sand which carries molasses as a binder is called *molasses sand.* It is very useful for making moulds of small castings of intricate shapes and their sections. The cores are also made from this sand.

12·8 Grain Shape and Size of Sand

The shape and size of the grains of sand have a large effect on its strength and the general properties. The shape of foundry sand grains may vary from *round to angular*. Some sands may consist almost entirely of grains of one shape while others have a mixture of various shapes. According to the shape, foundry sands are classified as follows :

---

* Refer Art.12.4. on sand binders.

1. Round, as shown in Fig. 12.1(*a*),
2. Sub-angular, as shown in Fig. 12.1(*b*),
3. Angular, as shown in Fig. 12.1(*c*), and
4. Compound, as shown in Fig. 12.1(*d*).

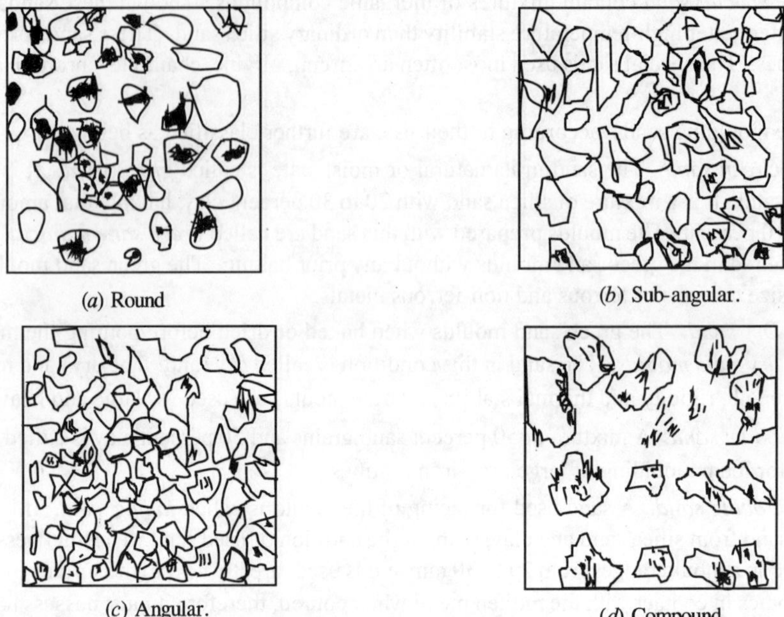

(*a*) Round (*b*) Sub-angular.

(*c*) Angular. (*d*) Compound.

Fig. 12.1. Shape of sands.

The moulding sand with rounded grains gives much better compatibility under the same conditions than the sands with angular grains. In other words, the sand with rounded grains has the greatest degree of close packing of particles while the sand with angular grains can not pack together closely. Since the compactibility increases with rounded grains, therefore the permeability or porosity of sand decreases with rounded grains and increases with angular grains. The sand with rounded grains have higher green strength than the sand with angular grains.

The moulding sands with angular grains require higher amounts of binder and moisture because of their greater specific surface area. The compound grains also require higher amounts of binder and moisture. The compound grains are least desirable in sand mixtures because they have a tendency to disintegrate at high temperatures. The round shape grain sands are best as moulding sand. The sub-angular grains may be taken as a best compromise as its characteristics lies between angular and round shape grains.

The size of the sand grains may be *fine*, *medium* or *coarse*. The fine sand is used for small and intricate castings so that all the details of the mould are brought out sharply. The medium sand is used in bench work and light floor work. As the size of the casting increases, the sand grains should be coarser to permit the easy escape of gases that are generated in the mould.

12·9 Preparation of Moulding Sand

The preparation of sand is required in order to obtain good castings. It consists of mixing of sand, tempering of sand and conditioning of sand. The efficient mixing of sand ingredients in proper ratio is very important. It may be done manually or mechanically. The natural sand does not possess all the qualities which the moulding sand should possess. It is, therefore, necessary that the natural sand should be mixed with other sands or other substances (such as clay, coal dust etc.) which possess that characteristic in a high degree. The clay is a necessary bond to the moulding sand and coal dust is the most widely used substance. The process of mixing consists of removing all foreign particles such fins, nails etc. from the sand grains by passing it through a sieve or by magnetic separators. The sand ingredients are mixed either by hand or by sand mixing machines. The mixing is continued till there is a uniform distribution of ingredients. The more uniform is this distribution, the better is the quality of sand.

The preparation of moulding sand by adding water in proper proportions is known as tempering of sand. The exact amount of water to temper a given amount of sand is impossible to state, because the sand conditions always vary. It can be mastered only through practical experience.

The preparation of moulding sand so that it becomes suitable for moulding purposes is called conditioning of sand. The proper conditioning of sand accomplishes uniform distribution of binder around the sand grains, controls the moisture content, eliminates foreign particles and aerates the sand so that it flows readily around and takes up the detail of the pattern.

12·10 Testing of Moulding Sand

We have discussed in the previous article that the moulding sand must possess certain properties for producing good moulds and castings. In modern foundries, the moulding sand is periodically tested in a sand testing shop to maintain the composition and properties. The successful production of good casting is dependent upon the uniform quality of materials used in their manufacture. Following are the various tests to be carried out in a sand testing shop :

1. *Moisture content test.* The moisture content may be determined by the loss of weight after evaporation or by means of a chemical reaction as discussed below :

(*a*) *By the loss of weight after evaporation.* This test is performed as follow :

(*i*) A 50 gm sample of tempered sand is weighed, dried from 105°C to 110°C, cooled to room temperature in a desiccator and reweighed.

(*ii*) The difference in the weights of the sample (in gm) is divided by the total weight of the sample taken (*i.e.*, 50 gm) to obtain percentage moisture content.

The modern foundries, a device known as moisture teller is widely used. This instrument blows pre-heated air through the moist sand sample, for about 5 minutes, contained in a moisture teller pan. The pan is equipped with a 500 mesh metal screen in the bottom. The sand sample is weighed directly in the moisture teller pan and determines the loss in weight of the sample.

(*b*) *By chemical reaction.* The moisture content may be determined by means of a chemical reaction as discussed below :

(*i*) A 10 gm sample of sand is placed in a special container along with a measure of powdered calcium carbide. The reaction will produce a measurable amount of acetylene gas proportional to the amount of moisture in the sand.

(*ii*) The moisture content is read directly from a calibrated scale on the instrument.

The instrument used for this purpose is known as *speedy moisture tester.* The test made with this device is not affected by other volatile materials. The test is fast, reasonably accurate and required a minimum of skill.

2. *Grain size or Fineness test.* This test is carried out to determine the percentage distribution of grain sizes in the sand. The test is performed on a dried-sand sample (50 gm) from which all clay substance is removed. It is conducted by screening sand grains by means of a set of graded sieves that are numbered according to the fineness of their mesh. The standard set of sieves according to American Foundrymen's Society (AFS) consists of eleven sieves and a pan. The meshes of these sieves are 6, 12, 20, 30, 40, 50, 70, 100, 140, 200 and 270.

The sand is placed on the coarsest sieve at the top and is vibrated by motor driven shakers for about 15 minutes. The weight of sand retained on each sieve is obtained and converted to a percentage basis. In order to obtain the fineness number, each percentage is multiplied by a factor (called multiplier) and added to obtain the total product. The AFS grain-fineness is calculated by using the following formula, *i.e.*,

$$\text{AFS grain-fineness number} = \frac{\text{Total product}}{\text{Total percentage of sand retained in screen}}$$

The determination of grain fineness number provides a rapid method of expressing the average grains size of sand. The following table shows a clear understanding how the AFS fineness number is calculated.

Table 12.1. Calculation of AFS grain fineness number.

S.No.	*Sieve No.*	*Openings (mm)*	*Amount retained on sieve gm*	*%*	*Multiplier*	*Product*
1	6	3.353	0	0	3	0
2	12	1.651	0	0	5	0
3	20	0.8533	0	0	10	0
4	30	0.590	0	0	20	0
5	40	0.414	1.50	3.00	30	90
6	50	0.295	0.30	0.60	40	24
7	70	0.208	2.50	5.00	50	250
8	100	0.147	0.50	1.00	70	70
9	140	0.104	8.20	16.40	100	1640
10	200	0.074	13.00	26.00	140	3640
11	270	0.053	12.80	25.60	200	5120
12	Pan		5.60	11.20	300	3360
	Total		44.40	88.80		14 194

$$\therefore \quad \text{AFS grain fineness number} = \frac{14\,194}{88.8} = 160$$

3. *Clay content test*. The clay content is determined by the loss of weight after washing of sample. The test is carried out as discussed below :

(*i*) A 50 gm sample of previously dried sand in weighed. It is placed in a mixing device and treated with a standard sodium hydroxide solution under controlled conditions.

(*ii*) After thorough washing, residue is dried and reweighed. The loss of weight is a measure of clay substance and may be expressed in percent by weight.

4. *Permeability test*. The permeability is expressed as the volume of air in cu. cm. that will pass per minute under a pressure of 1 gm/cm^2 through a specimen of sand 1 cm^2 in cross-sectional area and 1 cm in height. There are following four recognised conditions of permeability :

(*a*) *Basic permeability*. It is the permeability measured with a specimen of packed dry sharp sand.

(*b*) *Green permeability*. It is the permeability measured with a specimen made of moist moulding sand.

(*c*) *Dry permeability*. It is the permeability measured with a specimen made of moulding sand and dried at 105ºC to 110ºC.

(*d*) *Baked permeability*. It is the permeability measured with a specimen made of sand with thermosetting binders and baked at some temperature above 110°C.

The test is carried out as follows :

(*i*) A specimen of size 5.08 cm by 5.08 cm is placed in an instrument known as *permeability meter*.

(*ii*) A selected amount of air is forced through the specimen under controlled conditions.

(*iii*) The rate of flow of air is timed to calculate the permeability number.

(*iv*) The permeability number is calculated by the following formula :

$$P = \frac{vh}{p.a.t}$$

where
P = Permeability number to be determined,
v = Volume of air passing through the test specimen,
h = Height of specimen in cm,
p = Pressure of air in gm/cm^2,
a = Cross-sectional area of specimen in cm^2, and
t = Time taken by air to pass in minutes.

(*v*) The standard procedure requires 2000 cu cm of air to be forced through a specimen of sand 5.08 cm in height and 20.268 cm^2 in area. The drum in the permeability meter is designed to produce a constant pressure of 10 gm/cm^2. The formula may now be simplified by substituting these values.

$$\therefore \quad P = \frac{2000 \times 5 \cdot 08}{10 \times 20\ 268 \times t} = \frac{50 \cdot 12}{t}$$

$$= \frac{50 \cdot 12 \times 60}{t} = \frac{3007 \cdot 2}{t} \qquad \text{...(}t\text{ being in seconds)}$$

5. *Strength test*. The strength tests are carried out to test the holding power of various bonding materials in green, dry and core sand. Following are the various strengths to which a sand specimen may be tested :

(*a*) *Compressive strength*. It is measured when the specimen is caused to break under a compressive load.

(*b*) *Shear strength*. It is measured when the specimen is caused to break under a shearing load.

(*c*) *Tensile strength*. It is measured when the specimen is caused to break under tensile load.

(*d*) *Transverse strength*. It is measured when the specimen is caused to break under a bending load.

The compression and shear tests are conducted to determine the strength of green sand. These tests are carried on a universal testing machine, with sand specimen of size 5.08 cm by 5.08 cm, as discussed below :

(*i*) The compressive strength is obtained by applying a uniformly increasing load (at the rate of 2.5 N/s) on a sand specimen until it fails. The compression value in N/mm^2 is read directly on the scale.

(*ii*) The shear strength is obtained in the same way by changing the holding devices so that the load is applied on the upper half of the specimen at one end and on the lower half at the other end.

The tensile and transverse strength are often measured with a specially prepared specimens of core sand and baked.

6. *Mould and core hardness test.* The hardness of the mould achieved by ramming the sand can be measured by a mould hardness tester, which is like a pocket watch. The hardness is obtained by pressing a spring loaded steel ball of 5 mm diameter into the surface of mould. The depth of penetration is indicated on the dial in hundredths of a millimetre. The reading will not change, even with excessive pressure on the tester. The soft rammed moulds give a value around 40, medium rammed moulds 50 and hard rammed moulds 70.

12·11 Moulding Processes

The moulds may be made either by hand or moulding machines. The hand moulding is used for small work while the machine moulding is preferred for mass production work. The moulding processes in the ordinary commercial foundry may be classified into the following two groups :

1. According to the type of material used for the mould, and
2. According to the method used in making the mould.

These classifications are discussed, in detail, in the following pages.

12·12 Types of Moulds

The various types of moulds, according to type of material used, are as follows :

1. *Green sand mould.* The sand mould prepared from natural moulding sand in its green state is called *green sand mould.* The green sand mixture is prepared by mixing thoroughly silica sand (80 percent), clay (10 to 15 percent) and water (4 to 6 percent). This mixture should be properly tempered before it is used. The green sand moulds may be used for small and medium castings of ferrous and non-ferrous alloys. Since the green sand moulds contain moisture, therefore certain defects like blow holes may occur in casting. The surface finish of castings made in green sand moulds is also not good.

2. *Dry sand mould.* The sand mould prepared from fine grained sand mixed with suitable binder is called *dry sand mould.* These moulds must be baked in an oven. The dry sand moulds are stronger and may be handled more easily with less damage. These moulds are mostly used in small and medium-sized casting. For large castings, the dry sand moulds may be made in sections. These moulds may be used for many alloys but are more common in casting of steel.

3. *Skin-dried mould.* The sand mould with a dry sand facing and a green sand backing is called *skin-dried mould.* These moulds may be used for casting all ferrous and non-ferrous alloys. These moulds are more commonly used for large moulds. The skin-dried moulds are less expensive to construct than dry sand moulds but more expensive than green sand moulds. It requires less floor space and time for comparable sizes of dry sand moulding. The skin dried moulds are not so strong as dry sand moulds.

4. *Loam mould.* The mould made with loam sand (a mixture of sand and clay) is called *loam mould.* The loam sand also contains fire clay or ganisters. The loam moulds are similar to pit moulds and are used for large work where the pattern required will be too expensive to make. The mould is first built up with bricks or large iron parts. These parts are then plastered with a thick loam sand mixture consisting of silica sand (22 volumes), clay (5 volumes), coke (10 percent) and moisture (18-20 volumes). The shape of mould is obtained by either with sweep or skeleton patterns. The mould is then allowed to dry thoroughly so that it resist the heavy rush of molten metal and withstand heavy temperature. Since much time is required to make a loam mould, therefore these are not extensively used.

5. *Metal mould.* The mould made of metal is called *metal mould.* The metal moulds are used in casting of low melting temperature alloys. Since the castings produced by metal moulds have a

smooth finish, therefore much of the machine work is eliminated.

6. *Shell mould.* The mould made of two or more thin shell like parts consisting of thermosetting resin-bonded sand, is called *shell mould.* The resins include phenol formaldehydes, urea formaldehydes and poly esters. It is mixed with sand as a powder from 3 to 10 percent by weight. It may be applied as a liquid and then dried on the sand grains. The mixture for moulding must be dry and free flowing. The shell moulds are used for a special moulding process known as shell moulding (also known as croning process).

12·13 Methods of Moulding

The various methods used in making the mould are as follows :

1. *Bench moulding.* The moulding done on a bench of a convenient height to the moulder, is called *bench moulding.* It is usually used for small castings which are light in weight.

2. *Floor moulding.* The moulding done on the foundry floor is called *floor moulding.* In this method, the foundry floor itself acts as a drag and it may be covered with a cope or the mould may be cast open. It is used for all medium-sized and large castings.

3. *Pit moulding.* The moulding done in a pit instead of a flask is called *pit moulding.* It is used for extremely large castings. The pit acts as a drag part of the flask and a separate cope is used above it. The sides of the pit are brick lined and on the bottom there is a thick layer of cinders with connecting vent pipes to the floor level. Since pit moulds can resist high pressures developed by hot gases, therefore it greatly saves pattern expenses.

4. *Machine moulding.* The moulding done by a machine is called *machine moulding.* The *moulding machines perform the number of operations which the moulder does by hand. The ramming of sand, rolling the mould over, forming the gate and drawing out of pattern can be done by these machines much better and more efficiently than by hand.

5. *Sweep moulding.* The moulding done by using a sweep pattern is called *sweep moulding.* It is used for making moulds of large size and symmetrical in shape particularly of circular sections. The method of making a mould by using a sweep pattern is discussed in Chapter 10 (Art. 10.12).

12·14 Methods of Green Sand Moulding

Following are three methods commonly used for green sand moulding :

1. *Open-sand method.* It is the simplest form of green sand moulding, particularly used for solid patterns. In this method, the entire mould is made in the foundry floor or in a bed of sand levelled off in a wooden frame or moulding box (drag) raised above floor level for convenience in working and pouring. The moulding box is not necessary as the molten metal finds its own level. It is, therefore, essential that the bed of sand should be horizontal so as to obtain a plate of uniform thickness. The mould cavities are made deeper than the intended thickness of the casting. The overflow channels are cut, at the sides of the cavity, at the exact height from the bottom face of the mould to give the desired thickness. The sand must be rammed lightly, just hard enough to support the weight of metal, otherwise there is a danger that boiling may cause blow holes in the casting.

This method is only used for simple castings, floor plates, moulding boxes, weights etc. In other words, it is used for casting with flat tops in which it does not matter if the upper surface is decidedly rough.

2. *Bedded-in method.* When the upper surface of the casting is not flat or must be smoother than the rough surface produced by the open sand moulds, then it is necessary to provide a sand cover (or cope) for the open-sand bed. This is known as *bedded-in moulding.*

In this method, for small moulds, the pattern may be pressed or hammered down to bed into the sand to form the mould cavity. After the pattern is drawn, wires are thrust into the sand below the mould cavity and the vent holes connected to the atmosphere by another series of sloping vents which

* Refer Chapter 11 (Art. 11.5).

emerge in the sand surface outside the top box.

For large castings, from a solid pattern, the pattern is placed on sand bed, pressed or hammered into position and sand is filled in at the side upto the level of the mould joint. The pattern is then lifted away and facing sand is spread over the impression. The pattern is again placed and rammed up. It is sometimes necessary to remove and replace the pattern several times to ensure that the face of the mould has been properly prepared. This bedding-in method of moulding entails a high degree of moulding skill to produce an evenly rammed mould. After the joint is smoothed and parting sand spread over, a cope (top part of the moulding box) is placed over the pattern. The cope is rammed up, runners and risers cut and the cope is lifted leaving the solid impression in the floor. The pattern is drawn out, the surfaces of the parts of the mould finished, and the cope replaced in its correct position to complete the mould.

3. *Turn-over method*. This method is widely used for split patterns and for solid patterns. The complete process of making a green sand mould using split pattern and turn over method is discussed, in detail, in the following pages.

12·15 Making a Green Sand Mould by Turn Over Method

The procedure of making a green sand mould using a split pattern and turn over method is discussed in the following steps :

1. First of all, a flat board, known as *moulding board*, is placed on the floor with the lower half of the pattern placed in position.
2. The bottom half of the moulding box or flask known as *drag*, is placed on the moulding board, as shown in Fig. 12.2(*a*). The moulding box should be of such a size that it accommodates the pattern easily and some space is left around it for ramming the sand.
3. The tempered moulding sand is riddled on the pattern. The sand is pressed around the pattern with fingers.
4. The sand is then rammed in the drag by means of a hand rammer. It should be lightly rammed over the pattern and more heavily away from the pattern. In ramming the sand around the sides of the flask, the peen end of the rammer should be used first and then additional sand is placed into the drag as the sand is packed down. The inside area of the drag is then packed down with butt end of the rammer. It may be noted that if the sand is not sufficiently rammed, it will not hold together when handled or when the molten iron strikes it. On the other hand, if it is rammed too hard, it will not permit the steam and gas to escape when the molten iron comes into the mould.
5. After ramming, the excess sand is levelled off with a strike-off bar.
6. A vent wire is thrust into the sand in a number of places to reach the pattern so that small vent holes are made to escape of the gases when molten metal is poured.
7. A loose sand is now sprinkled over the mould and a bottom board is clamped in position, as shown in Fig. 12.2(*b*).
8. The drag is now rolled over and the moulding board is removed, exposing the pattern.
9. The surface of the sand is first smoothed over with a trowel and then covered with a dry parting sand. This is done to prevent the sand in the cope (upper part of moulding box) from sticking to the sand in the drag when the mould is separated to remove the pattern.
10. The top half pattern is now fixed in position with a cope on the drag, as shown in Fig. 12.2(*d*).
11. The facing sand is riddled over the pattern and the sprue and riser pegs fixed in position, the cope is filled with green sand. The operations of filling, ramming and venting of the cope is similar to that of drag.

12. The sprue and riser pegs are withdrawn, the top part lifted from the drag and rolled over to horizontal position.
13. Any loose sand is brushed from the faces of the mould and the joints smoothed.
14. Water from a swab is sprinkled on the sand around the pattern so that the edges of the mould hold firmly when the pattern is withdrawn.
15. A draw spike is driven into the patterns and rapped lightly in all directions in order to loosen the pattern. The patterns from both parts of the mould are withdrawn by lifting up the draw spike.

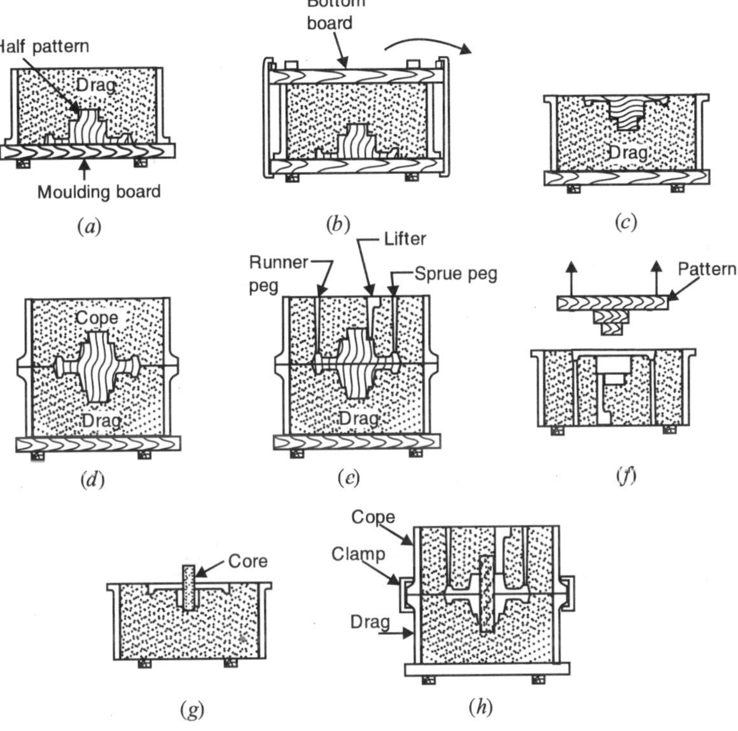

Fig. 12.2. Turn over method of green sand moulding.

16. A small passage known as *gate* must be cut from the mould at the bottom of the sprue opening which is made like a funnel to facilitate pouring of molten metal.
17. The centre core is now fixed in the print left by the pattern in the drag.
18. The mould surfaces may be sprayed, swabed or dusted with a prepared coating material containing silica flour and graphite. The composition of the coating material depends upon the type of material being cast. This coating material improves the surface finish of the casting and reduce possible surface defects.
19. The top part is rolled back, carefully lowered into position on the drag, and clamped or weighted. The mould is now ready for pouring, as shown in Fig. 12.2 (*h*).

In order to make a mould which requires more than one parting surface, such as a grooved pulley we may use two moulding boxes with dry sand cores as shown in Fig. 12.3 (*a*) or three moulding boxes to obtain two parting lines as shown in Fig. 12.3 (*b*).

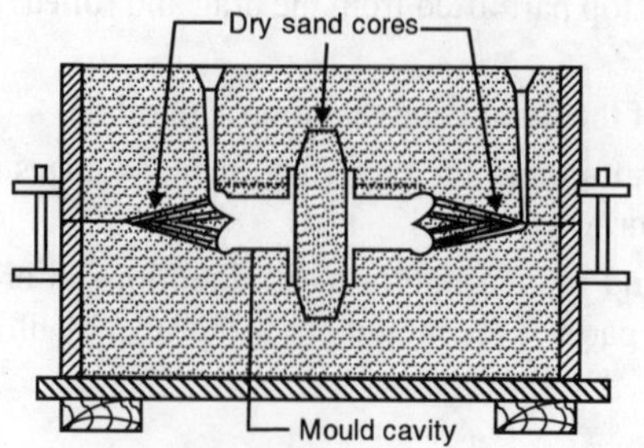

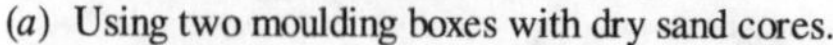
(*a*) Using two moulding boxes with dry sand cores.

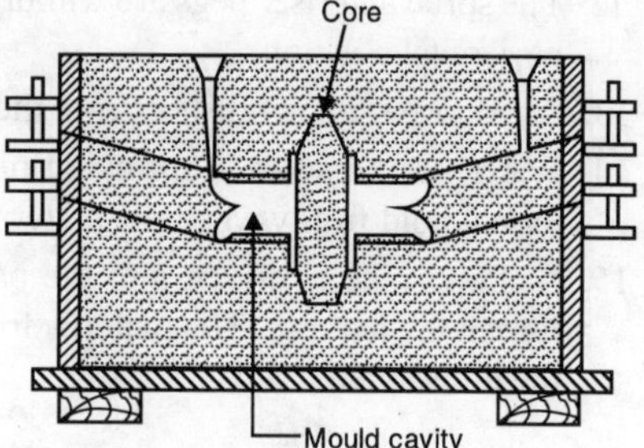

(*b*) Using three moulding boxes.

Fig. 12.3. Moulding of a grooved pulley.

12·16 Gates and Risers

The passage-way which serves to deliver the molten metal into the mould cavity is known as *gating system*. A gating system, as shown in Fig. 12.4. consists of the following parts :

1. A *pouring cup* which is a funnel-shaped opening in the upper surface of the cope above the sprue. It minimises the splash and turbulence and promote the entry of the clean metal only into the down sprue. In order to prevent the entry of dirt or slag into the down sprue, the pouring basin is provided with a *skim core*, *strainer core*, *delay screen* or a *sprue plug*.

2. A *down gate* or *sprue* which is a vertical opening (usually tubular) through the cope.

3. A *runner* which receives the metal from the down sprue and distributes to several gate passage ways (*i.e.*, ingates) around the mould cavity. A runner may be used in large castings.

4. An *ingate* is an opening (usually horizontal) which carries the metal from the runner to the mould cavity.

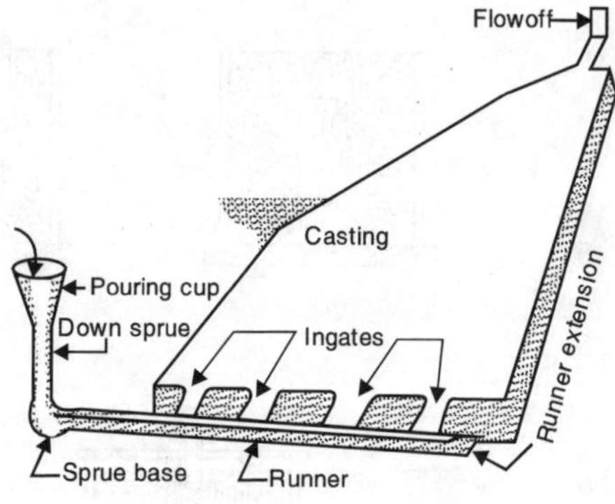

Fig. 12.4. Gating system.

A *riser* (or feed head) is an opening through the cope. Its main purpose is to feed the molten metal to the casting as it solidifies *i.e.*, to compensate for the shrinkage. The risers also serve as a large vent for generated steam and gases and afford a place for collecting loose sand or slag. The castings of heavy section and of high shrinkage alloys commonly require a riser. The riser, depending upon its location, may be described as *top riser* or *side riser* and may be either an *open riser* or a *blind riser*. Since risers are designed to stay liquid while the casting solidifies, *riser height* and *riser neck* are important dimensions as are those of the body of the riser itself. The riser distance and the shape of the *riser base* are additional important details that pertain only to side risers.

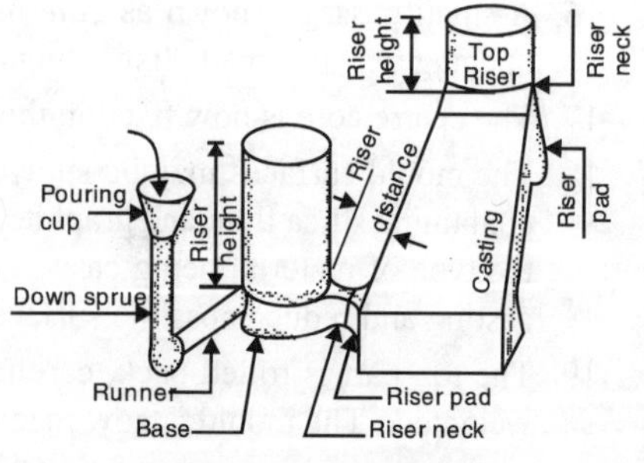

Fig. 12.5. Riser details.

The gates and risers are often designed to provide *directional solidification*. This requires that the solidification should progress from the mould surfaces to the hottest metal so that there is always hot metal available to compensate for shrinkage. To accomplish this, castings are riser-gated with metal entering the riser through a down sprue and runner, heating both the riser base and neck while flowing into mould cavity, as shown in Fig. 12.5.

12·17 Types of Gates

The molten metal can be directed into mould cavity in various ways. The each gating system depends upon its primary objectives. Thus, a gate may be designed for ease of moulding, to avoid turbulent flow, to prevent washing of sand from the mould walls and to avoid inclusion of slag with the metal entering the mould. The various types of gates are discussed below :

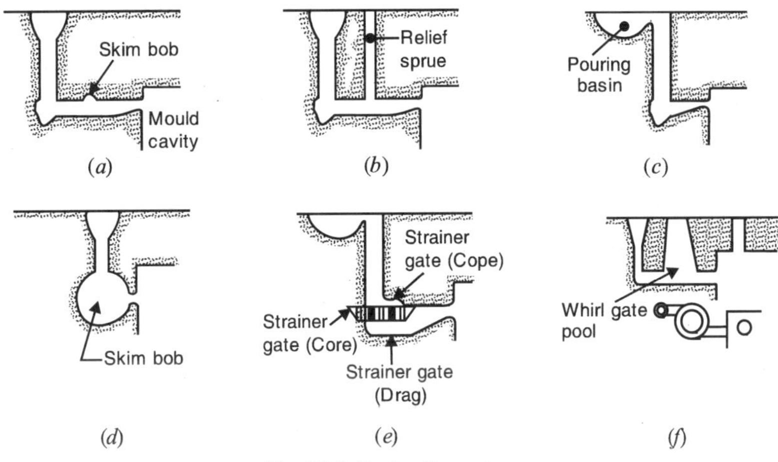

Fig. 12.6. Parting line gates.

1. *Parting line gates*. The runners and gates which are formed along the parting line separating the cope and drag portions of the mould, are called *parting line gates*. They may contain devices such as *skim bobs* or *relief sprues*, as shown in Fig. 12.6 (*a*) and (*b*), to collect slag or relieve pouring pressure. The use of a *pouring basin* to serve this function is shown in Fig. 12.6 (*c*). A *shrink bob*, as shown in Fig. 12.6 (*d*) serves the dual function of slag collector and metal reservoir to feed the casting as it shrinks. The use of *core inserts* to filter the metal and to prevent erosion of the mould is shown in Fig. 12.6 (*e*). The *whirlgate runner*, as shown in Fig. 12.6 (*f*) prevents the slag to enter into the mould.

2. *Bottom gates*. The gates which enter into the mould cavity near the bottom of the drag are called *bottom gates*. A bottom gate, as shown in Fig. 12.7 (*a*), is particularly used for steel castings or reduce erosion and gas entrapment and to prevent splashing.

3. *Horn gate*. A horn gate, as shown in Fig. 12.7 (*b*), is a type of bottom gate. There is one disadvantage in its use that it has a tendency for producing a fountain effect in the casting. However, it is a means of bottom gating without the necessity of a core for the gate.

4. *Branch gate*. A branch gate, as shown in Fig. 12.7 (*c*), is designed either to feed a single casting at several points or a number of individual castings.

5. *Top gate*. A top gate with strainer core, as shown in Fig. 12.7 (*d*), opens directly into the cavity from above. These gates are not used so often but occasionally used to feed the molten metal more efficiently. The *wedge gates*, *finger gates* and *pencil gates* are all types of top gating system and these are shown in Fig. 12.7 (*e*), (*f*) and (*g*) respectively.

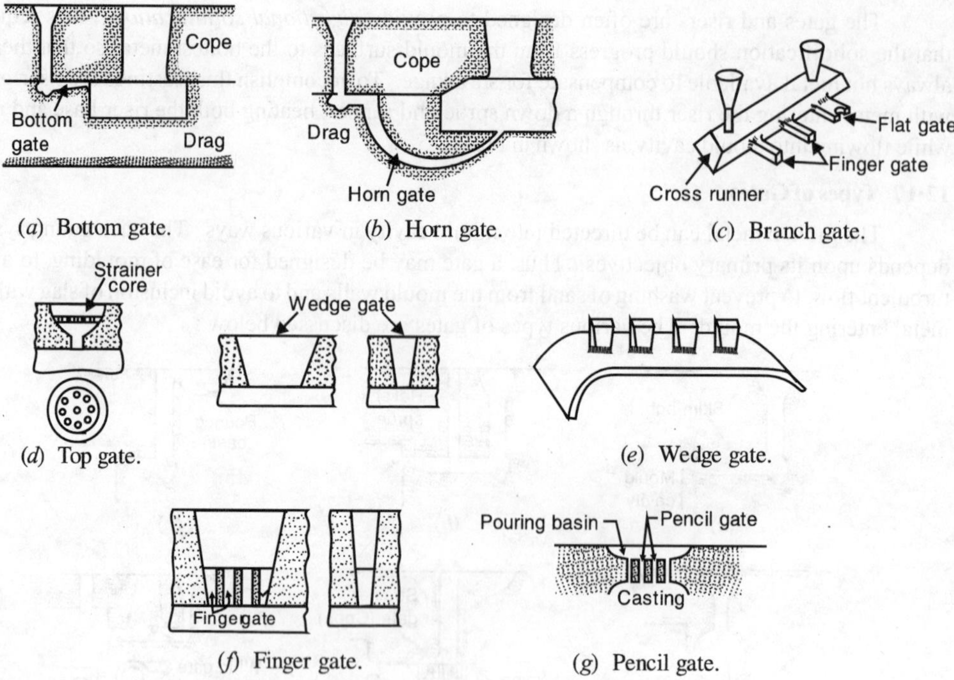

(a) Bottom gate. (b) Horn gate. (c) Branch gate.

(d) Top gate. (e) Wedge gate.

(f) Finger gate. (g) Pencil gate.

Fig. 12.7. Types of bottom gates.

12·18 Moulding Methods with Typical Patterns

Following are the various moulding methods employed specially for moulding with the patterns of typical shapes.

1. *Temporary top part or false cope method*. When the pattern is irregular, it will not be possible to put it on a flat board to begin with moulding. In order to prepare a mould for such a pattern, a temporary top part or false cope method may be used. It is very similar to bedded-in moulding as discussed in Art. 12.14. In this method, a top part is filled with sand and strickled off level. The shape of the pattern is dug out of the sand and the pattern is bedded-in to the proposed joint line. The face of the temporary top part is smoothed, parting powder sprinkled over it and moulding proceeds in the usual way as discussed earlier. A drag is rammed up , the two boxes clamped together and turned over. The temporary top part is removed and discarded at this stage, if only one casting is required. The sand is knocked out and the empty box is used to prepare a real cope with gating system in the usual manner.

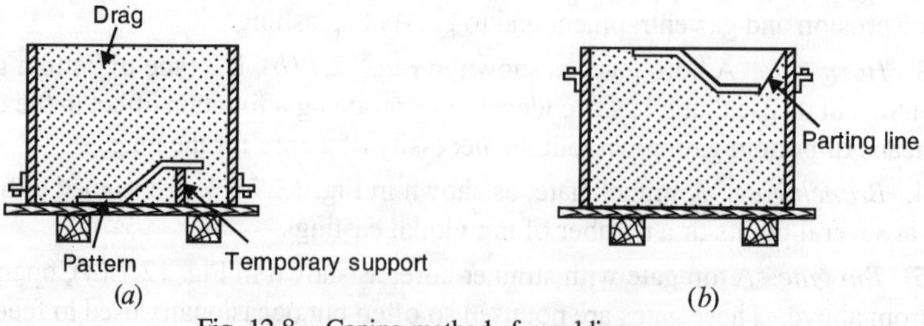

Fig. 12.8. Coping method of moulding.

2. *Coping down method*. A coping down method is used for irregular pattern having the parting line not in one plane. Such a pattern is first placed on a moulding board with some temporary

support, as shown in Fig. 12.8 (*a*). The drag is then placed over the moulding board and rammed as usual. The drag is now rolled over and a new irregular parting surface is formed to the parting line of the pattern, as shown in Fig. 12.8 (*b*). This new parting surface sloping gradually is called *coping down*. The mould is completed as usual.

3. *Green-sand match method.* A green-sand match method is used when castings in large quantities are required from an irregular pattern as used in the above method. It is a quicker method than the coping down method. A match is nothing but a hard support for the pattern during ramming of the drag.

In this method, a match similar to the slope of the pattern is made either of wood, plaster, quick setting compounds of gypsum, portland cement or green sand, as shown in Fig. 12.9. When sand is used for preparing a match, it should be strengthened by an extra addition of 4 to 5 percent of linseed oil. The match acts as a support for the pattern during ramming. The mould is prepared as usual.

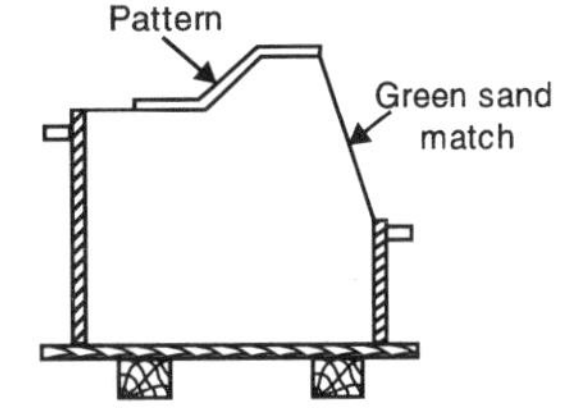

Fig. 12.9. Green sand match method of moulding.

This method saves considerable time to be spent in coping down of individual drags when the match is not used.

4. *Draw-back method.* The patterns which have projections and are made in a single piece, as shown in Fig. 12.10 (*a*), are difficult to remove from the sand. In order to overcome this difficulty, a draw-back method is used.

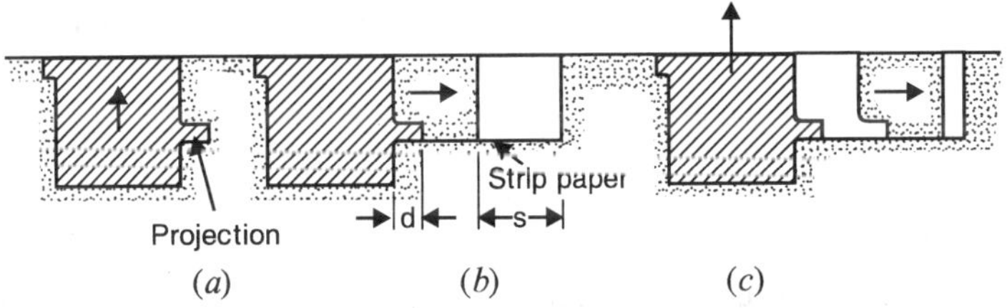

Fig. 12.10. Draw-back method of moulding.

In this method, a stiff paper is rammed up in the moulding sand which forms a parting surface and cavity for shifting the draw back, as shown in Fig. 12.10 (*c*). Fig. 12.10 (*b*) shows that when the draw back is shifted, the pattern can be easily withdrawn. After the pattern is removed, the draw back is again shifted to its correct position and the space behind it is filled with sand.

5. *Loose piece method.* Sometimes the projections of the patterns are made in loose pieces in order to facilitate its removal from the sand. A pattern with a loose piece is shown in Fig. 12.11 (*a*). The loose piece is fixed to the pattern with a pin. The preparation of the mould and removal of the

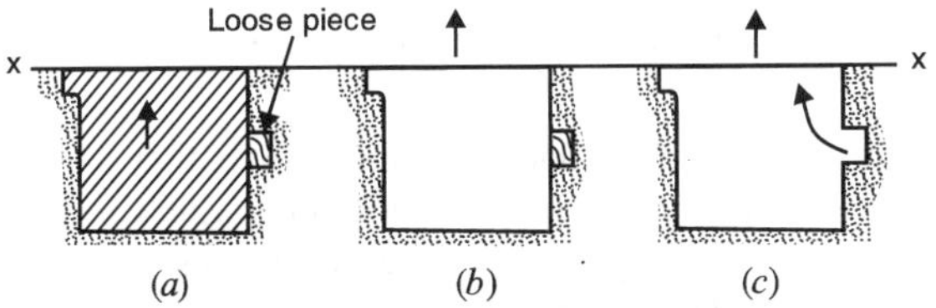

Fig. 12.11. Loose-piece method of moulding.

major portion of the pattern is carried out in the usual way, as shown in Fig. 12.11 (*b*). The loose piece is taken out of the sand by means a tool called skewer, as shown in Fig. 12.11 (*c*).

6. *Dry sand core method.* This method of moulding is widely used for small wheels, or pulleys with grooved rims and for shrouded gearing. The groove in the rim of a pulley, as shown in Fig. 12.12,

may be produced by placing dry sand cores (usually dried oil sand) in the groove while moulding.

7. *Cover core method.* A cover core is a flat core used to cover the cavity in the mould. Fig. 12.13 (*a*) shows the pattern of a steam hammer tup or block which has to be cast face downwards.

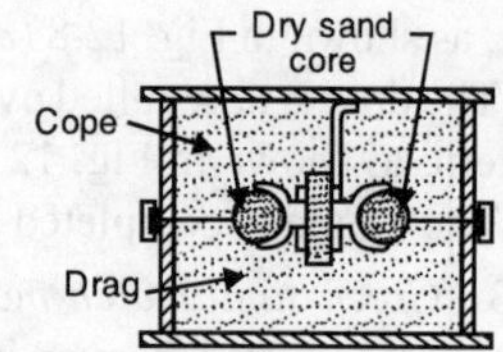

Fig. 12.12. Dry sand core method of moulding.

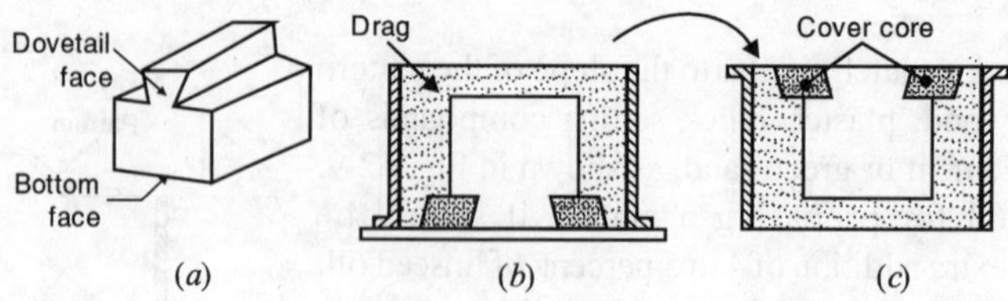

Fig. 12.13. Cover core method of moulding.

First of all, the drag is rammed up with dovetail face of the pattern downward and with a cover placed at each dovetail recesses, as shown in Fig. 12.13 (*b*). The drag is now rolled over, cover removed and the pattern is withdrawn. The cover cores are then replaced in position, as shown in Fig. 12.13 (*c*).

12·19 Cores

The cores are defined as sand bodies used to form the hollow portions or cavities of desired shape and size in a casting. These are not moulded with the pattern, but separately in a *core box. In order to achieve the utmost of intricacy in castings, cores must be collapsible after the metal is poured. The sand cores possesses this property and therefore most frequently used to produce almost any desired degree of casting intricacy. In addition to forming internal cavities in the casting, the cores serve a number of purposes as follows :

1. The cores may be used to construct a complete mould as in pit moulding.
2. The cores may be used to form a part of a green-sand mould. The patterns having projections, which can not be moulded, may be formed by placing a core in the mould after the pattern is drawn.
3. The cores strengthen or improve a mould surface.
4. The cores may be used as a part of the gating system. The strainer cores to prevent foreign matter in the molten metal, *pouring-basin cores* to provide a cup or basin on the top of the mould and *splash cores* to prevent metal erosion in gates and runners are the examples of cores used in gating system.

Since the cores are subjected to intense heat of the molten metal when it is poured into the mould, therefore a good core should have high strength, good permeability, high refractoriness, and sufficient collapsibility.

When cores are made of moulding sand in the moist state, they are called *green-sand cores*. When cores are made with a mixture of sand and special binders which develop strength when baked, they are called *dry sand cores*. The type of sand used depends upon the size of the core and temperature of molten metal poured into the mould. The cores made of smaller grain size of sand will give smoother surface of casting in contact with the cores.

---

* The core boxes are discussed in Chapter 10 (Art 10.19).

12·20 Types of Cores

We have discussed in the previous article that the cores are made in core boxes and placed in the core seat in moulds formed by core prints. A *core print* is an extra projection on a pattern and forms a seat to support and hold the core in the mould. The various types of cores, depending upon their position, are as follows :

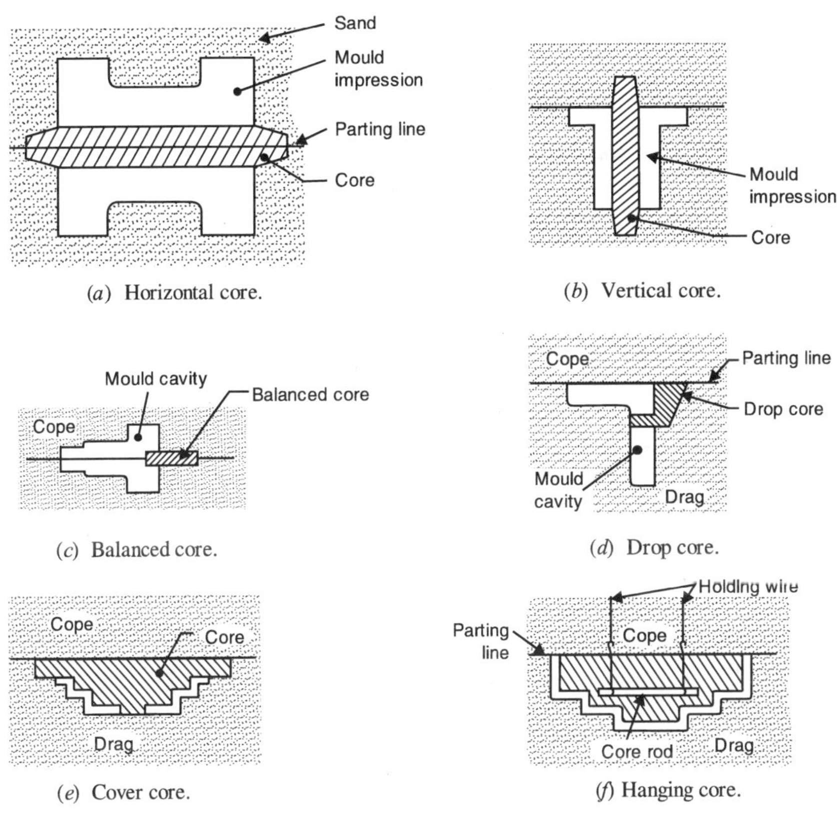

(*a*) Horizontal core. (*b*) Vertical core. (*c*) Balanced core. (*d*) Drop core. (*e*) Cover core. (*f*) Hanging core.

Fig. 12.14. Types of cores.

1. *Horizontal cores*. These cores are placed horizontally in the core prints at the parting line of the mould such that one-half remains in the cope and the other half in the drag, as shown in Fig. 12.14 (*a*). This type of core is often used on a split or two-piece pattern.

2. *Vertical cores*. These cores are placed vertically in the mould, as shown in Fig. 12.14 (*b*). The core print is provided on the cope and drag sides of the pattern. The upper end of the core requires a considerable taper so as not to tear the sand in the cope when the flask is assembled.

3. *Balanced cores*. These are similar to horizontal cores, but these cores are supported at one end only, as shown in Fig.12.14 (*c*). Therefore the core print should be made of sufficient length in order to prevent the core from falling into the mould. The balanced cores are used to produce blind holes along a horizontal axis in a casting.

4. *Drop cores*. The drop cores (also known as *wing core* or *stop off core*), as shown in Fig. 12.14 (*d*), are used when a hole or a cavity to be cored is not in line with the parting surface *i.e.*, it is either above or below the parting line.

5. *Cover cores or hanging cores*. The cover cores, as shown in Fig. 12.14 (*e*), are used when the entire pattern is rammed in the drag and the core is required to be supported from top of the mould. This type of core usually requires a hole through the upper part to permit the metal to reach the mould. If the core hangs from the cope and does not have any support at the bottom in the drag, then it is called *hanging core*, as shown in Fig. 12.14 (*f*).

12·21 Core Binders

The core sand may be simply defined as sand mixtures suitable for cores. The core-sand mixtures consist of sand grains, binders for green and cured strength and other additives used for special purposes.

The core binders serve to hold the sand grains together and impart strength, resistance to erosion and to breakage and degree of collapsibility. They may be classified as organic and inorganic binders. The organic binders are combustible and are destroyed by heat. Hence they contribute a degree of collapsibility to the core sand mixture. The inorganic binders are not combustible and may have considerable strength at high temperatures, may have resistance to erosion and may be relatively non-collapsible, depending upon their nature.

The organic binders include core oils, resins, cereal binders, sulphite binders, pitch, wood flour and plastics. The core oils such as linseed oil, from 0.5 to 3 percent by weight, is used in many core-sand mixtures (also called *oil sand* for cores). A common mixture uses 40 parts of river sand and 1 part of linseed oil. A core oil with 45 percent linseed oil, 30 percent resin oil and remaining kerosene oil has proved very successful. The core oils contribute little to the green strength of core sands, but are most commonly used for baked strength. The oil sand cores should be baked for 2 hours at 170°C to 190°C. The driers such as ammonium nitrate (0.10 percent or more) is used in core-oil bonded sands to hasten the curing process. The baking time may be reduced by 20 to 80 percent by driers.

The cereal binders (starches, molasses, dextrin, casein), soluble in water, are commonly used for green strength. The percentage of cereal binders must be less than three times the percentage of core oil in the mixture.

The sulphite binders (lignin) are water soluble compounds of wood sugars produced as a by-product of paper making. They provide strength on evaporation of water and produce very high surface hardness during baking. These binders are used upto 1.5 percent and are useful in green sand and dry sand core mixtures.

The pitch is a solid-coal-tar by product of the making of coke. It is used as a ground powder (also called rosin) upto 3 percent or sometimes more. It develops hot strength to resist metal.

The wood flour (ground hardwood cellulose) usually less than 1 percent is used in core sand for decreasing cracking and increasing collapsibility of cores.

The thermosetting plastics including urea and phenol formaldehyde are being successfully used as core binders. These are mixed with other ingredients like silica flour, cereal binder, water, kerosene and a parting liquid. The urea resin binders are baked at 160° to 170°C and the phenolic binders at 180° to 200°C. Both these binders respond to dielectric heating and are completely combustible under the heat of the metal. The binders have high adhesive strength, moisture resistance, burn out characteristic and ability to provide a smooth surface to the core.

The inorganic binders include fire clay, bentonites, silica flour and iron oxide. They are used to obtain green strength, baked strength, hot strength or a smooth finish. Since they are all finely pulverised materials, therefore they greatly increase the amount of oil which is necessary in oil-sand mixtrures.

The simplest core-sand mixture commonly used comprises sand, 1 percent core oil, 1 percent cereal and 2.5 to 6 percent water.

Note : The cores which have high hot strength and do not collapse sufficiently can cause tears in a casting as it shrinks around the core.

12·22 Core Making

The cores are mostly made of a core sand mixture consisting of sand grains and organic binders which provide green strength, cured strength and collapsibility. The green strength is required so that the core sand may be moulded to shape *i.e.*, for core making. The core obtains its real strength and hardness when it is cured to develop the bonded strength. The simplest core sand mixture commonly used comprises sand, 1 percent core oil, 1 percent cereal and 2.5 to 6 percent water. The other ingredients like kerosene oil, parting liquid and inorganic binders varies according to the type of metal being cast and its temperature, size and shape of the casting, method of making the core etc. The core making procedure consists of the following steps :

1. *Mixing of core sand.* The first consideration in making the dry sand cores is to mix and prepare the sand properly in order to obtain best cores. If the binder is dry, it should be thoroughly mixed with the sand before any moisture is added. The mixture must be homogeneous so that the core will be of uniform strength throughout. The mixing of sand is performed in paddle mixers or mullers.

2. *Ramming of core sand.* The cores are usually made in core boxes. The core box is filled with core sand, rammed and struck off. The ramming may be done manually or with machines. Small cores are usually made at core benches.

3. *Venting of core.* Since the cores are often surrounded by hot metal after the casting is poured, therefore the cores are required to be sufficiently permeable to allow core gases to pass through the core and into the mould or atmosphere. The vent holes are provided for the escape of gases. These vent holes are usually made with wires or rods.

4. *Reinforcing of core.* Some cores require internal reinforcing to prevent breakage or shifting when the casting is poured. The wires or arbors within the sand serve this purpose. Fig. 12.15 shows the wires being used to support the barrel of the cylinder core. The wires should be surrounded with the sand to keep them from fusing to the casting and making their removal difficult.

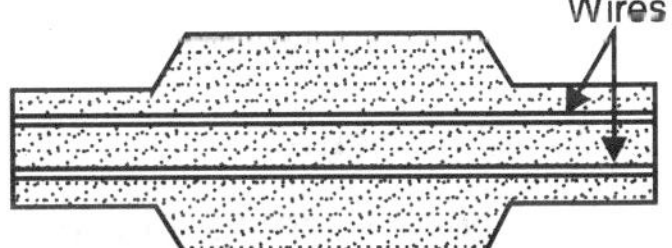

Fig. 12.15. Reinforcing of core.

5. *Baking of core.* The core is transferred to a cope plate or drier for baking. This is done by placing the plate over the core box, inverting both and removing the core box away from the core. The core which is to be baked on a core plate, must have a flat surface resting on it. The cores with no flat surfaces must be supported on a core drier until they are baked.

The cores are baked at temperatures upto about 260°C to develop strength obtainable from the organic binders in the core sand. During baking, moisture is driven off first, holding the core temperature to 100°C. Then the core oil or other binder changes chemically and molecularly from a liquid to a solid by oxygen absorption and polymerization as temperature rises from 200°C to 260°C. The baking cycles of 2 to 6 hours at 200°C to 240°C are quite commonly used. The proper baking is essential for a core to perform satisfactorily when the metal is poured. The under baked cores give off much gas and can cause a number of defects. The overbaked cores may collapse too soon and break or erode before the casting is solidified. The proper baking depends upon the good baking equipment if the core sand mixture is correct. The baking may be done in ovens, dielectric bakers and radiant bakers.

6. *Cleaning of cores.* The cleaning of cores consists of trimming, brushing, coating and muddling. The *trimming* is done to remove fins arising from loose joints or loose pieces in the core box or other sand projections by rubbing them or filing with an emery stone, core file or other abrasive tool. The *brushing* of cores is performed with a brush to remove loose sand. The brushing is required

on cores which, before baking, have been bedded in sand rather than on a core drier. The *coating* of cores is done with a refractory or protective materials which improve their resistance to molten metal or produce a better casting. The coating may be applied by spraying, dipping or swabbing. The *muddling* is a localised coating to make the core completely smooth. The graphite or red talc moistened to a putty with water may be used to fill up any cavities, rough spots, soft rammed areas or the joint lines of assembled cores. Another mud mixture consisting of 94 percent silica flour, 3 percent western bentonite and 3 percent dextrin moistened to a putty with water may also be used. The mud is applied by hand, rubbed into the cavities and smoothed off. The cores which have been coated or rubbed may cause trouble with defects in casting because of steam evolution when the metal is poured. This is avoided by drying in a core oven for 20 minutes to 1 hour.

7. *Sizing of cores*. The sizing usually involves gauging the core to see if its size is correct and then removing material. The gauges may be used to check critical dimensions. The templates, pieces of sheet metal cut out to the contour required of the core, may be used to determine whether the core has sagged or slumped during baking. The cores may be inserted in a gauging fixture which will check a number of dimensions at once. Some cores are made slightly oversize and ground to correct height. This may be done by hand filing with a scraper or on a core grinding machine.

8. *Joining of cores*. Sometimes, cores are made of two or more pieces before they can be used. They are held together by pasting, bolting or leading.

The pasting is commonly used on small work. A core paste, usually a mixture of talc, dextrine, flour, molasses, water or other ingredients, is applied to the surfaces to be joined.

The bolting is useful in larger core work. The recessed holes are left in the cores so that they may be bolted together. The bolt and nut heads are covered with a core plug or mud.

Sometimes, leading is used where a strong joint is desired in small cores. The cavities are left in the core parts to be joined. These are filled with molten lead. When the lead freezes, it holds the core together, just as in bolting.

12·23 Core Setting

The core setting is the operation of placing cores in moulds. The cores must be of correct size and positioned properly with respect to the mould cavity so that cored-out cavities are in their required location in casting. The cores are positioned in the mould by core prints. When the metal is poured, the cores may shift and rise unless they are securely anchored. A good core setting, therefore, requires advance planning so that the cores may be correctly positioned and firmly held when the metal is poured. The small cores are placed in the mould by hand whereas larger cores may require hoist or crane service.

12·24 Core Shifting and Chaplets

A little consideration will show that due to the upward thrust of the molten metal, the long thin cores tend to float more easily than short chunky cores. If the metal is molten for a long time, the core has greater opportunity to float, shift or crack. If the core gets shifted from its position in the mould, it will result in a displaced cavity and hence a defective casting is produced. In order to overcome the vertical movement of the core due to upward thrust of the molten metal, chaplets are used. The chaplets are metal forms placed between the mould and core surfaces. The various types of chaplets are shown in Fig, 12.16 (*a*) to (*f*).

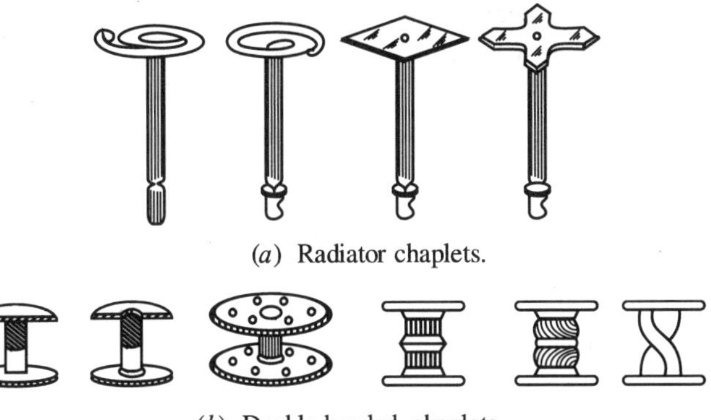

(*a*) Radiator chaplets.

(*b*) Double headed chaplets.

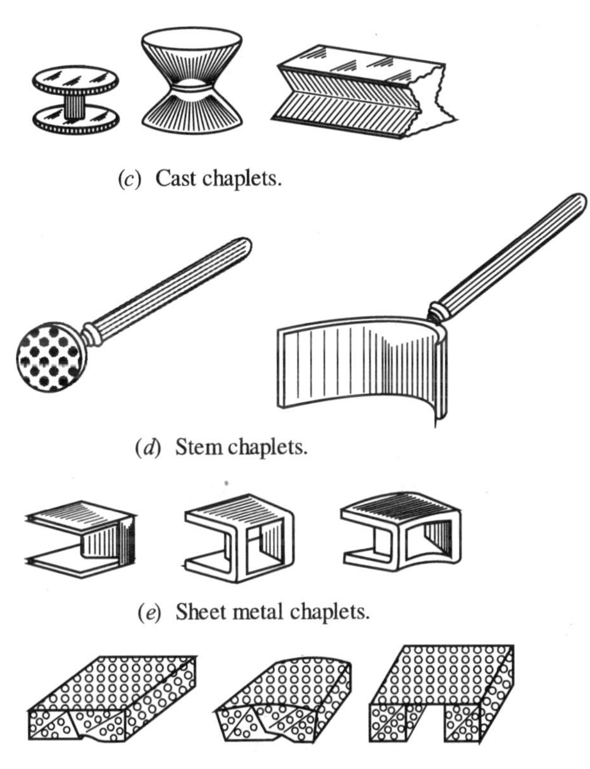

(*c*) Cast chaplets.

(*d*) Stem chaplets.

(*e*) Sheet metal chaplets.

(*f*) Perforated chaplets.

Fig. 12.16. Various types of chaplets.

QUESTIONS

1. What material is used for making a mould ? Write the composition of a good moulding sand.

2. Describe the different types of binders used in a silica sand to improve its binding quality.

3. Explain the following :

(*a*) Permeability ; (*b*) Cohesiveness ;

(*c*) Adhesiveness ; and (*d*) Collapsibility.

4. Discuss the various types of moulding sands used in a foundry.
5. Describe the following with the help of suitable sketches, whereas necessary :
   (*a*) Floor moulding ; (*b*) Bench moulding ;
   (*c*) Pit moulding, and (*d*) Machine moulding.
6. State the advantages of the following :
   (*a*) Dry sand moulds over green sand moulds ; and
   (*b*) Loam moulds over dry sand moulds.
7. Describe, with neat sketches, the procedure of making a green sand moulds by turn-over method.
8. What do you understand by gating system ? Discuss, with neat sketches, the various types of gates.
9. Sketch and describe the following methods of moulding :
   (*a*) False cope method, (*b*) Coping down method,
   (*c*) Draw back method, (*d*) Loose piece method,
   (*e*) Dry sand core method, and (*f*) Cover core method.
10. How the cores are defined ? Why they are used ? Explain with sketches, the different types of cores ?
11. Write the different stages in core making ? What do you understand by core setting ?
12. What are chaplets ? Why they are used ? Sketch the various types of chaplets.

OBJECTIVE TYPE QUESTIONS

1. The property of sand due to which it evolves a great amount of steam and other gases is called
   (*a*) collapsibility (*b*) permeability (*c*) cohesiveness (*d*) adhesiveness
2. The property of sand due to which the sand grains stick together is called
   (*a*) collapsibility (*b*) permeability (*c*) cohesiveness (*d*) adhesiveness
3. The adhesiveness is the property of a sand due to which
   (*a*) it evolves a great amount of steam and other gases
   (*b*) the sand grains stick together
   (*c*) it cling to the sides of a moulding box
   (*d*) none of the above
4. Green sand is a mixture of
   (*a*) 30% sand and 70% clay (*b*) 50% sand and 50% clay
   (*c*) 70% sand and 30% clay (*d*) 90% sand and 10% clay
5. The binder commonly used in making oil sand, is
   (*a*) clay-type binder (*b*) organic-type binder
   (*c*) inorganic-type binder (*d*) any one of these
6. The sand used for making cores is
   (*a*) green sand (*b*) dry sand (*c*) loam sand (*d*) oil sand
   (*e*) parting sand (*f*) none of these
7. A sand employed on the faces of the pattern before moulding is called
   (*a*) green sand (*b*) dry sand (*c*) loam sand (*d*) oil sand
   (*e*) parting sand (*f*) none of these
8. The purpose of a gate is
   (*a*) deliver molten metal into the mould cavity

(*b*) act as a reserviour for the molten metal

(*c*) feed the molten metal to the casting in order to compensate for the shrinkage

(*d*) deliver molten metal from pouring basin to gate

9. The purpose of a riser is to

(*a*) deliver molten metal into the mould cavity

(*b*) act as a reserviour for the molten metal

(*c*) feed the molten metal to the casting in order to compensate for the shrinkage

(*d*) deliver molten metal from pouring basin to gate

10. Cores are used to

(*a*) form internal cavities in the casting

(*b*) improve mould surface

(*c*) form a part of a green sand mould

(*d*) all of the above

(*e*) none of the above.

ANSWERS

1. (*b*)	2. (*c*)	3. (*c*)	4. (*c*)	5. (*b*)
6. (*d*)	7. (*e*)	8. (*a*)	9. (*c*)	10. (*d*)

13

Special Casting Processes

1. Introduction. 2. Permanent Mould Casting. 3. Semi-permanent Mould Casting. 4. Slush Casting. 5. Die Casting. 6. Centrifugal Casting. 7. Investment Casting. 8. Shell Moulding Process. 9. Continuous Casting. 10. Defects in Casting. 11. Cleaning of Castings. 12. Inspection of Castings. 13. Design of Castings.

13·1 Introduction

The sand moulds may be used for casting ferrous and non-ferrous metals, but these moulds can be used only once, because the mould is destroyed after the metal has solidified. This will increase the cost of production. The sand moulds also, can not maintain better tolerances and smooth surface finish. In order to meet these requirements, following casting methods may be used :

1. Permanent mould casting.
2. Semi-permanent mould casting,
3. Slush casting,
4. Die casting,
5. Centrifugal casting,
6. Investment or lost wax casting, and
7. Shell moulding process.

We shall now discuss these methods, in detail, in the following pages.

13·2 Permanent Mould Casting

A casting made by pouring molten metal into a mould made of some metallic alloy or other material of permanence is known as *permanent mould casting.* The term *gravity die casting* is another name for this type of permanent mould casting. Since the permanent moulds are costly, therefore it is recommended for mass production only. The permanent moulds are impractical for large castings and alloys of high melting temperatures, but they can be used advantageously for small and medium sized non-ferrous castings. The mould is not destroyed after removing the casting and may be reused many times.

The iron and steel moulds are suitable for non-ferrous castings. The steel moulds coated with refractory material, such as graphite, are successfully used for production of iron castings. The castings produced by these moulds require less skill, limited floor space, have improved surface finish, high dimensional accuracy and less number of rejections than sand castings. The bronze moulds are also used for low melting metals.

13·3 Semi-permanent Mould Casting

It is similar to permanent mould casting except that sand cores are used at some points, instead of metal cores. The use of sand cores results in lower cost and also allows cored openings of irregular shape or undercuts of such type where metal cores will be too costly and difficult to handle on production basis.

13·4 Slush Casting

The slush casting is a special application involving the use of a permanent mould. The process is based on the solidification of molten metal by the chilling effect. It is used for casting low-melting temperature alloys. The metals like gold, silver, aluminium, zinc, lead and their alloys have low melting point. It is also used for making hollow castings without the use of cores. In this process, the molten metal is poured into a metallic mould. The metal is retained in the mould long enough for the outer skin to solidify. Finally, the mould is turned over to remove metal still in molten condition. Due to the chilling effect, only a thin layer of metal sticks to the mould surface which is taken out by opening the halves of the mould. This method is only adopted for ornaments and toys of non-ferrous alloys.

13·5 Die Casting

The die casting (also known as *pressure die casting*) may be defined as that casting which uses the permanent mould (called die) and the molten metal is introduced into it by means of pressure. The castings produced by die casting method require very little machining. The dies are usually made in two parts which must be locked securely before molten metal is forced into them under high pressures of 7 to 700 MPa. The pressure may be obtained by the application of compressed air or by hydraulically operated pistons. The ferrous alloys are not yet commercially die-casted because of their high pouring temperature. Following are the two types of die casting machines commonly used for die casting.

1. *Hot chamber die casting machine.* The hot chamber die casting machine of the submerged type is shown in Fig. 13.1. The molten metal is forced in the die cavity at pressures from 7 to 14 MPa. The pressure may be obtained by the application of compressed air, as shown in Fig. 13.1(*a*) or by a hydraulically operated plunger, as shown in Fig. 13.1(*b*).

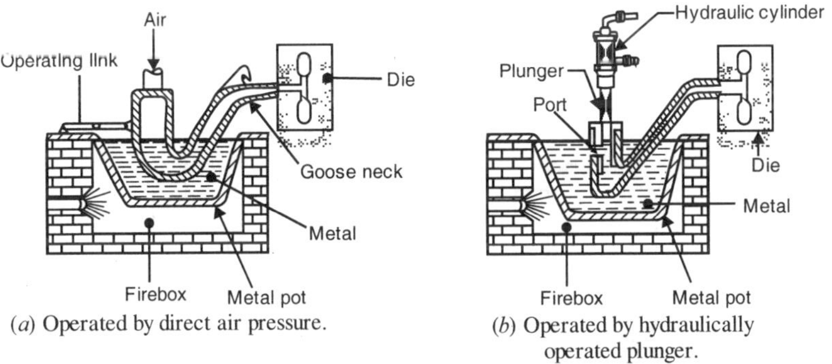

(*a*) Operated by direct air pressure.

(*b*) Operated by hydraulically operated plunger.

Fig. 13.1. Hot chamber die casting machine.

In the first method, the goose neck is lowered into the molten metal for filling it. It is then raised and connected to the die neck. A suitable mechanism is provided to raise and lower the goose neck. The compressed air at a pressure of about 2.5 to 5 MPa is now injected into the goose neck to force the molten metal into the die.

In the second method, the plunger acts inside a cylinder formed at the end of the goose neck, which is immersed in a pot of molten metal. A port is provided near the top of the cylinder to allow the entry of the molten metal. The downward stroke of the plunger pushes the molten metal through the goose neck into the die.

The hot chamber die casting machine is used for casting zinc, tin, lead and other low melting alloys.

2. *Cold chamber die casting machine.* In cold chamber die casting machine, the melting unit is usually separate and molten metal is transferred to injection mechanism by ladle. The pressure on the casting metal in cold chamber die casting machine may vary from 21 to 210 MPa and in some

cases may reach 700 MPa. The greater pressures are required for semi- molten alloys to compensate for reduced fluidity resulting from low pouring temperatures. This process is used for casting aluminium, magnesium, copper base alloys and other high melting alloys. The cold chamber die-casting machine, as shown in Fig. 13.2(*a*), consists of a pressure chamber of cylindrical shape fitted with a piston or ram that is usually operated by hydraulic pressure. A measured quantity of molten metal is brought in a ladle from the melting pot to a chamber and forced into the closed die sections by applying hydraulic pressure upon the piston. The complete cycle is completed in the following four steps :

1. The metal is loaded in the chamber.
2. The plunger forces the metal into the die cavity, as shown in Fig. 13.2(*b*).
3. After the metal solidifies, the die is opened, as shown in Fig. 13.2(*c*).
4. The casting, together with the slag of the excess metal, is ejected from the die, as shown in Fig. 13.2(*d*).

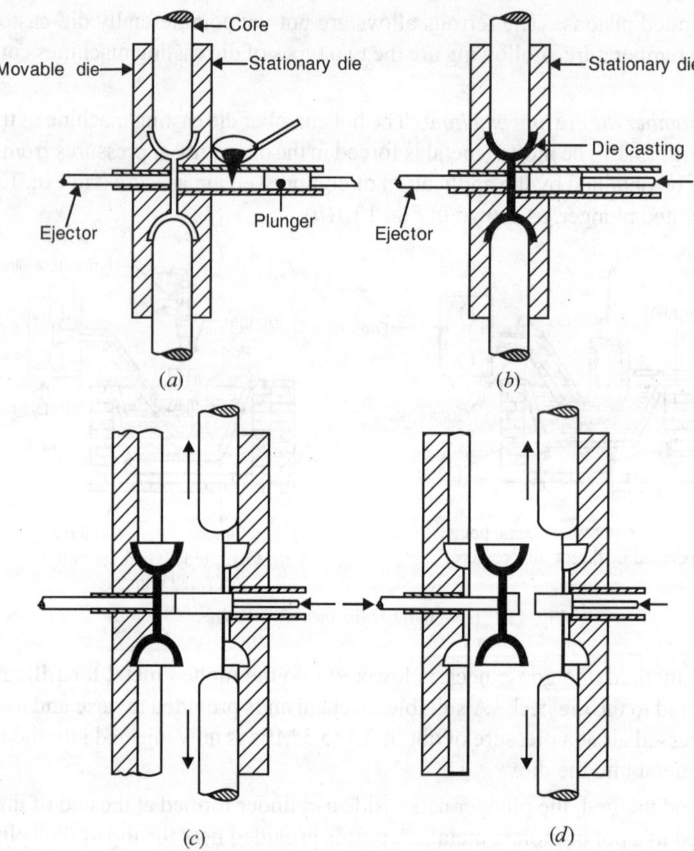

Fig. 13.2. Cold chamber die casting machine.

Following are the advantages and disadvantages of die casting :

Advantages

1. The rapid and economical production of large quantities of identical parts can be achieved.
2. The parts having smooth surfaces and close dimensional tolerances may be produced, very little machining is required.

3. The parts having thin and complex shapes can be casted accurately and easily.
4. The die casting requires less floor area than is required by other casting processes.
5. The castings produced by die-casting process are less defective, owing to increased casting soundness.
6. The rapid cooling rate produces high strength and quality in many alloys.
7. The inserts, if required, can be casted easily in desired places.
8. The die retains its trueness and life for longer periods. For example, the life of a die for zinc base castings is upto one million castings, for copper base alloys upto 75 000 castings, and for aluminium base alloys upto 50 000 castings.

Disadvantages

1. The cost of equipment and die is high.
2. There is a limited range of non-ferrous alloys which can be used for die castings.
3. The die castings are limited in size.
4. It requires special skill in maintenance.

13·6 Centrifugal Casting

A casting process in which molten metal is poured and allowed to solidify while the mould is revolving, is called *centrifugal process*. The castings produced under this centrifugal force are called *centrifugal castings*. The centrifugal casting may be classified into the following three types :

1. *True centrifugal casting.* This process is especially employed for casting articles of symmetrical shape *e.g.*, cast iron pipes, sleeves, steel gun barrels and other castings of cylindrical form.

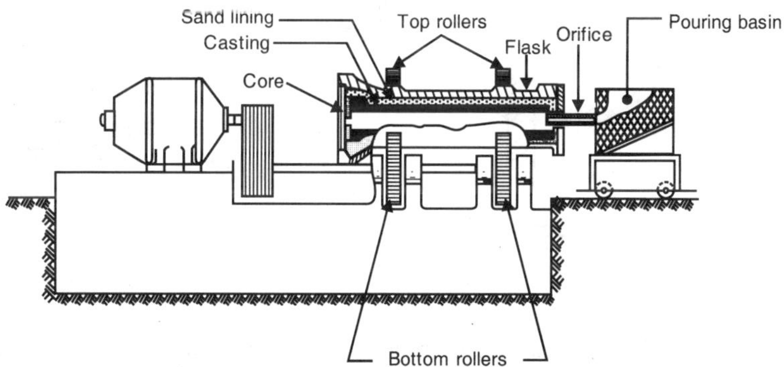

Fig. 13.3. Centrifugal casting.

In this process, as shown in Fig. 13.3, the mould is made of metal and lined with refractory material or sand. The molten metal is poured by ladle into the cavity of rapidly rotating mould. The centrifugal force directs the fluid metal to the inner surface of the mould with considerable pressure where solidification occurs forming hollow castings. The outside of mould is covered by water bath for quick cooling of metal.

The main advantages of a true centrifugal casting process are :

(*a*) It is a quick and economical than other methods.
(*b*) It eliminates the use of risers, feed heads, cores etc.
(*c*) The ferrous and non-ferrous metals can be casted by this process.
(*d*) The castings produced by this process have dense and fined grained structure with all impurities forced back to the centre where they can be frequently machined out.

2. *Semi-centrifugal casting.* This process is employed for making large size castings which are symmetrical about their own axis such as pulleys, spoked or disc wheels, gears, propellers etc.

In this process, as shown in Fig. 13.4, the mould is rotated about its vertical axis in a properly balanced state and the metal is poured into a central sprue from where it enters the hub. From the hub, it is forced outward to the rim by the centrifugal force. It may be noted that several moulds can be cast at one time.

If a central hole is required in the casting, a dry sand central core may be used in the centre. The speed of rotation of these moulds is lower than that of true centrifugal casting. This is due to the reason that the moulds are of larger size and there is a tendency of the molten metal to flow out of the mould joint. Because of the lower speed, the pouring pressure produced is low and thus the impurities are not effectively separated from the metal.

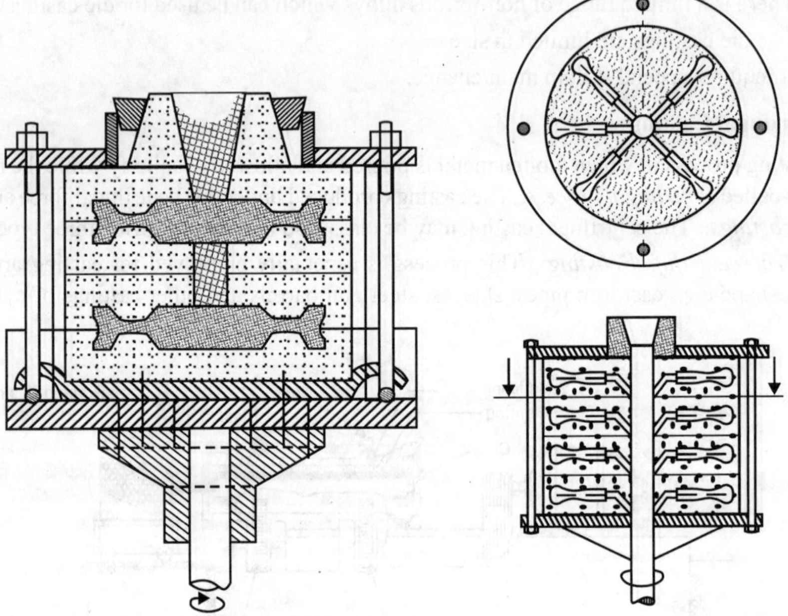

Fig. 13.4. Semi-centrifugal casting.

Fig. 13.5. Centrifuging.

3. *Centrifuging.* The castings of irregular shape can be produced by this method and a large number of small size castings can be produced at one time.

In this process, as shown in Fig. 13.5, a number of small mould cavities are made symmetrically around a common central sprue and the metal is fed to them by radial gates. In this case, the mould cavities are not rotated about their own axis, but they are rotated about the central sprue which acts as the axis of rotation. Since the centrifugal force used to fill the mould cavities is not same for all the castings, therefore it is not a purely centrifugal process.

13·7 Investment Casting

It is also known as *lost wax process* or *precision casting.* The castings produced by this method are within very close tolerances and do not require subsequent machining. The procedure adopted for investment casting is as follows :

1. First of all, a metal die for casting the wax patterns is made.
2. The wax patterns and gating systems are produced from the metal dies by injection. The wax is injected into the mould at 50°C to 80°C and at a pressure of 3.5 to 7MPa.

3. The wax assembly is dipped into a slurry of a refractory coating material. A typical slurry consists of silica flour suspended in ethyl silicate solution of suitable viscosity to produce uniform casting after drying. After dipping, the assembly is coated by sprinkling it with silica sand and allowed to dry.

Note : Sometimes pre-coating is not used and the wax pattern is directly invested in the moulding material. In this case, the moulding mixture must be vacuumed to remove air bubbles which may lodge next to the pattern.

4. The coated wax assembly is now invested in the mould. This is done by inverting the wax assembly on a table, surrounding it with a paper-lined steel flask and pouring the investment moulding mixture consisting of either sand 95%, water 27 to 31% and 5% alumina cement or sand with 3% or more ethyl silicate or sodium silicate around the pattern. The mould material settles by gravity and completely surrounds the pattern as the work table is vibrated. The moulds are then allowed to dry in air for 2 to 3 hours.
5. The wax is melted out of the hardened mould by heating it in an inverted position at 90°C to 150°C. The melted wax may be collected and re-used.
6. The mould is again heated at the rate of 40°C to 70°C per hour from about 150°C to 1000°C for ferrous alloys and 650°C for aluminium alloys. The temperature is controlled so that the mould is at a temperature desirable for pouring the particular alloy. The investment moulds may be poured under simple gravitational force or under the force of applied air pressure or by centrifugal force.

Following are the advantages and disadvantages of investment casting :

Advantages

1. The close tolerances (± 0.05 mm) are easily maintained in average work.
2. It produces extremely smooth surfaces.
3. It eliminates most machining operations including thread cutting and gear tooth forming.
4. It is adaptable to all metallic alloys.

Disadvantages

1. The investment moulds as well as the materials from which they are made are single purpose, therefore they can not be reused. This increases cost of production.
2. The larger objects are impractical for investment casting due to equipment size limits.

13·8 Shell Moulding Process

The shell moulding process (also called croning process) consists of making a mould that has two or more thin, shell like parts consisting of thermosetting resin-bonded sand. The resins include phenol-formaldehydes, urea formaldehydes and poly esters. It is mixed with sand as a powder from 3 to 10 percent by weight. It may be applied as a liquid and then dried on the sand grains. The mixture for moulding must be dry and free flowing. The procedure for shell moulding is as follows :

1. A metal pattern, heated to about 175°C to 350°C, is clamped over a box containing sand mixed with thermosetting resin, as shown in Fig. 13.6(*a*).
2. The box and pattern are inverted for a short time, as shown in Fig. 13.6(*b*). The mixture when comes in contact with hot pattern, it causes an initial set and builds up a coherent sand shell next to the pattern. This takes 5 to 20 seconds only. The thickness of this shell is about 6 mm to 18 mm and is dependant on the pattern temperature, dwell time on the pattern and the sand mixture.
3. The box and pattern are brought in its original position, as shown in Fig. 13.6(*c*). The shell of resin bonded sand is retained on the pattern surface, while the unaffected sand falls into box. The shell, still on the pattern is cured by heating it in an oven from 250°C to 350°C for 1 to 3 minutes. The time depends upon the type of resin and shell thickness.

4. The assembly is removed from the oven and the shell is stripped from the pattern by ejector pins as shown in Fig. 13.6(*d*). In order to obtain clean stripping, a silicon parting agent may be sprayed on the pattern.
5. The shell halves are assembled with clamps and supported in a flask with backing material. The shell mould is now ready for pouring. The complete process of shell moulding is shown in Fig. 13.6(*e*).

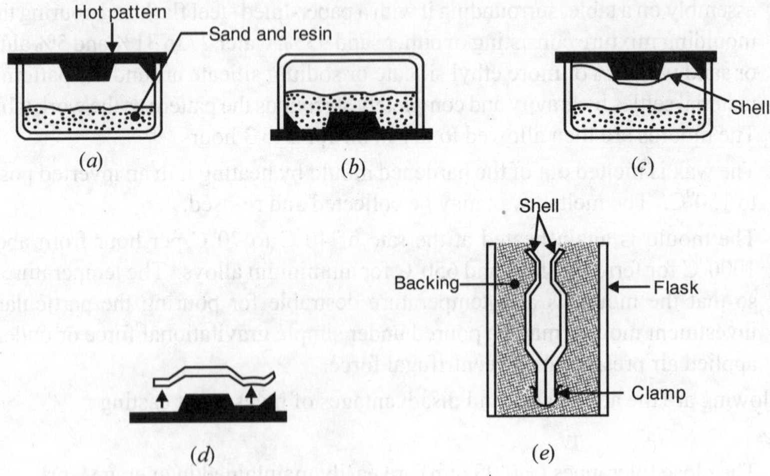

Fig. 13.6. Shell moulding process.

Following are the advantages and disadvantages of shell moulding.

Advantages

1. The shell cast parts can be produced with dimensional tolerances of ± 0.2 mm.
2. A very smooth surface is generally obtained.
3. The dimensional accuracy and smooth surfaces reduce cleaning and machining costs.

Disadvantages

1. The initial cost of metal patterns and other specialised equipment is high.
2. The resin binder is more expensive than other binders.

13·9 Continuous Casting

This process is generally employed for the production of rods, squares, pipes, tubes, sheets etc. It consists of pouring the molten metal into the upper end of a long vertical metal mould (open at both ends) and cooling rapidly by water. The solidified product is withdrawn by gravity in a continuous length from the lower end of the mould. The following two processes are commonly used for continuous casting of various metals and alloys :

1. *Asarco process.* It is one of the most popularly used process. In this process, as shown in Fig. 13.7(*a*), the molten metal is fed through gravity pull, into the mould lined with graphite.

The molten metal is solidified by the circulating water in the jacket around the open mould. The pulling rolls are provided below the mould to withdraw the solidified metal from the mould at a uniform speed. A saw is provided below the rolls to cut the product to desired length. This process is widely used for copper and bronzes.

2. *Reciprocating process.* In this process, as shown in Fig. 13.7(*b*), the copper mould reciprocates up and down with the surrounding water jacket. The molten metal is poured into the

mould at a uniform speed, through a pipe connected to the furnace. A valve controls the flow of molten metal. The solidified portion of the casting is withdrawn by the rolls at a constant speed. The movement of the rolls and the reciprocating motion of the rolls are properly controlled by means of cams.

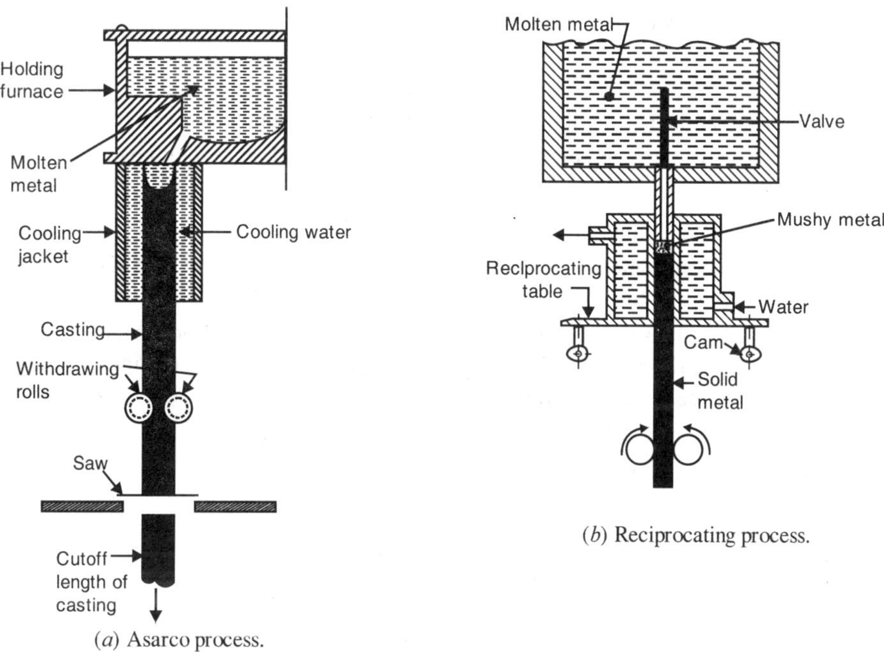

(*a*) Asarco process.

(*b*) Reciprocating process.

Fig. 13.7. Continuous casting.

13·10 Defects in Casting

The defects in a casting may be due to pattern and moulding box equipment, moulding sand, cores, gating system or molten metal. Some of the defects and their reasons are discussed below :

1. *Mould shift.* It results in a mismatching of the top and bottom parts of a casting, usually at the parting line. It occurs due to following reasons :

(*a*) misalignment of pattern parts, due to worn or damaged patterns, and

(*b*) misalignment of moulding box or flask equipment.

This defect can be prevented by ensuring proper alignment of the pattern, moulding boxes, correct mounting of pattern on pattern plates etc.

2. *Core shift.* It is an abnormal variation of the dimensions which are dependent on core position. It is caused by

(*a*) misalignment of cores in assembling cored-moulds,

(*b*) undersized or oversized coreprints, and

(*c*) by using incorrect size of chaplet.

This defect can be eliminated by providing the core at the proper place and must be gripped properly in the sand

3. *Swell.* It is an enlargement of the mould cavity by molten metal pressure resulting in localised or general enlargement of the casting.

It is due to the following reasons :

(*a*) insufficient ramming of sand,

(*b*) insufficient weighting of the mould during casting, and

(*c*) pouring of molten metal too rapidly or too hard.

The swells are avoided by the proper ramming of sand and uniform flow of molten metal into the mould.

4. *Fins and flash.* These are thin projections of metal not intended as a part of casting. These usually occur at the parting line of the mould or core sections. These are caused by

(*a*) excessive rapping of the pattern before it is withdrawn from the mould,

(*b*) insufficient weight on the top part of the mould, and

(*c*) loose clamping of the mould.

In order to avoid this defect, sufficient weight should be placed on the top part of the mould so that the two parts fit tightly together.

5. *Sand wash.* It usually occurs near the ingates as rough lumps on the surface of a casting. The sand that has been washed away appears on the upper surfaces of the casting as rough holes or depressions. This is due to the following reasons :

(*a*) soft ramming of sand,

(*b*) weak sand,

(*c*) poor pattern, and

(*d*) insufficient draft.

This defect is avoided by the proper ramming of sand.

6. *Shrinkage.* It is a crack in the casting or dishing on the surface of a casting which results from unequal contraction of the metal during solidification. This is due to the following reasons :

(*a*) improper location and size of gates and runners,

(*b*) inadequate risers,

(*c*) lack of directional solidification,

(*d*) incorrect metal composition, and

(*e*) incorrect pouring temperatures.

This defect can be eliminated by the use of feeders and chills at proper locations to promote directional solidification.

7. *Hot tear.* It is an internal or external ragged discontinuity in the metal casting resulting from hindred contraction occurring just after the metal has solidified. This defect is due to the following reasons :

(*a*) abrupt changes in section, inadequate filleting of inside corners, and improper placement of chills.

(*b*) poor collapsibility of mould and core materials which will place extra stress on certain details.

(*c*) improper pouring temperature.

In order to eliminate this defect, abrupt changes in section should be avoided. The pouring temperature should be correct and there should be even rate of cooling.

8. *Sand blow or blow hole.* It is an excessively smooth depression on the outer surface of a casting. This defect is also called *blow holes.* This defect is due to the following reasons :

(*a*) high moisture content in moulding sand,

(*b*) low permeability of sand,

(c) hard ramming of sand,

(d) defective gating system, and

(e) improper venting of sand.

This defect can be removed by proper venting, completely drying up the mould, selecting proper sand with required permeability and proper in-gate system for the flow of molten metal.

9. *Core blow.* It is an excessively smooth depression on the inner surface of a cored cavity or a gas pocket immediately above a cored cavity. This defect is caused by using insufficient baked cores.

Thus the cores should be sufficiently baked before using.

10. *Honeycombing or slag holes.* These are smooth depressions on the upper surfaces of the casting. They usually occur near the ingates. This defect is due to imperfect skimming of the metal or due to poor metal.

This defect can be avoided by preventing the slag from entering along with the molten metal.

11. *Scabs.* These are patches (*i.e.*, slightly raised areas) of sand on the upper surface of casting. This defect is due to the following reasons :

(a) uneven ramming of sand, and

(b) slow or intermittent running of metal.

The proper ramming of sand and uniform flow of the molten metal into the mould can eliminate this defect. Another method to remove this defect is to mix additives such as wood flour, sea coal or dextrine into the sand.

12. *Cold shuts and misruns.* These occur when the mould cavity is not completely filled and an incomplete casting results. This defect is due to the following reasons :

(a) too small gates,

(b) too many restrictions in the gating system,

(c) pouring head is too low,

(d) faulty venting of the moulds, and

(e) metal lacking in fluidity.

In order to eliminate these defects, the casting should be designed keeping in mind the fundamental principles of gating and risering. The thin sections should be preheated and the molten metal should be poured at the correct temperature.

13. *Pour short.* It occurs when the mould cavity is not completely filled because of insufficient metal. It is due to the following reasons :

(a) interruptions during pouring operation, and

(b) insufficient metal in the ladles being used to pour the metal.

In order to avoid this defect, the ladle should have sufficient molten metal at the correct temperature.

14. *Metal penetration.* It occurs when the alloy being cast tends to penetrate into sand grains and causes a fused aggregate of metal and sand on the surface of the casting. It is due to the following reasons :

(a) soft rammed sand,

(b) moulding sand and core sand being too coarse,

(c) improper use of mould and core washes will cause penetration, and

(d) excessive metal temperature or increased fluidity of metal.

This defect can be eliminated by removing the above mentioned reasons.

15. *Run-outs and bust-outs.* These permit drainage of the metal from the cavity and result in incomplete castings. These are due to the following reasons :

(*a*) A pattern that is too large for a given flask or pattern placed too close to the flask edge results in a weak spot and cause run-out,

(*b*) the match plate surfaces that are out of parallel or uneven results in a poorly formed parting line and cause a run-out,

(*c*) inadequate mould weights or clamps will permit the cope to lift which results a run-out,

(*d*) improper sealing of mould joints causes run-outs,

(*e*) excessive pouring pressures may cause run-out, and

(*f*) misalignment of cope and drag may promote a run-out.

The corrective measures taken in respect of the above reasons will prevent this defect.

16. *Rough surface finish.* It is merely a lack of sufficient smoothness in the casting. It is due to the following reasons :

(*a*) soft ramming of sand,

(*b*) coarser sand,

(*c*) hard pouring or too high metal fluidity, and

(*d*) improper use of mould and core washes often promote rough casting surfaces.

This defect can be avoided by using a proper mould and ramming of sand.

17. *Crush.* It is an irregularly shaped cavity or projection on the castings caused by the displacement of the sand at the mould joints or core prints, which usually occurs when the mould is being closed. It occurs due to the following reasons :

(*a*) badly made mould joints,

(*b*) excessive pressure on the sand surface, and

(*c*) too large cores or too small core prints.

This defect can be eliminated by taking proper care in placing the cope over the drag. The sand in the cope should be rammed properly.

18. *Warpage.* It is unintentional and undesirable deformation that occurs during or after solidification. It is due to the following reasons :

(*a*) continuous large flat surface on castings, indicating a poor design, and

(*b*) no directional solidification of casting.

This defect can be eliminated by modifying the casting design and proper directional solidification.

13·11 Cleaning of Castings

The cleaning of castings generally refers to all operations involved in the removal of adhering sand, gates, risers, or other metal not a part of the casting. The cleaning operations may also include a certain amount of metal finishing or machining to obtain the required casting dimensions. Following are the various cleaning operations usually performed on a casting :

1. *Rough cleaning.* It is that part of cleaning operation which includes removal of gates and risers. The following points may be noted in connection with rough cleaning :

(*i*) If the casting is of brittle material, the gating system may be broken off by impact when the castings are dumped and vibrated in shake-out or knock-out devices. Flogging with a hammer or sledge is a positive means of rough cleaning by impact.

(*ii*) If the casting is of ductile material, rough cleaning may be done with mechanical cut-off machines employing band saws, abrasive cut-off wheels and metal shears.

(*iii*) If the risers and sprues on steel castings are very large, these may be removed by cutting torches.

(*iv*) If the risers are large and cast of oxidation-resistant alloys, they can not be cut well with oxygen cutting torch alone. In such cases, *powder cutting* (in which a stream of iron powder is introduced into the oxygen torch flame) is employed.

Note : *Gouging* and *powder washing* are processes allied to torch and powder cutting which assist in cleaning up castings. Gouging is used to clean out or remove surface defects on steel castings. It is performed with the cutting torch and involves preheating the defective area and cutting it or washing it with a low velocity, large bore oxygen stream. The slag inclusions, blow holes, cracks, fins and wires may be cleaned out by this process. The powder washing is a similar process, but is used on more oxidation resistant alloys. In this process, iron powder is required as in powder cutting. Most of the trimming of riser pads, heavy fins, chills and other metal protuberances on medium and large steel castings may be done in this way.

2. *Surface cleaning.* It includes cleaning of interior and exterior surfaces when sand, scale and other adhering materials are involved. Following are the procedures used for surface cleaning :

(*i*) *Tumbling.* The sand, scale and some fins and wires may be removed by tumbling in a barrel-like machine known as tumbling mill. The tumbling action can be combined with sand, grit or similar materials to aid the process. The wet tumbling using caustic water solution is employed to help in cleaning the casting and to suppress the dust.

(*ii*) *Blasting.* The blasting is the most rapid means of removing sand and scale. The abrasives employed are sand, metal grit and metal shot. The sand blasting is performed by using coarse sand as abrasive and air as the carrying medium. The grit or shot blasting is done by throwing the metallic particles by centrifugal force from a rapidly rotating wheel.

(*iii*) *Other surface cleaning methods.* A number of methods of casting surface cleaning other than those mentioned above are in use. The other methods such as wire brushing, buffing, pickling and various polishing procedures may be used to aid in the surface cleaning.

3. *Trimming.* The trimming is used either after, before or during the initial surface cleaning. It involves the removal of fins, gate and riser pads protruding beyond the casting surface, chaplets, wires, parting-line, flash and other similar unwanted appendages to the casting which are not a part of its final dimensions. The following are the procedures used for trimming :

(*i*) *Chipping.* It may be done by hand with a hammer and chisel or by pneumatic chipping hammers. It is used to remove fins, gate and riser pads, wires etc., and to remove cores.

(*ii*) *Grinding.* The grinding or snagging of castings is used to remove excess metal. It may be done by portable grinders, stand grinder and swing-frame grinders.

4. *Finishing.* The latter stage of cleaning is often referred to as finishing. In some cases, cleaning is complete after trimming operations, but others may require additional surface finishing such as machining, chemical treatment, polishing, buffing, painting to put them into a suitable appearance.

13·12 Inspection of Castings

The inspection of castings after cleaning is necessary to determine the presence of any defects not readily visible. It comprises those operations which check the quality of castings and result in their acceptance or rejection. Following are the various methods of inspection :

1. *Destructive inspection method.* In destructive inspection method, the sample of casting is destroyed during inspection. This test is used for testing mechanical properties such as tensile strength, percentage elongation, hardness etc. These tests are performed on special test bars or on pieces cut from a sample casting.

2. *Non-destructive inspection method.* In non-destructive inspection methods, the inspection

is done without destroying the casting.

Following are the various methods of non-destructive inspection :

(*a*) *Visual inspection.* It is the most frequently used method for detecting various casting defects, *e.g.,* cracks, tears, dirt, blow holes, scabs, metal penetration, shifts, run-outs, poured short, swells and many others. The ultimate aim of visual inspection is to ensure that the outward appearance of the casting looks good.

(*b*) *Dimensional inspection.* The dimensional inspection of casting involves the principles of gauging. It is necessary to check whether the certain details are within tolerances or not. The surface plates, height and depth gauges, layout tables, dividing heads, go and not-go gauges, snap and plug gauges etc. may be applied for dimensional inspection of castings.

(*c*) *Pressure testing.* The pressure testing is used to locate leaks in a casting or to check the overall strength of a casting in resistance to bursting under hydraulic pressure. The pipes and tubes are pressure tested.

(*d*) *Radiographic inspection.* The radiographic inspection (X-ray or gamma-ray technique) is used for inspecting internal defects of a casting. The rays are passed through the test piece and the intensity of transmitted rays is recorded on a photographic film. The internal cracks and defects will show on the film as darker areas.

3. *Magnetic particle inspection.* This method of inspection is used on magnetic ferrous castings for detecting invisible surface or slightly sub-surface defects.

In this method, the casting surface or areas to be inspected is magnetised and finely ferromagnetic particles are applied on it. When a current or magnetising force is passed through the metal, the very small surface and sub-surface cracks and porosity will interrupt the magnetic field which cause ferro-magnetic particles to concentrate around the defect.

4. *Fluorescent penetrant.* This method of inspection is used to find minute pores and cracks on the surface of castings that may be missed even under magnification. In this method, a fluorescent penetrating oil mixed with whiting powder is applied to the casting surface by dipping, spraying or brushing. After the surface has been wiped dry, the oil will creep out of cracks or other defects and become visible at those places.

13·13 Design of Casting

An engineer must know how to design the castings so that they can effectively and efficiently render the desired service and can be produced easily and economically. In order to design a casting, the following factors must be taken into consideration :

1. The function to be performed by the casting,
2. Soundness of the casting,
3. Strength of the casting,
4. Ease in its production,
5. Consideration for safety, and
6. Economy in production.

In order to meet these requirements, a design engineer should have a thorough knowledge of production methods including pattern making, moulding, core making, melting and pouring etc. The best design will be achieved only when one is able to make a proper selection out of the various available methods. However, a few rules for designing castings are given below to serve as a guide.

1. The sharp corners and frequent use of fillets should be avoided in order to avoid concentration of stresses.
2. All sections in a casting should be designed of uniform thickness, as far as possible. If, however, variation is unavoidable, it should be done gradually.

3. An abrupt change of an extremely thick section into a very thin section should always be avoided.
4. The castings should be designed as simple as possible, but with a good appearance.
5. Large flat surfaces on the casting should be avoided because it is difficult to obtain true surfaces on large castings.
6. In designing a casting, the various allowances must be provided in making a pattern.
7. The ability to withstand contraction stresses of some members of the casting may be improved by providing the curved shapes *e.g.*, the arms or pulleys and wheels.
8. The stiffening members such as webs and ribs used on a casting should be minimum possible in number, as they may give rise to various defects like hot tears, and shrinkage etc.
9. The casting should be designed in such a way that it will require a simpler pattern and its moulding is easier.
10. In order to design cores for casting, due consideration should be given to provide them adequate support in the mould.
11. The deep and narrow pockets in the casting should be invariably avoided to reduce cleaning costs.
12. The use of metal inserts in the casting should be kept minimum.
13. The markings such as names or numbers etc. should never be provided on vertical surfaces because they provide a hindrance in the withdrawl of pattern.
14. A tolerance of ± 1.6 mm on small castings (below 300 mm) should be provided. In case more dimensional accuracy is desired, a tolerance of ± 0.8 mm may be provided.

QUESTIONS

1. State the difference between permanent mould casting and sand casting. What are the common materials used for permanent mould casting ?
2. Write short notes on 'semi-permanent mould casting' and 'slush casting'.
3. What is the difference between gravity die casting and pressure die casting ?
4. Sketch and describe the working of a hot chamber and cold chamber die casting machines.
5. What are advantages and disadvantages of die casting ?
6. Explain, with the help of a neat sketch, the process of centrifugal casting.
7. What is investment casting ? What are the main materials used for making the investment pattern ? Write the advantages and disadvantages of investment casting.
8. Describe step by step, the process of shell moulding. What are its advantages and disadvantages ?
9. Explain the defects in casting with their reasons. How these defects can be prevented ?
10. Discuss the different methods by which the castings can be inspected.
11. Write a short note on the design of castings.

OBJECTIVE TYPE QUESTIONS

1. When the molten metal is fed in the cavity of a metallic mould by gravity, the method of casting is known as

(*a*) die casting method (*b*) slush casting method
(*c*) permanent mould casting method (*d*) centrifugal casting method

2. The casting method adopted for ornaments and toys of non-ferrous alloys, is
   (*a*) permanent mould casting (*b*) slush casting
   (*c*) die casting (*d*) centrifugal casting
3. In a hot chamber die casting machine
   (*a*) melting pot is separate from the machine
   (*b*) melting pot is an integral part of the machine
   (*c*) melting pot may have any location
   (*d*) high temperature and pressure is used
4. In a cold chamber die casting machine
   (*a*) melting pot is separate from the machine
   (*b*) melting pot is an integral part of the machine
   (*c*) melting pot may have any location
   (*d*) high temperature and pressure is used
5. Cast iron and steel pipes are produced by
   (*a*) slush casting (*b*) investment casting
   (*c*) true centrifugal casting (*d*) die casting
6. In a centrifugal casting method
   (*a*) core is made of sand (*b*) core is made of ferrous metal
   (*c*) core is made of non-ferrous metal (*d*) no core is used
7. The centrifugal casting method is used for casting articles of
   (*a*) symmetrical shape about vertical axis
   (*b*) symmetrical shape about horizontal axis
   (*c*) irregular shape (*d*) non-ferrous metal only
8. A casting defect which results in general enlargement of a casting is known as
   (*a*) shift (*b*) sand wash
   (*c*) swell (*d*) blow hole
9. Scabs are casting defects which
   (*a*) results in a mismatching of the top and bottom parts of a casting
   (*b*) results in general enlargement of the casting
   (*c*) occurs near the ingates as rough lumps on the surface of a casting
   (*d*) occurs as sand patches on the upper surface of a casting.
10. A casting defect which occur near the ingates as rough lumps on the surface of a casting is known as
    (*a*) shift (*b*) sand wash (*c*) swell (*d*) scab

ANSWERS

1. (*c*)	2. (*b*)	3. (*b*)	4. (*a*)	5. (*c*)
6. (*d*)	7. (*b*)	8. (*c*)	9. (*d*)	10. (*b*)

14

Smithy and Forging

14·1 Introduction

The smithying and forging is a process of heating the metal to a plastic state and then shaping it with a hand hammer or power hammer. The *smithying* is an oldest form of forging and is largely used where small parts are shaped by heating them in an open fire or hearth. It is also called *hand forging* or *smith forging* and is done by the blacksmith.

The *forging* is a process of heating the metal in a closed furnace to a desired temperature in order to acquire sufficient plasticity and shaping it in dies under the pressure of heavy hammers, forging machines and presses. The term forging is usually referred to the production of heavy parts.

The forging processes are extremely important in any machine building industry such as in the manufacturing of tractors, automobiles, agricultural machinery, ship building, locomotive building, rail road equipments, cutting tools etc.

14·2 Advantages and Disadvantages of Forging

Following are the advantages and disadvantages of forging.

Advantages

1. It refines the structure of the metal.
2. It renders the metal stronger by setting the direction of grains.
3. It results in considerable saving in time, labour and material as compared to the production of a similar item by cutting from a solid stock and then shaping it.
4. The reasonable degree of accuracy may be obtained.
5. The forgings may be welded.
6. It increases the ductility and resistance to impact loading.

Disadvantages

1. The cost of initial set up is heavy.
2. The close tolerances under forging operations is difficult to maintain.
3. The metal gets cracked or distorted if worked below a specified temperature limit.

14·3 Forgeable Materials

Any metal or alloy which can be brought to plastic stage through heating can be forged. The extent to which a material can be forged is governed by its composition as well as temperature of forging. The selection of a forging material depends upon certain desirable mechanical properties such as strength, malleability, durability, machinability, etc. The metals and alloys which can be forged include carbon steels, alloy steels including stainless steel, wrought iron, copper base alloys, nickel and nickel-copper alloys, aluminium alloys and magnesium alloys.

14·4 Forgeability

The forgeability of a material is defined as the capacity of a material to undergo deformation under compression without rupture. It mainly depends upon the composition and purity of the material, number of phase present, and grain size. The pure metals have good malleability and thus good forging properties. The forgeability increases with temperature upto a point at which a second phase *e.g.*, from ferrite to austenite in steel appears or if grain growth becomes excessive. The fine grained materials containing insoluble compounds having lower concentration of impurities on grain boundaries have better forgeability than coarse grained materials.

The forgeability also depends upon the strain rate and strain distribution. The metals possessing low ductility at cold working temperatures have reduced forgeability at higher strain rates whereas metals possessing high ductility at cold working temperatures have good forgeability.

14·5 Heating Devices (Hearths and Furnaces)

The heating of metal is done either in a smith's hearth or in a furnace. The hearths (commonly known as forges) are used for heating metals for hand forging. It is a very old method of heating, but still it is used. The furnaces are used for heating metals for heavy forging. The hearths and furnaces are discussed, in detail, as follows :

1. *Hearths.* The hearths may be classified as open or closed hearths. The blacksmith's forges may have one or two hearths which are called single hearth or double hearth respectively.

A common form of the hearth, as shown in Fig. 14.1(*a*), is a single hearth open type forge. It consists of a shallow dish or tray made of heavy gauge mild steel or cast iron sheets. Its size varies from 1.5 m to 2.5m square and 0.8 m to 1.2 m deep. It is provided with a lining of fire clay or other refractory material to withstand the excessive heat produced due to the combustion of fuel.

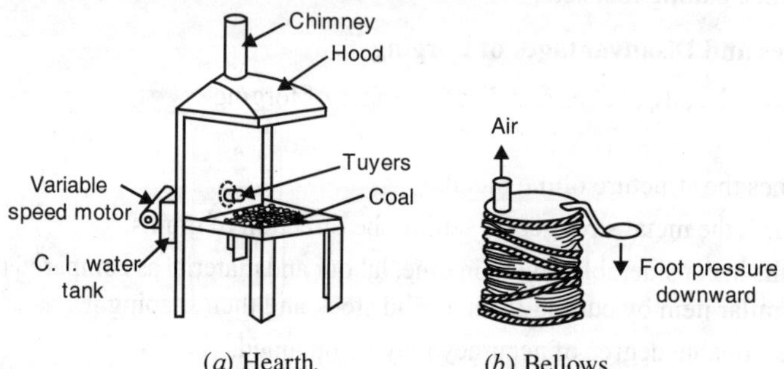

(*a*) Hearth. (*b*) Bellows.

Fig. 14.1. Single hearth (open type).

The fuel used in a hearth may be coke, coal or charcoal. For general work, a low-sulphur coke, is used. The low sulphur fuels are preferred because sulphur is detrimental to steel. There is an inlet for blowing air either through the back or bottom. The air is generally applied through a motorised fan blower. The motorised fan is generally fitted with a series-wound variable speed motor.

In small smithy shops, foot operated bellows are used for blowing air as shown in Fig. 14.1(*b*). The smoke and gases produced due to the burning of fuel is escaped through hood and chimney. The water tank is provided for the purpose of quenching the work.

The closed type hearths are employed in mass production for heating small parts. The hearths may be made of bricks especially for heavy work.

2. *Furnaces.* The furnaces of reverberatory type are mostly used for heating large work to be forged under power hammers. Since the work, in the furnaces, is heated by the flames produced from the combustion of fuel, therefore these furnaces may be called as flame furnaces. The gas and oil are mostly used as fuels as these are economical and easily controlled. The work does not come in direct contact with the fuel. The following are the various types of furnaces used for heating steel.

(*a*) *Box or batch type furnaces.* These are widely used in forging shops for heating small and medium size work. These furnaces are usually made of steel frame lined with insulating and refractory bricks.

(*b*) *Continuous type furnaces.* These furnaces are provided with mechanical pusher and are used for mass production of articles. In these furnaces, the pieces of steel are charged at one end and pushed to the furnace for heating at correct temperature. These heated pieces are moved to the other end for forging.

(*c*) *Slot type furnaces.* These furnaces are commonly used for heating bars at one end for drop forging or other forging operation. In these furnaces, a slot is provided at the front through which the bar is inserted for heating. When the desired forging temperature is reached, the bar is removed.

(*d*) *Rotary hearth furnaces.* These furnaces are sometimes used for heating large number of pieces of steel for forging. In these furnaces, the speed of rotation is adjusted in such a way that the work is heated to the required temperature after one or two revolutions.

(*e*) *High frequency induction furnaces.* These furnaces are quite popular with the availability of cheap electric power. The work produced by induction heating is free from oxide scale, have uniform temperature and takes less time.

(*f*) *Resistance furnaces.* These furnaces are faster than induction furnaces are often automated. In these furnaces, the work is connected to the circuit of a step down transformer. The simple fixtures are made to hold the work of different shapes and sizes.

14·6 Blowers

The supply of air at proper pressure is always necessary for the combustion of fuel in the hearth or furnace. A bellow, fan or blower may be used for this purpose. A *bellow* is generally used for portable fires to supply small quantity of air. A *fan* is best adopted for supplying a large volume of air at a small pressure. The *centrifugal fan* driven by an electric motor is commonly used for forging work. It consists of a fan mounted on a shaft which runs on two ball bearings. It first draws air from the atmosphere towards the centre and delivers it under pressure to the hearth or furnace through an outlet pipe located below the fan. The *Root's blower* is more efficient and economical when air has to be delivered at any appreciable pressure. It consists of two properly shaped rollers or impellers housed in a casing. These are mounted on two parallel shafts and are rotated by an electric motor through a set of reducing gears. The shafts revolve in opposite directions inside the casing. The air is drawn through the inlet pipe on the top and forced out under desired pressure through an outlet pipe.

14·7 Open Fire and Stock Fire

The fuel used in a hearth may be coke, coal or charcoal. For general work, a low sulphur coke is used. Since the work comes in direct contact with the fuel in a smith's hearth, therefore the condition of fire plays an important part in the efficient heating of the work. Following are the two methods commonly used for firing a smith's hearth :

1. *Open fire.* The open fire (also called loose fire), as shown in Fig. 14.2(*a*), is used for all general work. It is made in the hollow space with coke left from the last fire, covered with green coal. As the fire burns away, coke from the top and sides is thrust into the centre and its place is taken by

green coal from the supply maintained on the front plate of the forge. The ash and clinkers are cleared at regular intervals of time.

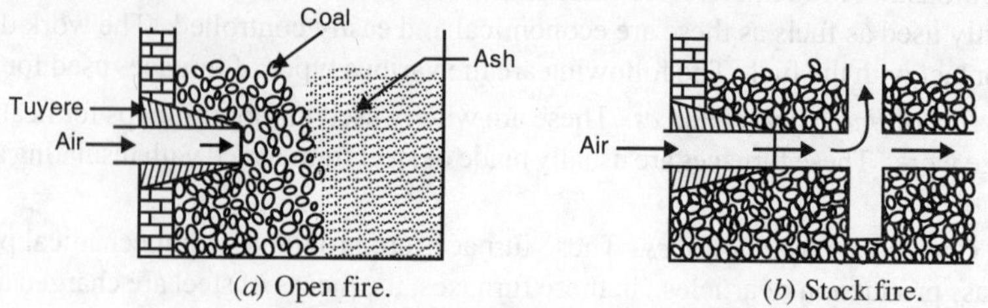

(a) Open fire. (b) Stock fire.

Fig. 14.2. Open fire and stock fire.

2. *Stock fire.* The stock fire, as shown in Fig. 14.2(b), is made to give more space for the work. It is especially useful in dealing with larger pieces, when heat may have to be kept up for some time. Before commencing the work, the fire is made up by placing a desired size of a round block of steel or wood over the tuyere upon which fresh coal is closely built into the form of a mound or stock. The block is then withdrawn from the bed of the hearth and the fire is lighted on the hollow space left. Here the work is heated, being carefully covered with freshly coked fuel from time to time as the fire burns away.

14·8 Forging Temperatures

Any metal or alloy heated to such a temperature which brings it to plastic stage can be forged. Wrought iron, steel and its alloy require a higher temperature depending upon their composition while non-ferrous metals and alloys need comparatively lower temperature to render them easily workable. The following table shows the temperature ranges for forging some common metals :

Table 14.1. Forging temperature for common metals.

Material	*Forging temperature, °C*	*Material*	*Forging temperature, °C*
Wrought iron	900-1300	High carbon and alloy steels	800-1150
		Stainless steel	940-1180
Low carbon steels	750-1300	Copper, brass and bronze	600-950
Medium carbon steels	750-1250	Aluminium and magnesium alloys	350-500

In case of hand forging, the forging temperature is judged by the colour of the light emitted by the hot metal. In case of furnaces used for mass production, the forging temperature is determined by the optical pyrometers.

14·9 Heating of Steel for Forging

When steel is heated in a furnace, oxidation takes place on its surface and scale is formed. The thickness of the layer of scale increases with the rise of temperature due to more oxidation. Some part of the scale falls away itself from the metal and combines with the lining material of the furnace to form a slag. The part of the scale that remains on the surface of the metal is pressed into the interior during forging. This lowers the quality of steel. At high temperatures, the oxidation of steel is accompanied by simultaneous decarborisation due to the burning of carbon. This results in reduction of carbon contents at the surface of the metal, hence reduces the mechanical properties of the steel. The degree of decarburisation depends upon the rise in temperature and chemical composition of the steel.

14·10 Overheating and Burning of Steel

When steel is heated, a wide range of changes takes place in its internal structure depending upon its carbon content and heating temperature. The temperature at which the change starts on heating is called *lower critical point* and the temperature at which the change ends in heating is called *upper critical point.* The lower critical point is same for all steels but the upper critical point varies according to the carbon content in steel. There is no change in grain size when steel is heated upto the lower critical point (about 723°C) and at this temperature birth of new grains take place. At the upper critical point, the average grain size is minimum. But when steel is heated beyond the upper critical point, there is an increase in the size of grains and when it is further heated to a higher temperature, the grain size increases rapidly. When the size of these grains become too large, the steel is said to be *overheated steel.* The overheated steel will tear and crack when forged and its fracture will have an extremely coarse structure. This steel can be improved by annealing (*i.e.,* by heating it to 30°C to 35°C above the upper critical point and then cooling slowly).

If the overheated steel is allowed to remain in the furnace for a long time, it will be burnt due to excessive oxidation of the inner grain boundaries. Such a steel is called *burnt steel.* As a result of burning, the steel becomes weak and large cracks appear on the surface. The further heating will cause fusing and destruction of the metal completely. The burnt steel cannot be rectified and has to be rejected as scrap.

14·11 Forging Processes

The forging processes, according to the method used, are classified into the following two groups :

1. Flat die forging, and 2. Closed die forging.

The *flat die forging* is also known as *smith forging.* This may be done by hand or power. In *hand forging,* the work is heated to the required temperature, placed on an anvil and then shaped with suitable hand hammer. It is employed only for shaping small number of light forgings. When the work is shaped under power hammers, it is then known as *power forging.*

The *closed die forging* includes such processes as drop forging, press forging and machine or upset forging. In the closed die forging, the work is heated to the required temperature, placed in a two part or multipart hollow mould called die (depending upon the shape of work required) and then pressed.

14·12 Black-Smith's Hand Tools and Appliances

The hand tools and appliances used by the blacksmith in various forging operations are as follows :

1. *Anvil.* The anvil is an important smith's tool. It is used for supporting the work while it is struck with hammer. It is made in several shapes but two most common forms are the English and

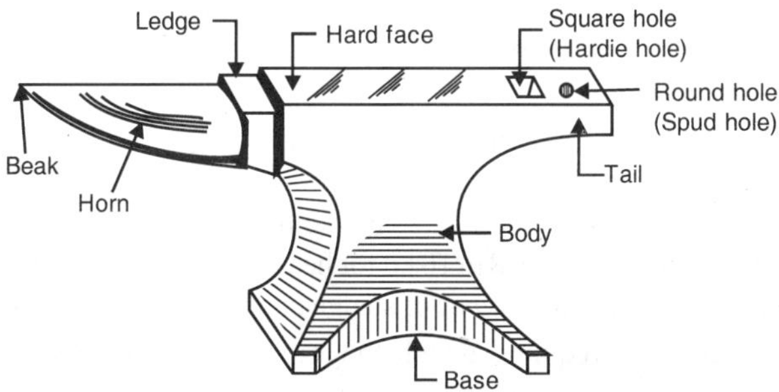

Fig. 14.3. Anvil.

French types. An English anvil has a single beak, as shown in Fig. 14.3, while the French anvil has a beak at each end. The body of the anvil is made of mild steel to which a piece of high carbon steel about 2.5 mm thick is welded on the top to give a hard top face. The beak is soft like the anvil body and its shape makes it useful for bending round section to different diameters. The ledge between the beak and the anvil face is also soft and may be used for resting metal when cutting through with a chisel. The soft underneath metal does not damage the chisel edge.

In the top of the anvil, is a square and a round hole. The square hole (also known as Hardie hole) is used for holding square shank shaping tools like bottom fullers, swages and hardie (bottom chisel) etc. The round hole is used for admitting the ends of the punches or drifts. This hole is also used for bending round bars of different curvatures.

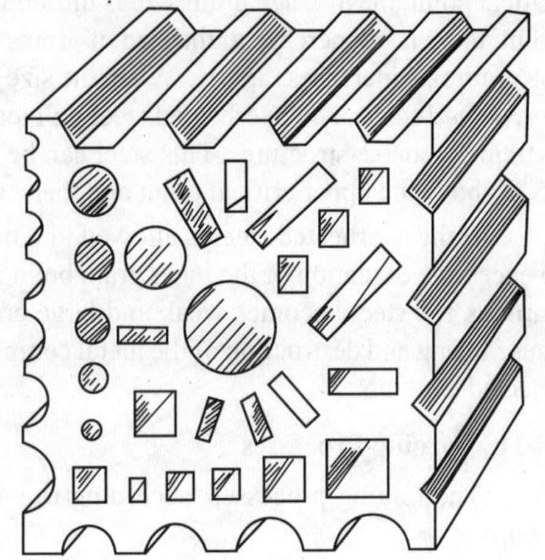

Fig. 14.4. Swage block.

The most common anvil has a mass of about 150 kg. The anvil is placed either on a wooden block or on cast iron stand such that the top face is about 600 mm to 700 mm from the floor.

2. *Swage block.* The swage block, as shown in Fig. 14.4, is generally made of cast iron and has round, square, rectangular and half round grooves. In addition to this, it is provided with holes which are useful for holding bars while bending and knocking up heads. The swage block is usually supported at a suitable height on a stand which is adaptable to hold it flatwise or on its edge.

3. *Hammers.* The hammers commonly used in hand forging are of the following two types :

(*a*) Hand hammer, and (*b*) Sledge hammer.

The *hand hammers* are generally ball peen hammers as shown in Fig. 14.5(*a*). They usually have a slightly convex striking face and should have mass from 1 to 1.5 kg. The hammer heads should be of cast steel, the ends are hardened and tempered with the portion round the eye soft.

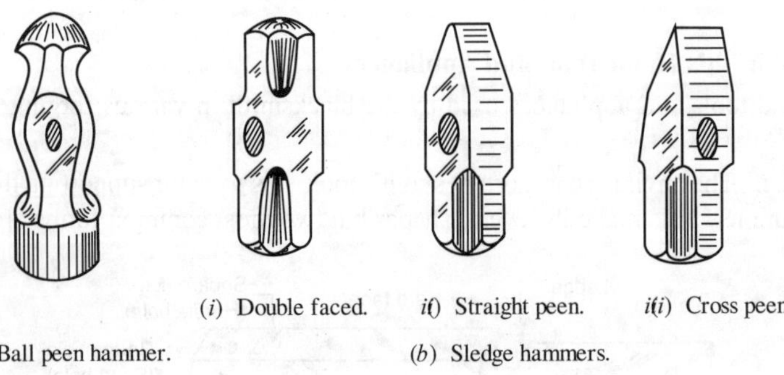

(*i*) Double faced. *ii*) Straight peen. *ii*(*i*) Cross peen.

(*a*) Ball peen hammer. (*b*) Sledge hammers.

Fig. 14.5. Types of hammers.

The *sledge hammers* are used by the smith's helper or striker. The mass of the sledge hammer varies from 4 to 6 kg for ordinary work and 8 to 10 kg for heavy work. The shaft is about one metre long and it is mostly made of wood. The three types of sledge hammers are shown in Fig. 14.5(*b*).

4. *Tongs.* The blacksmith requires suitable tongs to handle the work while forging. These are

made of mild steel. The various types of tongs commonly used for holding work are as follows :

(*a*) *Closed mouth tong.* A closed mouth tong, as shown in Fig. 14.6 (*a*), is used for holding thin sections.

(*b*) *Open mouth tong.* An open mouth tong, as shown in Fig. 14.6(*b*), is suitable for holding heavier stock.

(*c*) *Round hollow tong.* A round hollow tong, as shown in Fig 14.6 (*c*), is used for holding round, hexagonal and octagonal section.

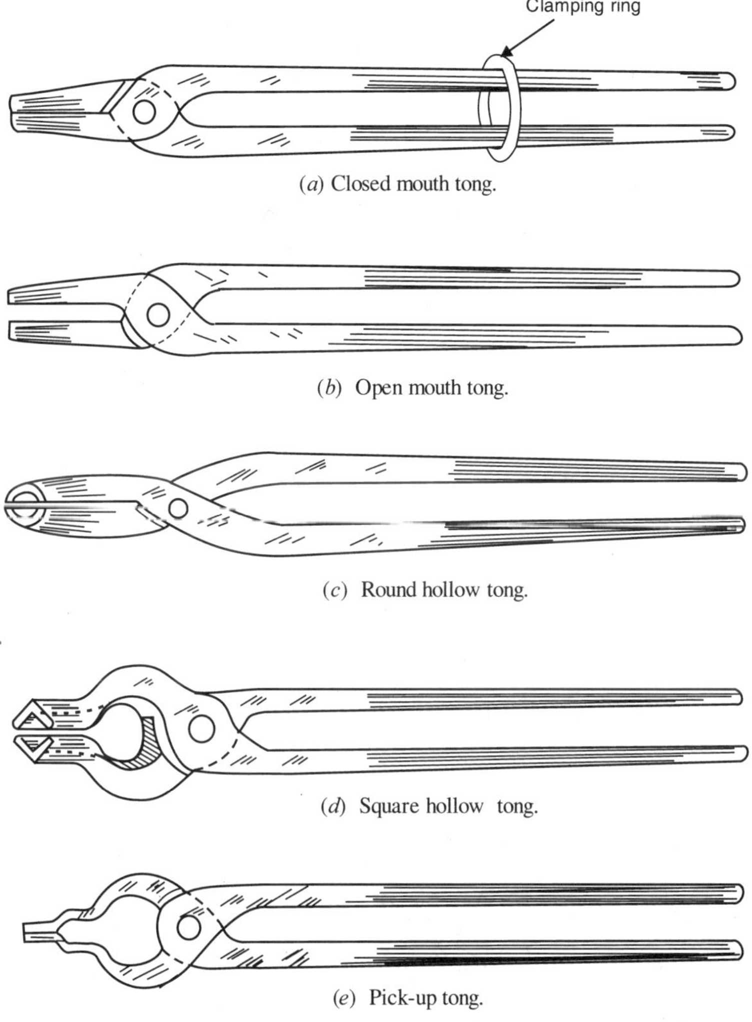

(*a*) Closed mouth tong.

(*b*) Open mouth tong.

(*c*) Round hollow tong.

(*d*) Square hollow tong.

(*e*) Pick-up tong.

Fig. 14.6. Types of tongs.

(*d*) *Square hollow tong.* A square hollow tong, as shown Fig. 14.6(*d*), is used for holding square, hexagonal and octagonal work.

(*e*) *Pick-up tong.* A pick-up tong, as shown in Fig. 14.6(*e*), is used for picking up round bars, but not for holding work during forging.

In order to relieve the smith of the strain of holding the tongs during a long forging operation, the tongs are provided with clamping rings as shown in Fig. 14.6(*a*).

5. *Chisels.* The chisels are used for cutting metals and for nicking prior to breaking. They may be cold or hot depending upon whether the metal to be cut is cold or hot. The main difference between the two chisels is in the edge.

The *cold chisel* is made of tool steel with a cutting angle of about 60 degrees and its edge is hardened and tempered as shown in Fig. 14.7(*a*). The handle of this tool is usually formed of an iron rod bent around a fullered recess made in the body of the chisel. The edge of a chisel should not be quite straight but slightly rounded.

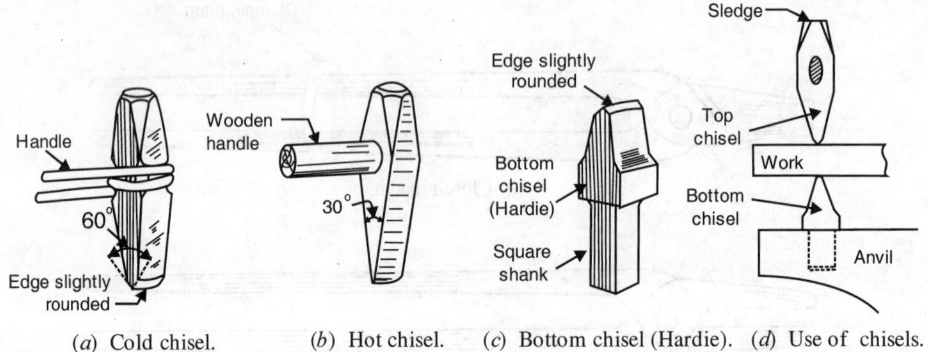

(*a*) Cold chisel. (*b*) Hot chisel. (*c*) Bottom chisel (Hardie). (*d*) Use of chisels.

Fig. 14.7. Chisels and their uses.

The *hot chisel* is made of low carbon steel and has a cutting angle of 30 degrees, as shown in Fig. 14.7(*b*). The hardening of the edge is not necessary as the hot metal would soften it. This type of chisel is fitted with wooden shafts.

The chisels are generally used in pairs. The top chisel which is held by the smith may be hot or cold chisel. The bottom chisel (often called *hardie*), as shown in Fig. 14.7(*c*), has a square shank and fits in the square hardie hole in the anvil face. Fig. 14.7(*d*) shows the use of top and bottom chisels.

6. *Swages.* The swages are used for work which has to be reduced and finished to round or hexagonal form and are made with half grooves of dimensions to suit the work being reduced. The

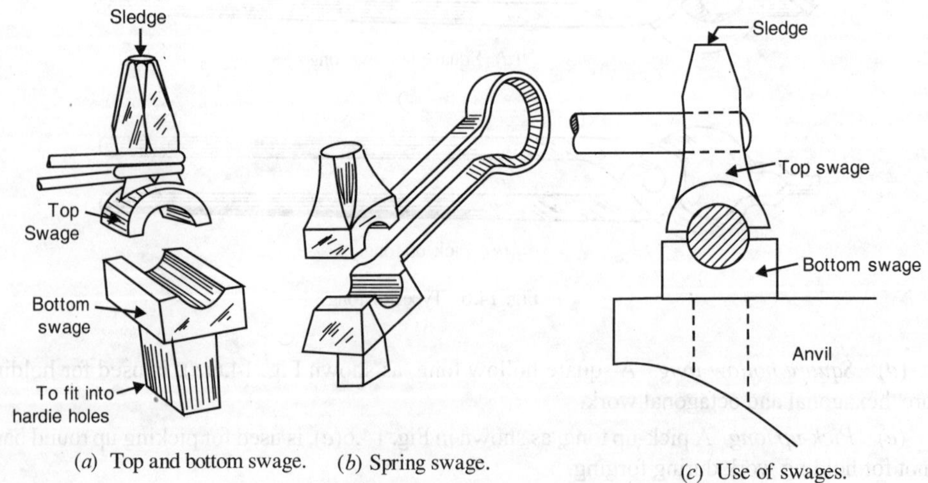

(*a*) Top and bottom swage. (*b*) Spring swage. (*c*) Use of swages.

Fig. 14.8. Swages.

swages may be in separate top and bottom halves, as shown in Fig. 14.8(*a*), or the two halves may be connected by strip of spring steel, as shown in Fig. 14.8(*b*), thus enabling them to be used by the smith single-handed. The use of swages is shown in Fig. 14.8(*c*).

7. *Fullers.* The fullers, as shown in Fig. 14.9(*a*), are used for necking down a piece of work. It is mainly used to reduce the metal while making shoulders. They are made of tool steel. The fullers are made in pairs. The top fuller is held by the blacksmith and struck by the striker while the bottom fuller is fitted into the hardie hole of the anvil. The fullers are made in various sizes according to the requirement. The use of fullers to neck before setting down is shown in Fig. 14.9(*b*).

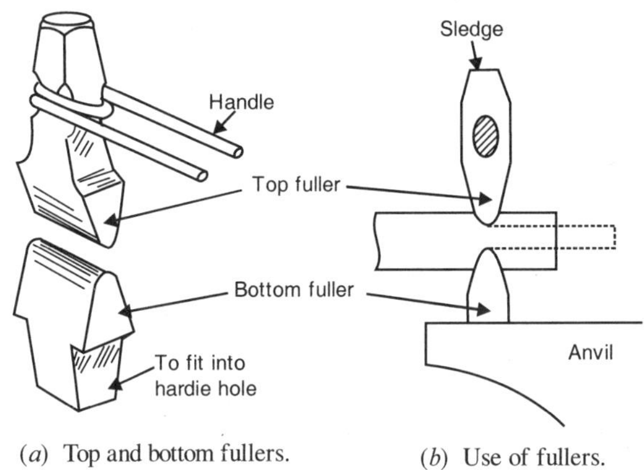

(*a*) Top and bottom fullers. (*b*) Use of fullers.

Fig. 14.9. Fullers and their use.

8. *Flatters.* The flatter, as shown in Fig. 14.10(*a*), is used for finishing flat surfaces and are made with a perfectly flat face. It gives smoothness and accuracy to articles which have already been finished by swages and fullers. Its size is about 7.5 cm square or round. It is made of tool steel. It is struck by a hammer on the head. The use of flatter is shown in Fig. 14.10(*b*).

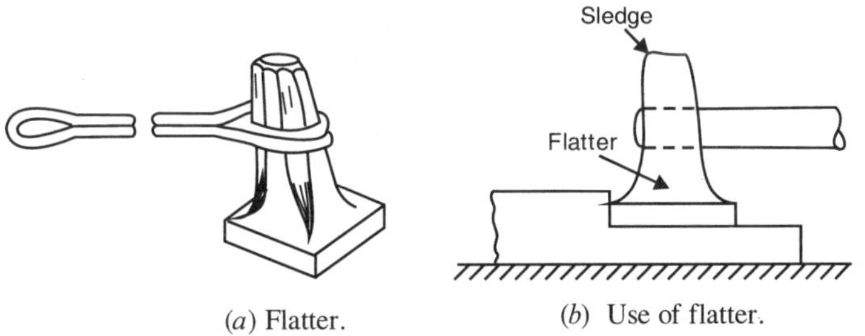

(*a*) Flatter. (*b*) Use of flatter.

Fig. 14.10. Flatters and its use.

9. *Set hammer.* The set hammer, as shown in Fig. 14.11, is similar to flatter but is a smaller tool. It is made up of forged steel and is hardened. It is generally available in 4 cm square size. It is used to set down shoulders or smooth out fuller marks near the shoulders before the blacksmith finishes with the flatter.

Fig. 14.11. Set hammer.

10. *Punches.* The punches, as shown in Fig. 14.12(*a*) are used for making holes when the metal is at forging heat. These are made in various shapes and sizes such as square, round, oval etc. It is made of tool steel, hardened and tempered. The round hole in the tail of the anvil is provided for facilitating the punch to pass without getting damaged after punching in a certain section kept on the anvil. Generally the punching of hole is

carried out from both sides, as shown in Fig. 14.12(*b*), to avoid driving the punch onto the hard anvil face.

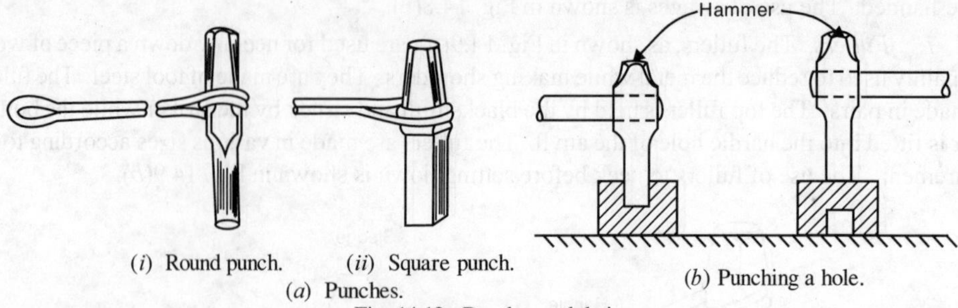

(*i*) Round punch. (*ii*) Square punch.
(*a*) Punches.
(*b*) Punching a hole.

Fig. 14.12. Punches and their use.

11. *Drifts.* The drift, as shown in Fig. 14.13, is used to finish the size of the punched holes. They are made taper and bulging in the centre. These are made in various shapes and sizes like punches. They are also made of tool steel, hardened and tempered.

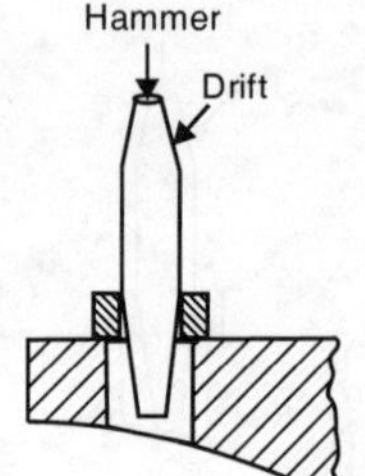

Fig. 14.13. Use of drift

14·13 Smith Forging Operations

The number of operations are performed in a smithy shop for the formation of desired shape of object by forging. The commonly used operations are as follows :

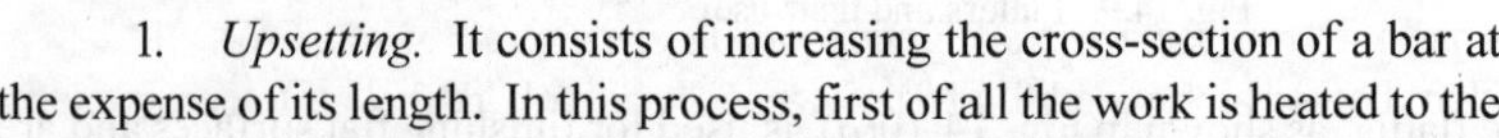

1. *Upsetting.* It consists of increasing the cross-section of a bar at the expense of its length. In this process, first of all the work is heated to the required temperature and then the heavy blow is given by hand hammer. The swelling of work takes place at the heating portion. The position and nature of upsetting depends upon the heating and upon the type of blow delivered. In general, the increase in lateral swelling will be greatest at the parts where the metal is hottest. Fig 14.14(*a*), (*b*) and (*c*) shows the effect of heavy blow when the heating is done at one end, at both ends and at the centre respectively.

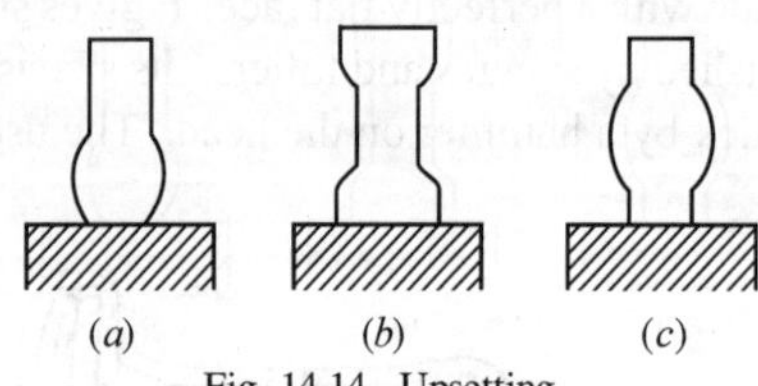

(*a*) (*b*) (*c*)

Fig. 14.14. Upsetting.

2. *Drawing down.* It is the process of reducing the cross-section of a bar and increasing its length. The operation of drawing down is carried out in the following two steps :

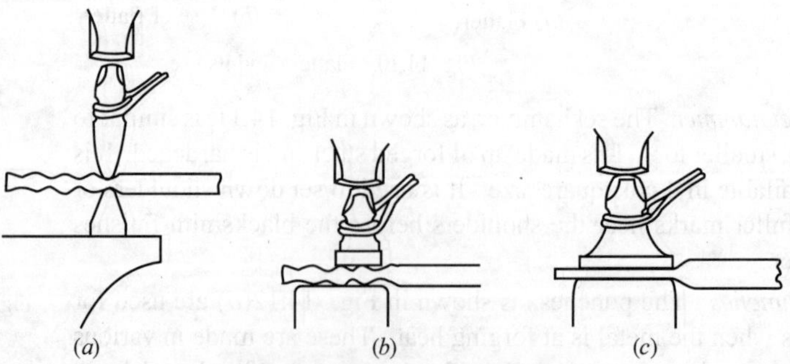

(*a*) (*b*) (*c*)

Fig. 14.15. Drawing down operation.

(*a*) First of all, the hammering is done with a straight peen hammer by keeping the work at the edge or beak of the anvil. The bar is turned through 90° if the thickness in both directions is to be reduced. A pair of fullers as shown in Fig. 14.15(*a*), are also used for this process.

(*b*) Now the curved top of the work is levelled off with a set hammer as shown in Fig. 14.15(*b*). and finally finished with a flatter as shown in Fig. 14.15(*c*).

Note : When a round bar is to be drawn down a considerable amount, it is first brought to a square by the above method and then squared up again. It is then reduced to octagonal shape by taking off the corners and finally rounded and finished between swages.

3. *Setting down.* It is a process of local thinning down effected by the set hammer. The work is usually fullered at the place where the setting down commences.

Fig. 14.16(*a*) shows the process of setting down both ends of a bar by using the top and bottom fuller and Fig. 14.16(*b*) shows how the flatter or set hammer is used close to a shoulder.

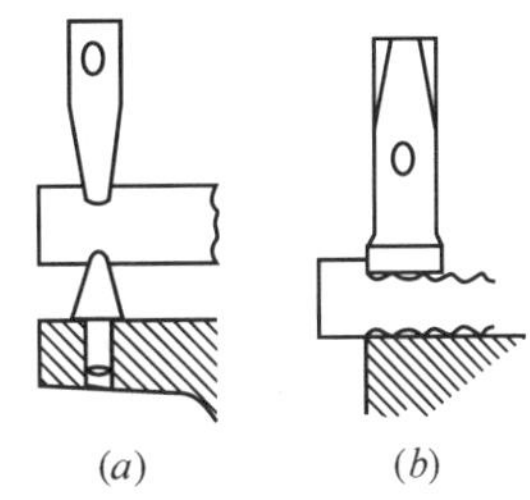

(*a*) (*b*)

Fig. 14.16. Setting down operation.

4. *Punching.* It is the process of producing holes by forcing a punch through the work. The punching operation is performed as follows :

The work is heated to nearly welding heat (*i.e.,* white heat) and then laid flat on the anvil. The punch is now driven about half way through one side as shown is Fig. 14.17(*a*). The work is then turned over and placed upon a tool called bolster which rests on the anvil as shown in Fig. 14.17(*b*). The punch is again completely driven through the other side to give a clean cut hole on both sides. In case a punch is driven right through from one side, the metal will get bulged as shown in Fig. 14.17(*c*). In order to prevent the bulging and to produce a hole in a single operation, a die is used with the punches shown in Fig. 14.17(*d*). The die is made of hard steel with a hole tapering downwards to give clearance. The diameter of the hole being the same as that of the holc to bc punched. The punch should also be made taper and slightly larger at the centre to permit its falling through the hole when punched.

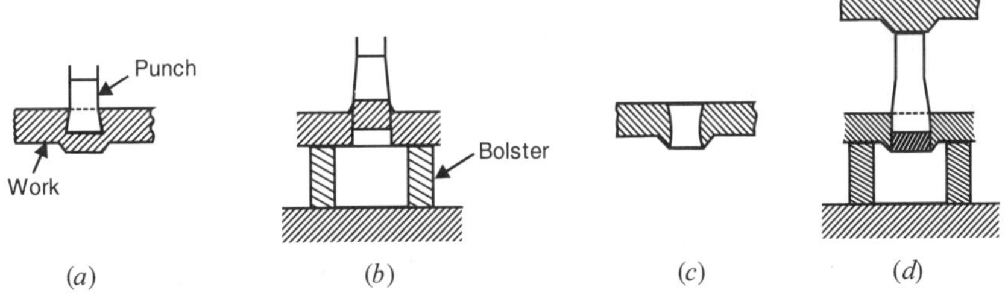

(*a*) (*b*) (*c*) (*d*)

Fig. 14.17. Punching operation.

5. *Bending.* It is an important operation in forging and is very frequently used. The bends may be either sharp cornered angle bends or they may be composed of a more gradual curve. The angle bends may be made by hammering the metal over the edge of the anvil as shown in Fig. 14.18(*a*), over a block of metal held in the hardie hole as shown in Fig. 14.18(*b*), or over the vice jaw while the metal is being gripped as shown in Fig. 14.18(*c*).

A little consideration will show that when the metal is bent, the layers of metal on the inside are shortened and those on the outside are stretched, as shown in Fig. 14.18(*d*). This causes a bulging of the sides at the inside and a radius on the outside of the bend. A sharp corner on the inside of a bend should be avoided as it constitutes a weakness which may lead to a fracture of the corner. In case double bends are to be made, a simple bending fixture should be used to save time and to enable more uniform results.

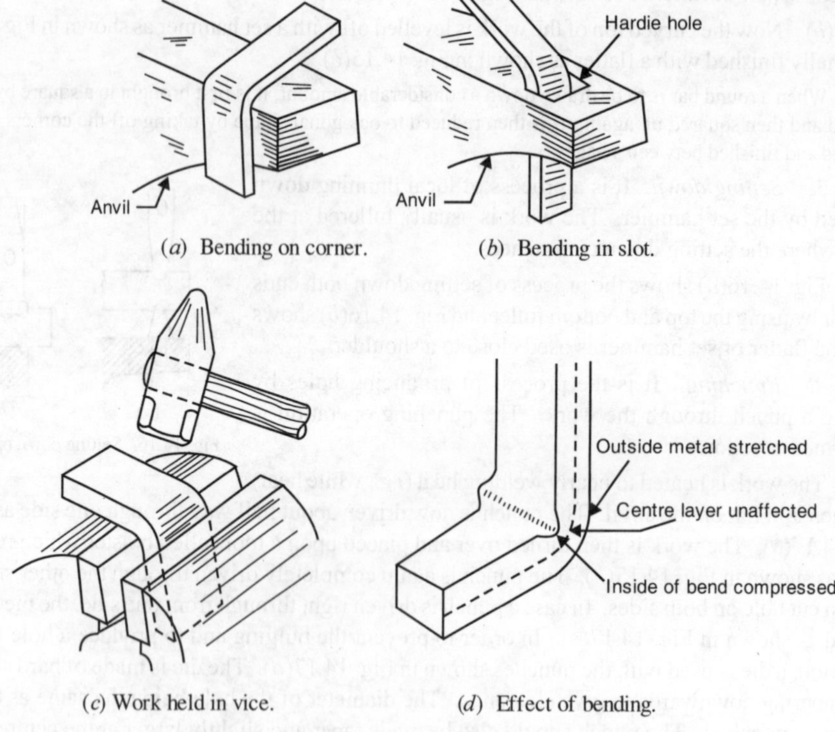

(*a*) Bending on corner. (*b*) Bending in slot.

(*c*) Work held in vice. (*d*) Effect of bending.

Fig. 14.18. Methods of making angle bends.

The gradual bends may be made by using the beak of the anvil as a former as shown in Fig. 14.19(*a*), or the metal may be bent round a bar of the correct radius held in a vice, as shown in Fig. 14.19(*b*). When a quantity of bending is to be done as shown in Fig. 14.19(*c*), some type of bending fixture should be used to save time and to produce better and uniform bends.

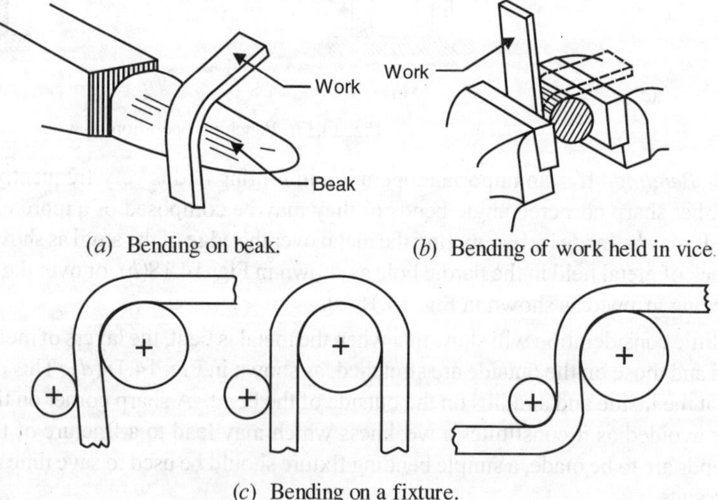

(*a*) Bending on beak. (*b*) Bending of work held in vice.

(*c*) Bending on a fixture.

Fig. 14.19. Methods of making gradual bends.

6. *Welding.* It is the process of joining the two surfaces of metal under pressure after they are heated to the correct temperature. The correct temperature for wrought iron is about 1350°C whereas for mild steel the temperature is a little lower than this. It may be noted that if the temperature is low, it will not cause the weld to take place. On the other hand, if the temperature is too high, it will ruin the metal by burning it.

Before the weld is made, the ends of the pieces for welding are formed to correct shape *i.e.,* they should not be flat but are rounded. This is due to the reason that the union of the metals commences at the centre first. The process of preparing the ends ready for the welding is known as *scrafing.*

Another essential point for a good weld is that the surfaces to be welded should be perfectly clean. The flux must be used to prevent it from oxidation. Sand is the suitable flux for wrought iron while calcined borax is useful for mild steel. The following four methods of forge welding are commonly used :

(*a*) *Scarf weld.* It is the most reliable and straight forward method. Before welding, the ends of the work piece are rounded as shown in Fig. 14.20(*a*).

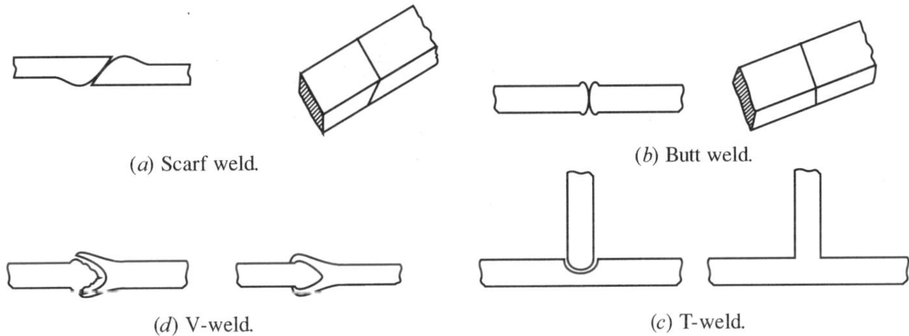

Fig. 14.20. Methods of forge welding.

(*b*) *Butt weld.* It is difficult to make by hammering because in this case the faces are to be joined and the hammering is always done perpendicular to the joint line. This type of weld is usually made by a welding machine in which the bars are gripped and their ends are forced together by hand or power operation. When they are so held, the sides of the weld may be smoothed up by hammering. The butt weld is shown in Fig. 14.20(*b*).

(*c*) *V-weld.* This is employed for joining heavy cross-section parts and should be used where the thickness of metal permits the preparing of vee. In this method, the scarfed end of the bar is roughened, as shown in Fig. 14.20(*c*), so that the weld may become solid and strong. This joint is also called *splice joint* or *split joint.*

(*d*) *T-weld.* This method is used where two parts are joined forming Tee. In this case, the end of one of the bar is upset while the other bar is upset in the centre. These two bars are placed at right angle and then welded as shown in Fig. 14.20(*d*). This is also called *jump weld.*

7. *Cutting.* It is a form of chiseling whereby a long piece of stock is cut into several required lengths or a forging is cut-off from its stock. In performing the cutting operation, first of all a notch is made with chisel about one-half the thickness or diameter of the stock. It is then turned through an angle of 180° and the chisel is placed exactly opposite to the notch. The required length of the metal is now cut-off by giving blows to the chisel with a sledge hammer.

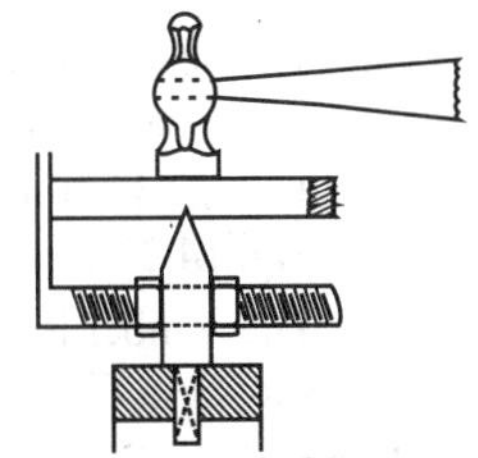
Fig. 14.21. Hardie with adjustable cutting-off gauge.

A cold chisel if used on hot work will have its temper drawn and if a hot chisel is used for cutting cold metal it will be turned and knocked up. These remarks apply equally to the corresponding bottom chisels or hardie. Fig. 14.21 shows a hardie with adjustable cutting-off gauge which is an extremely useful appliance when a number of pieces of the same length are required.

14·14 Applications of Forging

In the previous article, we have discussed the forging operations in detail because most of the jobs are produced by a combination of these processes. We shall now discuss the steps involved in producing parts of every day use in the following examples.

Example 1. To make a hexagonal bolt head from a round bar.

Procedure. The tools required are anvil, hand hammer, tong, bush or size plate, flattener, swage, cupping tool, scale and caliper. The sequence of operations are as follows :

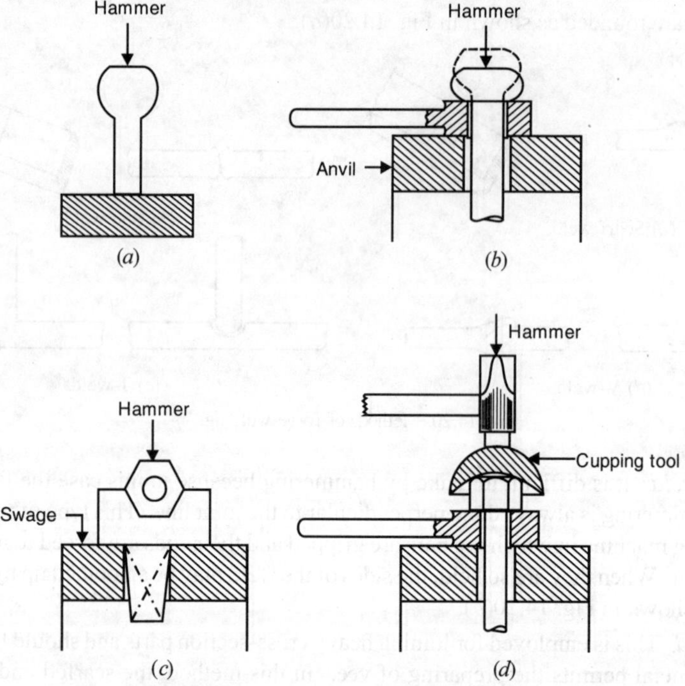

Fig. 14.22. Process of making a hexagonal bolt head.

1. First of all, one end of the bar is heated up for a sufficient length to make the head.
2. The heated end is jumped up on the anvil, as shown in Fig. 14.22(*a*).
3. The head is flattened by hammering against the end of a bush through which the shank passes, as shown in Fig. 14.22(*b*).
4. The head is formed to hexagonal shape by keeping it on the bottom swage, as shown in Fig. 14.22(*c*).
5. The camber is given on the head by means of a cupping tool using the bush as support as shown in Fig. 14.22(*d*).

Example 2. To make a chisel from an octagonal bar of mild steel.

Procedure. The tools required are anvil, hammer, tong, flattener and scale. The sequence of operations are as follows :

1. First of all, one end of the bar is heated and camber is given on the head as shown in Fig. 14.23(*a*).
2. The other end of the bar is heated and flattened as shown in Fig. 14.23(*b*).
3. The excess metal is cut-off as shown in Fig. 14.23(*c*).

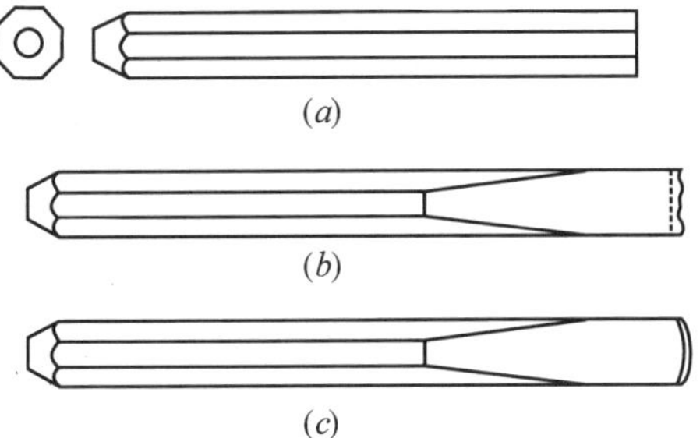

Fig. 14.23. Process of making a cold chisel from an octagonal bar

4. The cutting edge is hardened and tempered.
5. Now the cutting edge is ground to shape and the required angle.

Example 3. To make a flat drill from an octagonal bar of mild steel.

Procedure. The tools required are anvil, hammer, tong, flattener and scale. The sequence of operations are as follows :

1. First of all, one end of the bar is heated in a forge fire.
2. The heated end is flattened and the excess metal is cut off as shown in Fig. 14.24(*a*).

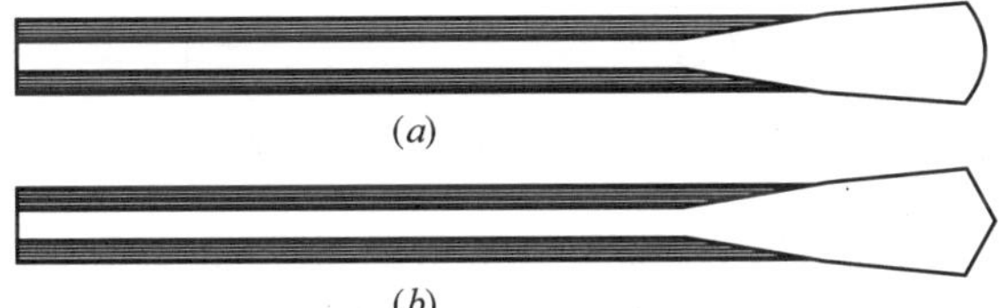

Fig. 14.24. Process of making a flat drill.

3. The cutting edge is hardened and tempered.
4. Now the cutting edge is ground to shape and the required angle, as shown in Fig. 14.24(*b*).

14·15 Power Hammers and Presses for Smith Forging

The hand forging is limited to forging of small machine parts only because the intensity of blows by hand hammering is small. For forgings of medium and large machine parts, the hand hammering is not enough for producing a great degree of deformation in the metal being forged. Moreover, hand forging is a lengthy process and requires repeated heating of the metal. Therefore, the power hammers and presses are employed in forgings. The machines which work on forgings by blow are called *hammers* while those working by pressure are called *presses.*

The capacity of the hammers is determined by the mass of their upper falling parts, *i.e.*, tup or ram and die. It may be noted that more the mass of these parts and greater the height from which they fall, the higher will be the intensity of blows. The anvil block, which serves as a rigid support for forging, is built on a foundation separate from the main frame so that the shock of the hammer blows is cushioned by the foundation and is not transmitted to the frames.

The power hammers in common use are lever-spring hammers, pneumatic hammers, steam or air hammers and drop hammers. These hammers are discussed, in detail, in the following articles.

14·16 Lever Spring Hammers

The lever spring hammer, as shown in Fig. 14.25, is mechanically driven hammer and is best suited for drawing out and flattening small forgings produced in large numbers. It consists of a leaf spring which oscillates in a bearing provided at the top of a heavy rigid frame. The front end of this spring is connected by links to the tup which carries the mass and moves vertically up and down between the fixed guides. The rear end of the spring goes up and down with the help of a connecting rod which receives motion from a crankshaft driven by belt. The throw of the crankpin or eccentric may be adjusted and thus the blows of varying intensity may be delivered.

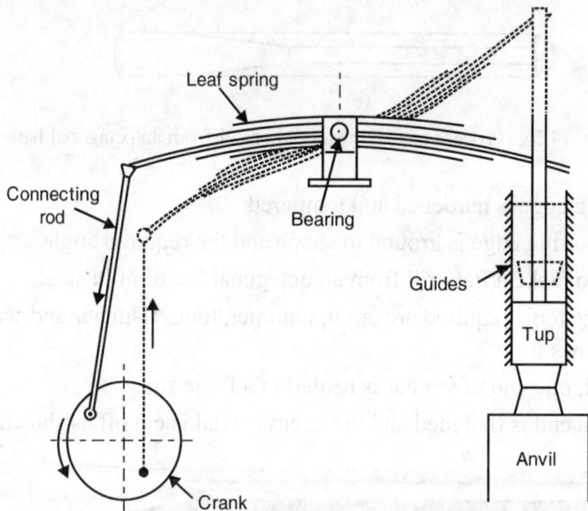

Fig. 14.25. Lever spring hammer.

In its operation, let the tup is in contact with the anvil and crank end of the spring is just commencing to move downwards. Before the tup beings to rise, both ends of the spring will bend downwards until the inertia due to the mass of the tup is overcome. The rear or crank end of the spring moves slowly towards the end of the stroke, allowing the front end of the spring to straighten itself, thus increasing the velocity of tup. When the rear end of the spring starts its upward stroke, the tup still moves upwards and is retarded and finally brought to rest by bending upwards its end of the spring. When whole of the energy previously stored up in the moving mass has been transferred to the spring, the tup is free to fall and a powerful blow is delivered. These hammers have the disadvantage of frequent breaking of springs due to rapidly repeated and alternate bending to which it is subjected. These hammers are made in various capacities having the tup mass from 30 to 250 kg and the speed of blows upto 300 per minute.

14·17 Pneumatic Hammers

The pneumatic hammer, as shown in Fig. 14.26, is widely used and operated by electric power. It consists of two cylinders *A* and *B*. The piston of the cylinder *A* is connected to a motor by a crank and connecting rod. The piston of cylinder *B* carries a tup which slides between the inside fixed guides. There are two air valves through which air passes into the cylinder *B* below and above the piston alternately. This drives the tup up and down respectively. These valves are operated by a hand lever or treadle.

In its operation, suppose the tup is resting upon the anvil. When the piston in cylinder *A* moves downwards, a partial vacuum is created above the piston of cylinder *B* and the external pressure of the atmosphere raises the tup. A little before the end of upstroke of the piston of cylinder *B*, the air is forced from cylinder *A* to cylinder *B* where it is powerfully compressed due to the upward stroke of the piston. This in turn drives tup downwards with considerable velocity thus giving a heavy blow. This operation is repeated again and again and the required blows are obtained. The intensity of blows are regulated by controlling the opening of an air valve which supplies air from cylinder *A* to cylinder *B*. When the valve is closed, the full pressure is maintained and the heaviest blows are obtained. When the valve is fully open, it allows the air to escape into the atmosphere and thus the hammer ceases to work. Thus the blows of any degree of intensity may be obtained by merely opening the valve to the desired extent.

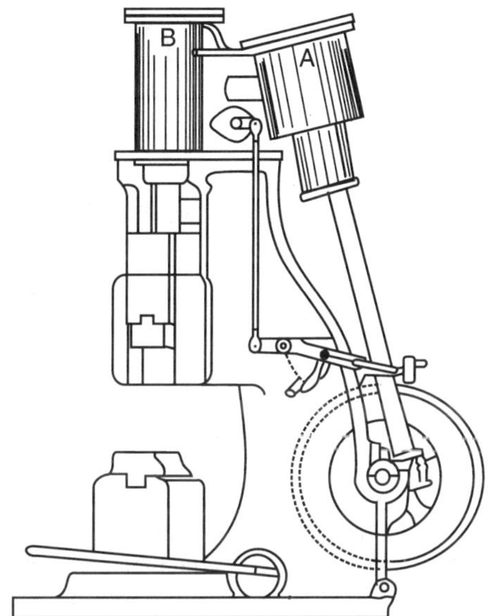

Fig. 14.26. Pneumatic hammer.

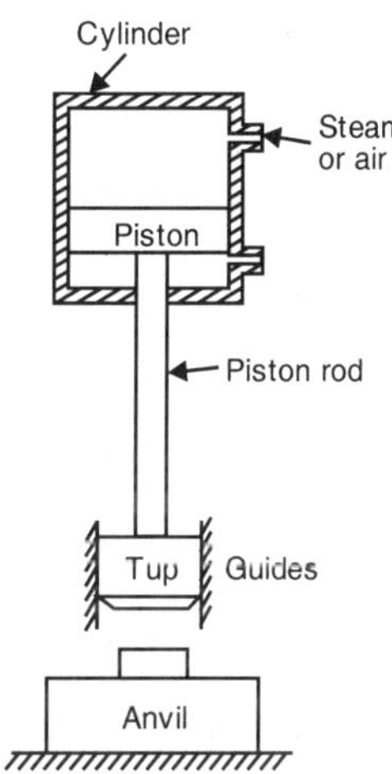

Fig. 14.27. Steam or air hammer.

14·18 Steam or Air Hammers

The steam or air hammers may be operated either by steam or compressed air. It consists of a vertical cylinder in which the piston moves up and down by the steam or air pressure. The tup is connected to the lower end of the piston rod which moves between the fixed guides, as shown in Fig. 14.27. The opening and closing of the steam or air passage on either side of the piston is controlled by a central device.

When the stem or air is admitted into the space below the piston, the tup is lifted and when the steam or air escapes from the cylinder, the tup falls freely. In case the steam or air is admitted from the top of the piston while the tup is moving down, it gives higher intensity of blows.

14·19 Forging Press

The blow of a heavy steam hammer produces a harsh sound which causes frequent breakage of parts, considerable wear and tear of the hammer itself and rest of the plant. Also, the effect of heavy blow is almost entirely confined to the surface and does not penetrate into the interior of the forging. Due to these reasons, hydraulic forging presses are used in place of very large hammers for heavy or large forgings.

There are several types of forging presses in use, but the one which works with water pressure derived from an accumulator is widely used. The principle of such a forging press is shown in Fig. 14.28. In its operation, the water from a large tank is pumped continuously and stored up in the accumulator until required by the press. The water from the accumulator is supplied to the main cylinder at a high pressure (about 20 to 35 N/mm^2) which forces the main hydraulic ram and press head down upon the work to be forged. After each stroke of the press head, the main hydraulic ram is returned to its initial position by the action of working fluid below the pistons of the pull back cylinders. In order to accelerate the return motion, it requires a relatively low water pressure but a large volume of water per unit time.

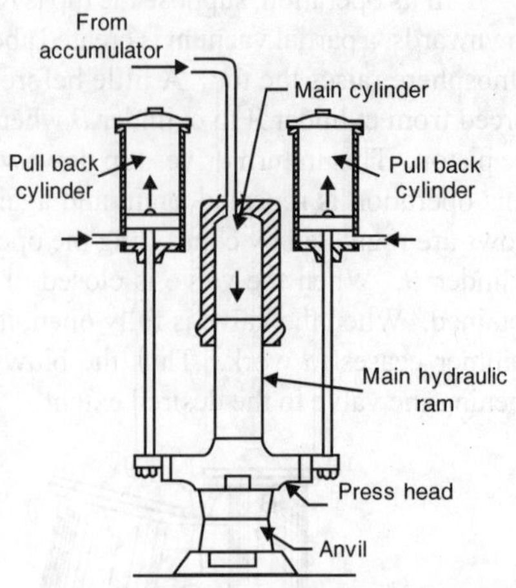

Fig. 14.28. Principles of forging press.

14·20 Press Forging

The forging of parts done in presses is called *press forging*. It provides a method for forming various sized shapes which may not be practicable by applying repeated severe blows by hammers (*i.e.,* by hammer forging). Though in press forging, the action of the press is slow in comparison to hammer action, but the reduction in size of heavy parts is comparatively rapid. The following two types of presses are usually used in press forging.

1. Hydraulic presses, and 2. Mechanical presses.

The hydraulic presses are used for heavy work whereas mechanical presses for light work. Though the mechanical presses operate faster than the hydraulic presses, yet the hydraulic presses provide greater squeezing force. It may be noted that the production of forgings by presses is faster than the hammer forging because the whole operation is completed in a single squeezing action. Moreover, the action goes deep into the centre of the part because it gives the metal enough time to flow.

Since the dies used in press forging have less draft, therefore more complicated shapes can be forged with better dimensional accuracy. In press forging, the speed, pressure and travel of the die are automatically controlled. There are less vibrations and noise than in hammer forging.

The capacity of presses varies from 500 to 8000 tonnes and the speeds from 35 to 90 strokes per minute. The presses are mostly used for producing rivets, screws and nuts where a high operating speed is desired.

14·21 Machine or Upset Forging

Some forgings are produced with equipment known as forging machines or upsetters. It consists of applying pressure longitudinally on a hot bar gripped firmly between grooved dies, to upset a desired portion of its length.

Fig. 14.29(*a*) shows a hexagonal head produced by upsetting on a cylindrical bar. In its operation, the cylindrical portion of a bar is securely gripped by the gripper die while its end, as shown dotted in Fig. 14.29(*a*), is upset into a hexagonal-shaped die cavity. The operation of the dies to produce this part is shown in Fig. 14.29(*b*). The bar is placed in a closely fitting groove of the stationary gripper die as its end is located against a bar stop. When the movable gripper die closes, a hexagonal shaped die impression is formed. Then while the bar is securely held, the heading die moves forward to press the protruding end of the bar, upsetting it and filling the die impression. In most cases, more than one forging step will be necessary, and the gripper dies will have two or more forging positions arranged vertically as shown in Fig. 14.29(*c*). In order to correspond with these gripper die positions, two or more heading dies are mounted on the heading die slide. The bar is placed in each die position in proper sequence until forging is completed.

Originally, this process was developed to form heads on bolts, but now-a-days it is widely used to produce a variety of useful forgings. The operation is usually hot working but cold working may also be done. The volume of metal which can be upset on a particular bar in a single die is limited. This limitation is determined by the ratio of the length to be upset and the diameter of the bar. It this ratio exceeds the certain limit, the head will not upset properly but will buckle and thus cause cold shuts, folds, laps or reenteering fibres.

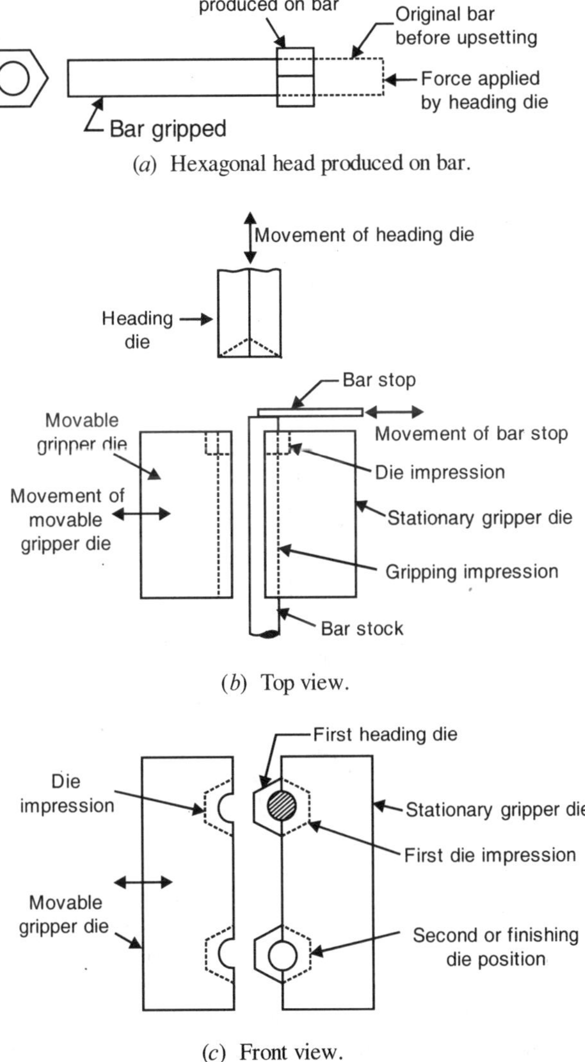

(*a*) Hexagonal head produced on bar.

(*b*) Top view.

(*c*) Front view.

Fig. 14.29. Machine or upset forging.

14·22 Drop Forging

The smith forging, as discussed in Art. 14.13, is a useful and indispensable process, but is not suitable for the production of large number of identical forgings. When a large quantities of a certain component with sound and high quality structure is required, a method known as drop forging is used. The drop forged parts are used in automobiles, airplanes, railroad equipment, engines, agricultural implements and house hold-appliances.

The drop forging process consists of hammering heated bars or billets of steel or other metals into closed impression dies. The closed impression dies compress the metal, causing it to flow and completely fill the die impression. A complete closed impression die consists of two alloy-steel blocks. One of these blocks is called the top or upper die which is fastened to the ram of a steam drop hammer, board drop hammer or other type of power hammer. The other block is the bottom or lower die and it is fastened to the anvil. The impressions for producing the desired shape to be forged are machined in the working surfaces of each die half. When these two die halves come together in alignment, a completely closed die cavity or die impression is formed. Since machined die impressions are employed, a much larger variety of shapes can be forged with greater accuracy and better detail than is possible with smith forging.

14·23 Steam Drop Hammer

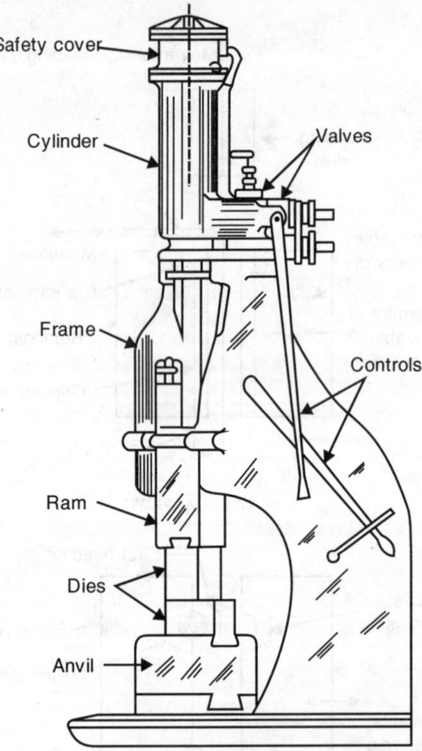

Fig. 14.30. Steam drop hammer.

The steam drop hammers, as shown in Fig. 14.30, are similar to the steam hammers used for smith forging with the added provision of larger and more accurate guides for the vertical movement of the ram. These larger and more accurate bearing surfaces, between the ram and the frame, guide the top die so that it will meet the bottom die in close alignment. A steam drop hammer is not truly a drop hammer, since steam is often admitted above the piston to increase the impact of the falling ram. The steam pressure is used for raising the ram and for controlling the impact of the striking blows, which can be varied from a light tap to a full blow. The energy supplied by the impact of a steam hammer is given by

$$\text{Energy} = m.v^2/2 \text{ (in N-m)}$$

where m = Total mass of all falling parts including upper die, ram, piston rod and piston (in kg), and

v = Velocity at impact (in m/s).

14·24 Board Drop Hammer

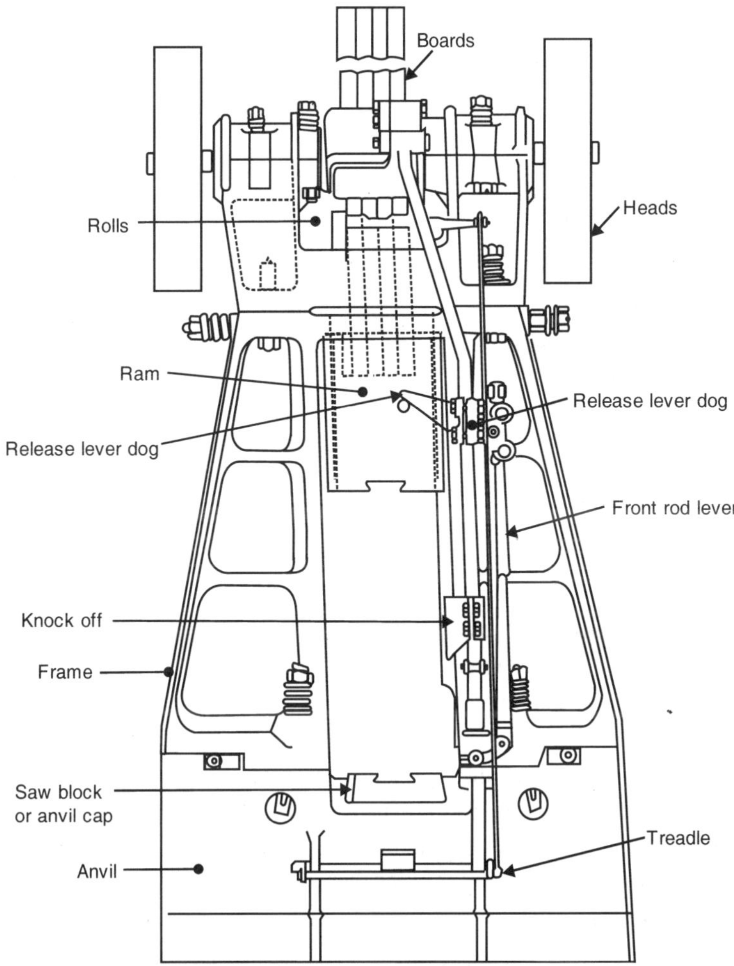

Fig. 14.31. Board drop hammer.

A board drop hammer (also called gravity drop hammer) is shown in Fig. 14.31. The ram is raised by a pair of steel rolls which rotate continually. When these rolls are moved together, they grip and raise the hard wood boards fastened to the top of the ram. The ram continues to be raised until the release lever is actuated by a pin on the ram. This causes the rolls to move outward, releasing the boards and allowing the ram to fall by gravity.

Fig. 14.32 shows the action of the rolls and the clamps in operation. The distance of fall can be adjusted by changing the vertical position of the release lever, but once it is set, it will remain constant while the hammer is in operation. Thus it is not possible to regulate the impact of the striking blows during operation. When the operator takes his foot off the treadle, a set of holding clamps locks the ram in its uppermost position. The energy supplied by the impact of a board drop hammer is determined as follows :

$$\text{Energy} = m.g.h \text{ (in N-m)}$$

where m = Total mass of all falling parts including upper die, ram, piston rod and piston (in kg), and

h = Height of fall (in meters), and

g = Acceleration due to gravity (in m/s^2).

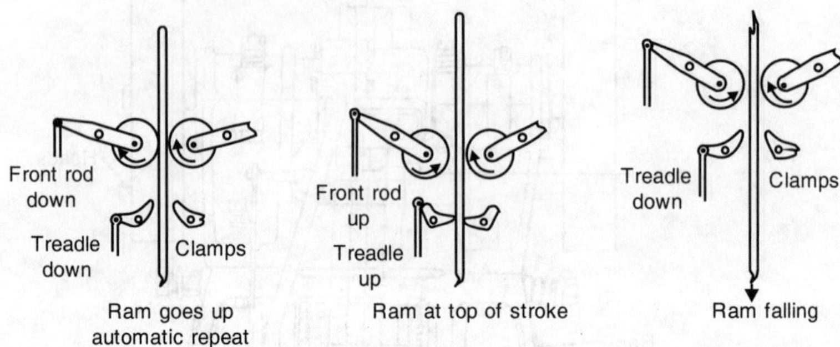

Fig. 14.32. Action of the rolls and clamps in operation.

14·25 Advantages and Disadvantages of Drop Forging

The following are the advantages and disadvantages of drop forging :

Advantages

1. There is little wastage of material.
2. The operations are relatively faster.
3. No sand-casting moulds are required.
4. The strength of material is enhanced.

Disadvantages

1. The closed impression dies normally cost more.
2. The maintenance of the dies is expensive.
3. The extra cost of removing the flash *i.e.,* a thin, irregular extra metal that is forced out between the parting surfaces of the dies.

14·26 Defects in Forgings

The following are some of the reasons for forging defects :

(*i*) Faulty original metal,

(*ii*) Incorrect die design,

(*iii*) Improper heating, and

(*iv*) Improper forging operation.

Some of the defects found in forgings are as follows :

1. *Cold shuts or laps.* This defect is usually found at corners and at right angles to the surface. It appears like cracks. It is due to the result of the metal surface folding against itself during forging. The possible causes are misplacing the metal in the die, poor forging design and incorrect die design.

2. *Cracks.* The cracks are longitudinal or transverse in nature and occur on the forging surface. It is due to the poor quality of metal, improper heating, incorrect cooling and low temperature of the forging.

3. *Scale pits.* These are shallow surface depressions caused by scale that was not removed from the die and was worked into the surface of the forging. When the scale is removed by cleaning, the pits remain in the forging surface. It can be avoided by proper control of forging temperature and frequent cleaning of dies.

4. *Unfilled sections.* This defect is caused when a closed-impression die cavity is not completely filled with metal. The possible causes for this defect are the use of wrong amount of metal, misplacing the metal in the die, not heating the metal sufficiently, poor forging design and incorrect die design.

5. *Mismatched forgings.* This defect results when the upper and lower die impressions are out of alignment as the final blow is struck.

6. *Ruptured fibre-flow lines.* This defect is usually caused by too rapid flowing of the metal. This can be revealed only by observing the macrostructure.

7. *Fins and rags.* These are small projections or pieces of loose metal driven into the forging surface. It can be controlled by proper die design and workmanship.

8. *Burnt and overheated metal.* This is caused by heating the metal at a high temperature or for a too long time. This defect can be avoided by the proper heating of metal.

14·27 Design of Forging

In designing a forging, the following points should always be considered :

1. The forged components should ultimately be able to achieve a radial flow of grains or fibres.
2. The forgings which are likely to carry flash such as drop and press forging, should preferably have the parting line in such a way that the same will divide them in two equal halves.
3. The parting line of a forging should be, as far as possible, in one plane.
4. Sufficient draft on surfaces should be provided to facilitate easy removal of forgings from dies.
5. The sharp corners should always be avoided to prevent concentration of stress and to facilitate ease in forging.
6. The pockets and recesses in forgings should be minimum in order to avoid increased die wear.
7. The ribs should be high and thin.
8. Too thin sections should be avoided to facilitate easy flow of metal.

QUESTIONS

1. What is the difference between smithing and forging ?

2. Explain, with a neat sketch, the smith's forge. What do you understand by 'open fire' and 'stock fire' ? Which of the two is more advantageous and why ?

3. Sketch and describe the following :

(*a*) Anvil, and (*b*) Swage block.

4. Describe, with neat sketches, the smithy hand tools.

5. Write short notes on the following :

(*a*) Swages, (*b*) Fullers, (*c*) Flatters, (*d*) Set hammer

(*e*) Punches, (*f*) Drifts, and (*g*) Hardie.

6. Explain with sketches, the following operations :

(*a*) Upsetting, (*b*) Drawing down,

(*c*) Setting down, and (*d*) Bending.

7. Describe in detail the process of hand welding.

8. What do you understand by drop forging ? Explain its process.

9. Sketch and describe the steam and board drop hammers.

10. What are the common forging defects and why they occur.

11. What are the important factors to be taken into consideration while designing a forging ?

OBJECTIVE TYPE QUESTIONS

1. The cutting edge of a cold chisel is about

(*a*) 30° (*b*) 45° (*c*) 60° (*d*) 75°

2. The cutting edge of hot chisel is about

(*a*) 30° (*b*) 45° (*c*) 60° (*d*) 75°

3. The fullers are used

(*a*) for finishing flat surfaces (*b*) for necking down a piece of work

(*c*) for punching a hole (*d*) to finish the punched hole

4. The upsetting is a process of

(*a*) increasing the cross-section of a bar

(*b*) reducing the cross-section of a bar

(*c*) joining the two surfaces of metal under pressure after heating

(*d*) bending of a bar

5. The drawing down is a process of

(*a*) increasing the cross-section of a bar

(*b*) reducing the cross-section of a bar

(*c*) joining the two surfaces of metal under pressure after heating

(*d*) bending of a bar

ANSWERS

1. (*c*) 2. (*a*) 3. (*b*) 4. (*a*) 5. (*b*)

15

Welding

15·1 Introduction

The welding is a process of joining two similar or dissimilar metals by fusion, with or without the application of pressure and with or without the use of filler metal. The fusion of metal takes place by means of heat. The heat may be obtained from blacksmith's fire, electric arc, electrical resistance or by chemical reaction.

The process of joining similar metals by melting the edges together, without the addition of filler metal, is called *autogenous welding*.

The process of joining similar metals with the help of filler rod of the same metal is called *homogeneous welding*.

The process of joining dissimilar metals using filler rod is called *heterogenous welding*. The filler rod material is such that its melting point is less than the parent metals.

The welding is extensively used in fabrication as an alternative method for casting or forging and as a replacement for bolted and riveted joints. It is also used as a repair medium *e.g.*, to reunite metal at a crack to build up a small part that has broken off such as gear tooth or to repair a worn surface such as bearing surface.

15·2 Weldability

The term 'weldability' may be defined as the property of a metal which indicates the ease with which two similar or dissimilar metals are joined by fusion with or without the application of pressure and with or without the use of filler metal.

Strictly speaking, a metal has good weldability if it can be easily welded in a fabricated structure. The various factors affecting the weldability of a metal are :

1. Composition of the metal.
2. Brittleness and strength of the metal at elevated temperatures.

3. Thermal properties of the metal.
4. Welding techniques, fluxing material and filler material.
5. Proper heat treatment before and after the deposition of the metal.

The common metals having weldability in the descending order are iron, carbon, steel, cast iron, low alloy steels and stainless steel.

15·3 Advantages and Disadvantages of Welded Joints

Following are the advantages and disadvantages of welded joints over other joints such as riveted or bolted joints.

Advantages

1. The welding structures are normally lighter than riveted or bolted structures. This is due to the reason, that in welding, gussets or other connecting components are not used.
2. The welded joint provides maximum efficiency (may be 100%) which is not possible in other type of joints.
3. The alterations and additions can be easily made in the existing structures.
4. As the welding structure is smooth in appearance, therefore it looks pleasing.
5. In welded connections, the tension members are not weakened as in the case of riveted joints.
6. A welded joint has a great strength. Often a welded joint has the strength of the parent metal itself.
7. Sometimes, the members are of such a shape (*i.e.*, circular steel pipes) that they afford difficulty in riveting. But they can be easily welded.
8. The welding provides very rigid joints. This is in line with the modern trend of providing rigid frames.
9. It is possible to any part of a structure at any point.
10. The process of welding takes less time than other type of joints

Disadvantages

1. Since there is an uneven heating and cooling during fabrication, therefore, the members may get distorted or additional stresses may develop.
2. It requires a highly skilled labour and supervision.
3. The edge preparation is generally required before welding.
4. Since no provision is left for expansion and contraction in frame, therefore there is a possibility of cracks developing in it.
5. The jigs and fixtures are required to hold the parts in position.
6. The inspection of welding work is more difficult than riveting or bolted work.

15·4 Types of Welding

The welding is broadly divided into the following two groups :

1. *Forge or pressure welding*. In forge or pressure welding (also known as plastic welding), the workpieces are heated to plastic state (except for cold pressure welding) and then the workpieces are joined together by applying pressure on them. In this case no filler material is used. The forge or pressure welding is further classified as follows :

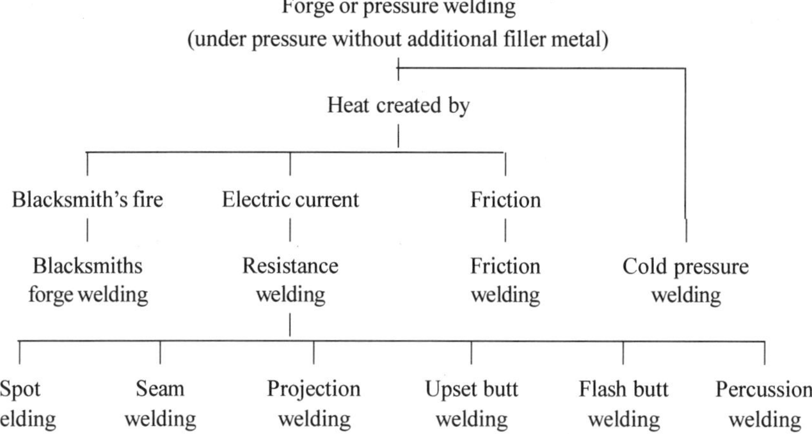

2. *Fusion or non-pressure welding.* In fusion or non-pressure welding, the edge of workpieces to be joined and the filler material are heatd to a temperature above the melting point of the metal and then allowed to solidify. The fusion or non pressure welding is further classified as follows :

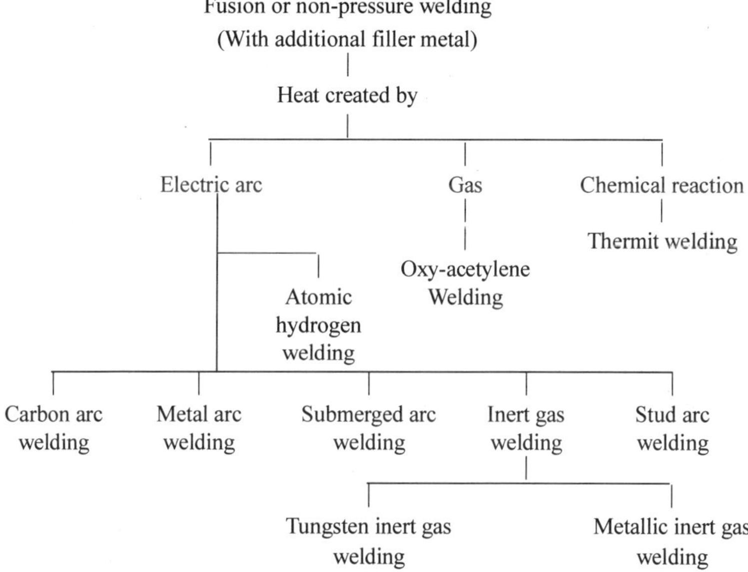

15·5 Cold Pressure Welding

This is a modern process and has been developed during recent years. In this process certain similar and dissimilar metals are joined without any source of heat. The two parts to be welded are subjected to high pressure which results in inter-surface molecular fusion of the parts to be joined. This process is mainly used for welding non-ferrous metals particularly aluminium and copper.

15·6 Types of Welded Joints

The relative positions of the two pieces being joined determine the type of joint. The following are the five basic types of joints commonly used in fusion welding.

1. *Lap joint.* The lap joint is obtained by over lapping the plates and then welding the edges of the plates. These joints are employed on plates having thickness less than 3 mm. The lap joints may be

(*a*) Single transverse,

(*b*) Double transverse, and

(*c*) Parallel lap joints.

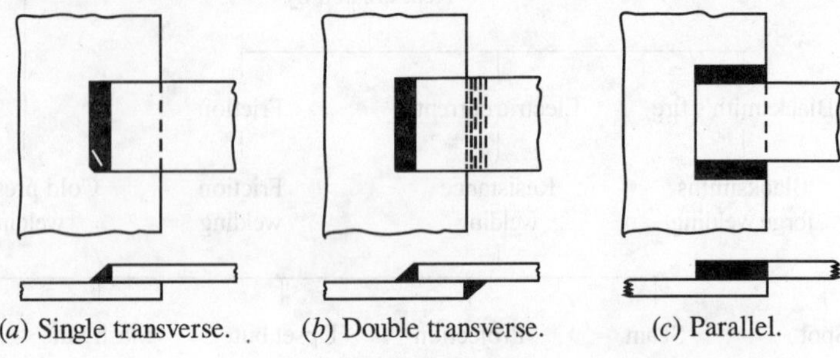

(*a*) Single transverse. (*b*) Double transverse. (*c*) Parallel.

Fig. 15.1. Lap joints.

The lap joints are shown in Fig. 15.1. A single transverse lap joint has the disadvantage that the edge of the plate which is not welded can buckle or warp out of shape.

2. *Butt joint*. The butt joint is obtained by welding the ends or edges of the two plates which are approximately in the same plane with each other as shown in Fig. 15.2. In butt welds, the plate edges do not require bevelling if the thickness of plate is less than 5 mm. On the other hand, if the plate thickness is 5 mm to 12.5 mm, the edges should be bevelled to *V* or *U*-groove and plates having thickness above 12.5 mm should have a *V* or *U*-groove on both sides.

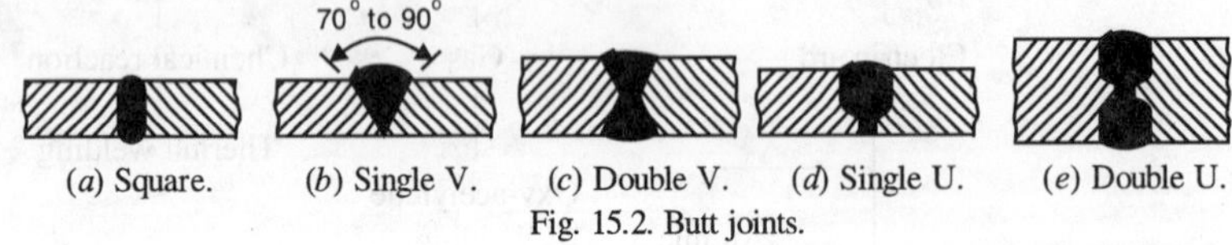

(*a*) Square. (*b*) Single V. (*c*) Double V. (*d*) Single U. (*e*) Double U.

Fig. 15.2. Butt joints.

The butt joints may be

(*a*) Square butt joint, (*b*) Single *V*-butt joint, (*c*) Double *V*-butt joint, (*d*) Single *U*-butt joint, and (*e*) Double *U*-butt joint.

These joints are shown in Fig. 15.2.

3. *Corner joint*. The corner joint, as shown in Fig. 15.3 (*a*), is obtained by joining the edges of two plates whose surfaces are at an angle of approximately 90° to each other. It is used for both light and heavy gauge sheet metal. In some cases corner joint can be welded, without any filler metal, by melting off the edges of the parent metal.

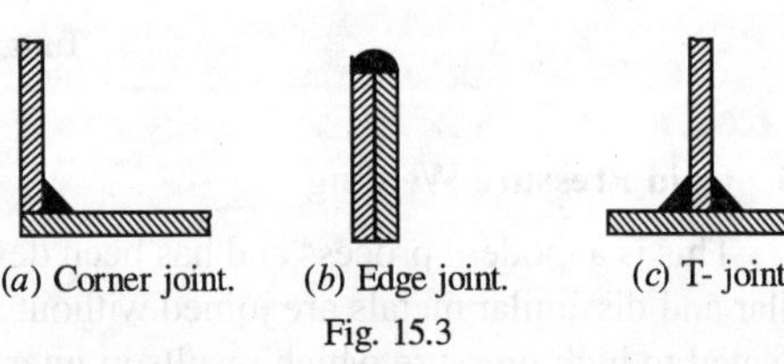

(*a*) Corner joint. (*b*) Edge joint. (*c*) T- joint.

Fig. 15.3

4. *Edge joint*. The edge joint, as shown in Fig. 15.3 (*b*), is obtained by joining two parallel plates. It is economical for plates having thickness less than 6 mm. This joint is unsuitable for members subjected to direct tension or bending.

5. *T-joint*. The T-joint, as shown in Fig. 15.3 (*c*) is obtained by joining two plates whose surfaces are approximately at right angles to each other. It is widely used to weld stiffeners in air craft and other thin walled structures. These joints are suitable upto 3 mm thickness.

15·7 Fillet Welded Joints

The lap joints, corner joints and T-joints are the fillet welded connections which are generally used. The rounding of a corner is known as filleting. The cross-section of the fillet is approximately triangular. The flush fillet, convex fillet and the concave fillet are the three types of fillet welds as shown in Fig. 15.4 (*a*), (*b*), and (*c*) respectively.

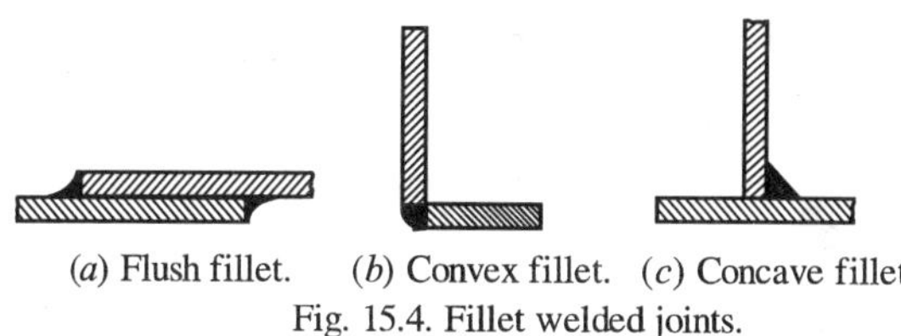

(*a*) Flush fillet. (*b*) Convex fillet. (*c*) Concave fillet

Fig. 15.4. Fillet welded joints.

15·8 Edge Preparation and Applications

The edge preparation is very essential in order to obtain sound welded joint. It consists of bevelling the edges and carefully cleaning the faces to be welded from dust, sand grit, oil etc. The edge preparation used in fusion welding for various types of butt welded joints are as follows :

1. The square butt welded joints are used when the thickness of the plates is from 3 to 5 mm. It does not require bevelling of the edges. The edges should be spaced about 3 mm apart as shown in Fig. 15.5 (*a*).
2. The single-V butt welded joints are used when the thickness of the plates is between 8 mm and 16 mm. The edges of the plates are bevelled to form an included angle of about 70° to 90° as shown in Fig. 15.5 (*b*), depending upon the method of welding used.
3. The double V butt welded joints are used when the thickness of the plates is more than 16 mm and where welding can be performed on both sides of the plate. The both edges of each plate are bevelled to form a double-*V* as shown in Fig. 15.5 (*c*).

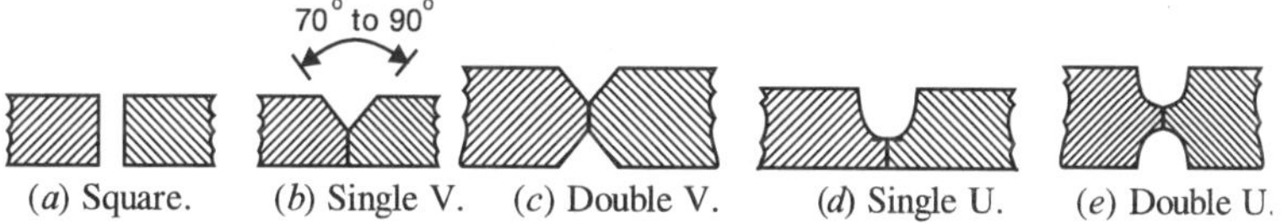

(*a*) Square. (*b*) Single V. (*c*) Double V. (*d*) Single U. (*e*) Double U.

Fig. 15.5. Edge preparation.

4. The single and double-*U* butt welded joints are used when the thickness of the plate is more than 20 mm. These joints are more satisfactory and require less filler rod, but the edge preparation as shown in Fig. 15.5 (*d*) and (*e*) is difficult.

15·9 Welding Positions

The welding positions are classified as follows :

1. *Flat position.* In this position, the filler metal is deposited from the upper side of the joint with the face of the weld horizontal, as shown in Fig. 15.6 (*a*).

2. *Horizontal position.* In this position, the weld is deposited upon the side of a horizontal and against a vertical surface, as shown in Fig. 15.6 (*b*).

3. *Vertical position.* In this position, the line of welding is in a vertical plane and the weld is deposited upon a vertical surface, as shown in Fig. 15.6 (*c*).

4. *Overhead position*. In this position, the weld is deposited from the under side of the joint and the face of the weld is horizontal, as shown in Fig. 15. 6 (*d*). It is the reverse of flat welding.

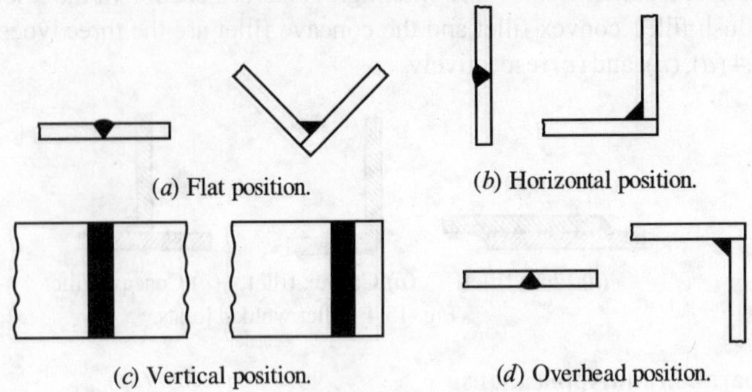

Fig. 15.6. Welding positions.

15·10 Blacksmith's Forge Welding

It is the oldest method of all the welding processes. The metal parts to be welded are heated in blacksmith's furnace to a plastic state (about 1300°C) and then pressed together by hand or power hammer.

Though the process is very slow, yet it is conveniently used for joining parts of mild steel, wrought iron and low carbon steels. When wrought iron or steel at high temperature is exposed to the air, it oxidises and is covered with a film of scale (*i.e.*, oxide). If this film scale and ashes from the fire are allowed to remain on the surfaces to be welded, the result will be failure. In order to avoid this, the fluxes are used. In case of wrought iron, sand in the form of flux is sprinkled on the work piece surface while for mild steel calcined borax is used as flux.

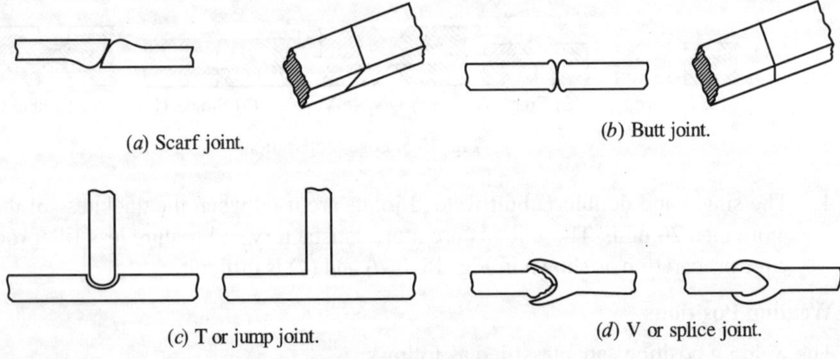

Fig. 15.7. Forged welding.

The common joints used in forge welding may be scarf or lap joint, butt joint, T or jump joint and V or splice joint as shown in Fig. 15.7 (*a*), (*b*), (*c*) and (*d*) respectively.

15·11 Electric Resistance Welding

This is also a type of pressure welding. It is used for joining pieces of sheet metal or wire. It is usually adopted for mass production. In this case, welding heat is obtained at the location of the desired weld by the electrical resistance through the metal pieces to a relatively short duration, low voltage (from 6 to 10 volts only) high amperage (varying from 60 to 4000 amperes) electric current. When the area to be welded is sufficiently heated, the pressure varying from 25 to 55 MPa is applied to the joining area by suitable electrodes until the weld is solid. The amount of heat generated by the resistance of the metal pieces at the points of contact is given by

$$H = I^2 R t$$

where H = Total heat generated in the work, in watt-second,

I = Current, in amperes,

R = Resistance of the metal being welded, in ohms, and

t = Time during which the electric current flows, in seconds.

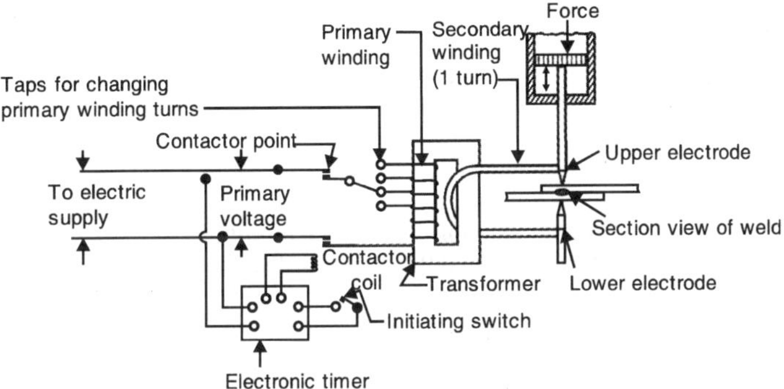

Fig. 15.8. Electric resistance welding.

Since the amount of current (I) is squared in the above relation, therefore, it has the greatest influence upon the amount of heat generated. Thus its control is most important.

A common method for regulating the amount of current is to change the number of functioning primary turns of the transformer. The total resistance through a joint depends upon the two metal pieces to be welded, the cleanliness of the surfaces to be joined and the pressure applied by the electrodes. The time may be controlled by an electronic timer, as shown in Fig. 15.8, which can control the time within ± 0.001 second. Such accurate time control is required only for metals like aluminium which is more difficult to weld by resistance welding.

The resistance welding has the following advantages and disadvantages :

Advantages

1. It is well suited for mass production.
2. It is economical in operation, because nothing is consumed except electrical power.
3. The welds are quickly made.
4. The skilled operators are not required.
5. It is possible to weld dissimilar metals.

Disadvantages

1. The resistance welding equipment has high initial cost.
2. Certain resistance welding processes are limited to lap joints. A lap joint has an inherent device between the two metal pieces, which causes stress concentration in applications where fatigue is present. The device may also cause trouble when corrosion is present.

Notes : 1. When both metal pieces to be welded together are similar and have the same thickness, then the weld will ordinarily penetrate both pieces to about the same extent.

2. When the pieces of dissimilar metals or of different thicknesses are to be welded, then care must be taken to see that the weld penetrate both metal pieces about equally. In other words, heat should be generated in both pieces equally to produce a good resistance weld of proper oval section.

3. The electrodes are of a high-heat conductivity copper alloy and they are water cooled to protect them from softening or melting.

4. In electric resistance welding, either A.C. or D.C. may be used, but A.C. is generally used.

15·12 Types of Electric Resistance Welding

The electric resistance welding may be classified as follows :

1. Spot welding,
2. Roll spot and seam welding,
3. Projection welding,
4. Butt welding, and
5. Percussion welding.

We shall now discuss these types of electric resistance welding, in detail, in the following articles.

15·13 Spot Welding

The spot welding is used for welding lap joints, joining components made from plate material having 0.025 mm to 1.25 mm in thickness. The plate thickness is limited by the pressure and current capacity of the available spot welding machine. The plates to be joined together are placed between the two electrode tips of copper or copper alloy as shown in Fig. 15.9. A low voltage and high amperage current is passed between electrodes, causing the metal temperature to be raised to welding temperature. The heat generated by the electric current is rapidly transferred from the outer faces of the plates to the electrodes which are water cooled. The pressure is applied which squeezes the metal together and thus completes the weld. A spot welding machine is shown in Fig. 15.10.

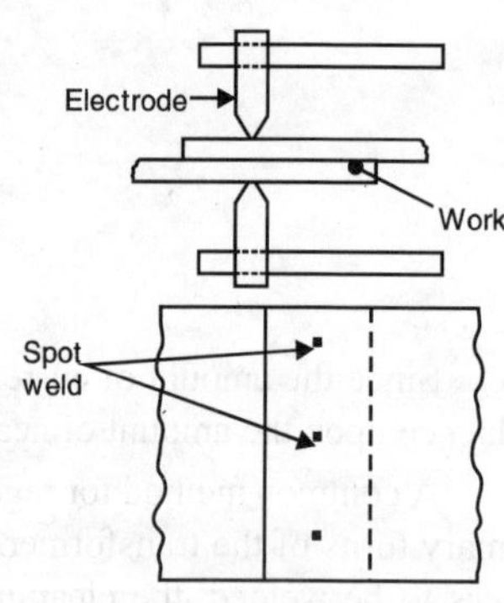

Fig. 15.9. Spot welding.

The spot welding is used for welding most of the ferrous and non-ferrous alloys. It is widely used in manufacturing automobiles, refrigerators and other metal stamping assemblies.

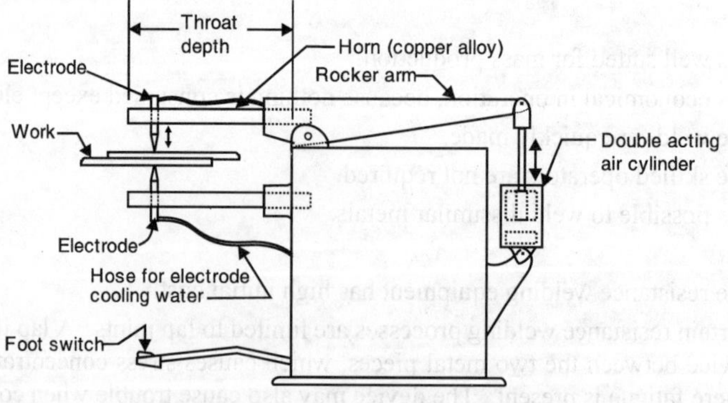

Fig. 15.10. Spot welding.

Notes : 1. The current density at the point to be welded depends upon the electrode tip diameter (d), which should be $d = \sqrt{t}$

where t = Thickness of plate material to be welded.

2. The distance between the nearest edge of the plate and centre of weld should be at least 1.5 d. This will help to prevent expulsion of molten metal and a low strength joint.

3. The spacing between two spot welds should not be less than 3 d in order to avoid shunting effect between neighbouring welds.

15·14 Roll Spot and Seam Welding

When the spot welds on two over-lapping pieces of metal are spaced, the process is known as *roll spot welding*. If the spot welds are sufficiently made close so that they overlap and make a leakproof seam (*i.e.*, joint), the process is called *seam welding*. The typical applications of seam welding include sheet metal radiator and drums where both longitudinal and circumferential lap joints are used.

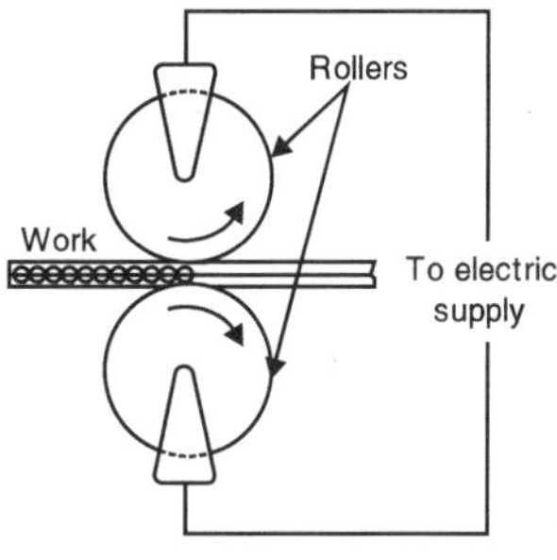

Fig. 15.11. Seam welding.

The work pieces to be welded are passed between the rotating disc shaped electrodes (of copper alloy) which are in the form of rollers as shown in Fig. 15.11. These electrodes are not separated after each individual weld as in spot welding, but maintain a continuous pressure on the work. The current passing from roller to roller through the workpiece heats up the parts to be joined and the pressure on the roller electrodes forms the weld. The cooling is accomplished by flooding the electrodes and metal pieces with water. This process is best adopted for metal thickness ranging from 0.025 mm to 3 mm.

15·15 Projection Welding

It is similar to spot welding except that one of the metal pieces to be welded has projections on its surface at the points where the welds are to be made. These projections should extend about 1 mm above the surface to be welded. The projections are readily made with punch presses by embossing. The contact between the two metal pieces to be joined is made at the projections, as shown in Fig. 15.12. When the welding current passes through these projections, they are melted and flattened, allowing the two surfaces to come together. The melted projections become the welds.

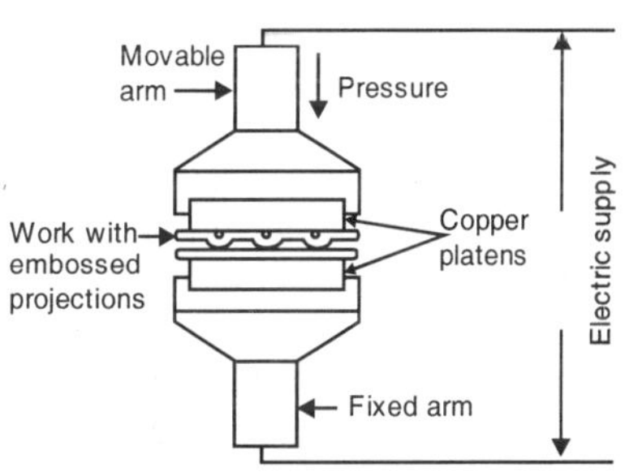

Fig. 15.12. Projection welding.

This process is used for cross-wise welding and for parts where the ridges are produced by machining. It is most suitable for large quantity production. The main advantages of projection welding over spot welding are as follows :

1. It is a quick process and number of welds can be made simultaneously. The ideal number of projections is three and the maximum number of projections which can be used satisfactorily is about six.
2. The welds may be more closely spaced than is possible with spot welding.
3. There is no thickness limitation.
4. The life of the electrodes used is much longer as compared with the life of spot welding electrodes.

Notes : 1. When two pieces of different metals are to be welded by projection welding, then the projections should be made on the metal piece with the higher conductivity.

2. When the two metal pieces are of different thicknesses, the projections should be made on the thicker metal piece.

15·16 Butt Welding

The butt welding is of the following two types :

(*a*) Upset butt welding, and (*b*) Flash butt welding.

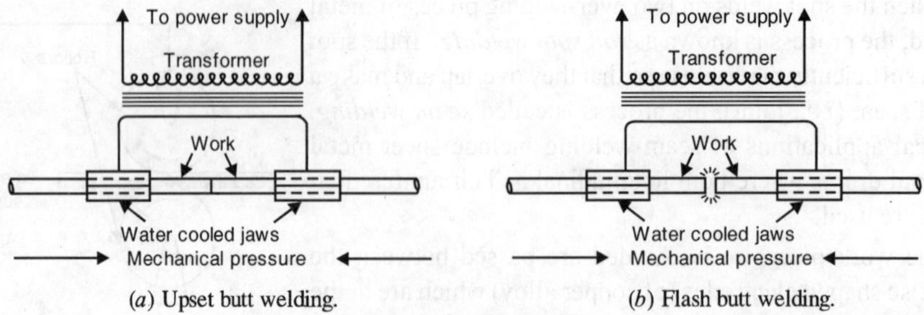

(*a*) Upset butt welding. (*b*) Flash butt welding.

Fig. 15.13. Butt welding.

In *upset butt welding*, the two parts of metal having the same cross-section are held by two heavy water cooled electrode jaws and forced together as shown in Fig. 15.13 (*a*). The heat is generated in the contact surface by the electrical resistance when a suitable electric current flows across the joint. When the metal at the joint is heated sufficiently, the pressure is applied by the two jaws in order to force the two pieces together. The upset butt welding is especially adopted to rods, pipes, small structural shapes and many other parts of uniform section.

In *flash butt welding*, the heat required for the welding operation is derived from the arcing that occurs in the space between the parts to be joined, as shown in Fig. 15.13 (*b*). In this process, the ends to be welded are pushed together and immediately after contact is made, they are separated by a short distance. In separating, an arc is formed which heats the ends of the pieces to the fusion temperature. After the ends are fused, the current is shut off and the pieces are forced together. In this process, a slight burr is produced when the two sides are forced together, forming what is called a *flash* (*i.e.,* a ragged line of extruded material which can be machined or ground away). This process is used extensively in the manufacture of steel containers and in the welding of mild steel shanks to high speed steel drills and reamers.

15·17 Percussion Welding

It is a recent development in welding. The welding heat is obtained from the arc produced by a rapid discharge of stored electrical energy. The electrostatic capacitors are used to store the electrical energy.

In this process, the pieces to be welded are held apart, one in a stationary holder and the other in a clamp mounted in a slide and backed up against heavy spring pressure. When the movable clamp is released, it moves rapidly carrying with it the piece to be welded. When the pieces are about 1.5 mm apart, there is a sudden discharge of electrical energy, causing intense arcing over the surface and bringing them to high temperature. The arc is extinguished by the percussion blow of the two parts coming together with sufficient force to effect the weld.

The process is so rapid (takes about 0.1 second) that there is little heating effect in the material adjacent to the weld. It may be noted that this process is limited to only small areas (upto 3.2 sq cm) of nearly regular sections. Thin sheets of equivalent area cannot be joined by this process.

This process is used for welding stellite tips to tools, copper to aluminium or stainless steel, silver contact tips to copper, cast iron to steel, lead-in wires on electric lamps and zinc to steel.

15·18 Arc Welding

The arc welding is a fusion welding process in which the welding heat is obtained from an electric arc between the work (or base metal) and an electrode. The electric arc is produced when two conductors of an electric circuit are touched together and then separated by a small distance, such that there is sufficient voltage in the circuit to maintain the flow of current through the gaseous medium (air). The temperature of heat produced by the electric arc is of the order of 6000°C to 7000°C.

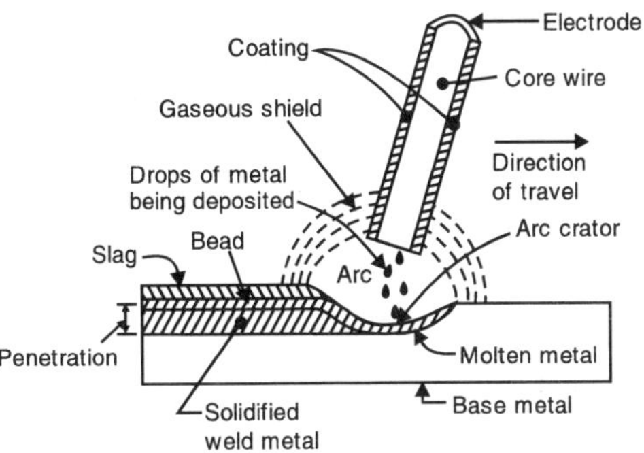

Fig. 15.14. Arc welding.

The most common method of arc welding is with the use of a metal electrode which supplies filler metal. The welding is done by first making contact of the electrode with the work* and then separating the electrode to a proper distance to produce an arc. When the arc is obtained, intense heat so produced quickly melts the work under the arc forming a pool of molten metal which seems to be forced out of the pool by the blast from the arc, as shown in Fig. 15.14. A small depression is formed in the work and the molten metal is deposited around the edge of this depression, which is called the *arc crator*. The slag is brushed off after the joint has cooled. The arc, once started, should be advanced at a uniform speed along the desired line of welding. The melting should reach to a sufficient depth below the original surfaces of the metal pieces to be joined to obtain the desired weld. This is known as obtaining *proper penetration.*

Both the direct current (D.C.) and alternating current (A.C.) arc used for arc welding. The direct current supply for arc welding is usually obtained from a generator driven by either an electric motor or a petrol or diesel engine. The alternating current supply for arc welding is obtained from a step down transformer which receives current from the supply mains at 200 to 440 volts and transforms it to the voltage actually required (*i.e.*, 80 to 100 volts) for striking the arc only. In order to maintain the arc, a still lower voltage say about 30 to 40 volts is required. The voltage required in case of D.C. welding is 60 to 80 volts for striking the arc and 15 to 25 volts for maintaining the arc.

15·19 Polarity in Arc Welding

We have already discussed that both the direct current and alternating current may be used for arc welding, but the direct current is preferred for most purposes. When the work is connected to the positive terminal of a D.C. welding machine and the negative terminal to an electrode holder, the welding set up is said to have *straight polarity*. On the other hand, when work is connected to negative and the electrode to a positive terminal, then the welding set up is said to have *reversed polarity*. The straight polarity is preferable for some welds while for other welds reversed polarity should be used. When it is desired to have more heat liberated at or near the work, straight polarity is used because the heat developed at the positive terminal is higher (about 60 to 75% of the total heat evolved) than the negative terminal. The temperature of the arc produced is 3500°C to 4000°C.

When alternating current is used, there is no fixed polarity at the terminals and they interchange in every cycle. The alternative current also acquires zero value twice in each cycle. Thus at these particular moments, the potential difference between the terminals is zero and hence a high voltage is required to maintain the arc at this moment.

*The electrode acts as one conductor and the work as the second conductor.

15·20 Comparison Between A.C. and D.C. Arc Welding

Following is the comparison between A.C. and D.C. arc welding.

A.C. arc welding	*D.C. arc welding*
1. The A.C. welding transformer has no moving parts and is simpler.	1. The D.C. welding generator has rotating parts and is more complicated.
2. The transformer costs less and its maintenance cost is low.	2. The generator costs more and its maintenance cost is high.
3. Since the distribution of heat is equal, therefore there is no need for changing the polarity. Hence only ferrous metals are usually welded by A.C.	3. Heat distribution is different in two poles, *i.e.*, two-third in positive and one-third in negative. By changing the polarity, all types of metals can be welded by D.C.
4. All types of electrodes can not be used in A.C. arc welding because the current constantly reverses with every cycle. Only coated electrodes can be used.	4. All types of electrodes, bare or coated can be used in D.C. arc welding because the polarity can be changed to suit the electrode.
5. The problem of 'arc blow' does not arise as it is very easy to control.	5. In D.C., the 'arc blow' is severe and cannot be controlled easily.
6. The arc is never stable.	6. The arc is more stable.
7. It can be used only when A.C. supply from the mains is available.	7. In the absence of A.C. mains supply, an engine driven D.C. generator set can be used.
8. A.C. is more dangerous.	8. D.C. is comparatively less dangerous.

15·21 Types of Arc Welding

Following are the two types of arc welding depending upon the type of electrode :

1. *Un-shielded arc welding*. When a large electrode or filler rod is used for welding, it is said to be *un-shielded arc welding*. In this case, the deposited weld metal while it is hot will absorb oxygen and nitrogen from the atmosphere. The effect of oxygen is to produce inclusions and blow holes in weld and nitrogen promotes hardness and brittleness. This decreases the strength of weld metal, lower its ductility and resistance to corrosion.

2. *Shielded arc welding*. When the welding rods coated with fluxing material are used, then it is called *shielded arc welding*. The fluxing material is applied to the rod by dipping or extrusion. These rods are sometimes called *coated electrodes*. The coated rods not only supply a fluxing agent but it also fuses at a slower rate than the rod itself. This property acts to exclude the atmosphere in the vicinity of the atmosphere and prevents the absorption of large amounts of harmful oxygen and nitrogen. The welds made with the coated rods or by shielded arc welding method are free from blow holes and produce much stronger welded joints.

15·22 Electrodes for Arc Welding

The electrodes for arc welding are generally of the following two types :

1. Bare or non-coated electrodes, and 2. Coated electrodes.

When the *bare* or *non-coated electrodes* are used, then the deposited weld metal while it is hot will absorb oxygen and nitrogen from the atmosphere. The effect of oxygen is to produce inclusions and blow holes in the weld and nitrogen promotes hardness and brittleness. This decreases the strength of the weld metal ; lowers its ductility and resistance to corrosion. The bare electrodes are rarely used in modern welding practice.

When the core of the bare metallic wire is provided with a covering or coating of some fluxing material, it is then known as *coated electrodes*. The fluxing material is applied to the wire cores either by dipping in a bath or during extrusion.

The coated electrodes are widely used in metal arc welding and perform the following functions:

1. It gives off an inert or protective gas when heated by the arc, which partially envelops and shields the arc and molten metal from the atmosphere. Thus the molten metal is protected from oxygen and nitrogen of the atmosphere.
2. It provides a flux, which helps to remove oxides and other impurities from the molten weld metal. This then produces a slag, which floats on the top of the molten metal, providing additional protection against the surrounding air.
3. It stabilizes the arc by providing certain chemicals which have this ability.
4. It provides means for adding various alloying elements into the weld metal.
5. Since the electrode coating is consumed at a slower rate than the metal core of an electrode, therefore, the coating, except at the start, extends beyond the metal core, as shown in Fig. 15.14. This helps to direct the arc and the molten metal being deposited.
6. It concentrates the arc stream and reduces thermal losses. This increases the temperature concentrated at the electrode tip. Thus, it increases the rate of metal deposition and the penetration obtained.

The coating may consists of two or more ingredients and there are different kinds of coatings to suit the different types of welds and metals to be joined. The principal constituents used for coating the electrode are as follows :

(*a*) The slag forming constituents are silicate, manganese oxide and iron oxide. Sometimes aluminium oxide is used, but it makes the arc less stable.

(*b*) The constituents to improve the arc characteristics are sodium oxide, calcium oxide, magnesium oxide and titanium oxide.

(*c*) The deoxidising constituents are graphite, aluminium and wood flour.

(*d*) The alloying constituents to improve the strength of weld are vanadium, cobalt, molybdenum, aluminium, chromium, nickel, manganese and vanadium.

(*e*) The binding materials are sodium silicate, potassium silicate and asbestos.

The coated electrodes, according to the thickness of coating, may be divided into the following three groups :

1. Lightly coated electrodes ; 2. Medium coated electrodes, and 3. Heavy coated electrodes.

The light and medium coated electrodes may have coating thickness varying from 10 to 50 percent of the total diameter of the coated electrode, whereas in some heavy coated electrodes it may be as high as 100 percent.

The standard diameter of the electrode varies from 1.6 mm to 9 mm and length from 250 mm to 450 mm.

Notes : 1. When iron powder is used in relatively large amounts in the coating of an electrode, then it is said to be a *contact electrode*. The arc welding with contact electrode is called *contact arc welding* and has the following advantages :

(*i*) The penetration is deep at the centre.

(*ii*) The welding speeds are high.

(*iii*) The welds are readily made in all positions.

2. In contact arc welding, the coating at the end of the electrode is kept in contact with the work at all times.

15·23 Arc Welding Equipment

The equipment commonly used for welding consists of the following :

1. Alternating current (A.C.) or direct current (D.C.) machines ; 2. Electrode ; 3. Electrode holder (insulated) ; 4. Cables, 5. Cable connectors ; 6. Cable lugs ; 7. Chipping hammer ; 8. Earthing clamps ; 9. Wire brush ; 10. Helmet or hand shield ; 11. Safety goggles ; 12. Hand gloves ; 13. Aprons, sleeves etc.

15·24 Precautions in Arc Welding

The following precautions should be strictly observed while performing electric arc welding.

1. The hand, face and eyes should be protected from the heat and flare of the electric arc. For this, the operator should wear hand gloves to protect hands and a hand shield or helmet with a window of coloured glass should be used to protect face and eyes.

2. The space for the electric arc welding should be screened off from the rest of the building so that other workmen are safe from the glare of the arc which is very harmful for eyes.

15·25 Arc Welding Processes

The following are the various arc welding processes commonly used in engineering practice.

1. Carbon arc welding,
2. Metal arc welding,
3. Metallic inert gas (MIG) arc welding,
4. Tungsten inert gas (TIG) arc welding,
5. Atomic hydrogen welding,
6. Stud welding,
7. Submerged arc welding,
8. Plasma arc welding,
9. Flux cored arc welding.
10. Electro-slag welding,
11. Electro-gas welding.

We shall now discuss these arc welding processes, in detail, in the following articles.

15·26 Carbon Arc Welding

In carbon arc welding, the welding heat is obtained from an electric arc between a carbon electrode and the work. Thus the two parts to be joined are fused as shown in Fig. 15.15 (*a*). In welding heavy plates, the additional metal is deposited in the weld from a filler rod.

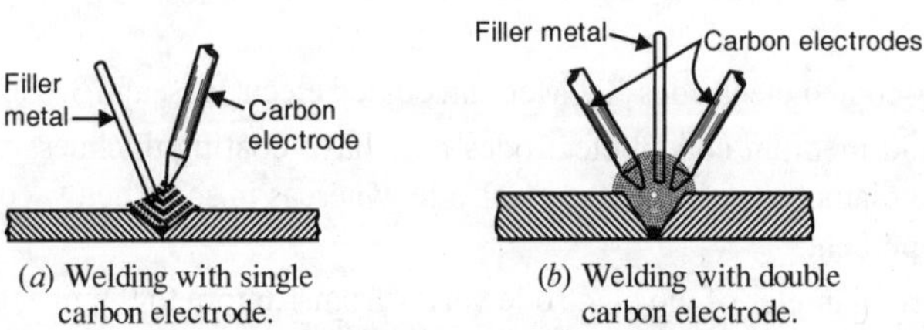

(*a*) Welding with single carbon electrode.

(*b*) Welding with double carbon electrode.

Fig. 15.15. Carbon arc welding.

In another method, known as *twin carbon arc welding*, two carbon electrodes are employed between which the arc is formed, as shown in Fig. 15.15 (*b*). In this way, the arc generates heat which is applied close to the parts to be welded, causing the edges to fuse. In heavy welds, a filler rod is melted by the arc and deposited in the weld. The twin carbon arc welding is very similar to gas welding in which the heat is generated above the work and is played upon it. The differences between the two are in the source of heat and differences in the atmosphere around the work.

In carbon arc welding, a current of 20 to 800 amperes is used which depends upon the thickness of metal to be welded and diameter of electrode. Most of the carbon arc welding is done with automatic welding equipment where the arc voltage and current, rate of travel and rate of feeding the filler rod are all properly controlled. The carbon arc welding is used in all modern industries for welding cast iron, steel, copper, bronze, galvanised iron, aluminium and for rough cutting of metal.

Note : In carbon arc welding, either carbon or graphite electrodes are used. The element carbon exists in three forms which are amorphous, graphite and diamond. The carbon electrodes are of amorphous form which are considerably cheaper. If amorphous carbon is heated to a sufficiently high temperature, it is converted to graphite carbon. The graphite electrodes have a longer life and approximately 400 percent greater electrical conductivity as compared to carbon electrodes. The graphite and carbon electrodes are consumed slowly during welding because of a slow oxidation of carbon.

15·27 Metal Arc Welding

In metal arc welding, the arc is produced between the metal electrode (also called filler rod) and the work piece. During the welding process, the metal electrode is melted by the heat of the arc and fused with the work piece. The temperature produced by the heat is about 2400°C to 2700°C. The electrodes used in this process are straight pieces of metal wire about 45 cm long and diameters ranging from 1.5 to 10 mm. The entire electrode (except for the bare end where it is held by the electrode holder) is covered with a* coating material. The basic set up for the metal arc welding is shown in Fig. 15.16.

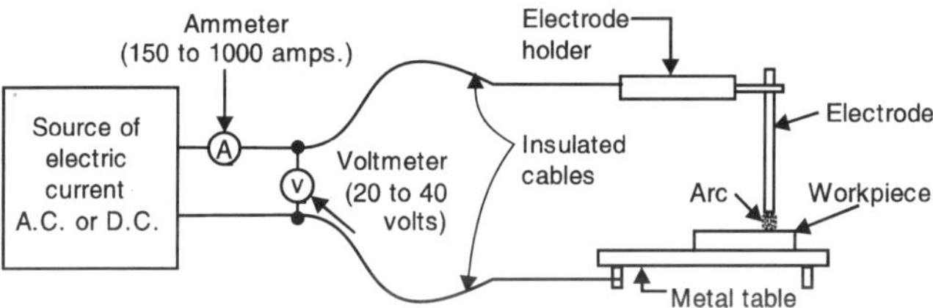

Fig. 15.16. Basic set up for metal arc welding.

The metal arc welding is normally a manual process. When the metal pieces to be joined are relatively thick, then a number of individual passes are required to complete the weld, as shown in Fig. 15.17. The line of metal deposited during a single pass is known as *bead*. By moving the electrode side to side, a wider bead may be obtained and more metal is deposited during a pass.

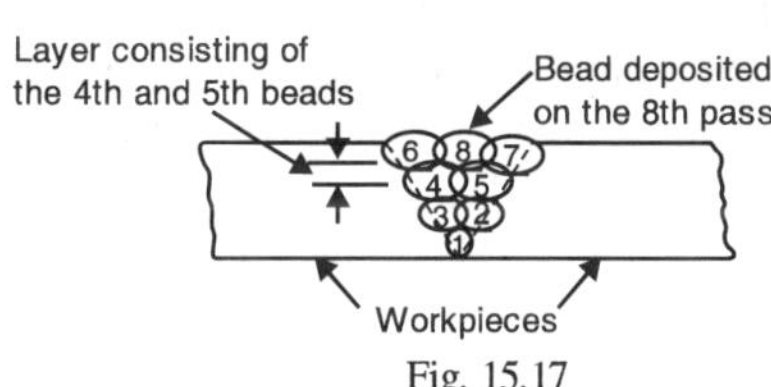

Fig. 15.17

15·28 Metallic Inert-gas (MIG) Arc Welding

In metallic inert-gas arc welding (commonly known as *MIG welding*) is similar to tungsten inert gas arc welding, except that the electrode is consumable. In this process, filler metal wire (from 0.75 mm to 1.5 mm diameter) of desired composition is automatically and continuously fed from a reel at a speed ranging from 250 to 750 cm per minute. The filler metal is deposited by the arc which is completely surrounded by an inert gas.

This process is used for the same applications as the tungsten inert-gas arc welding, but it is widely used for welding thick plates (above 4 mm in thickness). The process may be done either manually or with automatic equipment.

15·29 Tungsten Inert-gas (TIG) Arc Welding

In tungsten-inert gas arc welding (commonly known as *TIG welding*), the heat is produced from an arc between the non-consumable tungsten electrode and the work piece. The welding zone is

---

* Refer Art. 15.22.

shielded by an atmosphere of inert gas (such as helium or argon) supplied from a suitable source. The gas nozzle surrounds the tungsten electrode, as shown in Fig. 15.18. When the gas leaves the nozzle, it completely envelops the tip of the electrode and the work under it.

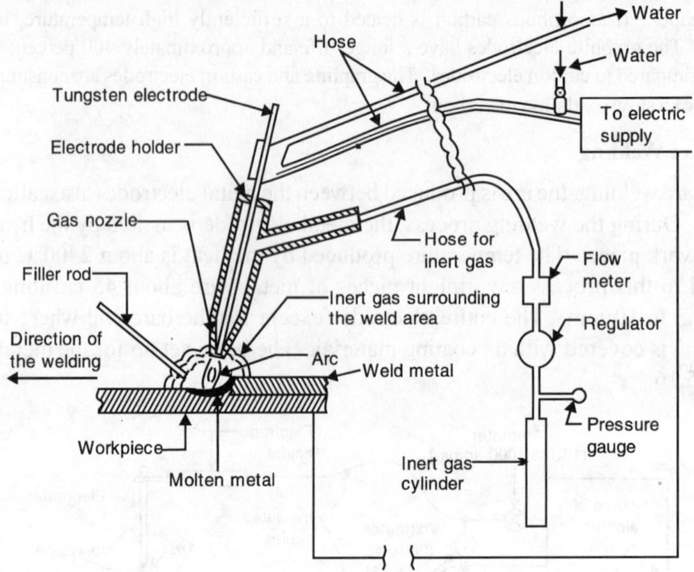

Fig. 15.18. Inert-gas arc welding.

In this process, either direct current or alternating current may be used, the selection depends upon the type of metal to be welded. The direct current with straight polarity is used for welding copper alloys and stainless steel, whereas the reversed polarity is used for magnesium. The alternating current is more versatile in its application and is used for steel, cast iron, aluminium and magnesium. The process is capable of making clean sound welds in aluminium without the use of corrosive fluxes. Though any metal can be welded by this process, yet it is best suited for the welding of relatively thin pieces of metal. The tungsten inert gas welding may be done either by a manual process or by the automatic machine welding. When it is manual, the welds can be made in any position.

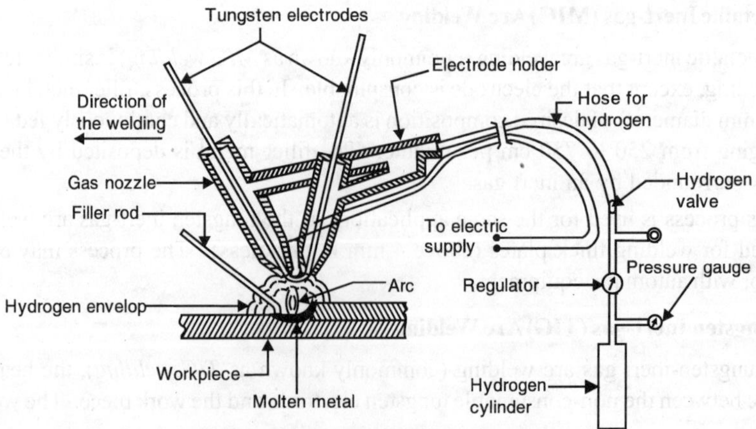

Fig. 15.19. Atomic hydrogen welding.

15·30 Atomic Hydrogen Welding

In atomic hydrogen welding, the arc is obtained between two tungsten electrodes (non-consumable), while a stream of hydrogen passes by the arc and envelops the welding zone as shown in Fig. 15.19. As the hydrogen enters the arc, the high temperature breaks up the molecules of hydrogen (consisting of two atoms) into single atoms. In dosing so, a large amount of heat is required and absorbed by the hydrogen. These single hydrogen atoms then have a strong tendency to recombine, especially when they approach the relatively cool surface of the work piece. When the atoms recombine, the heat which was absorbed is liberated, producing a temperature of about 3000°C at the work surface. Since the arc is independent of the work, the heat input to the weld can be controlled by varying the distance of the arc from the weld. The filler metal may be used when needed.

The main advantage of this process is its ability to provide high heat concentrations. In addition to this, the hydrogen also acts as a shielding gas to protect the electrodes and molten metal from oxidation. Since under the atomic-hydrogen arc, the molten metal becomes extremely fluid, therefore, the metal should be confined and welds should be made in the flat position. In addition to welding of thin pieces of aluminium, stainless steels etc., this process is widely used in the repair of steel moulds and dies.

15·31 Stud Arc Welding

The stud arc welding is a direct current arc welding process and is used for welding metal studs to the flat metal surfaces. A machine known as *stud welding gun* is used for this purpose. The stud is placed in the stud welding gun and the front end of the gun is held against the surface where the stud is to be fastened. When the trigger of the gun is pressed, it automatically goes through its welding cycle. The welding cycle starts by causing the welding current to flow between the end of the stud and the flat metal surface. The end of stud is then withdrawn by a small distance to create an arc. When this arc has properly melted the end of the stud and a molten metal pool on the metal surface, the melted end of the stud is forced against the molten pool on the metal surface. The entire operation takes only a fraction of a second and is controlled by the electronic timer which is present according to the size of the stud being welded. The welding may be shielded by an inert gas or the ends of the studs may be recessed slightly and filled with a suitable flux material which creates a protective shield of gas.

15·32 Submerged Arc Welding

In submerged arc welding, the arc is produced between a bare metal electrode and the workpiece. However, in this process, the region of the arc is completely submerged in a granulated material

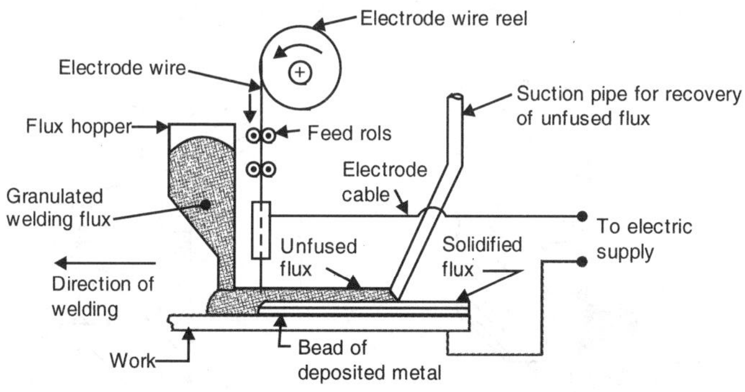

Fig. 15.20. Submerged arc welding.

acting as a flux. The entire process takes place within this fluxing material. During welding, the intense heat of the arc produces a pool of molten metal in the joint and at the same time melts some of the granular flux. This molten flux acts as a cleanser, absorbing impurities from the molten metal and producing a slag, which floats on the top of the molten metal. The slag solidifies with the metal and is easily removed. After welding, the excess granular material which has not been melted by the arc is picked up and used again.

In order to prevent molten metal running out of joint, it should be backed up by water cooled copper plates, steel plates or some refractory material. The equipment used for submerged arc welding is ordinarily mechanised and the electrode wire is fed automatically from a reel, as shown in Fig. 15.20. The process uses high currents upto 1200 amperes. Since in this process a considerable amount of metal is melted, therefore, it is used for flat welding. In this case, deep penetration is obtained and most commercial thicknesses of plate material can be welded with one pass. The thin plates upto 15 mm thickness can be welded without any bevelling of edges whereas a small vee is required on most others. The submerged arc welding is mostly done on low carbon and alloy steels, but it may be used on many of the non-ferrous metals.

15·33 Plasma Arc Welding

The plasma is a gas of positive ions and free electrons with an approximately equal positive and negative charge. In plasma arc welding, the coalescence is produced by striking an arc between a tungsten electrode and a water cooled nozzle. When gases are passed through the arc and nozzle, they get ionised and become plasma. The gases generally used are argon, hydrogen or helium. The plasma stream of the ionised particles is passed through a small orifice of a welding torch and made to impinge on to the area to be welded. The torch is held almost in vertical position to the base metal and filler metal wire is fed at a suitable angle. The plasma arc welding provides a deep penetration, faster welding, cleaner welds and narrow beads. It is used for welding stainless steel tubes, rockets, nuclear submarine plates, nickel and titanium alloy upto 8 mm thickness.

15·34 Flux Cored Arc Welding

The flux cored arc welding uses a continuous hollow electrode wire containing flux within its hollow core. The equipment required is a constant voltage direct current machine, a welding gun with cooling arrangement and a continuous wire feeder mechanism. Sometimes additional shielding is provided with a gas and then the process resembles gas metal arc welding.

The advantage of this process is that the metal can be deposited at a faster rate with deep penetration. This process is useful where there is large volume of work in flat or horizontal position only. The welds produced are strong and tough. It is primarily used for welding steel.

15·35 Electro-slag Welding

This process is particularly suitable for welding thick plates and structures for turbine shafts, boiler parts and heavy press. In this process, the plates to be welded are put in a vertical position with

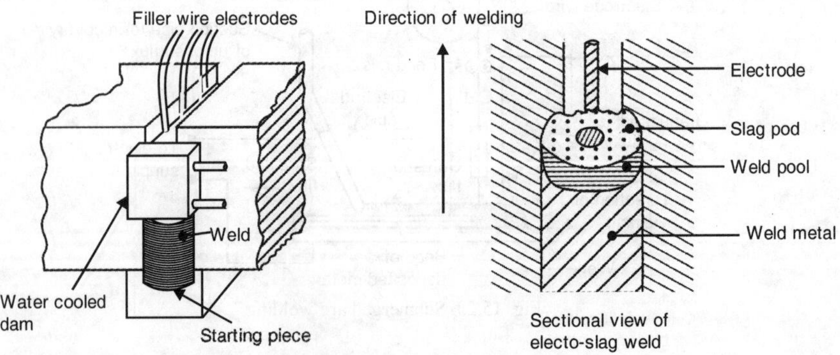

Fig. 15.21. Electro-slag welding.

a gap of about 20 to 30 mm, as shown in Fig. 15.21. The filler wires and the flux are kept in this gap. The filler wires are used as electrodes. Initially, an arc is struck between the electrode and the workpiece. The heat of the arc melts the flux. The arc is stopped and the slag is maintained in the molten state by the resistance of the electric current passing between the electrode and workpiece. The inside temperature of the molten metal pool is approximately 1950°C and the outside temperature is about 1650°C. This temperature is sufficient to melt the workpiece and filler wires to form a weld after solidification.

The molten metal pool so created is contained in the joint between the two water cooled copper shoes (dams), as shown in Fig. 15.21. Since the molten metal pool is large and the welding speed is slow, therefore, the cooling rate is quite low. This results in a coarse grain size and a follow up heat treatment is normally required to restore the strength.

15·36 Electro-gas Welding

The electro-gas welding is the development of electro-slag welding. In this method, either a solid electrode or a flux cored electrode is used. The electrodes are fed automatically by a slow speed motor. The main difference between the electro-slag welding and electro-gas welding is that in electro slag welding no flux is fed into the joint and the heat is created by an electric arc and an inert gas (carbon dioxide or argon) is fed into the joint to provide shield to the arc and the molten weld puddle. This process is used for welding workpieces of 20 to 80 mm thickness in a vertical position like fabrication of ships, bridges, large tanks and pressure vessels.

15·37 Thermit Welding

In thermit welding, the ends of the pieces to be welded are cut to provide a parallel-sided gap. This gap is filled up with wax which becomes the pattern for the weld. The mould sand is rammed, in the moulding box, around the wax pattern. A gate sprue, risers and a heating gate are provided in the sand as shown in Fig. 15.22. A preheating flame is used at the heating opening to melt and burn out the wax, to dry the mould and to bring the ends of the metal pieces to a red heat. After this, the heating opening is properly plugged with a core, which is backed up with moulding sand. Now a suitable

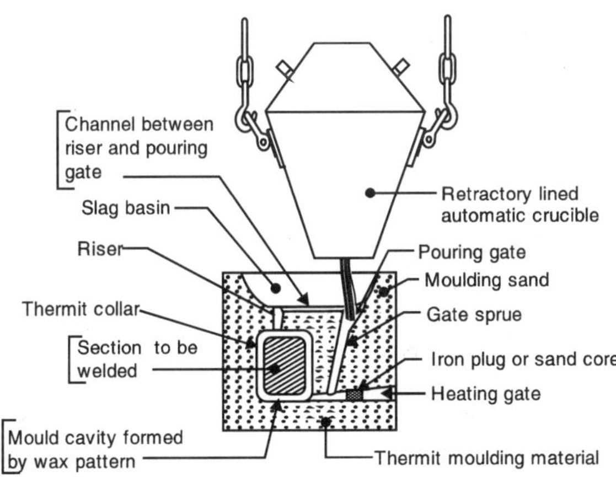

Fig. 15.22. Thermit welding.

refractory-lined crucible, containing a mixture of iron oxide and aluminium known as *thermit* is suspended above the gate spure. This mixture is ignited only at a temperature of about 1500°C. A special ignition powder is used to start the reaction. The temperature of the molten iron produced by this reaction is above 3500°C. The mixture reacts according to the following chemical equation :

$$8Al + 3Fe_3O_4 = 9Fe + 4Al_2O_3$$

The chemical reaction requires only about 30 seconds to obtain one tonne of molten iron (actually steel). The bottom plug of the crucible is removed and the steel is allowed to flow into the mould between the ends of the metal pieces to be welded. The temperature of the molten steel entering the mould is about 1500°C. The mild steel pieces may be added to the thermit powder to lower the temperature for pouring. The alloying elements may also be added to the thermit powder to obtain any desired analysis of the weld metal. A major advantage of the thermit welding is that all parts of the weld section are molten at the sametime and the weld cools almost uniformly. This results in a minimum problem with internal residual stresses.

The thermit welding is often used in joining iron and steel parts that are too large to be manufactured, such as rails, truck frames, locomotive frames, other large sections used on steam and rail roads, for stern frames, rudder frames etc. In steel mills, thermit electric welding is employed to replace broken gear teeth, to weld new necks on rolls and pinions and to repair broken shears.

15·38 Solid State Welding

The solid state welding depends upon the theory that if two perfectly clean metallic surfaces are placed in an intimate contact, the cohesive force between the atoms of the two surfaces is sufficient to hold or weld them together. The various solid state welding processes are friction welding, ultrasonic welding, diffusion welding, and explosive welding.

These welding processes are discussed, in detail, as follows :

1. *Friction Welding*. In friction welding, the heat is obtained by the rotating friction. The two pieces to be welded together are mounted in chucks on the same horizontal axis, as shown in Fig. 15.23. One of the chucks rotates while the other is held stationary. The pressure is applied to generate enough heat to reach welding temperature. As soon as the welding temperature is reached, the rotation of chuck is stopped. The pressure is maintained or increased until the completion of weld which usually takes from 10 to 25 seconds. This process is mainly used for joining round bars.

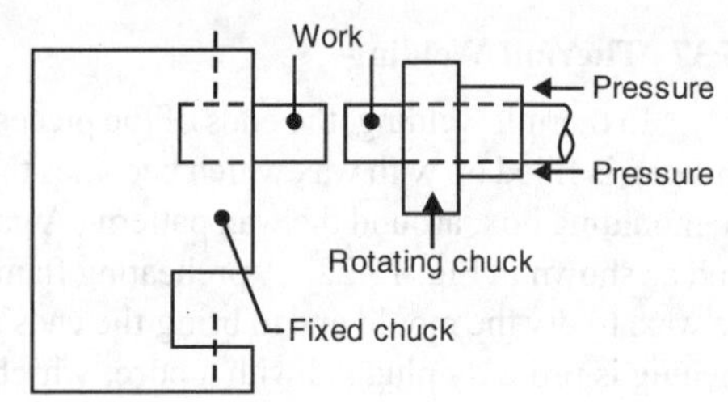

Fig. 15.23. Friction welding.

2. *Ultrasonic Welding*. In ultrasonic welding, the high frequency (ultrasonic) vibratory energy is applied to the weld area of the metal pieces to produce coalescence. The pieces to be welded are clamped rigidly in proper position and a high frequency vibtratory energy (frequency being 20 kHz to 60 kHz) is then transmitted into the weld area for about one second or less. It causes vibrations in metal molecules to raise the temperature sufficient to form a sound union between the faces of the material in contact. It is a very rapid process and produces a solid state bond of the two workpieces without applying external heat, filler metal or fluxes.

The ultrasonic welding process is used only for welding of thin strips and foils. The maximum thickness may vary from 0.38 mm to 2.5 mm depending upon the metal.

3. *Diffusion Welding*. The diffusion welding is a solid state welding process in which clean flat metal parts with fine surface finish are brought together in a vacuum or in an inert gas atmosphere under pressure. The pressure required varies from 35 MPa to 65 MPa and may be applied to the parts for hours. The heating is not essential, but if the temperature is raised, the diffusion rate is cut sufficiently. The basic requirement for this process is that the surfaces must be flat and have a fine surface finish. The process creates a metallic bond as a result of the pressure causing micro deformation of the surfaces into each other and the diffusion of atoms between the two materials being joined. The process is slow and costly because the use of an inert atmosphere requires a containment vessel and the time required varies from seconds to 4 to 20 hours.

This process makes it possible to join metal to metal, metal to ceramic and metal to metal with intermediate bonding materials. The diffusion bonding of steel is done by placing an intermediate fusion material between the surfaces and by heating upto about 850°C.

The principal applications of diffusion welding are in high technology particularly in the atomic, aerospace and electronic industries.

4. *Explosive welding.* The explosive welding (often called *cladding*) is the process of welding by high velocity movement. It is used to join two plates face to face. In this method, one of the plates (called the target plate) is held fixed while the other plate (called the flyer plate) is kept at an angle to the target plate, as shown in Fig. 15.24. The minimum distance between the two plates is of the order of ¼ to ½ of the flyer plate thickness. An explosive charge is kept on the top of the flyer plate with an intervening layer of rubber spacers. When the explosive charge is detonated, the flyer plate comes and hits the target plate with a huge velocity and the two plates are welded face to face. This process can be used to join dissimilar materials and the weld interface is seen to be wavy as shown in Fig. 15.24. It is used for the fabrication of structures used in critical space, nuclear applications and aerospace industries which can not be welded by any other process.

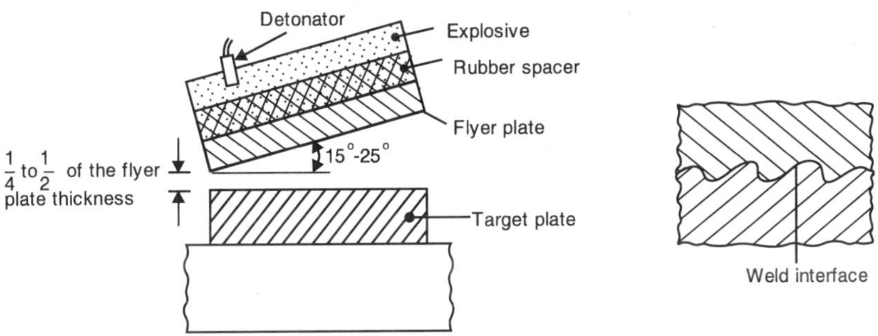

Fig. 15.24. Explosive welding.

The major advantage of this process is the simplicity with which extremely large surfaces can be joined. The heat treated metals can be welded without affecting the heat treated process. The incompatible materials can also be welded and the thin foils can be joined to heavier plates.

15·39 Modern Welding Processes

The following two processes have been recently developed using latest techniques :

1. *Electron beam welding.* The electron beam welding is a fusion welding process in which the coalescence is obtained by the heat generated from the kinetic energy of a dense beam of high velocity electrons. The electron optical system used is shown in Fig. 15.25. The high velocity electrons are emitted by a high voltage current which heats a tungsten filament called cathode. The

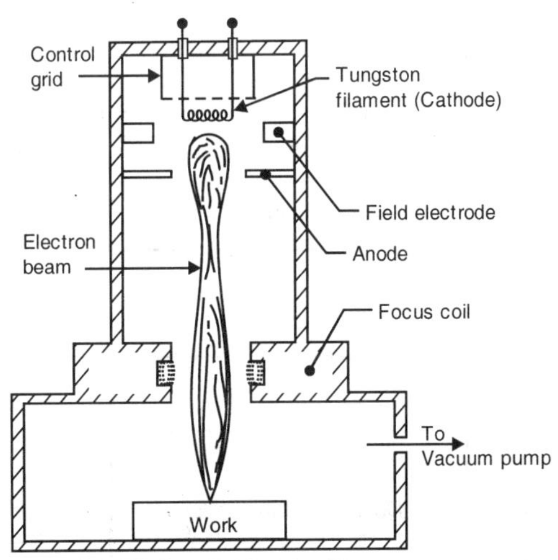

Fig. 15.25. Electron beam welding.

electrons are concentrated into a beam by means of a control grid and accelerated by a ring shaped anode and are focussed on the workpiece with the help of a focussing coil. The entire operation is carried out in a high vacuum with a pressure of 10^{-3} to 10^{-5} mm of Hg. The accelerating voltage is in the range of 20 to 200 kV with a welding current of the order of milliamperes. The heat generated is about 2500ºC. This is sufficient to melt and vaporize the workpiece material and thus fills a narrow weld gap even without a filler rod.

This process can be used to join dissimilar metals as effectively as similar metals and small thin parts can be joind to heavy sections. The materials (which are difficult to weld by other process) like tungsten, molybdenum, berylium, titanium, zirconium and columbium can be welded successfully by this process. Now-a-days, the electron beam process is used in automobile, airplane, aerospace, farm and other equipment including ball bearings over 100 mm diameter.

2. *Laser beam welding*. The laser is a device that generates an intense beam of coherent monochromatic radiation in the infra-red, visible or ultraviolet region of the electromagnetic spectrum, by stimulated emission of photons from an excited source. The word laser is an acronym from *light amplification by stimulated emission of radiation*.

In a laser beam welding, a laser beam is focussed on the workpiece and the heat generated is sufficient to form a coalescence on cooling. A laser beam is capable of producing a power density as high as 100 kW/mm^2.

This process is used for both cutting and welding of metals. It can be used to join dissimilar metals. The materials such as copper, nickel, tungsten, aluminium, stainless steel, titanium, molybdenum, zirconium and columbium can be welded conveniently by this process. It is also used for welding lead wires without removing the polyurethene insulation.

15·40 Basic Weld Symbols

The basic weld symbols according to IS : 813-1961 (Reaffirmed 1991) are shown in the following table :

Table 15.1. Basic weld symbols.

S.No.	*Form of weld*	*Sectional representation*	*Symbol*
1.	Fillet		
2.	Square butt		
3.	Single-V butt		
4.	Double-V butt		
5.	Single-U butt		
6.	Double-U butt		
7.	Single bevel butt	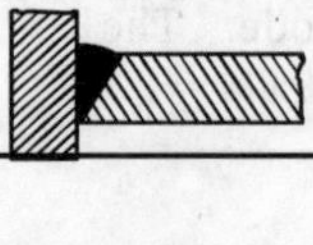	

S.No.	*Form of weld*	*Sectional representation*	*Symbol*
8.	Double bevel butt	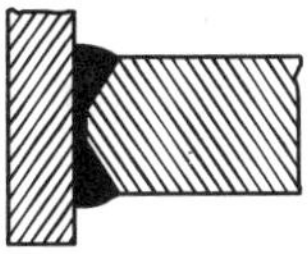	
9.	Single-J butt	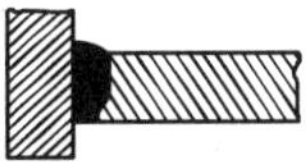	
10.	Double-J butt	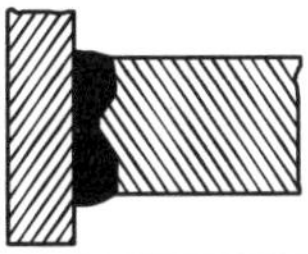	
11.	Bead (edge or seal)		
12.	Stud		
13.	Sealing run		
14.	Spot		
15.	Seam		
16.	Stitch		
17.	Mashed seam	 Before After	
18.	Plug	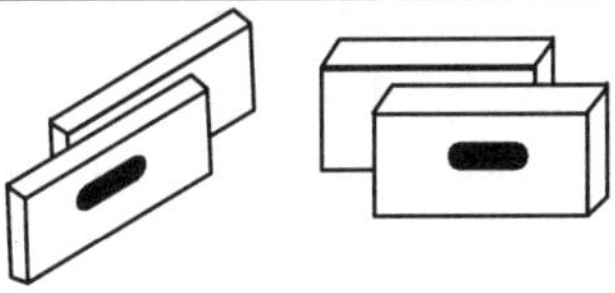	
19.	Backing strip		
20.	Projection	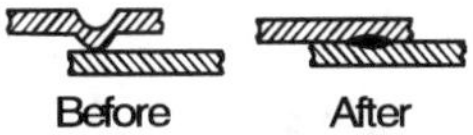 Before After	
21.	Flash	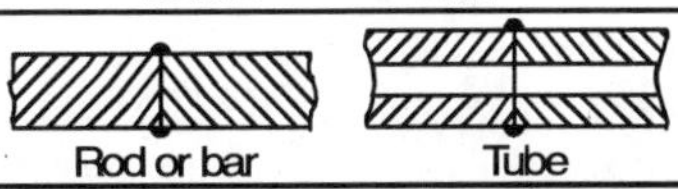 Rod or bar Tube	
22.	Butt resistance or pressure (upset)	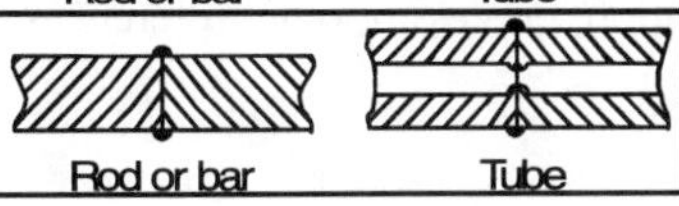 Rod or bar Tube	

15·41 Supplementary Weld Symbols

In addition to the above symbols, some supplementary symbols are also used as shown in the following table.

Table 15.2. Supplementary weld symbols.

S.No.	*Particulars*	*Drawing representation*	*Symbol*
1.	Weld all round		
2.	Field weld		
3.	Flush contour		
4.	Convex contour		
5.	Concave contour		
6.	Grinding finish	G	G
7.	Machining finish	M	M
8.	Chipping finish	C	C

15·42 Elements of a Welding Symbol

A welding symbol consists of the following eight elements.

1. Reference line,
2. Arrow,
3. Basic weld symbols,
4. Dimensions and other data,
5. Supplementary symbols,
6. Finish symbols,
7. Tail, and
8. Specification, process or other references.

15·43 Standard Location of Elements of a Welding Symbol

The *elements of a welding symbol shall have standard locations with respect to each other.

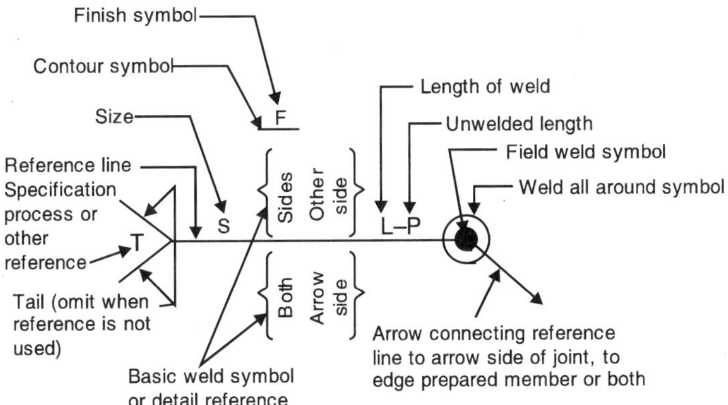

Fig. 15.26

The arrow points to the location of weld, the basic symbols with dimensions are located on one or both sides of reference line. The specification if any is placed in the tail of arrow. Fig. 15.26 shows the standard locations of welding symbols represented on drawing.

15·44 Gas Welding

It is a type of fusion welding, in which the heat for welding is obtained by the combustion of a fuel gas. In a gas welding, the edges or surfaces to be joined are melted by the heat of a gas flame. The molten metal is allowed to flow together and a solid continuous joint is obtained as the molten metal solidifies. The gas fusion welding is particularly suitable for joining metal sheets and plates having thickness from 2 to 50 mm. In order to weld materials having thickness more than 15 mm, additional metal (called filler metal) from the welding rod is melted into the gap between the two parts to be joined. The material of the welding rod is usually the same as that of the parts being welded. A flux is used during welding to remove impurities and oxides present on the surfaces of the metal to be joined and to obtain a good weld result.

The most widely used gas combination for producing a hot flame for welding metals is oxygen and acetylene (C_2H_2). When acetylene is mixed with pure oxygen, it burns to produce the highest flame temperature (about 3200°C). Such a flame is known as *oxy-acetylene flame*. The hydrogen gas or some other gas may also be used, but the temperatures obtained are lower than the oxy-acetylene flame. The approximate flame temperatures produced by differnt combination of gases are as follows:

1. Oxy-acetylene, 3200°C ;
2. Oxy-hydrogen, 2400°C ;
3. Oxy-propane, 2200°C ;
4. Oxy-town gas, 2100°C ;
5. Air-acetylene, 2400°C ;
6. Air-town gas, 1800°C ; and
7. Air propane, 1750°C.

Since oxy-acetylene flame is mostly used in gas welding, therefore we shall only discuss the oxy-acetylene welding in detail.

---

* For further description, please refer to IS : 813-1961 (Reaffirmed 1991).

Note : The oxy-hydrogen flame may be used for welding low-melting temperature metals and for welding thin pieces of steel.

15·45 Equipment for Oxy-acetylene Gas Welding

The basic equipment required to carry out oxy-acetylene gas welding is as follows :

1. *Welding torch.* The welding torch (also known as blow pipe) as shown in Fig. 15.27(*a*), is a tool for mixing the oxygen and acetylene in the desired volumes and burning the mixture at the end

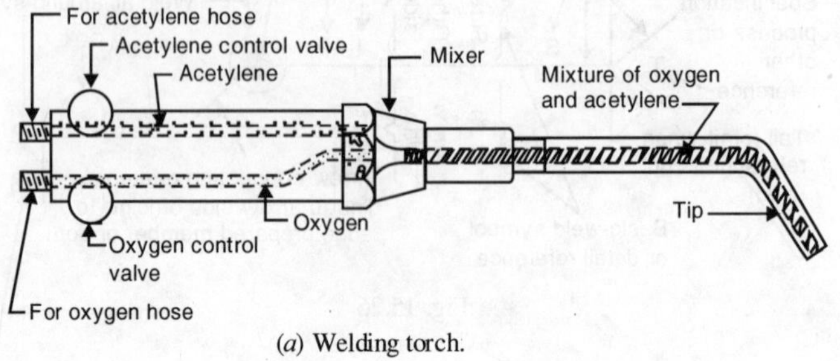

(*a*) Welding torch.

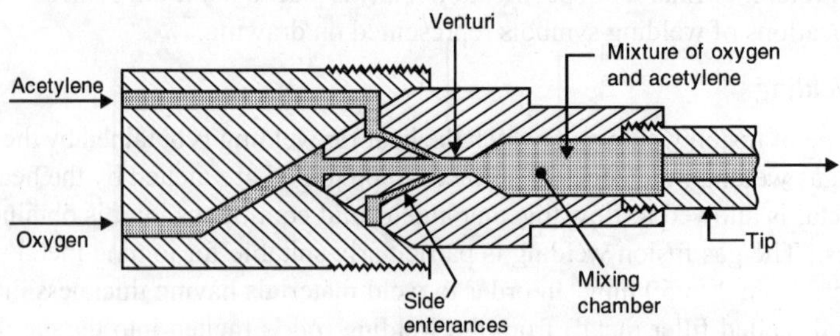

(*b*) Injector-type gas mixing chamber.

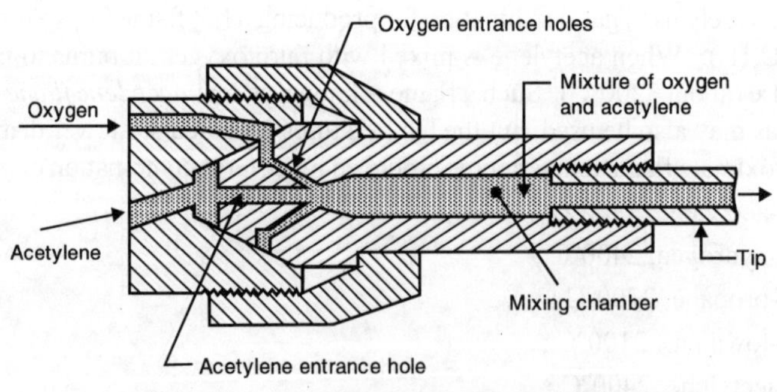

(*c*) Positive or equal pressure type gas mixing chamber.

Fig. 15.27.

of a tip, which produces a high temperature flame. The welding torches have a handle at one end with two inlet connections for gases. Each inlet has a valve that controls the volume of oxygen or of acetylene passing through them. The desired proportions of oxygen and acetylene are thoroughly mixed, in the mixer, before issuing from torch to the tip or nozzle. The flame is produced by igniting the mixture at the torch tip. The welding torches are commercially available in the following two types :

(*i*) Injector or low pressure type ; and

(*ii*) Positive or equal-pressure (also known as high pressure) type.

The principal difference between these types of torches is in the gas mixing chambers or gas mixer. The gas mixing chambers are shown in Fig. 15.27(*b*) and (*c*). A gas mixing chamber must be able

(*a*) to mix gases thoroughly for proper combustion,

(*b*) to arrest flash backs that might occur through improper operation,

(*c*) to stop any flame, travelling back through the tip at the mixer, and

(*d*) in some designs, to permit a range of tip sizes to be operated by the one size of mixer.

The *injector* or *low pressure type* welding torch operates at an acetylene pressure of less than 7 kN/m^2. This low pressure acetylene is produced at the welding site by the chemical reaction between water and calcium carbide (CaC_2). The oxygen is, however, supplied at a pressure ranging from 70 to 280 kN/m^2 depending upon the tip size. The high pressure oxygen passing through the venturi creates a low pressure which draws in low pressure acetylene and mix with it before both gases pass into the tip.

In case of *positive* or *equal pressure* type welding torch, the gases must be delivered to the torch at pressures generally above 7 kN/m^2. In the case of acetylene, the pressure must be between 7 kN/m^2 and 105 kN/m^2 while oxygen is generally supplied at approximately the same pressure.

Notes : (*a*) A high pressure welding torch should never be connected to a low pressure acetylene line.

(*b*) The high pressure acetylene is usually stored in strong steel cylinders.

2. *Welding torch tip.* The welding torch tip (also called nozzle) is that portion of torch through which the gases pass just prior to their ignition and burning. The tips are made of high thermal conductivity material, such as copper or copper alloy.

The tip size is measured by the diameter of the central hole through which the gaseous mixture passes. The tip size depends upon the thickness of the material to be welded. The interchangeable tips for the various thicknesses are usually provided with each welding torch.

3. *Pressure regulators.* The pressure regulators perform the following functions :

(*i*) It reduces the source pressure (from generator, cylinder, manifold or pipe line) to a workable pressure.

(*ii*) It maintains the constant delivery pressure and gas volumetric rate regardless of pressure variations at the source.

(*iii*) It permits adjustment to deliver gas at a certain desired pressure within its rated pressure range.

The mechanical details of a regulator may differ among the different manufacturers, but the fundamental operating principles are the same for all regulators. Generally, a pressure regulator has a union nipple to attach it to the cylinder and an outlet connection for the hose leading to the torch. There are two gauges on the body of the regulator, one showing pressure in the cylinder while the other shows pressure being supplied to the torch. The working pressures are adjusted by the operator according to the requirements of the job. The desired pressure at the welding torch for oxygen is between 70 and 280 kN/m^2 and for acetylene it is between 7 and 103 kN/m^2.

4. *Hose and hose fittings.* The hose for welding torches should be strong, durable, non porous, light and flexible. Two lengths of hose (one for oxygen and one for acetylene) are required to connect the regulator and the torch. The standard colour for oxygen hose is black and for acetylene hose it is red. The special hose fittings or connections are provided for attachments to the torch and pressure regulators.

5. *Gas cylinders*. The oxygen and acetylene are usually obtained from strong steel cylinders, as shown in Fig. 15.28, which are filled by commercial suppliers of these gases. The oxygen cylinders are usually filled at a pressure of 12.5 to 14 MN/m^2 and have capacity of 6.23 cubic metre. The mass of oxygen cylinder is approximately 66 kg when it is full. The standard colour for the oxygen cylinder is black. The acetylene is stored in cylinders under an initial pressure of 15 atmospheres (1.545 kN/m^2) and have the capacity of 7.6 cubic metre. Its mass is about 86 kg when it is full. The acetylene cylinders are equipped with fusible plugs at the bottom in order to save the cylinder from explosion in case of fire. The standard colour for the acetylene cylinder is maroon.

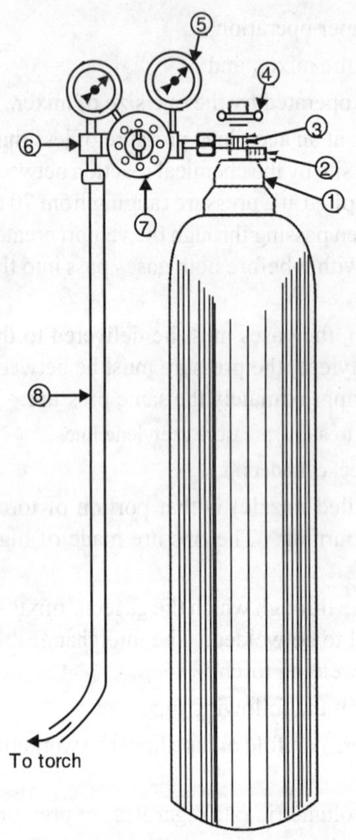

(*b*) Acetylene cylinder.

1. Fusible plugs. 2. Acetylene cylinder valve. 3. Valve wrench. 4. Cylinder pressure gauge. 5. Regulator outlet pressure gauge. 6. Acetylene regulator. 7. Adjusting screw. 8. Acetylenehose to torch.

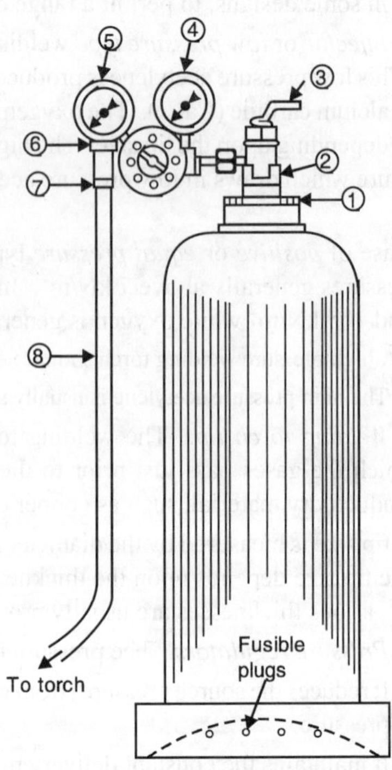

(*a*) Oxygen cylinder.

1. Cap. 2. Bursting disc. 3. Oxygen valve. 4. Handwheel. 5. Pressure gauge. 6. Oxygen regulator. 7. Adjustment screw. 8. Oxygen hose to torch.

Fig. 15.28. Parts of gas cylinder.

The pure acetylene is not stable at pressures above atmospheric pressure (*i.e.*, 103 N/m^2) and may decompose explosively. Therefore, in order to store acetylene at 15 atmospheres, it is dissolved in *acetone which is capable of absorbing a large volume of acetylene and of releasing it as the pressure falls. One volume of acetone at 15°C and at atmospheric pressure dissolves about 25 volumes

* The acetone is a chemical compound of hydrogen, oxygen and carbon ($CH_3.CO.CH_2$). It is a liquid with a pungent smell and is slightly poisonous.

of acetylene. The dissolving capacity grows in proportion to the pressure so that at the same temperature but at 15 atmospheres acetylene pressure, one volume of acetone absorbs $25 \times 15 = 375$ volumes of acetylene. The cylinder is first filled with a dry porous material (*e.g.*, charcoal or asbestos) which keeps the acetylene well separated. Then the tank is partially filled with acetone and acetylene is dissolved in it under a pressure of 15 atmospheres.

6. *Goggles and gloves*. The goggles, fitted with coloured glasses, should be worn by the operator when welding. The goggles protect the eyes from the glare of the flame cone and molten metal. The gloves should also be worn to protect the hands.

7. *Spark lighter*. It provides a convenient and instant means for lighting the welding torch.

15·46 Welding Rods

The welding rods (also called filler rods) are used in gas welding to provide extra metal to the weld by melting the end of the rod. Mostly, the composition of the welding rod material is same as that of the material being welded. In gas welding, the following types of welding rods are, usually, used.

1. For welding ferrous metals, steel rods with high carbon contents and more silicon and manganese (commonly known as silicon manganese rod) are widely used.
2. For welding alloy steels, the rods containing chromium and vanadium or molybdenum are used.
3. For welding stainless steels, the rods containing titanium, molybdenum or niobium (commonly known as 18/8 chromium-nickel rod) are used to prevent weld decay.
4. For welding copper and its casting, drawn copper rods are used.
5. For welding sheets of aluminium and its alloys, pure aluminium rods are used, but for aluminium castings, the rods containing upto 12 percent silicon are used.
6. For welding magnesium alloys, the rods having same composition as that of the material to be welded, are used.

15·47 Fluxes

When metals are heated, the oxygen from the air combines with them and form oxides. These oxides produce poor quality and low strength welds. In some cases, it makes the welding impossible. Thus in order to prevent oxidation and other unwanted chemical reactions during welding, fluxes are used. The fluxes are chemical compounds and form a fusible slag with the oxides of most metals at the welding temperature. The slag floats to the top of the molten metal and do not interfere with the deposition and fusion of filler metal. A good flux protects the molten metal from atmospheric oxygen and prevent the absorption and reaction of other gases in the flame. In addition, it helps to clean and protect the surfaces of the base metal.

The fluxes are available in several forms, such as a dry powder, a paste or thick solution or as a preplaced coating on the welding rod. The dry fluxes are applied by heating the end of the welding rod and dipping it into the powdered material. The flux adheres to the heated surface and is consumed with the rod during welding operation. The fluxes in the paste form are usually applied on the base metal with a brush, while the welding rod is either painted or dipped.

A single flux is not suitable for welding all metals. The ordinary grades of mild steels are successfully welded without the use of a flux. The commonly used fluxes for welding different metals are as follows :

1. For ferrous metals, the fluxes used are borax, sodium carbonate, sodium bicarbonate and sodium silicate.
2. For copper and its alloys, the fluxes used are mixtures of sodium and potassium borates, chlorides, carbonates, sulphates and boric acid.

3. For aluminium, the fluxes used are alkaline fluorides, chlorides and bisulpahtes.
4. For magnesium alloys, the fluxes used are similar in composition to those used for aluminium and its alloys.

Since all the fluxes are chemically active and very corrosive, therefore it is very essential to remove all traces of flux from the finished weld. The flux residue may be removed by hot water, steam or by chipping hammer.

15·48 Gas Flame

The combustion of acetylene with pure oxygen at the tip of a gas welding torch takes place in two stages. In the first stage, the carbon from the acetylene combines with oxygen to form carbon monoxide, the hydrogen of the acetylene being freed. The following chemical reaction occurs :

$$C_2H_2 + O_2 = 2CO + H_2$$

This reaction takes place within the inner, brilliant white cone flame, close to the tip opening, as shown in Fig. 15.29. This provides the most concentrated heat with the highest temperature for welding.

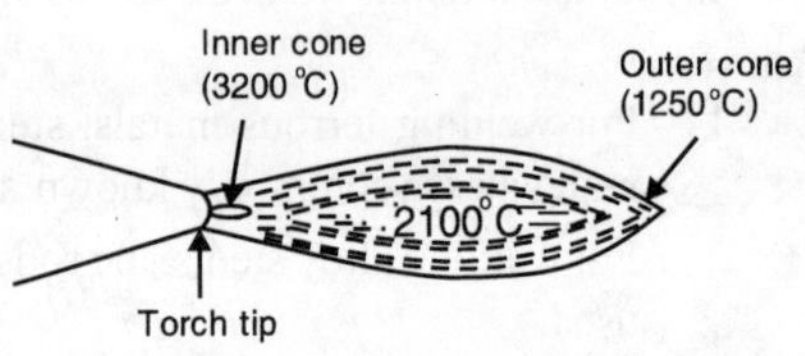

Fig. 15.29. Gas flame.

In the second stage, the carbon monoxide and hydrogen (produced in first stage), together with oxygen from the surrounding air, form carbon dioxide and water (steam). The reaction is represented by the following chemical equation :

$$4CO + 2H_2 + 3O_2 = 4CO_2 + 2H_2O$$

These reactions take place in the larger blue flame which surrounds whitish cone flame. This

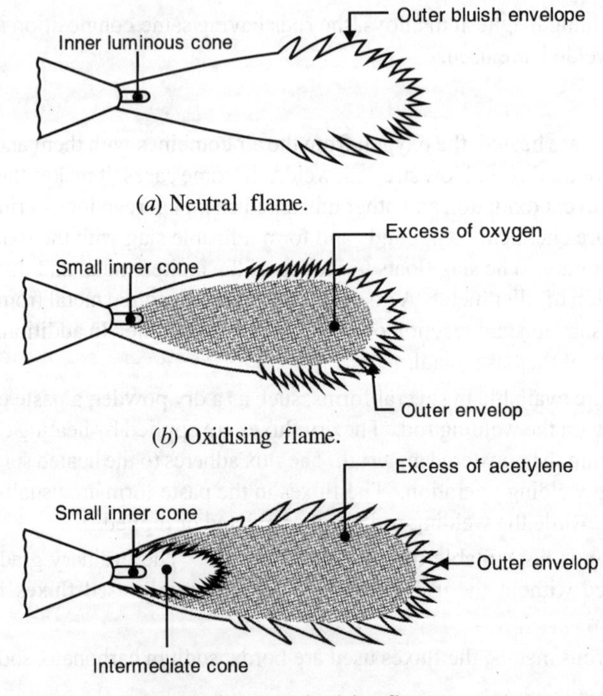

(*a*) Neutral flame.

(*b*) Oxidising flame.

(*c*) Reducing or carbonising flame.

Fig. 15.30. Types of flames.

larger blue flame contributes only a preheating effect for welding. It also protects the molten metal from oxidation.

The flame can be adjusted, to suit the welding conditions, by regulating the supply of acetylene and oxygen. The following three types of flames are used for gas welding:

1. *Netural flame.* This type of flame is obtained by supplying equal volumes of oxygen and acetylene. The neutral flame, as shown in Fig. 15.30 (*a*), has the following two sharply defined zones:

(*i*) An inner luminous cone, and

(*ii*) An outer cone or envelope of bluish colour.

The most of the oxy-acetylene welding (*e.g.*, welding of steel, cast iron, copper, aluminium etc.) is done with the neutral flame.

2. *Oxidising flame.* This type of flame is obtained where there is an excess of oxygen. It is similar to neutral flame but the inner cone is less luminous and shorter, as shown in Fig. 15.30 (*b*). It is used for welding brass and bronze.

3. *Reducing or carborising flame.* This type of flame is obtained by supplying an excess of acetylene. This flame, as shown in Fig. 15.30(*c*), has the following three zones :

(*i*) An inner cone,

(*ii*) An intermediate cone of whitish colour, and

(*iii*) An outer cone of bluish colour.

This flame is used where it is required to keep oxidation to a minimum. It is used for welding of molten metal, a certain alloy steels, many of the non-ferrous, hard surfacing materials such as satellite.

15·19 Gas Welding Technique

In a gas welding (or oxy-acetylene welding), the acetylene is, first of all, turned on by using the control valve on the torch and then it is ignited with a friction spark lighter. The flame is adjusted by supplying the oxygen with the oxygen control valve on the torch. The pieces to be welded are properly prepared and positioned. The weld is started by preheating and melting a small puddle of molten base metal. In case a welding rod is used to provide filler metal, it should be held in the flames so that its end melts at about the same time as the base metal. In order to obtain proper penetration and to produce a good weld, the tip of the torch should be moved with a side-to-side motion, slowly and uniformly. The usual techniques in oxy-acetylene welding are as follows :

1. *Left-ward or fore-hand welding.* In this method, the welding torch is held in the operator's right hand, the tip pointing towards the left and the weld is made from right to left. The torch makes an angle of 60°–70° with the plate and the welding rod makes an angle of 30°– 40°, as shown in Fig. 15.31. The flame is given a circular, rotational or side-to-side motion to obtain uniform fusion on each side of the plate. This method is more efficient for butt welding on plates upto 2 mm thickness. When plates over 3 mm thickness are to be butt welded, then the plate edges are bevelled to produce a 'V' of 80° to 90°. The larger volume of this 'V' requires sufficient quantity of filler material. If the 'V' is reduced, the welding torch flame pushes the molten metal from the pool towards the unmelted sides of the 'V'.

Note : The plates above 6 mm thickness are not economical to weld with this method.

2. *Right-ward or back-hand welding.* In this method, the welding torch is held in the right hand and the filler rod in the left hand. The welding begins at the left hand end of the joint and proceeds towards the right. The filler rod is given circular motion while the welding torch moves in straight line. In this case, the torch makes an angle of 40°–50° with the plate and the welding rod makes an angle of 30°– 40°, as shown in Fig. 15.32. This method is better and economical for plates

over 6 mm thickness. A good fusion can be obtained with this method without preparing 'V' upto 8 mm thick plates. The plate edges are bevelled to produce 'V' of 60° for plates above 8 mm thick.

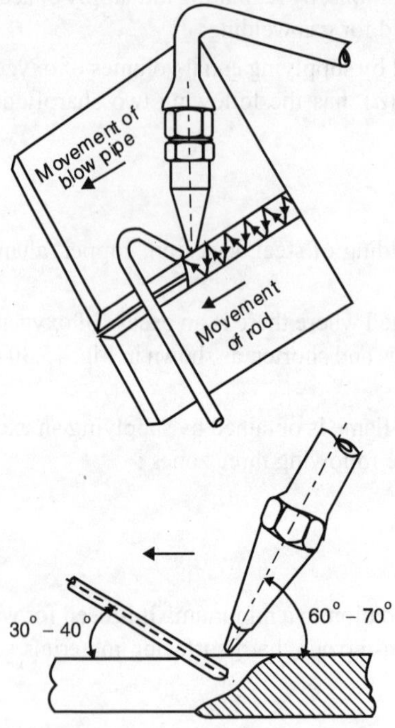

Fig. 15.31. Left-ward or fore-hand welding.

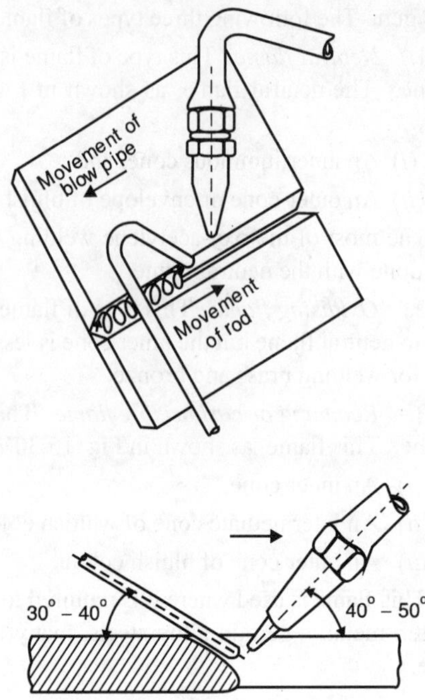

Fig. 15.32. Right-ward or back-hand welding.

This process uses a larger flame for a given plate thickness. This larger flame gives greater welding speed and less filler rod consumption. Since the welding torch moves in a straight line, the molten metal is agitated very little and excess oxidation is prevented. The plates upto 15 mm thick may be welded in one pass and above 16 mm thickness more runs must be deposited but the number is kept to a minimum.

3. *Vertical welding*. The vertical welding, as shown in Fig. 15.33, may be carried out either by the left-ward technique or by the right-ward technique. The great advantage of the vertical welding is that the plate edge preparation is not required upto 16 mm thickness. In vertical welding for plates above 5 mm thickness, two operators are required. These operators use the identical welding torches, same nozzle size, pressure settings, angles of welding torch and rod, speed of travel etc. The fusion is more easily achieved by this process, particularly when the two operators are employed. The both sides of the weld then have a uniform convex bead of weld metal. When the plates to be welded exceeds 6 mm, the combined power of two torches is much less than a single torch required to weld an equal thickness even if the right-ward technique is used.

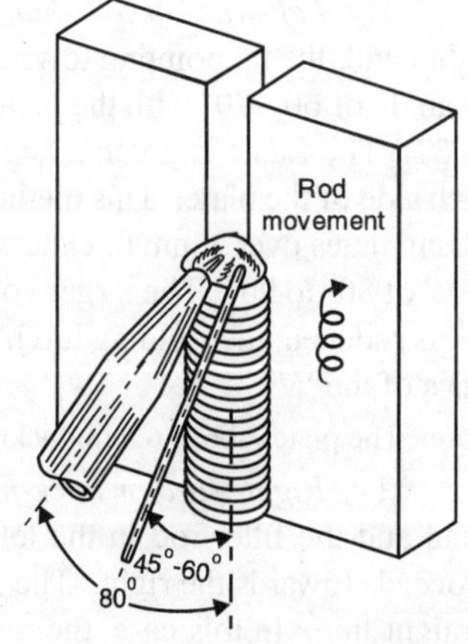

Fig. 15.33. Vertical welding.

4. *Linde welding*. It is a special welding technique which is basically used for butt welding of steel tubes at a temperature

below the melting point of the parent metal. In this method, the use of oxy-acetylene flame with excess acetylene is made and the right-ward technique is used.

The edges of the tube should be bevelled to have an included angle of 70° and butted together with a gap of approximately 2.5 mm. They are then tacked at regular intervals. The welding is carried out in the flat position and the tubes are being rotated to achieve this.

15·50 Gas or Oxygen Cutting of Metals

The cutting of iron and steel by using oxygen is extensively used now-a-days in industry. It is economically done by means of the chemical reaction of oxygen with the base metal at elevated temperatures. The process consists of preheating the material to be cut to its kindling or ignition temperature (about 1350°C). When this temperature is attained, a jet of a high pressure oxygen from a special cutting torch is directed on to the hot metal. The metal is rapidly oxidised. The molten oxides formed in the process are washed out by the stream of oxygen. As the molten metal runs down into the *kerf, it heats the cooler metal and makes the cutting operation continuous. During cutting operation, when the torch is directed against the metal ahead of completed kerf, the oxides are continuously removed from the kerf faces.

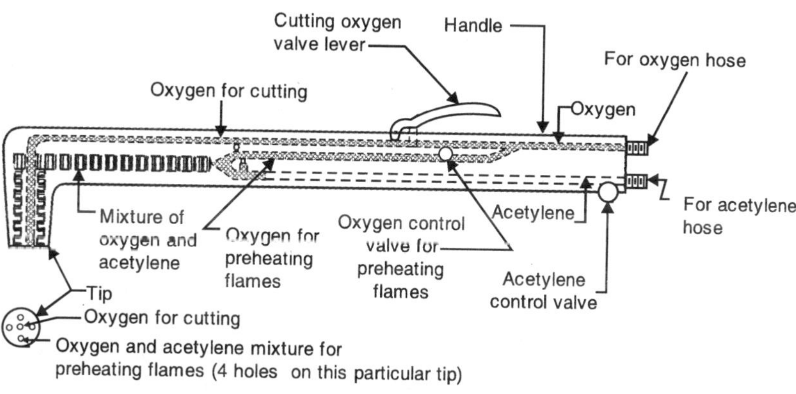

Fig. 15.34. Cutting torch.

It may be noted that only those metals can be cut by oxygen cutting whose oxidisation temperature is below their melting point, otherwise the metal would melt away before oxidisation and no clean cut edge will be obtained.

The oxygen cutting is well suited for the preparation of edges for welding. The cut edges are as smooth as rough machining cuts, and fairly close tolerances can be held.

The oxygen cutting is performed by means of ordinary gas welding equipment, except that the welding torch is replaced by a cutting torch.

The cutting torch tip, as shown in Fig. 15.34, consists of four openings for the preheating flame and in the centre of these openings there is a larger opening from which a jet of oxygen comes to cut the metal. The jet of oxygen is controlled by means of a spring loaded valve.

15·51 Cutting Machines

A very large part of the cutting is performed now-a-days by oxygen cutting machines. These machines hold the cutting torch and guide it along the work at an absolutely uniform rate of speed. The machine-guided torch cuts may be smooth and square with sharp edges, so that no further finishing is required. The machine cutting may be used for cutting straight lines, bevels, circles and other cuts

* Since only the metal in the direct path of the oxygen stream is oxidised or melted, therefore a narrow path with fairly smooth and parallel sides (known as kerf) is obtained.

of varied shape.

The oxygen cutting machines may be semi-automatic or automatic. In semi-automatic oxygen cutting machines, the torch is traversed by power but is guided by hand with the help of various guiding systems. The guiding mechanism used may be tracks, guide bars, templates etc.

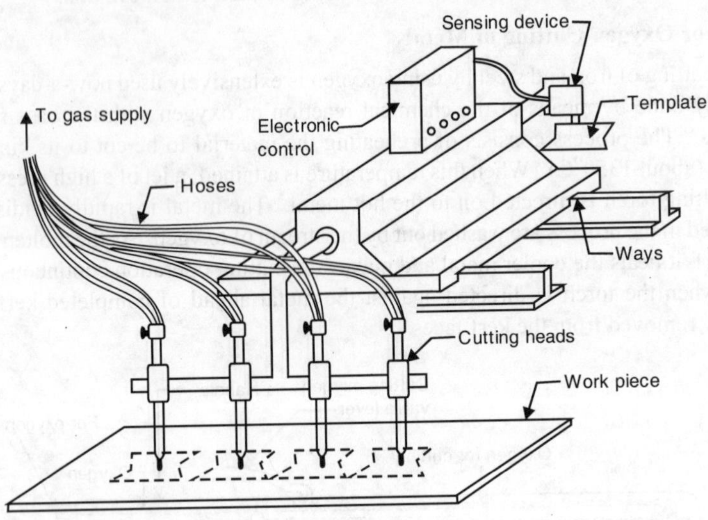

Fig. 15.35. Cutting machine.

In automatic machines, the movement of the torch to cut any desired shape is automatically controlled. Some machines carry a number of torches which are used to cut out a number of identical shapes during one complete operation. Such a machine cutting several parts simultaneously is shown in Fig. 15.35. All these machines are provided with a controlling or sensing device to guide the torches along a predetermined path. This control may be hand guided pointer following a drawing or it is held against a template. These machines are usually electrically driven and are provided with a knurled, magnetised drive spindle which follows a steel template and controls the movement of the machine at the proper cutting speed. There is also an electronically controlled sensing device which is provided with an electric eye capable of following a line drawing, thus eliminating the need for template construction.

15·52 Oxygen-lance Cutting

The oxygen-lance cutting is used for cutting heavy sections which are difficult to cut by ordinary oxygen cutting torch. The lance used is merely a suitable length of black iron pipe filled with a valve at one end to which an oxygen hose is connected. The pipe size may vary from 6 mm to 9 mm. In order to start the cut, it is necessary to preheat the starting spot with a suitable oxy-acetylene welding torch. After preheating, the welding torch is removed and the end of the lance is brought against the heated area and the oxygen valve on the lance opened. The oxygen pressure for this operation is about 500 to 700 kN/m^2. Once the operation is started sufficient heat is liberated to continue the cut. The lance is slowly consumed during the operation and must be periodically replaced.

The oxygen lance cutting is useful for piercing holes in heavy thickness of steel, for cutting of large risers in the foundry and for cutting tap holes in steel furnaces.

15·53 Arc Cutting

The processes of arc cutting are purely a melting by the electric arc of metal along a desired line of cut. It is not possible to do work of the same quality and smoothness as with gas cutting. However, the arc cutting is used for cutting of cast iron, scrap, non-ferrous metals, stainless steels, hole piercing, rivet cutting, nick breaking and other special uses. Its primary advantage is that it can be used on all types of metals. The arc cutting can be done with the carbon arc and with the metal arc.

15·54 Oxygen-arc Cutting Process

This new method, known as 'Arcos Leetrox' process is used for cutting of those metals which are impossible to cut by standard methods. Such metals are high chrome-nickel stainless steels, bronze, copper, brass, aluminium and cast iron. The mild steels can also be cut by this method.

The cutting is accomplished by directing a stream of oxygen into a pool of molten metal. The pool is made and kept molten by an arc established between the base metal and a tubular coated rod which is consumed during the cutting operation. The tubular rod also provides an oxidising flux and conveys oxygen to the surface being cut. In operation, the rod is held in a holder (similar to a welding electrode holder). The current and oxygen are fed to the rod through the holder. The flow of oxygen is easily controled by means of valve built into the holder.

The cutting action with this method is fast and no preheating is required. The cut material is not contaminated in any way and the finished cuts require very little grinding or machining.

15·55 Welding of Various Metals

Since a large number of metals are used in engineering practice, therefore, it is necessary to have an idea of weldability of these metals. Some welding methods as applied to different metals are discussed as below :

1. *Welding of low carbon steels or mild steel.* The low carbon steel or mild steel is widely used for heavy steel structural work where welding plays an important role in its fabrication. Therefore a careful study of welding technique for this metal is required. Carbon steels can be easily welded by forge welding, resistance welding, arc welding and gas welding.

Mild steel melts in the temperature range of 1450°C to 1540°C. In welding mild steel, no flux is required and the welding rods are made of pure iron or mild steel. No preheating is required even for welding thicker sections of mild steel. When mild steel is heated, it oxidises at full red heat. This oxide melts much earlier than mild steel and floats on the upper surface from where it can be removed. In gas welding, a neutral flame is used in order to minimise oxidation of steel. The properties of the weld are greatly improved by light hammering or peening. This is done after the weld has solidified but before it is cold.

2. *Welding of medium carbon steels.* The steels with carbon content between 0.3 to 0.5 percent are satisfactorily welded by arc welding, resistance welding, gas welding or thermit welding. The chilling effect in these steels is avoided by preheating the work. It may be noted that the preheating temperature (varying from 100°C to 400°C) depends upon the carbon contents in the steel. Since this steel is harder and brittle than mild steel, therefore, it is necessary to normalise the components after welding in order to relieve the residual stress present in them. In gas welding, a slight carbonising flame is used for welding these steels.

3. *Welding of high carbon steels.* In welding steels containing high percentage of carbon, preheating to about 400°C is essential. When the welded joint is cooled suddenly, cracking along the weld metal may occur. This is due to the fact that high carbon content in steel forms a very hard and brittle structure of martensite when cooled slowly. Therefore sudden cooling should be avoided. In order to relieve residual stresses set up during welding, heat treatment of these steels after welding is necessary.

4. *Welding of alloy steels.* These steels contain small amount of nickel, chromium, molybdenum, manganese, silicon, copper in addition to carbon. These alloy steels may be low alloy steels, medium alloy steels and high alloy steels. The welding of low alloy steels is similar to the welding of medium or high carbon steels. In welding such steels, preheating and slow cooling after

welding is essential in order to obtain crack free welds. The use of proper *welding rod relative to the composition of steel is very important to give desired mechanical properties to the weld metal. Shielded-metal arc and submerged arc welding is oftenly used. For most of low alloy steels, preheating is generally not necessary except for chromium-molybdenum steels which are highly air-hardening, produce crack sensitive zones and weld deposits.

5. *Welding of stainless steels*. This high alloy steel of composition 18% chromium and 8% nickel of austentic type is very common in use now-a-days. These high alloy steels can be welded by oxy-acetylene and metal-arc welding methods but the best method is electric butt welding followed by prompt annealing at 740°C to 800°C. Annealing is less effective in all types of stainless steels than in carbon steels because at elevated temperatures the austentic stainless steels have a relatively high yield strength. There is always a possibility of loss of chromium during welding. This can be prevented by using an electrode rich in chromium either in its core or in electrode coating. A wide range of electrodes is available in metal arc welding suitable for welding stainless steel of different composition. A suitable electrode should be selected as per manufacturer's advice. These electrodes prevent the weld decay which is the worst defect occured when welding stainless steels. In order to obtain a sound weld, cleaning of edges to be welded and removal of slag after each run is essential.

In oxy-acetylene welding of stainless steel, there should be a slightly excess acetylene in the blowing flame in order to avoid the risk of having an oxidising flame. It may be noted that an oxidising flame will result porous weld and too much of acetylene will produce brittle welds. To prevent oxidation of molten metal, it should be covered by the gas flame and the end of the welding rod should always be kept inside the pool of molten metal. Flux may or may not be used in welding stainless steel.

6. *Welding of cast steel*. In general, the weldability of cast steels is similar to wrought steels. Due to higher carbon and silicon contents, the mechanical properties and characteristics of the weld may be varied as per requirement. This term is used particularly for those components which have been produced by casting steel and have certain defects like blow holes etc. The welding operation for the repair of defective casting is termed as welding of cast steel. Preheating of heavy casting is essential to minimise straining of metal due to local heating at the time of welding. The surfaces to be welded should be bevelled before welding. Gas welding with a neutral flame is usually used for such a repair work.

7. *Welding of cast iron*. The welding of cast iron is mostly done by metal-arc welding, oxy-acetylene welding and braze welding. Carbon in cast iron is present in two forms, *i.e.*, as combined carbon in the form of iron-carbide and as free carbon in the form of graphite flakes. The weldability of cast iron usually decreases as the amount of free carbon in cast iron increases. Cast iron welding is used to make repairs which must restore the strength of the casting, to join together separately cast sections and to repair the surfaces of the castings to give a satisfactory appearance. In welding cast iron, the joint is prepared in the form of vee by milling, drilling, chipping or grinding. Since cast iron is weak and brittle, therefore great care is needed in welding this material. Care should be taken to prevent the development of contraction stresses during welding. Cast iron parts are generally preheated to a dull red heat and then welded. It may be noted that once the cast iron has melted, it should be cooled slowly as to allow time to carbon to change into free form. If the cooling is not slow, the carbon in the metal will be retained in the combined state to form white cast iron, which is hard, brittle and difficult to machine. To overcome this difficulty, preheating upto 600°C and slow cooling is recommended.

In welding cast iron with oxy-acetylene gas welding, a neutral flame and a clean and smooth welding rod is used.

During the welding operation, the end of the welding rod should always be kept inside the

* It is better to use low hydrogen electrode of chemical composition as that of parent metal.

molten metal. A good quality of flux will prevent oxide formation inside the pool of molten metal and help to float the impurities on the upper surface from where it can be removed easily. The quantity of flux used should be just sufficient as the higher quantity will tend to increase hardness.

Before welding cast iron, indirect preheating up to a temperature of 200°C is sufficient to prevent uneven expansion and contraction of casting. After preheating the entire casting, a crack to be welded is locally preheated upto 600°C and then finally welded. After welding, annealing of casting is very important.

8. *Welding aluminium and its alloys*. Aluminium and its alloys, for welding purposes are classified into two forms, *i.e.*, cast and wrought forms.

These can be welded by arc welding of all types, including inert-gas and atomic hydrogen welding. But the inert gas-tungston arc welding is extensively used than other arc welding processes. Other methods used for welding these metals are oxy-acetylene and oxy-hydrogen welding, brazing, spot welding, seam welding and flash welding.

In welding aluminium and its alloys with arc welding, direct current and reversed polarity is preferred. In order to remove oxide, the use of flux in welding is essential. The flux may either be applied dry or by forming a paste with water. The fluxes for aluminium usually contain chloride and fluoride salts. The edge preparation is required only for welding of thicker sections.

In welding wrought or sheet aluminium with oxy-acetylene welding, a neutral flame and a welding rod of the same material is used. Edge preparation upto 2 mm thickness is not necessary but for thicker sections the edges should be bevelled to an included angle of 80°. The flux is sparingly used by heating the rod and dipping it in the flux. In welding cast aluminium preheating is necessary and after welding it should be cooled slowly.

It is very important for aluminium welding to clean the work by brushing or filing and to impart degreasing treatment to the component taken for welding. The completed weld is dipped for about ten minutes in a 5% solution of nitric acid heated to about 80°C. Then the weld is washed in hot running water.

Note : Copper bearing aluminium alloys are unsuitable for arc and gas welding because of loss of mechanical properties and tendency to crack in the heat-effected zone. Butt welding is most suitable for these alloys.

9. *Welding copper and its alloys*. The copper having oxygen when heated forms cuprous oxide which offers difficulty in welding. Deoxidised or oxygen free copper can only be welded satisfactorily. In gas welding of copper, a neutral flame and a filler rod of copper and silver alloy is used. The brass, nickel-silver and phosphor bronze alloys should be welded with a slightly oxidising flame. Preheating to about 450°C should be done and this temperature is usually maintained during welding specially for heavy sections. Commercial rods are flux coated. When bare filler rods are used, a suitable flux paste is applied on the filler rod and the work to be welded. The principal constituent of copper welding fluxes is borax but it is not advisable to use borax alone because it forms a hard scale which is difficult to remove. Due to high thermal conductivity of copper, a great amount of heat is required at the base metal to produce fusion desirable for most welding processes. Thus, preheating should be done, otherwise cracking may occur where severe stresses are introduced into the weld.

The copper and its alloys can be successfully welded by metal arc or carbon arc welding processes. The direct current with straight polarity is usually employed for welding these metals.

10. *Welding nickel and its alloys*. These metals may be easily welded by most of the major welding processes like metal-arc welding, resistance welding and oxy-acetylene welding. In welding these metals, pre-heat treatment and post-heat treatment are rarely used except for special cases. No surface preparation is required in cold rolled sections. On the other hand, nickel-chromium heat resisting alloys in forged or hot-rolled conditions require surface treatments like grinding, blasting, and pickling before welding. Nickel, monel metal and inconel are readily welded by gas welding. A

neutral or slightly reducing flame should be used. Flux is not required for welding nickel. Silver brazing and soldering are other useful methods used for joining these metals.

In shielded metal arc welding, the direct current with reversed polarity is generally used. In tungsten inert-gas arc welding, although direct current with straight polarity is preferred but alternating current may also be used. Nickel alloys can also be successfully welded by submerged-arc welding. This process is not applicable to age-hardening alloys because the arc under the flux bed changes the composition of the weld deposit.

15·56 Testing of Welded Joints

In order to have good quality of welding, it is very essential to have careful examination of the component at each stage of manufacture. Thus by inspection or quality control, the welding defects are located and preventive measures are devised to reduce or eliminate them. The common defects which occur in welding are as follows :

1. The improper welding conditions and faulty welding methods, may result in poor fusion and the welded joint produced will be unsound.
2. The presence of oxides in the metal, uneven melting of the electrode coating and production of highly viscous slag may result slag inclusions in the weld.
3. The presence of gases in the metal and moisture in the flux may result formation of blow holes, gas pockets or roughness on the surface of the weld.
4. The non-uniform heating and cooling of the metal and gas forming constituents like phosphorus, sulphur and zinc vapours trapped inside the weld may produce cracks.
5. The non-uniform feeding of the welding rod, improper position of the electrode and excessive heating may result in pitting or under cutting of the weld. In other words, a cavity is formed into the base metal adjacent to the toe of the weld.

One of the most difficult problem in welding is to determine whether the finished weld is up to the expected standard or not. The welded portion may be inspected visually or examined with instruments to locate the defects and hidden flaws. The welding joints may be subjected to destructive testing like tensile test, bending test, impact test, hardness test etc. The non-destructive tests include spark test, X-ray test, magnetic test, hydraulic test and air-pressure test for pressure vessels.

15·57 Braze Welding

It is also known as *bronze welding*. The process of braze welding is intermediate between welding and brazing. In welding, the edges of the metal to be joined are melted and the molten metal pool is allowed to run between the edges which solidifies there to form a weld. Additional filler metal may or may not be used. In brazing, a low melting alloy is introduced between the edges to be joined and the joint is formed by adhesion. The flow of metal between the edges is due to capillary action.

In braze welding, the edges of the metal to be joined are heated to a temperature which corresponds to the melting point of the bronze filling rod used. The filler rod usually consists of 60% copper and 40% zinc with suitable amounts of silicon and tin which act as deoxidisers. The edge preparation of metals, for braze welding, is made in the similar way as for gas welding process. After applying the flux, the edges are heated either with oxy-acetylene flame or any other suitable source of heat. The temperature of heat should be corresponding to the fusion point of the filler rod material. When filler rod is brought in contact with the edges, it starts melting and the tinning action on the metal surfaces produces the adhesion on solidification to form a strong braze weld.

The metals with high melting point such as steel, cast iron, copper, brass and bronze are braze welded. Following are the advantages of braze welding :

1. The welding operation is carried out at low temperature.
2. It requires less heat to form the joint.

3. The operation of making the joint with braze welding is quicker than fusion welding.
4. Two different metals can also be joined by braze welding.

Notes : 1. The melting point of the metal to be joined should be higher than the bronze filler rod used for joining.

2. The braze welding process should not be used for service at over 250°C and for higher dynamic loads.

15·58 Soldering

The joining of two metal pieces by means of heat and a filler metal whose melting point is lower (below 400°C) than the melting point of metals to be joined, is called *soft soldering.* The filler metal, known as solder, is an alloy of lead and tin. The solders containing 45 to 62 per cent tin are general purpose solders of fine quality and have relatively rapid solidification properties. A solder containing 50 percent lead and 50 percent tin is recommended for stainless steel, though a solder upto 90 percent tin may be used. The higher tin solders are preferred because of greater strength, more silvery appearance (as compared to darker colour of high lead solders) and improved corrosion resistance. A suitable flux is always used in soft soldering in order to remove oxides already present on the work surface and to prevent further oxidation while the joint is being heated. The fluxes include rosin, rosin plus alcohol, zinc chloride and ammonium chloride.

There are many methods of heating the base metal and melting the solder and flux, but the oldest and the most widely used method of heating for soldering is by means of soldering iron. It is a simple, efficient and inexpensive tool consisting of a heat insulated handle at one end and a copper tip at the other end. The bit of the soldering iron may be heated electrically or in gas flames or in coke-fired furnaces.

The parts to be joined are, first of all, cleaned to remove foreign matter such as oil, grease, scale and dirt. The cleaning may be done by scrapping, filing or with emery cloth. It can be done chemically also and the most pickling chemicals used for cleaning are as follows :

1. For iron and steel — Solution of 10% sulphuric acid.
2. For brass — —do—
3. For copper — Dilute sulphuric acid.
4. For stainless steel — Mixture of sulphuric acid (3 parts), nitric acid (1 part) and water (10 parts).

After cleaning, the parts are arranged in proper alignment and position. The bit of the soldering iron is heated, cleaned, dipped in flux and then rubbed on the solder to pick up a small quantity of solder with it and then deposited gently along the joint.

The soft soldering is extensively used for making electrical connections, hermetic sealing of metal containers, joining lead pipe and copper tubing, filling casting imperfections and repair work. The most engineering metals which can be joined by soft solders are iron, steel, copper, brass, nickel and monel metal.

When silver alloyed with tin is used as a solder to join the parts, it is known as *hard soldering.* The temperatures of various hard solders vary from 600 to 900°C. In hard soldering, a blow torch is used.

15·59 Brazing

The joining of two metal pieces by means of heat and a special filler metal (known as spelter) having a melting point above 400°C but lower than the melting point of the parts of be joined, is called *brazing.* The copper base and silver alloys are commonly used as filler metals for brazing.

In brazing, the two metal pieces to be joined are, first of all, cleaned to remove all grease and oxide. The parts are fitted together along the line of joint and held in that position by some clamp.

The flux (usually borax) is applied on the joint and then heated to temperature above the melting point of the spelter. When the spelter is applied to the joint, it flows throughout the joint by capillary action.

The heat for brazing may be applied by several methods, some of which are discussed below :

1. *Torch brazing*. When the heating is done by means of an oxy-acetylene torch, the brazing is said to be torch brazing. It is used for brazing a small number of parts.

2. *Furnace brazing*. When the heating is done in a furnace, the brazing is said to be furnace brazing. It is employed in mass production of small parts.

3. *Electric brazing*. When the heating is done by an arc, resistance or induction heating techniques, the brazing is called electric brazing. It is used for large quantity production.

4. *Dip brazing*. When the heating consists of dipping the pieces with the spelter preplaced at the joint, into a bath of hot molten salt, the brazing is called dip brazing. It is used for large scale production.

QUESTIONS

1. What is welding ? State the difference between autogeneous welding and hetrogeneous welding.

2. What are the advantages and disadvantages of welded joints over other joints ?

3. How the welding is classified ? Explain them briefly.

4. Explain the various types of joints commonly used in welding.

5. Write short notes on the following :
(*a*) Cold pressure welding, (*b*) Smith's forge welding.

6. Discuss the method of electric resistance welding. What are its advantages and disadvantages ?

7. Describe, with the help of a neat sketch, the principle of spot welding.

8. What is the difference between roll spot welding and seam welding ? Explain with a neat sketch the principle of seam welding ?

9. State the difference between upset butt welding and flash butt welding. Explain their principles.

10. What do you understand by percussion welding ?

11. Discuss, with the help of neat sketch, the principle of electric arc welding. What is straight polarity and reversed polarity ?

12. What type of arc welding is obtained when the electrode is bare and coated ? Why the coating is provided on the elctrodes ?

13. Explain, with the help of a neat sketch, the principles of carbon arc and metal arc weldings.

14. Sketch and describe the principle of submerged arc welding.

15. Discuss, with a neat sketch, the principle of atomic hydrogen welding.

16. Describe, with the help of neat sketches, the following arc welding processes :
(*a*) TIG welding, and (*b*) MIG welding.

17. Write short notes on the following :
(*a*) Plasma arc welding ; (*b*) Flux cored arc welding ;
(*c*) Electro-slag welding ; and (*d*) Electro-gas welding.

18. What is solid state welding ? Explain the following solid state welding processes :
(*a*) Friction welding ; (*b*) Ultrasonic welding ;
(*c*) Diffusion welding ; and (*d*) Explosive welding.

19. Write short notes on the following :
(*a*) Stud arc welding, and (*b*) Thermit welding.

20. What do you understand by gas welding ?

21. List out the equipment required for oxy-acetylene welding. Explain each with sketches wherever necessary.

22. Describe, with the help of sketches, the three types of oxy-acetylene flames.

23. Write short notes on the following :

(*a*) Left-ward welding, (*b*) Right-ward welding, (*c*) Vertical welding and (*d*) Linde welding.

24. Sketch and describe the torch used in cutting of metals.

25. Explain the various processes used in cutting of metals.

26. Describe the process of welding used for :

(*a*) Cast irons, and (*b*) Aluminium.

27. How the welding of high speed steel is carried out ?

28. What is braze welding ? Describe its advantages and disadvantages.

29. What do you understand by soldering and brazing ? Describe the principle of each.

30. How you will braze a carbide tip to the mild steel shank of a turning tool ?

OBJECTIVE TYPE QUESTIONS

1. For welding plates of thickness less than 5 mm, its edges

(*a*) do not require bevelling

(*b*) should be bevelled to a single V or U groove

(*c*) should have a double V or U groove on one side

(*d*) should have a double V or U groove on both sides

2. In electric resistance welding, voltage required for heating is

(*a*) 1 to 5 volts (*b*) 6 to 10 volts (*c*) 11 to 20 volts (*d*) 50 to 100 volts

3. Spot welding is used for welding

(*a*) lap joints in plates having 0.025 mm to 1.25 mm thickness

(*b*) lap joints in plates having thickness above 3 mm

(*c*) butt joints in plates having 0.025 mm to 1.25 mm thickness

(*d*) butt joints in plates having thickness above 3 mm

4. Seam welding is a

(*a*) arc welding process (*b*) multi-spot welding process

(*c*) continuous spot welding process (*d*) process used for joining round bars

5. Seam welding is best adopted for metal thickness ranging from

(*a*) 0.025 to 3 mm (*b*) 3 to 5 mm (*c*) 5 to 8 mm (*d*) 8 to 10 mm

6. The welding process used in joining mild steel shanks to high speed drills, is

(*a*) spot welding (*b*) seam welding (*c*) projection welding

(*d*) upset butt welding (*e*) flash butt welding (*f*) any one of the above

7. In arc welding, the electric arc is produced between the work and the electrode by

(*a*) voltage (*b*) flow of current (*c*) contact resistance (*d*) all of these

8. In shielded arc welding

(*a*) large electrode is used

(*b*) welding rod coated with fluxing material is used

(*c*) welding rod coated with slag is used

(*d*) none of the above

9. The consumable electrode is used in
(*a*) carbon arc welding (*b*) submerged arc welding
(*c*) TIG arc welding (*d*) MIG arc welding
(*e*) thermit welding (*f*) none of these

10. In TIG arc welding, the welding zone is shielded by an atmosphere of
(*a*) hydrogen gas (*b*) oxygen gas (*c*) helium gas
(*d*) argon gas (*e*) either (*a*) or (*b*) (*f*) either (*c*) or (*d*)

11. Thermit, used in thermit welding, is a mixture of
(*a*) charcoal and iron oxide (*b*) charcoal and aluminium
(*c*) iron oxide and aluminium (*d*) charcoal, iron-oxide and aluminium

12. The oxy-acetylene gas used in gas welding produces a flame temperature of
(*a*) 1800ºC (*b*) 2100ºC (*c*) 2400ºC (*d*) 3200ºC

13. Acetylene gas is stored in cylinders in
(*a*) solid form (*b*) gaseous form (*c*) liquid form (*d*) any one of these

14. Most of the oxy-acetylene welding is done with
(*a*) neutral flame (*b*) oxidising flame (*c*) carbursing flame (*d*) all of these

15. The maximum flame temperature occurs
(*a*) at the outer cone (*b*) at the inner cone
(*c*) between the outer and inner cone (*d*) at the torch tip

16. A neutral flame is obtained by supplying
(*a*) equal volumes of oxygen and acetylene
(*b*) more volume of oxygen and less volume of acetylene
(*c*) more volume of acetylene and less volume of oxygen
(*d*) none of the above

17. Carburising flame is used to weld
(*a*) steel, cast iron, copper, aluminium etc. (*b*) brass and bronze
(*c*) hard surfacing materials such as stellite (*d*) all of the above

18. In fore-hand welding, the weld is made
(*a*) from left to right (*b*) from right to left
(*c*) first from left to right and then from right to left
(*d*) either from left to right or from right to left

19. Linde welding uses
(*a*) neutral flame and right-ward technique
(*b*) oxidising flame and left-ward technique
(*c*) carburising flame and right-ward technique
(*d*) neutral flame and left-ward technique
(*e*) oxidising flame and right-ward technique

20. The flux commonly used in brazing is
(*a*) zinc chloride (*b*) ammonium chloride
(*c*) rosin plus alcohol (*d*) borax

ANSWERS

1. (*d*)	2. (*b*)	3. (*a*)	4. (*c*)	5. (*a*)
6. (*e*)	7. (*c*)	8. (*b*)	9. (*d*)	10. (*f*)
11. (*c*)	12. (*d*)	13. (*c*)	14. (*a*)	15. (*b*)
16. (*a*)	17. (*c*)	18. (*b*)	19. (*c*)	20. (*d*)

16

Bench Work and Fitting

1. Introduction. 2. Fitting Tools. 3. Holding Tools (Vices). 4. Care of Vice. 5. Striking Tools (Hammers) 6. Cutting Tools. 7. Cold Chisels. 8. Types of Chisels. 9. Chipping. 10. Files. 11. Filing. 12. Methods of Filing. 13. Care of Files. 14. Scraper. 15. Method of Scraping. 16. Care of Scraper. 17. Grinding. 18. Hacksaw. 19. Power Hacksaw. 20. Sawing 21. Measuring, Marking and Testing Tools. 22. Method of Marking. 23. Drilling Tools. 24. Various Angles of Twist Drills. 25. Drilling Machines and Appliances. 26. Drilling. 27. Reamers. 28. Reaming. 29. Taps. 30. Tapping. 31. Dies and Stocks. 32. Dieing.

16·1 Introduction

The bench work and fitting plays an important role in every engineering workshop to complete and finish the job to the desired accuracy. Though today in industries most of the work is done on automatic machines and finished to a fairly good degree of accuracy, still they require some operations to be done by hand to finish them to the desired accuracy. The work carried out by hand at the bench is called *bench work,* whereas *fitting* is the assembling of parts together by filing, chipping, sawing, scraping, tapping etc., necessary after the machine operation. This may or may not be carried out at the bench.

16·2 Fitting Tools

The tools used in fitting practice may be classified into the following groups :

1. Holding tools,
2. Striking tools,
3. Cutting tools,
4. Scraping tools,
5. Drilling tools, and
6. Measuring, marking and testing tools.

We shall now discuss these tools, in detail, in the following pages.

16·3 Holding Tools (Vices)

The holding tools or vices are required to hold the work firmly. Following are the various types of vices for different purposes.

1. *Bench vice.* It is used for holding the work and is firmly fixed to the bench with bolts and nuts. It consists of an iron or steel cast body, square threaded screw, nut, handle (made of mild steel), two jaws and jaw plates (made of cast steel) as shown in Fig. 16.1. One of the two jaws is stationary, while the other jaw is movable which can be brought close on the work by rotating the square threaded screw in the nut. The jaws plates are knurled on one side and fixed in the jaws with the help of screws and can be replaced when worn out. The knurling is done to help in gripping the work. For holding soft material or finished work, soft plate liners or coverings made of lead, fibre, copper, brass or tin plate are fitted over the jaws for giving protective grip in order to prevent the surface of the finished work from damaging.

The height of the bench should be such that the top of the vice jaws is at about the same height as the operator's elbow. The size of the vice is given by the width of the jaws and the maximum opening between the jaws.

Some vices are provided with a swivelling base which enables them to be rotated and fixed at angle so that the work can be located in any desired position.

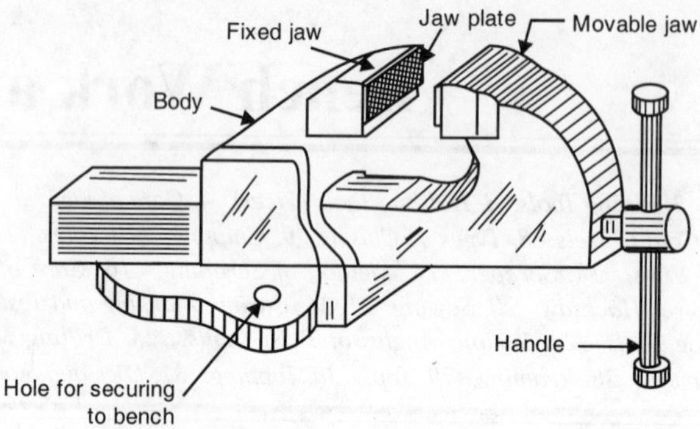

Fig. 16.1. Bench vice.

2. *Hand vice.* It consists of two jaws and hands which are hinged together, on a pivot, on the opposite ends as shown in Fig. 16.2. The work is held in between the serrated faces of the jaws with a screw and a wing nut. A flat spring, fixed between the two hand jaws, helps open them out by its pressure. The vice is held in one hand and the required operation is performed by the other hand. It is used for gripping small objects such as screws, rivets, small drills, keys of locks and other similar objects which are too small to be conveniently held in bench vice.

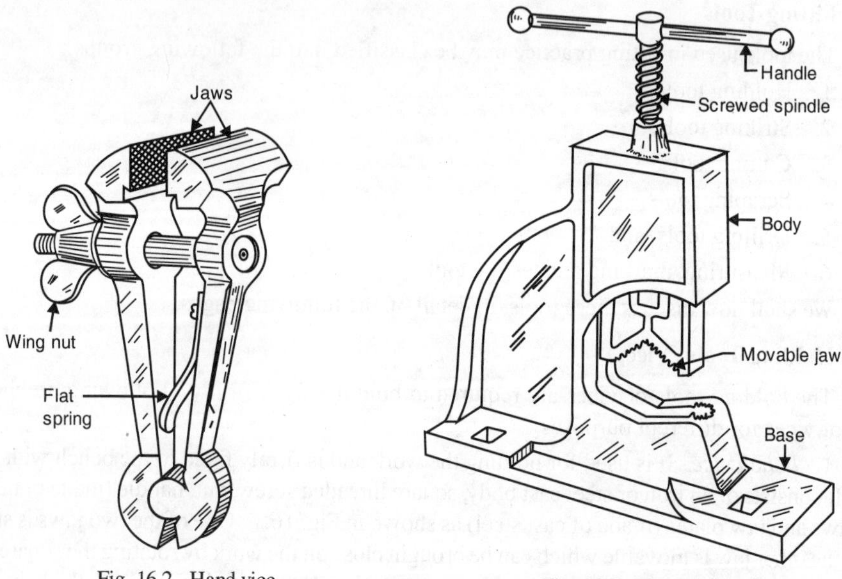

Fig. 16.2. Hand vice.

Fig. 16.3. Pipe vice

3. *Pipe vice.* It is used for holding pipes, shafts or round jobs. It works on the same principle as the bench vice. It consists of two jaws, one of which is fixed and the other is movable, as shown in Fig. 16.3. The shape of the jaws is like a V and it grips the work at four points or its surface.

4. *Leg vice.* It is generally used in a smithy shop, but it is also suitable for heavy hammering, chipping and cutting in fitting shop. The vice is secured to the top of the bench by an iron strap which is fastened to a plate bolted to the bench top. It differs from the bench vice in that the movable jaw is

pivoted about a point near to the middle of the leg of the vice. This prevents the faces of the two jaws from being parallel to each other in every position except one position.

5. *Pin vice.* This is a holding device for small diameter work, such as wires and pins. It consists of a fine self centering chuck on a handle which may be of wood or metal, as shown in Fig. 16.4. The work is gripped between the jaws of the chuck by rotating the handle.

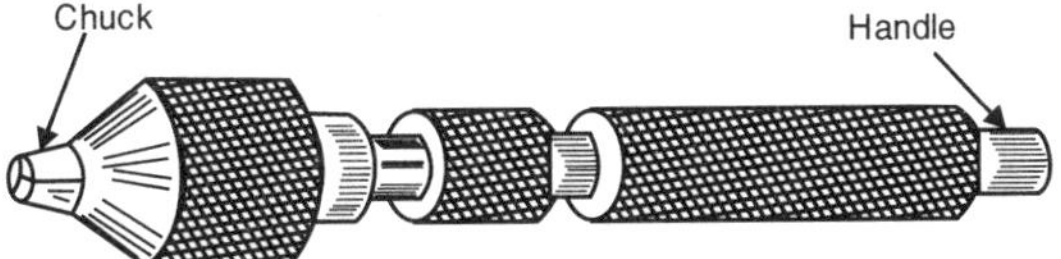

Fig. 16.4. Pin vice.

6. *Tool maker's vice.* It is particularly used for holding small work for filing or drilling. It is made of mild steel. It consists of a body with a solid jaw, a movable jaw, a screw and a handle for the control of movement of movable jaw, as shown in Fig. 16.5.

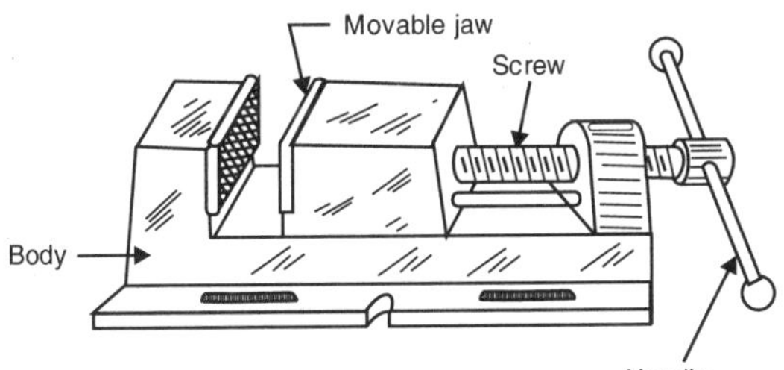

Fig. 16.5. Tool maker's vice.

16·4 Care of Vice

The Following points should be kept in mind while using a vice :

1. The vice should be kept clean and free from dust and metal chips by using a brush.
2. The threads and the nut should be occasionally oiled.
3. The vice should never be used as an anvil.
4. The serrated jaws should be covered with soft metal clamps when finished work is held.
5. For holding tubes, temporary wooden 'V' blocks should be used.

16·5 Striking Tools (Hammers)

The striking tools or hammers are used to strike the job or tool. A hammer consists of a head, striking face, peen and a shaft or handle as shown in Fig. 16.6. The hammer heads are made from a plain steel of about 0.6% carbon and are shaped by stamping or forging. The two ends (*i.e.,* face and peen) must be hardened and tempered and the centre of the head, with the eye, is left soft. The eye has an oval through recess, enlarged at one end, to accommodate the end of the shaft which is secured by inserting a hard wood or metal wedge into it. The shaft or handle is made of well seasoned, straight grained wood and it should be so shaped and sized that when gripped, it gives an easy feel to the hand. The size of the hammer is indicated by its weight. The various types of hammers in common use are as follows :

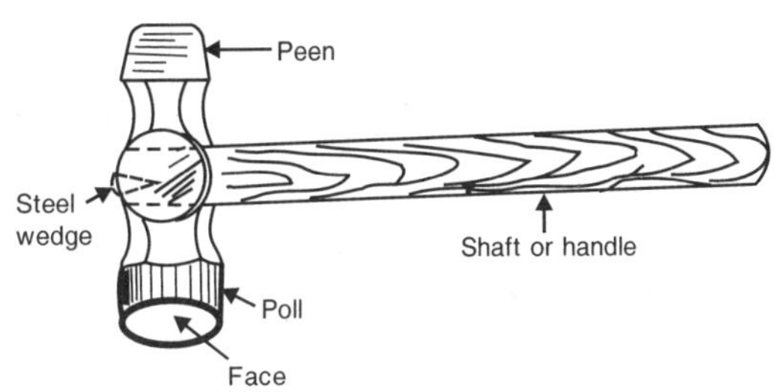

Fig. 16.6. Hammer.

1. *Ball-peen hammer.* The ball-peen hammer, as shown in Fig. 16.7(*a*), has the peen of the shape of a ball. It is the most common type of the hammers and is mostly used for riveting (*i.e.,* for forming a snap head on the tail of a rivet) and chipping. According to Indian standards (IS : 841-1957), the size of ball peen hammer varies from 0.11 kg to 0.91 kg.

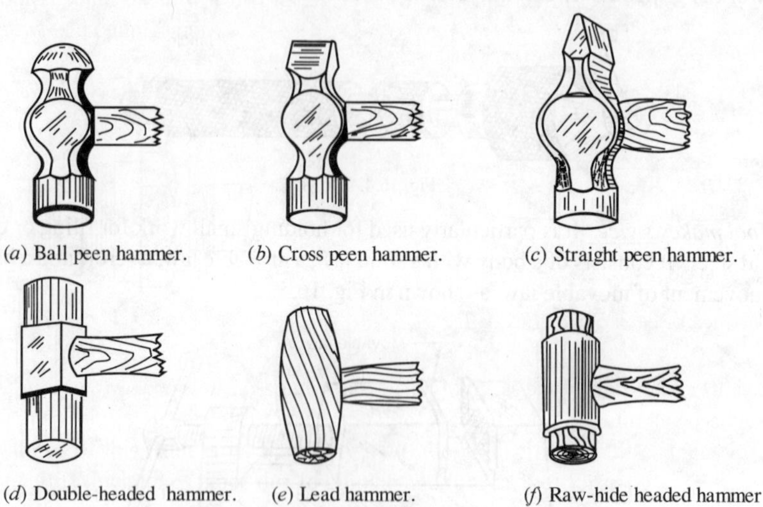

Fig. 16.7. Types of hammers.

2. *Cross-peen hammer.* The cross-peen hammer, as shown in Fig. 16.7(*b*), is similar to ball-peen hammer except that its peen is like a wedge which is perpendicular to the handle. It is used for bending, stretching, hammering into shoulders, inside curves etc. According to Indian standards (IS : 841 - 1957), the size of cross-peen hammer varies from 0.22 kg to 0.91 kg.

3. *Straight-peen hammer.* The straight-peen hammer, as shown in Fig. 16.7(*c*), has its peen straight or parallel to the handle. The width of peen is usually equal to the diameter of the face. It is used for stretching or peening the metal by hammering or fullering. According to Indian standards (IS : 841 - 1957), the size of straight-peen hammer varies from 0.11 kg to 0.91 kg.

4. *Double-faced hammer.* The double-faced hammer, as shown in Fig. 16.7(*d*), consists of a head which has two identical faces at both ends. It is used for striking cup tools for riveting.

5. *Soft hammer.* The soft hammers are used for striking finished surface which would be damaged by the hardened face of a hammer. The two types of soft hammers are shown in Fig 16.7(*e*) and (*f*). The *lead hammer,* as shown in Fig. 16.7(*e*), has a lead head cast on to a length of steel tube. The *raw hide hammer*, as shown in Fig. 16.7(*f*), has a cast iron head into which raw-hide plugs are inserted. These plugs can be replaced when they are worn out. The soft hammers may also be made of wood, hard rubber, copper and brass.

16·6 Cutting Tools

The chief cutting tools used in fitting are cold chisels, hacksaws and files. These are discussed, in detail, in the following pages.

16·7 Cold Chisels

The cold chisel, as shown in Fig. 16.8, is an important cutting tool used by the fitter.

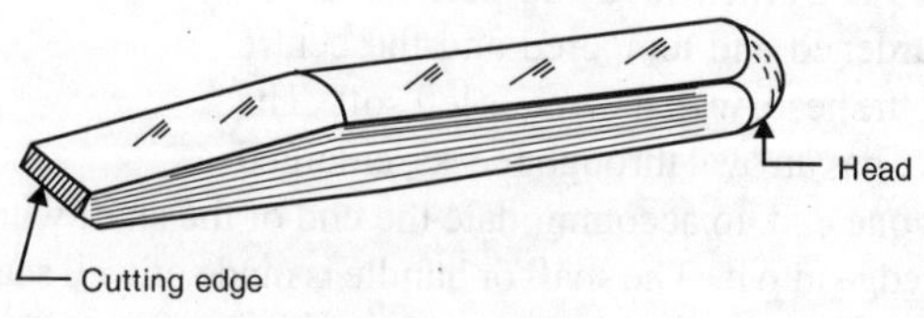

Fig. 16.8. Cold chisel.

It is distinguished from other chisels by the fact that it has no wooden handle. These are used to cut the cold metal and are made by forging from cast tool steel of octagonal cross-section. The length of the chisels is about 15 cm to 20 cm and the tapered part is from 5 cm to 8 cm long. The chisels are being made from 10 mm to 25 mm thick material.

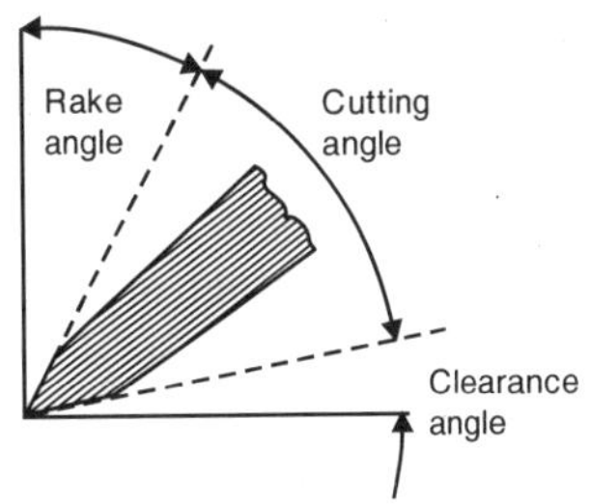

Fig. 16.9. Cutting angles of chisel.

The cutting edge is ground to an angle suited to the material being worked upon. After forging to shape and roughly grinding, the cutting edge should be hardened and tempered. The various angles formed on the chisel are cutting angle, clearance angle and rake angle as shown in Fig. 16.9. The most commonly used cutting angle is 60°, but this varies according to the type of material cut. The following table shows the least cutting angles for different materials.

Table 16.1. Cutting angles for different materials.

Material	*Cutting angles*
Aluminium	30°
Copper	45°
Brass	50°
Wrought iron and mild steel	55° to 60°
Cast iron	60° to 65°
Cast steel	70°

16·8 Types of Chisels

The various type of chisels commonly used for fitting are as follows :

1. *Flat chisel.* A flat chisel, as shown in Fig. 16.10(*a*), is the most common of all the chisels. It has a wide cutting edge about 16 mm to 32 mm and is slightly rounded to prevent the corners from digging into the metal. The length of the chisel varies from 100 mm to 400 mm. It is used for chipping flat surfaces, cutting off sheet metal, cutting bars and rivets.

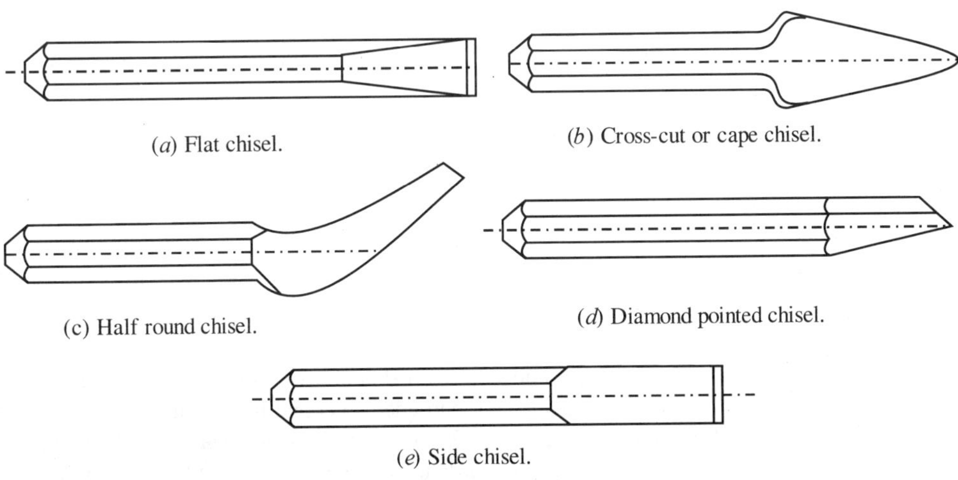
(*a*) Flat chisel.
(*b*) Cross-cut or cape chisel.
(c) Half round chisel.
(*d*) Diamond pointed chisel.
(*e*) Side chisel.
Fig. 16.10. Type of chisels.

2. *Cross-cut chisel or cape chisel.* A cross-cut or cape chisel, as shown in Fig. 16.10(*b*), has a cutting edge about 4 mm to 12 mm wide. From the edge, the metal thickness tapers off slightly. This is to permit the chisel to clear when a groove is being cut. It is widely used for cutting grooves in large surfaces before using the flat chisel. It is also used for cutting keyways in wheels and shafts.

3. *Half-round chisel.* A half-round chisel, as shown in Fig. 16.10(*c*) has a semi-circular cutting edge. The width of the cutting edge varies from 2 mm to 16 mm and length varies from 150 mm to 250 mm. It is used for cutting round or semi-circular oil grooves in bearings, bosses and pulleys.

4. *Diamond pointed chisel.* A diamond pointed chisel, as shown in Fig. 16.10(*d*) has a cutting edge shaped like a diamond. The width of the cutting edge varies from 6 mm to 16 mm and length varies from 100 mm to 400 mm. It is used to cut V-grooves and to chip square corners.

5. *Side chisel.* The side chisel, as shown in Fig. 16.10(*e*), has the cutting edge bevelled on one side only. It is particularly used for chipping and removing surplus metal in cotter ways and slots which are to be cut by hand after drilling.

16·9 Chipping

The operation used for cutting or removing thick layers of metal by means of cold chisels, is known as *chipping.*

The chipping operation is carried out as follows :

1. The work is firmly held in a vice.
2. The chisel is firmly gripped in the left hand leaving some distance above the thumb of the hand as shown in Fig. 16.11.
3. The hammer is grasped near the end of the hammer in the right hand. In operation, it should be brought upto the shoulder to ensure more power in the blows.

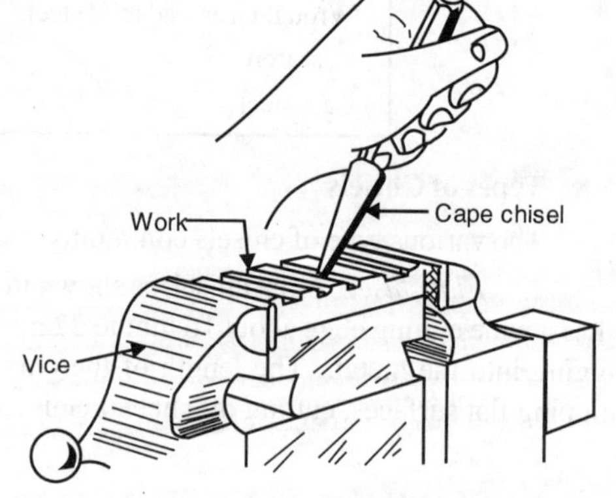

Fig. 16.11. Method of chipping.

The following points should be kept in mind while chipping the surface.

(*a*) The same cutting angle of the chisel must be maintained during the chipping operation.

(*b*) If the surface to be chipped is very large, it is advisable to cut grooves along the whole surface by means of a cross-cut or cape chisel and then chip off the remaining metal.

(*c*) When removing large amount of metal, it is necessary to lubricate the cutting edge frequently in order to have longer life of the chisel and to make chipping easier and smoother.

(*d*) The operator who performs the chipping operation, should always keep his eyes on the cutting edge of the chisel and not at the head.

16·10 Files

A file is a hardened piece of high grade steel with slanting rows of teeth. It is used to cut, smooth or fit metal parts. It cuts all metals except hardened steel and it cuts only on the forward stroke. It consists of a *blade* or *body* with a *tang* for fixing into the *wooden handle,* as shown in Fig. 16.12. The teeth are cut on the blade which are hardened and tempered. The tang is tempered to make it soft and tough. The various parts of the blade are shown in Fig. 16.12. The metal ring on the file handle is called *ferrule,* in order to prevent splitting of the handle. The files are classified according to the size, cut of teeth, grade and shape or cross-section of the file as discussed below :

(*a*) *Size of file.* The *size* of the file is indicated by its length. It is the distance from the *point* to the *heel* without tang. The length of the file, in general use, is 200 mm to 450 mm and 100 mm to 200 mm for finer work.

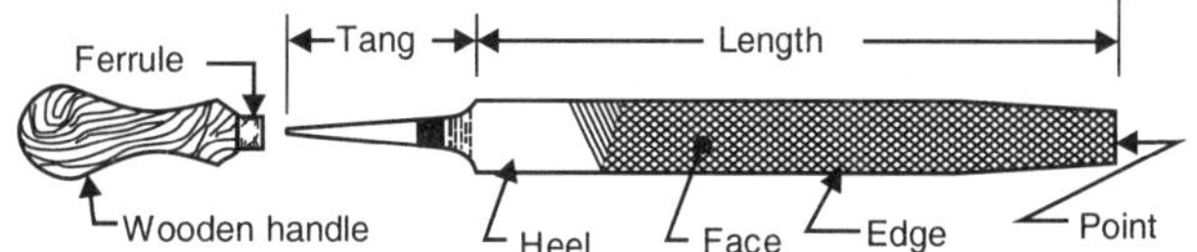

Fig. 16.12. Parts of a file.

(*b*) *Cut of teeth.* The files, according to the *cut of teeth,* are divided into two groups, *i.e., single cut* and *double cut.*

In *single cut files,* the teeth are cut parallel to each other running across the faces and at angle of 60° to the centre line of the file, as shown in Fig. 16.13(*a*), These files are frequently termed as *flats* and are particularly used for very hard metals.

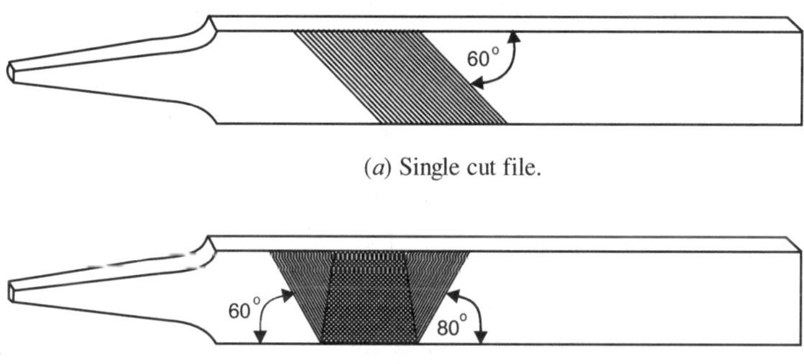

(*a*) Single cut file.

(*b*) Double cut file.

Fig. 16.13. Single cut and double cut file.

In *double cut files,* there are two sets of teeth. The first set of teeth are similar to those of single cut files (*i.e.,* at 60° to the centre line of the file) while the second set of teeth are cut diagonally across the first set of teeth at an angle of about 80° to the centre line of the file as shown in Fig. 16.13(*b*). All the teeth have negative rake *i.e.*, sloping backwards by which they cut only on the forward stroke. It removes metal faster and is used for general work.

(*c*) *Grade of cut of file.* The single cut and double cut files, depending upon the pitch of the teeth (*i.e.,* spacing between the rows of teeth) may be classified as rough (R), bastard (B), second cut (SC), smooth (S), dead smooth (DS) and superfine or supersmooth (SS). The number of teeth per 10 mm length and their uses for each of these grades of files are given in the following table.

Table 16.2. Grades of files and their uses.

S.No.	*Grade of file*	*Effective length of file in mm*								*Uses*
		100	*150*	*200*	*250*	*300*	*350*	*400*	*450*	
		Number of cuts per 10 mm of length								
1.	Rough (R)	10	8	7.1	6.3	5.5	5.3	4.8	4.5	It is used for heavy reduction and filing casting after chipping.

S.No.	Grade of file	*Effective length of file in mm*								*Uses*
		100	*150*	*200*	*250*	*300*	*350*	*400*	*450*	
		Number of cuts per 10 mm of length								
2.	Bastard (B)	18	13	11	10	9	8	7	6	It is used for reducing bulky work.
3.	Second cut (SC)	21	17	16	15	14	13	12	11	It is used for general work.
4.	Smooth (S)	30	24	20	20	19	18	16	15	It is used for smoothening surfaces and filing delicate work.
5.	Dead Smooth (DS)	35	33	31	30	28	–	–	–	These are used by toolmakers and silver smiths for high accuracy and finish.
6.	Supersmooth (SS)	63	49	45	40	–	–	–	–	

In may be noted that coarseness or pitch of the file varies directly as the length of the file. Thus larger the length of the file, coarser will be the pitch and smaller the file, finer will be the pitch.

(*d*) *Shape of files*. The files, according to their shape or cross-section, are classified as discussed below :

1. *Flat file.* A flat file, as shown in Fig. 16.14(*a*), is parallel for about two-thirds of its length and then tapers in width and thickness. It has double cut on both faces and single cut on both edges. It is mostly used for general work and for filing flat surfaces.

2. *Hand file.* A hand file, as shown in Fig. 16.14(*b*), has its width parallel throughout, but its thickness tapers. The both faces are double cut and one edge single cut. The uncut edge is called the *safe edge* and prevents cutting into one face of a square corner while the other face is being filed. It is used for general surfacing work, more particularly for filing steps or shoulders to square work without touching and spoiling the adjacent surface, already made true.

3. *Square file.* A square file, as shown in Fig. 16.14(*c*), is parallel for two-thirds of its length and then tapers towards the tip. It is double cut on all sides and is used for filing square corners and slots.

4. *Pillar file.* A pillar file, as shown in Fig. 16.14(*d*), is similar to hand file, but is narrower and thicker than hand file. It has one or both uncut edges (*i.e.*, safe edges) and is used for narrow work such as keyways, slots and grooves.

5. *Triangular file.* A triangular file, as shown in Fig. 16.14(*e*), has width either parallel throughout or upto middle and then tapered towards the tip. Its cross-section is triangular (equilateral) and the three faces are double cut and edges single cut. It is used for filing square shoulders or corners and for sharpening wood working saws.

6. *Round file.* A round file, as shown in Fig. 16.14(*f*), has round cross-section. A file with width parallel throughout is called *parallel round* and the file with width parallel upto middle and then tapering towards the tip is called *rat file.* The round files are usually double cut on the Rough and Bastard qualities over 15 cm long while the Rough and Bastard under 15 cm, together with second cut and smooth, are single cut. The round files are used for opening out holes, producing round corners, round-ended slots etc.

7. *Half round file.* The section of a half round file, as shown in Fig. 16.14(*g*), is not a true half circle but is only about one-third of a circle. The width of the file is either parallel throughout or upto middle and then tapered towards the tip. The flat side of this file is always double cut and curved side has single cut. It is used for filing curved surfaces.

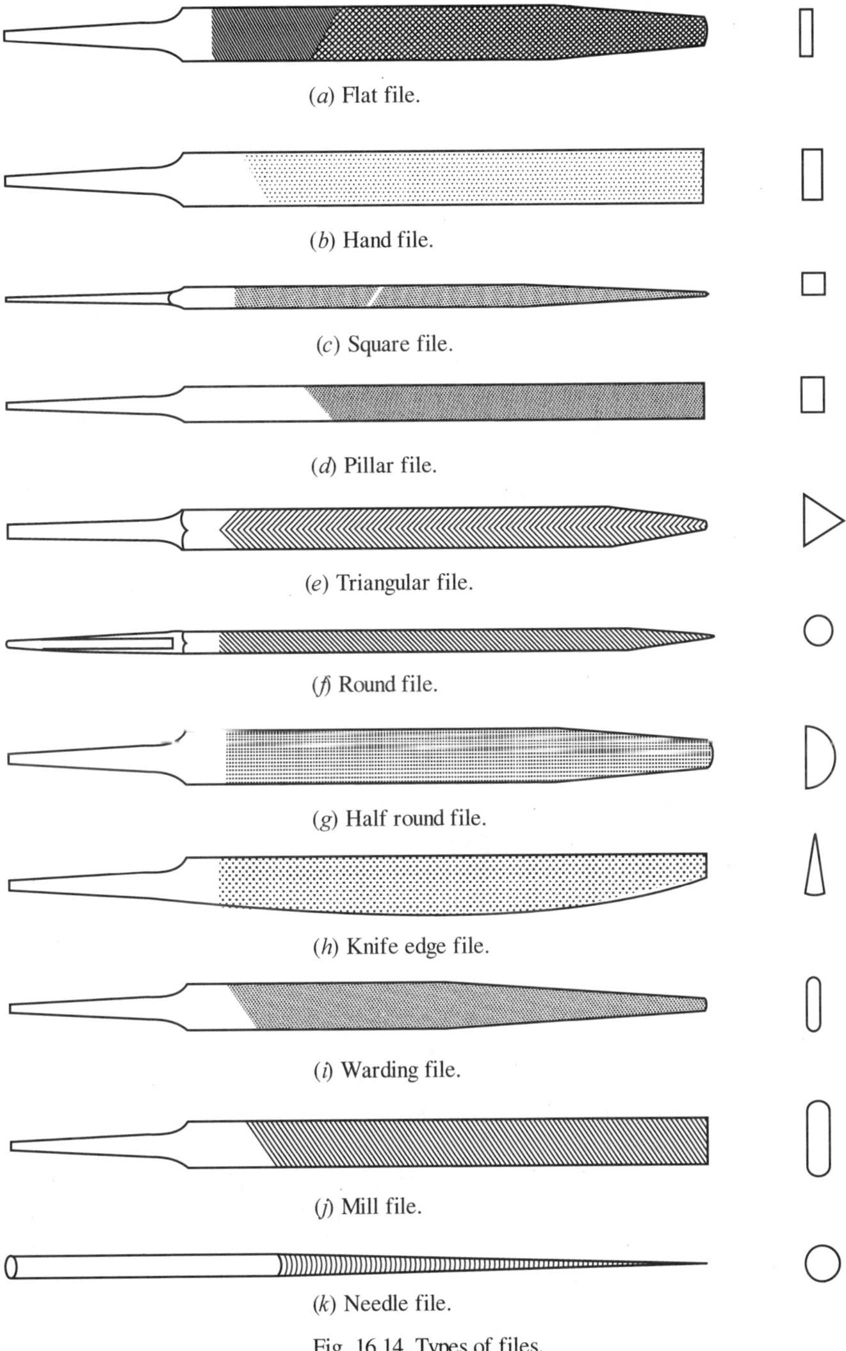
(a) Flat file.

(b) Hand file.

(c) Square file.

(d) Pillar file.

(e) Triangular file.

(f) Round file.

(g) Half round file.

(h) Knife edge file.

(i) Warding file.

(j) Mill file.

(k) Needle file.

Fig. 16.14. Types of files.

8. *Knife edge file.* A knife edge file, as shown in Fig. 16.14(*h*), has a width tapered like a knife blade. It is also tapered towards the tip and thickness. It is double cut on both flat faces and single cut on both edges. It is used for finishing sharp corners of grooves and slots.

9. *Warding file.* A warding file, as shown in Fig. 16.14(*i*), is similar to flat file but it is thinner and parallel on its thickness. It is used for filing narrow slots.

10. *Mill file.* A mill file, as shown in Fig. 16.14(*j*), is similar to flat file but parallel on both width and thickness and have both edges round. It is single cut and is suitable for forming the radius on saw teeth and in slots.

11. *Needle files.* The needle files, as shown in Fig. 16.14(*k*), are available in sizes from 100 mm to 200 mm of various shapes and cuts. These files are used for fine work such as pierced design in sheet metal.

16·11 Filing

Filing is required after chipping or cutting operation to remove burr and clean the face of the cuts and to finish the final shape ofthe work piece. It may be noted that not more than 0.6 mm should be left for filing. The filing operation consists of the following steps :

1. The work should be held firmly in the vice with the minimum amount of projection and with the surface to be filed truly horizontal.
2. The file handle is grasped in the right hand, as shown in Fig. 16.15, and the end of the file handle pressing against the palm of the hand. The pressure on the work is applied by holding the end of the blade with the left hand. The worker should take his position on the left side of the vice with the feet firmly planted, slightly apart.

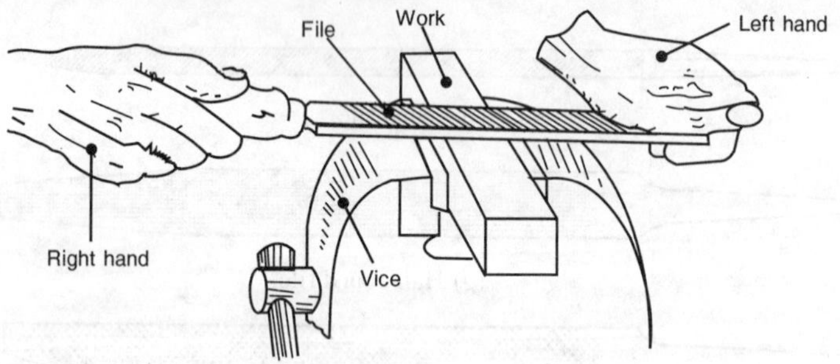

Fig. 16.15. Filing operation.

3. A stroke should be made by a slight movement of the right arm from the shoulder and by a sway of the body towards the work, each of these movements being about equal.
4. The file must remain horizontal throughout the stroke (which should be long, slow and steady) with pressure only applied on the forward motion.
5. The file on the return stroke remains in contact with the work but the pressure is relieved from it.
6. In order to test the surface of work during filing, a straight edge should be placed on it occasionally and the line of contact is viewed for day-light. When any considerable amount of metal has to be removed, the bulk of it should be removed by a rough or bastard-cut file and the surface progressively brought to a finish by second-cut and smooth files.

16·12 Method of Filing

The following three methods are commonly used for filing :

1. *Cross-filing.* It is the most common method of filing for general shaping work. It is used for efficient removal of maximum amount of metal in the shortest possible time. In this method, the filing strokes run alternatively from left to right and from right to left as shown in Fig. 16.16, so that whole of file surface moves across whole of work surface in one stroke. In cross-filing the possibility

of rounding of surface is less and the scratch marks of the file teeth cross each other so that maximum amount of metal is removed.

2. *Straight filing.* It is specially useful on long and narrow pieces of work whose width is less than that of the file. In this method, the file stroke is made nearly at right angles to the length of work.

3. *Draw filing.* The draw filing is used to remove file marks and for finishing operation. During this operation, the file is gripped as close to the work as possible between two hands as shown in Fig. 16.17. It is moved lightly to and fro over the work. For this purpose, a fine cut file with a flat face (*e.g.*, mill file) should be used.

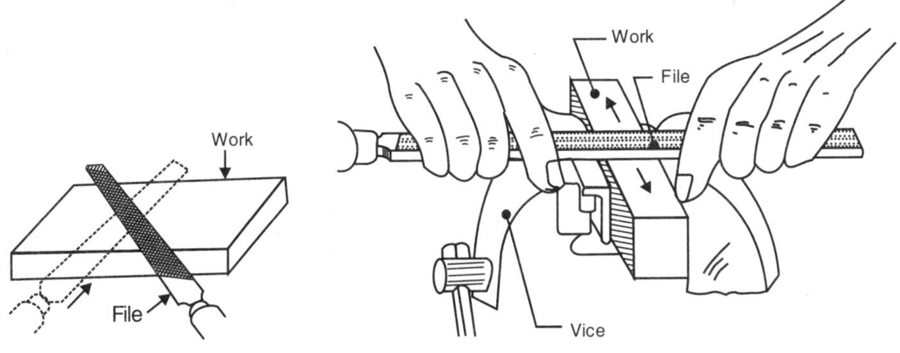

Fig. 16.16. Cross-filing.

Fig. 16.17. Draw filing.

16·13 Care of Files

The following points should be kept in mind while using files :

1. The file should not be used without handle or with loose fitting handle.

2. A file should never be used on hardened steel, hard surface scale or allowed to strike against the hardened vice jaws.

3. The new file should be used first on copper, brass and then on wrought iron and mild steel.

4. Sometimes when soft metals are filed, they tend to clog the file teeth with minute particles of metal (called pins). A little consideration will show that these pins will not only stop the efficient cutting action but they also spoil the work with scratches. These pins are removed and the file is cleaned with a thin hard wire brush or a file card as shown in Fig. 16.18. After cleaning the file, chalk should be rubbed into the file teeth to prevent clogging of pins.

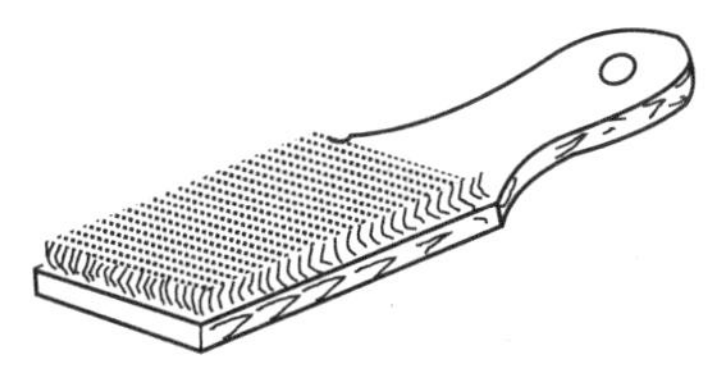

Fig. 16.18. File card.

5. The file should not be allowed to rust and to prevent it, the file is coated with machine oil. The oil should be removed before the file is put to use by carbon tetrachloride or caustic soda.

6. The worn-out files may be re-used by dipping it in hydrochloric acid. The worn-out files are best suited for making scrapers, punches, chisels etc.

16·14 Scraper

A scraper is a tool with sharp edge by means of which very small amount of metal can be removed. The process of removing metal with a scraper is known as *scraping.* Occasionally, it is carried out to improve the look or finish of surfaces.

The scrapers are made of tool steel or sometimes made from old worn-out files and provided with a wooden handle. The cutting part of the scraper is hardened but usually not tempered and then ground with at least one sharp cutting edge. The scrapers are made in a variety of lengths from 100 mm onwards and in many shapes. The various types of scrapers used are as follows :

1. *Flat scraper.* The flat scraper, as shown in Fig. 16.19(*a*), is the most common and similar to flat file. It has a cutting edge at the end which is slightly rounded. Its length vary from 100 mm to 250 mm and width from 20 mm to 30 mm for rough work, 16 mm to 20 mm for accurate work and 5 mm to 10 mm for extra accurate work. The thickness at the cutting end varies from 1 mm to 3.5 mm. It is used for producing a perfectly flat surface.

2. *Triangular scraper.* The triangular scraper, as shown in Fig. 16.19(*b*), is made from worn-out triangular file. It has three cutting edges. It is used for finishing small holes and for removing sharp corners and burrs. The length of this scraper is about 150 mm.

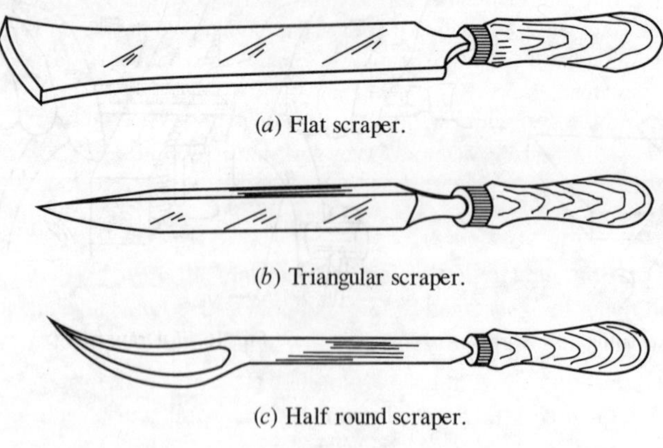

(*a*) Flat scraper.

(*b*) Triangular scraper.

(*c*) Half round scraper.

Fig. 16.19. Types of scrapers.

3. *Half-round scraper.* The half round scraper, as shown in Fig. 16.19(*c*), is made of half round file. It is used for scraping halves of bearings and curved surfaces. Its length is also about 150 mm.

16·15 Method of Scraping

We have already discussed that the process of removing metal with a scraper to obtain a true flat surface, is known as *scraping.* Before scraping, it is necessary to establish the high spots or impressions on the work which require scraping. This is done by moving the surface of the work to and fro on the surface plate, the top of which is covered with a thin film of Prussian blue or red lead. The high spot marked on the surface of the work are now scrapped down by a scraper.

The method of holding a scraper is at approximately 30° to the work piece as shown in Fig. 16.20. The blade should be held in the left hand, close to the cutting edge. The handle is grasped by the right hand, with the thumb and first finger pointing in the direction of cutting edge.

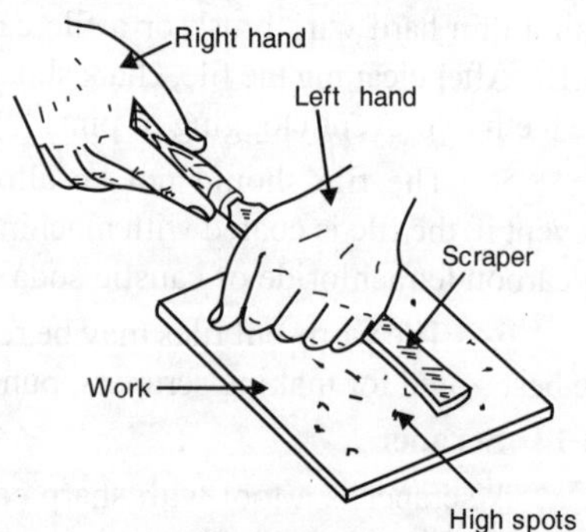

Fig. 16.20. Method of scraping.

16·16 Care of Scraper

Since the scrapers have a very sharp cutting edges, therefore, when these are not in use it is advisable to keep them lightly oiled and individually wrapped in a piece or cloth or felt or similar material, to protect the extremely hard cutting edge.

16·17 Grinding

The grinding is the process of removing metal by the use of grinding wheels. The work for grinding is pressed against the grinding wheel which rotates at a comparatively high speed and the surplus metal is removed by abrasion. It produces extremely smooth finish due to the small cutting edges on the wheel. It can finish work to very accurate dimensions in a short time.

The grinding wheel is made of abrasive grains which form the cutting edges in a wheel. These abrasive grains are held together by some form of bond such as clay, shellac or rubber. There are two types of abrasives *i.e.*, natural abrasives and artificial abrasives. The natural abrasives are *corundum* and *emery* and the artificial abrasives are *silicon carbide* (the trade name is Carborundum) and *aluminium oxide* (the trade name is aloxite). The natural abrasives are tougher than artificial ones but not as hard. The artificial abrasives break with sharp clear fracture and thereby present new cutting edge to the work. The silicon carbide wheels are recommended for materials of low tensile strength such as cast iron, brass, leather, cemented carbide etc. The aluminium oxide wheels are best for materials of high tensile strength like hardened steel, high speed steel, alloy steel etc. The wheels made from coarse grained abrasives are used for fast removal of materials where as fine grained wheels are used for finish of work. The coarse-grained wheel may be used for grinding soft materials and fine-grained wheels for hard and brittle materials. The hardness or softness of the wheel depends upon the ability of the bond used to hold the abrasive grains in wheel against the tearing force. In order to dissipate the heat which is normally generated during grinding, a large amount of coolant over the work, wheel face and sides is generally used. This is known as *wet grinding* and it gives longer wheel life and high finish of work.

When the wheel becomes loaded or *glazed, it should be trimmed or dressed immediately to remove dull grains of abrasives and metal particles and to restore a sharp crystalline surface for efficient cutting.

16·18 Hacksaw

The hacksaw is the chief tool used by the fitter for cutting rods, bars and pipes into desired

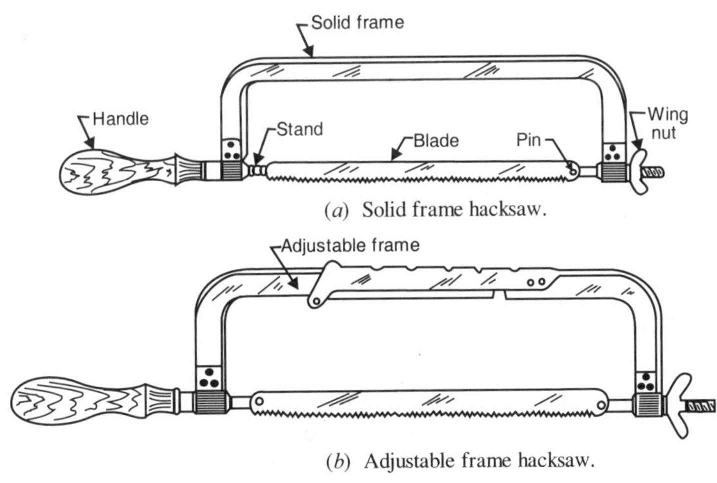

(*a*) Solid frame hacksaw.

(*b*) Adjustable frame hacksaw.

Fig. 16.21. Types of hacksaws.

---

* The wheel is said to be loaded when the spaces between the abrasive grains become clogged with particles of material being ground so that cutting edges do not project sufficiently to do their work.

The wheel is said to be glazed when grains which are worn down to the bond level have lost their sharpness are still detained by the bond.

lengths. It consists of a metal frame, which may be solid, as shown in Fig. 16.21(*a*), or adjustable, as shown in Fig. 16.21(*b*). The blade fits over two pegs which project from the pins sliding in the ends of the frame. The wing nut at the front end to the frame is for tensioning the blade. The blades are made of carbon or high-speed steel and may be finished with the cutting edge only hardened or they may be hard all over. The soft-backed blades are tougher and less liable to break than the all-hard blades. The blades are specified by its length and the point or pitch. The length of the blade is the distance between the outside edges of the holes which fit over the pins. The most usual blade for hand work is 250 mm long and 12.5 mm wide. The point or pitch is measured by the number of teeth per 25 mm length.

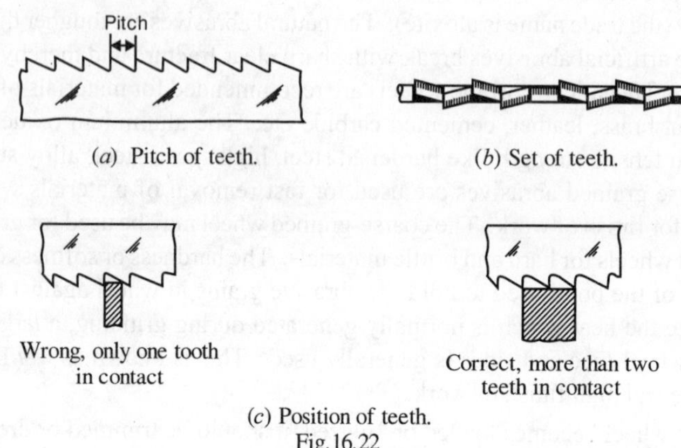

(*a*) Pitch of teeth. (*b*) Set of teeth.

(*c*) Position of teeth.
Fig.16.22

The points of the teeth are bent, as shown in Fig. 16.22(*a*), to cut a wide groove and prevents the body of the blade from rubbing or jamming in the saw cut. This bending of the teeth to the sides is called the *setting* of the teeth as shown in Fig. 16.22(*b*). Usually alternate teeth are set to right and left, every third or fifth tooth left straight to break up the chips and help the teeth to clear themselves. The fine-toothed blades for cutting thin metal are sometimes made with a wavy set to minimize stripping of the teeth from the blade.

The choice of the blade for any particular class of work depends upon the pitch of the teeth and at least two or three teeth should be in contact with the surface being sawn, as shown in Fig. 16.22(*c*). If this is not attained, the teeth will be stripped from the blade and sawing too sharply over a corner will also result in teeth being torn off.

The best all-round blade for hand use is one with 16 to 18 teeth per 25 mm. For other special classes of sawing, the following blades should be used :

(*a*) 14 teeth per 25 mm, for solid brass, copper and cast iron.

(*b*) 24 teeth per 25 mm, for silver, steel and thin cast steel rods.

(*c*) 32 teeth per 25 mm, for sheet metal and tubing *e.g.,* steel, copper and conduit tubing.

The following should be kept in mind while using a hacksaw :

(*i*) The blade must be strained tightly in the frame and steady strokes (about 50 per minute) should be used.

(*ii*) The breakage of blades may be due to the following reasons :

(*a*) rapid and erratic strokes,

(*b*) too much pressure,

(*c*) blade held too loosely in the frame, and

(*d*) work not held firmly in the vice.

(*iii*) Solid metals should be cut with a good pressure and thin sheets and tubes with light pressure.

16·19 Power Hacksaw

It is similar to ordinary hacksaw, but it has a motor driven attachment which moves the hacksaw blade to and fro. The power hacksaw has a special blade which is thicker and wider than ordinary blade. According to Indian standards (IS: 2594-1964), the length of the blade varies from 300 mm to 600 mm, width from 20 mm to 50 mm, thickness from 0.8 mm to 2.5 mm and pitch from 1.4 mm to 6.3 mm. It is used for cutting bars of diameter upto 200 mm or more. The cutting fluid should be used during cutting operation. The advantages of a power hacksaw are as follows :

1. It requires less time.
2. An unskilled worker can operate.
3. It automatically stops after cutting.
4. The accuracy can be maintained.

16·20 Sawing

It is the cutting and slotting operation and performed as follows :

1. The work is held tightly in the vice.
2. A suitable blade is fixed with its teeth facing forward so that it cuts the material in the forward stroke. The blade is tensioned sufficiently by a wing nut.
3. The course of the desired cut is marked by a guide line on the work and a nick or notch is made with a file.
4. The frame of the hacksaw is held, as shown in Fig. 16.23(*a*).

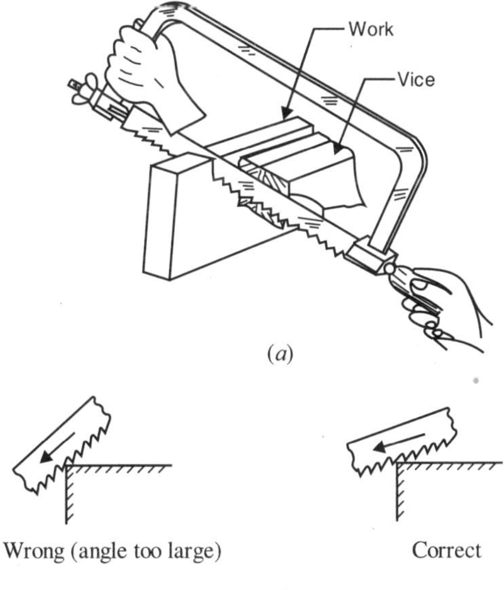

(*b*)
Fig. 16.23. Sawing.

5. Now the cutting is started at the nick by keeping the blade slightly inclined to the horizontal, as shown in Fig. 16.23(*b*). The pressure is applied during the forward stroke and it should be relieved during the return stroke.

Note : The tubes are not sawn in one go. First of all, the notch or nick portion is sawn upto the inner wall. Now it is rotated to such an extent that the saw blade is easily approachable to the saw cut. The process is repeated till whole of the tube is cut.

16·21 Measuring, Marking and Testing Tools

The following measuring, marking and testing tools as used in fitting shop are discussed as follows :

1. *Engineer's steel rule.* It is a stiff, straight steel strip having all the faces machined true. On one of the flat faces, graduations are marked in inches and centimeters as shown in Fig. 16.24. The least count in the former case is 1/64 of an inch and in the latter 1/10 of a centimeter. It is used to set out dimensions.

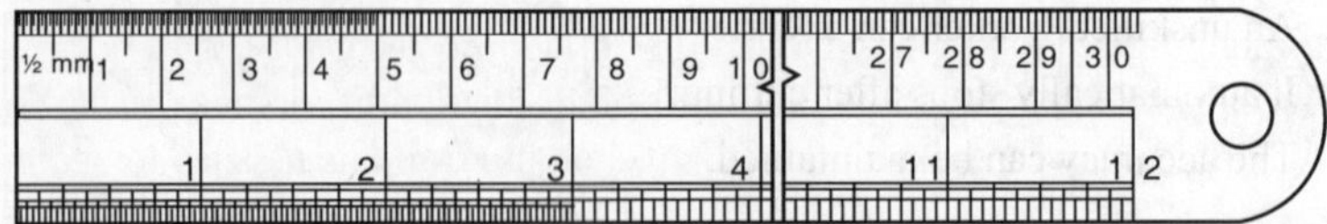

Fig. 16.24. Steel rule.

2. *Outside calipers.* The outside calipers are of two types *i.e.*, ordinary or spring type. It has two steel legs which are bent inward as shown in Fig. 16.25. It is used to measure the outside diameter of round objects and to measure width and thickness. This is done by opening the legs and then measuring the distance between the legs with a steel rule. The ordinary calipers have the two legs hinged between washers for free movement. The spring callipers have a flat curved spring instead of hinge and the legs are moved with a screw which helps to keep the set dimension unaltered.

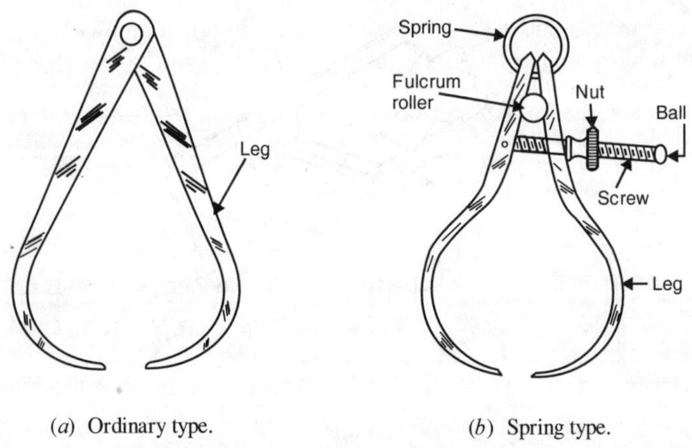

(*a*) Ordinary type. (*b*) Spring type.

Fig. 16.25. Outside calipers.

3. *Inside calipers.* The inside calipers are also of two types *i.e.*, ordinary and spring type. These are similar to outside calipers, the only difference being that the legs are bent outwards as shown in Fig. 16.26. These are used to set internal dimensions, to transfer them to work or to check with standards.

4. *Hermaphrodite, Jenny or Oddleg caliper.* It is made of steel tapered strip which is hinged between washers at one end as shown in Fig. 16.27. One leg is bent at the tip inwardly and the other has a straight pointed end. It is used to find out the centre of a round bar by holding a bent leg against a curved face of the bar and the pointed leg on the end face. It is also used to scribe lines parallel to the straight edges or for taking distances from the edges of the holes or plates and similar jobs.

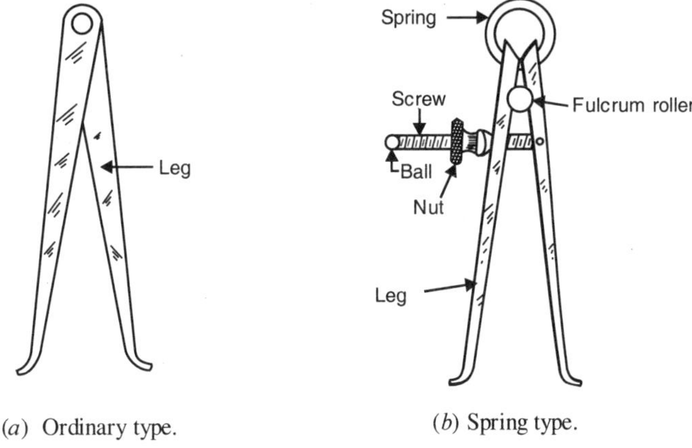

(*a*) Ordinary type. (*b*) Spring type.

Fig. 16.26. Inside calipers.

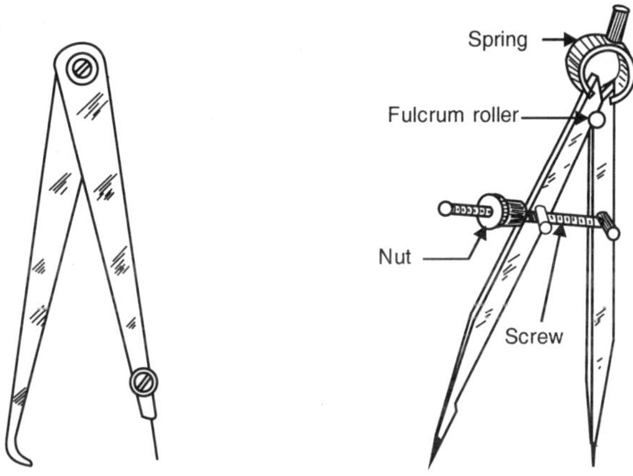

Fig. 16.27. Hermaphrodite caliper. Fig. 16.28. Divider.

5. *Divider.* A divider is an important instrument used for marking work. It is similar to calipers but its legs have sharp points as shown in Fig. 16.28. The most common type of the divider used in fitting have spring arrangement. The dividers are used for measuring the distance between two points, dividing a given length in a definite ratio, drawing circles and arcs and transferring dimensions from scales to objects.

6. *Surface plate.* It is made of grey cast iron in various sizes. It is usually square or rectangular, as shown in Fig. 16.29, and has the top and the adjacent four edges very accurately machined and scraped to form a true flat surface and are brought in square. The casting is ribbed to prevent it from deformation. It is placed horizontally on a firm support whose working height is about 800 mm from the floor.

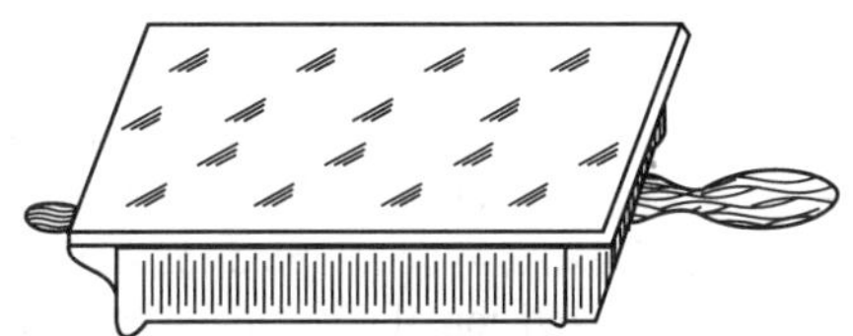

Fig. 16.29. Surface plate.

It is used to check the trueness of flat surfaces and to copy the master surface on a work. A thin coating of red lead mixed with oil is spread evenly over the surface plate. The work to be tested is cleaned, placed on the plate and gently moved along it. The parts which come in contact with red lead are marked and show where more metal has to be removed. The surface plates are made in two grades of accuracy *i.e. A* and *B*, depending upon scraping of within 0.005 mm and 0.02 mm respectively. It should be oiled and covered with a wooden cover to protect from dust, when not in use.

7. *Scriber.* It is a round hardened steel piece of about 150 mm to 300 mm long and 3 mm to 5 mm in diameter. Its one end is sharp pointed and the other end is also sharp pointed but bent as shown in Fig. 16.30. The bent end is used to scratch lines in places where the straight end can not reach.

Fig. 16.30 Scriber.

8. *Scribing blocks.* These have round or rectangular base blocks made of cast iron. Following are the two types of scribing blocks :

(*a*) *Ordinary scribing block.* It consists of a cast iron sliding base on the centre of which a steel rod (spindle) is fixed vertically as shown in Fig. 16.31. A steel scriber, held in the holder, is made to slide, tilt and rotate, on the rod after clamping it with a set screw. It is normally used in conjunction with either a surface plate or marking table. Its specific use is in locating centers of round rods held in V-block.

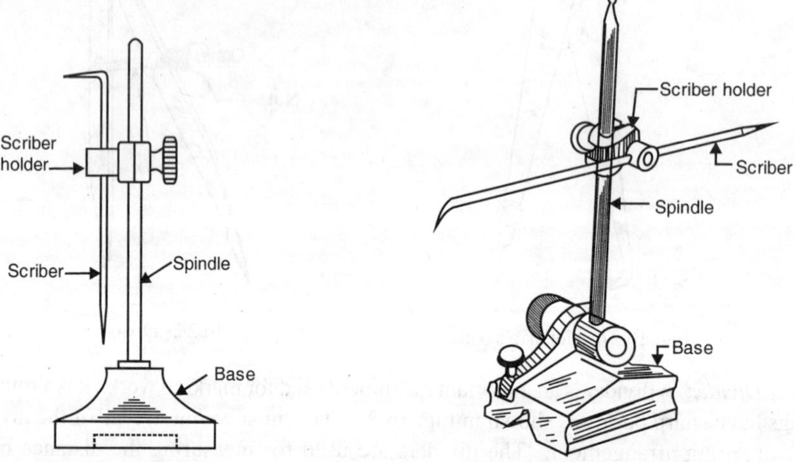

Fig. 16.31. Ordinary scribing block.

Fig. 16.32. Universal surface gauge.

(*b*) *Universal surface gauge.* It consists of a cast iron base perfectly machined and planed at the top, bottom and all sides as shown in Fig. 16.32. It carries a spindle which may be set at any angle. A scriber, which may be set at any angle or at any height is clamped to the spindle. A fine adjustment is provided by a screw (height adjusting screw) and movement of this screw swings the scriber through a small arc and thus varies the height. It is essentially used for marking on vertical or nearly vertical surfaces, but it can also be adopted for scribing horizontal surfaces. A further advantage of this type of gauge is the provision of a Vee in the base for resting the block on a round bar, so that the dimensions may be set off from the bar to some other part of the work. It is used to find out the centre of a round work, to align work in centre, and to layout markings on a work of any shape. The size of the gauge depends upon the job in hand but one with interchangeable spindles about 225 mm to 300 mm long is suitable.

9. *Punches.* A punch is primarily a marking tool used in bench work and is known as *centre*

punch, as shown in Fig. 16.33(*a*). It is generally made from 10 mm octagonal cast steel about 100 mm long. One end of the punch carries a point ground to an included angle of 90° and is tempered to a pale straw colour while the other end (called head) is slightly chamfered to prevent it from burring. The body is knurled for gripping it by hand. It is used for marking the ends of work to be centred for turning in lathe as well as for all centres of holes for drilling.

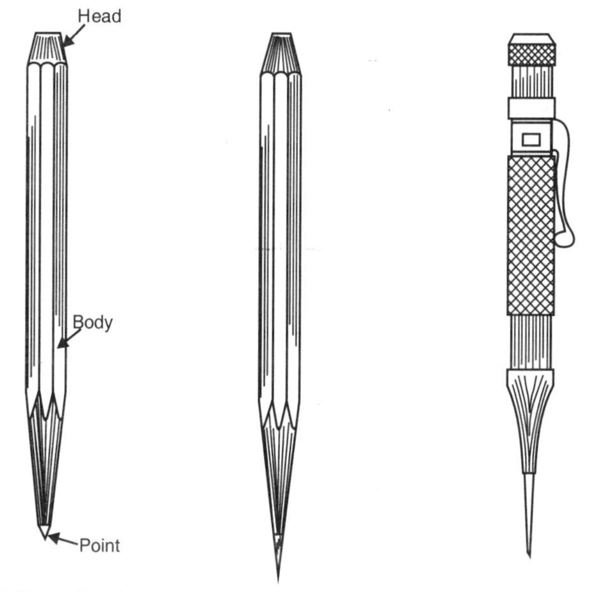

(*a*) Centre Punch. (*b*) Prick or dot Punch. (*c*) Automatic centre punch

Fig. 16.33. Types of punches.

The *prick or dot punch,* as shown in Fig. 16.33(*b*), is similar to centre punch but small in size. It is made from 6 mm diameter cast steel with a sharper point say 60 degrees. It is often used for marking position of lines and centres of circles to be drawn with the dividers.

The automatic centre punch, as shown in Fig. 16.33(*c*), is a hollow cylindrical body and has a spring loaded plunger by which a set blow is applied. The point of the punch is pressed against the work to be marked which compresses the spring inside. When the hand pressure is suddenly released, the spring in regaining its original length pushes the plunger downwards which applies a blow to the point. It is used for marking precision work.

10. *V-block.* The V-block is made of cast iron and all the faces are machined true. The two opposite side faces have V-grooves and the two have slots. The round bars are held firmly by a U-clamp engaging in slots, as shown in Fig. 16.34. The V-blocks are used for holding round bars during marking and centre drilling their end faces which are to be held between centres on lathe. It is also suitable for holding round

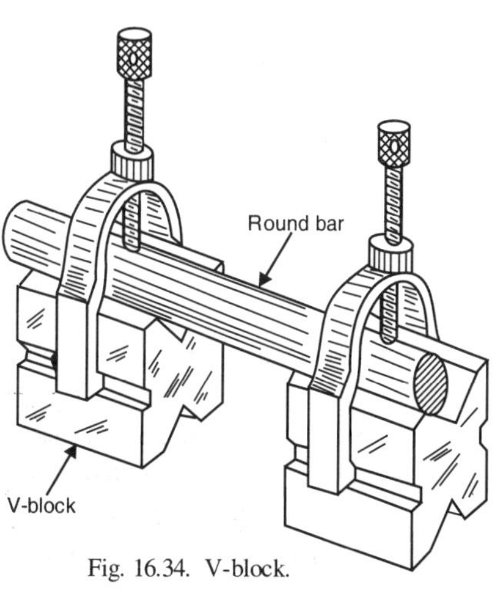

Fig. 16.34. V-block.

bars while they are drilled at right angles to their axis.

11. *Angle plate.* The angle plate, as shown in Fig. 16.35, is made of cast iron in different sizes. It has two planed surfaces at right angles to each other and has various slots in both the surfaces to hold the work firmly by means of bolts and clamps. The angle plate is used in conjunction with the surface plate for supporting work at right angles for marking.

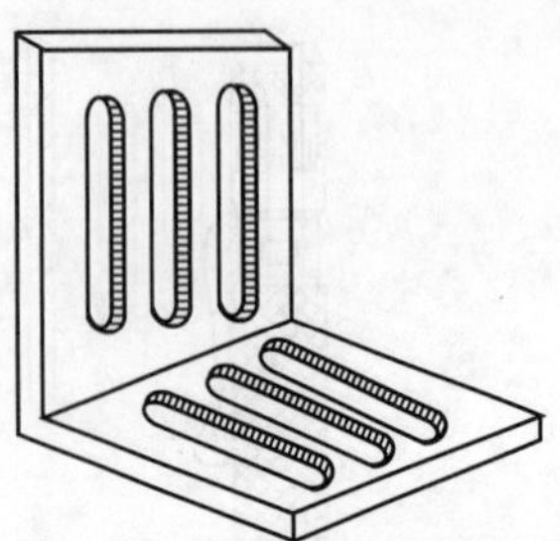

Fig. 16.35. Angle plate.

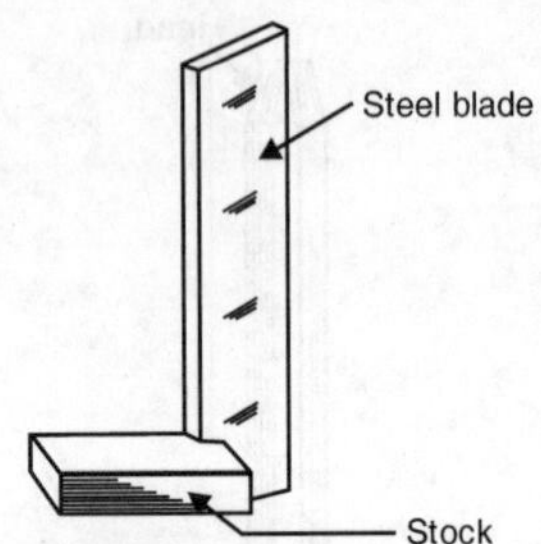

Fig. 16.36. Try square.

12. *Try square.* It consists of a steel blade fixed at right angles to the edge of a machined stock of cast iron as shown in Fig. 16.36. It is used for testing the squareness of surfaces and to set off lines at right angles to a given edge.

13. *Combination set.* It is an extremely useful too. It incorporates all the essential features of the try square, protractor, spirit level, rule and scriber. The combination set, shown in Fig. 16.37., consists of four parts *i.e.* square head, protractor head, centre square head and a blade or rule on which the other three parts slide. They are located in a groove in the blade and a tongue in the head to which the clamping screw is attached. The protractor and square heads are usually provided with spirit levels. The square head has a small scriber concealed in the end.

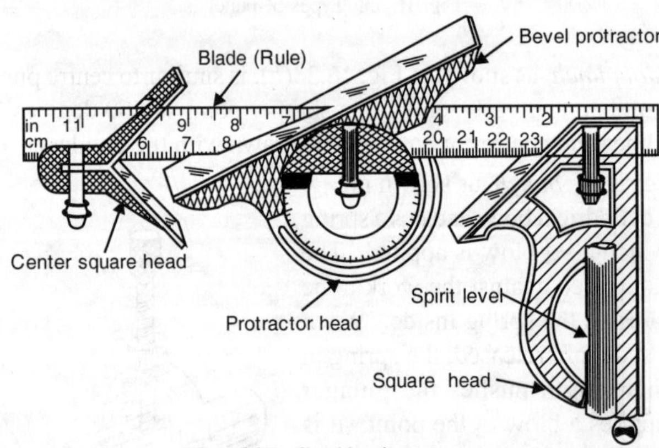

Fig. 16.37. Combination set.

The square head can be used, by moving the blade, as a try square, as a metre square or even as a depth gauge.

The centre square head may be used to extend a line around a corner, to find centre of a round piece or to find the centre at the end of a shaft. The protractor head is used for measuring angles and can be clamped in any desired position. The scale is graduated in degrees.

14. *Trammel.* A trammel, as shown in Fig. 16.38, is an alternative to dividers and it is used where a large circle having a large radius is to be marked. Since it has two adjustable vertical legs mounted on a common rod (called beam), therefore it is also known as *beam compass* or *beam*

trammel. It may also be used to measure distances in the same way as a divider is used.

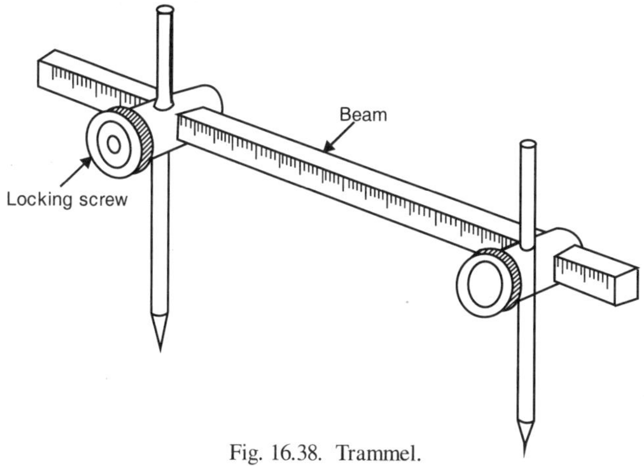

Fig. 16.38. Trammel.

15. *Straight edge.* The straight edge, as shown in Fig. 16.39, is like a solid bow. It is made of cast iron, ribbed at intervals and has machined straight flat edges made true. The ribbing is done to prevent warping. These are made of varying widths and lengths. It is used for testing the straightness and flatness of plane surfaces, particularly to test the accuracy of the ways of a lathe, planer, shaper etc.

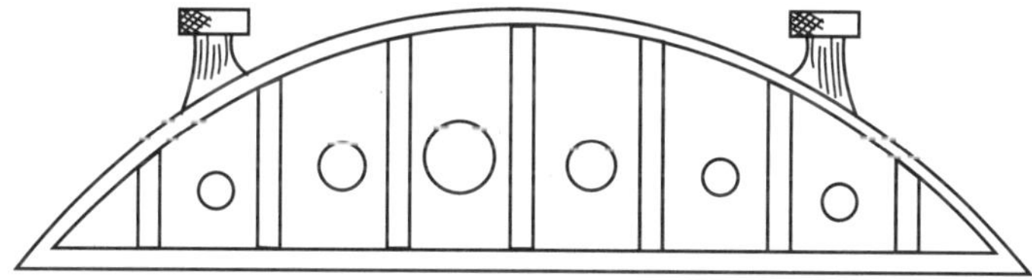

Fig. 16.39. Straight edge.

16·22 Method of Marking

The marking on a work is the most important operation in the bench work, in order to obtain an accurately finished product. It consists of setting out the dimensions on a work from the working drawing or transferring them from a similar part. The following procedure is adopted for marking on a work :

1. First of all, the surface of the work is coated with a paste of chalk or red lead and allowed to dry.
2. If the work is flat, it is laid with its uncoated surface on the face of the surface plate and if it is round, it is held in a V-block and clamped as shown in Fig. 16.34. In case the work is too thin, it is normally supported against an angle plate keeping the surface to be marked at right angles to the surface of surface plate.
3. Now the horizontal lines are first scribed by means of scribing block or surface gauge. The vertical lines are drawn by turning the work through 90° and by using the scriber. If a true surface is available on the edge of the work to be marked, then the horizontal and vertical lines may be marked with the help of a try-square.
4. The centre on the end of a round bar may be located by using a hermaphrodite caliper, surface gauge or a combination set.
5. The circles and arcs on a flat surface are marked by means of a divider.

6. After the scribing work is over, light indentations are made along the scribed lines, centres of circles and arcs, by means of a dot punch and a hammer. These punch marks serve as a guide for further operations like filing, chipping, drilling etc. During these operations, the metal is removed upto half the circumference of these punch marks to ensure that the job has gone undersize.

16·23 Drilling Tools

The drilling tools called drills are used for making round holes in the work. The following three types of drills are commonly used :

1. *Flat drill.* It is a simple type of drill and has two cutting edges bevelled at 60°, as shown in Fig. 16.40.

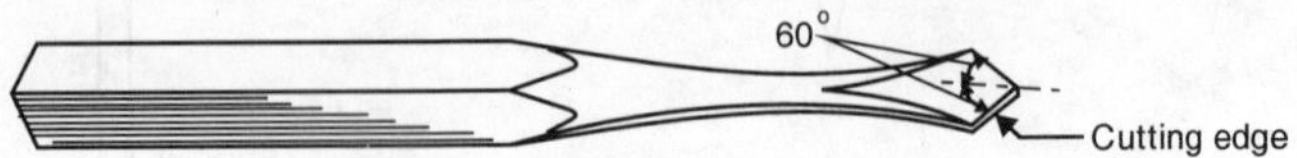

Fig. 16.40. Flat drill.

It is easily made by flattening a round bar of tool steel at one end by forging. It is then ground, hardened and tempered. The flat drill can not be used to drill deep holes because the chips (*i.e.* cut material) do not come out of the hole during the operation. The cutting edge of the drill wear quickly due to the metal chips remaining in the hole and by grinding every time, the diameter gets reduced. Strictly speaking, the flat drill can never be relied upon to give accurate straight hole.

2. *Straight fluted drill.* The straight fluted drill, as shown in Fig. 16.41, is used for making holes in brass, copper and other soft metals. Its only disadvantage is that the chips pack in the flutes.

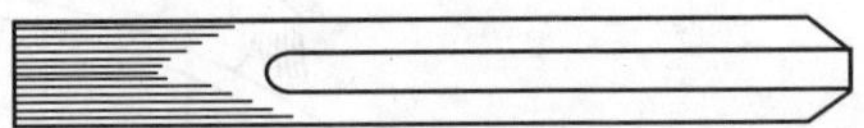

Fig. 16.41. Straight fluted drill.

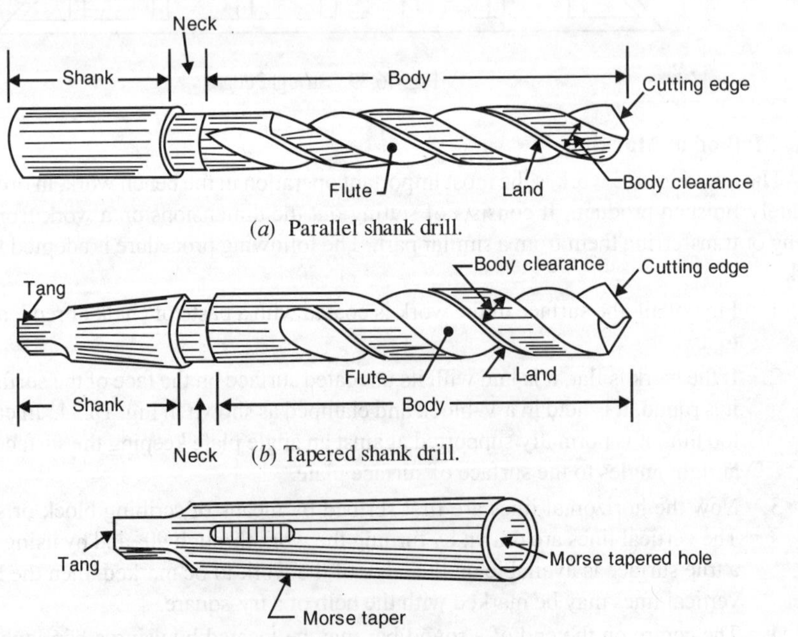

(*a*) Parallel shank drill.

(*b*) Tapered shank drill.

(*c*) Drill sleeve.

Fig. 16.42. Twist drills and drill sleeve.

3. *Twist drill.* It is the most commonly used variety of drill and is made with two, three or four cutting lips. The two lip drill is used to drill holes into solid metal while the three and four lip drill is used for enlarging holes previously drilled. The twist drills, as shown in Fig. 16.42(*a*) and (*b*), has a cylindrical body on which the grooves are cut.

These grooves are called *flutes.* The flutes of a twist drill are always made with a right hand helix. This forces the chips up the flutes and away from the cutting edges. While drilling, the drill is held by the shank. The shank may be parallel or tapered, as shown in Fig. 16.42(*a*) and (*b*) respectively. The parallel shank is provided on small sized drills (*i.e.*, upto 12.5 mm). The tapered shank drills have a taper called the *morse taper.* The size of taper on any particular drill depends upon the drill diameter. These drills are provided with a *tongue* (also called *tang*) at the end of the tapered shank. The drill is held by the shank while drilling. The shank fits into the slot in a drill sleeve, as shown in Fig. 16.42(*c*) or in the machine spindle. It drives the drill and keeps the shank from slipping especially on large drills.

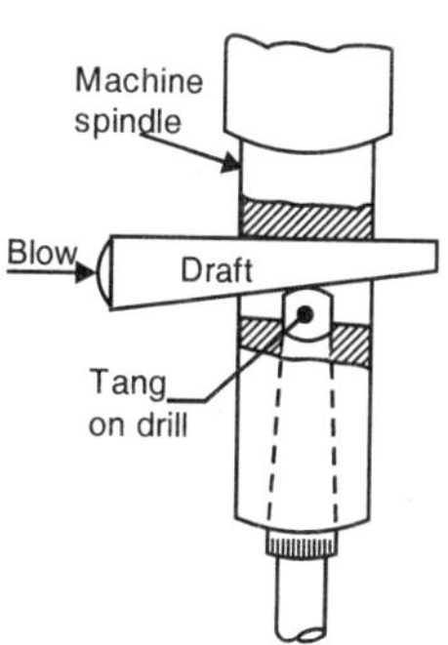

Fig. 16.43. Use of a drift

The drills are taken out from the machine spindle with the help of a drift, as shown in Fig. 16.43. The drills are made of high speed steel containing 18 percent tungsten, 4 percent chromium and 1 percent vanadium.

16·24 Various Angles of Twist Drills

The various angles of twist drills, as shown in Fig. 16.44, are discussed as follows :

1. *Cutting angle or angle of lip.* The two lips must be of the same length and the angles of both lips must be equal. The cutting angle for ordinary work is 59° and varies with the metal to metal.

Notes : **(*a*)** If the cutting angle is more, the drill will not cut the metal easily and will not hold its position centrally because of being too flat.

(*b*) If the cutting angle is less, more power will be needed to turn the drill and the drill will cut at a slower rate due to the longer cutting edges.

(*c*) If the angles on the cutting edges are different, then one cutting edge will wear quickly and the hole will be larger than the drill. This results in the wobbling of spindle and the drill wears out quickly.

2. *Lip clearance angle.* It is the angle formed by grinding away the heel (it is the part of the point) behind the cutting edge. The cone shaped cutting end is the point from the lips and varies from 12 to 15 degrees. This angle may be increased for drilling soft materials under heavy feeds. In cases of drilling the hardest material, this angle is 9 degrees and if it is reduced further, the drill cannot cut into the metal and may break in the centre along the web.

3. *Rake angle.* It is the angle between the flute and the work. It is usually 70 to 75 degrees. This helps to secure the lip over the correct space to curl the chips. If this angle is more, then there will be no edge for cutting and if it is less, the cutting edge is too thin and may break under the strain of the work.

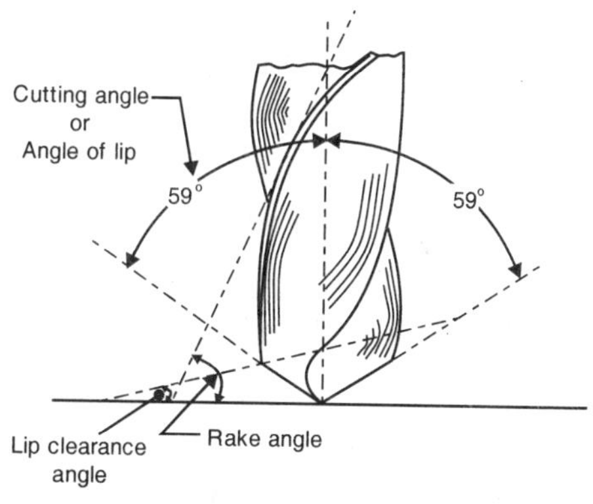

Fig. 16.44. Various angles of twist drills.

16·25 Drilling Machines and Appliances

The most common drilling machine used in fitting shop is a sensitive drilling machine which may be either bench or pedestal mounted type. A vertical drilling machine, as shown in Fig. 16.45, consists of a base, adjustable table electric motor, drill spindle, pulleys, belt and column. The machine is usually provided with four to six speeds. The speed can be varied by changing the position of Vee-belt on the pulleys. The driving motor is generally of 0.5 H.P.

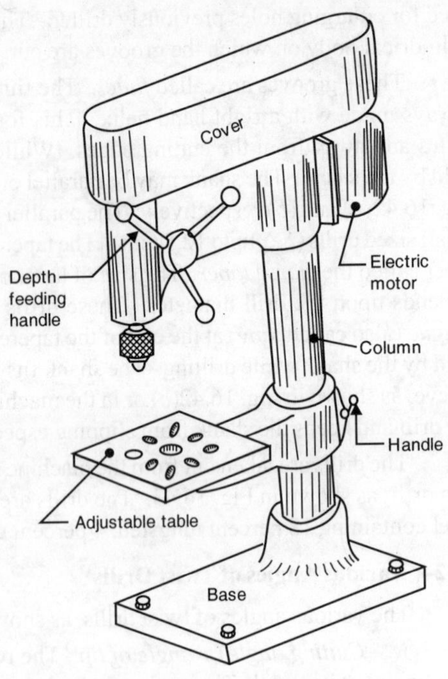

Fig. 16.45. Vertical drilling machine.

The hand-drilling appliances for holes are used where it is not possible to bring the work up to a machine. The following four types of hand-drilling appliances are mostly used.

1. *Hand or wheel brace.* A hand brace, as shown in Fig. 16.46(*a*), is operated by hand and is used only for small drilling operations (*i.e.,* for drilling upto 6 mm diameter). It has a main framework of metal with a wooden top. The centre of the frame carries a bevel wheel and a handle. This bevel wheel is meshed with a small pinion mounted on a spindle in the lower part of the frame. The screw-threaded end of the spindle carries a tapered chuck or drill holder with an opening at the end to accommodate a drill. The chuck contains three sliding jaws held apart by spiral springs. These jaws are tapered on the outside to correspond with the inside of the chuck. When the chuck is screwed further on to the

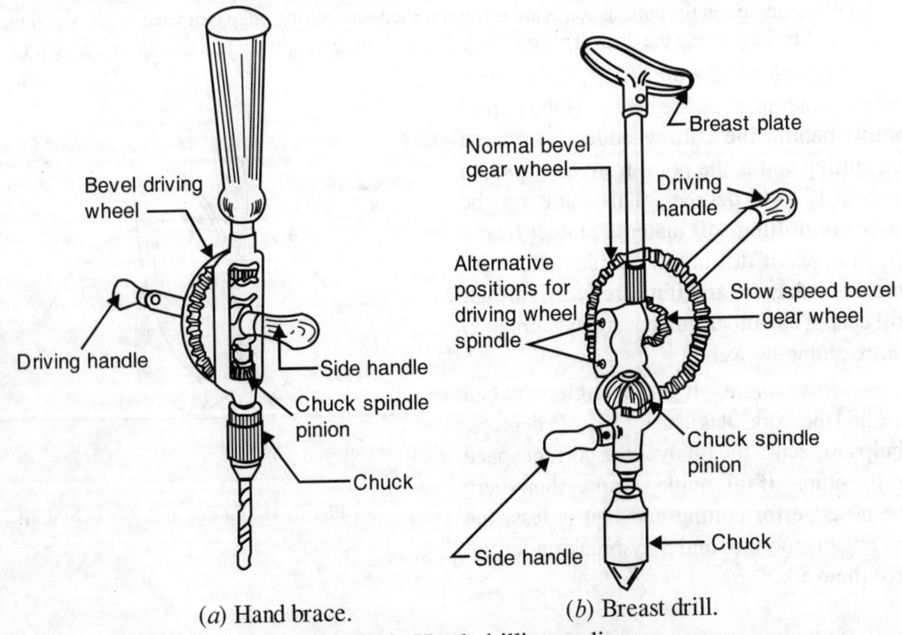

(*a*) Hand brace. (*b*) Breast drill.

Fig. 16.46. Hand drilling appliances.

spindle the jaws are forced towards the opening at the end of the chuck and due to the taper the jaws are made to close and grip the drill. In order to open the jaws, the chuck is turned in the opposite direction releasing the pressure and allowing the springs to open the jaws.

2. *Breast drill.* A breast drill, as shown in Fig. 16.46(*b*), resembles the hand brace and works on the same principle, but it is larger and more sturdy in construction. It may be used for holes upto 12.5 mm diameter. The top of the frame is fitted with a T-shaped metal plate which is curved to fit the chest. The pressure is given by the body, leaving the left hand free to steady the drill by the small handle fitted to the lower part of the frame.

Some breast drills are fitted with an additional smaller bevel wheel to lower the speed at which the spindle rotates. This is an advantage when holes have to be drilled in metals which require slower cutting speeds. In order to change the speed, the spindle of the bevel driving wheel is fitted in the lower hole in the drill frame bringing the small bevel wheel in contact with the chuck spindle pinion.

3. *Ratchet brace.* A ratchet brace, as shown in Fig. 16.47(*a*), has the following advantages, though the drilling a hole with ratchet brace is a slow process :

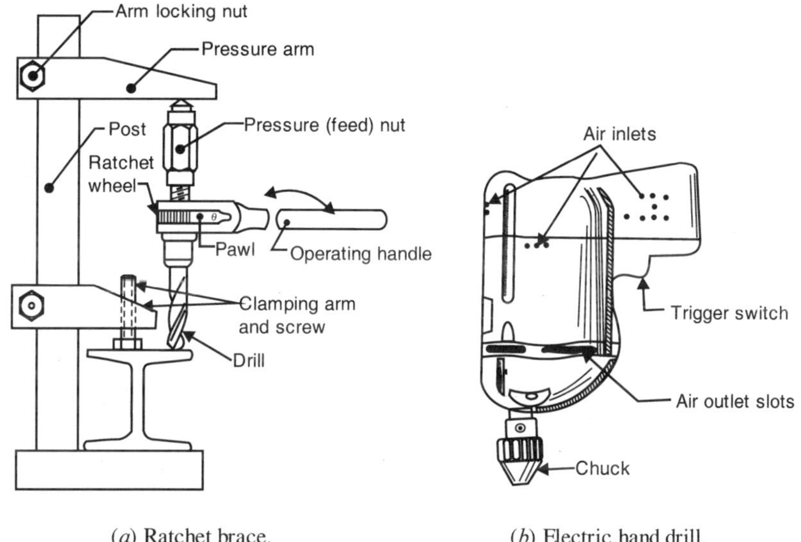

(*a*) Ratchet brace. (*b*) Electric hand drill.

Fig. 16.47. Hand drilling appliances.

1. It can drill holes of large diameter which are beyond the range of a hand or breast drill.
2. It can be used where power is not available.
3. It can be clamped to any suitable support or part of a machine at any suitable angle.

In its operation, the operating handle is moved in a short arc clockwise, which engages the pawl in the ratchet and rotates the drill. The handle is then drawn back to make the next stroke and the pawl slips over the ratchet. The pressure is gradually applied as the drilling proceeds by turning the large hexagonal nut about the ratchet. The taper or square shank drills may be used depending upon the type of drill holder fitted.

4. *Portable electric hand drill.* The portable electric hand drills are used for quickness of drilling operation. The small types of portable electric drills are usually pistol shaped as shown in Fig. 16.47(*b*). The current is passed to a small electric motor contained in the main casing and the armature is extended to drive the spindle and gear assembly carrying the chuck. A trigger or button switch is provided to control the current. The chuck principle is similar to that of hand brace except

that in some types a toothed key is used which engages with the teeth on the chuck to open and close the jaws. A small portable electric drill is used to drill holes upto 6 mm diameter.

The large portable electric drills are similar in construction except that power of the motor is passed through additional gears to reduce the chuck spindle speed. This is necessary because low cutting speeds are required for larger diameter drills.

16·26 Drilling

The operation of making round holes in metal pieces is known as drilling. It is done with the help of a drilling machine. There are many types of drilling machines, but the most common drilling machine used in fitting shop is a vertical drilling machine, as shown in Fig. 16.45.

The drilling operation is carried out as discussed below :

1. First of all, the work is marked with a centre punch at the centre of the hole to be drilled.
2. It is then firmly held in a vice or other suitable clamping device and placed on the adjustable table of the drilling machine.
3. The drill is fitted in the chuck.
4. The spindle with the chuck is lowered by the hand lever and it is ensured that the point of the drill is in exact alignment with the previously marked centre of the hole.
5. The motor is now started and the rotating drill is gradually pressed into the work to produce the desired hole.
6. The pressure should be frequently relieved during the drilling operation, otherwise the cutting edges of the drill will be strained and the drill is damaged.
7. In order to avoid the spoiling of cutting edges of the drill, a coolant such as oil or soap water should be used constantly during the drilling operation. The coolant is not required while drilling in brass and cast iron.

16·27 Reamers

In order to produce a hole having sufficiently good qualities of finish and accuracy for many purposes, and when accurate holes are required, a *reamer* must be used for finishing to size. The reamer does not originate the hole in the same way as the drill, but it merely imparts the necessary smoothness, to the previously drilled hole, parallelism, roundness and accuracy in size. The reaming, however, will not correct any errors which may be in the hole with regard to its position, or direction because the reamer merely follows the previously drilled hole. The various types of reamers commonly used for finishing the drilled hole are as follows :

1. *Hand reamers.* The hand reamers are operated by hand with a *tap wrench fitted on the square end of the reamer, with the work held in the vice. These reamers cut on the flutes which may be straight or helical and are tapered at the end for about one-fourth of its length. The extreme front end diameter is about 0.25 mm less than the reamer size. The shank is slightly less than the diameter of the flutes. A straight fluted hand reamer is shown in Fig. 16.48(*a*). The reamers are supposed to remove the minimum amount of metal *i.e.,* from 0.05 mm to 0.1 mm. If too much metal is allowed to remove by the reamer, then the edges of the reamer rapidly wear and lose their size. The finished hole may also be of over size and inaccurate. The hand reamers are usually made of cast steel.
2. *Machine or spiral fluted reamers.* The solid fluted reamer is similar to the hand reamer but it has a tapered shank. These reamers may have straight flutes, but more often they are cut with left-hand helix, as shown in Fig. 16.48(*b*). The angle made by the flute with the length of the reamer, varies from 4 degrees to 8 degrees. The flutes are made left-hand to counter act the tendency, the reamer would have to screw itself into the hole if the flutes are right-hand and even the straight-fluted reamers

* See Fig. 16.49 (*d*).

have a slight tendency to draw in. The machine reamers are usually made of high speed steel.

3. *Expanding reamers.* When a solid reamer becomes worn, it must either be ground down to the next smaller size or the teeth slightly enlarged by softening the teeth, expanding them slightly by hammering their fronts with a punch, re-hardening and grinding up again to size. In order to avoid this, expanding reamers are used. These are of various types, a common one is shown in Fig. 16.48(*c*). When the reamer has lost its size the diameter of the flutes is increased slightly by springing them open with a tapered pin operated by an adjustable screw, after which they are re-ground to the correct size.

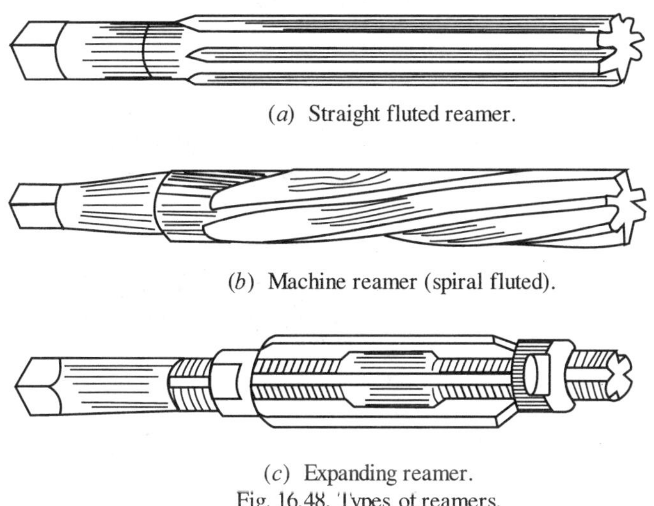

(*a*) Straight fluted reamer.

(*b*) Machine reamer (spiral fluted).

(*c*) Expanding reamer.

Fig. 16.48. Types of reamers.

16·28 Reaming

The reaming is a process of imparting necessary smoothness, parallelism, roundness and accuracy in size, to the previously drilled hole by using a reamer. The reaming does not correct any errors which may be in the hole with regard to its position or direction because the reamer merely follows the previously drilled hole. The hand reaming is mostly done when exactness is required. In its operation, the work is held in a vice and the reamer is placed into the hole which is operated by a tap wrench fitted on the square end of the reamer. Before starting the reaming, it must be ensured that the reamer is square with the axis of the hole being reamed. In order to obtain best results, reamer should be given a slow and steady motion and lubrication may be used. The reamer is turned only in clockwise direction even when it is removed from the hole.

16·29 Taps

A tap is used for cutting internal threads into a hole. The tap is provided with cutting edges and hardened, so that when it is screwed into a hole it cuts an internal thread to fit an external thread of the same size. The taps are made of high carbon or high speed steel by turning a screw and then cutting flutes along it in order to provide cutting edges which is used to screw it into the hole. After hardening and tempering, the flute at the front of the threads is ground up to provide the necessary sharpness on the threads for cutting. The shank of the tap is left plain and the end is squared to accommodate the tap wrench, as shown in Fig. 16.49(*d*).

The taps are usually made in sets of three to cut any particular size. These are called taper, intermediate (or second) and bottoming (or plug). The *taper tap* as shown in Fig. 16.49(*a*) has its leading end tapered off for a length of 8 to 10 threads and this is the first tap to use. The tapered end permits the tap to enter the hole and cut to the full thread gradually. The *second tap,* as shown in Fig.

16.49(*b*), has only two or three threads chamfered and follows the taper tap. If a hole is open at both ends, this tap is suitable for finishing the thread in it. The thread on the *bottoming or plug tap,* as shown in Fig. 16.49(*c*), runs to its extreme end and this tap should be used as the final tap, when a full thread must be cut to the bottom of a blind hole (*i.e.*, only open at one end).

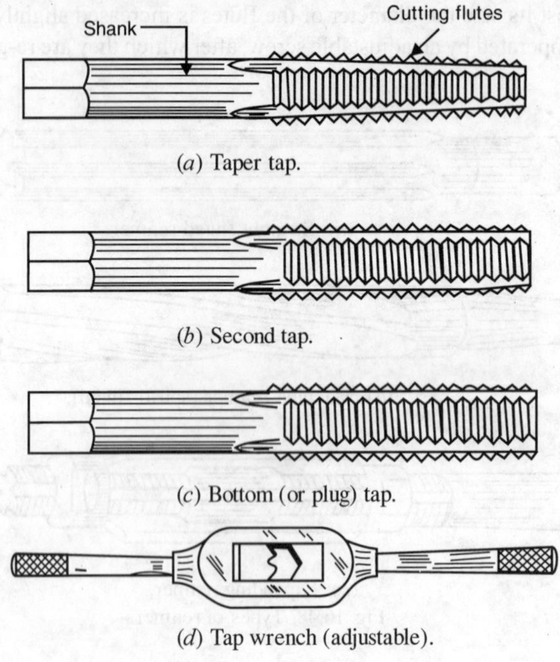

(*a*) Taper tap.

(*b*) Second tap.

(*c*) Bottom (or plug) tap.

(*d*) Tap wrench (adjustable).

Fig. 16.49. Taps and tap wrench.

16·30 Tapping

The process of cutting internal threads into a drilled hole by using a tap is known as tapping. For tapping, the following procedure is adopted :

1. First of all, a hole is drilled to a diameter smaller than the outside diameter of the thread on the tap *i.e.*, equal to the core diameter. The diameter of the hole to be drilled is called *tapping size.* Mathematically,

$$\text{Tapping size} = \text{Outside diameter} - 2 \times \text{Depth of thread}$$

where

$$\text{Depth of thread} = 0.64 \times \text{Pitch of thread}$$

2. After drilling the tapping hole, the taper tap is fixed in the tap wrench and screwed in the hole. Before commencing to screw it round for cutting the thread, the tap position must be adjusted until it stands square with the tap surface of the work and it must be maintained square.

3. For all materials except cast iron a little whale oil on the tap is applied to lubricate its action and improve the finish of the threads.

4. When the taper tap is felt to have started its work and its squareness has been checked, the cutting of threads proceeds. It should be kept in mind that the tap is not turned continuously, but after every half-turn it should be reversed slightly to clear the threads.

5. When the hole is straight through, the reduction of resistance indicates the taper tap is cutting a full thread and it may be removed from the hole which may be finished with the second tap.

6. When a blind hole is being tapped, the tap should be withdrawn from time to time and the

metal cleared from the bottom of the hole. When the end of the tap reaches the bottom of the hole, some resistance will be felt and no further force should be used at this point as the tap may be broken or the thread spoiled. Remove the taper tap and take the second one down as far as it will go and finally threads are cut at the bottom of the hole with a plug tap.

Note : When large number of holes are to be tapped, the hand tapping is a slow process. In such cases, the tapping is done by driving the tap with the drilling machine running on slow speed. This is called machine tapping.

16·31 Dies and Stocks

A tool used for cutting external threads on bars or tubes is called a *die*. It consists of a nut having portions of its thread circumference cut away and shaped to provide cutting edges to the remaining portions of the threads. After hardening and sharpening of the cutting edges, this is screwed on to the bar upon which the thread is to be cut. In order to hold and manipulate the die it is carried in the centre of a pair of operating handles called *stocks*.

A *solid die nut,* as shown in Fig. 16.50(*a*), is not usually employed for cutting threads from the solid but is used to rectify damage and knocks to existing screws such as on studs or bolts. They have usually five or more cutting edges and there is no means for adjustment.

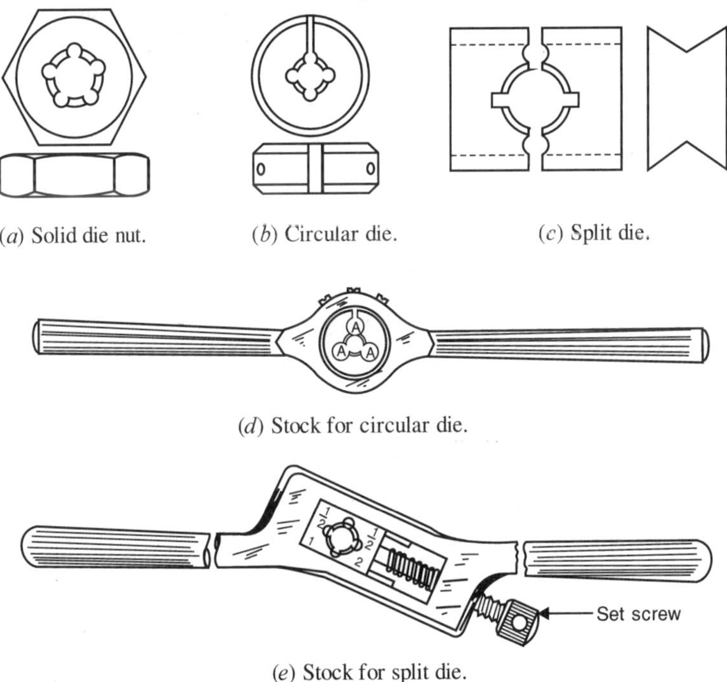

(*a*) Solid die nut. (*b*) Circular die. (*c*) Split die.

(*d*) Stock for circular die.

(*e*) Stock for split die.

Fig. 16.50. Dies and stocks.

The *circular dies,* as shown in Fig. 16.50(*b*), cut the threads in one operation and permits a certain amount of adjustment in the size that the die will cut. This is done by a small amount of opening or closing of the die with the help of screws in the stocks, as shown in Fig. 16.50(*d*).

The *split dies,* as shown in Fig. 16.50(*c*), consist of a pair of dies or jaws which fit into the stocks, as shown in Fig. 16.50(*e*), and are clamped by a screwed ring. These dies slide and may be adjusted by screws which bear against their outer faces. This permits the dies to be set a small amount open while the first cut is taken down a bar and closed into the correct size for the final finishing cut.

16·32 Dieing

The process of cutting external threads on the round stems of objects like bolts, studs and pipes by using a die and stock, is known as *dieing*. The following procedure is adopted for dieing :

1. First of all, the work is cleaned of hard scale (if any) and its end is slightly chamfered so that it can enter into a die easily.
2. The work is held in a vice or other suitable clamping device.
3. A die of suitable size is selected and held in a stock.
4. Before starting the cutting of thread, great care is necessary to ensure that the face of the die is square with the axis of the bar or work being threaded. The stock is now rotated with both hands and at the same time the die is pressed on to the end of the bar to help the commencement of cutting.
5. When the cutting has actually started, the die should be rotated backwards and forwards in a similar manner as for tapping. During operation, the threads should be supplied with some form of cutting oil.
6. The required depth of thread is not obtained in a single operation. Thus when sufficient length of the thread is made, the stock is screwed back and the jaws of the die are brought closer by tightening the adjusting screw. Now the process is repeated twice or thrice till the correct depth of thread is reached. However, if a solid die is used, the cutting operation is complete in one operation only, but it does not give satisfactory results, particularly in large work.

QUESTIONS

1. Sketch and describe the following hand tools :

(*a*) Bench vice, (*b*) Hand vice, and (*c*) Pipe vice.

2. Explain the following with the help of neat sketches :

(*a*) Ball peen hammer, (*b*) Straight peen hammer,

(*c*) Cross peen hammer, and (*d*) Soft hammers.

3. Draw the sketch of a cold chisel and name its various parts.

4. How are chisels classified ?

5. Explain, with the help of neat sketches, the various types of files used in fitting.

6. What are the methods commonly used for filing ?

7. What do you understand by pinning of files ?

8. Explain, with the help of neat sketches, the various types of scrapers. How are they used ?

9. Sketch the two types of hacksaw frames and explain them clearly.

10. Discuss the operation of sawing.

11. Write short notes on the following :

(*a*) Combination set, (*b*) Inside and outside calipers,

(*c*) Scribing blocks, (*d*) V-block,

(*e*) Surface plate, (*f*) Punches,

(*g*) Trammel, and (*h*) Straight edge

12. Sketch and describe flat drill and twist drill. Name the various angles of twist drill.

13. Why the reamers are used ? Explain, with neat sketches the different types of reamers.

14. Describe, with neat sketches, the taps and dies. How the internal threads are cut with taps ?

15. Discuss the procedure for cutting external threads on a pipe.

OBJECTIVE TYPE QUESTIONS

1. The cold chisels are made by
   (*a*) drawing (*b*) rolling (*c*) piercing (*d*) forging
2. The cold chisels are made from
   (*a*) cast iron (*b*) mild steel (*c*) cast tool steel (*d*) high speed steel
   (*e*) stainless steel (*f*) any one of the above
3. The cross-section of a chisel is usually
   (*a*) rectangular (*b*) square (*c*) hexagonal (*d*) octagonal
   (*e*) circular (*f*) triangular
4. A diamond pointed chisel is used for cutting
   (*a*) flat surfaces (*b*) grooves (*c*) keyways (*d*) V-shaped grooves
   (*e*) all of the above
5. A file with 20 teeth in 25 mm is called
   (*a*) rough file (*b*) bastard file (*c*) second cut file (*d*) smooth file
   (*e*) dead smooth file (*f*) none of these
6. When the file is moved to and fro over the work, it is known as
   (*a*) cross filing (*b*) draw filing (*c*) pull and push filing (*d*) none of these
7. When filing soft metals, the file teeth are clogged with minute particles of metal. The file should be cleaned by
   (*a*) washing it with water (*b*) rubbing on wood
   (*c*) washing it with dilute acid (*d*) using file card
8. A file removes the metal during
   (*a*) forward stroke (*b*) return stroke
   (*c*) both forward and return strokes (*d*) none of the above
9. A hacksaw blade cuts on the
   (*a*) forward stroke (*b*) return stroke
   (*c*) both forward and return strokes (*d*) cutting depends upon the direction of force
10. The teeth of hacksaw blade are bent
    (*a*) towards right (*b*) towards left
    (*c*) alternately towards right and left and every third or fourth left straight
    (*d*) may be bent in any direction
11. A hacksaw blade is specified by its
    (*a*) length (*b*) material (*c*) width (*d*) number of teeth
    (*e*) none of the above
12. To prevent the body of the blade from jamming in the saw cut, the teeth of blade are
    (*a*) strengthened (*b*) sharpened (*c*) set (*d*) all of the above
    (*e*) none of the above
13. The type of file used for a wood work is
    (*a*) single-cut file (*b*) double-cut file (*c*) rasp-cut file (*d*) any one of these
14. V-block is used to
    (*a*) check the trueness of flat surface (*b*) locate centres of round rods
    (*c*) check the surface roughness (*d*) none of the above
15. Scribing block is used to
    (*a*) hold the round bars during marking (*b*) check the trueness of flat surfaces
    (*c*) locate the centres of round bars (*d*) check the surface roughness

ANSWERS

1. (*a*)	2. (*c*)	3. (*d*)	4. (*d*)	5. (*a*)
6. (*b*)	7. (*d*)	8. (*a*)	9. (*a*)	10. (*c*)
11. (*a*)	12. (*c*)	13. (*c*)	14. (*d*)	15. (*c*)

17

Sheet Metal Work

1. Introduction. 2. Metals Used in Sheet Metal Work. 3. Sheet Metal Tools. 4. Sheet Metal Operations. 5. Marking. 6. Cutting. 7. Notching. 8. Bending. 9. Riveting. 10. Soldering. 11. Folding Edges. 12. Seam Making. 13. Hollowing or Blocking. 14. Sinking. 15. Raising. 16. Planishing. 17. Sheet Metal Machines. 18. Laying Out a Pattern. 19. Parallel Line Method of Laying Out a Pattern. 20. Radial Line Method of Laying Out a Pattern. 21. Triangulation Method of Laying Out a Pattern.

17·1 Introduction

The sheet metal shop is very important for every engineering concern. It deals with the working of metal sheets. It requires a thorough knowledge of projective geometry particularly the development of surfaces, because the laying out of pattern and cutting of metal sheets to correct sizes and shapes entirely depends upon the knowledge of the workman. The various operations performed in a sheet metal shop are cutting, shearing, bending etc. In this chapter, we shall discuss the sheet metal tools and the various processes.

17·2 Metals Used in Sheet Metal Work

The following metals are generally used in sheet metal work :

1. *Black iron sheet.* It has a bluish-black appearance and is often referred to as uncoated sheet. Since it is uncoated, therefore, it corrodes rapidly. The use of this metal is limited to articles that are to be painted or enameled such as stove pipes, tanks and pans.

2. *Galvanised iron.* It is a soft steel coated with molten zinc. The zinc coating resists rust, improves the appearance of the metal, and permits it to be soldered with greater ease. The galvanised iron sheet is used extensively in fabricated products such as pans, buckets, furnaces, heating ducts, cabinets, gutters and in many other articles.

3. *Copper.* It is a reddish coloured metal and is extremely malleable and ductile. It is used extensively in the electrical field. Since it does not deteriorate rapidly when exposed to the atmosphere, therefore, this metal is employed frequently in the building trades for water pipe, roofing, gutters and other parts of buildings.

4. *Aluminium.* It is a silvery white coloured metal and has many qualities like high ratio of strength to weight, corrosion resistant qualities and ease in fabrication. Now-a-days it is used in the manufacture of number of products such as household appliances, refrigerator trays, lighting fixtures, windows, duct work, in construction of airplanes, in the building trades, and in many electrical and transportation industries.

5. *Stainless steel.* It is an alloy steel possessing the ability to resist corrosion without any surface coating. One important type of stainless steel contains 18 percent chromium and 8 percent nickel. This steel is commonly known as 18/8 steel. It is widely used in building stream lined trains, food handling equipments, kitchenwares and in many other applications which require great strength and resistant to corrosion.

The stainless steel sheets can be formed, bent, drilled and soldered in the same manner as other types of sheet metal. Since stainless steel has greater tensile strength and hardness, therefore, sheets of this metal are slightly more difficult to work.

6. *Tin plate.* It is a steel sheet coated with pure tin. This metal has a very bright silvery appearance and is used principally in making food containers, cans and pans.

7. *Terne plate.* It is a thin steel coated with a mixture of molten tin and lead. This metal has dull appearance and is used for roofing and tanks. Since the lead coating is poisonous, therefore, it should not be used for containers that are to hold foods.

Note : The thickness of the metal sheets is indicated by series of numbers called *gauge numbers.* Each gauge designates a definite thickness. The thickness of sheet in inversely proportional to gauge number *i.e.*, larger the gauge number, lesser is the thickness.

17·3 Sheet Metal Tools

The tools commonly used in sheet metal work are as follows :

1. *Rules.* The rules are available in a variety of lengths and types, each of which is designed for measuring and laying out different work. The following rules are generally used in sheet metal shop :

(*a*) *Steel rule, folding rule and steel tape.* The steel rule (30 cm long), as shown in Fig. 17.1(*a*), is particularly useful in measuring and laying out small work. The folding rule (2 metres long) and the steel tape as shown in Fig. 17.1(*b*) and (*c*) respectively are very helpful in measuring and laying out large work. In order to measure accurate distance, the rule should be placed on its edge so that the graduations are in actual contact with the metal.

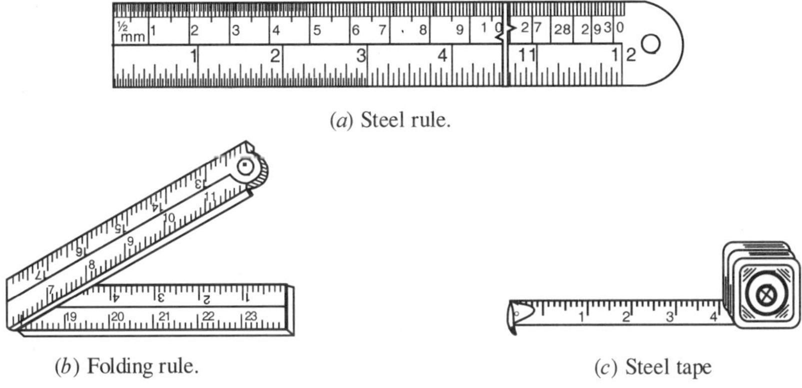

(*a*) Steel rule.

(*b*) Folding rule.

(*c*) Steel tape

Fig. 17.1. Rules.

(*b*) *Circumference rule.* The circumference rule is also used for measuring, laying out or as a straight edge. In addition to this, the rule also shows the circumference of circles. The top edge is graduated in centimeters for regular measuring and the lower half is graduated to indicate circumferences.

2. *Steel square.* It is a L-shaped piece of hardened steel with graduation marks on the edges for measuring. The narrow arm of the square is called the *tongue* and the wide part is known as the *body,* as shown in Fig. 17.2.

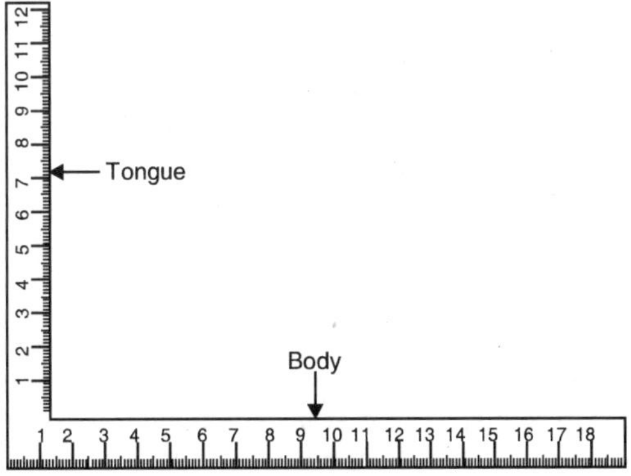

Fig. 17.2. Steel square.

Swinging blade protractor. The swinging blade protractor, as shown in Fig. 17.3, used in sheet metal work are made of steel. These are used for marking or measuring angles.

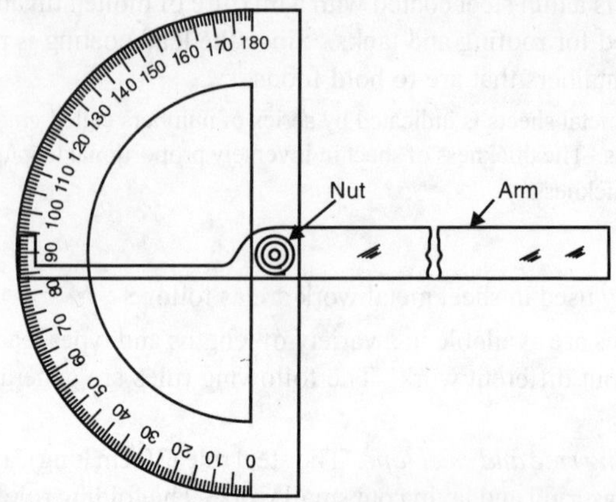

Fig. 17.3. Swinging blade protractor

4. *Straight edge.* It is a flat bar of steel with a bevelled edge. This bar comes in a variety of lengths ranging from 1 metre to 3 metre. It is useful for drawing long lines.

5. *Scratch awls or scribers.* They are used to scribe or mark lines on a metal surface for a variety of purposes in laying out patterns. The following are the three common types of scratch awls.

(*a*) *Ring scratch awl.* It is a solid steel rod about 5 mm in diameter and 150 mm to 200 mm long. It has a sharp tapered point at one end and a ring at the other end, as shown in Fig. 17.4(*a*).

(*b*) *Socket scratch awl.* It has a steel blade about 150 mm and provided with a replaceable wooden handle as shown in Fig. 17.4(*b*).

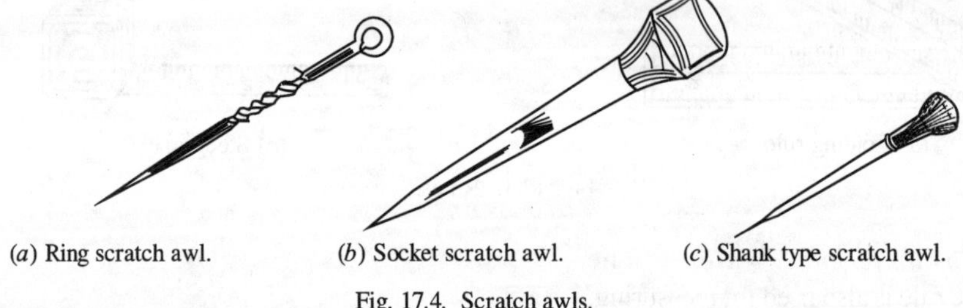

(*a*) Ring scratch awl. (*b*) Socket scratch awl. (*c*) Shank type scratch awl.

Fig. 17.4. Scratch awls.

(*c*) *Shank type scratch awl.* For general purposes, this type of scratch awl is preferred by most sheet metal workers because the steel blade passes through the handle which reinforces the top. The shank type scratch awl is shown in Fig. 17.4(*c*).

6. *Dividers.* The dividers are made with each leg tapered to a needle point. The two types of the dividers are the spring divider and the wing divider as shown in Fig. 17.5(*a*) and (*b*) respectively. The legs of the spring divider are adjusted by turning the knurled nut. The adjustment, on the wing divider is made by loosening the screw on the wing and then tightening the knurled nut on the end of the wing.

The dividers are available in number of sizes and types. These are used to space off equal distances, to divide lines into equal parts and to draw arcs and circles.

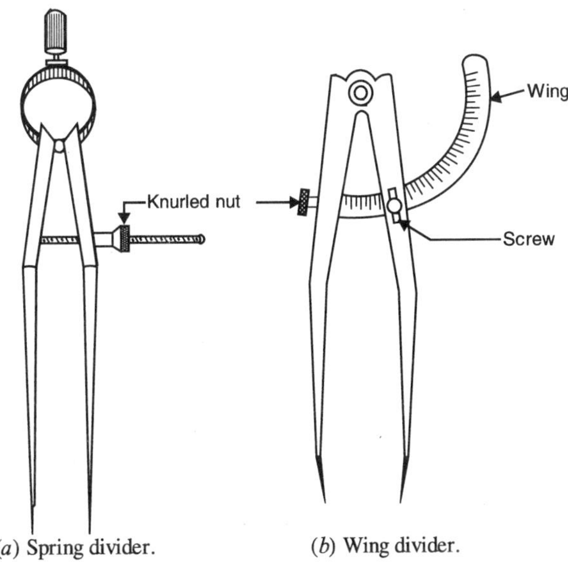

(*a*) Spring divider. (*b*) Wing divider.

Fig. 17.5. Dividers.

7. *Trammel points.* The trammel points (sometimes called a beam compass), as shown in Fig. 17.6, consists of two straight, removable legs tapered to needle points and attached to separate heads or holders. These heads or holders slide on wood or steel bars or beams and are held in place by thumb screws. A special clamp for a pencil can be attached to one of the points. The trammel points are used to draw large arcs and circles that are beyond the limit of dividers.

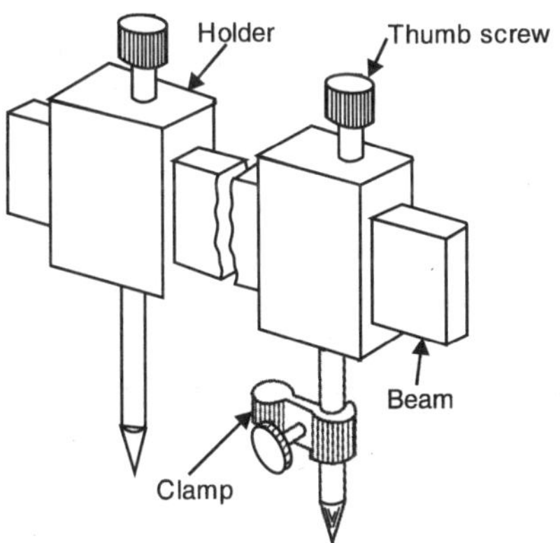

Fig. 17.6. Trammel points.

8. *Punches.* The various hand punches commonly used in sheet metal shop are as follows :

(*a*) *Prick punch.* The prick punch, as shown in Fig. 17.7(*a*), is a tool steel rod whose one end has a tapered point ground to approximately an included angle of 30°. It is used for making small indentations or establishing points for dividers and trammel points.

(*b*) *Centre punch.* The centre punch, as shown in Fig. 17.7(*b*), is similar to prick punch, but its point is ground to an angle of approximately 90°. It is used to mark the location of bend lines on heavy metal and to mark the centres of holes to be drilled.

(*c*) *Solid punch.* The solid punch, as shown in Fig. 17.7(*c*), is used for punching small holes in thin metal sheets.

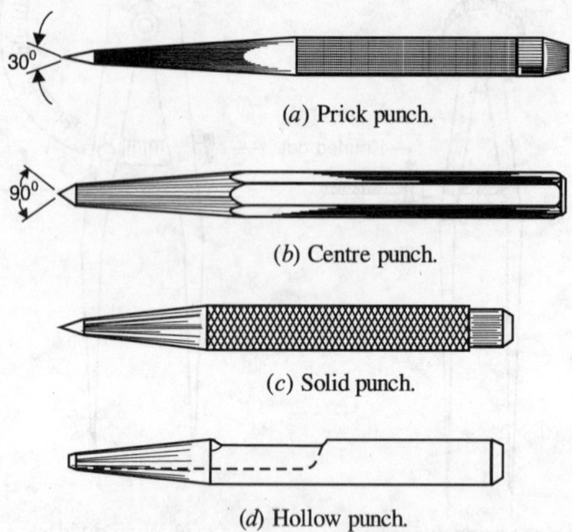

(*a*) Prick punch.

(*b*) Centre punch.

(*c*) Solid punch.

(*d*) Hollow punch.

Fig. 17.7. Various types of punches.

(*d*) *Hollow punch.* The hollow punch, as shown in Fig. 17.7(*d*) is used for punching holes upto 10 mm or above from metal sheets. The inner and outer faces of the punch meet at an angle of 40°. In order to avoid clipping the edges of the hollow punch, the metal sheets should be placed over a block of lead.

(*e*) *Hand lever punch.* A hand lever punch, as shown in Fig. 17.8, is sometimes used for making holes when too much holes are to be punched. The tool consists of a punch and die and is operated by hand. The die and punch may be replaced depending upon the size of the hole required.

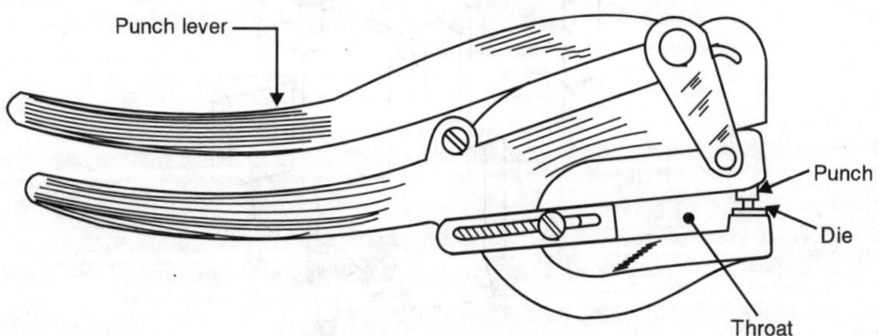

Fig. 17.8. Hand lever punch.

9. *Chisels.* The cold chisels are used to cut or shear metal. It is made from a piece of high carbon or alloy steel of hexagonal or octagonal in shape.

There are many different types of chisels, but the flat chisel, as shown in Fig. 17.9, is mostly used for cutting sheet metal, rivets, bolts and in chipping operations. The flat chisel has a tapered end with a sharp cutting edge which is properly hardened and tempered. The edge is bevelled to an angle of 40° to 45° for cutting sheet metal. The cutting edge is also ground to a slight curve to prevent the chisel from 'digging in' at the corners of metal being cut.

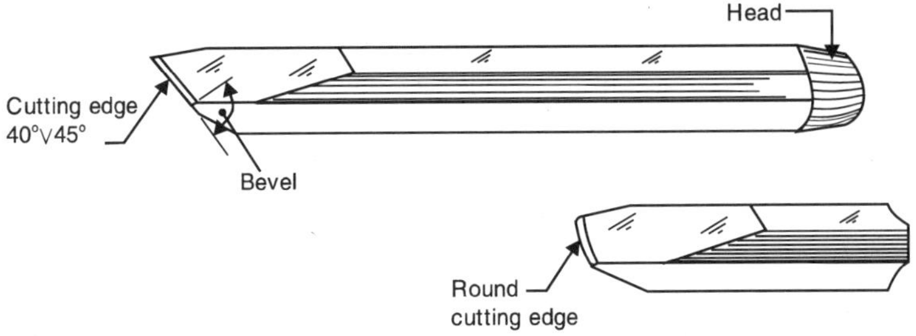

Fig. 17.9. Chisel.

10. *Snips.* The snips are somewhat similar to a pair of scissors but are considerably heavier. There are several types of snips available for making straight or circular cuts, but the most common are the straight snips and the curved snips.

The *straight snips,* as shown in Fig. 17.10(*a*), have straight blades for straight line cutting. These snips may be obtained in various sizes.

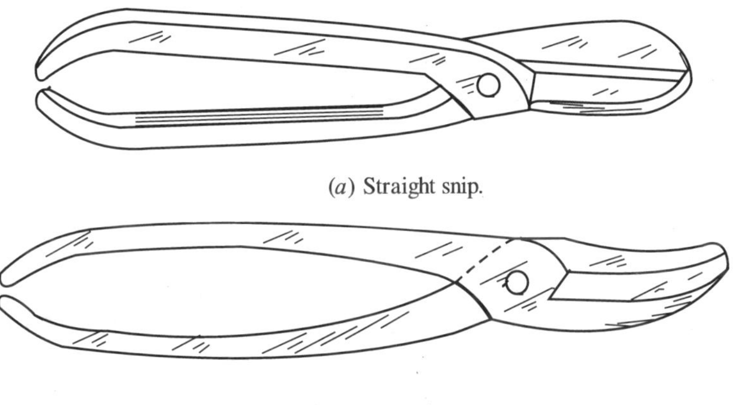

(*a*) Straight snip.

(*b*) Curved snip.

Fig. 17.10. Types of snips.

The *curved snips,* as shown in Fig. 17.10(*b*), have curved blades for making circular cuts. They are available for either right hand or left hand cuts.

11. *Hammers.* The hammers, in sheet metal work, are used for forming shapes by hollowing, raising, stretching or throwing off processes. There are many types of hammers, but the most commonly used hammers, in sheet-metal work, are as follows :

(*a*) *Ball peen hammer.* The ball peen hammer, as shown in Fig. 17.11(*a*), has a round, slightly curved face and a round head. It is a general purpose hammer.

(*b*) *Riveting hammer.* The riveting hammer, as shown in Fig. 17.11(*b*), has a square, slightly curved face with bevelled edges to prevent the head of the hammer from marking the metal. The peen side is double tapered and has a slightly rounded end. It is used for spreading rivets and for hammering a rivet set.

(*c*) *Setting hammer.* The setting hammer, as shown in Fig. 17.11(*c*) has a square, flat face and a tapered peen with bevelled end. The flat face is used for flattening seams without damage to the metal while peen end is used for peening operation.

(*d*) *Hollowing or blocking hammer.* The hollowing or blocking hammer, as shown in Fig. 17.11(*d*), has a dome face without any sharp corners. It is used for hollowing discs into bowl shapes.

(*e*) *Raising hammer.* The raising hammer, as shown in Fig. 17.11(*e*), has an oblong flat face with corners slightly rounded off. It is used in raising circular discs and many other raising and bumping operations.

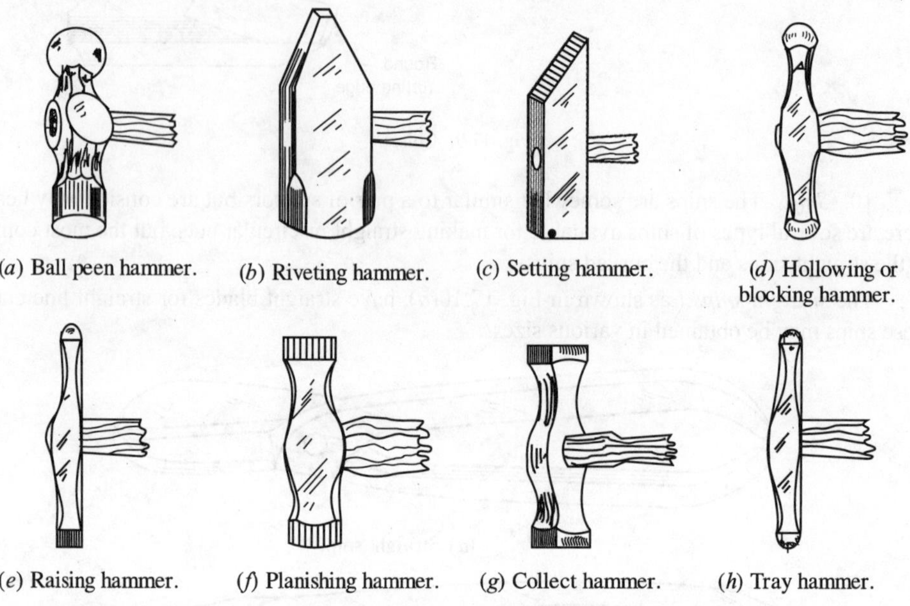

(*a*) Ball peen hammer. (*b*) Riveting hammer. (*c*) Setting hammer. (*d*) Hollowing or blocking hammer.

(*e*) Raising hammer. (*f*) Planishing hammer. (*g*) Collect hammer. (*h*) Tray hammer.

Fig. 17.11. Types of hammers.

(*f*) *Planishing hammer.* The planishing hammer, as shown in Fig. 17.11(*f*), has a round face. It is used on domed circular work.

(*g*) *Collect hammer.* The collect hammer, as shown in Fig. 17.11(*g*), has an oblong shaped faces. It is used on cylinders and in curved collects.

(*h*) *Tray hammer.* The tray hammer, as shown in Fig. 17.11(*h*), has oval shaped faces. It is used to sink the bottom when shaping a tray.

12. *Mallets.* The mallets may be made from hide, fibre or wood. The best size of mallet is 5 cm diameter. These may be obtained in various shapes to suit special work, as shown in Fig. 17.12.

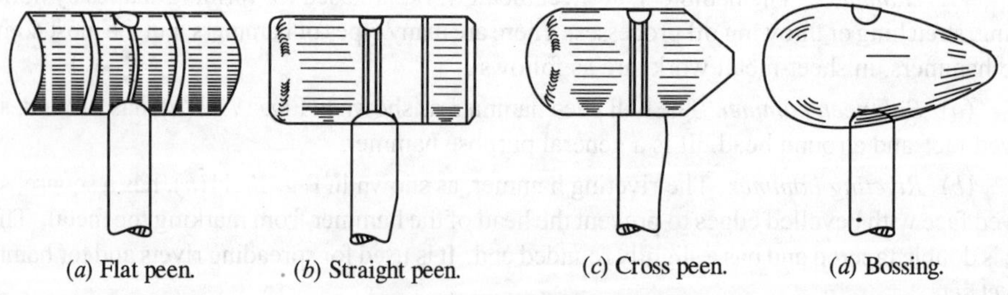

(*a*) Flat peen. (*b*) Straight peen. (*c*) Cross peen. (*d*) Bossing.

Fig. 17.12. Mallet forms.

13. *Pliers.* The pliers are used in sheet metal work for holding, cutting and bending work. The various types of pliers commonly used are as follows :

(*a*) *Flat nose plier.* The flat nose plier, as shown in Fig. 17.13(*a*), has flat jaws with small grooves. It is used for forming and holding work.

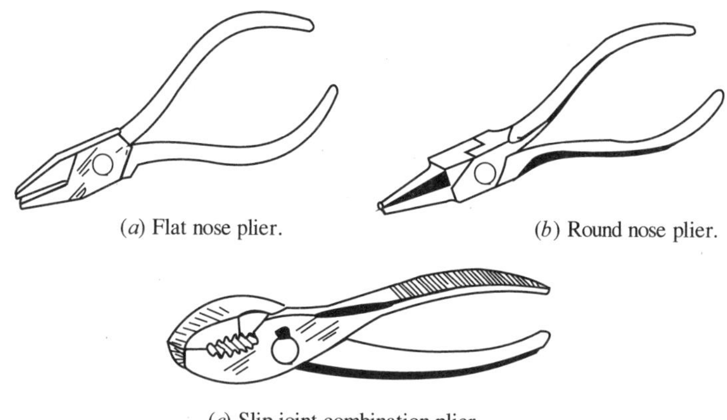

(*a*) Flat nose plier.

(*b*) Round nose plier.

(*c*) Slip joint combination plier.

Fig. 17.13. Types of pliers.

(*b*) *Round nose plier.* The round nose plier, as shown in Fig. 17.13(*b*), has long jaws rounded on the outside. It is used for holding and forming the various shapes and patterns.

(*c*) *Slip-joint combination plier.* The slip-joint combination plier, as shown in Fig. 17.13(*c*), has an adjustable jaw. It is a general purpose tool.

14. *Hacksaws and files.* The hacksaws and files used in sheet metal shop are similar to those used in bench work and fitting (Chapter 16).

15. *Bench stakes.* The stakes are steel anvils used to perform operations like bending, seaming or forming on sheet metal, when suitable machines are not available. The stakes are available in various shapes and sizes. The stakes consist of three parts, *i.e.,* the shank, the head and the horn as shown in Fig. 17.14. The shanks of the stakes are generally standard, the head and horn are available in various shapes and sizes. The different types of stakes and the operations performed by them are discussed in detail, as follows :

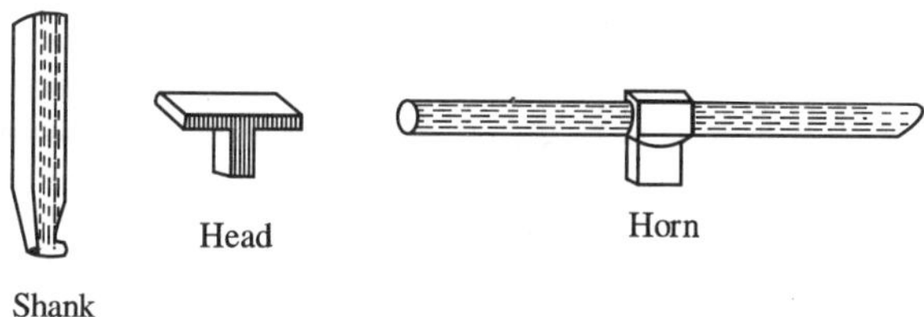

Fig. 17.14. Parts of a stake.

(*a*) *Blow horn stake.* The blow horn stake, as shown in Fig. 17.15(*a*), has a short tapered horn at one end and a long tapered horn at the other end. It can be held in a vice or in a square hole in a wooden block or bench top. It is used in forming, riveting or seaming tapered objects such as funnels, pitched covers etc.

(*b*) *Beak horn stake.* The beak horn stake, as shown in Fig. 17.5(*b*), has a round tapered horn at one end and a square tapered horn at the other end. The round tapered end may be used for many purposes like bending of sheet metal into cylinders and conical shapes, for setting down seams when used with a grooving punch or a seam set, for truing work to make circular when the seam is folded or soldered etc. The squared end of the stake is used for bending and truing straight sided work.

(*c*) *Candle mould stake.* The candle mould stake, as shown in Fig. 17.15(*c*), has two horns of different tapers. It is used in forming, riveting and seaming long, flaring articles.

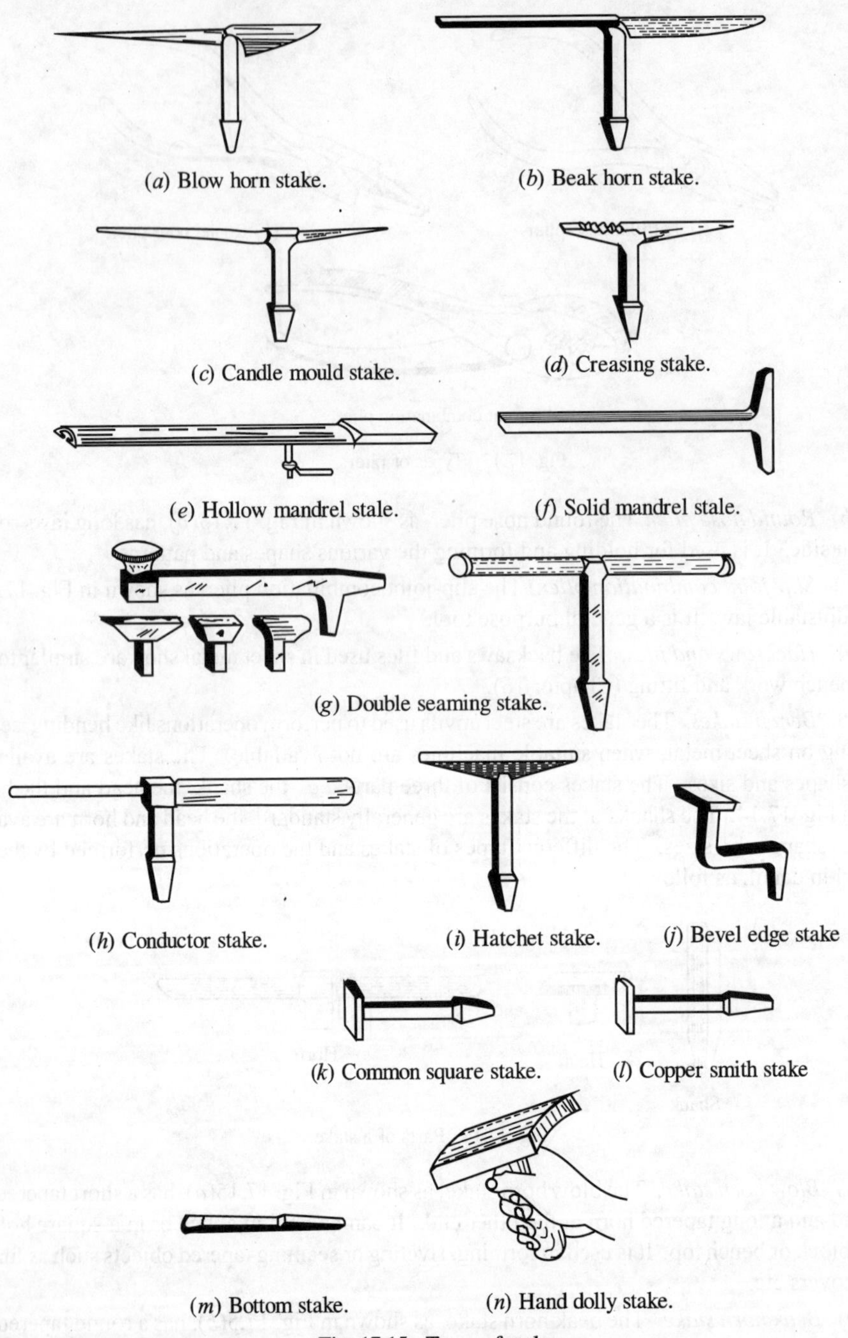

(*a*) Blow horn stake.
(*b*) Beak horn stake.
(*c*) Candle mould stake.
(*d*) Creasing stake.
(*e*) Hollow mandrel stale.
(*f*) Solid mandrel stale.
(*g*) Double seaming stake.
(*h*) Conductor stake.
(*i*) Hatchet stake.
(*j*) Bevel edge stake
(*k*) Common square stake.
(*l*) Copper smith stake
(*m*) Bottom stake.
(*n*) Hand dolly stake.

Fig. 17.15. Types of stakes.

(*d*) *Creasing stake.* The creasing stake, as shown in Fig. 17.15(*d*), has a square horn with grooved slots at one end and a tapered round horn at the other end. The slots in the square horn are used for wiring and beading and the round horn is used for forming conical shaped pieces.

(*e*) *Hollow mandrel stake.* The hollow mandrel stake, as shown in Fig. 17.15(*e*), has a single horizontal metal piece, one end of which has a rectangular surface while the other end has a round section. The rounded length has a slot which permits the stake to slide on a bolt so that it can be fastened in different positions on the bench. It is used for riveting and seaming pipes. The rectangular end is used for forming laps, riveting and double seaming corners of pans, boxes etc.

(*f*) *Solid mandrel stake.* The solid mandrel stake, as shown in Fig. 17.15(*f*), has a double shank so that the rounded or flat side can be used to perform operations similar to those of hollow mandrel stake.

(*g*) *Double seaming stake.* The double seaming stake with four heads, as shown in Fig. 17.15(*g*), has a horizontal bar with a hole at one end into which any of the four different heads may be inserted. The bar has a double shank so arranged that the stake may be used either horizontally or vertically. It is used for double seaming large work.

Another type of double seaming stake, as shown in Fig. 17.15(*g*), has two elliptical horns with two enlarged knobs at the ends. It is generally used for double seaming small cylindrical articles.

(*h*) *Conductor stake.* The conductor stake, as shown in Fig. 17.15(*h*), has two cylindrical horns of different diameters. It is used for forming , riveting and seaming tubes and small-sized pipes.

(*i*) *Hatchet stake.* The hatchet stake, as shown in Fig. 17.15(*i*), has a horizontal sharp straight edge bevelled along one side. It is used for making sharp bends, bending edges and forming pans and boxes by hand.

(*j*) *Bevel edge stake.* The bevel edge stake, as shown in Fig. 17.15(*j*), has a flat, square head with a bevel edge on the outside of the head for double seaming. It also has an offset shank which permits the work to clear the bench.

(*k*) *Common square stake.* The common square stake, as shown in Fig. 17.15(*k*), has a square head with a long shank. It is used for general operations.

(*l*) *Copper-smith stake.* The copper-smith stake, as shown in Fig. 17.15(*l*), has a rounded edge on one side of the head and a sharp rectangular on the other. It is used for general operations.

(*m*) *Bottom stake.* The bottom stake, as shown in Fig. 17.15(*m*), has a single vertical piece with a flared end. It is mostly used for dressing burred edges on a disc, for special double seaming and for turning small flanges.

(*n*) *Hand dolly stake.* The hand dolly stake, as shown in Fig. 17.15(*n*), has a flat face with two straight edges, one convex edge and one concave edge. It is used for all general purposes such as bucking rivets and double seaming. It is available in various shapes and sizes.

Note : The quality of the finished work mostly depends upon the condition of the stake in use. If the stake has punch or chisel marks, then the finished work will look rough. Thus a stake should never be used to back up the work directly for punching or cutting. A mallet should be used whenever possible when forming sheet metal by means of stakes.

16. *Stake holders.* The stake holder mostly used in steel metal shop is a rectangular bench plate, as shown in Fig. 17.16(*a*), which is fastened to the bench with bolts. The plate has tapered holes so that the stakes may be used in different positions.

Many sheet metal workmen prefer to use universal stake holder with a complete set of stakes, as shown in Fig. 17.16(*b*). It does not require a bench plate. One stake may be replaced for another very quickly by simply turning a swivel handle. The holder is clamped to any desired position on the bench.

17. *Hand groover.* The hand groover (also called seam set), as shown in Fig. 17.17, is used for grooving a seam by hand. The outer ends of the groover should be slightly rounded to prevent them from marking the work. It is available in various sizes.

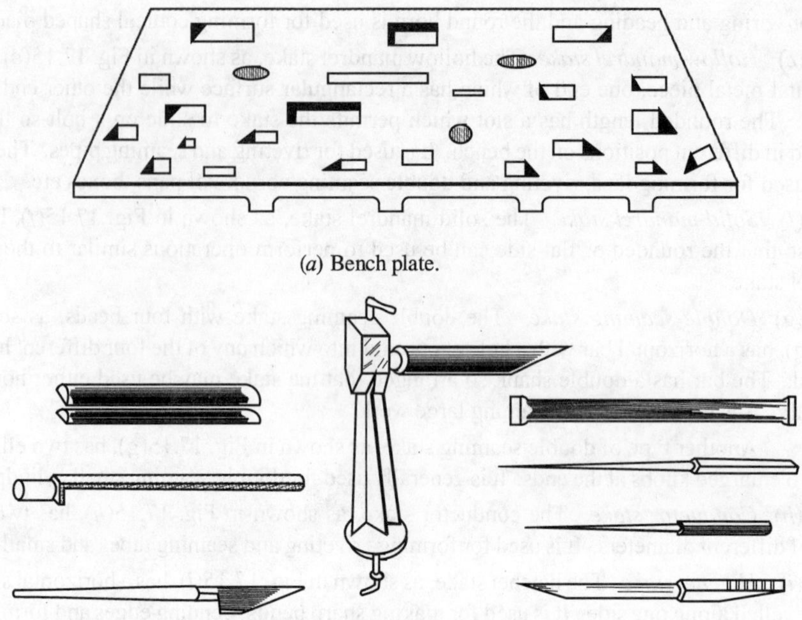

(a) Bench plate.

(b) Universal stake holder with a complete set of stakes.

Fig. 17.16

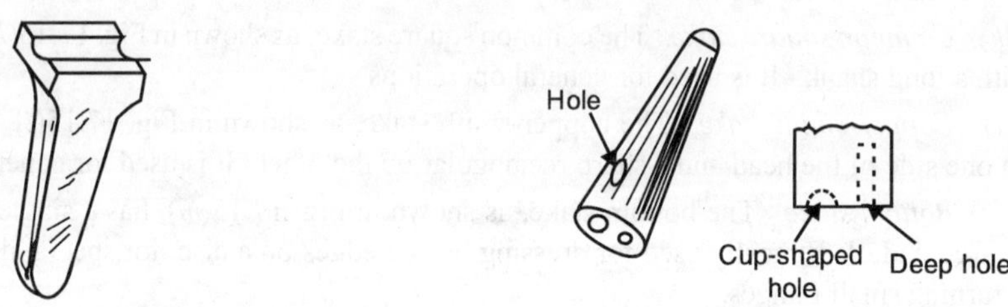

Fig. 17.17. Hand groover.

Fig. 17.18. Rivet set.

18. *Rivet set.* The rivet set, as shown in Fig. 17.18, is made of tool steel. It has a deep hole in the bottom to draw a rivet through metal and a cup shaped hole to form the finished head of a rivet. The hole in the side of a rivet set removes burr that is punched through the metal with the rivet. The rivet sets are available in various sizes.

19. *Hand seamer or Tongs.* The hand seamer or tongs, as shown in Fig. 17.19, is made of drop-forged steel with about 85 mm blades and an adjustable gage. It is used in seaming operations and bending in situations where it is impossible or inconvenient to bend metal on the stake.

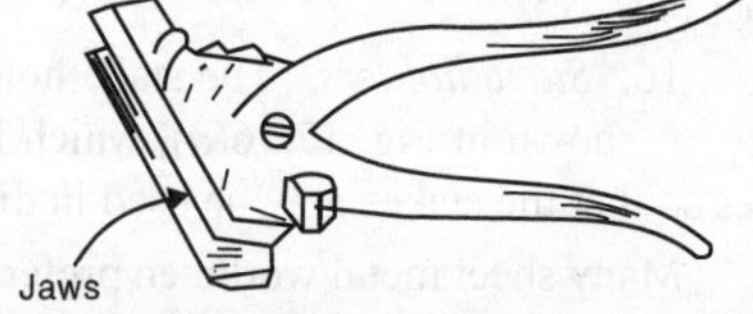

Fig. 17.19. Hand seamer or tongs.

20. *Soldering iron.* The soldering iron consists of a forged piece of copper fastened to an iron rod with a wooden handle on one end, as shown in Fig. 17.20. It is used to join two pieces of metal by means of an alloy of tin and lead (known as soft solder). The solder is usually transferred to the joint by means of a soldering iron.

The copper bit of the soldering iron may be straight pointed or of hatchet shape, as shown in Fig. 17.20(*a*) and (*b*) respectively. The straight pointed soldering iron is used for general-purpose work while the hatchet shaped soldering iron is used for soldering bottoms in all kinds of containers.

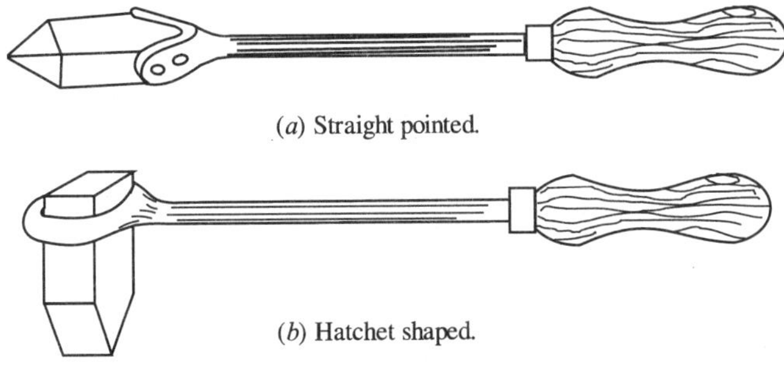

(*a*) Straight pointed.

(*b*) Hatchet shaped.

Fig. 17.20. Soldering irons.

21. *Folding bars.* These are used for bending and folding straight material. There are two kinds of folding bars in common use. The folding bar, as shown in Fig. 17.21(*a*), is made from rectangular mild steel. It is bent to bring the ends together to form a loop which springs the working faces together. The folding bar, as shown in Fig. 17.21(*b*), is made from two rectangular mild steel bars with holes drilled at the ends. These are held together with spring loaded nuts and bolts to give tension.

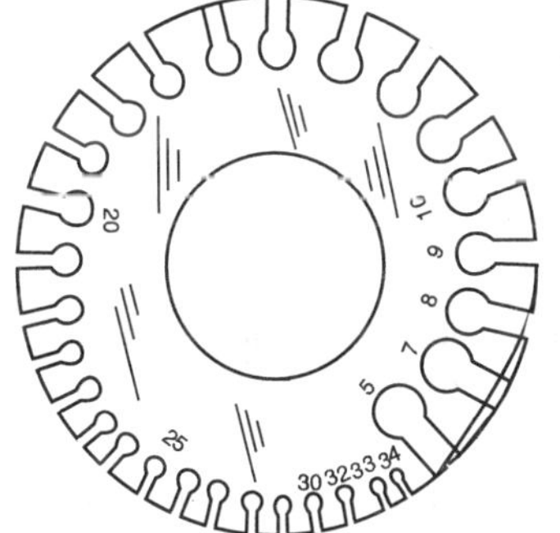

Fig. 17.21. Wire gauge.

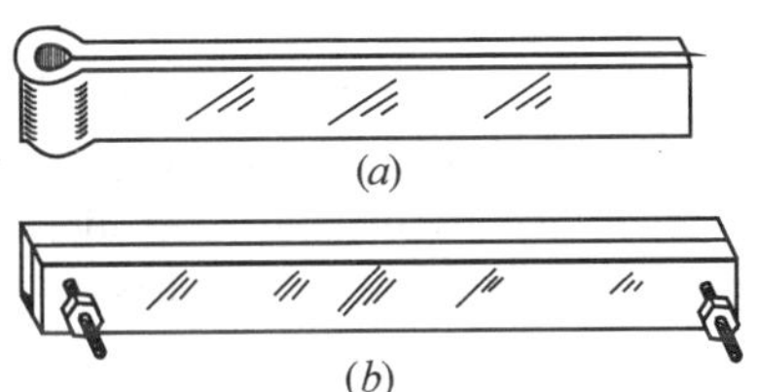

(*a*)

(*b*)

Fig. 17.22. Folding bars.

22. *Wire Gauge.* The wire gauge, as shown in Fig. 17.22, is used to check the diameter of wires or thickness of sheet metal. The commonly used wire gauge is Imperial Standard Wire Gauge (known as SWG). It has slots to indicate different width sizes. The most common wire gauge used in sheet metal has 21 slots with gauges varying from 4 to 24. The following table shows the imperial standard wire gauge sizes :

Table 17.1. Standard wire gauge sizes.

SWG No.	*Size in mm*	*SWG No.*	*Size in mm*	*SWG No.*	*Size in mm*
4	6.0	11	3.0	18	1.25
5	5.5	12	2.7	19	1.05
6	4.8	13	2.4	20	0.95
7	4.5	14	2.1	21	0.85
8	4.0	15	1.9	22	0.72
9	3.6	16	1.65	23	0.62
10	3.3	17	1.45	24	0.58

17·4 Sheet Metal Operations

The following are the various sheet metal operations :

1. Marking ; 2. Cutting ; 3. Notching ; 4. Bending ; 5. Riveting ; 6. Soldering ; 7. Folding edges ; 8. Seam making ; 9. Hollowing or blocking ; 10. Sinking ; 11. Raising ; and 12. Planishing.

We shall now discuss these operations, in detail, in the following pages.

17·5 Marking

The marking out operation consists of scratching of lines on the surface of a sheet metal. It is also called scribing operation. Before marking operation is carried out, the paper or metal pattern of the object is prepared. The metal pattern is desireable for repetitive work because it maintains accuracy for a long time, whereas the paper pattern tears quickly if used repeatedly. The pattern is now transferred to the flat sheet metal and marked as discussed below :

1. When a paper or metal pattern is transferred to the flat sheet metal, it should be held in place by weights to prevent the pattern from slipping.
2. A sharp pointed pencil is used to mark the outline of the paper pattern, whereas a scriber is used for metal pattern. The scriber should not be used to mark the paper pattern because it will tear the edges of the paper.
3. When a pattern is to be traced on aluminum and stainless steel sheets, a sharp pointed pencil should be used because the scriber will ruin the finish of the metal.
4. The best method of transferring the paper pattern to sheet metal is to mark the ends of all lines by a prick punch through the paper. Then remove the paper from the metal and join the all prick points by a scriber in the proper manner.

Many simple objects can be laid out directly on the sheet metal. Usually, it is better for the beginner to prepare a paper pattern for objects of complicated design rather than to layout directly on the sheet metal. Since the paper pattern may be checked by folding it into the required shape, therefore, unnecessary wastage of material is avoided.

In making a layout directly on the steel metal involves operations such as drawing straight lines or curved lines and making indentation marks to serve as guide for drilling holes. The procedure for marking straight lines, curved lines and indentation marks is as follows :

1. *Marking straight lines.* The straight edge (it is flat bar of steel with bevelled edge) or a steel square is used to draw straight lines as discussed below :

(*a*) When a straight edge is used, it is placed on the sheet metal in the correct position. A sharp pointed scriber is held in one hand at an angle with the point resting against the edge of the straight edge. Now the line is drawn by applying a little pressure to the scriber.

(*b*) When a steel square is used, the body or tongue of square is placed against the even side of the sheet metal. The scriber is run along the edge of the tongue or body of the square to mark the straight line.

2. *Marking circles and arcs.* The circles and arcs on the surface of a sheet metal are marked with a divider. One leg of the divider is placed at the centre of a circle or arc to be drawn while the other end is moved with a little pressure on the divider to mark the circle or arc. The divider is held in one hand. In order to prevent slipping, the divider is inclined in the direction in which the circle or arc is drawn. The large circles and arcs are drawn with trammel points.

3. *Marking irregular curves.* The irregular curves are marked by first laying out a series of points to indicate the shape of the curve. Now the french curve is set in such a way that atleast three points coincide with the curve and mark this with a scriber on the surface of the sheet metal. The french curve is now moved to coincide next three points and the curve is scribed connecting the first curve. This process is repeated till the whole curve is marked. The irregular curves may also be drawn by a flexible rule made of spring steel.

4. *Marking indentation marks.* The indentation marks are used as locations for drilling holes, centres for dividers and for transferring a pattern. The indentation marks for drilling are made with a centre punch, whereas marks for all other purposes are made with a prick punch. Both the centre punch and prick punch are used in the same manner. In using these tools, first of all the location of the hole is marked with two intersecting lines. The punch is held is one hand with its point directly on the intersection of the two lines. Now strike the head of the punch with a light blow of the hammer held in other hand.

17·6 Cutting

The sheet metals upto approximately 18 S.W.G. can be cut with hand snips. The following procedure is adopted for straight cutting.

1. Hold the snip in one hand and the nearest part of the metal in the other hand, as shown in Fig. 17.23(*a*).

2. Open the blades of a snip and place the upper blade on the line of cut to be followed. The blade should be kept perpendicular to the surface of the metal.

3. Start the cut at the edge of sheet by exerting pressure on handles of snips. When the cut is about 15 mm from the blade tips, open the blades again and push the snip forward. Repeat this until cut is completed.

4. As the metal is sheared off, it will curl up on the left hand side of the snips and curls down on the right hand side of the snips.

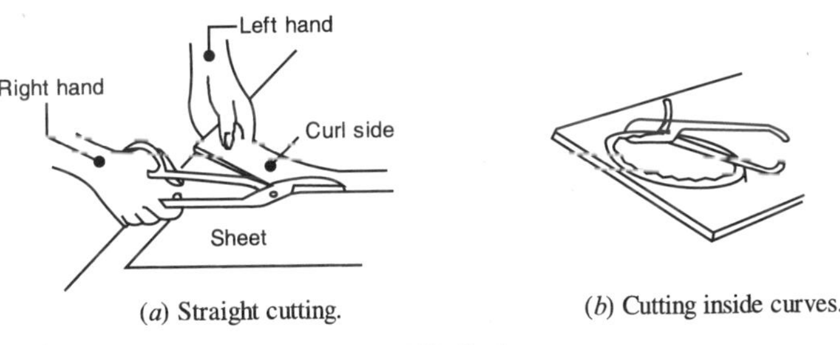

(*a*) Straight cutting. (*b*) Cutting inside curves.

Fig. 17.23. Cutting process.

5. When cutting a large sheet of metal, it is advisable to cut along the left hand edge of the sheet to permit the scrap or smaller section to curl away.

6. The inside curves such as circular holes are cut with curved snips. The holes are first cut roughly with a cold chisel and hammer. It is finally trimmed to the required size by the curved snip, as shown in Fig. 17.23(*b*).

7. The outside curves are easily cut with straight snips.

17·7 Notching

In bent sections that have folded edges, there should be some provision so that there is no overlapping of metal where the corners come together. In order to prevent bulging at such a place, it is necessary to slit or clip the metal or provide small openings. The openings left at the corners of seams and edges are known as *notches* and the operation is called *notching*, as shown in Fig. 17.24. The size, locations and types of notches depend upon the shape of object. The following are the

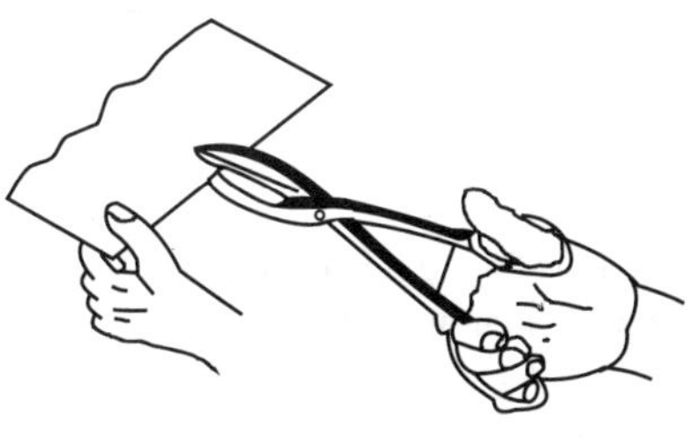

Fig. 17.24. Notching process.

different types of notches commonly used in sheet metal.

1. *Straight notch.* This notch is prepared by simply making a straight cut where the bend is to occur. This is used in dovetail seams. The method of laying out a straight notch is shown in Fig. 17.25.

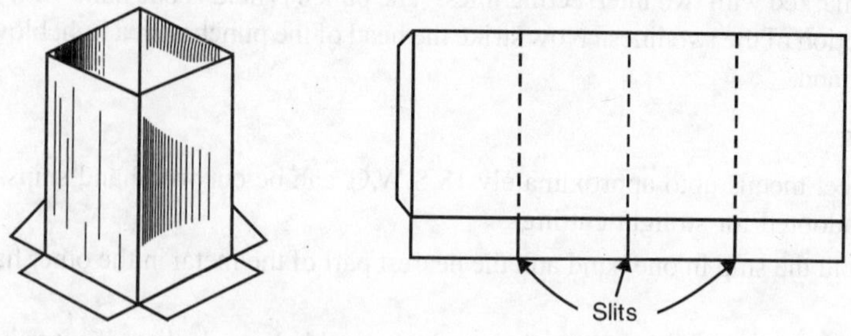

Fig. 17.25. Laying out a straight notch.

2. *Square notch.* The square notch is used for making a square or rectangular box. The squares on the corners, as shown in Fig. 17.26, must be removed to permit the sides to be bent properly.

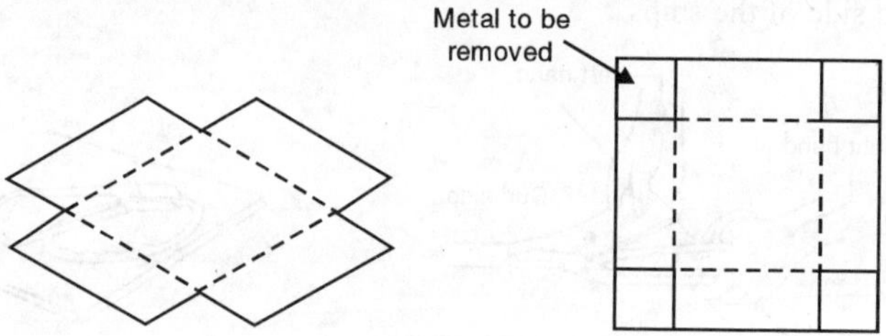

Fig. 17.26. Laying out a square notch.

3. *V-notch.* The *V*-notch is used when the corners of a flange must come together. If the flange forms a right angle, both sides must be cut at an angle of 45°, as shown in Fig. 17.27.

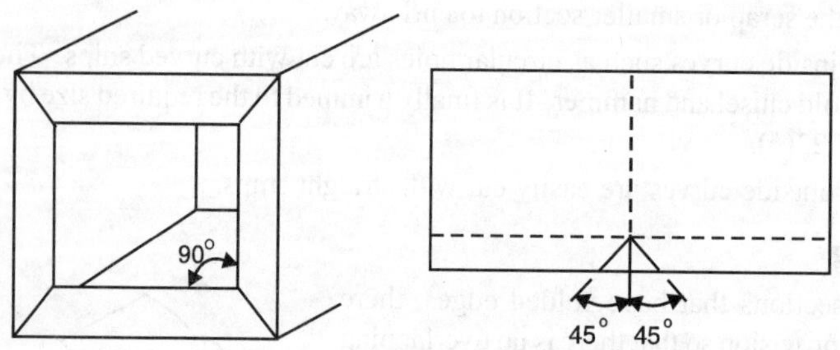

Fig. 17.27. Laying out a V-notch for 90°.

If the flange is bent at an angle that is less or more than a right angle, the notch must correspond to the particular angle that is required, as shown in Fig. 17.28.

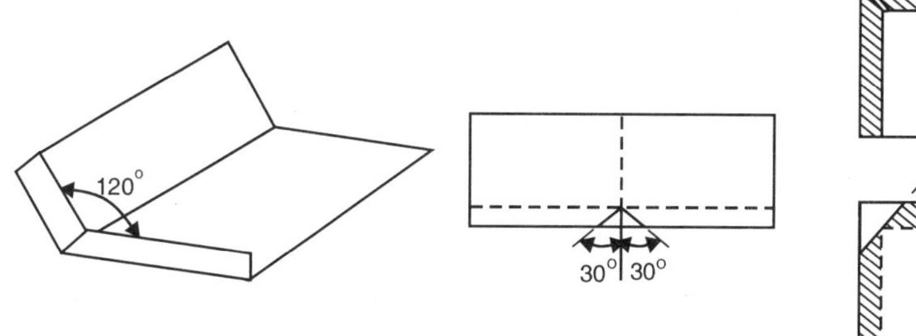

Fig. 17.28. Laying out a V-notch for 120°.

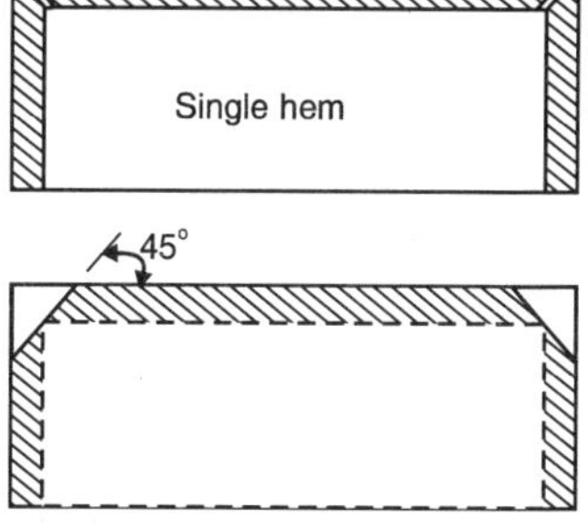

Fig. 17.29. Laying out a slant notch.

4. *Slant notch.* When single hems are to meet at right angles, the corners must be clipped at an angle of 45°, as shown in Fig. 17.29.

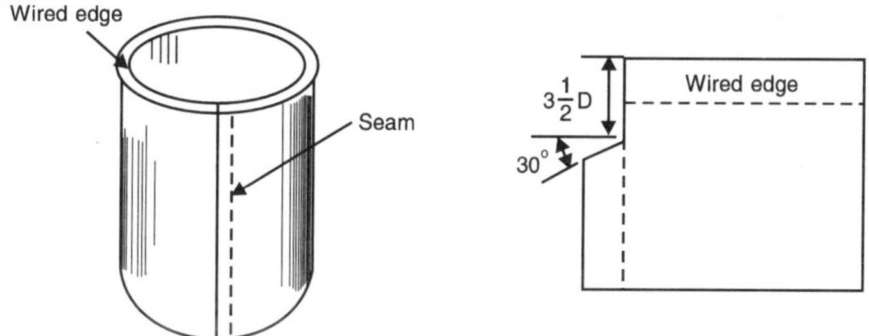

Fig. 17.30. Laying out a wire notch.

5. *Wire notch.* The wire notch is used on an article which has wired edges. This notch must be provided to prevent the wired edge from overlapping at the seam. The usual practice is to notch the seam at an angle of about 30°, as shown in Fig. 17.30. The angle is started at a point located at a distance from the upper edge of the pattern equal to approximately 3.5 times the diameter of wire.

17·8 Bending

The bending of sheet metal may be done over stakes, blocks of wood, pieces of angle iron, or the edge of a bench top. Sometimes the bends can be made in various bending machines. The following procedure may be followed while bending the sheet metal by hand :

1. First of all, a line is marked with pencil or scriber where the metal is to be bent. In order to make a right angled bend the sheet is placed on the bench so that the line is even with the edge of the bench. A block of wood is set on the metal with is edge directly over the bending line. It is then clamped to the bench with C-clamps. In order to bend a small piece of metal, the sheet is placed between two blocks of wood and then clamped in a vice, as shown in Fig. 17.31. The bend is now made by striking the metal with a mallet using very light blows. The work is started at one end and finished along the full length of the metal.

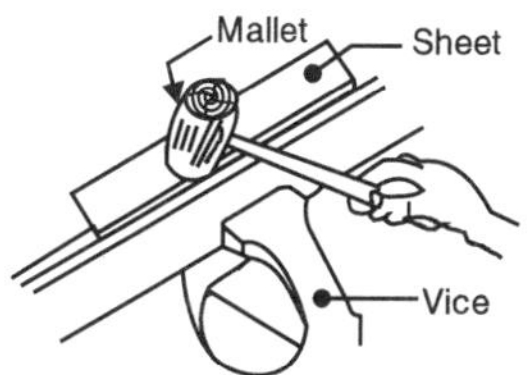

Fig. 17.31. Bending process.

2. In order to make a curved bend, the edge of the wooden block is planed to have the desired curvature. The wooden block is placed even with the edge of the bench. Another block of the same thickness is set on the bench top so that the sheet is in level. Then clamp a piece of angle iron or a

hardwood block over the sheet metal, as shown in Fig. 17.32. Now strike the metal with a mallet, gradually bending the sheet over the curved edge of the block.

3. A tool known as a hand seamer (see Fig. 17.19), can be used to make sharp bends on light-gauge sheet metal. Such a tool is especially helpful in bending narrow portions that are difficult to fold by other methods. This hand seamer has two adjustable screws which can be regulated for the width of the bend. The metal is simply placed between the jaws of the seamer and these jaws are clamped together. The flat surface of the metal is held firmly with one hand and at the same time the handle of the seamer is raised. If the metal is longer than the jaws, then small sections are bent at a time.

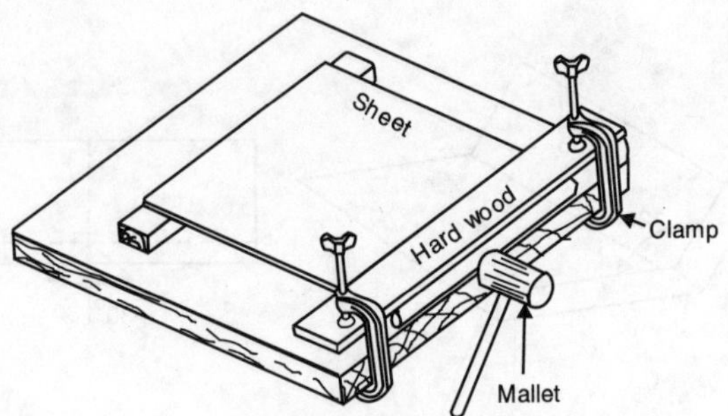

Fig. 17.32. Bending process.

17·9 Riveting

It is a process of joining permanently two pieces of sheet metal with rivets. The special rivets, called tinners' rivets, are used for such a purpose. These rivets are made of soft iron and are usually coated with tin. This prevents corrosion and makes them easier to solder. When riveting aluminium, special aluminium rivets should be used. The following procedure is adopted in making a riveted joint :

1. First of all, drill or punch the holes of correct size. Insert the rivet in the hole and place the head on some metal bar or stake.

2. Place the deep hole of the rivet set over the rivet and strike the rivet set with few sharp blows, as shown in Fig. 17.33(*a*). This draws the metal and rivet together. This process is called *drawing process.*

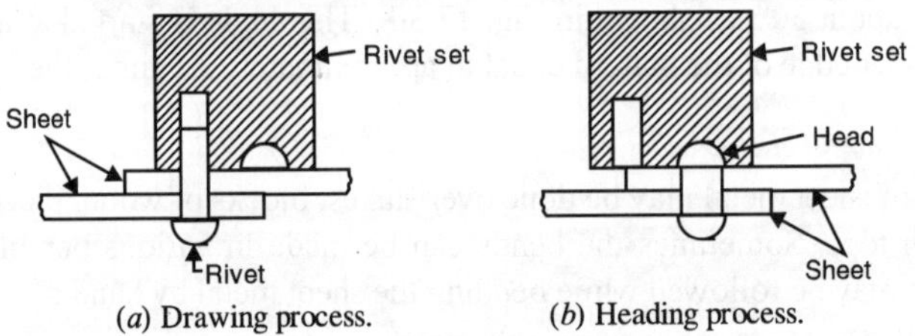

(*a*) Drawing process. (*b*) Heading process.

Fig. 17.33. Riveting process.

3. Remove the rivet set and flatten the end of the rivet with the face of a riveting hammer or ball peen hammer. It should be noted that each blow of hammer should hit the rivet shank squarely.

4. Place the cup shaped opening of the rivet set over the flattened rivet end, and strike the rivet set until the head is shaped properly, as shown in Fig. 17.33(*b*). This process is called *heading process.*

17·10 Soldering

It is the process of joining two or more pieces of metal by means of an alloy (known as solder) having lower melting point than the metal being joined. In order to solder a sheet metal we require a soldering iron, furnace or blow torch to heat the copper bit of soldering iron, a solder and a flux.

There are several types of solders, but the soft solder is mostly used in sheet metal work. The soft solder consists of 50 percent tin and 50 percent lead and has a melting point of approximately 190°C. A solder consisting of 60 per cent tin and 40 per cent lead is often used and melts at about 170°C. This solder flows more freely and produces a slightly stronger joint. Both these solders are available either in bar or wire form.

The metal is normally covered with a film of oxide and as the heating takes place, the amount of oxide increases. This oxide is removed by means of a flux. The fluxes commonly used for soft soldering are solutions or pastes containing zinc chloride. When the fluxes are applied to the surface, the heat of the soldering iron melts the flux which in turn dissolves the oxide. The soldered joint is made by the following procedure :

1. Clean the surface to be joined so that it is free from oxides or grease etc., and then apply the flux along the joint to prevent formation of oxides.

2. Heat the copper bit of soldering iron and when it is warm enough, it is dipped in the flux and rubbed against solder. The bit of soldering iron picks up a certain amount of solder.

3. The end of bit is now placed at the end of joint and the solder flows into the gap. When the solder ceases to flow into the joint, the bit is reheated and the procedure started again.

17·11 Folding Edges

The edges of the sheet metal may be folded to strengthen the edges and to eliminate the sharp edges. The three common types of folded edges are single hem, double hem and wired edges. The *single hem* is made by turning the edge over the hatchet stake, as shown in Fig. 17.34(*a*), and then completed it with a mallet over a flat stake, as shown in Fig. 17.34(*b*). The *double hem* is a single hem with its end bent under, as shown in Fig. 17.34(*c*).

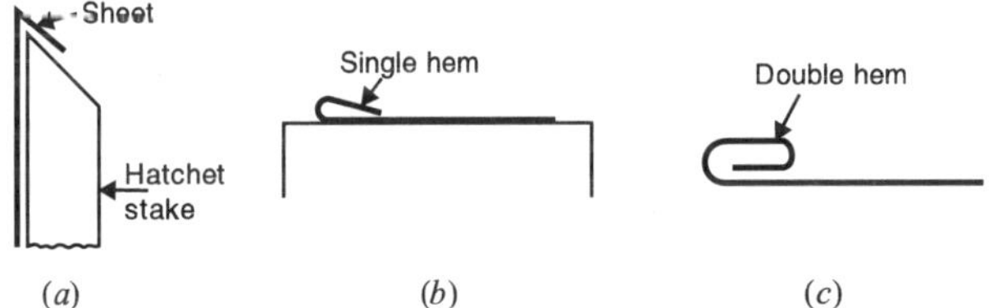

Fig. 17.34. Forming of single and double hem.

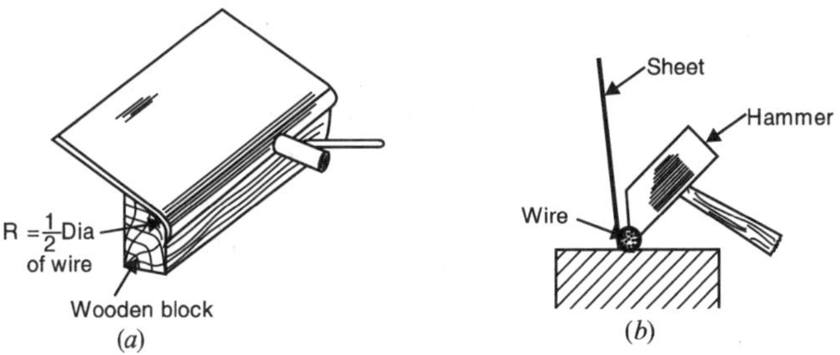

Fig. 17.35. Process of folding edges.

The *wired edges* are made by bending the sheet with a mallet over a length equal to 2½ times the diameter of wire, as shown in Fig. 17.35(*a*). This is done on a wooden block whose one edge is rounded to a radius equal to approximately one half the diameter of the wire. The wire is now placed in the turned edge and held in position with pliers. The metal is folded over the wire by striking it with a mallet. The final bend is made by striking the metal with the peen end of a setting hammer, as shown in Fig. 17.35(*b*).

17·12 Seam Making

A seam is a section where pieces of sheet metal are joined. There are many methods of making seams, but the type of seam is determined by the thickness of metal, the purpose for which the object is to be used. The seams commonly used in sheet metal are as follow :

1. *Lap seam.* It is the simplest type of seam in which one edge overlaps the other edge and is soldered or riveted as shown in Fig. 17.36. The width of lap depends upon the size of job. This is generally used on flat or cylindrical work, such as boxes, cylinders etc.

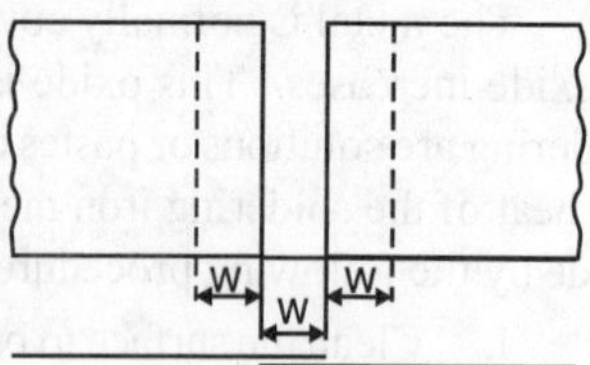

Fig. 17.36. Forming of lap seam.

2. *Grooved seam.* It is another method used to join pieces of sheet metal. This seam is made by hooking two folded edges together and then offsetting the folded edges. The offset is usually made on the outside when the inside surfaces are to be smooth as in case of a pipe. The grooved seam can be made by hand or by machine. The following procedure is adopted to make a grooved seam by hand.

(*a*) First of all, a line is marked at a distance of one-half the width of seam from the edge of the pieces and then draw another line equal to the width of grooved seam required.

(*b*) Now bend the two edges and hook them together, as shown in Fig. 17.37. This is done by placing the work on a hollow mandrel stake or a stake having a suitable curvature. If the pieces are flat, rest them on a square surface. Then flatten the seam slightly with a mallet.

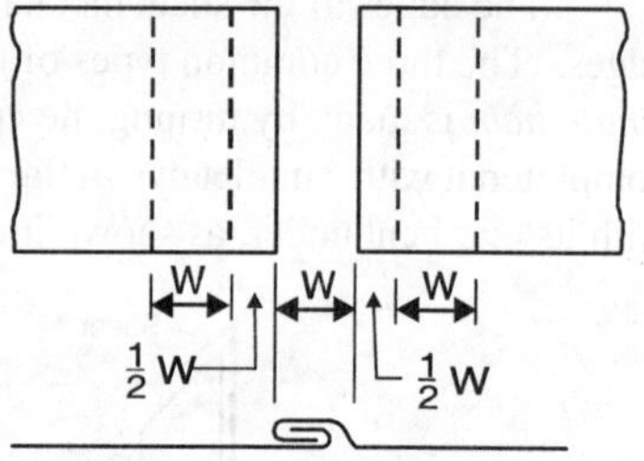

Fig. 17.37. Forming of grooved seam.

(*c*) The hand groover of the correct size (the width of the groove in the groover must be little larger than the width of seam) is fitted over one end of the seam and struck with a mallet. This process is repeated on the other end of the seam to prevent the edges from coming apart. The entire seam is grooved by striking the groover while moving it over the seam.

(*d*) The operation is completed by flattening the seam with a mallet.

3. *Single seam.* It is used to join bottoms to the bodies of vertical containers of various shapes. This seam makes a good joint for cylindrical, square or rectangular shaped vessels. The single seam may be made as follow :

(*a*) First of all, draw a line parallel to one edge of the body stretchout at a distance equal to the width of seam (*W*), as shown in Fig. 17.38(*a*).

(*b*) Draw two lines parallel to the edges of the bottom stretchout, as shown in Fig. 17.38(*b*) and (*c*). The plus and minus dimensions of 1 mm are used to prevent the folded edge of the bottom from interfering with the folded edge of the body.

(*c*) Now form the flange on the body and on the bottom piece by folding the metal with a mallet over a suitable stake or wooden block, as shown in Fig. 17.38(*d*).

(*d*) Insert the body of the container in the bottom piece and place the assembled parts on a flat stake.

(*e*) Bend the edge of bottom gradually over the edge of the body with a setting hammer.

4. *Double seam.* The double seam is used to connect the parts of a container when an exceptionally strong joint is required. This seam is actually a single seam which is bent up against the body of the container. The double seam may be used on cylindrical, square or rectangular shaped

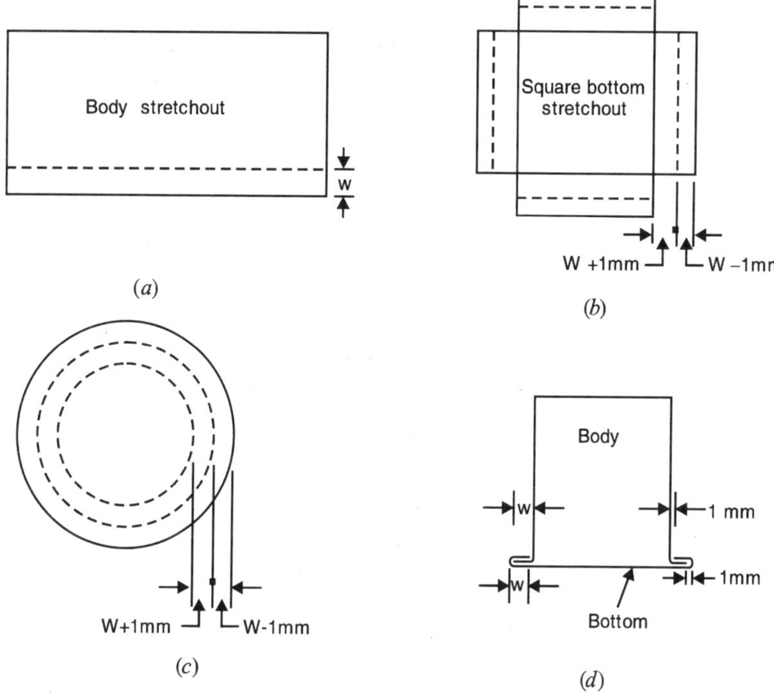

Fig. 17.38. Making of single seam.

jobs. The double seam is made as follow :

(*a*) First of all, make a single seam as discussed earlier.

(*b*) Place the container over a double seaming stake (or any other stake having the required curvature) with its inside edge against the edge of the stake.

(*c*) Now bend the single seam to approximately 45° with a mallet, as shown in Fig. 17.39(*a*). Continue to strike the partially bent seam with a mallet until it is flattened against the body of the container. The bend is made gradually and the container is kept turning as the seam is bent.

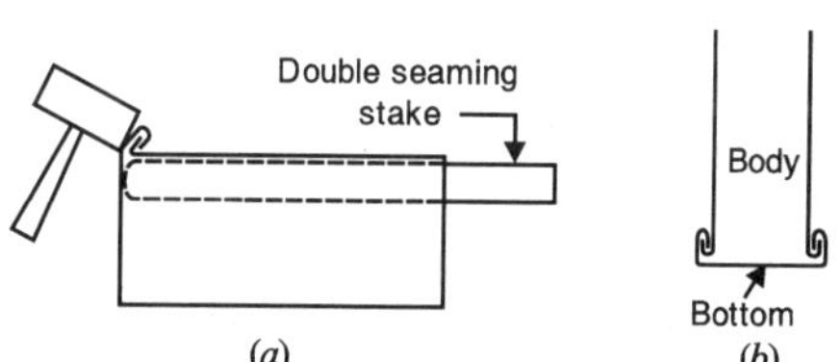

Fig. 17.39. Making of double seam.

(*d*) When the seam has been completely flattened, remove the container from the double seaming stake and place it on a square stake with its edge resting on the edge of the stake. Now tap the bottom lightly to straighten the seam. The double seam need not to be soldered unless the container is to be made water tight. The container with double seam is shown in Fig. 17.39(*b*).

5. *Dovetail seam.* It is used to join sections such as one pipe to another pipe or a flat plate to a pipe. The square as well as cylindrical objects can also be joined with this seam. The dovetail seam consists of a series of narrow strips of metal which are formed by slitting the end of pipe or collar. The narrow strips are bent in such a manner that they form a permanent joint. The procedure for making a dovetail seam to join flat plate to a pipe is as follow :

(*a*) Scribe a line on one end of the pipe to represent the depth of the seam. This depth may be

6 to 15 mm for small articles and may extend upto 25 mm for larger pieces. The line can be scribed with a suitable marking gauge. This can be easily made by notching a piece of sheet metal so that the point scribes a line at the desired distance from the edge.

(*b*) Lay out the width of strip (6 to 25 mm) and slit the metal along each vertical line with a pair of snips. Notch each strip slightly, as shown in Fig. 17.40(*a*).

(*c*) Bend every strip at right angle, beginning the strip at the seam of pipe as shown in Fig. 17.40(*b*). These bends may be made with pliers or by striking them with a hammer against the square stake.

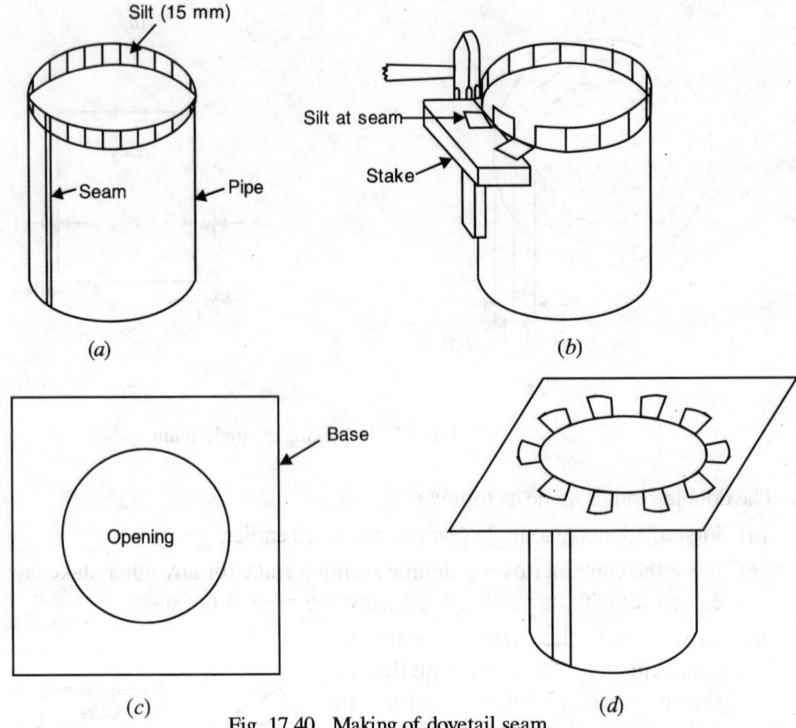

Fig. 17.40. Making of dovetail seam.

(*d*) Cut an opening to the flat piece equal to the size of pipe, as shown in Fig. 17.40(*c*). Now place the pipe with bent strips in the opening of the flat piece and bend the remaining upright strips to lock the two sections together.

(*e*) Set the work on a flat plate or stake, and strike the strips with a hammer. The final shape is shown in Fig. 17.40(*d*).

6. *Flanged or burred bottom seam.* It is used to join the bottom of a container to its body. The joint consists of a narrow flange which may be joined to the outside or inside of the vessel by soldering, as shown in Fig. 17.41(*a*) and (*b*). The flange, on cylindrical jobs, is often referred to as a *burr* and the process of making a narrow flange is known as *burring.* This operation is performed on a burring machine. The following procedure is followed in making a burred bottom seam :

(*a*) For a round container, draw a circle to represent the outline of the bottom, as shown in Fig. 17.41(*c*). In the case of a square or rectangular article, draw the bottom of the shape and size required, as shown in Fig. 17.41(*d*).

(*b*) Draw a second line to show the width of the flange or burr. The width of flange may range from 3 mm to 6 mm. If the flange is to be fastened on the inside of the body, the size of the bottom piece should be slightly smaller to allow it to slip in without interference.

(*c*) The flange or burr for the round container is made on a burring machine. If the article is square or rectangular in shape, the burr on the straight edges may be made on the bar folder or by hand.

(*d*) When the burr is completed, attach it to the bottom and solder the edges.

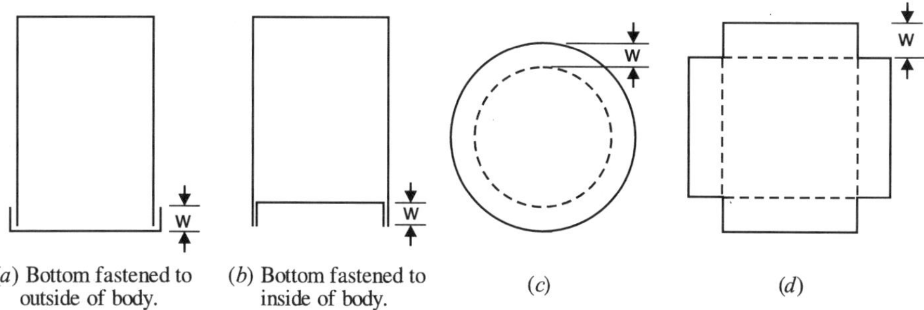

(*a*) Bottom fastened to outside of body. (*b*) Bottom fastened to inside of body. (*c*) (*d*)

Fig. 17.41. Making of flanged or burred bottom seam.

17·13 Hollowing or Blocking

It is the process of beating the sheet metal into a particular shape such as sauce pan lid or bowl. It is usually done on the hollowing block which is a wooden block with hollow cuts into it, as shown in Fig. 17.42(*a*). The metal to be worked is placed on the top of the hollowing block over a suitable hole and hammered with a bossing mallet or a hollowing hammer, as shown in Fig. 17.42(*b*). Since the metal is beaten down in concentric circles from the edge to the centre, therefore, the metal becomes thin towards the centre. After each completed set of circles, the work should be annealed and the process is repeated until the desired shape is attained. The hammering should be carefully done so that no crinkling edge is formed.

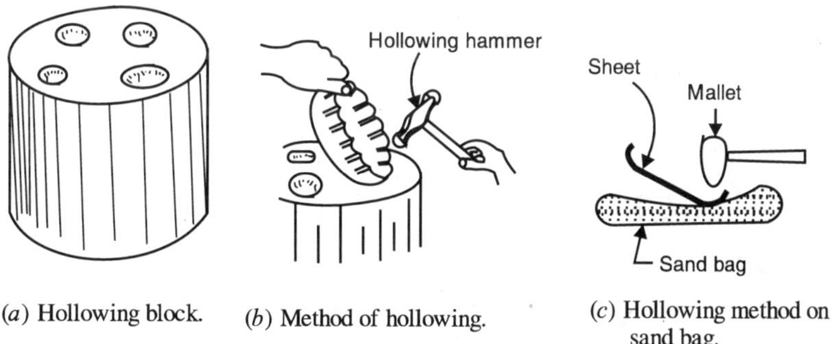

(*a*) Hollowing block. (*b*) Method of hollowing. (*c*) Hollowing method on sand bag.

Fig. 17.42. Hollowing or blocking process.

The hollowing process may also be done on a sand bag, as shown in Fig. 17.42(*c*). The main advantage of sand bag is that there are no indentation on the top.

17·14 Sinking

It is a process used to sink the bottom for forming a tray with a flat rim. The position of the corner of sinking is, first of all, scribed and then the depth of tray is scribed inside that. A piece of wood with a sharp corner is placed vertically in the vice and over the edge of this, the metal is beaten down with a tray hammer to shape the tray, as shown in Fig. 17.43.

If the tray is circular, two nails may be used as a gauge to give equal distance from the edge to the corner of the sinking. The rim is kept flat with a small mallet or by inverting the tray on a surface plate and using a piece of wood as a punch on the rim.

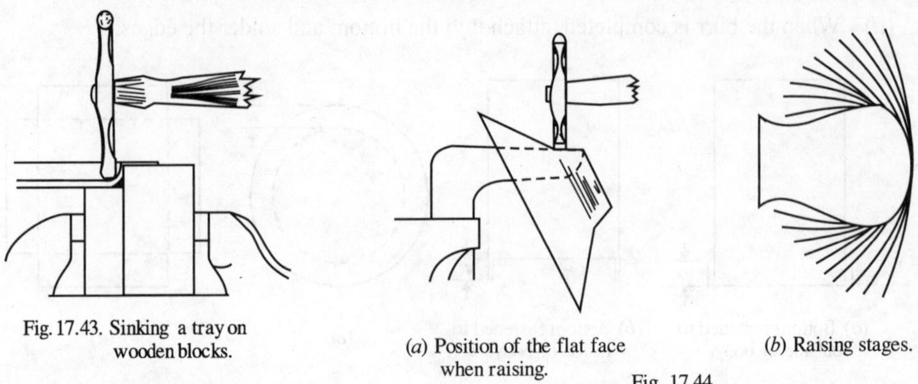

Fig. 17.43. Sinking a tray on wooden blocks.

(*a*) Position of the flat face when raising. (*b*) Raising stages.

Fig. 17.44

17·15 Raising

It is the process of dinting the metal down to shape over a tool with a raising hammer or mallet. The concentric circles are, first of all, drawn with pencil compass as a guide to keep the work true. The hammering is started from the centre and worked around in circles till the outer edge is reached. The final round of hammering is done with a mallet to keep the thickness of the edge. The metal must be annealed after each completed course and the process is repeated. The work must be held at an angle, as shown in Fig. 17.44(*a*), so that contraction can be made when hammered down to the tool. Fig. 17.44(*b*) shows the various stages that the metal would go through in raising.

17·16 Planishing

It is a process which is applied to a sheet with the main object of stretching the metal with an improved surface. The process brings the article to its final shape and imparts some degree of hardness to the metal. In planishing the bowl shapes, a round head stake, that fits the work at the point to be planished, is placed in the vice. The planishing is started at the centre with a planishing hammer and

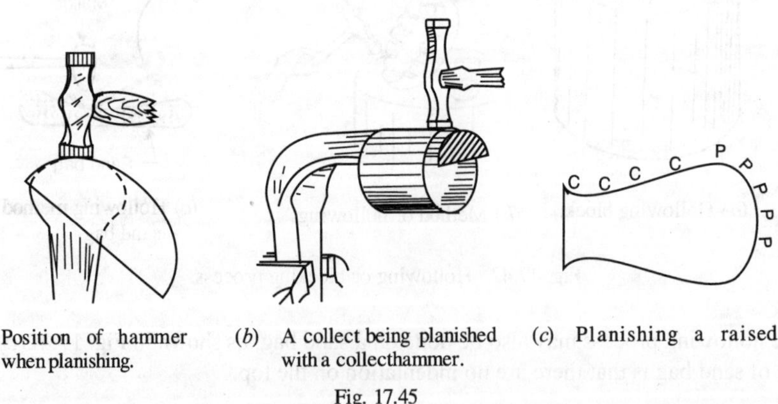

(*a*) Position of hammer when planishing. (*b*) A collect being planished with a collecthammer. (*c*) Planishing a raised

Fig. 17.45

is continued in concentric circles towards the edge. The planishing of bowl shapes is shown in Fig. 17.45(*a*). It may be noted that for planishing domed circular work, a round faced planishing hammer should be used and for planishing cylinders, collect hammer with oblong shaped faces, as shown in Fig. 17.45(*b*), is used.

The planishing of a raised article is shown in Fig. 17.45(*c*). The part P is planished with a planishing hammer, whereas the part C is planished with a collect hammer.

17·17 Sheet Metal Machines

A sheet metal shop must be provided with various machines in order to perform different operations. The machines may be hand operated for working on thin sheets and power operated for thick sheets. They are also available as either bench or floor models. Some of the machines commonly used are as follows :

1. *Shearing machine.* It is used for cutting metal sheets.
2. *Folding machine.* It is used for bending and folding the edges of the sheet metal.
3. *Bending machine.* It is used for shaping the metal sheets into cylindrical objects.
4. *Grooving machine.* It is used for grooving longitudinal seams in cylindrical objects.
5. *Forming machine.* It is used for forming sheet metal into cylindrical shapes of various diameters. There are two types of forming machines; *i.e.*, slip-roll forming machine and plain forming machine, but the former is generally preferred.
6. *Beading or swedging machine.* It is used to make depressions in metal such as in pipes, machine guards or wherever reinforcing is necessary.
7. *Burring machine.* It is used to turn burrs on circular discs such as bottoms and covers. It is also used for preparing edges for double seaming cylindrical articles.
8. *Turning machine.* It is similar to a burring machine but differs in the sharpness of the edge it makes. The burring machine produces a sharp edge, while the turning machine makes a rounded edge for wiring operations, for bodies of cylinders and for double seaming.
9. *Wiring machine.* It is used to complete the metal edge around the wire after the seat to receive the wire has been prepared by the turning machine.
10. *Crimping machine.* It is used for crimping (*i.e.*, reducing the diameter of a circular object) and beading in one pass. The machine may be used only for beading or crimping by changing the rolls of the machine.
11. *Setting-down machine.* It is used for setting down seams on containers of various shapes.
12. *Double seaming machine.* It is used for double seaming flat bottoms on straight or flared cylindrical pieces.

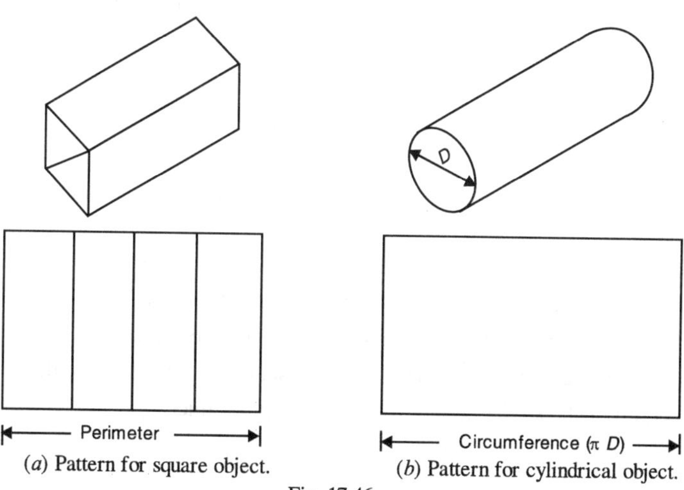

(*a*) Pattern for square object. (*b*) Pattern for cylindrical object.

Fig. 17.46

17·18 Laying Out a Pattern

All sheet metal articles are fabricated from flat pieces of metal cut according to the outlines that are drawn or traced on the sheets of metal. In order to obtain correct size and shape of the article, the patterns are used. A *pattern* is defined as a flat outline or stretch out or development of the article. The pattern may be drawn on paper first and then transferred to the metal or it may be laid out directly on the metal. If the pattern is to be used repeatedly, it is made of metal and is called *template* or *master pattern.* The knowledge of development of surfaces of the objects for a sheet metal worker is very important so as to enable him to cut the proper size of sheet and then form it into pattern of the desired article. Fig. 17.46 shows the pattern or stretchout or development of square and cylindrical objects.

All the sheet metal patterns may be laid out by one of the following three methods of development.

1. Parallel line method.
2. Radial line method, and
3. Triangulation method.

These methods of development are discussed, in detail, in the following pages.

17·19 Parallel Line Method of Laying Out a Pattern

This method of laying out a pattern is applied to those objects which are in the form of prisms or cylinders as in the case of patterns for ducts, elbows and Tee joints. The following examples show the patterns (or developments) of various objects by parallel line method.

Example 17.1. *Draw a pattern for a rectangular duct.*

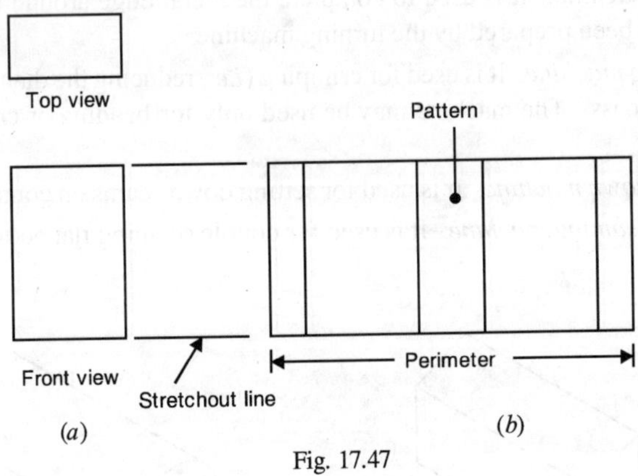

Fig. 17.47

Procedure. The following steps may be followed in laying out a pattern for the rectangular duct :

1. First of all draw the top view and front view of the duct, as shown in Fig 17.47(*a*).
2. Draw the stretchout line equal to the perimeter of the duct.
3. Mark the distances on the stretchout line from the top view.
4. Draw perpendicular lines at each point and set of the height of the duct.
5. Fig. 17.47(*b*) shows the pattern for the rectangular duct.

Example 17.2. *Draw a pattern for a truncated cylinder.*

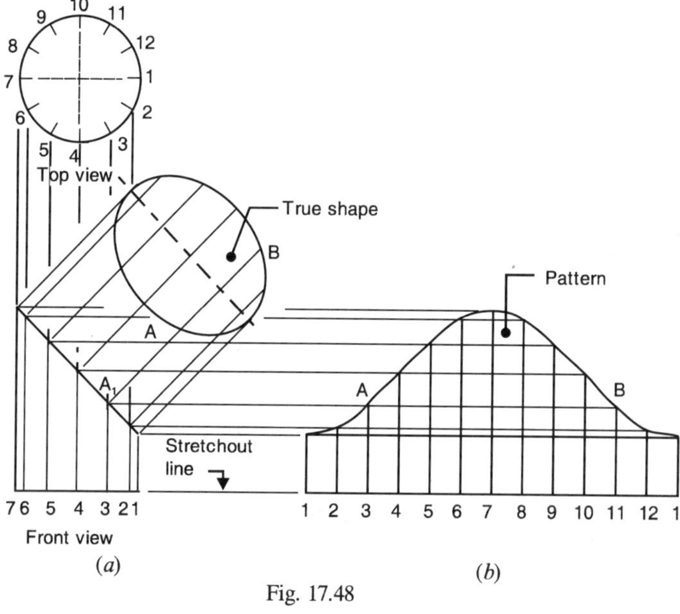

(*a*) (*b*)

Fig. 17.48

Procedure. Since the cylinder is truncated (*i.e.*, cut off at an angle), therefore, the bevelled end will develop into a curved line and the lower end into a straight line. The following steps may be followed in laying out a pattern for a truncated cylinder :

1. Draw the top view and front view. Divide the top view into twelve number of equal parts and project them in the front view, as shown in Fig. 17.48(*a*).
2. Draw the stretchout line equal to the circumference (πd) of the circle and divide it also into twelve equal parts.
3. Draw perpendiculars from each division point and locate the top points of these perpendicular lines by projecting the corresponding points from the front view.
4. Join the top points by a smooth curve and complete the pattern as shown in Fig. 17.48(*b*).
5. The true shape, if required, may be drawn as shown in Fig. 17.48(*a*).

Example 17.3. *Draw a pattern for the ventilating pipe and elbow which passes through an inclined roof, the front view of which is shown in Fig. 17.49* (*a*).

Procedure. The following steps may be followed for laying out a pattern for the pipe and elbow :

1. Draw the supplementary view of the elbow and divide it into twelve equal parts.
2. Project the points on the both *A* (pipe) and *B* (elbow) as shown in Fig. 17.49(*a*).
3. Draw the stretchout line equal to the circumference and divide it also into twelve equal parts.
4. Obtain the corresponding points of the given front view and join them with a smooth curve as shown in Fig. 17.49(*b*).

17·20 Radial Line Method of Laying Out a Pattern

This method of laying out a pattern is applied to those objects in which the slant edges (or elements or generators) radiate (like spokes of a wheel) from a point known as centre or apex as in the case of cones and pyramids. A little consideration will show that some elements in top view or front

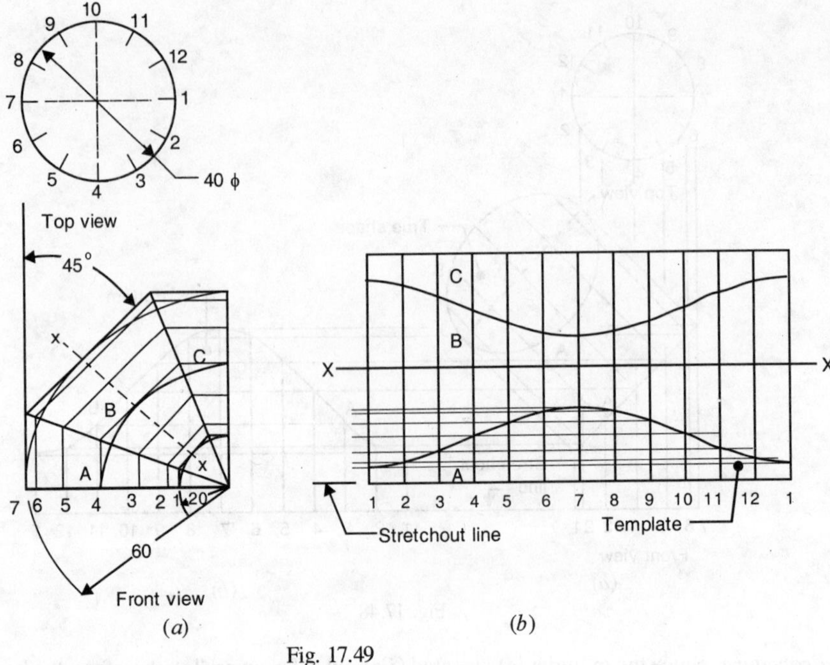

Fig. 17.49

view do not represent true lengths of the object. Therefore, their true lengths may be found out before laying out a pattern. The following example shows the laying out a pattern by radial line method.

Example 17.4. *Draw the pattern of a truncated cone by radial line method.*

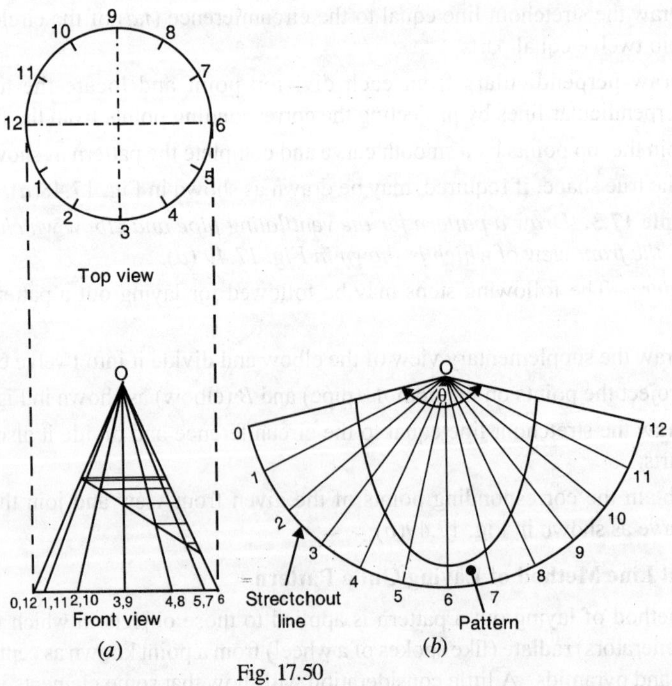

Fig. 17.50

Procedure. The following steps may be followed in laying out a pattern of a truncated cone.

1. First of all, draw the front and top view of the cone, as shown in Fig. 17.50(*a*).
2. The outer generators (*i.e.*, O-6 and O-12) represent the true length of all the elements (equal to *S*) and shall be the radius of the arc of the stretch out line.
3. The position of the cut points, in the front view, is obtained by drawing the lines parallel to the base and up to the slant side O-6.
4. Draw an arc with centre O and radius equal to O-6. This is the stretchout line for the pattern. The included angle (θ) may be calculated by the following relation :

$$\theta = \frac{R}{S} \times 360^{\circ}$$

where R = Radius of base of the cone.

5. Divide the arc into twelve equal parts and draw the lines O-0, O-1 etc. and cut these lines with corresponding radii measured from O along the line O-6.
6. The pattern is completed, as shown in Fig. 17.50(*b*).

17·21 Triangulation Method of Laying Out a Pattern

This method of laying out a pattern is applied to those objects which cannot be developed exactly as in case of *transition piece. The method consists of :

1. The surface to be developed is divided into a number of approximate triangles,
2. The true length of each side of triangle is found out, and
3. The true lengths are transferred in sequence to form the pattern.

Example 17.5. *Draw the pattern of a transition piece which connects two parallel circular and rectangular openings.*

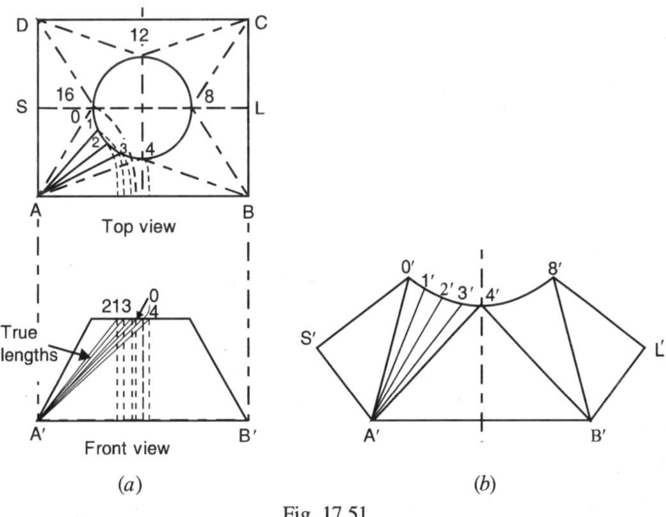

Fig. 17.51

Procedure. The pattern of a transition piece may be drawn as discussed in the following steps :

1. First of all, draw the front and top view, as shown in Fig. 17.51(*a*). From the top view, we see that the object is composed of four conical surfaces and forms isosceles triangles.

* The transition piece is that piece which connects openings of different shapes and sizes. In many cases, these are composed of plane surfaces and conical surfaces, the latter is developed by triangulation method. These pieces are used in ventilating, heating and other similar works.

2. Divide the circle into sixteen equal parts and draw the cone elements $A0, A1, A2$ etc.
3. In order to draw the pattern, as shown in Fig. 17.51(*b*), start on the line $A'S'$ being the base equal to AS and construct a right- angled triangle $A'S'0'$ in which $A'0'$ is equal to the true length of $A0$ *i.e.*, $A'0$ (measured from front view).
4. Now 0′ as centre and radius equal to the chord 01 (measured from top view) draw an arc. Again with A' as centre and $A'1$ as radius (true length of A1 measured from front view), cut the previous arc at 1′. In the similar way, points 2′, 3′, 4′, etc. are obtained.
5. By joining these points, a curve 0′8′ is obtained. Thus figure A′S′0′8′ L B′ A′ represents the half pattern of the transition piece.

QUESTIONS

1. What are the commonly used materials in sheet metal work ?
2. Write short notes on the following tools used in sheet metal workshop :
   (*a*) Steel square, (*b*) Semi-circular protractor,
   (*c*) Punches, and (*d*) Hand lever punch.
3. Why snips are used ? Explain, with the help of neat sketches, the straight and curved snips.
4. What do you understand by stakes ? Name the different types of stakes and explain two of them with neat sketches.
5. Discuss the various types of hammers used in sheet metal work.
6. Sketch and describe the use of folding bars.
7. Why a wire gauge is required in sheet metal shop ? Explain with a sketch.
8. How the cutting and notching of sheet metal is done with the help of snips ? Explain fully.
9. Describe the process of bending a sheet metal.
10. Explain the procedure for riveting two metal sheets.
11. What do you understand by soldering ? How the soldered joint is made ?
12. Write a complete process of folding edges of sheet metal.
13. What is a seam ? Describe the different types of seams.
14. Write short notes on the following sheet metal processes :
    (*a*) Hollowing or blocking, (*b*) Sinking.
    (*c*) Raising, and (*d*) Planishing.
15. Name the various types of machines used in sheet metal shop.
16. Discuss the various methods used in laying out a pattern.

OBJECTIVE TYPE QUESTIONS

1. Terne plate is a
   (*a*) steel sheet coated with pure tin
   (*b*) steel sheet coated with a mixture of molten tin and lead
   (*c*) soft steel coated with molten zinc
   (*d*) reddish coloured metal
2. A hand lever punch is used
   (*a*) for making small indentations
   (*b*) to mark the centres of holes to be drilled
   (*c*) for punching small holes in thin metal sheets
   (*d*) for making holes when too much holes are to be punched

3. A stake having a short tapered horn at one end and a long tapered horn at the other end, is called

(*a*) blow horn stake (*b*) beak horn stake

(*c*) creasing stake (*d*) bottom stake

4. A stake having two cylindrical horns of different diameters is known as

(*a*) candle mould stake (*b*) copper smith stake

(*c*) conductor stake (*d*) hatchet stake

5. Which of the following seam is used to join sections such as one pipe to another pipe or a flat plate to a pipe ?

(*a*) Lap seam (*b*) Dovetail seam

(*c*) Single seam (*d*) Grooved seam

6. In order to join the bottom of a container to its body, the seam used is

(*a*) flanged seam (*b*) single seam

(*c*) double seam (*d*) grooved seam

7. The process of beating the sheet metal into a particular shape, such as sauce pan or bowl, is called

(*a*) sinking (*b*) hollowing

(*c*) planishing (*d*) raising

8. A process applied to a sheet with the main object of stretching the metal with an improved surface, is known as

(*a*) sinking (*b*) hollowing

(*c*) planishing (*d*) raising

9. A beading or swedging machine is used

(*a*) to make depressions in metal such as in pipes, machine guards

(*b*) for shaping the metal sheets into cylindrical objects

(*c*) for cutting metal sheets

(*d*) none of the above

10. A sheet metal pattern is

(*a*) a three dimensional drawing of the object

(*b*) an isometric drawing of the object

(*c*) a full size drawing of the object

(*d*) none of the above

ANSWERS

1. (*b*)	2. (*d*)	3. (*a*)	4. (*c*)	5. (*b*)
6. (*a*)	7. (*b*)	8. (*c*)	9. (*a*)	10. (*c*)

18

Rivets and Screws

1. Introduction. 2. Material of Rivets. 3. Essential Qualities of a Rivet. 4. Manufacture of Rivets. 5. Methods of Riveting. 6. Types of Rivet Heads. 7. Types of Riveted Joints. 8. Lap Joint. 9. Butt Joint. 10. Important Terms used in Riveted Joints. 11. Caulking and Fullering. 12. Screw Thread. 13. Forms of Screw Threads. 14. Designation of Screw Threads. 15. Standard Dimensions of Screw Threads. 16. Screwed Joint. 17. Location of Screwed Joints. 18. Common Types of Screw Fastenings. 19. Locking Devices.

18·1 Introduction

A rivet is a short cylindrical bar with a head integral to it. The cylindrical portion of the rivet is called *shank or body* and lower portion of shank is known as *tail*, as shown in Fig. 18.1. The rivets are used to make permanent fastening between the plates such as in structural work, ship building, bridges, tanks and boiler shells. The riveted joints are widely used for joining light metals.

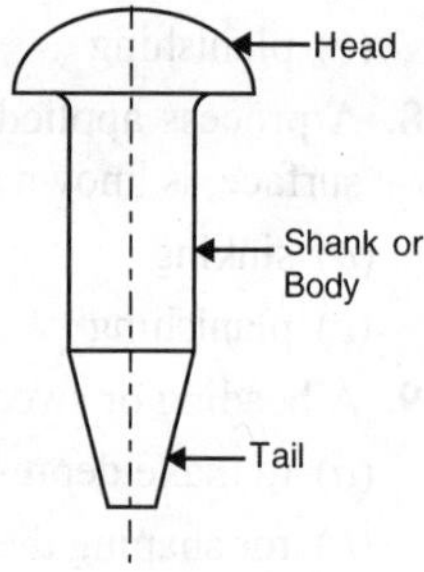

Fig. 18.1. Rivet parts.

The fastenings (*i.e.*, joints) may be classified into the following two groups :

1. Permanent fastenings, and
2. Temporary or detachable fastenings.

The *permanent fastenings* are those fastenings which can not be disassembled without destroying the connecting components. The examples of permanent fastenings in order of strength are soldered, brazed, welding and riveted joints.

The *temporary* or *detachable fastenings* are those fastenings which can be disassembled without destroying the connecting components. The examples of temporary fastenings are screwed, keys, cotters, pins and splined joints.

18·2 Material of Rivets

The material of the rivets must be tough and ductile. They are usually made of steel (low carbon steel or nickel steel), brass, aluminium or copper, but when strength and a fluid tight joint is the main consideration, then the steel rivets are used.

The rivets for general purposes shall be manufactured from steel conforming to the following Indian Standards :

(*a*) IS : 1148–1982 (Reaffirmed 1992) – Specification for hot rolled rivet bars (up to 40 mm diameter) for structural purposes ; or

(*b*) IS : 1149–1982 (Reaffirmed 1992) – Specification for high tensile steel rivet bars for structural purposes.

The rivets for boiler work shall be manufactured from material conforming to IS : 1990–1973 (Reaffirmed 1992) – Specification for steel rivets and stay bars for boilers.

Note : The steel for boiler construction should conform to IS : 2100–1970 (Reaffirmed 1992) – Specification for steel billets, bars and sections for boilers.

18·3 Essential Qualities of a Rivet

According to Indian standard, IS : 2998 – 1982 (Reaffirmed 1992), the material of a rivet must have a tensile strength not less than 40 N/mm^2 and elongation not less than 26 percent. The material must be of such quality that when in cold condition, the shank shall be bent on itself through 180° without cracking and after being heated to 650°C and quenched, it must pass the same test. The rivet when hot must flatten without cracking to a diameter 2.5 times the diameter of shank.

18·4 Manufacture of Rivets

According to Indian standard specifications, the rivets may be made either by cold heading or by hot forging. If rivets are made by the cold heading process, they shall subsequently be adequately heat treated so that the stresses set up in the cold heading process are eliminated. If they are made by hot forging process, care shall be taken to see that the finished rivets cool gradually.

18·5 Methods of Riveting

The function of rivets in a joint is to make a connection that has strength and tightness. The strength is necessary to prevent failure of the joint. The rightness is necessary in order to contribute to strength and to prevent leakage as in a boiler or in a ship hull.

When two plates are to be fastened together by a rivet as shown in Fig. 18.2 (*a*), the holes in the plates are punched and reamed or drilled. Punching is the cheapest method and is used for relatively thin plates and in structural work. Since punching injures the material around the hole, therefore, drilling is used in most pressure-vessel work. In structural and pressure vessel riveting, the diameter of the rivet hole is usually 1.5 mm larger than the nominal diameter of the rivet.

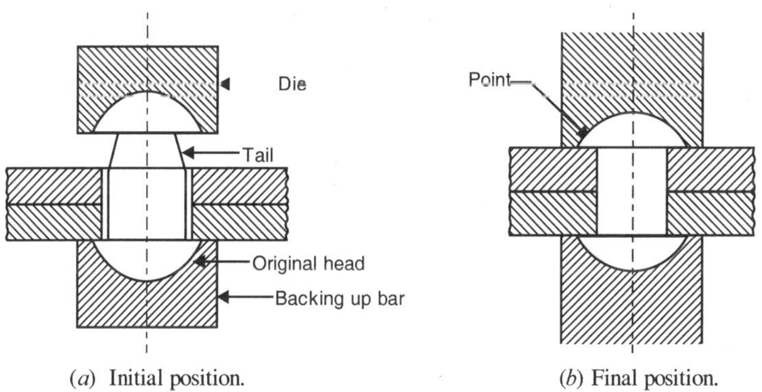

(*a*) Initial position. (*b*) Final position.

Fig. 18.2. Methods of riveting.

The plates are drilled together and then separated to remove any burrs or chips so as to have a tight flush joint between the plates. A cold rivet or a red hot rivet is introduced into the plates and the *point* (*i.e.*, second head) is then formed. When a cold rivet is used, the process is known as *cold riveting* and when a hot rivet is used, the process is known as *hot riveting*. The cold riveting process is used for structural joints while hot riveting is used to make leak proof joints.

The riveting may be done by hand or by a riveting machine. In hand riveting, the original rivet head is backed up by a hammer or heavy bar and then the die or set, as shown in Fig. 18.2 (*a*), is placed against the end to be headed and the blows are applied by a hammer. This causes the shank to expand thus filling the hole and the tail is converted into a *point* as shown in Fig. 18.2 (*b*). As the rivet cools, it tends to contract. The lateral contraction will be slight, but there will be a longitudinal tension introduced in the rivet which holds the plates firmly together.

In machine riveting, the die is a part of the hammer which is operated by air, hydraulic or steam pressure.

Notes : 1. For steel rivets upto 12 mm diameter, the cold riveting process may be used, while for larger diameter rivets, hot riveting process is used.

2. In case of long rivets, only the tail is heated and not the whole shank.

18·6 Types of Rivet Heads

According to Indian standard specifications, the rivet heads are classified into the following three types :

1. Rivets heads for general purposes (below 12 mm diameter) as shown in Fig. 18.3, according to IS : 2155 – 1982 (Reaffirmed 1996).

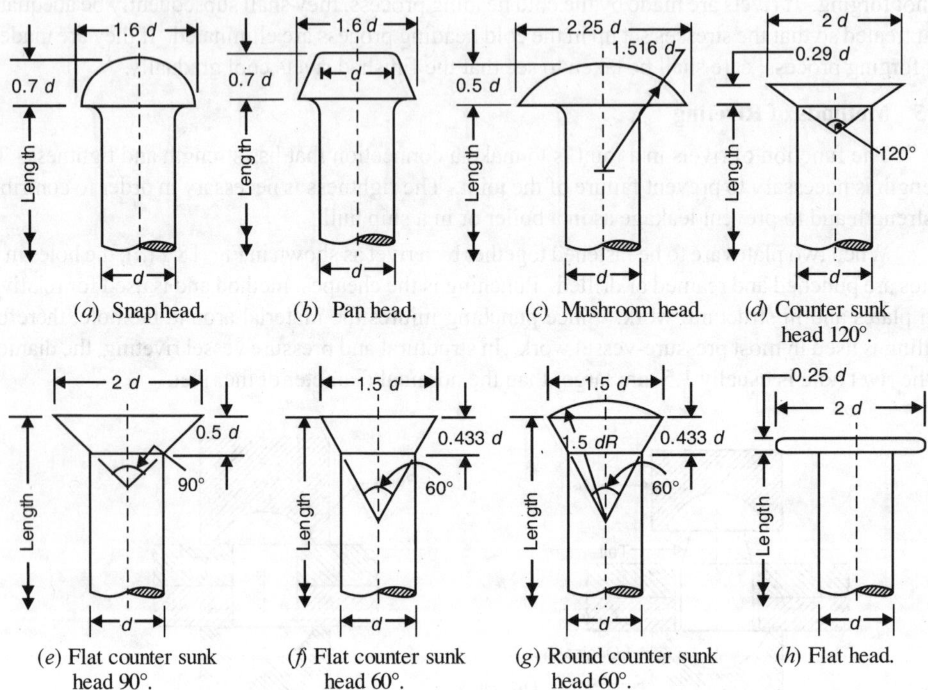

Fig. 18.3. Rivet heads for general purposes (below 12 mm diameter).

2. Rivet heads for general purposes (From 12 mm to 48 mm diameter) as shown in Fig. 18.4, according to IS : 1929 – 1982 (Reaffirmed 1996).

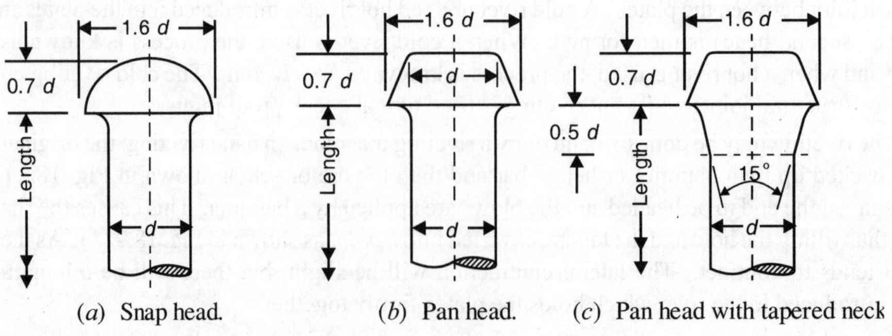

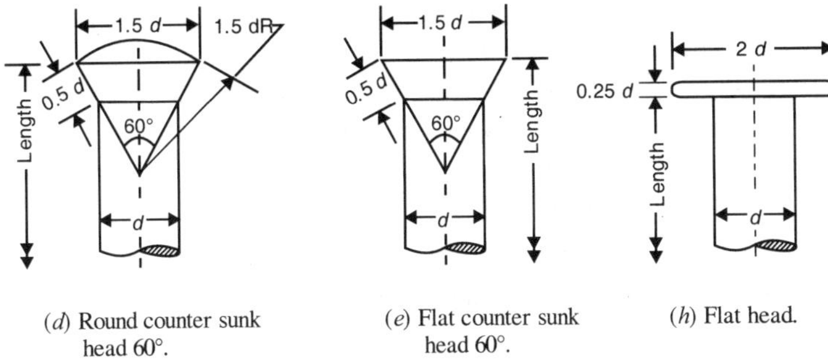

(*d*) Round counter sunk head 60°.

(*e*) Flat counter sunk head 60°.

(*h*) Flat head.

Fig. 18.4. Rivet heads for general purposes (from 12 mm to 48 mm diameter).

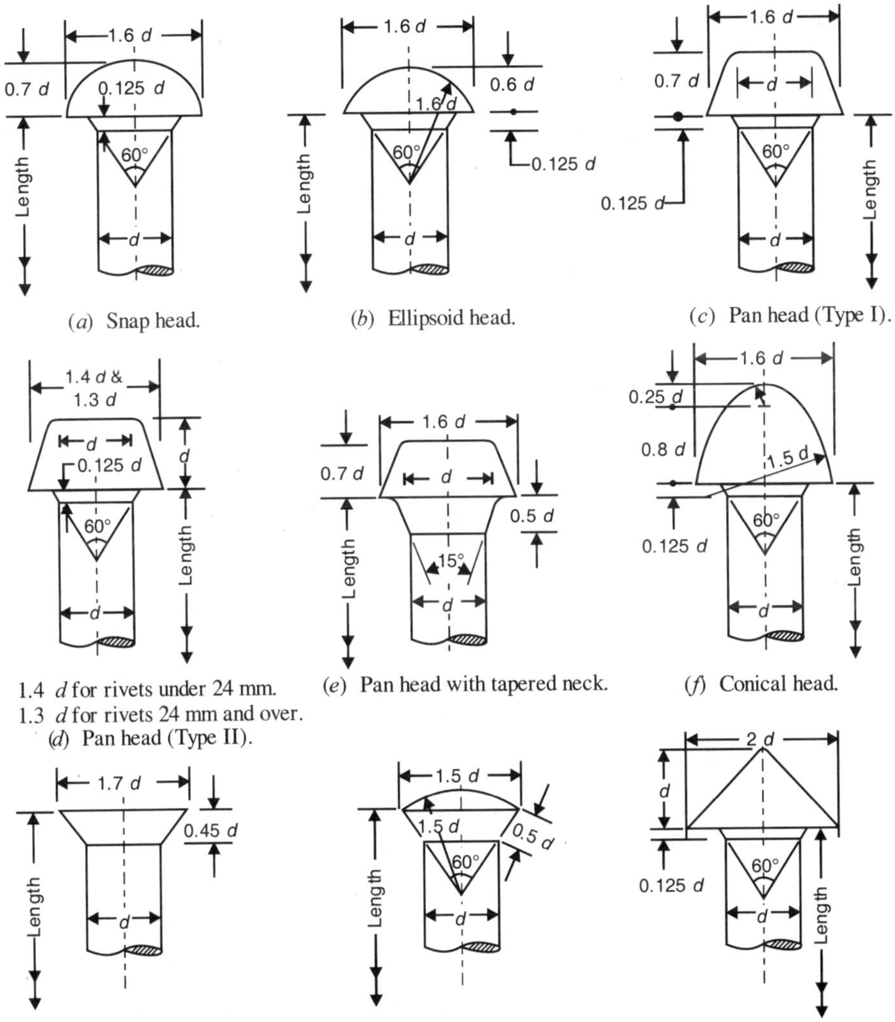

(*a*) Snap head.

(*b*) Ellipsoid head.

(*c*) Pan head (Type I).

1.4 *d* for rivets under 24 mm.
1.3 *d* for rivets 24 mm and over.
(*d*) Pan head (Type II).

(*e*) Pan head with tapered neck.

(*f*) Conical head.

(*g*) Counter-sunk head.

(*h*) Round counter sunk head.

(*i*) Steeple head.

Fig. 18.5. Rivet heads for boiler work.

3. Rivet heads for boiler work (from 12 mm to 48 mm diameter, as shown in Fig. 18.5, according to IS : 1928 – 1961 (Reaffirmed 1996).

The snap heads are usually employed for structural work and machine riveting. The counter sunk heads are mainly used for ship building where flush surfaces are necessary. The conical heads (also known as conoidal heads) are mainly used in case of hand hammering. The pan heads have maximum strength, but these are difficult to shape.

18·7 Types of Riveted Joints

Following are the two types of riveted joints, depending upon the way in which the plates are connected.

1. Lap joint, and 2. Butt joint.

18·8 Lap Joint

A lap joint is that in which one plate overlaps the other and the two plates are then riveted together.

18·9 Butt Joint

A butt joint is that in which the main plates are kept in alignment butting (*i.e.*, touching) each other and a cover plate (*i.e.*, strap) is placed either on one side or on both sides of the main plates. The cover plate is then riveted together with the main plates. Butt joints are of the following two types:

1. Single strap butt joint, and 2. Double strap butt joint.

In a *single strap butt joint*, the edges of the main plates butt against each other and only one cover plate is placed on one side of the main plates and then riveted together.

In a *double strap butt joint*, the edges of the main plates butt against each other and two cover plates are placed on both sides of the main plates and then riveted together.

In addition to the above, following are the types of riveted joints depending upon the number of rows of the rivets.

1. Single riveted joint, and 2. Double riveted joint.

A *single riveted joint* is that in which there is a single row of rivets in a lap joint as shown in Fig. 18.6 (*a*) and there is a single row of rivets on each side in a butt joint as shown in Fig. 18.8.

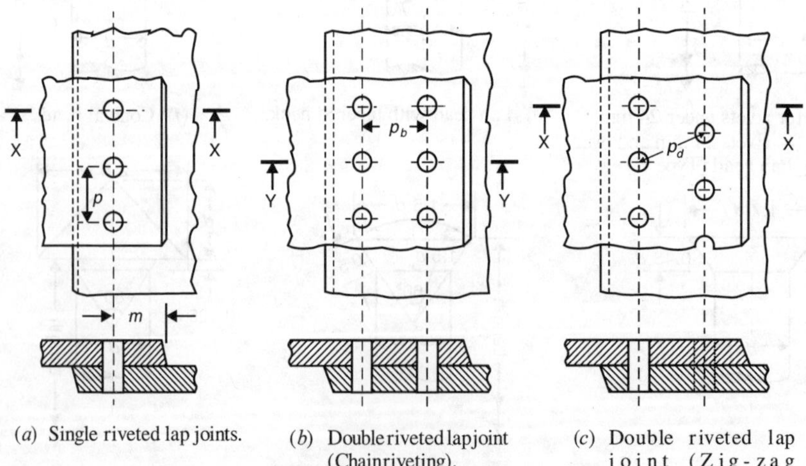

(*a*) Single riveted lap joints. (*b*) Double riveted lap joint (Chain riveting). (*c*) Double riveted lap joint (Zig-zag

Fig. 18.6. Single and double riveted lap joints,

A *double riveted joint* is that in which there are two rows of rivets in a lap joint as shown in Fig. 18.6 (*b*) and (*c*) and there are two rows of rivets on each side in a butt joint as shown in Fig. 18.9.

Similarly the joints may be *triple riveted* or *quadruple riveted*.

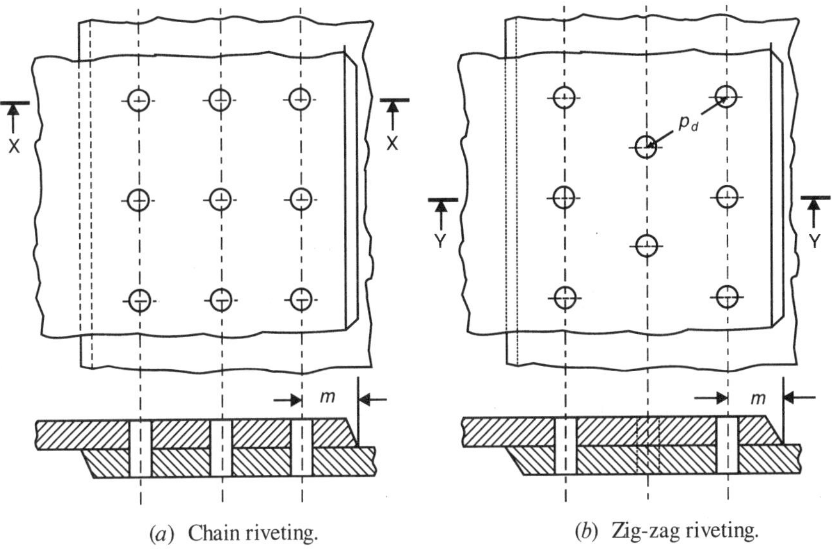

(*a*) Chain riveting. (*b*) Zig-zag riveting.

Fig. 18.7. Triple riveted lap joint.

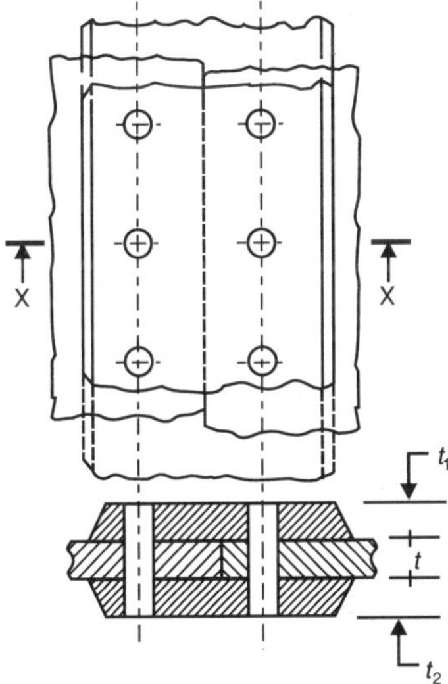

Fig. 18.8. Single riveted double strap butt joint.

Note : When the rivets in the various rows are opposite to each other, as shown in Fig. 18.6 (*b*), then the joint is said to be *chain riveted*. On the other hand, if the rivets in the adjacent rows are staggered in such a way that every rivet is in the middle of the two rivets of the opposite row as shown in Fig. 18.6 (*c*), then the joint is said to be *zig-zag riveted*.

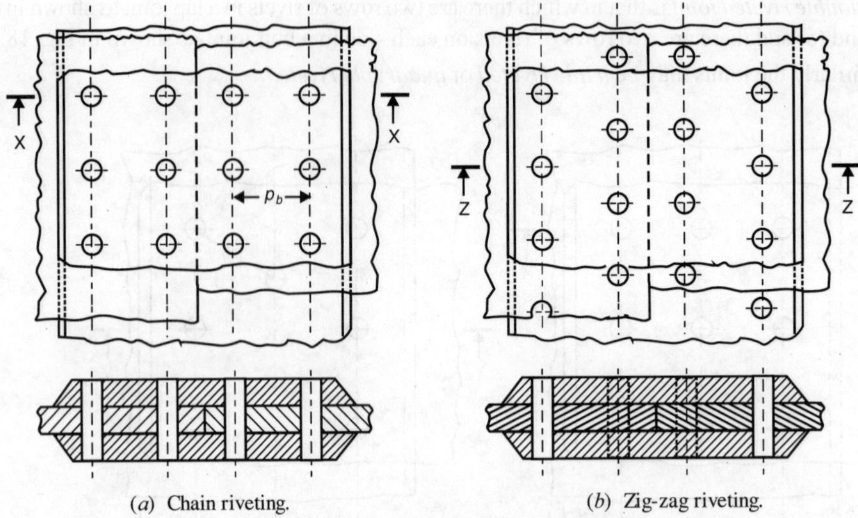

(a) Chain riveting. (b) Zig-zag riveting.

Fig. 18.9. Double riveted double strap (equal) butt joints.

18·10 Important Terms Used in Riveted Joints

The following terms in connection with the riveted joints are important from the subject point of view :

1. *Pitch.* It is the distance from the centre of one rivet to the centre of the next rivet measured parallel to the seam as shown in Fig. 18.6. It is usually denoted by *p*.

2. *Back pitch.* It is the perpendicular distance between the centre lines of the successive rows as shown in Fig. 18.6. It is usually denoted by p_b.

3. *Diagonal pitch.* It is the distance between the centres of the rivets in adjacent rows of zig -zag riveted joint as shown in Fig. 18.6. It is usually denoted by p_d.

4. *Margin* or *marginal pitch.* It is the distance between the centre of rivet hole to the nearest edge of the plate as shown in Fig. 18.6. It is usually denoted by *m*.

18·11 Caulking and Fullering

In order to make the joints leak proof or fluid tight in pressure vessels like steam boilers, air receivers and tanks etc., a process known as *caulking* is employed. In this process, a narrow blunt

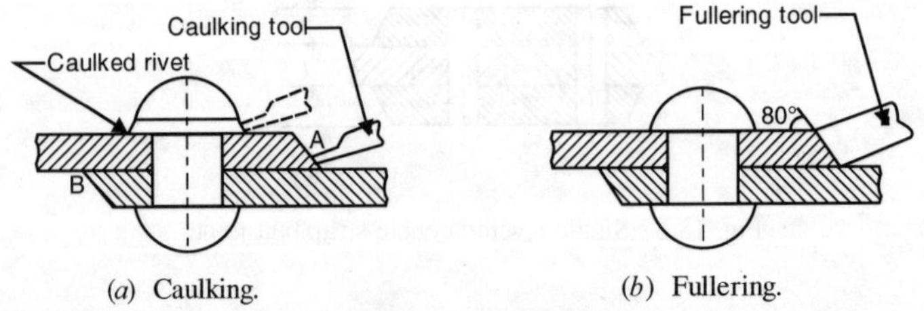

(a) Caulking. (b) Fullering.

Fig. 18.10. Caulking and fullering.

tool called caulking tool, about 5 mm thick and 38 mm in breadth, is used. The edge of the tool is ground to an angle of 80º. The tool is moved after each blow along the edge of the plate, which is planed to a bevel of 75º to 80º to facilitate the forcing down of edge. It is seen that the tool burrs down the plate at *A* in Fig. 18.10 (*a*) forming a metal to metal joint. In actual practice, both the edges at A and B are caulked. The head of the rivets as shown at C are also turned down with a caulked tool to make a joint steam tight. A great care is taken to prevent injury to the plate below the tool.

A more satisfactory way of making the joints staunch is known as *fullering* which has largely superseded caulking. In this case, fullering tool with a thickness at the end equal to that of the plate is used in such a way that the greatest pressure due to the blows occur near the joint, giving a clean finish, with less risk of damaging the plate. A fullering process is shown in Fig. 18.10 (*b*).

18·12 Screw Thread

A screw thread is formed by cutting a continuous helical groove on a cylindrical surface. A screw made by cutting a single helical groove on the cylinder is known as *single threaded* (or single-start) screw and if a second thread is cut in the space between the grooves of the first, a *double threaded* (or double-start) screw is formed. Similarly, triple and quadruple (*i.e.*, multiple-start) threads may be formed. The helical grooves may be cut either *right hand* or *left hand*.

The following terms used in screw threads, as shown in Fig. 18.11, are important from the subject point of view :

1. *Major diameter*. It is the largest diameter of an external or internal screw thread. The screw is specified by this diameter. It is also known as *outside* or *nominal diameter*.

2. *Minor diameter*. It is the smallest diameter of an external or internal screw thread. It is also known as *core* or *root diameter*.

3. *Pitch diameter*. It is the diameter of an imaginary cylinder, on a cylindrical screw thread, the surface of which would pass through the thread at such points as to make equal the width of the thread and the width of the spaces between the threads. It is also called an *effective diameter*. In a nut and bolt assembly, it is the diameter at which the ridges on the bolt are in complete touch with the ridges of the corresponding nut.

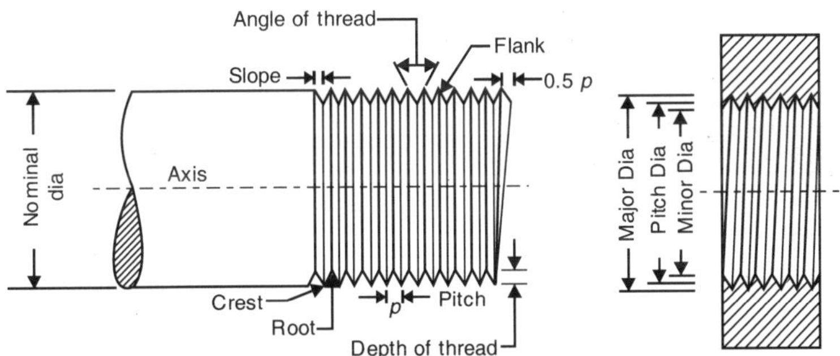

Fig. 18.11. Terms used in screw threads.

4. *Pitch*. It is the distance from a point on one thread to the corresponding point on the next. This is measured in an axial direction between corresponding points in the same axial plane. Mathematically,

$$\text{Pitch} = \frac{1}{\text{No. of threads per unit length of screw}}$$

5. *Lead*. It is the distance between two corresponding points on the same helix. It may also be defined as the distance which a screw thread advances axially in one rotation of the nut. Lead is equal to the pitch in case of single start threads, it is twice the pitch in double start, thrice the pitch in triple start and so on.

6. *Crest*. It is the top surface of the thread.

7. *Root*. It is the bottom surface created by the two adjacent flanks of the thread.

8. *Depth of thread.* It is the perpendicular distance between the crest and root.

9. *Flank.* It is the surface joining the crest and root.

10. *Angle of thread.* It is the angle included by the flanks of the thread.

18·13 Forms of Screw Threads

The following are the various forms of screw threads.

1. *British standard whitworth (B.S.W.) thread.* This is a British standard thread profile and has coarse pitches. It is a symmetrical *V*-thread in which the angle between the flanks, measured in an axial plane, is 55°. These threads are found on bolts and screwed fastenings for several purposes. The various proportions of B.S.W. threads are shown in Fig. 18.12.

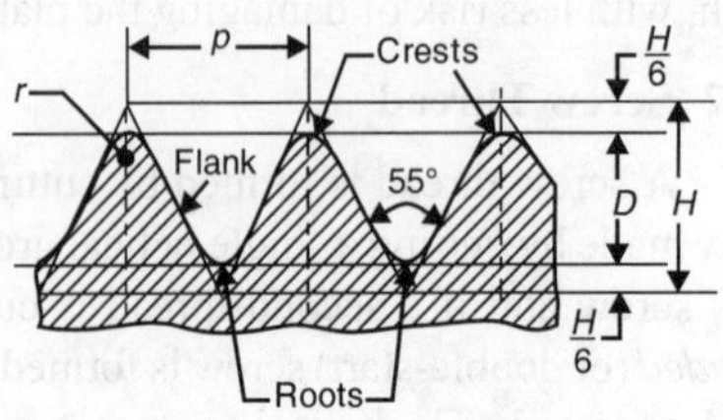

$H = 0.96\,p$; $D = 0.64\,p$; $r = 0.1373\,p$.

Fig. 18.12. British standard whitworth thread

The British standard threads with fine pitches (B.S.F.) are used where great strength at the root is required. These threads are also used for fine adjustments and where the connected parts are subjected to increased vibrations as in aero and automobile work.

The British standard pipe (B.S.P.) threads with fine pitches are used for steel and iron pipes and tubes carrying fluids. In external pipe threading, the threads are specified by the bore of the pipe.

2. *British association (B.A.) thread.* This is a B.S.W. thread with fine pitches. The proportions of the B.A. thread are shown in Fig. 18.13. These threads are used on screws for precision work.

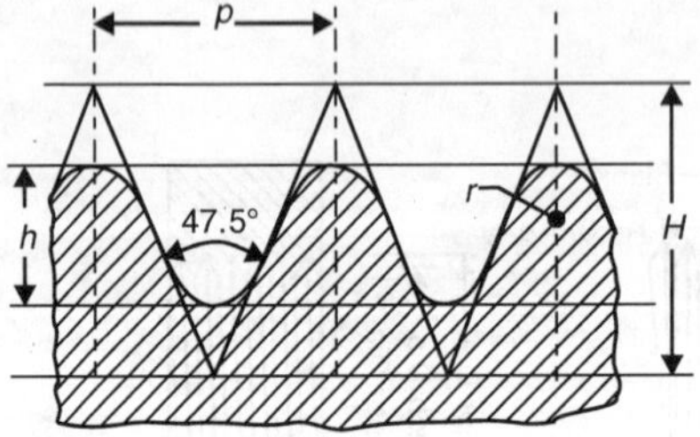

$H = 1.13634\,p$; $h = 0.6p$; $r = 0.18083\,p$.

Fig. 18.13. British association thread.

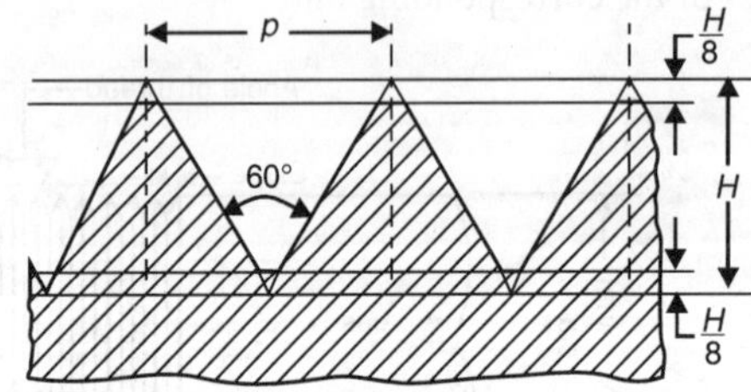

$H = 0.866\,p$.

Fig. 18.14. American national standard thread.

3. *American national standard thread.* The American national standard or U.S. or Seller's thread has flat crests and roots. The flat crest can withstand more rough usage than sharp *V*-threads. These threads are used for general purposes *e.g.*, on bolts, nuts, screws and tapped holes. The various proportions are shown in Fig. 18.14.

4. *Unified standard thread.* The three countries *i.e.*, Great Britain, Canada and United States came to an agreement for a common screw thread system with the included angle of 60° in order to facilitate the exchange of machinery. The thread has rounded crests and roots, as shown in Fig. 18.15.

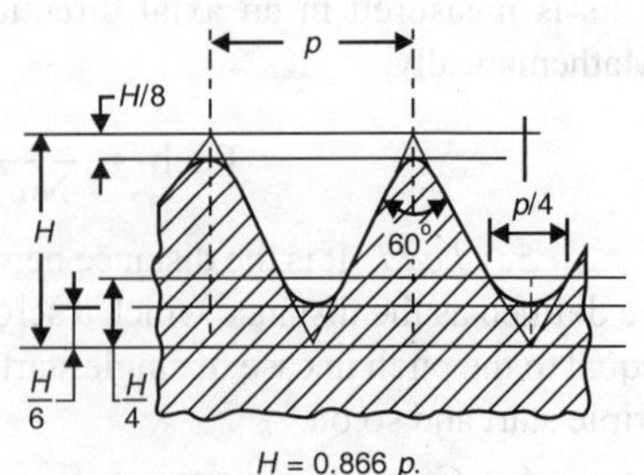

$H = 0.866\,p$.

Fig. 18.15. Unified standard thread.

5. *Square thread.* The square threads, because of their high efficiency, are widely used for transmission of power in either directions. Such types of threads are usually found on the feed mechanisms of machine tools, valves,

spindles, screw jacks etc. The square threads are not so strong as *V*-threads but they offer less frictional resistance to motion than Whitworth threads. The pitch of the squre thread is often taken twice that of a B.S.W. thread of the same diameter. The various porportions are shown Fig. 18.16.

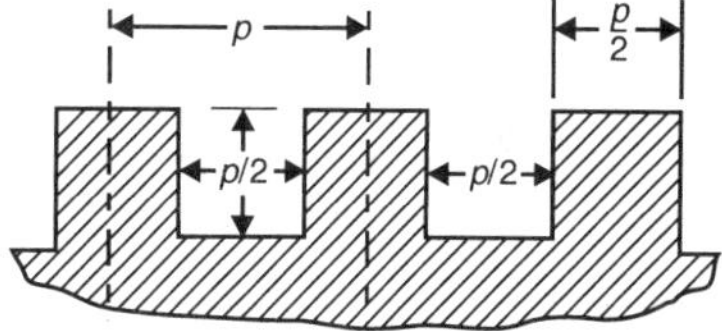

Fig. 18.16. Square thread.

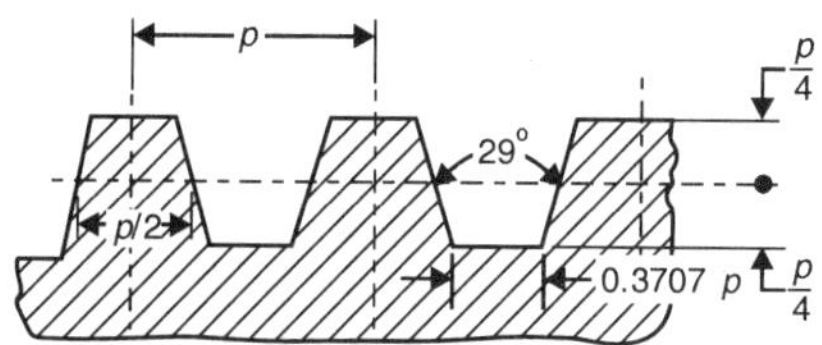

Fig. 18.17. Acme thread.

6. *Acme thread.* It is a modification of square thread. It is much stronger than square thread and can be easily produced. These threads are frequently used on screw cutting lathes, brass valves, cocks and bench vices. When used in conjuction with a split nut, as on the lead screw of a lathe, the tapered sides of the thread facilitate ready engagement and disengagement of the halves of the nut when required. The various proportions are shown in Fig. 18.17.

7. *Knuckle thread.* It is also a modification of square thread. It has rounded top and bottom, as shown in Fig. 18.18. It can be cast or rolled easily and cannot economically be made on a machine. These threads are used for rough and ready work. They are usually found on railway carriage couplings, hydrants, necks of glass bottles and large moulded insulators used in electrical trade.

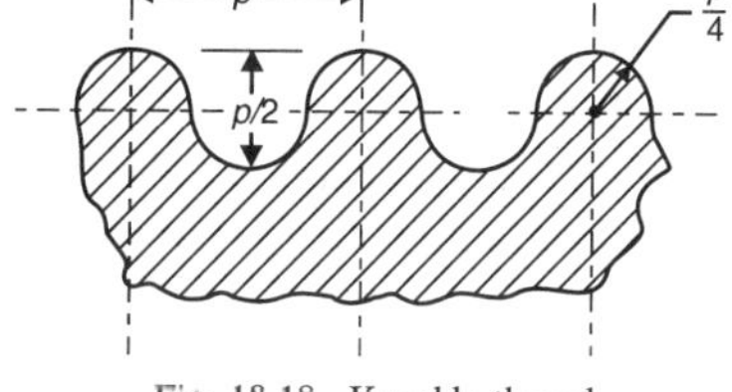

Fig. 18.18. Knuckle thread.

8. *Buttress thread.* It is used for transmission of power in one direction only. The force is transmitted almost parallel to the axis. This thread unites the advantage of both square and *V*-threads. It has low frictional resistance characteristics of the square thread and have the same strength as that of *V*-thread. The spindles of bench vices are usually provided with buttress threads. The various proportions of buttress thread are shown in Fig. 18.19.

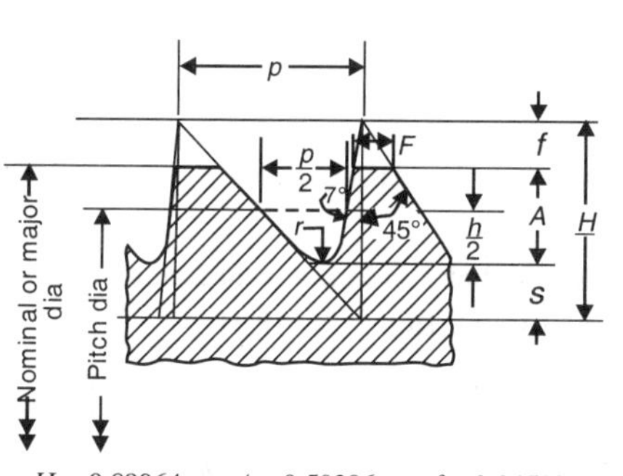

$H = 0.89064\,p$; $A = 0.50286\,p$; $f = 0.24532\,p$;
$s = 0.13946\,p$; $F = 0.27544\,p$; and $r = 0.12055\,p$.

Fig. 18.19. Buttress thread.

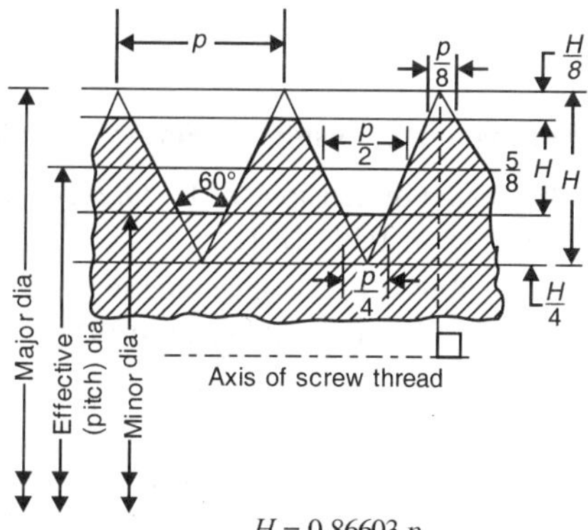

$H = 0.86603\,p$.

Fig. 18.20. Basic profile of the metric thread.

9. *Metric thread.* It is an Indian standard thread and is similar to B.S.W. threads. It has an included angle of 60° instead of 55°. The basic profile of the thread is shown in Fig. 18.20 and the design profile of the nut and bolt is shown in Fig. 18.21.

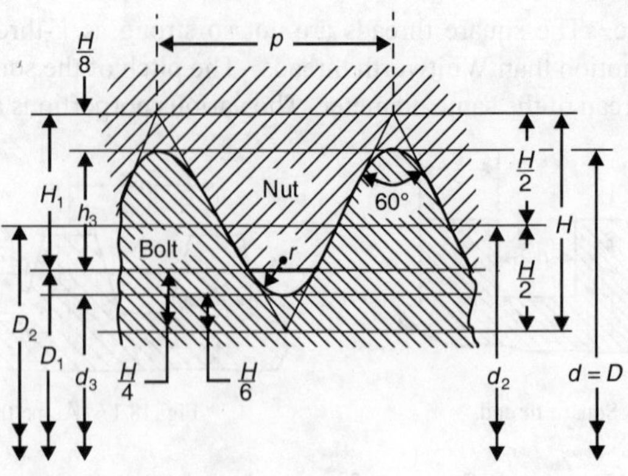

$$H = 0.86603\,p\,;\,D_2 = d_2 = d - 0.6495\,p\,;\,D_1 = d - 1.0825\,p\,;$$
$$d_3 = d - 1.2268\,p\,;\,H_1 = \frac{5}{8}H\,;\,h_3 = \frac{17}{24}H\,;\,r = \frac{H}{6}$$

Fig. 18.21. Design profile of the nut and bolt.

18·14 Designation of Screw Threads

According to Indian standards, IS : 4218 (Part IV) 1976 (Reaffirmed 1996), the complete designation of the screw thread shall include

1. *Size designation.* The size of the screw thread is designated by the letter '*M*' followed by the diameter and pitch, the two being separated by the sign ×. When there is no indication of the pitch, it shall mean that a coarse pitch is implied.

2. *Tolerance designation.* This shall include

(*a*) A figure designating tolerance grade as indicated below :
'7' for fine grade, '8' for normal (medium) grade, and '9' for coarse grade.

(*b*) A letter designating the tolerance position as indiciated below :
'*H*' for unit thread, '*d*' for bolt thread with allowance, and '*h*' for bolt thread without allowance.

For example, A bolt thread of 6 mm size of coarse pitch and with allowance on the threads and normal (medium) tolerance grade is designated as *M*6-8*d*.

18·15 Standard Dimensions of Screw Threads

The design dimensions of I.S.O. screw threads for screws, bolts and nuts of coarse and fine series are shown in Table 18.1.

Table 18.1. Design dimensions of screw thread, bolts and nuts according to IS : 4218 (Part III) 1976 (Reaffirmed 1996) (Refer Fig. 18.21)

Designation	*Pitch mm*	*Major or nominal diameter Nut and Bolt (d = D) mm*	*Effective or pitch diameter Nut and Bolt (d_p) mm*	*Minor or core diameter (d_c) mm*		*Depth of thread (bolt) mm*	*Stress area mm^2*
				Bolt	*Nut*		
(1)	(2)	(3)	(4)	(5)	(6)	(7)	(8)
Coarse series							
M 0.4	0.1	0.400	0.335	0.277	0.292	0.061	0.074

(1)	(2)	(3)	(4)	(5)	(6)	(7)	(8)
M 0.6	0.15	0.600	0.503	0.416	0.438	0.092	0.166
M 0.8	0.2	0.800	0.670	0.555	0.584	0.123	0.295
M 1	0.25	1.000	0.838	0.693	0.729	0.153	0.460
M 1.2	0.25	1.200	1.038	0.893	0.929	0.158	0.732
M 1.4	0.3	1.400	1.205	1.032	1.075	0.184	0.983
M 1.6	0.35	1.600	1.373	1.171	1.221	0.215	1.27
M 1.8	0.35	1.800	1.573	1.371	1.421	0.215	1.70
M 2	0.4	2.000	1.740	1.509	1.567	0.245	2.07
M 2.2	0.45	2.200	1.908	1.648	1.713	0.276	2.48
M 2.5	0.45	2.500	2.208	1.948	2.013	0.276	3.39
M 3	0.5	3.000	2.675	2.387	2.459	0.307	5.03
M 3.5	0.6	3.500	3.110	2.764	2.850	0.368	6.78
M 4	0.7	4.000	3.545	3.141	3.242	0.429	8.78
M 4.5	0.75	4.500	4.013	3.580	3.688	0.460	11.3
M 5	0.8	5.000	4.480	4.019	4.134	0.491	14.2
M 6	1	6.000	5.350	4.773	4.918	0.613	20.1
M 7	1	7.000	6.350	5.773	5.918	0.613	28.9
M 8	1.25	8.000	7.188	6.466	6.647	0.767	36.6
M 10	1.5	10.000	9.026	8.160	8.876	0.920	58.3
M 12	1.75	12.000	10.863	9.858	10.106	1.074	84.0
M 14	2	14.000	12.701	11.546	11.835	1.227	115
M 16	2	16.000	14.701	13.546	13.835	1.227	157
M 18	2.5	18.000	16.376	14.933	15.294	1.534	192
M 20	2.5	20.000	18.376	16.933	17.294	1.534	245
M 22	2.5	22.000	20.376	18.933	19.294	1.534	303
M 24	3	24.000	22.051	20.320	20.752	1.840	353
M 27	3	27.000	25.051	23.320	23.752	1.840	459
M 30	3.5	30.000	27.727	25.706	26.211	2.147	561
M 33	3.5	33.000	30.727	28.706	29.211	2.147	694
M 36	4	36.000	33.402	31.093	31.670	2.454	817
M 39	4	39.000	36.402	34.093	34.670	2.454	976
M 42	4.5	42.000	39.077	36.416	37.129	2.760	1104
M 45	4.5	45.000	42.077	39.416	40.129	2.760	1300
M 48	5	48.000	44.752	41.795	42.587	3.067	1465
M 52	5	52.000	48.752	45.795	46.587	3.067	1755
M 56	5.5	56.000	52.428	49.177	50.046	3.067	2022
M 60	5.5	60.000	56.428	53.177	54.046	3.374	2360
Fine series							
M 8 × 1	1	8.000	7.350	6.773	6.918	0.613	39.2
M 10 × 1.25	1.25	10.000	9.188	8.466	8.647	0.767	61.6
M 12 × 1.25	1.25	12.000	11.184	10.466	10.647	0.767	92.1
M 14 × 1.5	1.5	14.000	13.026	12.160	12.376	0.920	125
M 16 × 1.5	1.5	16.000	15.026	14.160	14.376	0.920	167
M 18 × 1.5	1.5	18.000	17.026	16.160	16.373	0.920	216
M 20 × 1.5	1.5	20.000	19.026	18.160	18.376	0.920	272

(1)	(2)	(3)	(4)	(5)	(6)	(7)	(8)
M 22 × 1.5	1.5	22.000	21.026	20.160	20.376	0.920	333
M 24 × 2	2	24.000	22.701	21.546	21.835	1.227	384
M 27 × 2	2	27.000	25.701	24.546	24.835	1.227	496
M 30 × 2	2	30.000	28.701	27.546	27.835	1.227	621
M 33 × 2	2	33.000	31.701	30.546	30.835	1.227	761
M 36 × 3	3	36.000	34.051	32.319	32.752	1.840	865
M 39 × 3	3	39.000	37.051	35.319	35.752	1.840	1028

Note : In case the table is not available, then the core diameter (d_c) may be taken as 0.84 *d*, where *d* is the major diameter.

18·16 Screwed Joint

A screwed joint is mainly composed of two elements *i.e.*, bolt and nut. Screwed joints are widely used where the machine parts are required to be readily connected or disconnected without damage to the machine or the fastening. This may be for the purpose of holding or adjustment in assembly or service inspection, repair, or holding or adjustment in assembly or service inspection, repair, or replacement or it may be for the manufacturing or assembly reasons.

The following are the advantages and disadvantages of the screwed joints :

Advantages

1. Screwed joints are highly reliable in operation.
2. Screwed joints are convenient to assemble and disassemble.
3. A wide range of screwed joints may be adopted to various operating conditions.
4. Screws are relatively cheaper to produce due to standardisation and highly efficient manufacturing processes.

Disadvantages

The main disadvantage of the screwed joints is the stress concentration in the threaded portions which are vulnerable points under variable load conditions.

Note : The strength of the screwed joints is not comparable with that of riveted or welded joints.

18·17 Location of Screwed Joints

The choice of type of fastenings and its location are very important. The fastenings should be located in such a way so that they will be subjected to tensile and/or shear loads and bending of the fastening should be reduced to a minimum. The bending of the fastening due to misalignment, tightening up loads, or external loads are responsible for many failures. In order to relieve fastenings of bending stresses, the use of clearance spaces, spherical seat washers, or other devices may be used.

18·18 Common Types of Screw Fastenings

Following are the common types of screw fastenings :

1. *Through bolt*. A through bolt (or simply a bolt) is shown in Fig. 18.22 (*a*). It is a cylindrical bar with threads for the nut at one end and head at the other end. The cylindrical part of the bolt is known as shank. It is passed through drilled holes in the two parts to be fastened together and clamp them securely to each other as the nut is screwed on to the threaded end. The through bolts may or may not have a machined finish and are made with either hexagonal or square heads. A through bolt should pass easily in the holes, when put under tension by a load along its axis. If the load acts perpendicular to the axis, tending to slide one of the connected parts along the other end thus subjecting it to shear the holes should be reamed so that the bolt shank fits snugly therein. The

through bolts according to their usage may be known as machine bolts, carriage bolts, automobile bolts, eye bolts etc.

2. *Tap bolt*. A tap bolt or screw differs from a bolt. It is screwed into a tapped hole of one of the parts to be fastened without the nut, as shown in Fig. 18.22 (*b*).

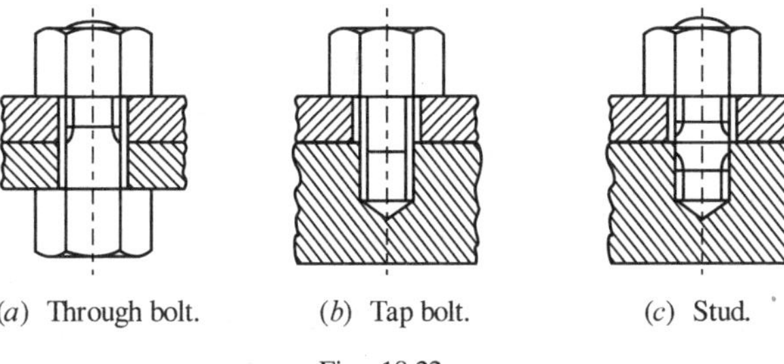

(*a*) Through bolt. (*b*) Tap bolt. (*c*) Stud.

Fig. 18.22

3. *Studs*. A stud is a round bar threaded at both ends. One end of the stud is screwed into a tapped hole of the parts to be fastened, while the other end receives a nut on it, as shown in Fig. 18.22 (*c*). Studs are chiefly used instead of tap bolts for securing various kinds of covers *e.g.*, covers of engine and pump cylinders, valves, chests, etc.

This is due to the fact that when tap bolts are unscrewed or replaced, they have a tendency to break the threads in the hole. This disadvantage is overcome by the use of studs.

4. *Cap screws*. The cap screws are similar to tap bolts except that they are of small size and a variety of shapes of heads are available, as shown in Fig. 18.23.

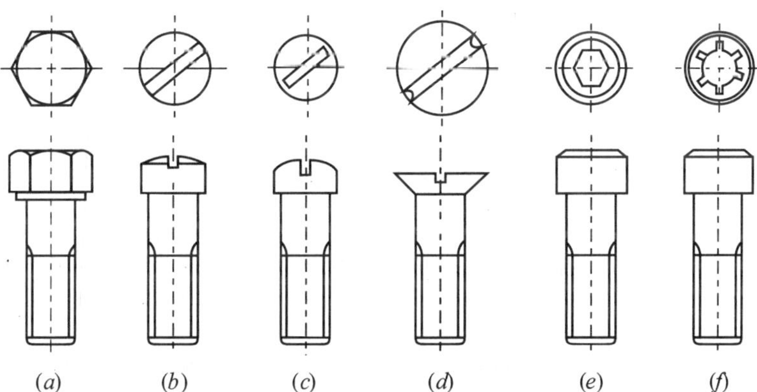

(*a*) (*b*) (*c*) (*d*) (*e*) (*f*)

(*a*) Hexagonal head ; (*b*) Fillister head ; (*c*) Round head ; (*d*) Flat head ; (*e*) Hexagonal socket ; (*f*) Fluted socket.

Fig 18.23. Cap screws.

5. *Machine screws*. These are similar to cap screws with the head slotted for a screw driver. These are generally used with a nut.

6. *Set screws*. The set screws are shown in Fig. 18.24. These are used to prevent relative motion between the two parts. A set screw is screwed through a thread hole in one part so that its point (*i.e.*, end of the screw) presses against the other part. This resists the relative motion between the two parts by means of

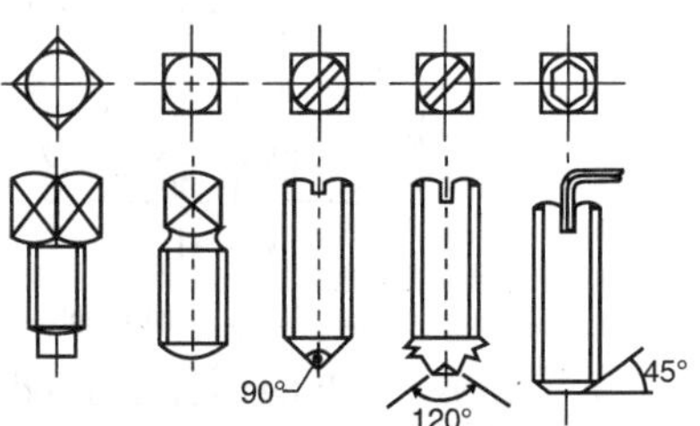

Fig. 18.24. Set screws.

friction between the point of the screw and one of the parts. They may be used instead of key to prevent relative motion between a hub and a shaft in light power transmission members. They may also be used in connection with a key, where they prevent relative axial motion of the shaft, key and hub assembly.

18·19 Locking Devices

Ordinary thread fastenings, generally, remain tight under static loads, but many of these fastenings become loose under the action of variable loads or when machine is subjected to vibrations. The loosening of fastening is very dangerous and must be prevented. In order to prevent this, a large number of locking devices are available, some of which are discussed below :

1. *Jam or lock nut*. A most common locking device is a jam, lock or check nut. It has about one-half to two-third thickness of the standard nut. The thin lock nut is first tightened down with ordinary force, and then the upper nut (*i.e.*, thicker nut) is tightened down upon it, as shown in Fig. 18.25 (*a*). The upper nut is then held tightly while the lower one is slackened back against it.

In slackening back the lock nut, a thin spanner is required which is difficult to find in many shops. Therefore, to overcome this difficultly, a thin nut is placed on the top as shown in Fig. 18.25 (*b*).

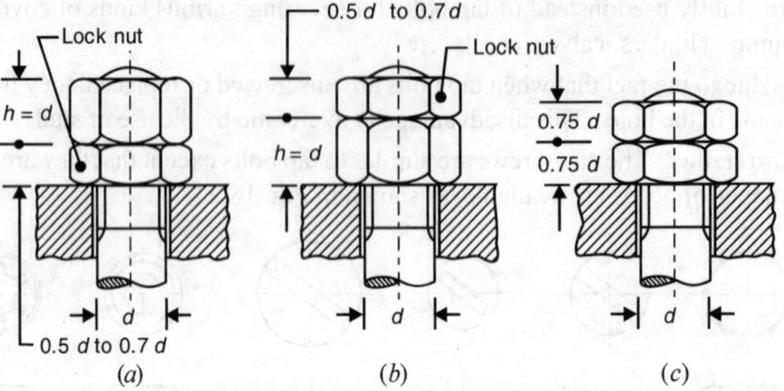

Fig. 18.25. Jam or lock nut.

If the nuts are really tightened down as they should be, the upper nut carries a greater tensile load than the bottom one. Therefore, the top nut should be thicker one with a thin nut below it because it is desirable to put whole of the load on the thin nut. In order to overcome both the difficulties, both the nuts are made of the same thickness as shown in Fig. 18.25 (*c*).

2. *Castle nut*. It consists of a hexagonal portion with a cylindrical upper part which is slotted in line with the centre of each face, as shown in Fig. 18.26. The split pin passes through two slots in the nut and a hole in the bolt, so that a positive lock is obtained unless the pin shears. It is extensively used on jobs subjected to sudden shocks and considerable vibration such as in automobile industry.

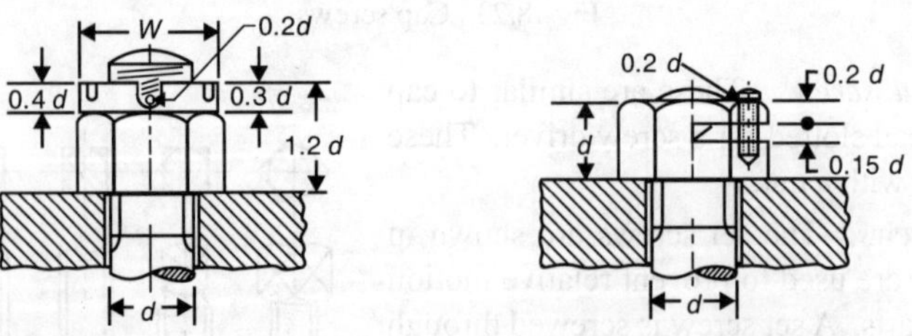

Fig. 18.26. Castle nut. Fig. 18.27. Sawn nut.

3. *Sawn nut*. It has a slot sawed about half way through, as shown in Fig. 18.27. After the nut is screwed down, the small screw is tightened which produces more friction between the nut and

the bolt. This prevents the loosening of nut.

4. *Penn, ring or grooved nut.* It has a upper portion hexagonal and a lower part cylindrical as shown in Fig. 18.28. It is largely used where bolts pass through connected pieces reasonably near their edges such as in marine type connecting rod ends. The bottom portion is cylindrical and is recessed to receive the tip of the locking set screw. The bolt hole requires counter-boring to receive the cylindrical portion of the nut. In order to prevent bruising of the latter by the case hardened tip of the set screw, it is recessed.

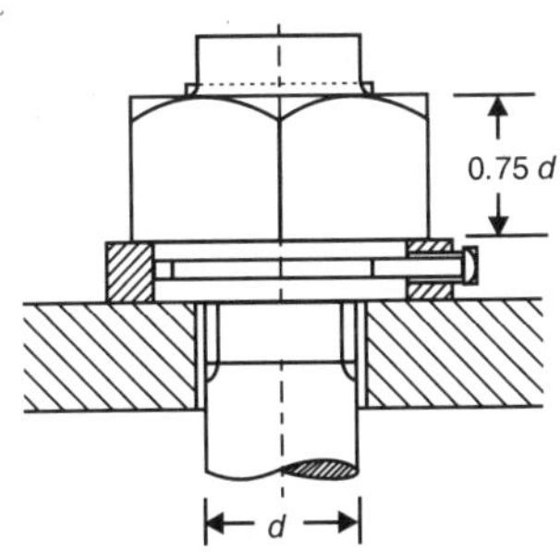

Fig. 18.28. Penn, ring or grooved nut.

5. *Locking with pin.* The nuts may be locked by means of a taper pin or cotter pin passing through the middle of the nut as shown in Fig. 18.29 (*a*). But a split pin is often driven through the bolt above the nut, as shown in Fig. 18.29 (*b*).

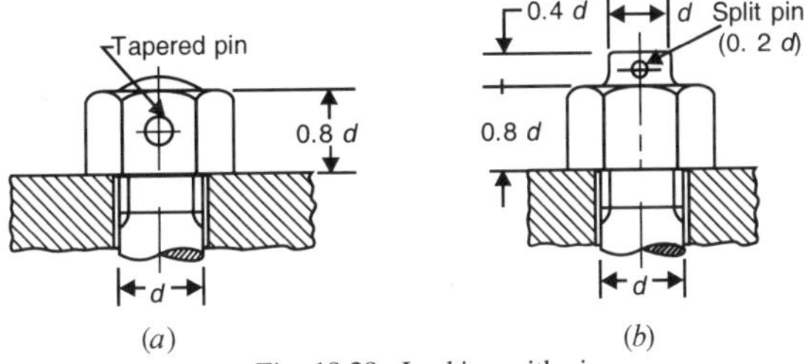

Fig. 18.29. Locking with pin.

6. *Locking with plate.* A form of stop plate or locking plate is shown in Fig. 18.30. The nut can be adjusted and subsequently locked through angular intervals of 30° by using these plates.

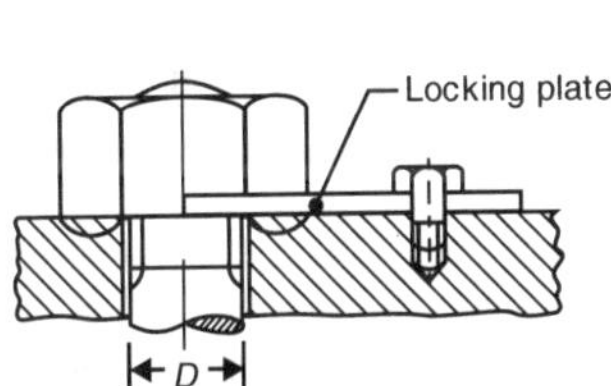

Fig. 18.30. Locking with plate.

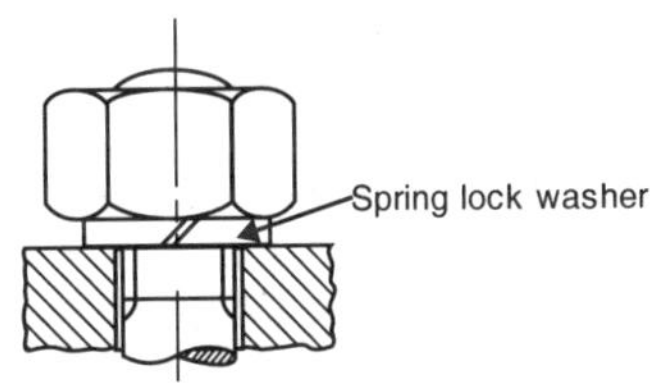

Fig. 18.31. Spring lock washer

7. *Spring lock washer.* A spring lock washer is shown in Fig. 18.31. As the nut tightens the washer against the piece below, one edge of the washer is caused to dig itself into that piece, thus increasing the resistance so that the nut will not loosen so easily. There are many kinds of spring lock washers manufactured, some of which are fairly effective.

QUESTIONS

1. Classify the rivet heads according to Indian Standard.
2. How the rivets are manufactured and state the material used ?
3. Explain the method of making a riveted joint.
4. Describe the common types of riveted joints. Illustrate your answer with neat sketches.
5. Why the caulking and fullering is employed in a riveted joint ?
6. What do you understand by single start and double start threads ?

7. Define the following terms :

(*a*) Major diameter ; (*b*) Minor diameter ; (*c*) Pitch ; and (*d*) Lead.

8. Describe, with neat sketches, the various forms of screw threads.

9. Discuss the advantages and disadvantages of a screwed joint.

10. Write short note on nut locking devices covering the necessity and various types. Your answer should be illustrated with neat sketches.

OBJECTIVE TYPE QUESTIONS

1. A rivet is specified by

(*a*) shank diameter (*b*) length of rivet (*c*) type of head (*d*) length of tail
(*e*) all of the above (*f*) none of the above

2. In rivet joints, the distance from the centre of one rivet to the centre of the next rivet measured parallel to the seam, is called

(*a*) margin (*b*) back pitch (*c*) diagonal pitch (*d*) pitch

3. The edges of the boiler plates are bevelled to an angle of

(*a*) 20° (*b*) 40° (*c*) 60° (*d*) 80°

4. The largest diameter of an external or internal screw thread is known as

(*a*) minor diameter (*b*) major diameter (*c*) pitch diameter (*d*) pitch

5. A screw is specified by its

(*a*) major diameter (*b*) minor diameter (*c*) pitch diameter (*d*) none of these

6. The included angle for the acme threads is

(*a*) 29° (*b*) 47.5° (*c*) 55° (*d*) 60°

7. The railway carriage couplings have

(*a*) square threads (*b*) acme threads (*c*) kunckle threads (*d*) buttress threads

8. The square threads are usually found on

(*a*) spindle of bench vices (*b*) railway carriage couplings
(*c*) feed mechanism of machine tools (*d*) screw cutting lathes

9. A bolt has

(*a*) threads for the nut at one end and head at the other end
(*b*) head at one end and the other end fits in a tapped hole of the parts of be fastened
(*c*) threads at both ends
(*d*) point threads

10. Which of the following has threads at both ends ?

(*a*) Bolt (*b*) Tap bolt (*c*) Stud

11. The machine screws are

(*a*) slotted for the screw driver and are generally used with a nut
(*b*) similar to tap bolts except that they are of small size and a variety of shapes of heads are available
(*c*) used to prevent relative motion between the two parts
(*d*) none of the above

12. A locking device in which the bottom cylindrical portion is recessed to receive the tip of the locking set screw, is called

(*a*) castle nut (*b*) jam nut (*c*) ring nut (*d*) sawn nut

13. A locking device extensively used in automobile industry is a

(*a*) jam nut (*b*) castle nut (*c*) sawn nut (*d*) ring nut

(*e*) any one of the above (*f*) none of the above

14. A bolt of M 24 × 2 means that

(*a*) the pitch of the thread is 24 mm and depth is 2 mm

(*b*) the cross-sectional area of the threads is 24 mm^2

(*c*) the nominal diameter of the bolt is 24 mm and the pitch is 2 mm

(*d*) the effective diameter of the bolt is 24 mm and there are 2 threads per cm

15. According to Indian standards, a bolt thread of 6 mm size of coarse pitch and with allowance on the threads and normal tolerance grade is designated as

(*a*) M 6 (*b*) M 8 (*c*) M 6-8d (*d*) M 8-6d

ANSWERS

1. (*a*)	2. (*d*)	3. (*d*)	4. (*b*)	5. (*a*)
6. (*a*)	7. (*d*)	8. (*c*)	9. (*a*)	10. (*c*)
11. (*a*)	12. (*c*)	13. (*b*)	14. (*c*)	15. (*c*)

19

Limit System and Surface Finish

1. Interchangeability. 2. Important Terms Used in Limit System. 3. Fits. 4. Types of Fits. 5. Basis of Limit System. 6. Indian Standard System of Limits and Fits. 7. Calculation of Fundamental Deviation for Shafts. 8. Calcualtion of Fundamental Deviation for Holes. 9. Surface Roughness and its Measurement. 10. Preferred Numbers.

19·1 Interchangeability

The term interchangeability is normally employed for the mass production of identical items within the prescribed limits of sizes. A little consideration will show that in order to maintain the sizes of the part within a close degree of accuracy, a lot of time is required. But even then there will be small variations. If the variations are within certain limits, all parts of equivalent size will be equally fit for operating in machines and mechanisms. Therefore, certain variations are recognised and allowed in the sizes of the mating parts to give the required fitting. This facilitates to select at random from a large number of parts for an assembly and results in a considerable saving in the cost of production.

In order to control the size of finished part with due allowance for error for interchangeable parts is called *limit system*.

It may be noted that when an assembly is made of two parts, which enters into the other, is known as *enveloped surface* (or shaft for cylindrical parts) and the other in which one enters is called *enveloping surface* (or hole for cylindrical part).

Notes : 1. The term *shaft* refers not only to the diameter of a circular shaft, but it is also used to designate any external dimension of a part.

2. The term *hole* refers not only to the diameter of a circular hole, but it is also used to designate any internal dimension of a part.

19·2 Important Terms Used in Limit System

The following terms used in limit system (or interchangeable system) are important from the subject point of view.

1. *Nominal size.* It is the size of a part specified in the drawing as a matter of convenience.

2. *Basic size.* It is the size of a part to which all limits of variation (*i.e.*, tolerances) are applied to arrive at final dimensioning of the mating parts. The nominal or basic size of a part is often the same.

3. *Actual size.* It is the actual measured dimension of the part. The difference between the basic size and the actual size should not exceed a certain limit, otherwise it will interfere with the interchangeability of the mating parts.

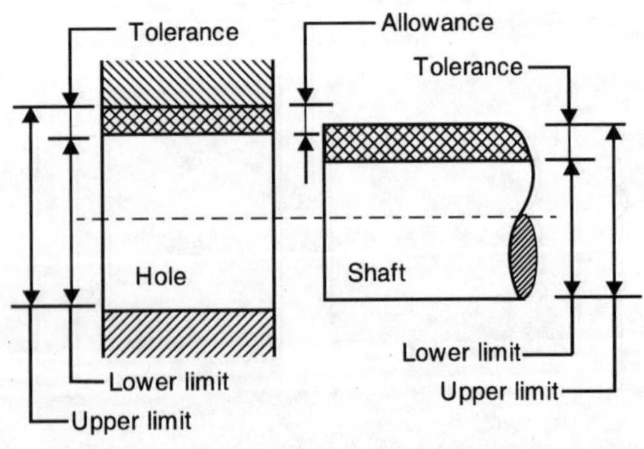

Fig. 19.1. Limits of sizes.

4. *Limits of sizes.* There are two extreme permissible sizes for a dimension of the part as shown in Fig. 19.1. The largest permissible size for a dimension is called *upper* or *high* or *maximum limit,* whereas the smallest size is known as *lower* or *minimum limit.*

5. *Allowance.* It is the difference between the basic dimensions of the mating parts. The allowance may be *positive* or *negative.* When the shaft size is less than the hole size, then the allowance is *positive* and when the shaft size is greater than the hole size, then the allowance is *negative.*

6. *Tolerance.* It is the difference between the upper limit and lower limit of a dimension. In other words, it is the maximum permissible variation in a dimension. The tolerance may be *unilateral* or *bilateral.* When all the tolerance is allowed on one side of the nominal size *e.g.*, $20^{+0.000}_{-0.004}$, then it is said to be *unilateral system of tolerance.* The unilateral system is mostly used in industries as it permits changing the tolerance value while still retaining the same allowance or type of fit.

When the tolerance is allowed on both sides of the nominal size, *e.g.*, $20^{+0.002}_{-0.002}$, then it is said to be *bilateral system of tolerance.* In this case + 0.002 is the upper limit and – 0.002 is the lower limit.

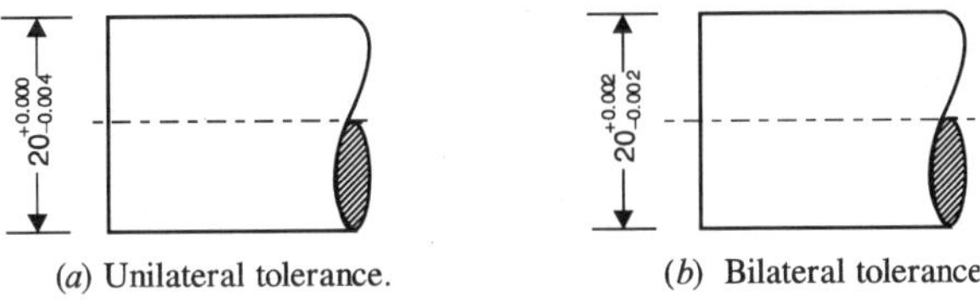

(*a*) Unilateral tolerance. (*b*) Bilateral tolerance

Fig. 19.2. Method of assigning tolerances.

The method of assigning unilateral and bilateral tolerance is shown in Fig. 19.2 (*a*) and (*b*) respectively.

7. *Tolerance zone.* It is the zone between the maximum and minimum limit size, as shown in Fig. 19.3.

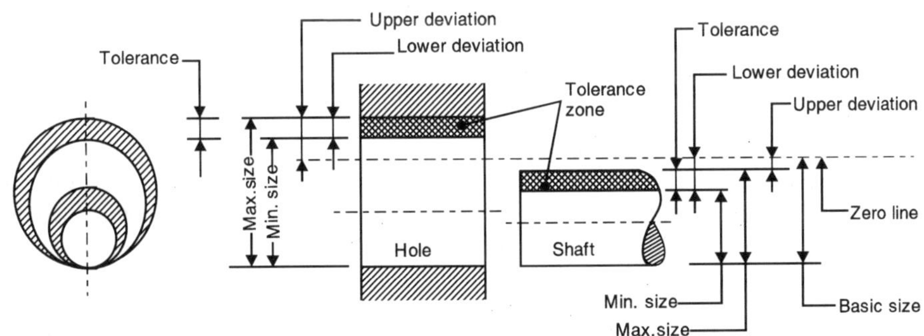

Fig. 19.3. Tolerance zone.

8. *Zero line.* It is a straight line corresponding to the basic size. The deviations are measured from this line. The positive and negative deviations are shown above and below the zero line respectively.

9. *Upper deviation.* It is the algebraic difference between the maximum limit and the basic size. The upper deviation of a hole is represented by a symbol *ES* (Ecart Superior) and of a shaft, it is represented by *es*.

10. *Lower deviation.* It is the algebraic difference between the minimum limit and the basic

size. The lower deviation of a hole is represented by a symbol *EI* (Ecart Inferior) and of a shaft, it is represented by *ei*.

11. *Actual deviation*. It is the algebraic difference between an actual size and the corresponding basic size.

12. *Mean deviation*. It is the arithmetical mean between the upper and lower deviations.

13. *Fundamental deviation*. It is one of the two deviations which is conventionally chosen to define the position of the tolerance zone in relation to zero line, as shown in Fig.19.4.

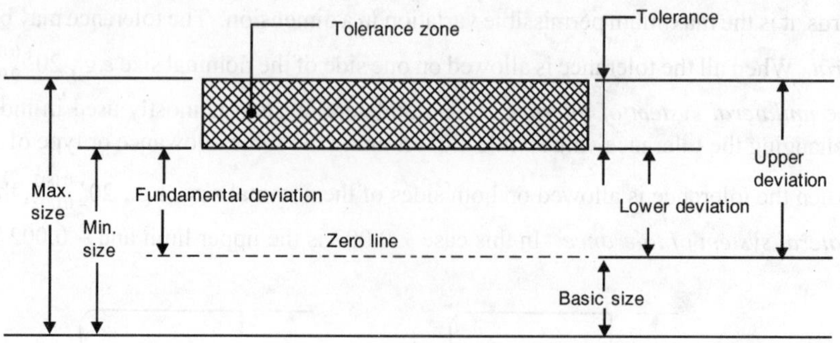

Fig. 19.4. Fundamental deviation.

In an example, $20^{+0.035}_{-0.025}$, the basic size is 20 mm, the upper deviation is 0.035 mm and the lower deviation in 0.025 mm. Therefore,

Maximum limit size = 20 + 0.035 = 20.035 mm

Minimum limit size = 20 – 0.025 = 19.975 mm

Since the tolerance is the difference between the maximum and minimum limit size, therefore, the tolerance in this case is 20.035 – 19.975 = 0.06 mm.

19·3 Fits

The degree of tightness or looseness between the two mating parts is known as a *fit* of the parts. The nature of fit is characterised by the presence and size of clearance and interference.

The *clearance* is the amount by which the actual size of the shaft is less than the actual size of the mating hole in an assembly as shown in Fig. 19.5 (*a*). In other words, the clearance is the difference between the sizes of the hole and the shaft before assembly. The difference must be *positive*.

The *interference* is the amount by which the actual size of a shaft is larger than the actual finished size of the mating hole in an assembly as shown in Fig. 19.5 (*b*). In others words, the interference is the arithmetical difference between the sizes of the hole and the shaft, before assembly. The difference must be *negative*.

19·4 Types of Fits

According to Indian standards, the fits are classified into the following three groups :

1. *Clearance fit*. In this type of fit, the size limits for mating parts are so selected that clearance between them always occur, as shown in Fig. 19.5 (*a*). It may be noted that in a clearance fit, the tolerance zone of the hole is entirely above the tolerance zone of the shaft.

In a clearance fit, the difference between the minimum size of the hole and the maximum size of the shaft is known as *minimum clearance,* whereas the difference between the maximum size of the hole and minimum size of the shaft is called *maximum clearance,* as shown in Fig. 19.5 (*a*).

The clearance fits may be slide fit, easy sliding fit, running fit, slack running fit and loose running fit.

2. *Interference fit.* In this type of fit, the size limits for the mating parts are so selected that interference between them always occur, as shown in Fig. 19.5 (*b*). It may be noted that in an interference fit, the tolerance zone of the hole is entirely below the tolerance zone of the shaft.

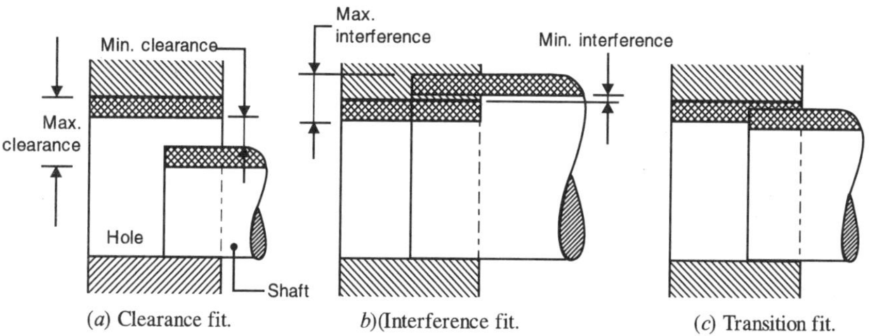

(*a*) Clearance fit. *b*)(Interference fit. (*c*) Transition fit.

Fig. 19.5. Types of fits.

In an interference fit, the difference between the maximum size of the hole and the minimum size of the shaft is known as *minimum interference*, whereas the difference between the minimum size of the hole and the maximum size of the shaft is called *maximum interference*, as shown in Fig. 19.5 (*b*).

The interference fits may be shrink fit, heavy drive fit and light drive fit.

3. *Transition fit.* In this type of fit, the size limits for the mating parts are so selected that either a clearance or interference may occur depending upon the actual size of the mating parts, as shown in Fig. 19.5 (*c*). It may be noted that in a transition fit, the tolerance zones of hole and shaft overlap.

The transition fits may be force fit, tight fit and push fit.

19·5 Basis of Limit System

The following are two bases of limit system :

1. *Hole basis system.* When the hole is kept as a constant member (*i.e.,* when the lower deviation of the hole is zero) and different fits are obtained by varying the shaft size, as shown in Fig. 19.6 (*a*), then the limit system is said to be on a hole basis.

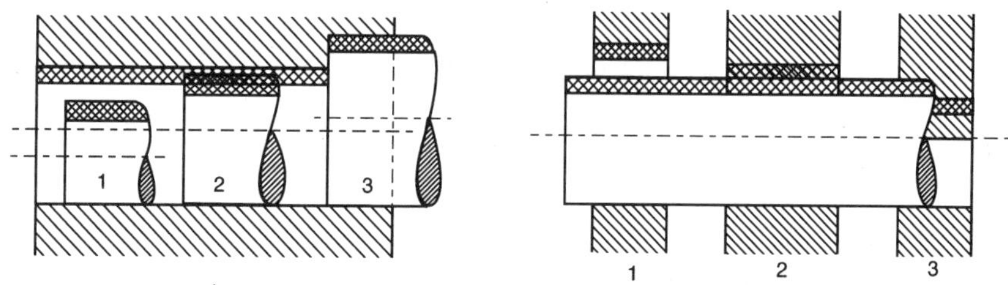

1. Clearance fit. 2. Transition fit. 3. Interference fit.

(*a*) Hole basis system. (*b*) Shaft basis system.

Fig. 19.6. Bases of limit system.

2. *Shaft basis system.* When the shaft is kept as a constant member (*i.e.*, when the upper deviation of the shaft is zero) and different fits are obtained by varying the hole size as shown in Fig. 19.6 (*b*), then the limit system is said to be on a shaft basis.

The hole basis and shaft basis system may also be shown as in Fig. 19.7, with respect to the zero line.

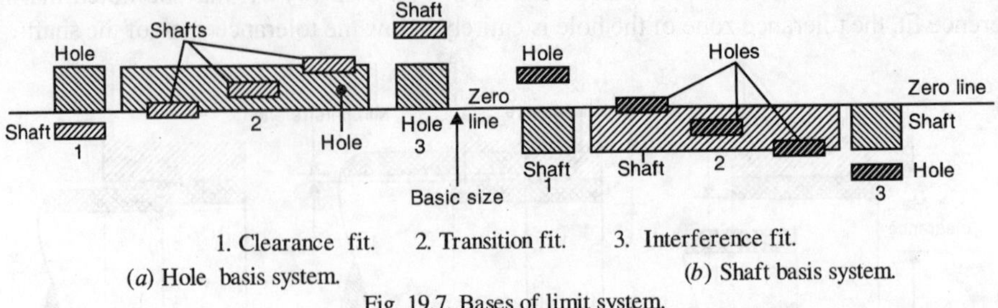

1. Clearance fit. 2. Transition fit. 3. Interference fit.

(*a*) Hole basis system. (*b*) Shaft basis system.

Fig. 19.7. Bases of limit system.

It may be noted that from the manufacturing point of view, a hole basis system is always preferred. This is because the holes are usually produced and finished by standard tooling like drill, reamers etc., whose size is not adjustable easily. On the other hand, the size of the shaft (which is to go into the hole) can be easily adjusted and is obtained by turning or grinding operations.

19·6 Indian Standard System of Limits and Fits

According to Indian standard [IS : 919 (Part I)–1993], the system of limits and fits comprises 18 grades of fundamental tolerances *i.e.*, grades of accuracy of manufacture and 25 types of fundamental deviations indicated by letter symbols for both holes and shafts (capital letters *A* to *ZC* for holes and small letters *a* to *zc* for shafts) in diameter steps ranging from 1 mm to 500 mm. A unilateral hole basis system is recommended but if necessary a unilateral or bilateral shaft basis system may also be used. The 18 tolerance grades are designated as *IT* 01, *IT* 0 and *IT* 1 to *IT* 16. These are called *standard tolerances*. The standard tolerances for grades *IT* 5 to *IT* 7 are determined in terms of standard tolerance unit (i) in microns, where

$$i \text{ (microns)} = 0.45 \sqrt[3]{D} + 0.001\, D,$$ where D is the size or geometric mean diameter in mm.

The following table shows the relative magnitude for grades between *IT* 5 and *IT* 16.

Table 19.1. Relative magnitude of tolerance grades.

Tolerance grade	*IT* 5	*IT* 6	*IT* 7	*IT* 8	*IT* 9	*IT* 10	*IT* 11	*IT* 12	*IT* 13	*IT* 14	*IT* 15	*IT* 16
Magnitude	7 *i*	10 *i*	16 *i*	25 *i*	40 *i*	64 *i*	100 *i*	160 *i*	250 *i*	400 *i*	640 *i*	1000 *i*

The values of standard tolerances corresponding to grades *IT* 01, *IT* 0 and *IT* 1 are given below :

For *IT* 01, i (microns) $= 0.3 + 0.008\, D$,

For *IT* 0, i (microns) $= 0.5 + 0.012\, D$, and

For *IT* 1, i (microns) $= 0.8 + 0.020\, D$

where D is the size or geometric mean diameter in mm.

The tolerance values of grades *IT* 2 to *IT* 4 are scaled approximately geometrically between *IT* 1 and *IT* 5. The fundamental tolerances of grades *IT* 01, *IT* 0 and *IT* 1 to *IT* 16 for diameter steps ranging from 1 mm to 500 mm are given in Table 19.2. The manufacturing processes capable of producing the particular *IT* grades of work are shown in Table 19.3.

The alphabetical representation of fundamental deviations for basic shaft and basic hole system is shown in Fig. 19.8.

Table 19.2 Fundamental tolerances of grades IT 01, IT 0 and IT 1 and IT 16, according to IS : 919 (Part I) – 1993.

Basic size (Diameter steps) in mm	Standard tolerance grades, in micron (1 micron = 0.001 mm)																	
	IT 01	*IT* 0	*IT* 1	*IT* 2	*IT* 3	*IT* 4	*IT* 5	*IT* 6	*IT* 7	*IT* 8	*IT* 9	*IT* 10	*IT* 11	*IT* 12	*IT* 13	*IT* 14	*IT* 15	*IT* 16
Over 1 To and inc. 3	0.3	0.5	0.8	1.2	2	3	4	6	10	14	25	40	60	100	140	250	400	600
Over 3 To and inc. 6	0.4	0.6	1	1.5	2.5	4	5	8	12	18	30	48	75	120	180	300	480	750
Over 6 To and inc. 10	0.4	0.6	1	1.5	2.5	4	6	9	15	22	36	58	90	150	220	360	580	900
Over 10 To and inc. 18	0.5	0.8	1.2	2	3	5	8	11	18	27	43	70	110	180	270	430	700	1100
Over 18 To and inc. 30	0.6	1	1.5	2.5	4	6	9	13	21	33	52	84	130	210	330	520	840	1300
Over 30 To and inc. 50	0.6	1	1.5	2.5	4	7	11	16	25	39	62	100	160	250	390	620	1000	1600
Over 50 To and inc. 80	0.8	1.2	2	3	5	8	13	19	30	46	74	120	190	300	460	740	1200	1900
Over 80 To and inc. 120	1	1.5	2.5	4	6	10	15	22	35	54	87	140	220	350	540	870	1400	2200
Over 120 To and inc. 180	1.2	2	3.5	5	8	12	18	25	40	63	100	160	250	400	630	1000	1600	2500
Over 180 To and inc. 250	2	3	4.5	7	10	14	20	29	46	72	115	185	290	460	720	1150	1850	2900
Over 250 To and inc. 315	2.5	4	6	8	12	16	23	32	52	81	130	210	320	520	810	1300	2100	3200
Over 315 To and inc. 400	3	5	7	9	13	18	25	36	57	89	140	230	360	570	890	1400	2300	3800
Over 400 To and inc. 500	4	6	8	10	15	20	27	40	63	97	155	250	400	630	970	1550	2500	4000

Table 19.3. Manufacturing processes and *IT* grades produced.

S.No.	*Manufacturing process*	*IT grade produced*	*S.No.*	*Manufacturing process*	*IT grade produced*
1.	Lapping	4 and 5	9.	Extrusion	8 to 10
2.	Honing	4 and 5	10.	Boring	8 to 13
3.	Cylindrical grinding	5 to 7	11.	Milling	10 to 13
4.	Surface grinding	5 to 8	12.	Planing and shaping	10 to 13
5.	Broaching	5 to 8	13.	Drilling	10 to 13
6.	Reaming	6 to 10	14.	Die casting	12 to 14
7.	Turning	7 to 13	15.	Sand casting	14 to 16
8.	Hot rolling	8 to 10	16.	Forging	14 to 16

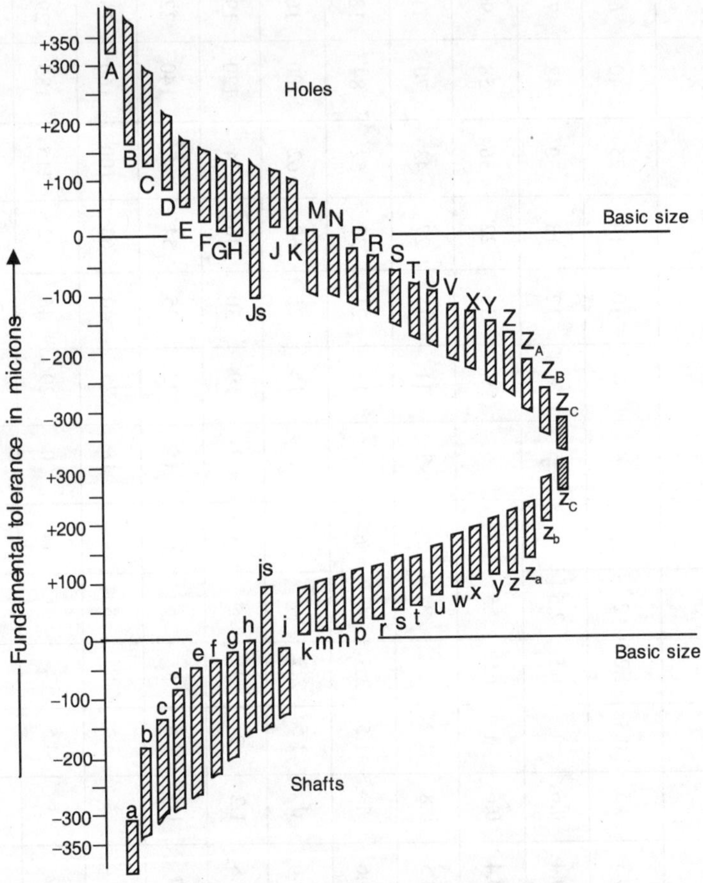

Fig. 19.8. Fundamental deviations for shafts and holes.

For hole, *H* stands for a dimension whose lower deviation refers to the basic size. The hole *H* for which the lower deviation is zero is called a *basic hole*. Similarly, for shafts, *h* stands for dimension whose upper deviation refers to the basic size. The shaft *h* for which the upper deviation is zero is called a *basic shaft*.

A fit is designated by its basic size followed by symbols representing the limits of each of its two components, the hole being quoted first. For example, 100 *H*6/*g*5 means basic size is 100 mm and the tolerance grade for the hole is 6 and for the shaft is 5. Some of the fits commonly used in engineering practice, for holes and shafts are shown in Tables 19.4 and 19.5 respectively according to IS : 2709–1982 (Reaffirmed 1993).

Table 19.4. Commonly used fits for holes according to IS : 2709–1982 (Reaffirmed 1993).

Types of fit	*Class of shaft*	*With holes*				*Remarks*
		H6	*H7*	*H8*	*H11*	
Clearance fit	*a*	—	—	—	*a*11	Large clearance fit and widely used.
	b	—	—	—	*b*11	
	c	—	*c*8	**c*9	*c*11	Slack running fit.
	d	—	*d*8	**d*8, *d*9, *d*10	*d*11	Loose running fit—used for plummer block bearings and loose pulleys.
	e	*e*7	*e*8	**e*8-*e*9	—	Easy running fit—used for properly lubricated bearings requiring appreciable clearance. In the finer grades, it may be used on large electric motor and turbogenerator bearings according to the working condition.
	f	**f*6	*f*7	**f*8	—	Normal running fit—widely used for grease lubricated or oil lubricated bearings where no substantial temperature differences are encountered —Typical applications are gear box, shaft bearings and the bearings of small electric motors, pumps etc.
	g	**g*5	**g*6	*g*7	—	Close running fit or sliding fit—Also fine spigot and location fit—used for bearings for accurate link work and for piston and slide valves.
	h	**h*5	**h*6	**h*7–*h*8	**h*11	Precision sliding fit. Also fine spigot and location fit—widely used for non-running parts.
Transition fit	*j*	**j*5	**j*6	**j*7	—	Push fit for very accurate location with easy assembly and dismantling—Typical applications are coupling, spigots and recesses, gear rings clamped to steel hubs etc.
	k	**k*5	**k*6	*k*7	—	True transition fit (light keying fit)—used for keyed shaft, non-running locked pins etc.
	m	**m*5	**m*6	*m*7	—	Medium keying fit.
	n	*n*5	**n*6	*n*7	—	Heavy keying fit—used for tight assembly of mating parts.
Interference fit	*p*	*p*5	**p*6	—	—	Light press fit with easy dismantling for non-ferrous parts. Standard press fit with esay dismantling for ferrous and non-ferrous parts assembly.
	r	*r*5	*r6	—	—	Medium drive fit with easy dismantling for ferrous parts assembly. Light drive fit with easy dismantling for non-ferrous parts assembly.
	s	*s*5	**s*6	*s*7	—	Heavy drive fit on ferrous parts for permanent or semi-permanent assembly. Standard press fit for non-ferrous parts.
	t	*t*5	*t*6	**t*7	—	Force fit on ferrous parts for permanent assembly.
	u	*u*5	*u*6	**u*7	—	Heavy force fit or shrink fit.
	v, *x*	—	—	—	—	Very large interference fits—not recommended for use.
	y, *z*	—	—	—	—	

*Second preference fits.

Table 19.5. Commonly used fits for shafts according to IS : 2709–1982 (Reaffirmed 1993).

Types of fit	*Class of hole*	*With shafts* *h5	*h6*	*h7*	*h8	*h9*	*h11*	*Remarks and uses*
Clearance fit	*A*	—	—	—	—	—	*A*11	Large clearance fit and widely used.
	B	—	—	—	—	—	*B*11	
	C	—	—	—	—	—	*C*11	Slack running fit.
	D	—	**D*9	—	*D*10	*D*10	**D*11	Loosing running fit.
	E	—	**E*8	—	*E*8*	*E*9	—	Easy running fit.
	F		**F*7	—	*F*8	**F*8	—	Normal running fit.
	G	**G*6	*G*7	—	—	—	—	Close running fit or sliding fit, also spigot and location fit.
	H	**H*6	*H*7	*H*8	*H*8	*H*8,*H*9	*H*11	Precision sliding fit. Also fine spigot and location fit.
	Js	**Js*6	*Js*7	**Js*8	—	—	—	Push fit for very accurate location with easy assembly and disassembly.
Transition fit	*K*	**K*6	*K*7	**K*8	—	—	—	Light keying fit (true transition) for keyed shafts, non-running locked pins, etc.
	M	**M*6	**M*7	**M*8	—	—	—	Medium keying fit.
	N	**N*6	*N*7	**N*8	—	—	—	Heavy keying fit (for tight assembly of mating surfaces).
Interference fit	*P*	**P*6	*P*7	—	—	—	—	Light press fit with wasy dismantling for non-ferrous parts. Standard press fit with easy dismantling for ferrous and non-ferrous parts assembly.
	R	**R*6	*R*7	—	—	—	—	Medium drive fit with easy dismantling for ferrous parts assembly. Light drive fit with easy dismantling for non-ferrous parts assembly.
	S	**S*6	*S*7	—	—	—	—	Heavy drive fit for ferrous parts permanent or semi-permanent assembly standard press fit for non-ferrous parts.
	T	**T*6	*T*7	—	—	—	—	Force fit on ferrous parts for permanent assembly.

*Second preference fits.

19·7 Calculation of Fundamental Deviation for Shafts

We have already discussed that for holes, the upper deviation is denoted by *ES* and the lower deviation by *EI*. Similarly for shafts, the upper deviation is represented by *es* and the lower deviation by *ei*. According to Indian standards, for each letter symbol, the magnitude and sign for one of the two deviations (*i.e.*, either upper or lower deviation), which is known as fundamental deviation, have been determined by means of formulae given in Table 19.6. The other deviation may be calculated by using the absolute value of the standard tolerance (*IT*) from the following relation :

$$ei = es - IT \quad \text{or} \quad es = ei + IT$$

It may be noted that for shafts *a* to *h*, the upper deviations (*es*) are considered, whereas for shafts *j* to *Zc*, the lower deviation (*ei*) is to be considered.

The fundamental deviation for Indian standard shafts for diameter steps from 1 mm to 200 mm may be taken directly from Table 19.9 (Page 419).

Table 19.6. Formulae for fundamental shaft deviations.

Upper deviation (es)		*Lower deviation (ei)*	
Shaft designation	*In microns (for D in mm)*	*Shaft designation*	*In microns (for D in mm)*
a	$= -(265 + 1.3\ D)$ for $D \leq 120$	j 5 to j 8	No formula
	$= -3.5\ D$ for $D > 120$	k 4 to k 7	$= +0.6 = \sqrt[3]{D}$
		k for grades ≤ 3 and ≤ 8	$= 0$
b	$= -(140 + 0.85\ D)$ for $D \leq 160$	m	$= +(IT\ 7 - IT\ 6)$
	$= -1.8\ D$	n	$= +5\ (D)^{0.34}$
	for $D > 160$	p	$= +IT\ 7 + 0$ to 5
c	$= -52\ (D)^{0.2}$ for $D \leq 40$	r	= Geometric mean of values of ei [for shaft p and s]
	$= -(95 + 0.8\ D)$ for $D > 40$	s	$= +(IT\ 8 + 1$ to $4)$ for $D \leq 50$ $= +(IT\ 7 + 0.4D)$ for $D > 50$
d	$= -16\ (D)^{0.44}$	t	$= +(IT\ 7 + 0.63\ D)$
e	$= -11\ (D)^{0.41}$	u	$= +(IT\ 7 + D)$
f	$= -5.5\ (D)^{0.41}$	v	$= +(IT\ 7 + 1.25\ D)$
		x	$= +(IT\ 7 + 1.6\ D)$
g	$= -2.5\ (D)^{0.34}$	y	$= +(IT\ 7 + 2\ D)$
		z	$= +(IT\ 7 + 2.5\ D)$
h	$= 0$	za	$= +(IT\ 8 + 3.15\ D)$
		zb	$= +(IT\ 9 + 4\ D)$
		zc	$= +(IT\ 10 + 5\ D)$

For js, the two deviations are equal to $\pm\ IT/2$.

19·8 Calculation of Fundamental Deviation for Holes

The fundamental deviation for holes for those of the corresponding shafts, are derived by using the rule as given in the following table.

Table 19.7. Rules for fundamental deviation for holes.

All deviation except those below			*General rule* Hole limits are identical with the shaft limits of the same symbol (letter and grade) but disposed on the other side of the zero line. EI = Upper deviation es of the shaft of the same letter symbol but of opposite sign.
For sizes above 3 mm	N	9 and coarser grades	$ES = 0$
	J, K, M and N P to ZC	upto grade 8 inclusive upto grade 7 inclusive	*Special rule* ES = Lower deviation ei of the shaft of the same letter symbol but one grade finer and of opposite sign increased by the difference between the tolerances of the two grades in question.

The fundamental deviation for Indian standard holes for diameter steps from 1 mm to 200 mm may be taken directly from the following table.

Table 19.8. Indian standard 'H' Hole

Limts for H5 to H13 over the range 1 mm to 200 mm as per IS : 919 (Part II) - 1993.

Diameter steps in mm		*Deviations in micron (1 micron = 0.001 mm)*									
		H5	*H6*	*H7*	*H8*	*H9*	*H10*	*H11*	*H12*	*H13*	*H5 – H13*
Over	To	High +	High +	High +	High +	High +	High +	High +	High +	High +	Low
1	3	5	7	9	14	25	40	60	90	140	0
3	6	5	8	12	18	30	48	75	120	180	0
6	10	6	9	15	22	36	58	90	150	220	0
10 14	14 18	8	11	18	27	43	70	110	180	270	0
18 24	24 30	9	13	21	33	52	84	130	210	330	0
30 40	40 50	11	16	25	39	62	100	160	250	460	0
50 65	65 80	13	19	30	46	74	120	190	300	390	0
80 100	100 120	15	22	35	54	87	140	220	350	540	0
120 140	140 160	18	25	40	63	100	160	250	400	630	0
160 180	180 200	20	29	46	72	115	185	290	460	720	0

Example 19.1. *The dimensions of the mating parts, according to basic hole system, are given as follows :*

Hole : 25.00 mm Shaft : 24.97 mm

25.02 mm 24.95 mm

Find the hole tolerance, shaft tolerance and allowance.

Solution. Given : Lower limit of hole = 25 mm ; Upper limit of hole = 25.02 mm ; Upper limit of shaft = 24.97 mm ; Lower limit of shaft = 24.95 mm

Hole tolerance

We know that hole tolerance = Upper limit of hole – Lower limit of hole

= 25.02 – 25 = 0.02 mm **Ans.**

Shaft tolerance

We know that shaft tolerance = Upper limit of shaft – Lower limit of shaft

= 24.97 – 24.95 = 0.02 mm **Ans.**

Allowance

We know that allowance = Lower limit of hole – Upper limit of shaft

= 25.00 – 24.97 = 0.03 mm **Ans.**

Table 19.9. Indian standard shafts for common use as per IS :919 (Part II) – 1993.

Shaft	*Limit*	*Values of deviations in microns for diameter steps 1 mm to 200 mm (1 micron = 0.001 mm)*																
		1 to 3	3 to 6	6 to 10	10 to 14	14 to 18	18 to 24	24 to 30	30 to 40	40 to 50	50 to 65	65 to 80	80 to 100	100 to 120	120 to 140	140 to 160	160 to 180	180 to 200
*s*6	High +	22	37	32	39	39	48	48	59	59	72	78	93	101	117	125	133	151
*s*7	High +	24	31	38	46	46	56	66	68	68	83	89	106	114	132	140	148	168
*s*6 & *s*7	Low +	15	19	23	28	28	35	35	43	43	53	59	71	79	92	100	108	122
*p*9	High +	16	20	24	29	29	35	35	42	42	51	51	59	59	68	68	68	79
	Low +	9	12	15	18	18	22	22	26	26	32	32	37	37	43	43	43	50
*k*6	High +	—	—	10	12	12	15	15	18	18	21	21	25	25	28	28	28	33
*k*7	High +	—	—	16	19	19	23	23	27	27	32	32	38	38	43	43	43	50
*k*6 & *k*7	Low +	—	—	1	1	1	2	2	2	2	2	2	3	3	3	3	3	4
	High +	7	9	10	12	12	13	13	15	15	18	18	20	20	22	22	22	25
	Low –	2	3	5	6	6	8	8	10	10	12	12	15	15	18	18	18	21
*h*6	High –	0	0	0	0	0	0	0	0	0	0	0	0	0	0	0	0	0
*h*7	High –	0	0	0	0	0	0	0	0	0	0	0	0	0	0	0	0	0

Shaft	Limit	Values of deviations in microns for diameter steps 1 mm to 200 mm (1 micron = 0.001 mm)															
		1 to 3	3 to 6	6 to 10	10 to 14	14 to 18	18 to 24	24 to 30	30 to 40	40 to 50	50 to 65	65 to 80	80 to 100	100 to 120	120 to 140	140 to 160	160 to 180	180 to 200
h6 & h7	Low –	7 & 9	8 & 12	9 & 15	11 & 18	11 & 18	13 & 21	13 & 21	16 & 25	16 & 25	19 & 30	19 & 30	22 & 35	22 & 35	25 & 40	25 & 40	26 & 40	29 & 46
g6	High –	3	4	5	6	6	7	7	9	9	10	10	12	12	14	14	14	15
	Low –	10	12	14	17	17	20	20	25	25	29	29	34	34	39	39	39	44
f7	High –	7	10	13	16	16	20	20	25	25	30	30	36	36	43	43	43	50
f8	High –	7	10	13	16	16	20	20	25	25	30	30	36	36	43	43	43	50
f7 & f8	Low –	16 & 21	22 & 28	28 & 35	34 & 43	34 & 43	41 & 53	41 & 53	50 & 64	50 & 64	60 & 76	60 & 76	71 & 90	71 & 90	83 & 106	83 & 106	83 & 106	96 & 122
e8	High –	14	20	25	32	32	40	40	50	50	60	60	72	72	85	85	85	100
e9	High –	14	20	25	32	32	40	40	50	50	60	60	72	72	85	85	85	100
e8 & e9	Low –	28 & 39	38 & 50	47 & 61	59 & 75	58 & 75	73 & 92	73 & 92	89 & 112	89 & 112	106 & & 134	106 7 & 134	126 & & 158	126 & & 158	148 & 185	148 & 185	148 & 185	172 & 215
d9	High –	20	30	40	50	50	65	65	80	80	100	100	120	120	145	145	145	170
	Low –	45	60	76	93	93	117	117	142	142	174	174	207	207	245	245	245	285
c9	High –	60	70	80	95	95	110	110	120	130	140	150	170	180	200	210	230	240
	Low –	85	100	116	138	138	162	162	182	192	214	224	257	267	300	310	330	355
b9	High –	140	140	150	150	150	160	160	170	180	190	200	220	240	260	280	310	340
	Low –	165	170	186	193	193	212	212	232	242	264	274	307	327	360	380	410	545

Example 19.2. *Calculate the tolerances, fundamental deviations and limits of sizes for the shaft designated as 40 H8 / f 7.*

Solution. Given : Shaft designation = 40 $H\,8\,/\,f7$

The shaft designation 40 $H\,8\,/\,f7$ means that the basic size is 40 mm and the tolerance grade for the hole is 8 (*i.e.*, IT 8) and for the shaft is 7 (*i.e.*, IT 7).

Tolerances

Since 40 mm lies in the diameter steps of 30 mm to 50 mm, therefore, the geometric mean diameter,

$$D = \sqrt{30 \times 50} = 38.73 \text{ mm}$$

We know that standard tolerance unit,

$$i = 0.45 \sqrt[3]{D} + 0.001\, D$$

$$= 0.45 \sqrt[3]{38.73} + 0.001 \times 38.73$$

$$= 0.45 \times 3.38 + 0.038\ 73 = 1.559\ 73 \text{ or } 1.56 \text{ microns}$$

$$= 1.56 \times 0.001 = 0.001\ 56 \text{ mm} \qquad ...(\because 1 \text{ micron} = 0.001 \text{ mm})$$

From Table 19.1, we find that standard tolerance for the hole of grade 8 (IT 8)

$= 25\, i = 25 \times 0.001\ 56 = 0.039$ mm **Ans.**

and standard tolerance for the shaft of grade 7 (IT 7)

$= 16\, i = 16 \times 0.001\ 56 = 0.025$ mm **Ans.**

Note : The value of IT 8 and IT 7 may be directly seen from Table 19.2.

Fundamental deviation

We know that fundamental deviation (lower deviation) for hole H,

$$EI = 0$$

From Table 19.6, we find that fundamental deviation (upper deviation) for shaft f,

$$es = -5.5\,(D)^{0.41}$$

$$= -5.5\,(38.73)^{0.41} = -24.63 \text{ or } -25 \text{ microns}$$

$= -25 \times 0.001 = -0.025$ mm **Ans.**

$\therefore$ Fundamental deviation (lower deviation) for shaft f,

$ei = es - IT = -0.025 - 0.025 = -0.050$ mm **Ans.**

The –ve sign indicates that fundamental deviation lies below the zero line.

Limits of sizes

We know that lower limit for hole

= Basic size = 40 mm **Ans.**

Upper limit for hole = Lower limit for hole + Tolerance for hole

= 40 + 0.039 = 40.039 mm **Ans.**

Upper limit for shaft = Lower limit for hole or Basic size – Fundamental deviation (upper deviation) ...($\because$ Shaft f lies below the zero line)

= 40 – 0.025 = 39.975 mm **Ans.**

and lower limit for shaft = Upper limit for shaft – Tolerance for shaft

= 39.975 – 0.025 = 39.95 mm **Ans.**

Example 19.3. *Give the dimensions for the hole and shaft for the following :*

(*a*) *A 12 mm electric motor sleeve bearing ;*

(*b*) *A medium force fit on a 200 mm shaft ; and*

(*c*) *A 50 mm sleeve bearing on the elevating mechanism of a road grader.*

Solution. (*a*) *Dimensions for the hole and shaft for a 12 mm electric motor sleeve bearing*

From Table 19.4, we find that for an electric motor sleeve bearing, a shaft *e* 8 should be used with *H* 8 hole.

Since 12 mm size lies in the diameter steps of 10 mm to 18 mm, therefore, the geometric mean diameter,

$$D = \sqrt{10 \times 18} = 13.4 \text{ mm}$$

We know that standard tolerance unit,

$$i = 0.45\sqrt[3]{D} + 0.001\,D$$

$$= 0.45\sqrt[3]{13.4} + 0.001 \times 13.4 = 1.07 + 0.0134 = 1.0834 \text{ microns}$$

∴ *Standard tolerance for shaft and hole of grade 8 (*IT* 8)

$$= 25\,i \qquad \text{...(From Table 19.1)}$$

$$= 25 \times 1.0834 = 27 \text{ microns}$$

$$= 27 \times 0.001 = 0.027 \text{ mm} \qquad \text{...(}\because \text{ 1 micron} = 0.001 \text{ mm)}$$

From Table 19.6, we find that upper deviation for shaft '*e*',

$$es = -11\,(D)^{0.41} = -11\,(13.4)^{0.41} = -32 \text{ microns}$$

$$= -32 \times 0.001 = -0.032 \text{ mm}$$

We know that lower deviation for shaft '*e*',

$$ei = es - IT = -0.032 - 0.027 = -0.059 \text{ mm}$$

∴ Dimensions for the hole (*H* 8)

$$= 12^{+0.027}_{+0.000} \quad \textbf{Ans.}$$

and dimensions for the shaft (*e* 8)

$$= 12^{-0.032}_{-0.059} \quad \textbf{Ans.}$$

(*b*) *Dimensions for the hole and shaft for a medium force fit on a 200 mm shaft*

From Table 19.4, we find that shaft *r* 6 with hole *H* 7 gives the desired fit.

Since 200 mm lies in the diameter steps of 180 mm of 250 mm, therefore, the geometric mean diameter,

$$D = \sqrt{180 \times 250} = 212 \text{ mm}$$

We know that standard tolerance unit,

$$i = 0.45\sqrt[3]{D} + 0.001\,D$$

$$= 0.45\sqrt[3]{212} + 0.001 \times 212 = 2.68 + 0.212 = 2.892 \text{ microns}$$

∴ Standard tolerance for the shaft of grade 6 (*IT* 6) from Table 19.1

$$= 10\,i = 10 \times 2.892 = 28.92 \text{ microns}$$

$$= 28.92 \times 0.001 = 0.028\,92 \text{ or } 0.029 \text{ mm}$$

and standard tolerance for the hole of grade 7 (*IT* 7)

$$= 16\,i = 16 \times 2.892 = 46 \text{ microns}$$

$$= 46 \times 0.001 = 0.046 \text{ mm}$$

We know that lower deviation for shaft '*r*' from Table 19.6,

$$ei = \frac{1}{2}[(IT\,7 + 0.4\,D) + (IT\,7 + 0 \text{ to } 5)]$$

* The tolerance values may be taken directly from Table 19.2.

$$= \frac{1}{2}[(46 + 0.4 \times 212) + (46 + 3)] = 90 \text{ microns}$$

$$= 90 \times 0.001 = 0.09 \text{ mm}$$

and upper deviation for the shaft r,

$$es = ei + IT = 0.09 + 0.029 = 0.119 \text{ mm}$$

∴ Dimension for the hole $H\,7$

$$= 200^{+0.046}_{+0.00} \quad \textbf{Ans.}$$

and dimension for the shaft $r\,6$

$$= 200^{+0.119}_{+0.09} \quad \textbf{Ans.}$$

(*c*) *Dimensions for the hole and shaft for a 50 mm sleeve bearing on the elevating mechanism of a road grader*

From Table 19.4, we find that for a sleeve bearing, a loose running fit will be suitable and a shaft $d\,9$ should be used with hole $H\,8$.

Since 50 mm size lies in the diameter steps of 30 mm to 50 mm or 50 mm to 80 mm, therefore, the geometric mean diameter,

$$D = \sqrt{30 \times 50} = 38.73 \text{ mm}$$

We know that standard tolerance unit,

$$i = 0.45\sqrt[3]{D} + 0.001\,D$$

$$= 0.45\sqrt[3]{38.73} + 0.001 \times 38.73$$

$$= 1.522 + 0.038\,73 = 1.560\,73 \text{ or } 1.56 \text{ microns}$$

∴ Standard tolerance for the shaft of grade 9 ($IT\,9$) from Table 19.1

$$= 40\,i = 40 \times 1.56 = 62.4 \text{ microns}$$

$$= 62.4 \times 0.001 = 0.0624 \text{ or } 0.062 \text{ mm}$$

and standard tolerance for the hole of grade 8 ($IT\,8$)

$$= 25\,i = 25 \times 1.56 = 39 \text{ microns}$$

$$= 39 \times 0.001 = 0.039 \text{ mm}$$

We know that upper deviation for the shaft d, from Table 19.6,

$$es = -16\,(D)^{0.44} = -16\,(38.73)^{0.44} = -80 \text{ microns}$$

$$= -80 \times 0.001 = -0.08 \text{ mm}$$

and lower deviation for the shaft d,

$$ei = es - IT = -0.08 - 0.062 = -0.142 \text{ mm}$$

∴ Dimension for the hole $H\,8$

$$= 50^{+0.039}_{+0.000} \quad \textbf{Ans.}$$

and dimension for the shaft $d\,9$

$$= 50^{-0.08}_{-0.142} \quad \textbf{Ans.}$$

Example 19.4. *A journal of nominal or basic size of 75 mm runs in a bearing with close running fit. Find the limits of shaft and bearing. What is the maximum and minimum clearance ?*

Solution. Given : Nominal or basic size = 75 mm

From Table 19.4, we find that the close running fit is represented by $H\,8/g\,7$ *i.e.,* a shaft $g\,7$ should be used with $H\,8$ hole.

Since 75 mm lies in the diameter steps of 50 mm to 80 mm, therefore, the geometric mean diameter,

$$D = \sqrt{50 \times 80} = 63 \text{ mm}$$

We know that standard tolerance unit,

$$i = 0.45\sqrt[3]{D} + 0.001\,D = 0.45\sqrt[3]{63} + 0.001 \times 63$$

$$= 1.79 + 0.063 = 1.853 \text{ micron}$$

$$= 1.853 \times 0.001 = 0.001\,853 \text{ mm}$$

$\therefore$ Standard tolerance for hole 'H' of grade 8 ($IT8$)

$$= 25\,i = 25 \times 0.001\,853 = 0.046 \text{ mm}$$

and standard tolerance for shaft 'g' of grade 7 ($IT7$)

$$= 16\,i = 16 \times 0.001\,853 = 0.03 \text{ mm}$$

From Table 19.6, we find that upper deviation for shaft g,

$$es = -2.5\,(D)^{0.34} = -2.5\,(63)^{0.34} = -10 \text{ micron}$$

$$= -10 \times 0.001 = -0.01 \text{ mm}$$

$\therefore$ Lower deviation for shaft g,

$$ei = es - IT = -0.01 - 0.03 = -0.04 \text{ mm}$$

We know that lower limit for hole

= Basic size = 75 mm

Upper limit for hole = Lower limit for hole + Tolerance for hole

= 75 + 0.046 = 75.046 mm

Upper limit for shaft = Lower lomit for hole – Upper deviation for shaft

...($\because$ Shaft g lies below zero line)

= 75 – 0.01 = 74.99 mm

and lower limit for shaft = Upper limit for shaft – Tolerance for shaft

= 74.99 – 0.03 = 74.96 mm

We know that maximum clearance

= Upper limit for hole – Lower limit for shaft

= 75.046 – 74.96 = 0.086 mm **Ans.**

and minimum clearance = Lower limit for hole – Upper limit for shaft.

= 75 – 74.99 = 0.01 mm **Ans.**

19·9 Surface Roughness and its Measurement

A little consideration will show that surfaces produced by different machining operations (*e.g.*, turning, milling, shaping, planing, grinding and superfinishing) are of different characteristics. They show a marked variations when compared with each other. The variation is judged by the degree of smoothness. A surface produced by superfinishing is the smoothest while that by planing is the roughest. In the assembly of two mating parts, it becomes absolutely necessary to describe the surface finish in quantitative terms which is a measure of micro-irregularities of the surface and expressed in microns. In order to prevent stress concentrations and proper functioning, it may be necessary to avoid or to have certain surface roughness.

There are many ways of expressing the surface roughness numerically, but the following two methods are commonly used :

1. Centre line average method (briefly known as CLA method), and
2. Root mean square method (briefly known as RMS method).

The *centre line average method* is defined as the average value of the ordinates between the surface and the mean line, measured on both sides of it. According to Indian standards, the surface finish is measured in terms of 'CLA' value and it is denoted by *Ra*. Mathematically

$$\text{CLA value or } Ra \text{ (in microns)} = \frac{y_1 + y_2 + y_3 + \ldots y_n}{n}$$

where $y_1, y_2, \ldots y_n$ are the ordinates measured on both sides of the mean line and n are the number of ordinates.

The *root mean square method* is defined as the square root of the arithmetic mean of the squares of the ordinates. Mathematically,

$$\text{R.M.S. value (in microns)} = \sqrt{\frac{y_1^2 + y_2^2 + y_3^2 + \ldots y_n^2}{n}}$$

According to Indian standard, following symbols are used to denote the various degrees of surface roughness :

Symbol	Surface roughness (*Ra*) in microns
∇	8 to 25
∇ ∇	1.6 to 8
∇ ∇ ∇	0.025 to 1.6
∇ ∇ ∇ ∇	Less than 0.025

The following table shows the range of surface roughness that can be produced by various manufacturing processes.

Table 19.10. Range of surface roughness.

S.No.	*Manufacturing process*	*Surface roughness in microns*	*S.No.*	*Manufacturing process*	*Surface roughness in microns*
1.	Lapping	0.012 to 0.016	9.	Extrusion	0.16 to 5
2.	Honing	0.025 to 0.40	10.	Boring	0.40 to 6.3
3.	Cylindrical grinding	0.063 to 5	11.	Milling	0.32 to 25
4.	Surface grinding	0.063 to 5	12.	Planing and shaping	1.6 to 25
5.	Broaching	0.40 to 3.2	13.	Drilling	1.6 to 20
6.	Reaming	0.40 to 3.2	14.	Sand casting	5 to 50
7.	Turning	0.32 to 25	15.	Die casting	0.80 to 3.20
8.	Hot rolling	2.5 to 50	16.	Forging	1.60 to 25

19·10 Preferred Numbers

When a machine is to be made in several sizes with different powers or capacities, it is necessary to decide what capacities will cover a certain range efficiently with minimum number of sizes. It has been shown by experience that a certain range can be covered efficiently when it follows a geometrical progression with a constant ratio. The preferred numbers are the conventionally rounded off values derived from geometric series including the integral powers of 10 and having as common ratio of the following factors :

$$\sqrt[5]{10},\ \sqrt[10]{10},\ \sqrt[20]{10} \text{ and } \sqrt[40]{10}$$

These ratios are approximately equal to 1.58, 1.26, 1.12 and 1.06. The series of preferred numbers are designated as **R*5, *R*10, *R*20 and *R*40 respectively. These four series are called *basic series*. The other series called *derived series* may be obtained by simply multiplying or dividing the

* The symbol *R* is used as a tribute to Captain Charles Renard, the first man to use preferred numbers.

basic sizes by 10, 100 etc. The preferred numbers in the series *R*5 are 1, 1.6, 2.5, 4.0 and 6.3. Table 19.11 shows basic series of preferred numbers according to IS : 1076 (Part I) – 1985 (Reaffirmed 1990).

Notes : 1. The standard sizes (in mm) for wrought metal products are shown in Table 19.12 according to IS : 1136–1990. The standard G. P. series used correspond to *R*10, *R*20 and *R*40.

2. The hoisting capacities (in tonnes) of cranes are in *R*10 series, while the hydraulic cylinder diameters are in *R*40 series and hydraulic cylinder capacities are in *R*5 series.

3. The basic thickness of sheet metals and diameter of wires are based on *R*10, *R*20 and *R*40 series. Wire diameter of helical springs are in *R*20 series.

Table 19.11. Preferred numbers of the basic series, according to IS : 1076 (Part I) – 1985 (Reaffirmed 1990).

Basic series	*Preferred numbers.*
*R*5	1.00, 1.60, 2.50, 4.00, 6.30, 10.00
*R*10	1.00, 1.25, 1.60, 2.00, 2.50, 3.15, 4.00, 5.00, 6.30, 8.00, 10.00
*R*20	1.00, 1.12, 1.25, 1.40, 1.60, 1.80, 2.00, 2.24, 2.50, 2.80, 3.15, 3.55, 4.00, 4.50, 5.00, 5.60, 6.30, 7.10, 8.00, 9.00, 10.00
*R*40	1.00, 1.06, 1.12, 1.18, 1.25, 1.32, 1.40, 1.50, 1.60, 1.70, 1.80, 1.90, 2.00, 2.12, 2.24, 2.36, 2.50, 2.65, 2.80, 3.00, 3.15, 3.35, 3.55, 3.75, 4.00, 4.25, 4.50, 4.75, 5.00, 5.30, 5.60, 6.00, 6.30, 6.70, 7.10, 7.50, 8.00, 8.50, 9.00, 9.50, 10.00

Table 19.12. Preferred sizes for wrought metal products according to IS : 1136 –1990.

Size ranges	*Preferred sizes (mm)*
0.01– 0.10 mm	0.02, 0.025, 0.030, 0.04, 0.05, 0.06, 0.08 and 0.10
0.10 – 1 mm	0.10, 0.11, 0.12, 0.14, 0.16, 0.18, 0.20, 0.22, 0.25, 0.28, 0.30, 0.32, 0.35, 0.36, 0.40, 0.45, 0.50, 0.55, 0.60, 0.63, 0.70, 0.80, 0.90 and 1
1 – 10 mm	1, 1.1, 1.2, 1.4, 1.5, 1.6, 1.8, 2.22, 2.5, 2.8, 3, 3.2, 3.5, 3.6, 4, 4.5, 5, 5.5, 5.6, 6, 6.3, 7, 8, 9 and 10
10 – 100 mm	10 to 25 (in steps of 1 mm), 28, 30, 32, 34, 35, 36, 38, 40, 42, 44, 45, 46, 48, 50, 52, 53, 55, 56, 58, 60, 62, 63, 65, 67, 68, 70, 71, 72, 75, 78, 80, 82, 85, 88, 90, 92, 95, 98 and 100
100 – 1000 mm	100 to 200 (in steps of 5 mm), 200 to 310 (in steps fo 10 mm), 315, 320, 330, 340, 350, 355, 360, 370, 375, 380 to 500 (in steps of 10 mm), 520, 530, 550, 560, 580, 600, 630, 650, 670, 700, 710 and 750 – 1000 (in steps of 50 mm)
1000 – 10 000 mm	1000, 1100, 1200, 1250, 1400, 1500, 1600, 1800, 2000, 2200, 2500, 2800, 3000, 3200, 3500, 3600, 4000, 4500, 5000, 5500, 5600, 6000, 6300, 7000, 7100, 8000, 9000, and 10 000

QUESTIONS

1. What do you understand by the term interchangeability ?
2. Write short note on the following :

   (*a*) Nominal size ; (*b*) Basic size ; (*c*) Tolerance ; (*d*) Allowance ; and (*e*) Fits
3. State briefly unilateral system of tolerances covering the points of definition, application and advantages over the bilateral system.
4. What are commonly used fits according to Indian Standards ?
5. What is meant by 'hole basis system' and 'shaft basis system' ? Which one is preferred and why ?
6. Discuss the Indian Standard system of limits and fits.

7. How the fundamental deviation for shafts and holes is calculated ?
8. How is surface roughness is measured and what are the symbols used for surface roughness ?
9. What do you understand by preferred numbers ? Explain fully.

OBJECTIVE TYPE QUESTIONS

1. The difference between the upper limit and lower limit of a dimension is called
   (*a*) nominal size (*b*) basic size (*c*) actual size (*d*) tolerance
2. In a unilateral system of tolerance, the tolerance is allowed on
   (*a*) one side of the actual size (*b*) one side of the nominal size
   (*c*) both sides of the actual size (*d*) both sides of the nominal size
3. The algebraic difference between the maximum limit and the basic size is called
   (*a*) upper deviation (*b*) lower deviation
   (*c*) actual deviation (*d*) mean deviation
4. A basic shaft is one whose
   (*a*) lower deviation is zero (*b*) upper deviation is zero
   (*c*) lower and upper deviations are zero (*d*) none of these
5. According to Indian standard specifications, 100 $H6/g5$ means that the
   (*a*) basic size is 100
   (*b*) actual size is 100 mm
   (*c*) difference between the actual size and basic size is 100 mm
   (*d*) none of the above
6. According to Indian standards, the total number of tolerance grades are
   (*a*) 8 (*b*) 12 (*c*) 16 (*d*) 20
7. According to Indian standards, the total number of fundamental deviations are
   (*a*) 10 (*b*) 15 (*c*) 20 (*d*) 25
8. According to Indian standards, 100$H6/g5$ means that
   (*a*) tolerance grade for the hole is 6 and for the shaft is 5
   (*b*) tolerance grade for the shaft is 6 and for the hole is 5
   (*c*) tolerance grade for the shaft is 4 to 8 and for the hole is 3 to 7
   (*d*) tolerance grade for the hole is 4 to 8 and for the shaft is 3 to 7
9. In a bilateral system of tolerance, the tolerance is allowed on
   (*a*) one side of the actual size
   (*b*) one side of the nominal size
   (*c*) both sides of the actual size
   (*d*) both sides of the nominal size
10. When the dimension is expressed as $20^{+0.035}_{-0.025}$, then the tolerance is
    (*a*) 0.035 mm (*b*) 0.025 mm (*c*) 0.01 mm (*d*) 0.06 mm

ANSWERS

1. (*d*)	2. (*b*)	3. (*a*)	4. (*b*)	5. (*a*)
6. (*c*)	7. (*d*)	8. (*a*)	9. (*d*)	10. (*b*)

20

Measuring Instruments and Gauges

20·1 Introduction

The parts manufactured on a mass production basis must be produced according to rigid engineering specifications in order to facilitate easy assembly. The measuring instruments, tools and gauges are used to measure, check and set the dimensions specified. They are used to detect inaccuracy in machining so as to eliminate defects and rejections. Their applications are numerous. In addition to measurements they are used to set, adjust tools, to align machines and layout the surfaces to be machined.

The *measuring instruments* are dimensional control instruments used to measure the exact size of object. These are adjustable devices and can measure with an accuracy of 0·001 mm or better.

The *gauges* are fixed dimension instruments and are not graduated. In most cases, these have no adjustable member for measuring varying dimensions of lengths and angles.

20·2 Classification of Measuring Instruments

The measuring instruments may be classified as follows :

1. *Precision or non-precision measuring instruments.* The precision instruments are those which have ability to measure parts with an accuracy of 0.001 mm or better. The non-precision instruments are limited to the measurements of parts to a visible line graduation on the instrument used, such as graduated rule or scale etc.

2. *Line measuring or end measuring instruments.* In line measuring instruments, the ends of a dimension being measured are aligned with graduations of the scale from which the length is read directly such as a steel rule. In end measuring instruments, the measurements are taken between the two ends as in micrometers, calipers, gauge block etc. These instruments are more useful and important for precision measuring work.

3. *Direct or indirect type measuring instruments.* The direct type measuring instruments are used to determine the actual size and dimension of a work piece. The indirect type measuring instruments are used to compare measurements of work piece with the direct measuring instrument.

20·3 Types of Measuring Instruments

Every measuring instrument has limited use and can be used for one or more purposes, *i.e.*, for measuring linear dimensions, angular dimensions and for measuring surface finish. The various measuring instruments, depending upon their use, are as follows :

1. *Linear measurements.* The following are the various instruments used for linear measurements :

(*a*) *Non-precision instruments*

(*i*) Steel rule, (*ii*) Calipers,
(*iii*) Dividers, (*iv*) Depth gauge, and
(*v*) Telescopic gauge.

(*b*) *Precision instruments*

(*i*) Micrometers, (*ii*) Vernier calipers,
(*iii*) Vernier height gauge, (*iv*) Vernier depth gauge,
(*v*) Vernier gear tooth caliper, and (*vi*) *Slip gauges.

2. *Angular measurements.* The following are the various instruments used for angular measurements :

(*a*) *Non-precision instruments*

(*i*) Protractor, (*ii*) Adjustable bevel,
(*iii*) Combination set, and (*iv*) T-square.

(*b*) *Precision instruments*

(*i*) Universal bevel protractor, (*ii*) Sine bar,
(*iii*) Dividing head (indexing head), (*iv*) Angle gauge blocks,
(*v*) Spirit level, and (*vi*) Auto-collimator.

3. *Taper measurements*

4. *Surface measurements.* The following are the various instruments used for surface measurements :

(*i*) Straight edge, (*ii*) Surface plate,
(*iii*) Surface gauge, (*iv*) Optical flat,
(*v*) Interferometer, (*vi*) Profilometer, and
(*vii*) Profilogram.

All the above mentioned measuring instruments are discussed, in detail, in the following pages.

20·4 Steel Rule

The steel rule (also known as *scale*) is the simplest and most common non-precision linear measuring instrument, and it is made of tempered steel. All the faces of a steel rule are machined true. On one of the flat faces, the graduations are marked in inches and in centimetres, as shown in Fig. 20.1. The steel rules in common use are either 150 mm or 300 mm long. It can be used for taking linear measurements of the workpiece to an accuracy of from 1 mm to 0.5 mm.

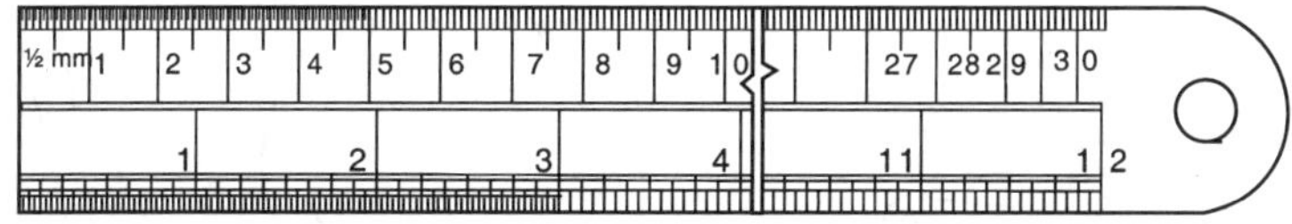

Fig. 20.1. Steel rule.

Notes : 1. Since the edges of a steel rule are important in measurements, therefore it should never be used as a substitute for the screw driver.

2. The sharpness of graduations must be maintained by the use of grease dissolving fluids.

* See Art. 20·41.

20·5 Calipers

These are non-precision linear measuring instruments used to transfer and compare a dimension from one object to another or from one part to a micrometer or to a scale. The various types of calipers, commonly used, are as follows :

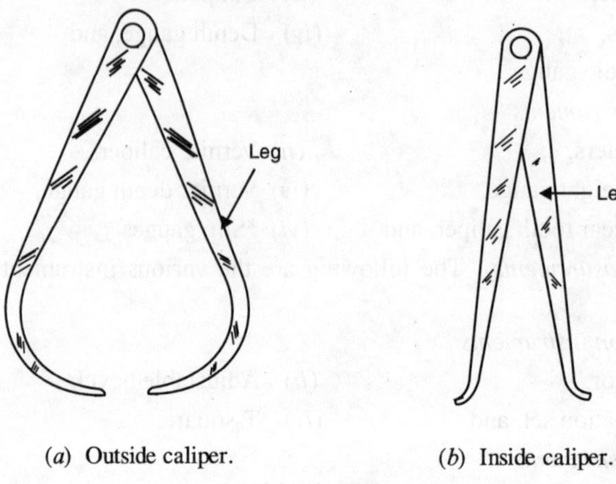

(*a*) Outside caliper. (*b*) Inside caliper.

Fig. 20.2

1. *Outside calipers.* The outside calipers, as shown in Fig. 20.2(*a*), have their legs bent inward and hinged at the top. These are used for checking or measuring outside dimensions or plate thickness. Care must be taken not to twist or slant the caliper, otherwise wrong reading will result. The accuracy in caliper measurement depends upon the sense of feel. A steel rule must be used with the caliper when direct reading is required.

2. *Inside calipers.* The inside calipers, as shown in Fig. 20.2(*b*), have their legs bent outwards. These are used for checking or measuring the internal dimensions of recesses, hole diameters and parallel surfaces of inside dimension. A steel rule must be used to obtain a specific reading, as in outside caliper.

3. *Hermaphrodite or odd leg calipers.* The hermaphrodite or odd leg caliper (also known as jennies or mophs caliper) has one of its leg pointed while the other leg is bent inwards, as shown in Fig. 20.3. It is chiefly used for marking out lines parallel to the edge of the work and for locating centre of cylindrical work.

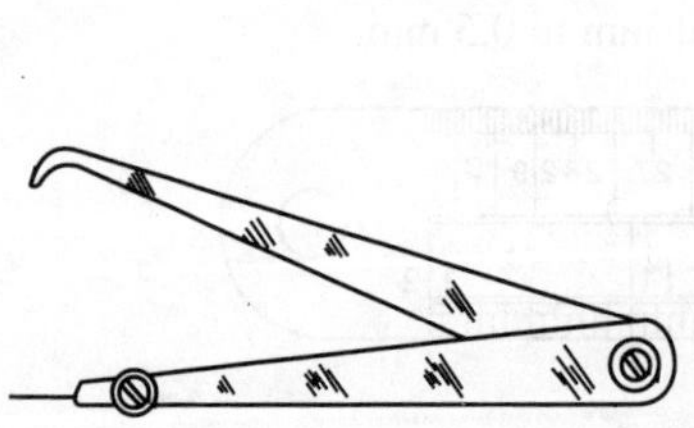

Fig. 20.3. Hermaphrodite or odd leg caliper

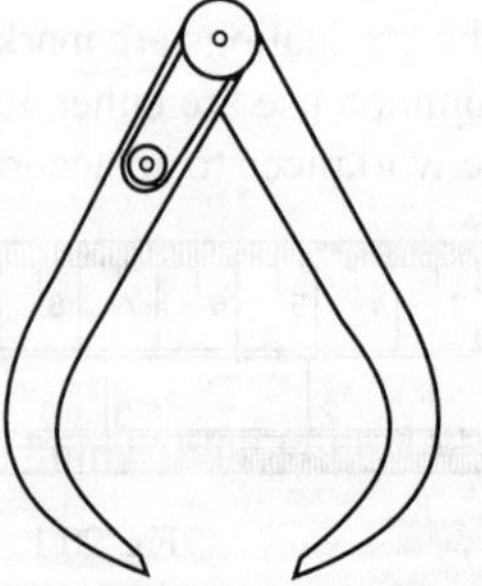

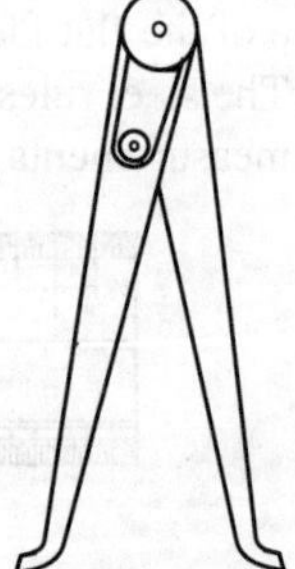

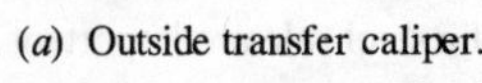

(*a*) Outside transfer caliper. (*b*) Inside transfer caliper

Fig. 20.4

4. *Transfer calipers.* The transfer calipers, may be outside or inside, as shown in Fig. 20.4(*a*)

and (*b*) respectively. The outside transfer calipers are used for measuring outside projections, and the inside transfer calipers are used for measuring inside cavities from which the legs of the caliper cannot be removed directly but are collapsed after the dimension is measured. A steel rule must be used with the caliper to obtain direct reading.

Note : The calipers, discussed above, are known as *firm joint calipers*.

5. *Spring calipers.* The spring calipers (also known as tool maker's calipers) are outside and inside calipers, as shown in Fig. 20.5(*a*) and (*b*) respectively. These calipers are expensive and have a fine screw adjustment to regulate or to set the position of the legs. A steel rule must be used as in ordinary calipers.

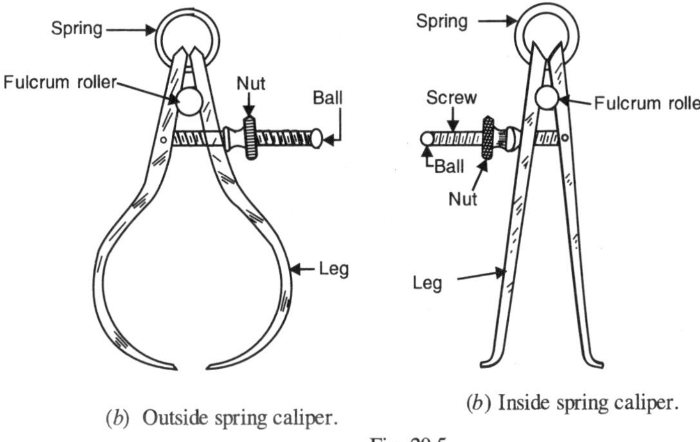

(*b*) Outside spring caliper. (*b*) Inside spring caliper.

Fig. 20.5

20·6 Divider

The dividers may be firm joint type or spring type, but the spring type dividers, as shown in Fig. 20.6, are preferred for smaller and more accurate work. The necessary adjustment of the legs is

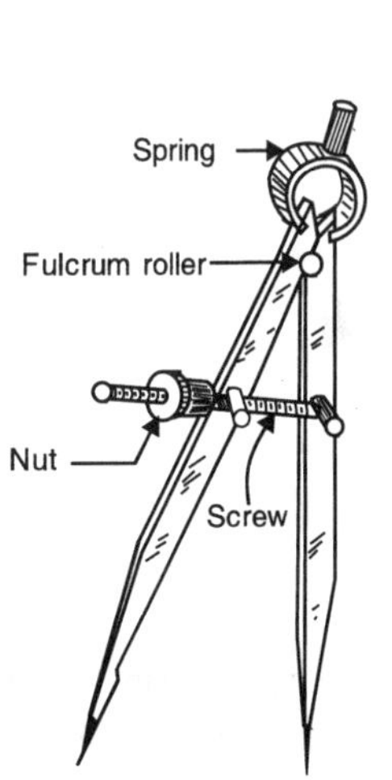

Fig. 20.6. Divider.

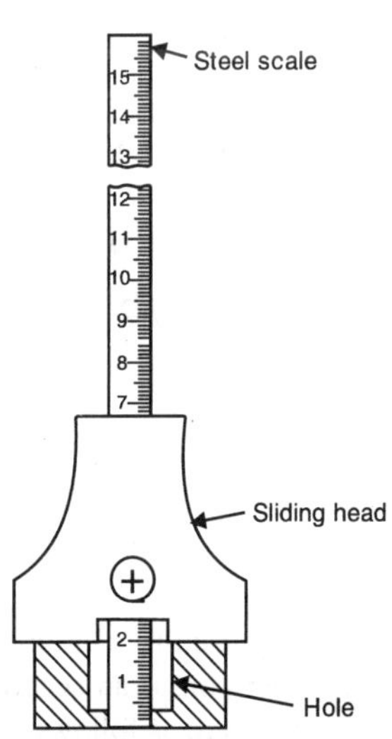

Fig. 20.7. Depth gauge.

obtained by a nut and a screw. This tool is used for transferring dimensions, marking out curves and circles and for doing general layout work.

20·7 Depth Gauge

It is used for measuring the depth of blind holes, grooves, slots and heights of shoulders in holes etc. The simple type of depth gauge, as shown in Fig. 20.7, consists of a normal steel rule to which a sliding head, at right angle, is clamped.

In using this instrument, the steel rule is lowered into the hole and the sliding head rests on the top surface of hole, as shown in Fig. 18.7. The reading is directly taken from the rule. The depth gauge does not measure as accurate as *depth gauge micrometers.

20·8 Telescopic Gauge

It is used for measuring internal size of holes, slots and grooves. A telescopic gauge, as shown in Fig. 20.8, consists of a handle, one end of which carries two rods, one on each side at right angles to the handle. These two rods are made telescopic *i.e.,* one of the rods (movable) can enter into the other fixed rod and both are under spring tension. The other end of the handle is provided with a knurled screw so that the rods may be locked in any desired position.

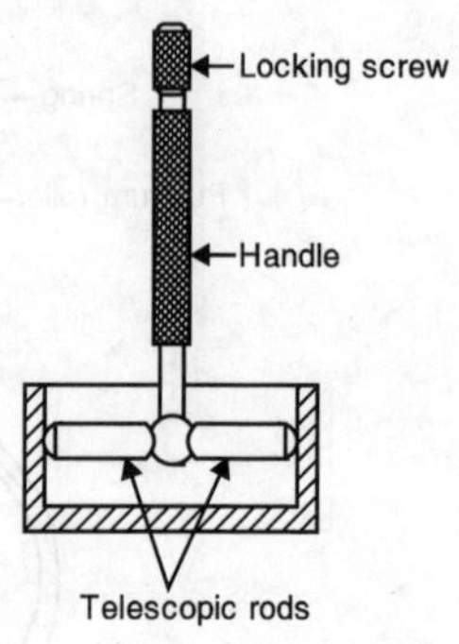

Fig. 20.8. Telescopic gauge.

In using this instrument to measure a hole diameter, the rods are first pressed and locked in position. Now the rods are inserted into the hole and unlocked in order to touch the ends of a hole. The instrument is locked in this position and taken out of the hole. The distance between the end faces of the rods is measured by an outside micrometer. This instrument cannot measure itself because it has no graduations of its own.

20·9 Micrometers

The micrometers are precision measuring instruments. These are available in English system or in metric system, but the metric system micrometers are mostly used. The various types of micrometers are as follows :

1. Outside micrometer,
2. Inside micrometer,
3. Screw thread micrometer, and
4. Depth gauge micrometers.

We shall now discuss these micrometers, in detail, in the following articles.

20·10 Outside Micrometer

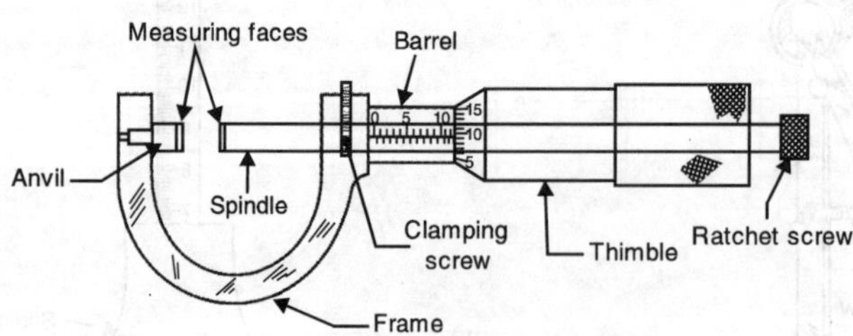

Fig. 20.9. Outside micrometer.

It is mainly used to measure the outside diameter of a job or length of a small part. It can measure the dimension to an accuracy of 0.01 mm. The main parts of an outside micrometer, as shown in Fig. 20.9, are as follows :

1. *Frame.* The frame may be U-shaped or C-shaped, in order to permit measurements of a

---

* A depth gauge micrometer is discussed in Art. 20.13.

cylinder of diameter equal to the range of micrometer. It is made of steel, cast steel, malleable cast iron or light alloy. The use of light alloy is made for sizes over 300 mm in order to facilitate the handling of large micrometers.

2. *Anvil and spindle.* The anvil of the micrometer is fixed and it is on the left hand side of the frame. It should protrude at least 3 mm from the frame in order to permit of the attachment of a measuring wire support. There is a screwed spindle on the right hand side of the frame and have threads of 0.5 mm pitch. The anvil and spindle are generally made of high grade tool steel. In some cases, the measuring faces of the anvil and spindle are tipped with tungsten carbide or some other suitable material. The diameter of the anvil and spindle should be equal.

3. *Spindle clamp or lock nut.* A lock-nut is provided on the micrometer to lock the spindle when the micrometer is at its maximum reading.

4. *Barrel and thimble.* The barrel (also known as sleeve) has datum or reference line and fixed graduations. The thimble is a tubular cover fastened with the spindle and moves with the spindle. The bevelled edge of the thimble is divided into 50 equal parts and every fifth part is numbered.

5. *Ratchet or friction stop.* The micrometer is provided with a ratchet or friction stop in order to enable satisfactory operation. When the spindle reaches near the work whose dimension is to be measured, then the ratchet is used to prevent excessive pressure on the spindle.

The barrel, thimble, ratchet and locking nut, for all sizes of micrometers, are made from wear resistant steel.

The least distance which can be measured accurately with micrometer is called *least count of micrometer.* It is the ratio of pitch of the spindle screw to the number of divisions on the thimble. Mathematically,

$$\text{Least count} = \frac{\text{Pitch of the screw}}{\text{Number of divisions on the thimble}} = \frac{0.5 \text{ mm}}{50} = 0.01 \text{ mm}$$

Reading a micrometer. The micrometer reading is taken as discussed below :

1. Before taking a micrometer reading, the range of the micrometer and zero error must be checked. The zero error is checked by contacting the faces of the anvil and spindle for micrometers from 0-25 mm, and by placing a master piece of 25 mm between the anvil and spindle faces for micrometers from 25-50 mm. When the zero on the thimble does not coincide with the zero on the reference line on the barrel, then there is a zero error in the micrometer. It may be noted that

(*a*) When the zero on the thimble is above the reference line, then the zero error in negative. Therefore it should be added to the reading.

(*b*) When the zero on the thimble is below the reference line, then the zero error is positive. Therefore it should be subtracted from the reading.

The zero error can be adjusted with the help of a special spanner provided with the micrometer.

2. The graduations on the barrel above the reference line are in 1 mm intervals and are numbered 0, 5, 10, 15, 20 and 25. The graduations below the reference line are also in 1 mm intervals but each graduation shall be placed at the middle of two successive upper graduations so as to read 0.5 mm, as shown in Fig. 20.10.

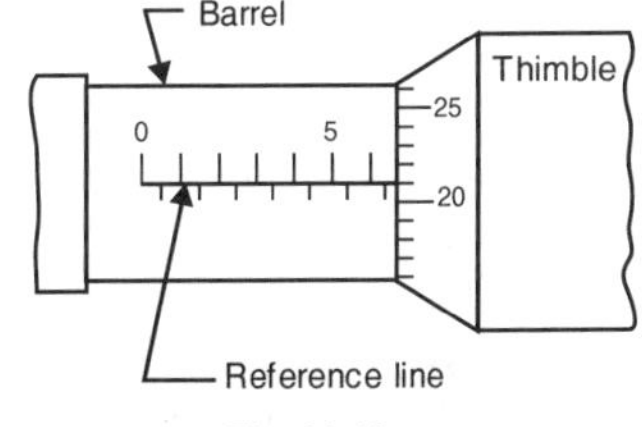

Fig. 20.10

3. Now hold the work between the faces of the anvil and spindle and note the number of divisions on the barrel (or main scale) above and below the reference line, and the number of divisions on the thimble (or circular scale) coinciding with the reference line.

4. The reading, as shown in Fig. 20.10, in obtained as follows :

Number of divisions above the reference line = 6

$= 6 \times 1 \text{ mm} = 6 \text{ mm}$

Number of divisions below the reference line = 1

$= 1 \times 0.5 \text{ mm} = 0.5 \text{ mm}$

Number of divisions on the thimble coinciding with the reference line = 21

$= 21 \times$ *$0.01 \text{ mm} = 0.21 \text{ mm}$

$\therefore$ Total reading $= 6 + 0.5 + 0.21 = 6.71 \text{ mm}$

Note : The outside micrometers are available in the measuring ranges from 0-25 mm, 25-50 mm, 50-75 mm, 75-100 mm and so on. The largest measuring range available is 575-600 mm. In using micrometers over the range of 0-25 mm, the initial distance between the measuring faces must be added with the reading obtained from the micrometer. Greater measuring ranges (over 300 mm) can be obtained with interchangeable and adjustable anvils.

20.11 Inside Micrometer

The inside micrometer is used for measuring large internal diameters (over 50 mm) to an accuracy of 0.01 mm. It works on the same principle as that of outside micrometer.

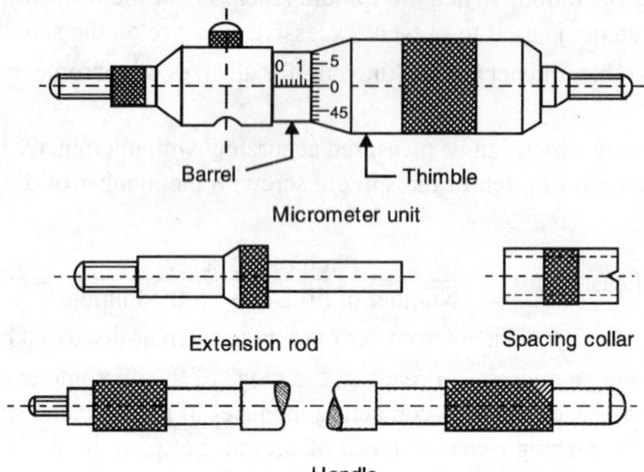

Fig. 20.11. Inside micrometer.

The inside micrometer, as shown in Fig. 20.11, consists of the following four parts :

1. Micrometer unit,
2. Extension rod,
3. Spacing collar, and
4. Handle.

The *micrometer unit* (also known as *measuring head*) consists of a barrel or sleeve and a thimble, similar to outside micrometer. It has no frame and spindle. The measuring points are at the extreme ends and the adjustment is made by advancing or withdrawing the thimble along the barrel with the help of a micrometer screw. The micrometer screw has a pitch of 0.5 mm. The barrel is provided with a scale of 13 mm long and is graduated into half- millimeter and millimeter division as in the outside micrometer. The bevelled edge of the thimble is divided into 50 divisions round the circumference.

* The number of divisions on the thimble coinciding with the reference line are multiplied by the least count of the micrometer.

A series of *extension rods* of different lengths, with or without spacing collars, are provided in order to obtain a wide measuring range. The measuring faces of the extension rods are tipped with tungsten carbide or some other hard material. The range of measurement, for which an extension rod is applicable, is always marked on it. These are usually available in the range of 25 mm.

A detachable *handle* of sufficient length is usually provided so that the micrometer unit can be easily lowered into deep holes.

Reading an internal micrometer. Since the micrometer unit has only 13 mm movement, therefore using one range of extension rods (say 50-75 mm), without the spacing collar, the distance between the measuring or contacting faces can be varied from 50 to 63 mm by rotating the thimble, In other words, we can measure the internal diameter from 50 to 63 mm with the extension rod having the range 50-75 mm. In order to measure more than 63 mm and upto 75 mm, a spacing collar (which is exactly 12 mm in length) is inserted on the extension rod. Thus we can measure the diameter from 63 to 75 mm, by using a spacing collar with the extension rod. In the similar way, other dimensions can be taken by using extension rods of different ranges.

20·12 Screw Thread Micrometer

It is designed to measure the pitch diameter of screw threads to an accuracy of 0.01 mm. In construction, the screw thread micrometer, as shown in Fig. 20.12, is similar to outside micrometer with the following differences.

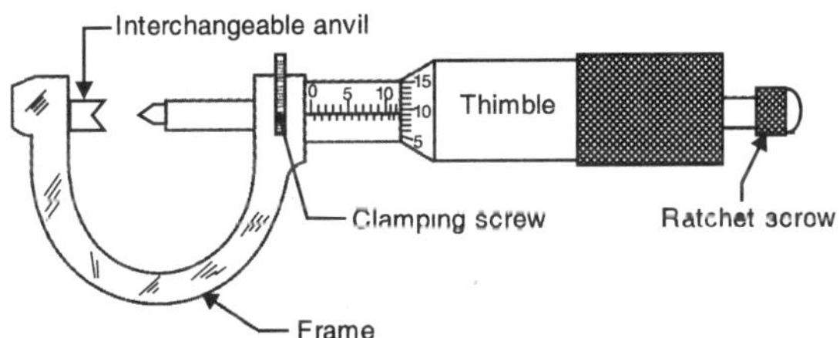

Fig. 20.12. Screw thread micrometer.

1. The movable spindle is pointed, and
2. The end of the anvil is of the same form as the screw thread to be measured.

The different pairs of interchangeable vee-anvils and spindle points are used with this micrometer. In order to measure the pitch diameter, the pointed end of the spindle and the sides of the vee-anvil should contact the surfaces of the thread. The reading on the micrometer is read in the similar way as in outside micrometer.

20·13 Depth Gauge Micrometer

The depth gauge micrometer (also known as *depth micrometer*), is used to measure the depth of holes, slots and recessed areas to an accuracy of 0.01 mm. The various parts of this micrometer are shown in Fig. 20.13.

The *micrometer head* acts as a reference surface and is held firmly and perpendicular to the centre line of the hole. The *screwed spindle* does the actual measuring. The *graduated sleeve* has the reference line and fixed graduations

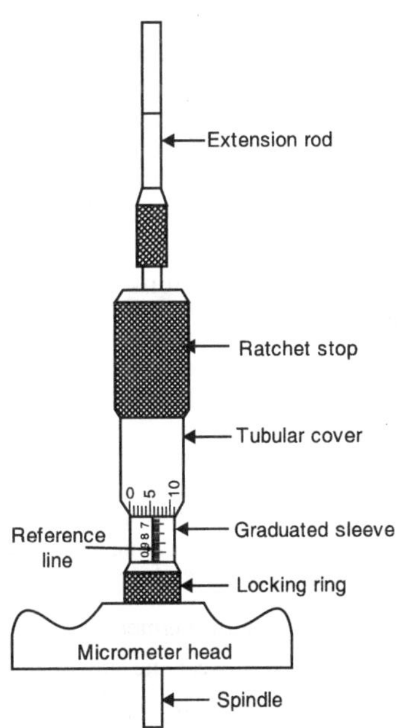

Fig. 20.13. Depth gauge micrometer.

in the reverse direction *i.e.,* in the order 0, 9, 8, 7, 6, etc. The thimble is a tubular cover fastened to the spindle and moves with the spindle. The bevelled edge of the thimble has graduations from 0 to 50, and every fifth is numbered. The *locking ring* is provided to lock the micrometer at any desired reading. The *ratchet stop* is a small extension to the thimble. The *extension rods* are used for larger ranges of measurement, as in inside micrometer. It is inserted through the top of the micrometer.

In using this instrument, it should be first ensured that the edge of the hole is free from burrs. Care should be taken when the spindle touches the bottom of the hole and the micrometer head rests on the surface. The accuracy in reading depends upon the sense of touch. The reading of this micrometer is taken in the similar way as outside micrometer.

20·14 Vernier Caliper

It is a precision instrument which is used for measuring external as well as internal diameters of shafts, thickness of parts etc., to an accuracy of 0.02 mm. It can also be used to measure the depth of slots and holes. The principle of vernier is that when two scales or divisions slightly different in size are used, the difference between them can be utilised to determine the accuracy of measurement.

The vernier calipers, as shown in Fig. 20.14, consists of a beam on which the fixed graduations (known as the *main scale*) is engraved. One end of the beam has fixed or stationary jaw. The vernier head, which carries the sliding or movable jaw, has a vernier scale engraved on it. This jaw moves along the guiding surfaces provided by the beam. When the measuring faces of the fixed and sliding jaws are brought in contact with each other, the scale shows a zero reading. The final adjustment of the movable jaw, which depends upon the sense of correct feel, is made by the adjusting screw, provided on the auxiliary head. After the final adjustment is made, the movable jaw is locked by the clamping screw on the vernier head and the reading is noted down.

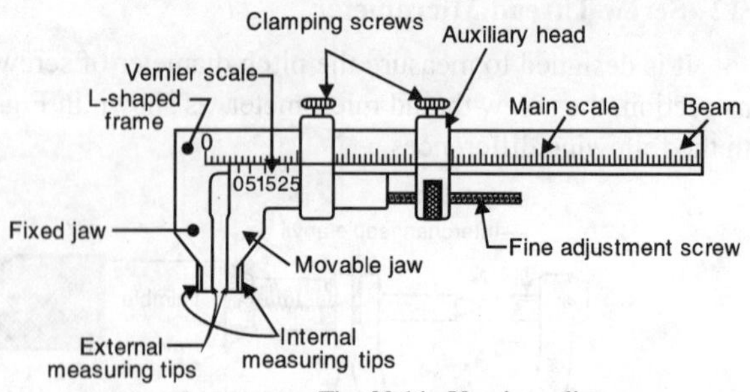

Fig. 20.14. Vernier caliper.

It may be noted that the main scale, shown in Fig. 20.14, is used for external measurements only. The readings for the internal measurements are obtained by adding the combined width of the internal measuring jaws to the scale reading.

Reading a vernier caliper. The smallest division on the main scale is either 0.5 mm or 1 mm and the vernier scale may have 25 or 50 divisions. The ratio of the smallest division on the main scale to the number of divisions on the vernier scale is known as *least count* of the vernier caliper. Mathematically,

$$\text{*Least count} = \frac{\text{Smallest division on main scale}}{\text{Number of divisions on vernier scale}} = \frac{0.5}{25} = 0.02 \text{ mm}$$

---

* The least count may also be obtained as follows :
Since 25 divisions on the vernier scale will coincide with 24 small divisions (each division = 0.5 mm) on the main scale, therefore

$$1 \text{ division on the vernier scale} = \frac{24}{2} \times \frac{1}{25} = \frac{24}{50} \text{ mm}$$

Difference between the smallest division on the main scale and 1 division on the vernier scale

$$= 0.5 - \frac{24}{50} = \frac{1}{2} - \frac{24}{50} = \frac{1}{50} = 0.02\text{mm}$$

This difference is called the *least count.*

The least count is usually marked on the vernier caliper.

The reading for the positions of the main scale and vernier scale shown in Fig. 20.15(*a*) is taken as follow :

1. First of all, note the centimetres, millimetres and half millimetres on the main scale which the zero of the vernier has moved from the zero of the main scale. From Fig. 20.15(*b*), we see that the zero of the vernier has moved the main scale by a distance 1 cm + 1 mm + 0.5 mm *i.e.* 1.15 cm or 11.5 mm.

Fig. 20.15

2. Now count the number of divisions on the vernier scale from zero to a line which exactly coincides with the line on the main scale. From Fig. 20.15(*b*), we see that the line of 16th division exactly coincides with the line on the main scale. This shows a reading of $16 \times 0.02 = 0.32$ mm.

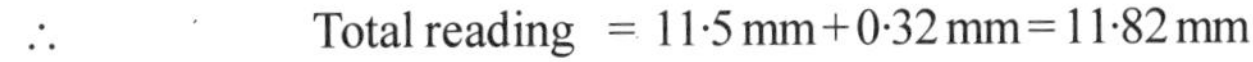

$$\therefore \quad \text{Total reading} = 11{\cdot}5\,\text{mm} + 0{\cdot}32\,\text{mm} = 11{\cdot}82\,\text{mm}$$

Notes : **(*a*)** Before using the instrument, it should be checked for zero reading. The zero line on the vernier scale should coincide with the zero on the main scale.

(*b*) The same procedure as discussed above can be used for measuring with 50 divisions vernier scale.

(*c*) The *dial vernier calipers,* in which the readings are taken directly from a dial indicator, are also available.

20·15 Vernier Height Gauge

It is mainly used for measuring heights of parts to an accuracy of 0.02 mm. It is also used for scribing lines in layout work.

The principle of this instrument is similar to a vernier caliper. The essential parts of the vernier height gauge, as shown in Fig. 20.16, are as follows:

1. A special base with accurately finished bottom,
2. A graduated beam, called the main scale,
3. A sliding vernier head carrying the vernier scale and clamping screw,
4. An auxiliary head having the fine adjustment and clamping screw, and
5. A measuring and scribing jaw.

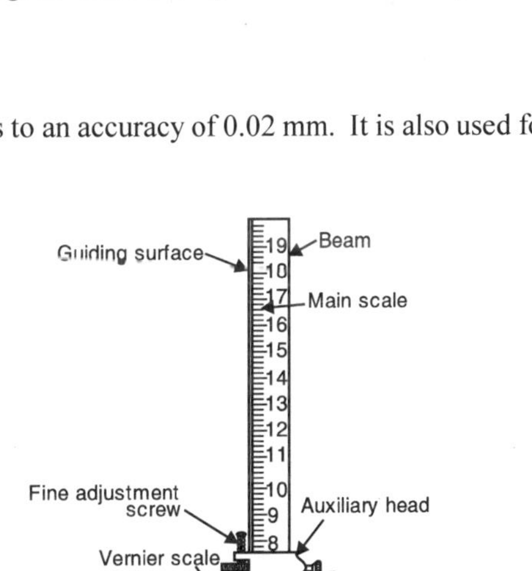

Fig. 20.16. Vernier height gauge.

The vernier height gauge, commonly used, have the measuring range from 0 to 300 mm. Before using this instrument, it should be checked for zero error. For this, the vernier height gauge is placed on a surface plate and the vernier head is brought down till the measuring jaw touches the surface plate. In this position, the zero on the main scale should coincide with the zero on the vernier scale. In order to measure the height, the work is placed between the surface plate and the measuring face. The final adjustment depending upon the sense of correct feel is made by the fine adjustment screw. The measurement is read in the similar way as in the vernier caliper. For scribing lines, the scriber is set at a specified height and moved along the work piece to be marked.

Note : The vernier height gauges with the lower and upper measuring limits, *e.g.*, 20 to 250, 30 to 300, 40 to 500, 60 to 800 and 60 to 1000 mm, are also available. In using such an instrument, the zero error is checked by placing a corresponding gauge block (having the height equal to the lower measuring limit) between the surface plate and the measuring jaw. The height of the work is obtained by adding the lower measuring limit to the reading obtained on the vernier height gauge.

20·16 Vernier Depth Gauge

It is used to measure the depth of holes, recesses, and distances from plane surface to a projection. It can measure to an accuracy of 0.02 mm. It works on the same principle as that of a vernier caliper. The essential parts of a vernier depth gauge, as shown in Fig. 20.17, are as follows:

Fig. 20.17. Vernier depth gauge.

1. A sliding vernier head having a vernier scale, clamping screw and a base at the bottom,
2. A graduated blade known as main scale, and
3. An auxiliary head with the fine adjustment and clamping screw.

In using this instrument, the base is held firmly against the upper surface and the blade is lowered into the hole until it contacts the surface of the hole. The final adjustment depending upon the sense of correct feel is made by the fine adjusting screw. The instrument, after tightening the clamping screws, is removed from the hole and read in the same way as the vernier caliper.

20·17 Vernier Gear-tooth Caliper

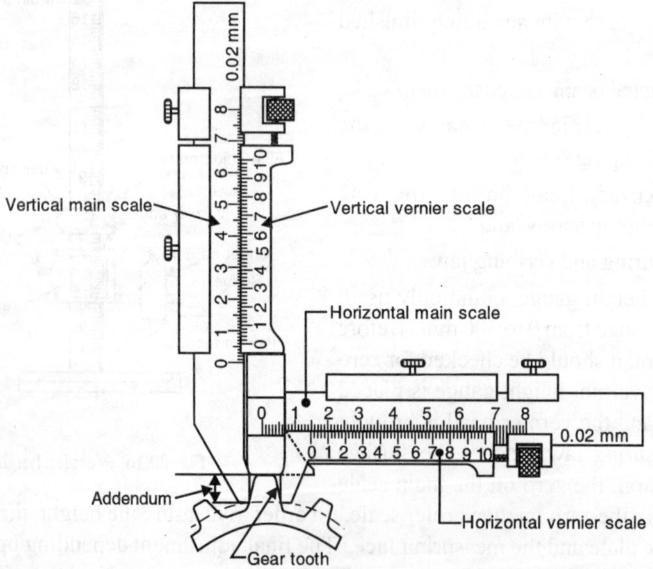

Fig. 20.18. Vernier gear-tooth caliper.

It is used for measuring the chordal thickness of a gear tooth at the pitch circle of a gear. The vernier gear-tooth caliper, as shown in Fig. 20.18, consists of two vernier calipers perpendicular to each other.

The horizontal vernier caliper which measures the tooth thickness is similar to an outside vernier caliper whereas the vertical vernier caliper is adjusted for measuring the distance from the top of a tooth (or addendum circle) to the pitch circle of a gear. The tables which give these distances and the chordal tooth thicknesses for gears of different diametral pitches and different number of teeth are available.

20·18 Protractor

It is the most simple device used to measure the angle between the two faces of a component. It consists of a base in the form of a steel flat bar and an angle measuring device graduated in degrees and minutes, as shown in Fig. 20.19. The base of the instrument is set along the edge of the work to be checked for angular measurement. The instrument can be read upto half a degree. A vernier scale can also be provided to the protractor to measure angle upto 5 minutes.

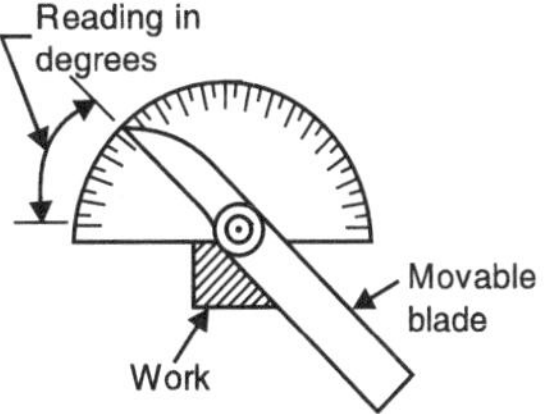

Fig. 20.19. Protractor.

20·19 Adjustable Bevel

It is an instrument used for checking, measuring and transferring angles and laying out work. The adjustable bevel, as shown in Fig. 20.20, consists of two blades which can be set and locked in any position by a locking screw. The angle is checked by some other angular measuring device as it is not a direct reading instrument.

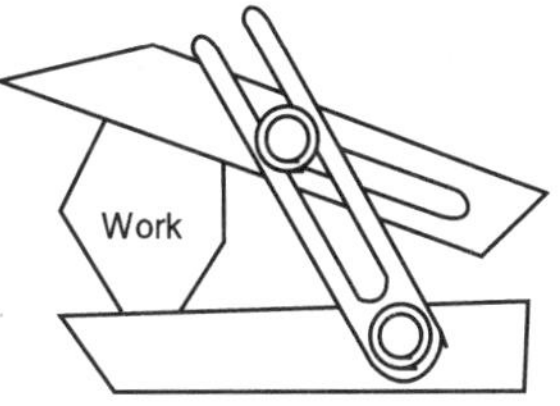

Fig. 20.20. Adjustable bevel.

20·20 Combination Set

It is an extremely useful instrument and has all the essential features of try square, bevel protractor, rule and scriber. The combination set, as shown in Fig. 20.21, consists of four parts *i.e.*, square head, protractor head, centre square head and a blade or rule on which the other three parts slide. They are located in groove in the blade and a tongue in the head to which clamping screw is attached. The protractor and square heads are usually provided with spirit levels. The square head has a small scriber concealed in the end for marking purposes.

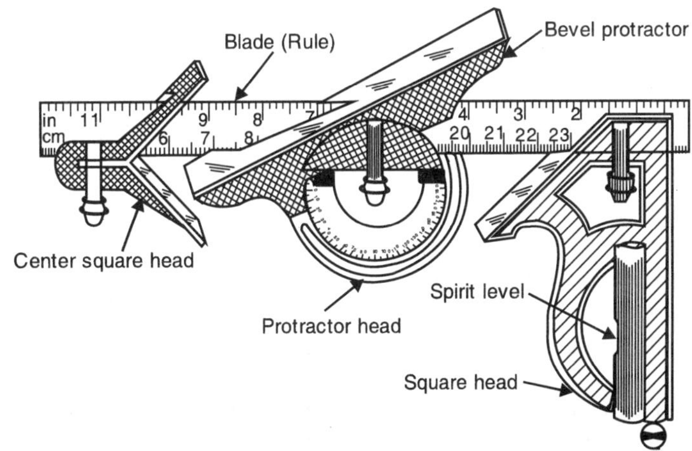

Fig. 20.21. Combination set.

The square head can be used, by moving the blade, as a try square, mitre square or even as a depth gauge. The centre square head may be used to extend a line around a corner or to find centre of a round object. The protractor head is used for measuring angles and can be clamped in any desired position. The scale is graduated in degrees.

20·21 Try Square

It is used for testing right angles. It consists of a steel blade fixed at right angles to the edge of a machined stock of metal, as shown in Fig. 20.22. The stock, when in use, acts as a fence and the blade as a guide for marking. It is available is various sizes like 100 mm, 150 mm and 300 mm.

Fig. 20.22. Try square.

20·22 Universal Bevel Protractor

It is also called *vernier bevel protractor.* It is used for measuring and testing angles. The accuracy of this instrument, to measure an angle, is within the limits of 5 minutes $\left(i.e., \frac{1}{12} \text{ of a degree} \right)$.

The universal bevel protractor, as shown in Fig. 20.23, consists of a base, sliding blade, graduated disc (or the main scale) and vernier scale. The protractor's disc is free to rotate at the pivot and can be clamped with a nut in the base assembly. The blade can slide both ways. The vernier scale is fitted with the disc to take readings during measurement.

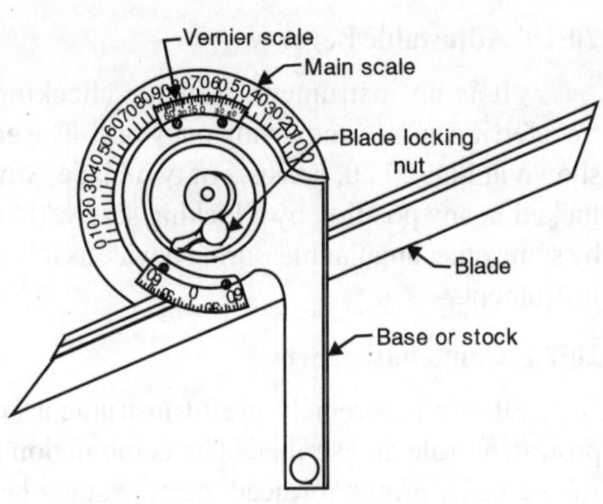

Fig. 20.23. Universal bevel protractor.

Reading a universal bevel protractor. The protractor's disc is graduated in degrees over an arc of 180°, reading 0 to 90° each way. The vernier scale has 12 divisions both to the right and left of the zero line. These 12 divisions coincides with 23 divisions on the main scale, as shown in Fig. 20.24(*a*).

∴ 1 division on the vernier scale

$$= \frac{23^\circ}{12}$$

and difference between the 2 divisions on the main scale and 1 division on the vernier scale

$$= 2 - \frac{23}{12} = \frac{1^\circ}{12} \text{ or 5 minutes.}$$

The reading on the instrument, as shown in Fig. 20.24(*b*), is taken as follows:

1. First of all, note the number of whole degrees between the zero on the main scale and zero on the vernier scale. From Fig. 20.24(*b*), we see that the whole degrees are equal to 52°.

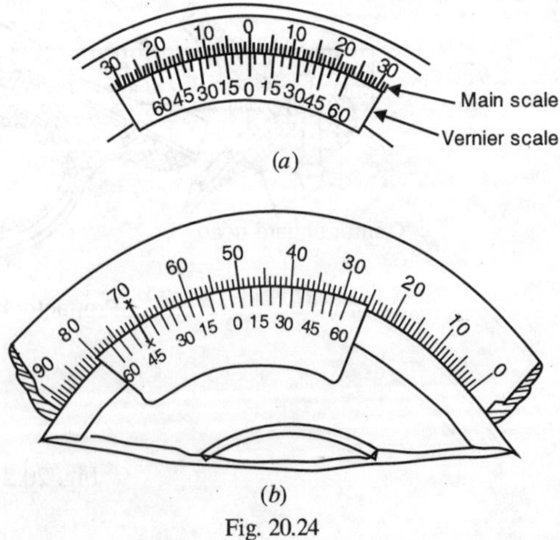

Fig. 20.24

2. Now count the number of divisions, in the same direction, from zero on the vernier scale to the line which coincides with the line on the main scale. Since each division represents 5 minutes, therefore the number of these divisions multiplied by 5 give the number of minutes to be added to the whole degrees. From Fig. 20.24(*b*) we see that the ninth division from zero on the vernier scale exactly coincides with the line on the main scale, therefore number of minutes = 9 × 5 = 45 minutes.

$$\therefore \quad \text{Total reading} = 52^\circ + 45' = 52^\circ 45'$$

20·23 Sine Bar

It is used either to measure angles more precisely than a bevel protractor or for locating any work to a given angle within very close limits.

It is generally used in conjunction with slip gauges. A sine bar, as shown in Fig. 20.25, consists of a hardened and ground steel bar which is stepped at the ends. A roller is fastened, in each step, with a screw. These rollers should touch the both faces of the step. A sine bar is specified by the distance between the centres of two rollers (L). The most commonly used sine bar has this distance as 100 mm. Some holes are drilled in the body of the sine bar in order to reduce the weight and to facilitate handling. It should satisfy the following conditions in order to measure accurately :

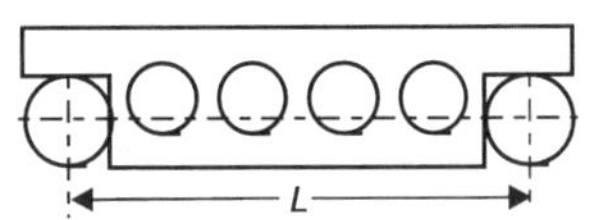

Fig. 20.25. Sine bar.

1. The diameter of the two rollers must be equal.
2. The centre lines of the two rollers must be parallel and separated by an accurate known distance.
3. One surface of the sine bar must be flat and lie in a plane which is parallel to the plane containing the centre lines of the two rollers.

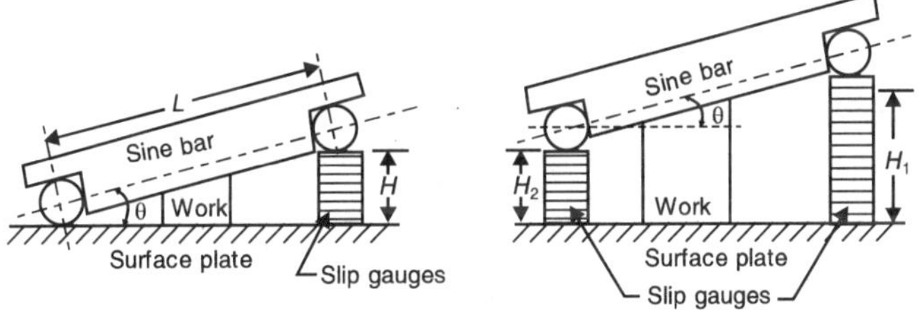

(*a*) One roller placed on slip gauges. (*b*) Both the rollers placed on slip gauges

Fig. 20.26. Use of slip gauges.

In measuring the angles or locating any work to a given angle, the use of sine bar is made as follows :

1. The work having tapered top and flat base is placed on the surface plate.
2. One of the rollers is placed on the surface plate, as shown in Fig. 20.26(*a*), while the other roller is made to rest on slip gauges of height H in such a way that the top of the work piece coincides with the sine bar.
3. Let the sine bar is set at an angle θ.

$$\therefore \quad \sin\theta = \frac{H}{L}$$

4. In order to obtain better results, both the rollers of the sine bar are placed on slip gauges of height H_1 and H_2, as shown in Fig. 20.26(*b*). In such cases

$$\sin\theta = \frac{H_1 - H_2}{L}$$

20·24 Dividing Head

The dividing head (also known as *indexing head*) was actually meant for use on milling machines, but its application is now extended to inspection work also. It is specially used for the measurement of angular dimensions about a common centre.

The dividing head, as shown in Fig. 20.27, mainly consists of a spindle, crank, indexing plate, worm and worm wheel. The spindle supports and rotates the work. A centre may be mounted in the tapered hole at the front end of the spindle for holding a work between centres. The crank is attached to a worm which is in mesh with a worm wheel. The worm and worm wheel have a gear ratio 40. In other words, one revolution of the crank will turn the spindle through $\frac{1}{40}$th of a revolution or 9°.

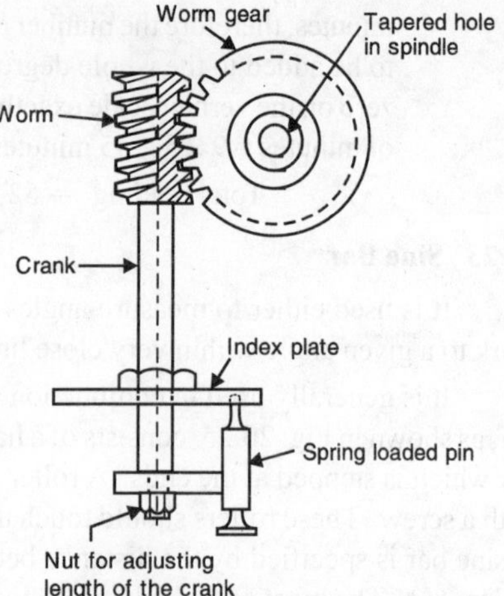

Fig. 20.27. Dividing head.

A dividing head is usually provided with two or more index plates by means of which any desired angle can be obtained with close accuracy. These various indexing methods for obtaining a wide range of angles are as follows :

1. Direct indexing,
2. Plain or simple indexing,
3. Differential indexing, and
4. Compound indexing.

The methods are beyond the scope of this volume.

20·25 Angle Gauge Blocks

These blocks are used to set up any angle to the nearest of 3 minutes. These are commonly used to measure the angle in the die insert. The angle gauge blocks are made of hardened steel and the measuring faces are lapped and polished to a high degree of accuracy and flatness. These blocks are about 75 mm long and 15 mm wide. There are thirteen separate angle gauge blocks in conjunction with one square block and one parallel straight edge. These are available in the following three series:

First series : 1°, 3°, 9°, 27° and 41°

Second series : 1′, 3′, 9′ and 27′.

Third series : 3″, 6″, 18″ and 30″.

All these angle gauge blocks in combination can be added or subtracted in order to get large

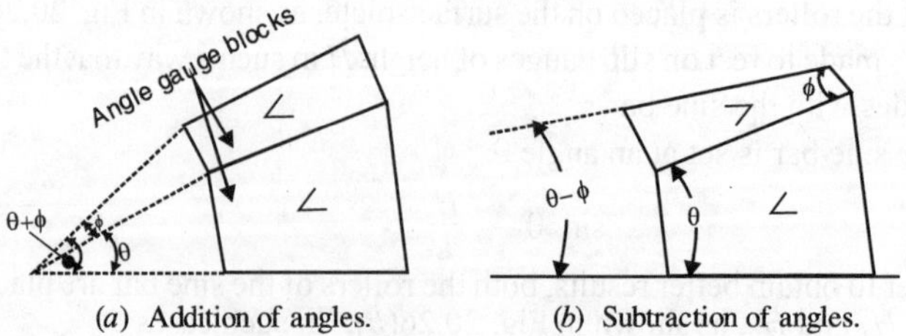

(*a*) Addition of angles. (*b*) Subtraction of angles.

Fig. 20.28. Use of angle gauge blocks.

number of combinations possible. All these blocks are marked V to indicate the direction of included angle. When the angles of the individual angle blocks are to be added up as shown in Fig. 20.28(a), then all the marked V should be in the same direction and when the angle is to be subtracted, as shown in Fig. 20.28(b), then marked V should be in opposite direction.

Note : Since the angle gauge blocks cannot always be directly applied to work, therefore these are being used as a reference and as an aid to other angle measuring devices.

20·26 Spirit Level

It is used for checking or testing the levelness and verticalness of surfaces. Its use is widely found for the static levelling of the machinery and other equipment, but it is based on precision angle measuring principle. These are available in various sizes. It is specified by the length *e.g.*, 200 mm spirit level.

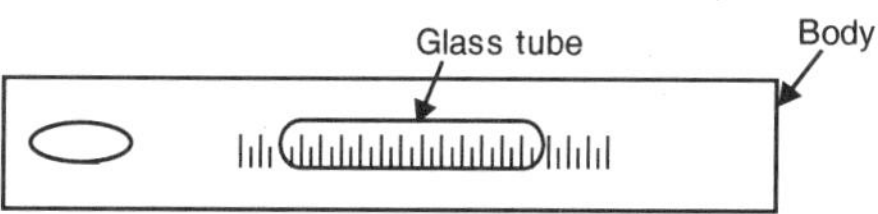

Fig. 20.29. Spirit level.

A spirit level consists of a glass tube ground to a large radius enclosed in a wooden or metal body, as shown in Fig. 20.29. The tube is partially filled with spirit so that a bubble of air is left, which comes to the top when placed horizontal. A scale is engraved on the glass at the top of the tube. When the spirit level is resting on the flat and levelled surface, the bubble will rest at the centre of the scale. If the base and hence the level is slightly tilted, the bubble will move along the scale a small distance depending upon the angle of tilt. If R is the radius of curvature of the inside surface of the glass tube and α is the angle of tilt in radians, then from Fig. 20.30, we find that the displacement of the bubble in the glass tube,

$$d = R.\,\alpha \qquad \ldots(i)$$

and $$\sin\alpha = h/L \quad \text{or} \quad \alpha \text{ (in radians)} = h/L \qquad \ldots(ii)$$

From equations (i) and (ii),

$$d = R \times h/L$$

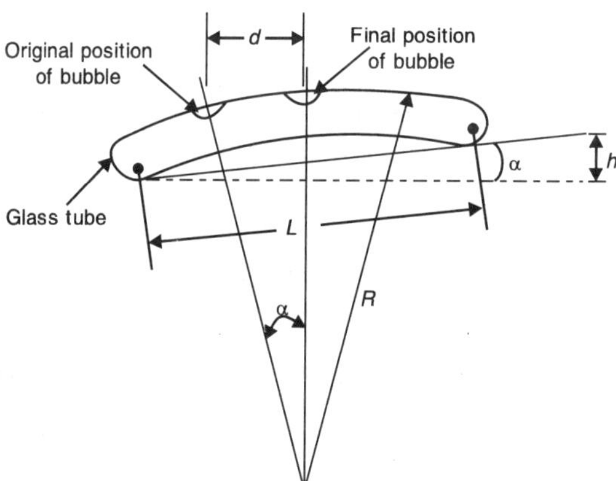

Fig. 20.30. Angle of tilt of spirit level.

The sensitivity of a spirit level is defined as the angle of tilt which will move the bubble through one division on the scale and is expressed in seconds per division. It increases as R increases. If α is expressed in seconds, then from equation (i),

$$d = \frac{R \times \alpha}{206\,265} \qquad \ldots(\because 1 \text{ rad} = 206\,265 \text{ seconds})$$

The scale spacing of a spirit level scale (d) is usually 2 mm.

Therefore for a radius of curvature, $R = 206$ m or 206×10^3 mm,

$$d = \frac{206 \times 10^3 \times \alpha}{206\ 265} = 0.9987\ \alpha$$

or $$\alpha = d/0.9987 = 2/0.9987 = 2 \text{ seconds}$$

Therefore an inclination of 2 seconds of the spirit level will cause a bubble movement of 1 division or 2 mm. Such a high sensitivity is required for spirit levels designed for laboratory research. For highly precise shop measurements, sprit levels with scale division values from 4 minutes to 10 minutes are employed.

20·27 Auto-collimator

It is an optical instrument used for the measurement of small angular differences. This instrument works on the following principle.

The point source of the light O is placed on the principal focus of a collimating lens as shown in Fig. 20.31(a). The rays of light from O incident on the lens will travel as a parallel beam of light. If this beam of light strikes a plane reflector which is normal to the collimator axis, then it will be reflected back along its own path and will focus at the same point O.

When the plane reflector is turned through a small angle θ, then the parallel rays will be reflected back through twice the angle *i.e.,* 2θ, as shown in Fig. 20.31(b). Now the reflected parallel rays will focus at point O' in the same focal plane. The angle θ may be calculated by measuring the distance OO'. The relation between θ and OO' is given by

$$\theta = \frac{OO'}{2f} \text{ radians}$$

where f = Focal length of the lens.

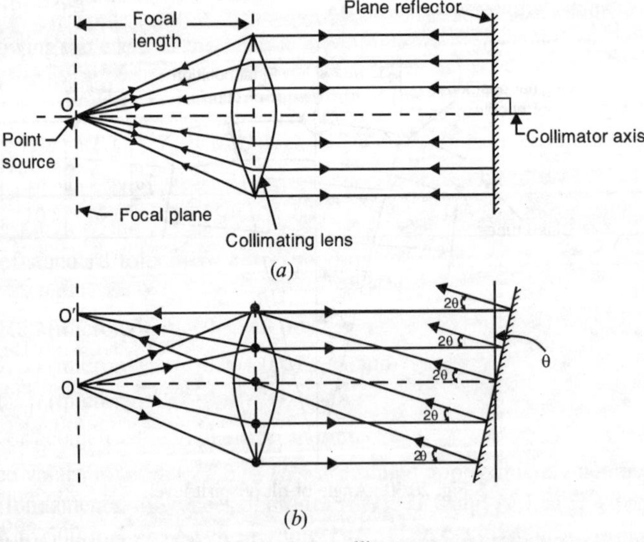

Fig. 20.31. Auto-collimator.

20·28 Straight Edge

It is used for testing the straightness and flatness of plane surfaces, particularly to test the accuracy of the ways of a lathe, planer, shaper etc.

The straight edge, as shown in Fig. 20.32, is like a solid bow. It is made of cast iron, ribbed intervals and have machined straight flat edges made true. The ribbing is done to prevent warping. These are made of varying widths and lengths.

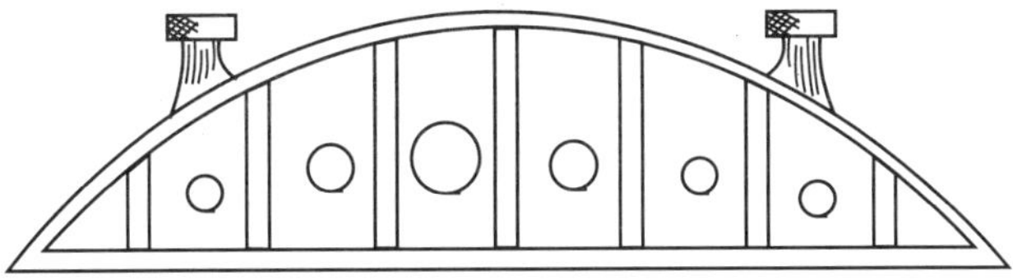

Fig. 20.32. Straight edge.

The accuracy in straightness and flatness is tested by placing the straight edge on the surface to be tested and watching the light between the surface and the straight edge. The light seen at one place or another shows the deviation from straightness.

20·29 Surface Plate

The *surface plate is used to check the trueness of flat surfaces and to copy the master surface on a work. It forms a reference surface for measurements and acts as a functional gauge when used in conjunction with other gauges.

20·30 Surface Gauge

The surface gauge (also known as scribing block), as shown in Fig. 20.33, is used for checking parallelism of work, for scribing lines at specified heights and for marking out parts that have to be fitted or machined. It consists of a cast iron base perfectly machined and planed at the top, bottom and all sides. It carries a spindle which may be set at any angle. A scriber, which may also be set at any angle or at any height, is clamped to the spindle. It is normally used in conjunction with a surface plate to scribe lines at specified heights. This tool has a V-shaped groove in the base for resting the block on a round bar, so that the dimensions may be set off from the bar to some other part of the work.

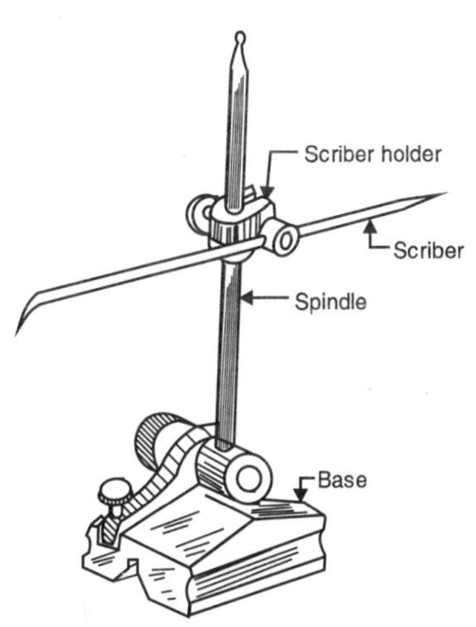

Fig. 20.33. Surface gauge.

20·31 Optical Flat

The optical flats are used to test the flatness of measuring surfaces of flats, slip gauges, gauge blocks, micrometer units etc. They are cylindrical in form and the working surfaces are flat. An optical flat is made of fused quartz and have parallel surfaces about 25 mm apart. The optical flats may have either one or both parallel surfaces ground and lapped to gauge block flatness. These instruments work on the principle of light wave interference as discussed below :

The optical flat is placed slightly inclined on the surface of the work piece to be tested. The angle θ, between the surface of work piece and optical flat is shown greatly exaggerated in Fig. 20.34. The optical flat is illuminated by a **monochromatic (*i.e.,* single colour) source of light. The incident

---

* See also chapter on 'Bench Work and Fitting'.

** Helium is most commonly used as a monochromatic source of light.

ray from the source of light S penetrates the optical flat and as it reaches the bottom surface at A, it is partially reflected along AB and is partially transmitted across the air space along AC. From C, the ray is reflected again along CD and passes out towards the eye.

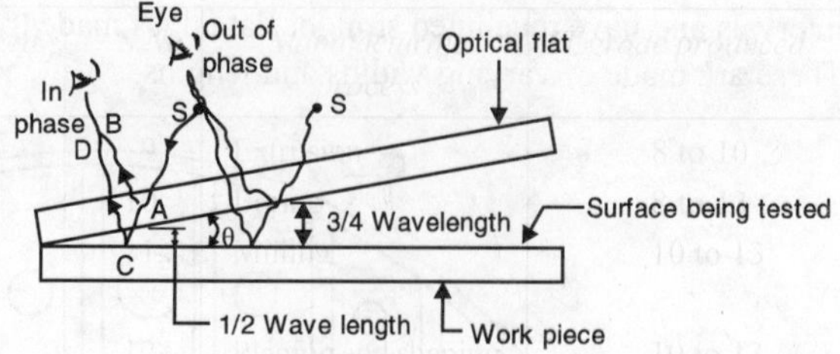

Fig. 20.34. Optical flat.

When the thickness of the air space, between the bottom surface of the optical flat and the surface to be tested, is equal to *one-half the wavelengh* of light, then a *dark band* which represents the absence of light, will appear. This is due to the fact that the two reflected rays are *out of phase* by 180° (*i.e.,* the crests of one ray falls on the troughs of the other ray) and their energies cancel to appear as an absence of light or dark band. Due to the same reasoning, other dark bands representing the absence of light, will appear when the thickness of air space between the surfaces of flat and work piece is equal to 0, $\frac{1}{2}$, 1, $\frac{3}{2}$, 2 etc. of wavelength of light.

When the thickness of the air space, between the bottom surface of the optical flat and surface to be tested, is equal to one-fourth wavelength, three-fourth wavelength and so on, then a *bright band* will appear. This is due to the fact that the two reflected rays are *in phase* (*i.e.* the crest of one ray falls on the crest of the other ray), their energies reinforce each other and they appear bright.

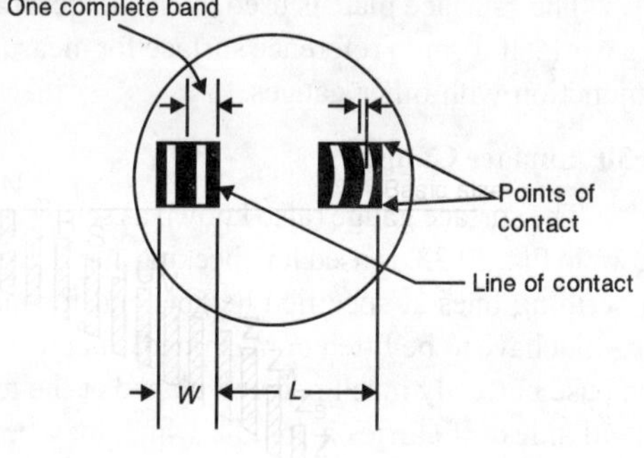

Thus, in case of perfectly flat surface, a pattern of alternate dark and light straight lines is obtained. Any deviation from this pattern will be a measure of the error in the flatness of the surface being inspected.

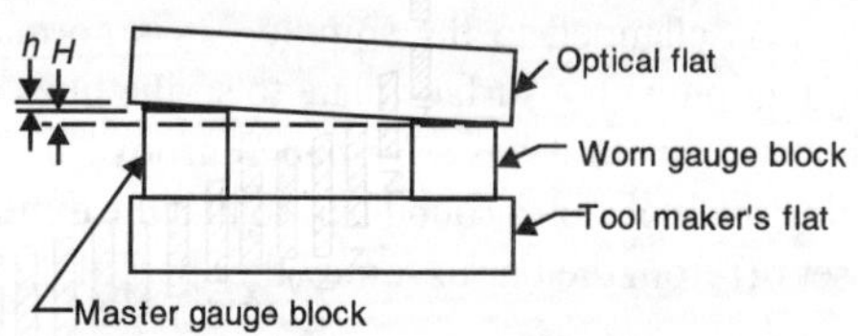

Fig. 20.35. Use of optical flat.

Fig. 20.35 shows the use of an optical flat to check the worn gauge block with a master gauge block. First of all, both the gauge blocks are placed on the tool makers' flat. Then the optical flat is placed on gently and pushed downward carefully until it rests. By observing the bands on the optical flat, the surface of the master gauge block is found to be flat. The difference in heights of the two gauge blocks (h) is obtained by the following relation :

$$h = \frac{L}{W} \times N \times \frac{\text{Wavelength of light}}{2}$$

where L = Distance between contact points,

W = Width of master gauge block, and

N = Number of complete bands.

The worn gauge-block surface, as observed in Fig. 20.35, is not flat but concave. The distance x indicates how much the edge is lower than the centre of the gauge block. One method of determining whether the surface is convex or concave is to push the optical flat downward gently at various points and observe the change at various points. Sometimes interference bands take odd shapes, but they may be interpreted as contour lines on a map. These contour lines connect points which are equidistant from the surface of the optical flat.

20·32 Interferometer

It is also an optical instrument used for measuring flatness and determining the length of slip gauges by making the use of light-wave interference. The use of this instrument is faster and easier than that of optical flats and is considered to be the most accurate measuring instrument now-a-days. There are various designs of interferometers, but the one, as shown in Fig. 20.36(*a*), was designed by National Physical Laboratory, New Delhi. It is mainly used for checking the flatness of flat surfaces. The flatness of any surface is judged by comparing with an optically flat surface which is generally the base plate of the instrument.

The instrument consists of a mercury vapour lamp light source. A mirror is used to deflect these light rays through 90°. These light rays, when passed through a green filter and focussed on pin hole, a monochromatic source of light is obtained. The pin hole is placed in the focal plane of the collimating lens which throws a parallel beam of light on the gauge and base plate via an optical flat. Since the optical flat is placed above the surface to be tested, at a slight inclination, therefore the interference fringes are formed due to the reflected rays from the bottom surface of the optical flat, base plate and gauge surface to be tested. These interference fringes are viewed by means of an eyepiece and a thick glass plate semi-reflector set at 45° to the optical axis.

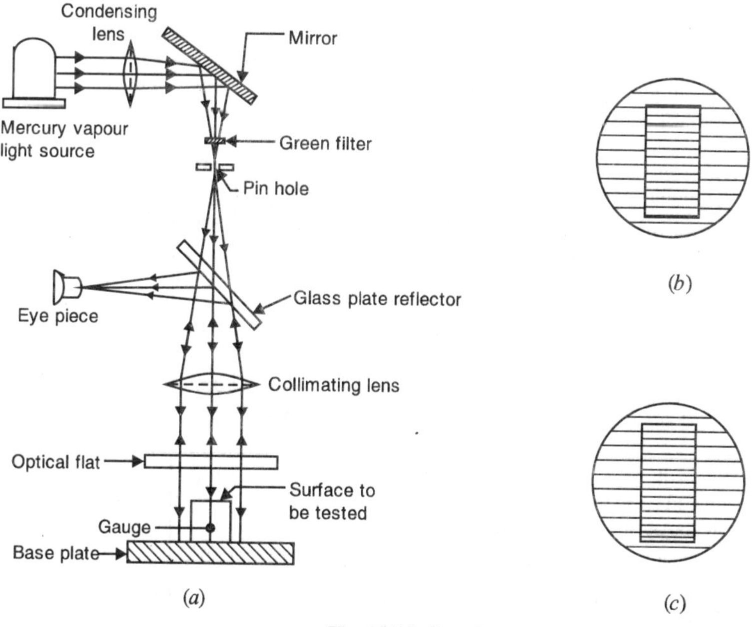

Fig. 20.36. Interferometer.

If the surface to be tested is flat and parallel to the base plate, then the optical flat being equally inclined on both the surfaces, the fringe pattern from both the gauge surfaces and the base plate will consist of straight, parallel and equally spaced fringes as shown in Fig. 20.36(*b*). When the gauge is flat but not parallel to the base plate, then the straight and parallel fringes of different pitch above the gauge surface as compared with those of the base plate are seen, as shown in Fig. 20.36(c).

20·33 Profilometer

It is most commonly used for direct measurement of surface roughness. It works on the same principle as a gramophone pickup. The instrument consists of a sharp pointed diamond stylus in the pick-up unit which moves over the surface. The stylus rises and falls depending upon the surface roughness. These movements of the stylus set up electrical impulses which are amplified and operates an indicating device graduated in microns. This instrument records the average height of the surface roughness.

20·34 Profilogram

The surface finish can also be evaluated by the tracer type profilogram, as shown in Fig. 20.37.

The surface to be tested is placed on a working table which can be moved to and fro. The stylus or tracer point is pivoted with a mirror, as shown in Fig. 20.37. The pivot point is fixed in space with the stand and its position can be changed initially. When the surface to be tested moves, the oscillations of the tracer are transmitted to the mirror. The beam of light received from the light source and through the lens and precision slit, strikes the oscillating mirror. From this oscillating mirror, the beam of light is directed to the revolving drum upon which sensitised film is arranged. After developing, the inspected surface magnified from 400 to 2000 times in the vertical direction and from 10 to 50 times in the horizontal direction, is obtained.

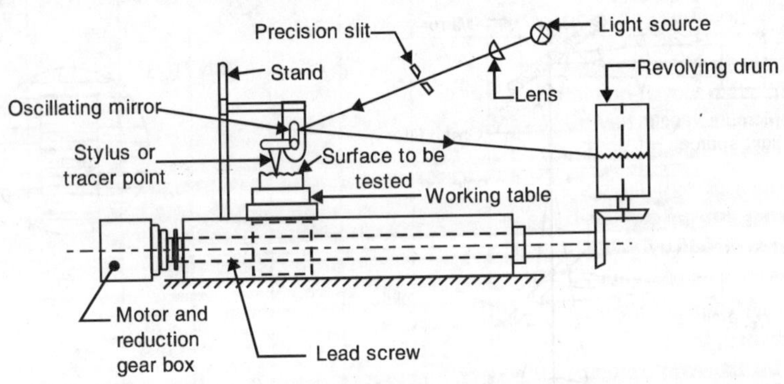

Fig. 20.37. Profilogram.

20·35 Gauges

The gauges are fixed-dimension instruments. These are generally used to check the particular dimensions of a work piece within their tolerance. The gauges should be made so that a minimum of time and skill is required in their use. The instruments differ from measuring instruments because they have no graduations to measure various lengths and angles and normally no adjustment can be made in their use. The gauges are specially made for some particular job which is to be produced in large quantities.

20·36 Classification of Gauges

The gauges are classified as follows :

1. *According to their type.*

   (a) Standard gauges, and

   (b) Limit gauges.

The *standard gauges* are mainly used as standards of reference. These gauges are made to the nominal size of the part to be tested and have the measuring member equal in size to the mean permissible dimension of the part to be checked.

The *limit gauges* are also known as '*Go*' and '*Not go*' gauges. These gauges are made to control the sizes of the work to a definite amount above and below the standard size. The limit gauges may be either *single ended, progressive* or *double ended.* The single ended limit gauges have separate 'Go' and 'Not go' members. In the progressive limit gauges, the 'Go' and 'Not go' members are on the same side of the gauge. The double ended limit gauge has 'Go' member at one end and 'Not go', member at the other end. In using the limit gauges, the 'Go' end of the gauge should pass over the acceptable work and the 'Not go' end must not pass over the work. If the work enters both the ends of the gauge, the work will be *under size* and if it does not enter either end of the gauge, the work will be *over size.*

The limit gauges may be *solid* or *adjustable.* The solid limit gauges are fixed for only one set of limit whereas in the adjustable limit gauges, the limit can be changed to suit the work. The limit gauges may, also, be *internal* or *external* for checking the internal or external sizes of the work.

2. *According to their purposes.*
   (*a*) Workshop or working gauges,
   (*b*) Inspection gauges, and
   (*c*) Reference or master gauges.

The *workshop* or *working gauges* are used in workshop to check the work while it is being made. The *inspection gauges* are used by inspection department to check the finished parts. The *reference* or *master gauges* are used only for checking the accuracy of other gauges. These gauges represent the duplicate of the work to be checked.

3. *According to the form of tested surface.*
   (*a*) Plug gauges for checking holes,
   (*b*) Snap and ring gauges for checking shafts,
   (*c*) Thread gauges for checking threads,
   (*d*) Thickness gauges (slip gauges and feeler gauges) for checking thickness, and
   (*e*) Form gauges (screw pitch gauges and radius or fillet gauges) for checking forms.

These gauges are discussed, in detail, in the following articles.

20·37 Ring Gauges

The ring gauges are used to check the diameter of shafts or studs. These are cylindrical in shape and has a hole of the exact size specified for the part to be checked. In using a ring gauge, it should fit over the part being checked without the use of force and without any noticeable side movement. A standard ring gauge commonly used is shown in Fig. 20.38. The limit ring gauges with 'Go' and 'Not go' ends are also available. The 'Go' and 'Not go' ended are identified by an annular groove on the periphery of the gauge.

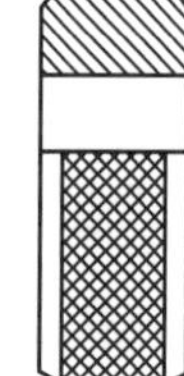

Fig. 20.38. Ring gauge.

20·38 Plug Gauges

The plug gauges are used to test the accuracy of holes. A *standard plug* gauge, as shown in Fig. 20.39(*a*), has its diameter finished to the standard size. It is used in general engineering workshop, tool rooms etc. where production quantities are not too great and accuracy is more important. The *limit plug gauges* are used where large quantities are to be produced and the variations in the standard size is allowed.

The *single-ended limit plug gauge,* as shown in Fig. 20.39(*b*), has separate 'Go' and 'Not go' members, the *progressive limit plug gauges,* as shown in Fig. 20.39(*c*), has 'Go' and 'Not go' members on the same side of a handle and the *double ended limit plug gauge* as shown in Fig. 20.39(*d*), has 'Go' member at one end and 'Not go' member on the other end of the handle. It may be noted that the 'Not go' end has less width than the 'Go' end. The 'Go' and 'Not go' ends are detachable so that they may be renewed separately when worn.

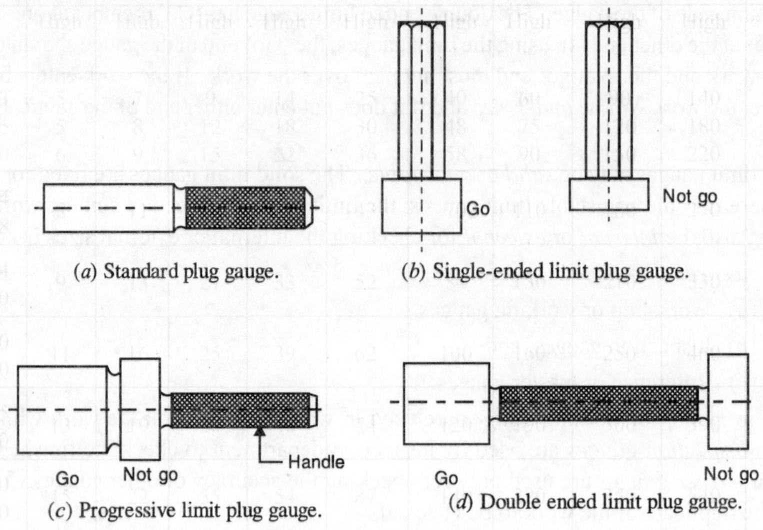

(*a*) Standard plug gauge. (*b*) Single-ended limit plug gauge.
(*c*) Progressive limit plug gauge. (*d*) Double ended limit plug gauge.

Fig. 20.39. Types of plug gauges.

Since the 'Not go' end will not enter the hole, therefore it is not subject to the same amount of wear as the 'Go' end.

In using a plug gauge the following points should be kept in mind :

1. It should engage the hole to be checked without using pressure.
2. It should be able to stand up in the hole without falling through.
3. It should just able to slide through slowly.

Note : In order to check a tapered hole, the standard and limit taper plug gauges are used.

20·39 Snap Gauges

The snap gauges are used for checking the external dimensions. They are made with openings to fit over a part to be checked. The part may be cylindrical or flat. The most commonly used types of snap gauges are as follows :

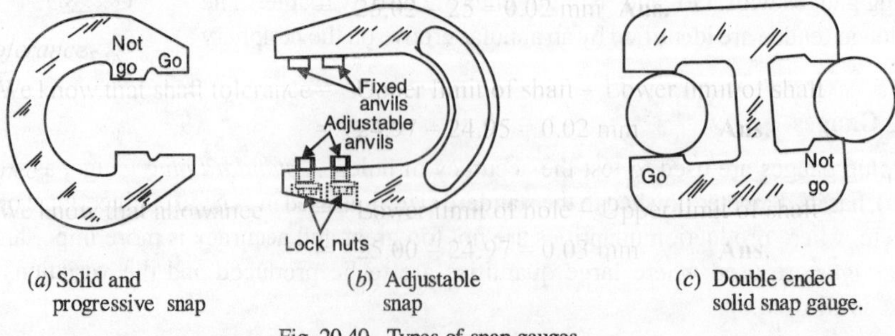

(*a*) Solid and progressive snap (*b*) Adjustable snap (*c*) Double ended solid snap gauge.

Fig. 20.40. Types of snap gauges.

1. *Solid or non-adjustable snap gauge.* A progressive type (*i.e.,* having 'Go' and 'Not go' members on the same side) solid or non adjustable snap gauge is shown in Fig. 20.40(*a*). Such gauges are used for checking larger sizes.

2. *Adjustable snap gauge.* An adjustable type of snap gauge is shown in Fig. 20.40(*b*). It has two fixed anvils and two adjustable anvils, one for the 'Go' and the other for the 'Not go'. The frame of the adjustable gauge is made of cast iron and the anvils are made of hardened steel. The frame is recessed for receiving the adjustable anvils secured by two screws. These gauges are also used for checking larger sizes.

3. *Double-ended solid snap gauge.* A double ended solid snap gauge is shown in Fig. 20.40(*c*). It has 'Go' member at one end and 'Not go' member at the other end. Such gauges are used for checking smaller sizes.

20·40 Thread Gauges

The thread gauges are used to check the pitch diameter of screw threads. The internal limit thread gauge, as shown in Fig. 20.41(*a*), is a plug gauge having screws on both ends. One end has a 'Go' member and the other has 'Not go' member. These gauges are used for checking the pitch diameter of internal threads.

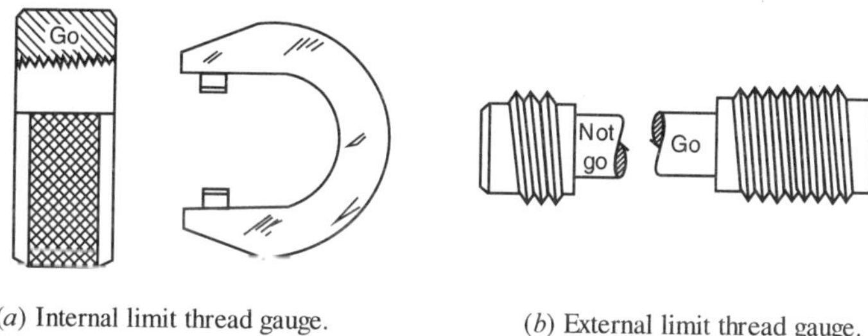

(*a*) Internal limit thread gauge. (*b*) External limit thread gauge.

Fig. 20.41. Thread gauge.

The external limit thread gauge, as shown in Fig. 20.41(*b*), may be of ring type or snap type. These gauges are used for checking pitch diameter of external threads.

20·41 Slip Gauges

The slip gauges (also known as precision gauge blocks) are used for checking the accuracy of measuring instruments such as micrometers, calipers, snap gauges, dial indicators etc. These are also used for setting the sine bars for angular measurements, for accurate measurements in die manufacture and in various precision measuring machines like tool room microscopes etc. The slip gauges consist of alloy steel blocks of section about 30 mm by 10 mm. These are hardened before being finished to size. In order to obtain a very high order of accuracy both as regard to flatness or parallelism of opposite faces, they are ground and lapped. The slip gauges are made in the following four grades.

1. Workshop or production grade,
2. Inspection grade,
3. Calibration grade, and
4. Reference grade.

The work shop or production grade slip gauges are used for actual production of components, tools and gauges whereas inspection grade is used for higher accuracy and inspection work. The calibration and reference grade slip gauges are used in laboratories and standard rooms.

The slip gauges are supplied in sets of suitable graded sizes so that numbers of combination may be obtained. In English system, there are five common sets containing 81, 49, 41, 35 and 28 pieces respectively. The size of the block is marked on its wide surface. In metric system, the commonly used sets are M 105, M 78, M 50, M 17 and M 9 pieces respectively. The range, steps and number of pieces of a M 105 set of slip gauges are as shown in the following table :

Table 20.1. Range of slip gauges.

Range (in mm)	*Steps (in mm)*	*Number of pieces*
1.01—1.49	0.01	49
0.50—24.50	0.5	49
25,50,75,100	25	4
1.0025	—	1
1.005	—	1
1.0075	—	1

In order to build a size, the blocks are brought together with a sliding action known as *wringing*, so as to clear air film from the junction and they are held by the atmospheric pressure on the opposite ends. The protective slips (or wear blocks) are usually supplied with a standard set of slip gauges. When the size is build up, these slips are used as outside members. These are made from tungsten carbide or sintered alumina which gives extremely hard and wear resistant surface. The slip gauge accessories are also available to increase its usefulness. The following points are important in using the slip gauges :

1. The slip gauges should be used in controlled atmosphere (20°C) and free from dust.
2. The faces of the slip gauges should be cleaned with chomise leather before being wrung together.
3. The slip gauges should not be wrung together after use. These should be kept in their marked position in the case or box provided with the set.

20·42 Feeler Gauge

The feeler gauge, as shown in Fig. 20.42, is used to check the clearances between two mating surfaces. It consists of a series of thin steel sheet strips (known as leaves) hardened and ground to various thicknesses. Each leaf is marked with its thickness which varies from 0.05 mm to 1 mm. The leaves are pivoted in holder of knife shape.

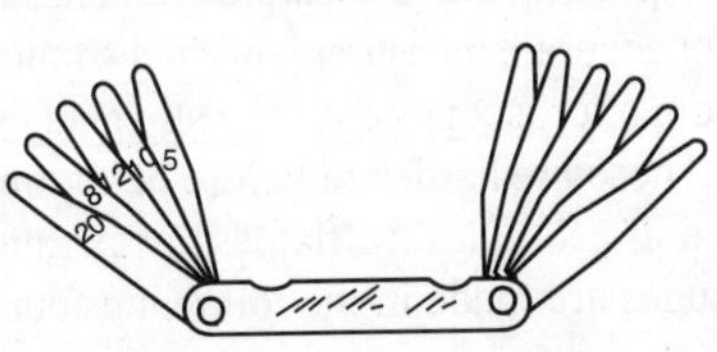

Fig. 20.42. Feeler gauge.

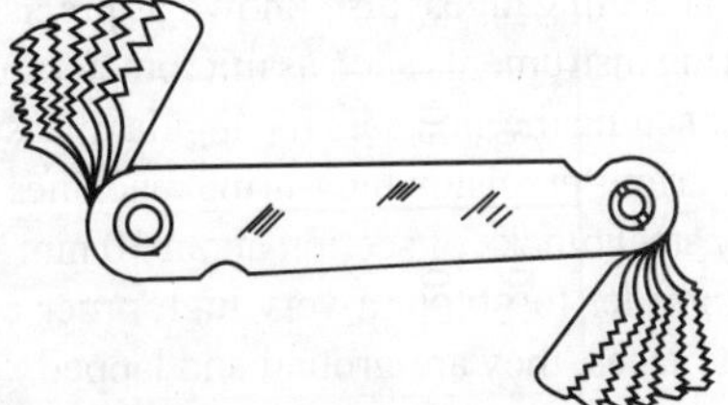

Fig. 20.43. Screw pitch gauge.

20·43 Screw Pitch Gauge

The screw pitch gauge, as shown in Fig. 20.43, is used to check the pitch of a screw. It is similar to a feeler gauge except that the leaves are notched on one edge according to the various pitches and contours of the specific threads. The free end of the leaf is generally stamped to indicate the pitch of the screw or number of threads per inch. These gauges are made for the American, British and the International standard or metric form of threads.

In using the screw pitch gauge, place successive leaves over the thread until one of them is found to coincide or mesh with the thread. The pitch of the thread (or the number of threads per inch) is read directly from the stamping on the leaf.

20·44 Radius or Fillet Gauge

The radius or fillet gauge, as shown in Fig. 20.44, is used to check the radii of curvature of concave or convex surfaces. It consists of a number of steel blades having standard radii ground and lapped on the ends and sides. The size of the radius of the curve is stamped on the blade for ready reference. The blades on one end of the gauge are used to check the concave surfaces and those on the other end are used for the convex surfaces.

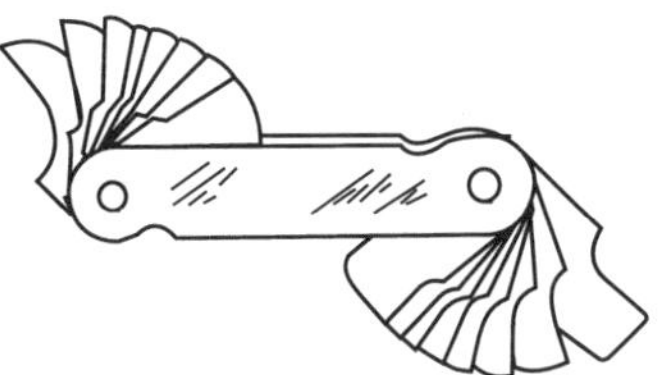

Fig. 20.44. Radius ro fillet gauge.

20·45 Comparators

A comparator is an indirect type of instrument with the help of which a known and unknown dimension of a work piece is compared with the working standard (usually slip gauges). The basic principle of operation of a comparator is as follows :

The comparator is first adjusted to zero on its dial or recording device with a gauge block in position. The gauge block is of dimension which the work piece should have. The work piece to be checked is placed in position and the comparator is used to check its dimension. The dimension of the work piece may be less than, equal to or greater than the standard dimension. If the dimension is less or greater than the standard, the difference will be shown on the dial or the recording device of the comparator. Thus, a comparator does not give the dimension of a work piece, but only gives the difference between the standard and the actual dimension of the work piece. In comparators, this difference is shown as magnified on the dial or recording device. The magnifications of 2000 times are quite common and some may have as high as 40 000. If the comparator has a magnification of 1000 and if the difference between the standard and the actual dimensions of a work piece is 0.02 mm, then it will result in pointer movement of 20 mm on the dial or recording device of the comparator.

Depending upon the method by which the difference in dimensions is magnified, the comparators are classified as follows :

1. Mechanical comparators,
2. Electrical comparators,
3. Optical comparators, and
4. Pneumatic comparators.

These comparators are discussed, in detail, in the following pages.

20·46 Mechanical Comparators

The mechanical comparators employ mechanical means to get the magnification *e.g.*, gear systems, levers etc. The magnification ranges from 250 to 5000. These comparators are relatively cheap and are used mainly for comparative measurement of external surface. Their main drawback is that any wear, play, backlash or dimensional faults in the mechanical devices used will also get magnified. Though there are many types of mechanical comparators, yet the following two types are important from the subject point of view :

1. Dial indicator or dial gauge; and 2. Lever type comparators.

These two comparators are discussed, in detail, as follows :

1. *Dial indicator.* The dial indicator (also known as dial gauge or indicating gauge) is a common type of mechanical comparator used on lathe. It is mainly used for checking parallelism and

concentricity of rods, holes and flatness of surfaces to an accuracy of 0.01 mm.

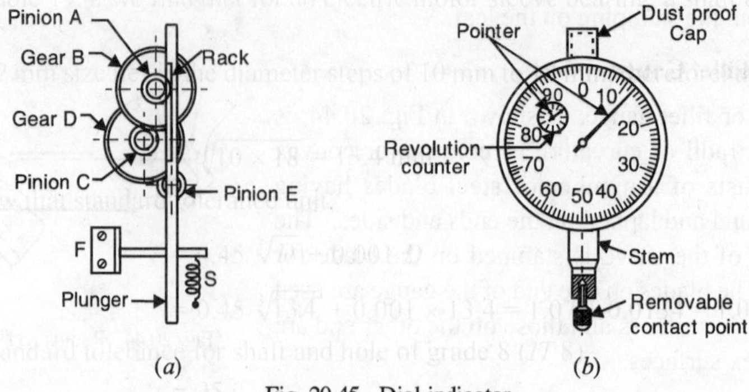

Fig. 20.45. Dial indicator.

A dial indicator, as shown in Fig. 20.45, is like a small clock with a plunger projecting at the bottom. The plunger is a perfect sliding fit in its own bearings. This carries a rack which accurately meshes with a pinion A. The rotation of the plunger about its own axis is prevented by a pin located in a slot in a rack guide F. In order to keep the plunger in an extended or normal position, a light coil spring S is employed. A very slight movement of the contact point causes the rack to turn the pinion A with which it is meshed. A large gear B is attached to the same spindle as pinion A. The gear B is further meshed with a pinion C, which thus magnifies the movement of pinion A. Another gear D (mounted on the same spindle as pinion C) meshes a pinion E which carries a indicator pointer.

The dial is graduated into 100 divisions. A full revolution of the indicator pointer represents 1 mm travel of the plunger. Thus, 1 division of the dial represents 1 mm travel of the plunger. A revolution counter is also incorporated in the gauge to indicate the travel of the plunger in whole millimetres.

The dial indicator with total plunger movement or lift of 3, 5 and 10 mm are available. The dial indicator with lift 3 mm has 50 divisions instead of 100.

2. *Lever type comparator.* A lever type mechanical comparator is shown in Fig. 20.46. In this type, a gauge block of standard dimension is wrung to anvil surface below the plunger and the

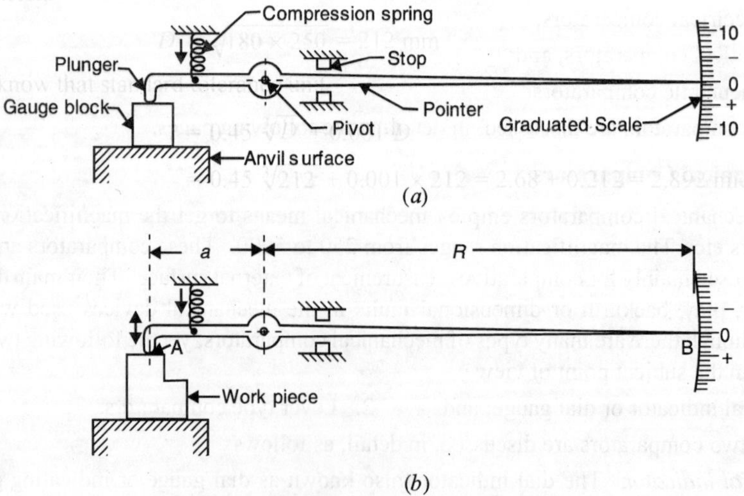

Fig. 20.46. Lever type comparator.

pointer set to zero. The gauge block is now removed and in its place, the work piece to be measured is placed on the anvil surface. The plunger moves up and down depending upon the difference in dimension of gauge block and the work piece. This movement of plunger gets magnified and shown on a graduated scale. A compression spring controls measuring pressure. The main draw back of this comparator is that the linear movement of point A on the plunger gets magnified into the movement of point B along an arc. The angular movement of point B must be small, otherwise, equal increments of plunger do not produce equal increments of the pointer. The magnification in this case is R/a.

20·47 Electrical Comparators

In electrical comparators, the movement of the measuring contact is converted into an electrical signal. This electrical signal is recorded by an instrument which can be calibrated in terms of plunger movement. For this, an alternating current wheat stone bridge circuit incorporating a galvanometer is used. The principle of an electrical comparator is shown in Fig. 20.47. An armature supported on thin steel strips is suspended between two coils A and B. When the distance of the armature surface from the two coils is equal, the wheat stone bridge is balanced and no current flows through its galvanometer. The slight movement of the measuring plunger unbalances the bridge resulting in the flow of current through the galvanometer. The scale of the galvanometer is calibrated to give the movements of the plunger. The comparators give a high degree of reliability because of minimum moving parts. The magnifications of the order of 40 000 or higher are obtainable with these comparators.

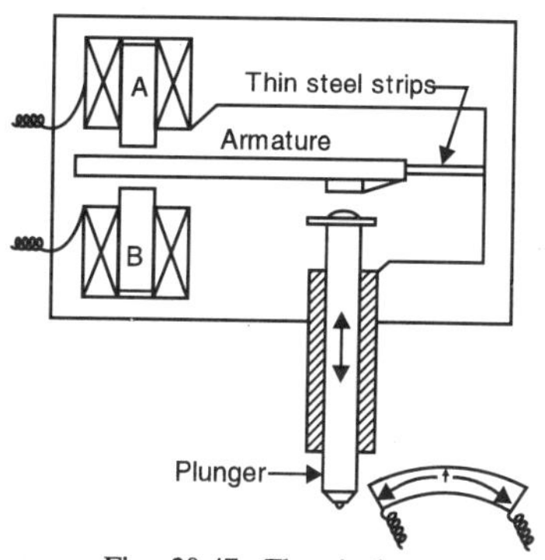

Fig. 20.47. Electrical comparator.

20·48 Optical Comparators

The optical comparators use a fundamental optical law and instead of a pointer, the edge of a shadow is projected on to a curved graduated scale to indicate the comparison measurement. The optical principle adopted is that of optical lever which is shown in Fig. 20.48. If a ray of light OA strikes a mirror, it is reflected as ray AB, such that ∠OAN = ∠NAB.

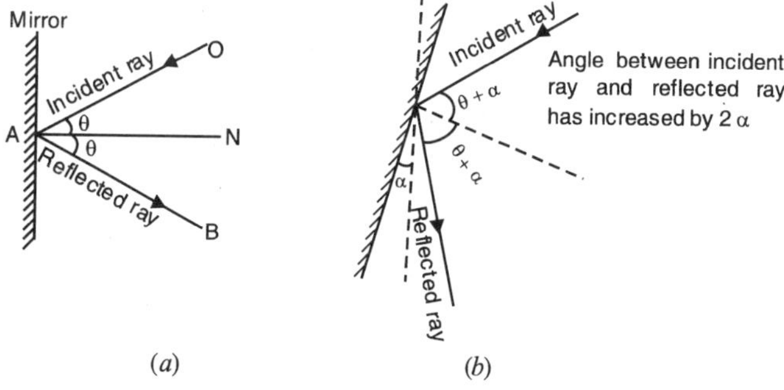

Fig. 20.48. Optical lever.

Now, if the mirror is tilted through an angle α, the reflected ray of light has moved through an angle 2α. In optical comparators, the mirror is tilted by the measuring plunger movement and the movement of the reflected light is recorded as an image on a screen. Fig. 20.49 shows the working principle of an optical comparator in which both mechanical and optical levers are used.

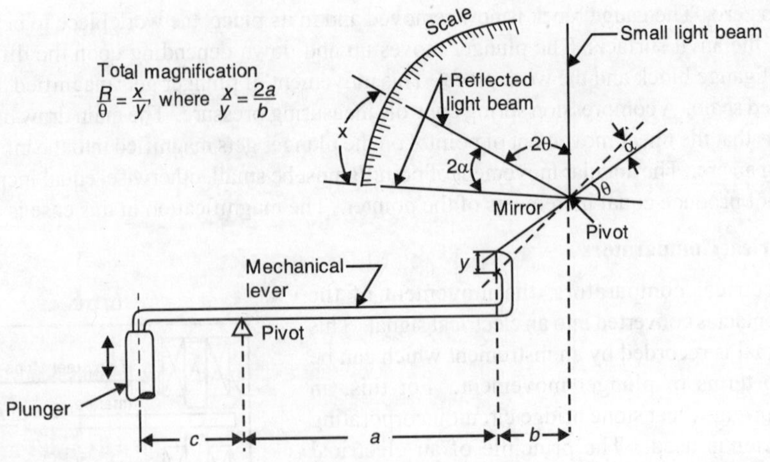

Fig. 20.49. Optical comparator.

20·49 Pneumatic Comparator

In pneumatic comparator, either air flow or air pressure is used for measurement. The response of the comparators working on air flow is quicker than those working on air pressure, but the latter are more versatile than the former. The pneumatic gauging is based on a double orifice arrangement and the drop in pressure flowing through the two orifices leads to measurement.

The working principle of a pneumatic comparator is shown in Fig. 20.50. A pressure reducer and regulator reduces the high pressure air supply to about 0.007 N/mm^2 and this pressure is maintained closely by including a restriction between the regulator and the dip tube *A*. Initially, the levels of the

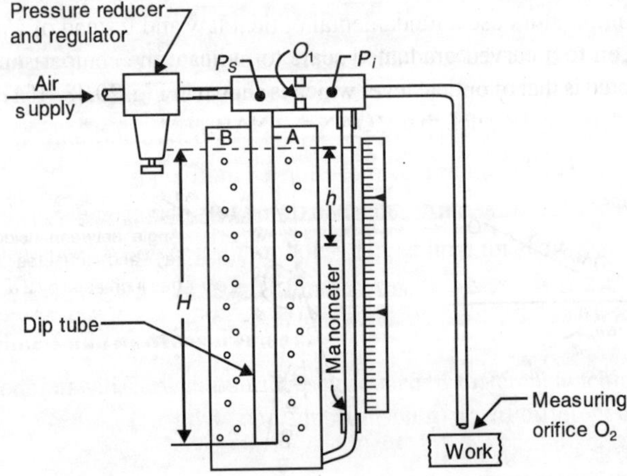

Fig. 20.50. Pneumatic comparator.

liquid in the tank *B* and the dip tube are same. When the air under pressure is supplied to the system, the liquid in the dip tube will be pushed out into the tank. Any pressure greater than needed to just clear the dip tube will escape through it into the tank as air bubbles. Thus the pressure between the regulator and the control orifice O_1, will always be exactly the same, irrespective of any variation in air pressure. Air after passing through the control orifice O_1, will reach the mouth of measuring orifice O_2. The back pressure will build up behind the measuring orifice if it is not able to pass the coming air flow. Due to this back pressure, the level of the liquid in the manometer tube will change. The back pressure is created due to the restriction at the mouth of the orifice O_2 caused by the variations in the dimensions of the work being checked. Thus, the manometer tube can be calibrated to give the measurement deviation from a standard.

The use of pneumatic comparators is quite extensive because they are cheap, simple to operate, accurate, flexible, reliable. They are best suited for checking multiple dimensions and have very high magnification (10 000).

QUESTIONS

1. Classify the measuring instruments. Write its various types.
2. What are the various instruments used for linear measurements ? Explain, with neat sketches, the use of the following :
   (*a*) Calipers, (*b*) Depth gauge, and (*c*) Telescopic gauge.
3. Describe, with the help of a neat sketch, the working of an outside micrometer.
4. Draw the sketches of the important parts of an inside micrometer and explain how the reading is taken ?
5. What dimension on a screw does a screw thread micrometer measure ?
6. How does the sleeve of a depth gauge micrometer differ from a common micrometer with respect to graduations ?
7. Discuss the principle of a vernier caliper. How a reading is measured by using such an instrument?
8. Draw the neat sketches of a vernier height gauge and depth gauge. Explain their working principles.
9. What dimensions of a gear tooth are checked by a gear tooth vernier ?
10. Describe, with the help of a neat sketch, a universal bevel protractor. What is the value of each graduation on the vernier of this instrument ?
11. Draw the neat sketch of a combination set. Explain the use of different parts.
12. What are the advantages of a sine bar over a bevel protractor ? How the angle is measured with the help of this instrument ?
13. What are angle gauge blocks and how these are used ?
14. Discuss the use of an optical flat to test the flatness of measuring surfaces.
15. What are the various instruments used for measuring surface roughness ? Describe one of them, in detail.
16. What are gauges ? Describe the following with neat sketches.
    (*a*) Plug gauges, (*b*) Snap gauges, and (*c*) Slip gauges.
17. What do you understand by comparators. Name the various types of comparators used.
18. Explain the working of a dial indicator and a pneumatic comparator.

OBJECTIVE TYPE QUESTIONS

1. The instrument used to measure external and internal diameter of shafts, thickness of parts and depth of holes, is
   (*a*) outside micrometer (*b*) inside micrometer
   (*c*) depth gauge micrometer (*d*) vernier caliper
   (*e*) combination set (*f*) any one of these

2. The instrument which has all the features of try-square, bevel protractor, rule and scriber, is

(*a*) outside micrometer (*b*) inside micrometer

(*c*) depth gauge micrometer (*d*) vernier caliper

(*e*) combination set (*f*) any one of these

3. The accuracy of micrometers, calipers, dial indicators can be checked by a

(*a*) feeler gauge (*b*) slip gauge (*c*) ring gauge (*d*) plug gauge

4. A progressive limit gauge has

(*a*) 'Go' and 'Not go' member on the same side of the gauge

(*b*) separate 'Go' and 'Not go' members

(*c*) 'Go' member at one end and 'Not go' member at the other end

(*d*) none of the above

5. A ring gauge is used to

(*a*) check the diameter of shafts or studs

(*b*) test the accuracy of holes

(*c*) check the clearance between two mating surfaces

(*d*) all of the above

ANSWERS

1. (*d*) 2. (*e*) 3. (*b*) 4. (*a*) 5. (*a*)

21

Quality Control

1. Introduction. 2. Inspection. 3. Inspection Methods. 4. Various Areas of Inspection. 5. Quality Control. 6. Total Quality Control. 7. Statistical Quality Control. 8. Measures of Central Tendency. 9. Measures of Dispersion. 10. Frequency Distribution. 11. Probability. 12. Normal Distribution Curve. 13. Variables and Attributes. 14. Variability in Manufacturing Process. 15. Control Charts. 16. Control Charts for Variables. 17. Control Charts for Attributes. 18. Sampling Inspection. 19. Types of Sampling Plan. 20. Operating Characteristic Curve.

21·1 Introduction

The term quality is a relative term and has different meanings to different people. When referring to a product, it generally signifies the degree of its excellence. The quality of a product consists of number of elements such as shape, size and finish etc. known as quality characteristics and these are specified in part drawings or manufacturing drawings.

The control is a system for measuring and checking a phenomenon. It suggests when to inspect, how often to inspect and how much to inspect. In addition, it incorporates a feedback mechanism which explores the cause of poor quality and helps to take corrective action.

21·2 Inspection

The inspection is a method to check materials or parts necessary for product production to see if it meets the quality standards of the next process or to check if the completed products meet the customer specifications. The inspection involves measurement of materials, parts or completed product by some or other method and the results are compared with relevant standards for judging if they are acceptable or not. Thus inspection helps to select the parts which will function satisfactorily for its life. It may be noted that the philosophy of inspection is only 'preventive' and not 'remedial'.

21·3 Inspection Methods

The following are two types of inspection methods :

1. *Total inspection.* It consists of measuring each quantity of a product lot and approving it.

2. *Sampling inspection.* It consists of taking some samples of each product lot for examination and based on the result of inspection, approving the lot. In sampling inspection, the number of samples for inspection and acceptability judgement criteria are calculated and determined using statistical methods. One of these method is selected taking full account the importance of each product quality characteristics and the economic aspect of the inspection.

21·4 Various Areas of Inspection

The following are three basic areas of inspection :

1. *Receiving inspection.* The receiving inspection is carried on incoming material, parts or partially fabricated items at production site. This is also known as *acceptance inspection.* The receiving inspection is essential in order to check whether the purchased parts confirm to quality characteristics given on drawings. The proper receiving inspection will prevent waste of time and money in production by identifying non-conforming items at the receiving stage itself.

2. *Process inspection.* The process inspection is carried out when delivering items from one

process to another in a series of processes. If the process inspection is carried out properly it would be possible to keep the rejections and rework at a low level and in many cases, it will help to make corrections before any rejections are produced, particularly in the case of dimensional control.

3. *Final inspection*. The final inspection is carried out on the finished product to ensure the quality of the delivered product. This is also known as *delivery inspection*. In this type of inspection the product is evaluated for both its appearance and performance, keeping in view the customer's requirement.

21·5 Quality Control

The quality control (briefly written as QC) is a system which uses the economical means to produce a quality product that meets the customer requirements. Thus, quality control is the total of a company's activities as a whole, which is to develop, design, produce, inspect and sell high quality products, providing the customers with long lasting satisfaction.

It may be noted that inspection is a part of quality control which determines whether the raw material supplies and finished products are acceptable whereas quality control on a broader aspect finds solution to reduce rejections and defective products. It is seen that when the control-becomes better, the need of inspection is reduced.

The basic idea of quality control is the prevention of defective products, rather than the 'elimination of defective products'. It is through quality control that some measures are taken to see that defective items are not produced at all. Even if they do occur, corrective actions must be taken to prevent further recurrence. Thus, quality control is economical in its purpose, objective in its procedure, dynamic in its operation and helpful in its treatment. It suggests when to inspect, how often to inspect and how much to inspect.

21·6 Total Quality Control

The total quality control is a system in which all in-house departments of an organisation together produce and sell high quality products at the lowest possible price, at the fastest possible speed and satisfies the customers. In order to provide the customer with the highest possible quality product, at the fastest possible speed, all departments including development, production, sales and staff members should produce high quality product with high efficiency at the upstream source of production. These activities are referred to as *total quality control* or *organisation-wide quality control*.

To promote total quality control, all employees should perform their own job, working towards the common target of satisfying the customer. This requires that each department classifies what kind of work it carries out for providing satisfaction to the customer and to promote that function.

21·7 Statistical Quality Control

A quality control system performs inspection, testing and analysis to judge whether the quality of each product is as per laid down quality standard or not. Since 100% inspection of parts is not found practicable in continuous production, therefore certain statistical techniques have been devised to evaluate machines, material and processes by observing capabilities and trends in variation so that continual analysis and predictions may be made to control the desired quality level or to solve quality control problems. The statistical techniques are also called as *statistical quality control methods*. These methods or tools are as follows :

1. The Shewart control charts for measurable quality characteristics. These are the charts for variables such as $\overline{X}$ and *R*, *i.e.*, the average and the range charts and the $\overline{X}$ and σ (sigma), *i.e.*, the average and the standard deviation chart.
2. The Shewart control chart for fraction defective, *i.e.*, the *p* chart.
3. The Shewart control chart for number of defects per unit, *i.e.*, the *c* chart.

4. Acceptance sampling.

Before discussing, the above charts and methods in detail, we shall first discuss the fundamental of statistics.

21·8 Measures of Central Tendency

Most frequency distributions exhibits a central tendency *i.e.*, a shape such that the most of the observations lie in the area between the two extremes. The measure of this central tendency is one of the fundamental measures in all statistical analysis. Following are the three principal measures of central tendency :

1. *Mean.* The mean is calculated by adding the observations and dividing by the number of observations. It is generally denoted as $\overline{X}$. Mathematically,

$$\overline{X} = \frac{X_1 + X_2 + X_3 + \ldots.. + X_n}{n} = \frac{\Sigma X}{n}$$

where X_1, X_2, X_3 ... etc. represents the observations and n represents the number of observations.

2. *Median.* It is the middle most or most central value when the observations (data) are arranged according to increasing value. The half of the observations lie above the median and the half of the observations lie below the median.

3. *Mode.* It is the value which occurs most often in a data set.

21·9 Measures of Dispersion

The measure of dispersion shows us how closely the individual observations cluster around the mean or some other averages. Following are three basic measures of dispersion :

1. *Range.* It is the difference between the largest observed value and the smallest observed value. It is sometimes known as *spread.* The range is denoted by R or ω.

2. *Standard deviation.* It is the measure of dispersion of a dimension about a mean dimension. It is denoted by σ. Mathematically, standard deviation,

$$\sigma = \sqrt{\frac{(X_1 - \overline{X})^2 + (X_2 - \overline{X})^2 + (X_3 - \overline{X})^2 \ldots.. + (X_n - \overline{X})^2}{n}}$$

$$= \sqrt{\frac{\Sigma (X - \overline{X})^2}{n}}$$

where $X_1, X_2, X_3, \ldots X_n$ are the observed values, $\overline{X}$ is the mean of all the observed values and n is the total number of observed values.

This is also called as *root mean square* (r.m.s.) deviation of the observed values from their mean.

3. *Variance.* It is defined as the square of standard deviation *i.e.*,

$$\text{Variance} = \sigma^2.$$

21·10 Frequency Distribution

The frequency distribution is a simple method of showing variation in manufactured parts. In frequency distribution, the first step is to count the number of times (frequency) each value occurs. These values are then arranged in ordered groups or classes with their frequency of occurence. The chart obtained as a result of this arrangement is known as *frequency distribution chart.* These charts are often used to judge the capability of a manufacturing process. By studying the pattern, the observer can make the amount of correction needed to bring the process back to best performance. A typical frequency distribution chart for 100 observations of diameter of a solid shaft is shown below :

Table 21.1. Frequency distribution chart.

Diameter of shaft in mm	*Tally marks*	*Frequency*
9.95	𝍷𝍷	2
9.96	𝍷𝍷𝍷𝍷	4
9.97	𝍸 𝍷	6
9.98	𝍸 𝍸	10
9.99	𝍸 𝍸 𝍷𝍷𝍷𝍷	14
10.00	𝍸 𝍸 𝍸 𝍸 𝍸 𝍸	30
10.01	𝍸 𝍸 𝍷𝍷𝍷𝍷	14
10.02	𝍸 𝍷𝍷𝍷	8
10.03	𝍸 𝍷	6
10.04	𝍷𝍷𝍷𝍷	4
10.05	𝍷𝍷	2

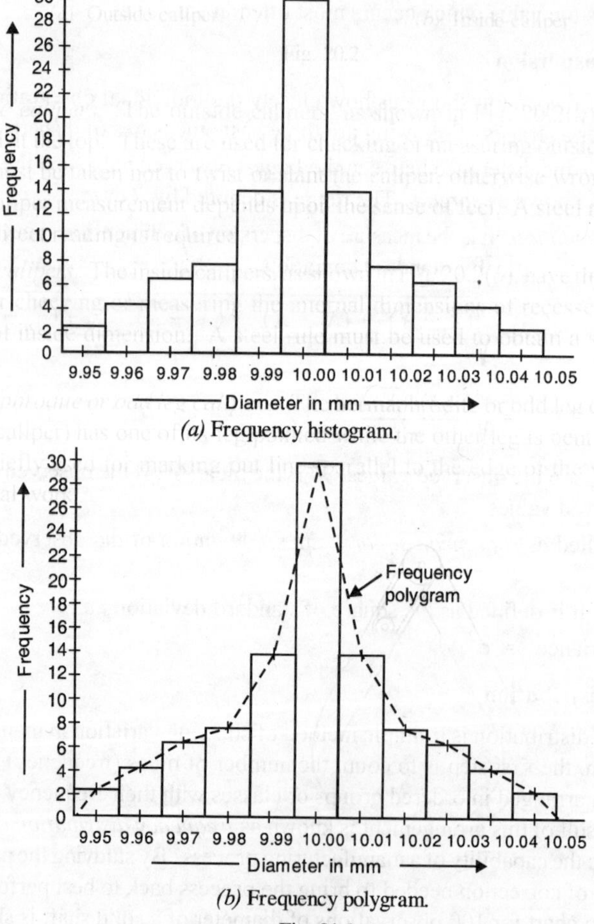

(a) Frequency histogram.

(b) Frequency polygram.

Fig. 21.1

There are number of ways of presenting the data shown in the above table. The following two methods are important from the subject point of view :

1. *Frequency histograms.* In histograms, the vertical axis represents the frequency of measurement and the horizontal axis represents the class interval or occurence of each dimension on a suitable scale. The difference between each value on the horizontal axis is called a class interval and each value is represented at the centre of the class-interval. A historgram for the data given in Table 21.1 is shown in Fig. 21.1(a).

2. *Frequency polygon.* If a smooth curve is drawn by joining the top centre points of the histogram, a frequency polygon or frequency distribution curve is obtained as shown in Fig. 21.1 (b).

21·11 Probability

The concept of probability is important for a quality control engineer as it forms the basis of statistical quality control. The understanding of the probability theory becomes necessary in order to follow sampling inspection and operation of control charts. The probability generally refers to totality of possible occurences. It is a mathematical measure of the likelihood of an occurence. Mathematically, probability,

$$p = \frac{\text{Number of favourable outcomes to the event}}{\text{Total number of possible events}}$$

From the above expression, it can be understood that the value of probability varies between zero and 1. If the value of probability is zero, then it represents absolute impossibility of occurence of an event and if the value of probability is 1, then it represents absolute certainty of occurence of an event. Also if p represents the probability of occurence of an event and q for non-occurence of an event then the sum of p and q will be unity *i.e.*, $p + q = 1$.

Example 21.1. *In a bag there are 5 black balls and 45 white balls. If a person picks up one ball randomly out of the bag, what is the probability of 1. getting a black ball ; and 2. getting a white ball.*

Solution. Given : Black balls (a) = 5 ; White balls (b) = 45

∴ Total number of balls in the bag $= a + b = 5 + 45 = 50$

1. *Probability of getting a black ball*

We know that probability of getting a black ball,

$$p_1 = \frac{\text{Number of favourable outcomes}}{\text{Total number of possible events}}$$

$$= \frac{a}{a+b} = \frac{5}{50} = 0.1 \text{ or } 10\% \quad \textbf{Ans.}$$

2. *Probability of getting a white ball*

We know that probability of getting a white ball,

$$p_2 = \frac{b}{a+b} = \frac{45}{50} = 0.9 \text{ or } 90\% \quad \textbf{Ans.}$$

21·12 Normal Distribution Curve

The frequency curves are of different shapes. The most important and useful of these curves is *normal distribution curve* or *probability curve*. The distribution has a symmetrical bell shaped form and tends to infinity in both directions as shown in Fig. 21.2.

The horizontal axis represents value of standard deviation (σ) above and below the mean $\overline{X}$. It may be noted that if a process follows normal distribution, then 99.73% of all the measured values will occur between $\overline{X} + 3\sigma$ and $\overline{X} - 3\sigma$ and the rest 0.27% are expected to lie outside the limits.

Similarly, 95.45 % of the measured values will occur between $\overline{X} + 2\sigma$ and $\overline{X} - 2\sigma$ and 68.27% will occur between $\overline{X} + \sigma$ and $\overline{X} - \sigma$.

For example, the diameter readings of a part manufactured in a process follows the normal distribution of $\overline{X} = 50$ mm and $\sigma = 0.02$ mm, then

$$\overline{X} + 3\sigma = 50 + 3 \times 0.02 = 50.06 \text{ mm}$$

and $$\overline{X} - 3\sigma = 50 - 3 \times 0.02 = 49.94 \text{ mm}$$

This means that, if the process is stable, then diameters of parts manufactured in this process lying between 49.94 mm and 50.06 mm is 99.73% and diameter readings outside this range accounts for 0.27%

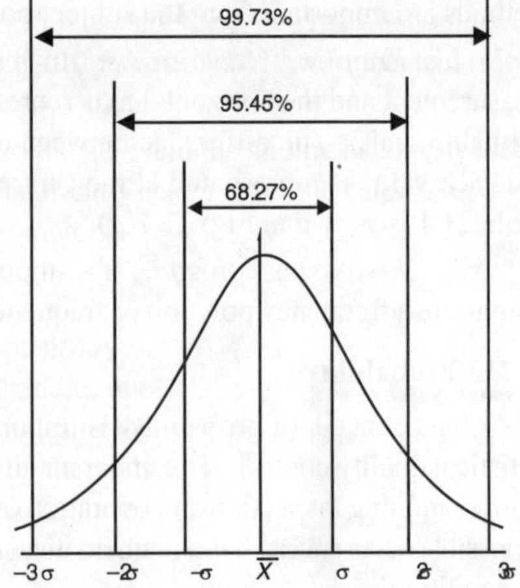

Fig. 21.2. Normal distribution curve.

21·13 Variables and Attributes

The quality characteristics of a product such as dimension, hardness, temperature etc. which are inspected by actual measurements are known as *variables*. Inspection using variables is mostly done on the shopfloor and is important for the control of process.

The quality characteristics of a product which are not measured directly but are gauged to determine whether they are acceptable or rejected are known as *attributes*. Attribute inspections finds applications where there is no need to measure exact dimensions. A very common type of attribute inspection is done by the GO and the NO GO gauges.

21·14 Variability in Manufacturing Process

Whenever a product is manufactured, it is observed that no two units of that product are exactly similar. This is because there occur some variations in the process due to inherent characteristic of that process, though the amount of these variations may differ from process to process.

Statistical analysis is used to study these variations in processes and to detect when and to what extent external factors enter into the process to change the variability pattern. There are two reasons for variation in quality characteristics of a product :

1. *Inherent sources*. These are caused by a number of chance factors and these mainly depend upon the machine tools. The inherent sources of variation occur even if the best methods have been employed. For example, a little play between a nut and screw may lead to backlash in a machine causing the cutting tool to operate differently and thereby producing different dimensions of a specified size. In quality control, the variation in dimension due to chance factors is not considered as it is generally very small compared to actual variation. It may be noted that if variation is only because of chance factors, the process will follow normal distribution.

2. *Assignable causes*. These lead to variation in dimensions due to tool wear, tool setting, tool getting loose, etc. These must be checked and corrected before the defective pieces are produced. The variation due to assignable causes have greater magnitude as compared to those due to chance causes.

21·15 Control Charts

The control charts are graphical presentation of observed values plotted as points, which are used to check if a process to produce some quality characteristics is in stable state (*i.e.*, process analysis) or to keep the process in stable state (*i.e.*, process control). The points are plotted in order of their occurence so that each value has its own identity relative to the time of observation. The control chart has control limit lines (normally upper and lower) plotted to indicate dispersion limits. It may be noted that the dispersion limits are not quality standards or tolerance specified on drawings but a statistically significant value of process dispersion method determined from probability. The control limits of the process are the limiting values of a dimension within which practically all values fall.

If the process follows normal distribution, then these control limits are positioned $\pm 3\sigma$ (sigma) away from the centre line. Thus, 3σ forms the basis of quality control. However, values which fall outside the control limits do not necessarily means rejections but only indicates that some necessary improvement and corrective action is to be taken to prevent manufacturing faulty parts.

21·16 Control Charts for Variables ($\bar{X}$ and R Chart)

The $\bar{X}$ chart (*i.e.*, the chart for mean values) and R chart (*i.e.*, the chart of ranges) are used to control the mean and range of a process in the case where inspection is done by measurements and not by gauging.

The $\bar{X}$ and R control charts can provide much information on a process by data of limited number of samples (usually about five) of parts that are continuously produced in large quantity. For this reason, it is known as an effective quality control method.

The procedure for preparing the $\bar{X}$ chart and R chart are discussed, in detail, as follows :

1. $\bar{X}$ *chart.* A good number of samples of parts manufactured are taken at random, over a period and their quality characteristics (say length and diameter) are measured. It may be noted that a sample consists of usually about five parts, thus for each sample, the mean is given as,

$$\bar{X} \text{ (Mean)} = \frac{X_1 + X_2 + X_3 + X_4 + X_5}{5}$$

The mean values are then plotted on the $\bar{X}$-chart instead of individual measurements because sample mean tends to approach normal distribution curve closely than the individual values. Thus, the total mean, $\bar{\bar{X}}$ is the mean of all the values of $\bar{X}$. Mathematically,

$$\bar{\bar{X}} = \frac{\Sigma \bar{X}}{N}$$

where N = Number of sample groups.

Now, for $\bar{X}$ chart,

$$\text{Upper control limit } (\text{UCL})_{\bar{X}} = \bar{\bar{X}} + 3\sigma_{\bar{X}} \quad ...(i)$$

and

$$\text{lower control limit } (\text{LCL})_{\bar{X}} = \bar{\bar{X}} - 3\sigma_{\bar{X}} \quad ...(ii)$$

where

$$\sigma_{\bar{X}} = \sqrt{\frac{\Sigma(\bar{X} - \bar{\bar{X}})^2}{N}}$$

The above equations (*i*) and (*ii*), can be written in a simplied form as :

$$(\text{UCL})_{\bar{X}} = \bar{\bar{X}} + A_2 \bar{R}$$

and

$$(\text{LCL})_{\bar{X}} = \bar{\bar{X}} - A_2 \bar{R}$$

where A_2 = Constant factor which depends upon the sample size and the value of it can be seen from Table 21.2, and

$\bar{R}$ = Average range.

A typical $\overline{X}$ chart is shown in Fig. 21.3 (*a*). The chart shows that a particular sample went above UCL which may be due to tool wear. After the replacement of tool the process was again under control.

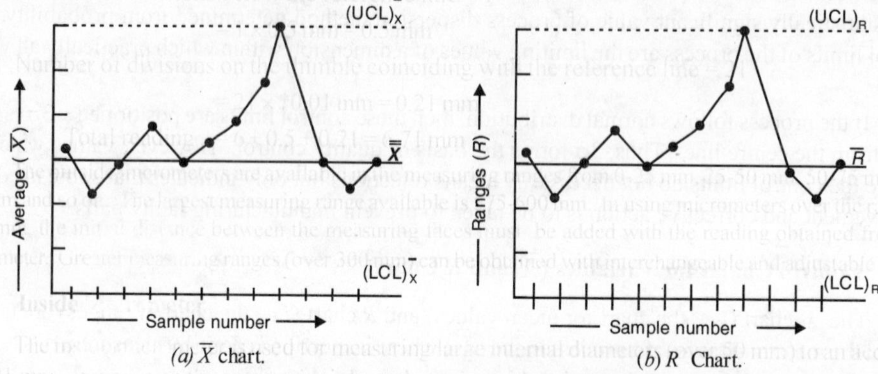

(*a*) $\overline{X}$ chart. (*b*) *R* Chart.

Fig. 21.3. Typical $\overline{X}$ and *R* control charts.

2. *R Chart.* Same sample groups are considered for *R* chart which were used in determining the values of $\overline{X}$. It may be noted that *R* which indicates dispersion in the group is the difference between maximum and minimum value in the group. Now, to determine range (*R*), for each group, $X_{max} - X_{min}$ is calculated.

The mean of ranges ($\overline{R}$) is the mean of all the values of *R*. Mathematically,

$$\overline{R} = \frac{\Sigma R}{N}$$

where N = Number of sample groups.

Now, for *R* chart

$$(\text{UCL})_R = D_4 \overline{R}$$

and

$$(\text{LCL})_R = D_3 \overline{R}$$

where D_4 and D_3 = Constant factor which depends upon the sample size and the values of which can be seen from Table 21.2.

A typical *R* chart is shown in Fig. 21.3(*b*). The chart shows that the entire process is under statistical control. Sudden drifting of two consecutive points above and below the centre line have been caused due to chance causes.

Table 21.2. Constant factors in $\overline{X}$ and *R* charts.

Sample size (Number of values in a sample)	*Limit average* (A_2)	*Range lower limit* (D_3)	*Range upper limit* (D_4)
2	1.88	0	3.27
3	1.02	0	2.57
4	0.75	0	2.28
5	0.58	0	2.11
6	0.48	0	2.00
7	0.42	0.08	1.92
8	0.37	0.14	1.86

Sample size (Number of values in a sample)	*Limit average* (A_2)	*Range lower limit* (D_3)	*Range upper limit* (D_4)
9	0.34	0.18	1.82
10	0.31	0.22	1.78
11	0.29	0.26	1.74
12	0.27	0.28	1.72
13	0.25	0.31	1.69
14	0.24	0.33	1.67
15	0.22	0.35	1.65

Notes : 1. If the values of $\overline{X}$ and R, for each sample is within the control limits, the process is said to be in *statistical control*. If the process is not under statistical control, it means that there is some external cause which throws the process out of control. Then that cause is to be detected and corrected so that the process is brought under the statistical conditions again.

2. In $\overline{X}$ and R charts, all values will occur within the control limits if the variations are due to chance causes only. If either $\overline{X}$ or R values fall outside the control limits, then these are due to assignable causes and corrective measures must be taken.

Example 21.2. *10 samples of 5 numbers each are taken from a manufacturing process at regular intervals. A particular quality characteristic is measured and $\overline{X}$ and R values are computed. After 10 samples, $\Sigma\overline{X} = 75$ and $\Sigma R = 25$. Compute the control limits.*

Solution. $N = 10$; $n = 5$; $\Sigma\overline{X} = 75$; $\Sigma R = 25$

We know that mean of all values,

$$\overline{\overline{X}} = \frac{\Sigma\overline{X}}{N} = \frac{75}{10} = 7.5$$

and mean of ranges,

$$\overline{R} = \frac{\Sigma R}{N} = \frac{25}{10} = 2.5$$

Control limits for $\overline{X}$

For $\overline{X}$ chart, we know that

$$(\text{UCL})_{\overline{X}} = \overline{\overline{X}} + A_2\overline{R}$$

$$= 7.5 + (0.58 \times 2.5) \quad ...(A_2 = 0.58 \text{ from Table 21.2 for } n = 5)$$

$$= 7.5 + 1.45 = 8.95 \quad \textbf{Ans.}$$

and

$$(\text{LCL})_{\overline{X}} = \overline{\overline{X}} - A_2\overline{R} = 7.5 - (0.58 \times 2.5)$$

$$= 7.5 - 1.45 = 6.05 \quad \textbf{Ans.}$$

Control limits for R

For R chart, we know that

$$(\text{UCL})_R = D_4\overline{R} = 2.11 \times 2.5 = 5.275 \quad \textbf{Ans}$$

$$...(\because D_4 = 2.11 \text{ from Table 21.2 for } n = 5)$$

and

$$(\text{LCL})_R = D_3\overline{R} = 0 \times 2.5 = 0 \quad \textbf{Ans.}$$

$$...(\because D_3 = 0 \text{ from Table 21.2 for } n = 5)$$

21·17 Control Charts for Attributes

We know that the quality characteristics which are not measured directly but are gauged to determine whether they are acceptable or not are known as *attributes*. Following are the different types of control charts for attributes.

1. *p-chart*. Some products or parts are subjected to defects that cannot be measured directly. The decision to accept or reject the parts is based on an attribute, not on a numerical value. For example, in a foundry, a casting has a crack. It is possible to measure it but not necessary because the casting is defective. Similarly, in checking non-critical parts with GO and NO GO gauges, we are interested to know whether they are good or bad, not how bad.

The *p*-chart is the quality control chart that looks at the number of defectives found in the total parts inspected. In other words, *p*-chart is a fraction defective chart that can be used in the following conditions :

1. When it is possible to obtain measurements but not necessary (*e.g.*, a crack in a casting), and

2. When it is desirable to know whether a product is good or bad (*e.g.*, checking with GO and NO GO gauges).

Thus, fraction defective (p) is defined as the ratio of number of defective articles found in any inspection or series of inspection to the total number of articles actually inspected. Mathematically,

$$p = \frac{\text{Number of defectives in subgroup}}{\text{Number inspected in subgroup } (n)}$$

and $$\bar{p} = \frac{\text{Total number of defectives during period}}{\text{Total number inspected during period } (\Sigma n)}$$

Fraction defective is always expressed as a decimal fraction.

Similar to $\overline{X}$ and R control charts, the control chart for p has a central line indicating average value $\bar{p}$, upper control limit and lower control limit. A typical *p*-chart is shown in Fig. 21.4.

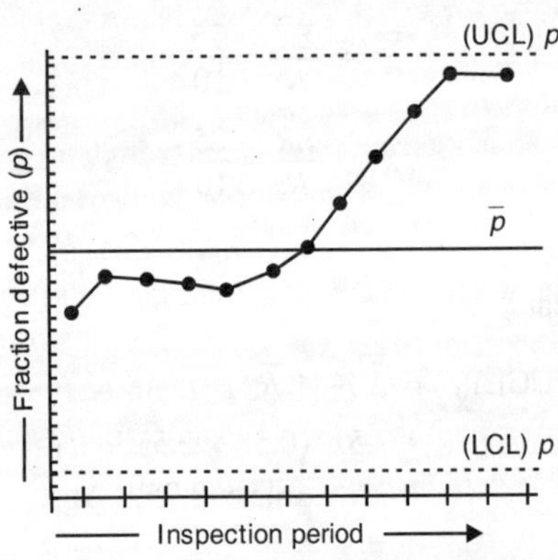

Fig. 21.4. Typical p control chart.

Now, formula for upper and lower control limits is given as

$$(\text{UCL})_p = \bar{p} + 3\sigma_p = \bar{p} + 3\sqrt{\frac{\bar{p}(1-\bar{p})}{n}}$$

and $$(\text{LCL})_p = \bar{p} - 3\sigma_p = \bar{p} - 3\sqrt{\frac{\bar{p}(1-\bar{p})}{n}} \quad \ldots \left(\because \sigma_p = \sqrt{\frac{\bar{p}(1-\bar{p})}{n}} \right)$$

2. *np-chart*. When the sample size is constant, then *np*-chart is used. After finding the value of p, it is then multiplied by n to get np, which represents the central line. The upper and lower limits in that case are :

$$(UCL)_{np} = n\bar{p} + 3\sigma_{np} = n\bar{p} + 3\sqrt{n\bar{p}\,(1-\bar{p})}$$

and
$$(LCL)_{np} = n\bar{p} - 3\sigma_{np} = n\bar{p} - 3\sqrt{n\bar{p}\,(1-\bar{p})} \qquad ...[\because \sigma_{np} = 3\sqrt{n\bar{p}\,(1-\bar{p})}]$$

The *np*-chart is similar to *p*-chart with the difference that the vertical axis represents the number of defectives *i.e.*, *np* and the centre line and control limits are magnified by a factor *n*.

Example 21.3. *Twenty sample of parts were taken from a production line for 100 per cent inspection, each sample containing 200 parts. The total number of defectives were 300. Compute upper and lower control limits.*

Solution. Given : Number of samples = 20 ; $n = 200$; Number of defectives = 300

We know that fraction defective during the period,

$$\bar{p} = \frac{\text{Total number of defectives during the period}}{\text{Total number inspected during the period}}$$

$$= \frac{300}{20 \times 200} = 0.075 \text{ or } 7.5\ \%$$

We know that upper control limit,

$$(UCL)_p = \bar{p} + 3\sqrt{\frac{\bar{p}\,(1-\bar{p})}{n}} = 0.075 + 3\sqrt{\frac{0.075\,(1-0.075)}{200}}$$

$$= 0.075 + 0.056 = 0.131 \quad \textbf{Ans.}$$

and lower control limit,
$$(LCL)_p = \bar{p} - 3\sqrt{\frac{\bar{p}\,(1-\bar{p})}{n}} = 0.075 - 3\sqrt{\frac{0.075\,(1-0.075)}{200}}$$

$$= 0.075 - 0.056 = 0.019 \quad \textbf{Ans.}$$

3. *c-Chart (Control Chart for defects).* The *c*-chart is also used for plotting attribute characteristics. The *p* or *np* chart applies to the number of defectives in a given sample of constant size, while the *c*-chart applies to the number of defects in each defective part or in a given sample of constant size. It may be noted that a defective refers to a part whereas a defect refers to lacking conformity to a specification. A defective part may contain more than one defect, for example a cast part may have blow holes and surface crackes at the same time.

The *c*-chart is generally preferred for large and complex parts, as a part having many dimensions cannot be rejected just because one of the dimension is not correct. Under such conditions, the incorrect dimensions can be expressed as the number of defects, over the number of dimensions measured. If $\bar{c}$ is the average number of defects, then

$$\bar{c} = \frac{\text{Total number of defects in all samples}}{\text{Total number of samples inspected}}$$

The three sigma (3σ) limits are :

$$(UCL)_c = \bar{c} + 3\sqrt{\bar{c}}$$

and
$$(LCL)_c = \bar{c} - 3\sqrt{\sigma}$$

where $\sqrt{\bar{c}}$ = Standard deviation (σ) for the number of defects per unit production.

The *c*-chart is plotted in the same manner as *p*-chart except that the control limits are based on Poisson's distribution which describes more appropriately the distribution of defects.

Example 21.4. *Twenty castings were inspected in order to check defects in them. After inspection 100 defects were found. Calculate c control limits.*

Solution. Given : Total number of defects = 100 ; Total number of samples inspected = 20

We know that average number of defects,

$$\bar{c} = \frac{\text{Total number of defects}}{\text{Total number of samples inspected}} = \frac{100}{20} = 5$$

Now, control limits are given as :

$$(UCL)_c = \bar{c} + 3\sqrt{\bar{c}} = 5 + 3\sqrt{5} = 11.7 \quad \textbf{Ans.}$$

and $$(LCL)_c = \bar{c} - 3\sqrt{\bar{c}} = 5 - 3\sqrt{5} = -1.7 = 0 \quad \textbf{Ans.} \quad ...(\text{–ve defect is not possible})$$

21·18 Sampling Inspection

Sampling is a very important tool of statistical quality control. When 100% inspection is not practicable, samples are taken from a lot and tested. The results of tests on the samples are then the representatives of the whole lot. Such inspection is known as *sampling inspection*. A well designed sampling plan gives greater assurance for quality than 100% inspection. In 100% inspection, the inspection becomes dull and boring because of the monotonous job of checking the same part, time and again, which may result in passing of defective parts also. But in sampling inspection there is more interest and job satisfaction for the inspectors.

Now a days, much of the inspection is done by sampling technique, which offers best protection against the acceptance of defective parts provided few defects are allowed among large number of parts. Depending upon the quality protection, a lot may be accepted if it contains less than the number of defectives called the *acceptance number*. The lot is rejected if it contains more than a certain number of defectives, called the *rejection number*. A 100% inspection is then made for a rejected lot, and the defectives are replaced by non-defectives.

It may be noted that the entire lot is either accepted or rejected on the basis of the inspection of the sample lot. If there is a sampling error then there is always a probability or risk that a bad lot is accepted and that a good lot is rejected. The former is called the *consumer's risk* and the latter is called the *producer's risk*.

21·19 Types of Sampling Plan

Following are three types of sampling plans :

1. *Single sampling plan*. In this type of sampling plan, a single sample is taken at random from the lot and lot is accepted or rejected on the basis of number of defects in the sample.

2. *Double sampling plan*. In this type of sampling plan, after the first sample, a second sample is taken and the number of defects are compared with the first sample and the decision is taken on the basis of the combined result of both the samples.

3. *Sequential sampling plan*. In this type of sampling plan, there is no specified limit on the total number of samples to be inspected before it is possible to reach a decision to reject or accept a lot.

21·20 Operating Characteristic Curve

The operating characteristic curve (briefly written as OC curve) is a graph, in which the horizontal axis shows the percentage of defectives (p%) in the lot and the vertical axis shows the probability of acceptance (Pa). The OC curve shows the ability of the sampling plan to distinguish between good and bad lots. The probability of the lot getting accepted varies according to the quality of lots. Therefore, it is always possible to draw OC curve for every sampling inspection plan or method. The sampling plan is selected in such a way that it passes through two important points on the curve *i.e.*, the producer's risk and the consumer's risk. The producer's interest is that most of the less defective units should be accepted and the consumer's interest is not to accept the more defective units. By seeing OC curve, it is possible to find out the percentage of accepted items of the lot with certain percentage of defects under the sampling method implemented.

An OC curve for a single sampling plan for sample size (n) = 50 and acceptance number (c) = 2 is shown in Fig. 21.5. This means that out of 50 samples taken randomly, and it is found that 2 or less than 2 are defective, then the lot will be accepted and the lot will be rejected if more than 2 defectives are found in the sample of 50.

From the figure, it is clear that the probability of acceptance of a lot which is worse than the lot with 10.5% defect is 0.1 (10%). Again, for a lot having 1.5% defective pieces, the probability of acceptance is 0.95 (95%). The 1.5% defective point (Pa = 95%) on the curve is called AQL (acceptance quality level) and 10.5% defective point (P = 10%) is called LTPD (lot tolerance percent defective). It may be noted that lot having 5% defective level (Pa = 50%) is called IQL (indifference quality level). It is a point where the producer's and consumer's risk are equal *i.e.*, the chances of accepting or rejecting a lot is 50%. If the percentage of defects is higher than IQL, the probability of rejection increases and if the percentage of defects is lower than IQL, the probability of acceptance of a lot increases.

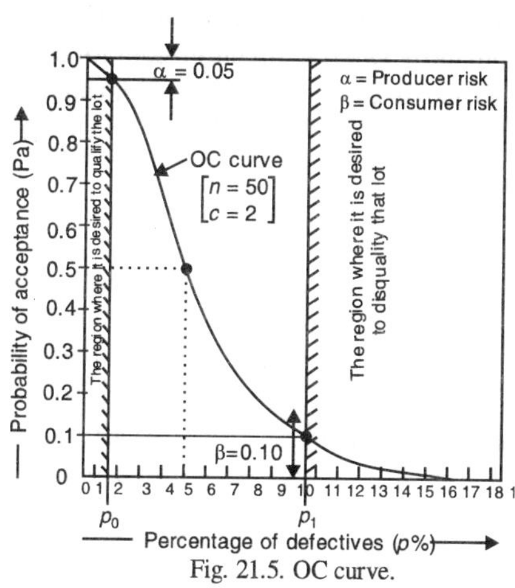

Fig. 21.5. OC curve.

Notes : 1. The shape of an OC curve depends upon the value of sample size (n) and acceptance number (c).

2. By changing the parameters (n and c) of an OC curve, an appropriate curve can be chosen which involves minimum purchaser's risk.

QUESTIONS

1. Define the term 'quality' as used in manufacturing science.
2. Define inspection and explain the two different methods of inspection.
3. Why is inspection of manufactured parts necessary ?
4. What do you understand by total quality control ?
5. Differentiate between inspection, quality control and statistical quality control.
6. Explain how statistical quality control is used to control the 'quality'.
7. State the significance of normal distribution curve in statistical quality control.
8. What is the difference between 'inspection by variables' and 'inspection by attributes'. Show why distinction is important ?
9. Define : (*i*) Mean ; (*ii*) Median ; (*iii*) Standard deviation ; and (*iv*) Range.
10. What is a control chart ? Name the various types of control charts.
11. What are $\overline{X}$ and R charts ? Under what conditions are they used.
12. What are 'p' and 'c' charts ? Why they are adopted ?
13. Briefly explain the sampling inspection. Explain its advantages over 100 per cent inspection.
14. What is an operating characteristic (OC) curve. Explain with an example.

OBJECTIVE TYPE QUESTIONS

1. The quality of part does not depend upon
   (*a*) how cheap the part is
   (*b*) how well the part perform its function
   (*c*) how reliable the performance of the part is
   (*d*) how far the performance can be maintained
2. Final inspection of the product is done at
   (*a*) receiving stage (*b*) process stage (*c*) delivery stage (*d*) after sales

3. The middle most or most central value of the data is

(*a*) mean (*b*) median (*c*) mode

4. The difference between the largest observed value and the smallest observed value is

(*a*) range (*b*) variance (*c*) standard deviation

5. If a process follows normal distribution curve, then the likely percentage of observed values, lying between $\overline{X} \pm 3\sigma$ is

(*a*) 99.52 (*b*) 99.73 (*c*) 99.65 (*d*) 99.82

where $\overline{X}$ = Mean of observed values, and

σ = Standard deviation.

6. The assignable cause of variation of dimension of any product is due to

(*a*) tool wear (*b*) tool setting (*c*) tool getting loose (*d*) all of these

7. Control charts are used for

(*a*) process control (*b*) production control

(*c*) material control (*d*) planning

8. The control chart used for a process where inspection is done by actual measurement is

(*a*) *p*-chart (*b*) $\overline{X}$ chart (*c*) *np*-chart (*d*) *c*-chart

9. For attributes, when sample size is constant, the control chart used for process control is

(*a*) *p*-chart (*b*) *c*-chart (*c*) *np*-chart

10. In *c*-chart, the control limits are based upon

(*a*) normal distribution (*b*) poisson's distribution

(*c*) mean value (*d*) none of these

ANSWERS

1. (*a*)	2. (*c*)	3. (*b*)	4. (*a*)	5. (*b*)
6. (*a*)	7. (*a*)	8. (*b*)	9. (*c*)	10. (*b*)

22

Powder Metallurgy

1. Introduction. 2. Production of Metal Powders. 3. Characteristics of Metal Powders. 4. Preparation of Metal Powders. 5. Processes Used for Manufacturing Parts from Metal Powders. 6. Primary Processes. 7. Secondary Processes. 8. Advantages of Powder Metallurgy. 9. Limitations of Powder Metallurgy. 10. Design Considerations for Powder Metallurgy. 11. Typical Applications of Powder Metallurgy.

22·1 Introduction

The powder metallurgy may be described as an art of manufacturing commercial products from the powdered metals and alloys. The process consists of compressing the irregularly shaped powdered metal particles in a die of the desired shape of the part to be produced. The part is then heated in an atmosphere-controlled furnace at a temperature below the melting point of one or more of the constituents of the powdered metals. This is known as *sintering*. The articles are subsequently heat treated to a required hardness.

The powder metallurgy technique is used

1. for producing parts from certain hard materials or refractory materials which are difficult to work with machines.
2. in fabrication of non-ferrous products like alloys of tungsten, carbides etc.
3. in lamination of intricate parts such as magnets, and self lubricating bearings.
4. in production of special electrical contacts and turbine blades having high thermal strength.
5. in producing friction materials used in clutch facings, brake linings etc.

Note : The parts made with powder metallurgy technique can be shaped to size with high precision (about 0.1 mm). This process has provided a practical solution to the problem of producing heat-resistant materials and cutting tools of extreme hardness.

22·2 Production of Metal Powders

A wide range of metal powders comprising of almost all metals and alloys are used in powder metallurgy. Sometimes, the manufacturing process include powder mixture of both metals and non-metals like copper and graphite as self-lubricating bearings, dynabrushes and ceramic-metal mixture for high temperature services. It may be noted that the use of powder metallurgy depends upon the production of metal powders of required degree of fineness and purity. The powders commonly used are copper-base and iron-base materials, stainless steel, nickel, aluminium alloys, chromium, titanium metal powders and other refractory materials including asbestos, graphite etc.

The powdered metals are produced by mechanical, physical and chemical means depending upon the particular application involved. The following are the methods commonly used for producing metal powders.

1. *Crushing.* This method is used for the brittle materials. The crushing equipment used for crushing brittle materials may be rapidly moving hammers, jaw crushers or gyratory crushes. The various ferrous and non-ferrous alloys can be heat-treated in order to obtain a sufficiently brittle material which can be easily crushed to powder form. The powders produced by this method are of angular shape which are subsequently reduced to powder by milling to attain the required fineness of the powder.

2. *Machining.* This method is employed to produce filings, turnings, scratchings, chips etc. which are subsequently pulverised by crushing and milling. Since relatively coarse and bulky powders entirely free from fine particles are obtained by this method, therefore it is particularly suitable in a very few special cases, such as for the production of magnesium powders for pyrotechnic applications. The powder particles produced by this method are of irregular shape. This method is higly expensive and therefore has a limited application.

3. *Milling or grinding.* This is the most important and widely used method of producing powders of the required grade of fineness. Though the milling or grinding process is best suited to brittle materials, yet it can be used for malleable and ductile materials. It involves the application of impact force on the material being powdered. The milling action is carried out by the use of ball mill, rod mill, impact mill, eddy mill etc.

4. *Shotting.* In this method, a fine stream of molten metal is poured through a screen into water. The molten metal on solidification produces large-sized spherical or pear-shaped particles. The powders produced by this method are of aluminium, lead and zinc.

5. *Atomisation.* In this method, the molten metal is forced through a nozzle into a stream of compressed air or water, as shown in Fig. 22.1. The metal on cooling solidifies into tiny particles of various sizes and irregular shapes. The size of these particles depend upon the size of nozzle, rate of flow of metal and temperature and pressure of water and air stream. This process is widely used in the production of powders from metals having low melting point such as zinc, tin and aluminium etc. The air is usually supplied at a pressure of about 0.2 to 0.3 N/mm^2.

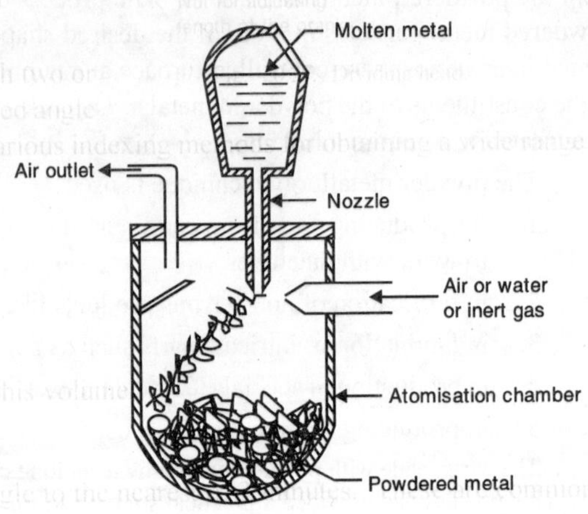

Fig. 22.1. Atomisation.

A stream of inert gas may also be used in atomisation of metal powders. The powders produced by this method are usually spherical or pear-shaped coated with a thin oxide layer. The formation of oxides on the individual particles during stirring is known as granulation.

6. *Electrolysis.* In this method, the metal plates are placed in a tank containing an acid solution *i.e.,* electrolyte. These metal plates act as anodes. The other metal plates are also placed in the tank to act as cathodes. When high current is passed, metal powders are deposited electrolytically on the cathodes under controlled conditions. The cathodes are taken out from the tank, washed with water to remove the acid solution and then dried. After drying, the deposit is scraped off, milled or ground for fineness and pulverised for desired grain size. This is further heat treated to remove the effect of work hardening and to drive out hydrogen. The electrolysis process produces fine and purest powders of tantalum, tin, silver, copper, zinc and iron, but the process is very costly. The powders produced by this process are very resistant to oxidation and have dendritic structure with low apparent density.

7. *Chemical reduction.* In this method, the metal powders are obtained by reducing refined ores or oxides in a current of hydrogen gas. The particles obtained by this process have softness and plasticity. The uniformity in size of the produced particles may be obtained by crushing or grinding. The metal powders, produced by this method are iron, copper, nickel, cobalt, molybdenum and tungsten.

8. *Vapour condensation.* This method, in fact, may be considered as a modification of the

usual distillation process employed for refining zinc. The powders of metals having low melting point such as zinc dust, is produced in very large quantities by this method. First of all, zinc oxide is mixed with powdered charcoal and heated until zinc vapour is formed by the reaction of zinc oxide with carbon monoxide. The zinc vapour is condensed to the dust like particles by cooling. The reoxidation of the zinc vapour to zinc oxide also occurs during cooling due to the formation of carbon dioxide. The presence of this oxide will prevent its coalescence into liquid causing blue powder as a normal by-product. Because of its high oxygen content, it is not considered suitable for powder metallurgy purposes.

22·3 Characteristics of Metal Powders

Following are the main characteristics of metal powders upon which the properties of finished product depend.

1. *Chemical composition and purity.* The chemical composition is an outstanding characteristic of the metal powders. It usually reveals the type and percentage of impurities present in the metal powders. The impurities greatly influence the physical and mechanical properties of the powdered metals.

It may also exert a decisive effect on pressing, sintering and other post-sintering operations which are essential for the production of finished product from powders. For example, the insoluble oxides (such as alumina, silica) present in the metal powders adversely affect the sintering and briquetting processes and thus reduce the life of dies and lower the mechanical properties.

2. *Size of particle and its distribution.* The size of particles is an important characteristic of metal powders. It has a great influence over density, porosity, compressibility and shrinkage during sintering. The size of the particle may be determined by many methods, but the simplest and the most popular method is a sieve method. In this method, the powdered metal is passed through screens having a definite number of meshes per linear centimetre. The majority of the metal powders employed in powder metallurgy industry vary in size from 4 to 200 microns.

The particle size distribution is also of fundamental importance as it affects pressing and sintering behaviour as well as physical and mechanical properties of the sintered material. The size distribution is given by the percentage of the coarse, medium and fine particles in a given sample of powdered metal. It may be noted that the porosity of metal increases if similar size of particles are used and decreases by the use of the dissimilar particles. The density of metal decreases if all the particles have coarse grains and increases by the use of fine grains. Thus, a balanced proportion of different particles should be worked out.

3. *Shape of particle.* The shape of particle may be flat, angular, ragged, spherical or dendritic. It has a great influence over porosity, compactness and strength of the part produced. It permits automatic feeding of powder flow, the maintenance of pre-determined tolerances and the control of deformation during sintering and cooling.

4. *Apparent density.* The apparent density (sometimes called packing density) of a powdered metal may be defined as the mass per unit volume of loose or unpacked powder. It is usually expressed in grams per cubic centimetre (g/cm^3). It depends upon the chemical composition, size of the particle and its distribution, shape of the particle and the method of manufacturing the metal powders. The apparent density helps in controlling the quantity of powder required for each part. When the apparent density decreases, it becomes necessary to use more powder to produce a part. It may be noted that the spherical particles have maximum apparent density whereas irregularly shaped and dendritic powders have reduced apparent density.

5. *Flowability.* The flowability is that property of a metal powder due to which it flows uniformly, readily and rapidly into a die or mould cavity. It governs the time required to fill a die or mould and the rate of production. The flowability of the metal powder must be good so that it flows

rapidly and uniformly into every corner of the die for obtaining rapid rate of production and consistent compacts. The flowability depends upon the shape of the particle, size and its distribution, coefficient of interparticle friction and the absorbed moisture. It may be noted that the coarse and spherical particles have good and maximum flowability respectively whereas the fine or dendritic and irregular particles have poor and reduced flowability respectively.

6. *Compressibility.* The compressibility of metal powders is defined as the ratio of initial volume (*i.e.,* volume of loose or unpacked volume) of the powder to the final volume after compression. It generally varies from 2 : 1 to 3 : 1. It controls the *green strength of the powder which is an important factor for easy handling of parts during operation. It is controlled by the size and shape of the particle.

7. *Sintering ability.* The sintering ability is that property of a metal powder due to which it provides and promotes adhesion between all the particles, on sintering at controlled temperature. The sintering ability of the metal powders should be high in order to produce the parts successfully. It may be noted that the irregularly and spherical shaped particles have good pressing and sintering properties.

22·4 Preparation of Metal Powders

The metal powders just after its production may not possess the desired properties for subsequent processes such as compacting and sintering processes. In powder metallurgy, it is very important that the powders is in dry form, free from impurities and has uniform properties. In order to maintain the properties of the metal powders, preliminary heat treatment (*i.e.,* annealing) is carried out usually in a reducing atmosphere (particularly hydrogen) and sometimes in vacuum. This has a considerable effect on the structure and other properties of the metal powders. It eliminates work-hardening, reduces the impurity content and alters the apparent density. It may be noted that the elimination of work hardening, decrease in impurity content because of the removal or reduction of oxides present and the change in apparent density improve the pressing quality of the metal powder.

The annealing process generally causes sintering of powder particles which forms a spongy mass. Thus, the powder should be pulverised and screened before its further use. This disintegration should be carried out carefully so that the ductility of the powder is not considerably reduced due to work hardening. When the powder is to be mixed with alloying elements, the powder should be used immediately after annealing.

22·5 Processes Used for Manufacturing Parts from Metal Powders

After preparing the powders, they must be stored under dry conditions before use. All the processes required for manufacturing parts through powder metallurgy are divided into the following two groups :

1. *Primary processes,* such as blending or mixing, briquetting or compacting, pre-sintering and sintering.

2. *Secondary processes,* such as sizing, coining, machining, impregnation, infiltration, plating and heat treatment.

These processes are discussed, in detail, in the following pages.

22·6 Primary Processes

The primary processes for manufacturing parts through powder metallurgy are discussed as follows :

1. *Blending or mixing.* The blending or mixing of powders is the first step in the manufacture

* The green strength of a powder is defined as the mechanical strength of a green compact (*i.e.,* compressed part before sintering) required to withstand, without damage to sharp edges, ejection from the die and transfer to the sintering furnace.

of parts from powdered metals. When one metal powder of the correct particle size is used, then no blending is required. But blending is essential when powders of the different metals with the same particle size or when powders of the same metal with different particle size are used. The main purpose of blending or mixing is to produce uniformity of particle size and shape throughout the large amount of powder. This helps in better control of subsequent pressing and sintering operations.

In blending or mixing, the various powders are weighed in the correct proportions and may be mixed either dry or wet in a ball mill. The wet mixing with water or some organic solvent is employed to produce a more fine and uniform mixture of powder particles. It should be done under favourable conditions to prevent surface oxidation and to keep dust to a minimum. The lubricants are added during blending operation in order to achieve required green strength and to facilitate pressing. It also reduces die wall friction and makes easy ejection of pressed parts. The common lubricants used are graphite, stearic acid and lithium stearate.

2. *Briquetting or compacting.* After mixing the metal powders thoroughly, the briquetting or compacting is the next process used in the manufacture of parts through powder metallurgy. It is the process of converting loose powder particles into a green compact of the desired shape and size. The process is carried out at room temperature in hardened steel dies and punches. It may be noted that due to the internal friction of the metal powders, the pressure applied from one side of the punch will not be uniformly distributed throughout the part. Thus in order to compress the powder properly, usually two punches (commonly known as upper punch and lower punch) are used, as shown in Fig. 22.2. These punches operate independently of each other and are located at each end of the die for applying uniformly distributed pressure throughout the part. The lower punch also acts as an ejector to remove the compacted part from the die. The die cavity must be smooth and should have a slight draft to facilitate the removal of the part. The ejected part is thus known as *green compact.*

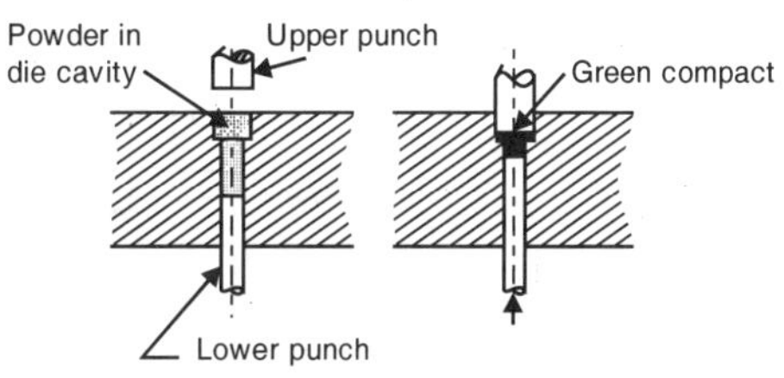

Fig. 22.2. Briquetting.

The soft and malleable metal powders are compressed at a very low pressure, whereas hard and refractory powders like tungsten require a very high pressure ranging from 800 to 1500 MPa. During the compacting process, the powder is compressed nearly one third of its original volume (*i.e.,* the ratio of loose volume of filled powder to the compressed volume is nearly 3 : 1). It may be noted that the density and hardness increase with pressure but improvement in these properties can not be obtained when the pressure increases beyond the maximum limit.

The presses used for the compacting process may be either mechanical or hydraulic or a combination of both. Though the mechanical presses are generally used for low pressures, yet they give high rate of production. The hydraulic presses are mostly used for high pressures and large components. These presses maintain a constant pressure at any point throughout the entire stroke and thus produce parts of uniform density which is not possible with mechanical presses.

3. *Pre-sintering.* It is the process of heating the green compact to a temperature below the sintering temperature. This is necessary to remove the lubricants and binders added during blending and to increase the strength of green compact. This process is not applied to all metals. Some metals like tungsten carbide can be easily machined after pre-sintering as they become very hard after sintering.

4. *Sintering.* It is the process of heating the briquetted component at an elevated temperature in a furnace under reduced atmospheric conditions. The atmospheric conditions are necessary to prevent any oxidation. The sintering temperature is kept below the melting point of the major constituent in the components. During the sintering process, bonding of the individual powder particles takes

place is the following three ways :

(*a*) melting of minor low temperature constituents,

(*b*) diffusion of particle to particle, and

(*c*) closing of voids present in the green briquettes.

The sintering process increases density, porosity, mechanical strength and ductility of the component. The furnaces used for sintering may be either of batch type or continuous type. A continuous type of furnace which has a mechanical conveyor to carry the compacts, is shown in Fig. 22.3. The furnace is electrically heated and the temperature is automatically controlled. The speed of the mesh belt is so adjusted that when the components leave the heating zone, the sintering is complete. Slow cooling takes place in a cooling zone so that no deformation takes place.

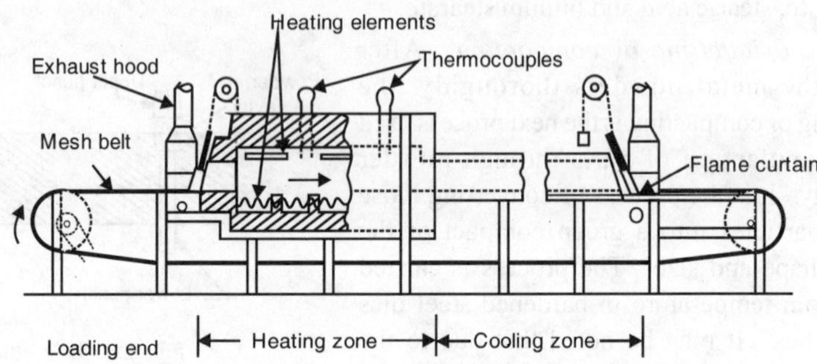

Fig. 22.3. Continuous type furnance for sintering.

It may be noted that the change in dimensions or shrinkage may occur during sintering. This depends upon the powder composition, shape and particle size variation of the powder, the briquetting pressure, and the sintering procedure. The accurate size is maintained by compensating for the change in making the green compact and then maintaining uniform conditions.

5. *Hot pressing.* It is a process which combines both the compacting and sintering processes. In a sintering process, the green compacts or briquettes are sintered. But in hot pressing, the powdered metal is placed in a die and the pressure is applied simultaneously with heat. The graphite dies are most commonly used because they are very cheap and have good thermal and electrical conductivity. The dies are usually heated, under protective atmosphere or in vacuum, by resistance wires wrapped around the die or by induction heating. The method of hot pressing is particularly used for large sized parts and for those powders (such as cemented carbide) which are difficult to press at room temperature.

The hot pressed parts possess higher tensile strength, increased hardness, improved tolerance and higher densities than the parts produced by the separate compacting sintering methods.

22·7 Secondary Processes

Many parts, manufactured through powder metallurgy, are suitable for use directly after sintering. However, when better surface finish and close tolerances are desired, then additional processes known as secondary or finishing processes, are necessary. The various secondary processes are sizing, coining, machining, impregnation, infiltration, plating and heat treatment. These processes are discussed, in detail, as follows :

1. *Sizing.* The sizing is a cold working operation, in which the sintered component is pressed in a die similar to that used for compacting. This operation is necessary to achieve dimensional accuracy.

2. *Coining.* The coining is similar to the sizing operation. The object of coining is to increase

the density by closing the pores, to improve the surface hardness and smoothness and to give additional strength to the part.

3. *Machining.* Most of the parts produced through powder metallurgy are finished to required dimensions and tolerances and no further machining is required. However, the parts which require small holes, grooves, undercuts, threads etc. cannot be produced through powder metallurgy. These shapes are generally produced by machining of the sintered part. The tungsten carbide tools are mostly used for machining. The high speed steel tools may also be used but because of its rapid wear, these tools are used when only a few parts are to be machined. It may be noted that the coolants with water should not be used for iron base parts as it results in corrosion.

4. *Infiltration.* It is the process of filling the pores of a sintered part with molten metal. The molten metal is introduced into the sintered part by capillary action. The metal used for infiltration should have low melting point than the part and the operation should be done as quickly as possible. This process reduces the porosity and increases the strength, hardness and density of the part.

5. *Impregnation.* It is a process of completely filling the voids and interconnected pores in a powdered metal component. This process is mostly used to provide self lubricant properties to the *porous bearings. The self lubricating properties are achieved by impregnating the sintered bearing with oil, grease or other lubricating materials. The usual practice is to immerse the bearing in a tank of oil, held in it at a temperature of 80° to 90°C for atleast 30 minutes. Since this method of impregnation with oil is time-consuming, therefore another method, known as vacuum impregnation, is always desirable. This method takes lesser time and fills all the interconnected pores. In the vacuum impregnation method, the bearing is introduced into the vacuum chamber and evacuation is continued until the pressure drops to 1/2 mm of Hg, to remove most of the air and moisture from within the pores. At this stage, oil is allowed to enter the vacuum chamber and the pressure is increased to atmospheric and finally air is introduced at atmospheric pressure.

6. *Plating.* The sintered parts are plated to provide a more pleasing appearance and to protect from corrosion. The high density parts can be readily plated by using convectional methods. But the low density parts, due to porosity, require special surface preparation. Such parts are first impregnated with plastic resin to eliminate the porosity and then regular plating procedure is adopted.

7. *Heat treatment.* The purpose of heat treatment of powder metal parts, particularly sintered ferrous parts, is to improve hardness, strength, wear resistance and grain structure.

22·8 Advantages of Powder Metallurgy

Following are the advantages of powder metallurgy :

1. There is no wastage of material throughout the entire manufacturing process, thus material is saved.
2. The parts produced have accurate dimensions and surface finish, thus the need for machining operations is eliminated.
3. The large quantity production is relatively economical to machine parts.
4. The skilled workers are not required to operate presses or other equipment, thus the labour costs are low.
5. It is possible to ensure uniformity of composition and structure as the exact desired proportion and purity of various metal powders can be used.
6. The parts with controlled porosity can be made.
7. The parts of higly refractory metals and high melting point such as tungsten carbide can be produced.
8. It is possible to make parts from a mixture of metals and non metals such as motor brushes.

* The porous bearings are usually made of bronze, with or without the addition of graphite or other dry lubricant.

9. It is possible to make parts of a mixture of metals which do not alloy with each other.
10. The impregnated parts as self lubricating bearings can only be produced by powder metallurgy technique.

22·9 Limitations of Powder Metallurgy

Though the parts produced by powder metallurgy technique are better and economical, yet it has the following limitations :

1. The production of high purity powders is highly expensive.
2. The cost of storing metal powders is very high as they oxidise when come in contact with air. Some powders such as aluminium, magnesium and titanium, in a finely divided state, may explode and are fire hazards.
3. It is used only for large scale production due to expensive tools and equipments such as dies, punches, presses and sintering equipment.
4. It is not possible to completely eliminate the porosity in the parts. This results in lower tensile and yield strengths, lower elongation and much lower impact values. In other words, a uniformly high density parts are difficult to produce by this process.
5. The size of the part is limited as it depends upon the compression ratio of the powders and the capacity of press used.
6. The powders when pressed do not flow around corners as do plastic materials and they do not transmit pressures like a liquid. Therefore intricate designs in products are difficult to attain.

22·10 Design Considerations for Powder Metallurgy

The metal powders do not flow like plastics and there is always some spring back after compression. Thus, in designing powdered metal parts, the following points should be kept in mind:

1. The sudden changes in section thickness should be avoided as far as possible.
2. The deep and narrow sections should be avoided.
3. The holes should be made in the direction of pressing.
4. The very close dimensional tolerances should be avoided in the direction of pressing.
5. The production of very small holes, knurling, threads and undercuts should be produced by machining.

22·11 Typical Applications of Powder Metallurgy

Following are some typical applications of powder metallurgy :

1. *Filament wire for electric bulbs.* The tungsten powder is used for making filament wires which have high melting point.

2. *Self lubricating bearings.* The iron powders, mixtures of copper and tin powder, iron and copper powders, all with or without graphite are used for self lubricating bearings. After sintering, the bearings are impregnated with oil, grease or wax.

3. *Magnets.* The powders of aluminium, nickel, cobalt and iron are used for permanent magnets.

4. *Cemented carbide tools and dies.* The powders of tungsten carbide, titanium carbide, boron carbide and tantalum carbide with cobalt are used for cemented carbide tools and dies. The cemented carbide tools are used for high speed machining of cast iron, bakelite, hard rubber etc.

5. *Clutch facings and brake linings.* The fine copper-base metal powder mixtures (copper, tin, silica, graphite, lead) are used for clutch facings and brake linings particularly for air-craft landing gear.

6. *Motor brushes.* The copper and graphite powders are used to make motor brushes.

7. *Pump rotors and gears.* The iron and graphite powders are used for making pump rotors and gears.

8. *Electrical contact parts.* Many combinations of powders such as tungsten and copper, tungsten and cobalt, tungsten and silver are used for electrical contact parts.

9. *Porous filters.* The powders of brass and bronze are used for manufacturing porous filters which require high porosity.

10. *Porous bearings.* The powders of copper, tin and iron with or without graphite are used for porous bearings.

11. *Metal coatings.* The powder metal coatings can be done on various parts to increase corrosion resistance or to repair worn out costly parts.

QUESTIONS

1. What is powder metallurgy ?

2. Name the various methods of manufacturing metal powders and give brief description of each.

3. What are the main characteristics of metal powders which control their processing ?

4. Describe briefly the primary processes required for manufacturing parts from powdered metals.

5. Name the various secondary operations which are required to be performed on powdered metal. parts ? Describe them briefly.

6. What are the advantages and limitations of powder metallurgy ?

7. What are the main points which should be kept in mind while designing powdered metal parts ?

8. Give the metal powders used for the following :

(*a*) Pump rotors and gears; (*b*) Self lubricating bearings;
(*c*) Porous bearings; (*d*) Permanent magnets; and
(*e*) Filament wire for electric bulbs.

9. Write short notes on the following :

(*a*) Briquetting; (*b*) Sintering;
(*c*) Hot pressing; (*d*) Sizing and coining;
(*e*) Infiltration; and (*f*) Impregnation.

OBJECTIVE TYPE QUESTIONS

1. The powder metallurgy technique is used

(*a*) for producing parts from hard materials
(*b*) in fabrication of non-ferrous products
(*c*) in producing friction materials
(*d*) all of the above (*e*) none of the above

2. The magnesium powders for pyrotechnic applications are produced by

(*a*) crushing (*b*) grinding (*c*) machining (*d*) atomisation

3. The compressibility of metal powders is defined as

(*a*) the rate of unpacked volume of the powder to the final volume after compression
(*b*) the mass per unit volume of loose or unpacked powder
(*c*) volume of powder after compression
(*d*) all of the above

4. The blending or mixing of powders is not required when

(*a*) powders of the same metal with different particle size is used

(*b*) powders of different metals with the same particle size is used

(*c*) powders of one metal of the correct particle size is used

(*d*) all of the above

5. The process of filling the pores of a sintered part with molten metal is called

(*a*) coining (*b*) infiltration (*c*) impregnation (*d*) sizing

ANSWERS

1. (*d*) 2. (*c*) 3. (*a*) 4. (*c*) 5. (*b*)

23

Plastics

1. Introduction. 2. Raw Material for Plastics. 3. General Properties of Plastics. 4. Types of Plastics. 5. Types of Thermosetting Resins. 6. Types of Thermoplastic Resins. 7. Synthetic Rubber or Elastomers. 8. Moulding Compounds. 9. Fabrication of Plastics. 10. Moulding. 11. Calendering. 12. Casting. 13. Slush Casting. 14. Laminating. 15. Machining of Plastics. 16. Joining of Plastics.

23·1 Introduction

The plastics are synthetic organic materials which can be moulded into any desired shape when subjected to heat and pressure. These materials are prepared from natural or synthetic resins with or without the moulding compounds. These moulding compounds are mixed with the granular resins before moulding.

The plastics, now-a-days, are being used in one form or the other in almost all fields of engineering. They are mostly used in the manufacture of aeroplane and automobile parts. They are also used for making safety glasses, laminated gears, pulleys, self lubricating bearings.

23·2 Raw Material for Plastics

In general, the plastics are compounds of carbon with other elements such as oxygen, hydrogen and nitrogen. The carbon combines with itself and other elements and forms more complicated compounds.

The raw material for plastic compounds are various agricultural products and numerous minerals and organic materials including petroleum, coal, gas, limestone, silica and sulphur. The natural resins are wax, shellac, pitch, bitumen resin, rubber etc. The synthetic resins, for the plastic materials, are produced by linking together a large number of small molecules of synthetic resin (known as *monomers* or *monoliths*) which forms large or long-chain molecules known as *polymers* (*poly* means many and *mers* means units or parts). The process by which monomers are combined to form polymers is known as *polymerisation.* The following two methods are used to achieve polymerisation.

1. Addition polymerisation, and
2. Condensation polymerisation.

In *addition polymerisation,* two or more similar monomers directly react to form polymers. The *condensation polymerisation* takes two or more dissimilar monomers that react to form polymers plus the by-product water.

23·3 General Properties of Plastics

The following are the general properties of plastics due to which it is widely used in engineering practice.

1. Lightness in weight. The specific gravity varies from 1 to 2.4.
2. Easy workability. The plastics can be easily casted, moulded, drilled, machined etc.
3. Highly resistant to corrosion.
4. Highly resistant to abrasion.
5. Good thermal and electrical insulators.
6. Good strength and rigidity.

7. Absorbent of vibrations and sound.
8. Good resistant to most of the chemicals.
9. Impermeable to water.
10. Low fabrication cost.
11. Good dimensional stability.
12. Can be made transparent or coloured.

23·4 Type of Plastics

The plastics are broadly classified in the following two groups :

1. Thermosetting plastics, and 2. Thermoplastic materials.

The thermosetting plastics (also known as heat-setting materials) are those plastics which are formed into shape under heat (in the temperature range of 127°C to 177°C) and pressure and results in a permanently hard product. The heat first softens the resins and when special chemicals are added, it becomes hard by a chemical change known as polymerisation. The reheating does not soften them again. The thermosetting plastics are soluble in alcohol and other organic solvents when they are in thermoplastic stage. This property is utilised in making paints and varnishes from plastics. These plastics are hard, strong and durable. These are manufactured in a variety of beautiful colours.

The *thermoplastic materials* (also known as cold-setting materials) are those plastics which do not become hard with the application of heat and pressure and no chemical change occurs. They remain soft at elevated temperatures and become hard on cooling. They can be remelted repeatedly by the successive application of heat. These plastics are better than thermosetting plastics because they can be used again and again. The scrap obtained from the moulding operations and rejected articles can be shaped into granules and reused to make first quality products. These plastics are used for makings toys and other ordinary purposes.

23·5 Types of Thermosetting Resins

The following are the various types of thermosetting resins :

1. *Phenolic resins.* The most important of the phenolic resins is *phenol formaldehyde.* It is obtained by condensing the phenol with formoldehyde in the presence of a catalyst. It is popularly known by its trade name of *Bakelite,* after the name of its inventor Dr. L. Bakelite.

It is a hard, rigid and scratch resistant material. It is highly resistant to heat, water, non-oxidising acids, salts and many organic solvents. It possesses excellent electrical insulating property. It is one of the cheapest material of all the thermosetting resins.

It is used in manufacturing handles for cooking pots, knobs, toilet seats, bottle caps, dials, telephone parts, cabinets for radio and television, electrical components like switches, plugs, switch-boards etc. It is also used as a binder in paints and varnishes and as an adhesives for grinding wheel.

Notes : **(*a*)** The phenol is carbolic acid obtained as a by-product during distillation of organic substances such as wood or coal.

(*b*) The phenolic resins are used with a variety of fillers to modify the properties. For example, wood flour for general purpose, asbestos for heat resistance and mica for electrical insulation.

(*c*) The phenolic resins may be moulded by compression or transfer moulding methods (See Art. 23.10).

2. *Amino resins.* The two important amino resins are *urea formaldehyde* and *melamine formaldehyde.* These are condensation products obtained by the reaction of urea or melamine with formaldehyde. The amino resins can be produced in a wide range of colours and are hard, rigid and durable. They possesses good electrical properties and are heat and scratch resistant.

The urea formaldehyde is widely used in domestic electrical fittings such as switch covers, plug tops, socket bases and lamp sockets. It is also used for cabinets, toilet seats, buttons and clock cases.

The melamine formaldehyde, due to good flowability of melamine, is principally used for moulded cups, plates, saucers, bowls etc.

Both the resins are used as coatings and adhesives.

Notes : (*a*) The urea is obtained by mixing liquid carbon dioxide and liquid ammonia under heat and pressure.

(*b*) The melamine is obtained from calcium carbide.

3. *Furane resins.* The furane resins are obtained when waste farm products such as cotton seeds, rice hulls, corncobs are processed with certain acids. These resins are dark in colour, water resistant and have good electrical properties. These are used as core sand binders and as hardening additives for gypsum plaster.

4. *Silicon resins.* The silicon resins differ from most other resins which are based on the carbon atom. The silicon resins have silicon and oxygen chain to which are linked various organic groups such as methyl side groups.

The silicon resins may be in the form of liquids, semi-solids (like greases), rubbers and solids. The liquid silicones or silicon oils possesses great wetting power for metals, low surface tension and show very small changes in viscosity with temperature. These are used as high temperature lubricants, anti-foaming agents, water-repellent finishes for leather and textiles heat transfer media, damping and hydraulic fluids. They are also used in cosmetics and polishes.

The silicones in the semi-solid form (*i.e.,* silicon greases) are modified silicon oils and are obtained by adding fillers like silica, carbon black etc. These are particularly used as lubricants in situations where very high and low temperatures are encountered.

The silicones in the rubber form have high abrasion resistance, stability at high temperatures and remain flexible even at very low temperatures. They are mostly used in gaskets, insulations and as additives in other rubbers.

The solid silicones possesses good electrical insulating properties and outstanding heat resistance. They are mostly used in high voltage insulators, high temperature insulating foams and mouldings which require high thermal stability.

Note : The silicon resins can be processed by compression or transfer moulding, extrusion and casting.

5. *Epoxy resins.* The epoxy resins are obtained from certain special types of organic chemicals, specifically epichlorohydrin and bisphenol (double phenol). These are cured or cross-linked by the addition of a hardner. The cured epoxy resins have low shrinkage, good flexibility, excellent chemical resistance and electrical insulating properties.

These are used for surface coatings, adhesives for glass and metals, and laminating materials used in electrical equipments. The moulds made from epoxy resins are employed for the production of components for aircrafts and automobiles.

6. *Polyester resins.* The polyester resins are obtained by the reaction between a dihydric alcohol and a dibasic acid. They are divided into the following three groups :

(*a*) Saturated polyesters, (*b*) Unsaturated polyesters, and (*c*) Alkyds.

The *saturated polyesters* are obtained by reacting glycol with saturated dibasic acid. They are a good fibre forming material and are converted into commercial fibres. Such fibres have high scratch resistance, high crease and wrinkle resistance. They are mostly used for making synthetic fibres like terylene, dacron etc.

The *unsaturated polyesters* are made by reacting glycol with unsaturated dibasic acid (like maleic anhydride). They have good flexural strength and can withstand temperatures upto 145°C. They are good resistance to water but possesses low resistance to acids and alkalies. These are generally used in safety helmets, air-craft battery boxes, motor-car body components etc.

The *alkyds* are produced by reacting polyhydric alcohol (like glycerol) with polybasic acid (like phthalic anhydride) in correct proportions in the presence of heat and catalyst (CO_2 gas). The alkyd resins are modified either by oil (drying or non-drying) or fatty acids.

The drying oil-modified alkyds are used as a coating material in numerous formulations. The acid or oil modified alkyds are hard, dimensionally stable and resistance to corrosion and acids. They are used for making good insulators, aircraft and automobile parts, sheets, rods, tubes, switches, gears, circuit-break insulators etc.

7. *Polyurethanes.* The polyurethanes are obtained, commercially, by treating di-isocynate and diol. They have excellent resistance to abrasion and solvents. These are used as coatings, films, foams, adhesives and elastomers.

23·6 Types of Thermoplastic Resins

The following are the various types of thermoplastic resins :

1. *Cellulose derivatives.* The cellulose derivatives are as follows :

(*a*) *Cellulose nitrate.* It is obtained by reacting cellulose with nitric acid in the presence of sulphuric acid which acts as a dehydrating agent. It is known by its trade name *celluloid.* It is a hard and brittle material. It is available in a wide variety of beautiful colours. It has good resistance to moisture but is highly inflammable. It is used for making spectacle frames, pen bodies, tooth brushes, radio dials, drawing instruments, table-tennis balls, toys and toilet articles. Since it is soluble in many solvents, therefore it is widely used for making lacquers.

(*b*) *Cellulose acetate.* It is obtained by reacting cellulose with acetic acid in the presence of sulphuric acid. It is stable, transparent, wear resistance and strong. It can be easily moulded and extruded. It is lighter than cellulose nitrate and has a tendency to absorb moisture. It is used for making toys, knobs, flashlight cases, radio panels, film for recording tape etc. The cellulose acetate in the extruded sheet form, is known as *cellophane.* It is used for packaging, curtains, drapers and wrapping.

(*c*) *Cellulose acetate-butyrate.* It is obtained by reacting cellulose with acetic acid and butoric acid. It is similar to cellulose acetate and can be produced in all colours. It is well suited for injection moulding and extrusion. It has low moisture absorption, toughness and good stability against heat and light. It is used for steering wheels, football helmets, goggle frames, insulating tapes, pipes and tubes for gas and water etc.

(*d*) *Ethyl cellulose.* It is obtained by reacting sodium cellulose with ethyl chloride under pressure and subsequent precipitation in water, followed by purification. It is strong, tough, moisture resistant and good insulators. It is used as a base for coating materials, jigs and fixtures, forming dies and moulded components.

(*e*) *Cellulose propionate.* It is a stable and tough thermoplastic. It has low tendency for moisture absorption and can be easily moulded. It is used for fountain-pens, flashlight cases, telephones and safety glasses.

2. *Polystyrene.* It is obtained by polymerisation of styrene (dissolved in ethyl benzene) in the presence of benzoyl peroxide. It is hard, brittle and low in impact resistance. It is available in colours from clear to opaque. It is resistant to water and most chemicals. It has dimensional stability and insulating ability. It is widely used in moulding of articles like toys, combs, buttons, radio and television parts, refrigerator parts, battery cases, high frequency electric insulators, lenses, indoor lighting panels etc.

3. *Polyethylene.* It is also known as polythene and is obtained by polymerisation of ethylene. It may be of low density or high density depending upon the process used in its manufacture. It is resistant to moisture and most chemicals. It is flexible at room and low temperatures and is capable of being heat sealed. Since it is one of the inexpensive plastics, therefore it is made in large amounts than any other plastics. The low density polythene is widely used as film, bags and sacks for packaging,

for houseware, dustbins and toys. The more rigid high density polythene is used for large moulded containers, bottles, milk crates and for better quality houseware.

4. *Polypropylene.* It is obtained by polymerising propylene in the presence of Zieglor-Natta catalyst. It is one of the lightest plastics available. It has excellent electrical properties, high impact and tensile strength. It is resistant to heat and chemicals. It is used for hospital and laboratory equipment, toys, furniture, television cabinets and electrical insulation.

5. *Acrylic resins.* It is light in weight and has good light transmitting power. It is moisture resistant and easy to fabricate. It can be cut, sawn or turned. It possesses excellent optical properties and acts as a good electric insulator. The plastics made from the acrylic resin are for safety glasses, roof lights, light fittings, signs, etc.

6. *Vinyl resins.* The most important plastics made from vinyl resins is *polyvinyl chloride.* The other types of plastics made from vinylresins are polyvinyl acetate, polyvinyl butyrates and polyvinylidene chloride.

The *polyvinyl chloride* (PVC) is obtained by heating a water emulsion of vinyl chloride in the presence of small amount of benzoyl peroxide or hydrogen peroxide in an autoclave under pressure. The PVC may be rigid or flexible. The rigid or unplasticised PVC is a hard thermoplastic, non-burning and weather resistant. It is mostly used for rain water pipes and fittings, roofing sheets, electrical conduit, safety helmets, refrigerator components, gramophone records, and bottles for edible oils etc. The flexible or plasticised PVC is a soft thermoplastic. It is used for garden hose, insulation of cables and wires, rain coats, table cloths and curtains etc.

The *polyvinyl acetate* (PVA) is obtained by heating vinyl acetate in the presence of benzoyl peroxide or acetyl chloride as catalyst. Since the articles made from PVA are distorted under the influence of compressive or tensile forces, so it is not used for moulding purposes. However, it is used for making paints, lacquers, plastic emulsions, coatings, wrapping paper etc.

The *polyvinyl butyrates* (PVB) is a clear tough resin and is used as adhesive for safety glass, sealing fuel tanks and flexible moulded products.

The *polyvinylidene chloride* (also known as Saran) is a stiff plastic like PVC and is used for film wraps.

7. *Polymides.* It is popularly known by its trade name Nylon. It is produced by the reaction of a diamine with an organic acid. There are various grades of nylons, but the most commonly used nylons is nylon 6/6, which is obtained by reacting hexamethylenediamine with adipic acid.

It is a crystalline thermoplastic, though and stiff. It is translucent and whitish in colour. It possesses high temperature stability and good abrasion resistance. It has excellent wear properties and can be used for applications without lubrication. It is widely used for light engineering components such as gears, bushes, bearings etc.

8. *Polytetrafluoroethylene (PTFE).* It is obtained by polymerisation of water emulsion of tetrafluoro ethylene, under pressure, in the presence of benzoyl peroxide as catalyst. It is popularly known by its trade name *Teflon* or *Fluon.* It has extreme toughness, high softening point (about 350°C), very low coefficient of friction, extremely good electrical and mechanical properties. It can be machined, punched and drilled. It is used for making gaskets, packings, pump parts, tank linings, non-lubricating bearings, chemical-carrying pipes and containers, electrical insulators, and for coating and impregnating glass fibres, asbestos fibres and cloths.

9. *Polymethyl methyacrylate (PMMA).* It is obtained by polymerisation of methyl methacrylate (ester of methyl acrylic acid), in the presence of acetyl peroxide or hydrogen peroxide. It is an acrylic polymer. It is popularly known by its trade name *Lucite* or *Plexiglass.* It is a hard, fairly rigid material with a high softening point of about 130-140°C, but it becomes rubber-like at a temperature above 65°C. This property of PMMA makes it easy to fabricate. It has good light transmitting power and possesses excellent optical properties. It is widely used for making lenses,

aircraft light fixtures, bomber noses, gun turrets, cockpit canopies, automotive appliances, wind screens etc.

10. *Acrylonitrile-butadiene-styrene (ABS).* It has high impact strength, resistance to low temperatures and chemicals. It can be produced with a high glass finish in a wide range of colours. It is used in place of PVC and polystyrene where greater toughness is required. The applications of ABS include instrument panels on cars, sailing and rowing boats, cameras, electrical hand tool housings, telephone hand sets, liners for refrigeration and for gas pipes. The sheets of ABS can be produced with a grained finish for use in the manufacture of suitcases, machine guards and furniture.

23·7 Synthetic Rubbers or Elastomers

The synthetic rubbers or elastomers are any vulcanisable rubber-like polymers which can be stretched atleast twice its length on loading in tension and returns to its original dimensions as soon as the load is released. Though there are many types of synthetic rubbers, yet the following are commonly used :

1. *Styrene rubber.* It is the most important type of synthetic rubber, which is produced by copolymerisation of butadiene (75% by weight) and styrene (25% by weight). It resembles natural rubber in processing characteristics as well as quality of finished products. It possesses high abrasion resistance, high load bearing capacity and resilience. It is mainly used for manufacturing motor tyres.

2. *Nitrile rubber.* It is a copolymer of butadiene and acrylonitrile. It possesses excellent resistance to heat, sun light, oils, acids and salts. It is used for making conveyor belts, aircraft components, tank-lining, hoses, gaskets, printing rollers, adhesives and automobile parts etc.

3. *Polychloroprene rubber.* It is made by polymerisation of chloroprene, a chlorinated butadiene. It is popularly known by its trade name *Neoprene.* It is resistant to oils, heat and light. Its general properties are improved by compounding it with metallic oxides such as ZnO or MgO. It is used for making hoses, gaskets, tubes for carrying corrosive gases and oils, conveyor belts, adhesives etc.

4. *Butyl rubber.* It is made by copolymerisation of iso-butylene with small amount (usually 1 to 5%) of isoprene. It has excellent resistance to heat, abrasion, and chemicals. It has also good electrical insulating properties. It is used for making cycle and automobile tubes, hoses, conveyor belts, tank linings, insulation for high voltage wires and cables etc.

5. *Polysulphide rubber.* It is made by the reaction between sodium polysulphide and ethylene dichloride. It possesses extremely good resistance to mineral oils, fuels, solvents and sunlight. It is used for making hoses, gaskets, cable coverings, oil tank linings etc.

23·8 Moulding Compounds

The various moulding compounds which are usually added with the granular resins before moulding, under heat and pressure, are as follows :

1. *Fillers.* The fillers are added to reduce the cost and to improve certain properties of the final plastic product, such as compressive strength or impact resistance. It is often essential for thermosetting plastics for reducing mould strinkage.

The commonly used fillers are : wood flour, silica sand, ground granite, diatomaceous earth, clays and powdered metal.

2. *Reinforcements.* The reinforcements (also known as reinforcing fillers) are added for increasing the mechanical strength of the plastics. The most widely used reinforcements are : fibre glass, asbestos, cellulose and synthetic fibre.

3. *Stabilisers.* The stabilisers are added to plastics to improve thermal stability during processing *i.e.,* to prevent their deterioration due to the action of heat and light. For example, zinc soap is used to vinyls and phenols to the styrenes.

4. *Plasticisers.* The plasticisers are added to plastics to make them soft, to improve toughness and flexibility of the final product. The commonly used plasticisers are camphore, paraffins, napathalanes, phosphates etc.

5. *Solvents.* The solvents are used for dissolving the plasticisers and to improve the flowability of the plastic in the mould. For example, alcohol is added in cellulose nitrate plastics to dissolve camphor.

6. *Pigments.* The pigments are used to impart decorative and colouring effect to the plastics. They also act as fillers for plastics. The commonly used pigments are zinc oxides, barytes etc.

7. *Catalysts.* The catalysts are used to accelerate the chemical reaction during the process of polymerisation of plastics. They act as *hardners* and *accelerators.* For example, ester is used as a catalyst for urea formaldehyde.

8. *Modifiers.* The modifiers are the chemicals added to plastics for changing the mechanical properties of the base resin. Some of them help in improving the mechanical properties also.

9. *Elastomers.* The elastomers are those materials which are added to the plastics in order to provide maximum property of elasticity.

23·9 Fabrication of Plastics

The process of making various products of plastics is known as fabrication of plastics. Though there are many methods used in the fabrication of plastics, yet the following methods are important :

1. Moulding, 2. Calendering, 3. Casting, and 4. Laminating.

The methods are discussed, in detail, in the following pages.

23·10 Moulding

It is the most common method used in the fabrication of plastics. The various methods of moulding of plastics are discussed as follows :

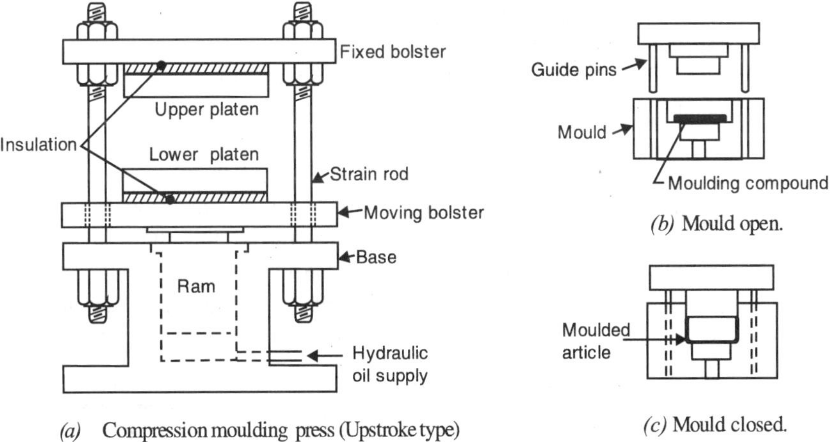

(a) Compression moulding press (Upstroke type) *(b)* Mould open. *(c)* Mould closed.

Fig. 23.1. Compression moulding.

1. *Compression moulding.* This method is mostly used for thermosetting plastics but can also be applied to some of the thermoplastic materials. In this method, the mould is opened and the predetermined amount of thermosetting compound either in powder form or granular form, is placed into the pre-heated mould. The mould is closed and heated in the temperature range of 120°C to 250°C, depending upon the material being moulded. The pressure, varying from 0.7 to 55 MPa depending upon the nature of the material and the size of the product, is applied hydraulically so that

the molten plastic flows into every corner of the mould. The mould is kept closed for sometime to allow the chemical change to complete so that the plastic product becomes permanently hard. The mould is now opened by releasing the pressure and the product taken out.

Some of the products made by compression moulding of thermosetting plastics are gears, knobs, handles for kitchenware, toilet seats, wall switch plates etc. When thermoplastic materials are made by compression moulding, the mould should be sufficiently cooled before the product is taken out of it. Thus in this case, the mould has to be heated and cooled rapidly which is difficult one.

Most of the compression moulding presses are of the up stroke type as shown in Fig. 23.1. In this type, a hydraulic ram moves the bed or bolster of the press upwards to close the mould. The upper part of the mould (also called plunger) is bolted to the upper platen while the lower part of the mould (also known as die) is bolted to the lower platen. These platens have heating channels and are backed with asbestos insulation. The heat may be supplied by steam or electricity.

Note : The downstroke presses (in which the lower bolster is fixed and the upper bolster is moving) are best suited to the moulding of large components which require a large stroke.

2. *Transfer moulding.* This process is mostly used for thermosetting plastics and is similar to compression moulding. We have discussed above that in compression moulding, the material is introduced into the mould in solid state and has to plasticise and flow within the mould. This requires very high pressures and even it is difficult to get the material to flow into intricate parts of the complex moulds. In order to overcome this difficulty, transfer moulding process is used.

In the transfer moulding process, as shown in Fig. 23.2, the thermosetting material in the powder or granular form, is plasticised in a chamber outside the mould by heating. It is forced or transferred into the mould cavity through one or more sprues (channels). The mould is held under pressure until the product is completely cured. After curing, the mould is opened and the product is removed. The curing time for transfer moulding is usually less than for compression moulding. The loading time is also shortened. This method is preferred to compression moulding for intricate parts with wide variations in section thickness.

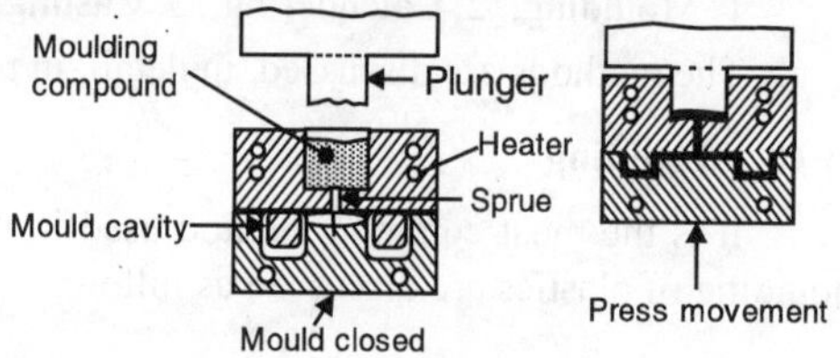

Fig. 23.2. Transfer moulding process.

The disadvantage of transfer moulding is that the material left in the chamber and in the runner and sprue leading to the mould, will be cured along with the material in the mould. This material must be extracted and discarded as scrap. Thus the wastage of material is rather higher than with compression moulding. The mould cost is also higher than for compression moulding.

3. *Injection moulding.* This method is mostly used for thermoplastic materials which plasticise when heated and solidify when cooled. The process consists of converting the thermoplastic material from powdered or granular form into a liquid and then injecting into the mould where it solidifies.

Fig. 23.3 shows a plunger type of injection moulding machine which is commonly used for producing parts from thermoplastic materials. In its operation, the plastic compound in the powdered or granular form is fed from a hopper through a metering device which supplies a constant amount of material for each moulding cycle. The hydraulic plunger forces

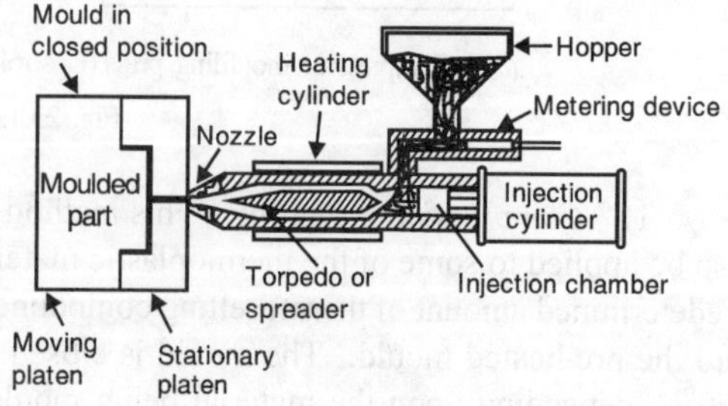

Fig. 23.3. Injection moulding machine (Plunger type).

the plastic material into the mould through the heating cylinder (where it melts), the nozzle and the sprue and runners. The temperature of heating cylinder ranges from 120°C to 250°C, depending upon the material and the mould size. After cooling, the mould is opened and the product is removed by the ejector system. The plastic that solidifies in the sprue and runners is removed from the product and reground to be fed into the moulding machine. The torpedo or spreader provides an increased surface-to-mass ratio by reducing the cross-section of the passages, thus improving the rate of heat transfer into the plastic.

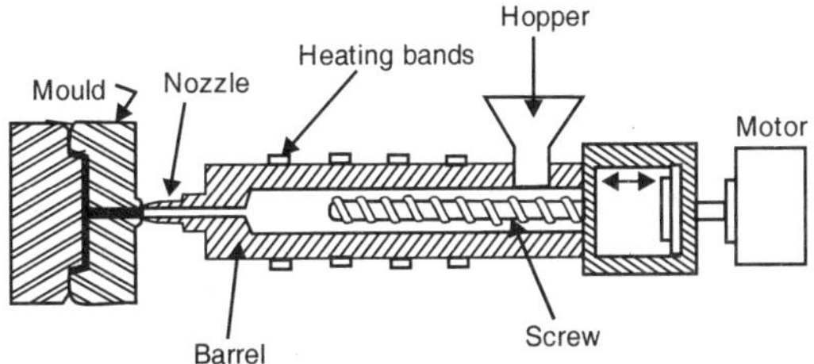

Fig. 23.4. Reciprocating screw injection moulding machine

Another type of injection moulding machine uses an extruder screw for melting the thermoplastic material, as shown in Fig. 23.4. The screw acts as a combination of injection and plasticising unit. It its operation, the powdered thermoplastic material is fed from a hopper to the rotating screw which discharges the material in front of the extruder barrel. The frictional forces developed by the rotating screw makes the plastic material into a semi-liquid state. The additional heat is supplied from the electric band heaters provided around the barrel surface. When the space in front of the screw becomes full with plasticised material, it forces the screw back tripping a limit-switch. This activates a hydraulic cylinder that forces the screw forward and injects the plasticised material into the mould, where it remains long enough to solidify. An anti-flow valve is provided to prevent any return of the plastic into the extruder barrel.

The injection moulding has the following advantages over compression moulding :

(*a*) It is a faster method than compression moulding.

(*b*) It is most economical method for mass production of single product.

(*c*) The articles of difficult shape and of thin walls are successfully produced.

(*d*) The metal inserts such as bearings, contacts or screws can be placed in the mould and cast integrally with the product.

(*e*) The material wastage is low as the runners and sprues can be ground up and reused.

4. *Jet moulding.* The thermosetting plastics may be moulded by a process known as jet moulding. It is similar to injection moulding. The standard injection moulding machines used for thermoplastic material, may be converted for jet moulding, with few changes. The nozzle, which is the important part of the machine, must be both heated and cooled during the moulding cycle. In jet moulding, the resin is first slightly heated in the cylinder and then allowed to pass through the nozzle

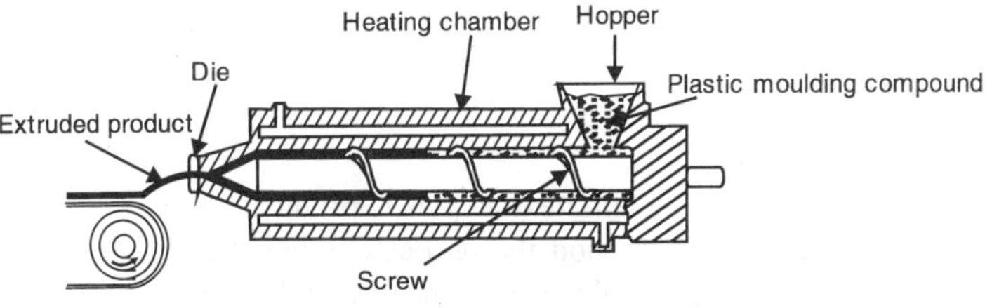

Fig. 23.5. Plastic extrusion machine.

(which is preheated to a high temperature) to the mould. When the mould is full, the nozzle is rapidly cooled by water to prevent polymerisation of the remaining material.

5. *Extrusion moulding.* The extrusion means the continuous flow of material through a die. The extrusion moulding is similar to injection moulding and is used for thermoplastic materials. This method is mostly used for making long tubes, rods, sheets, films, pipes, ropes etc. In its operation, the thermoplastic material in the powered or granular form is fed into a hopper and forced by a spiral screw into the heating chamber, as shown in Fig. 23.5. The material becomes soft in the heating chamber and the rotating screw forces this material out through a die. The finished product coming out of the die, is cooled by blow air or spraying water and it is taken away by the belt conveyor.

6. *Blow moulding.* This method is mostly employed for making thin walled hollow articles, especially bottles, from thermoplastic materials. In its operation, a cylinder of plastic material (known as *parison*) is extruded as rapidly as possible and cut to length. The hot parison is placed between the jaws of a split mould, as shown in Fig. 23.6.

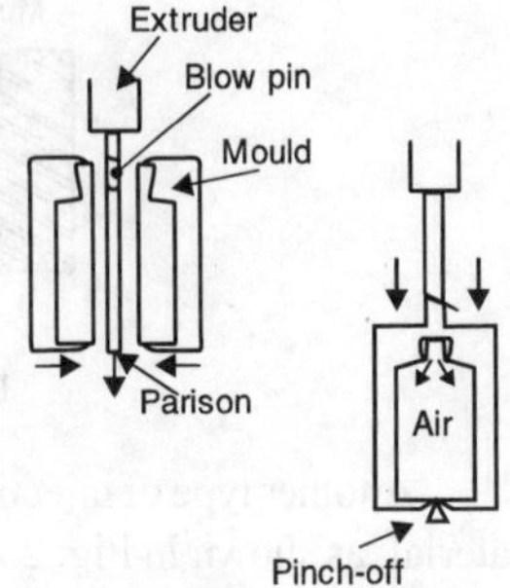

Fig. 23.6. Blow moulding.

When the mould is closed, it pinches off the parison and the product is completed by air pressure which forces the material against the mould surfaces. The mould should be adequately vented to eliminate poor surface finish. After the product is cooled sufficiently, the mould is opened and the product is removed.

23·11 Calendering

It is the process of producing film or thin sheets by squeezing a thermoplastic material between the revolving cylinders as shown in Fig. 23.7. In its operation, the material consisting of thermo-plastic resin, plasticiser, filler and colour pigments are compounded and heated before being fed into the calender. The thickness of the sheet produced depends upon the spacing between the cylinders that stretch the plastic. Before the film is wound, it passes through water cooled rolls. The products produced by calendering are vinyl, polyethylene, cellulose acetate films and sheets. The same process is used for rolling uncured rubber stock in the manufacture.

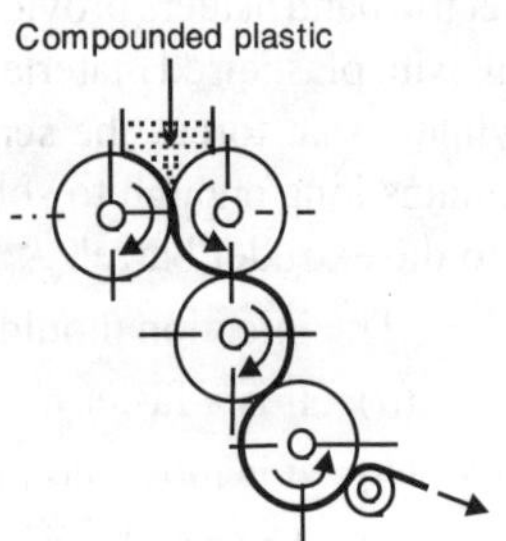

Fig. 23.7. Calendering.

23·12 Casting

It is the process of producing plastic products in the moulds without the application of pressure. This method is used only when the number of parts to be produced are limited. It is similar to metal casting process. In this method, the plastics are heated to a suitable temperature and the molten resin is poured into the open moulds which are usually made of lead and cured for several days. After due curing and hardening, the products are removed from the moulds.

The thermosetting materials which can be successfully casted are the phenolics, polyesters, epoxies, acrylics and silicones because they have greater fluidity in pouring than the thermoplastic resins.

23·13 Slush Casting

It is the process of producing thin flexible plastic products. It is similar to casting process in which no pressure is applied. In this method, the viscous solution of the thermoplastic resin is prepared and then poured into the preheated mould. Due to the heat of the mould, a uniform layer of resin sets all along the surface of the mould. The excess material, if any, is drained off. The material set in the mould is then cured by heating, followed by hardening of the product by chilling. The product is now removed from the mould. This process is used for making flexible toys and artificial flowers etc.

23·14 Laminating

It is the process of bonding together a variety of materials by heat and pressure to form a single piece (known as laminates). The most laminated plastics consist of sheets of wood, asbestos, fabric and other similar materials. These materials are first impregnated or coated with *thermosetting resin and then combined together under heat and pressure to form the commercial product. The laminated materials are hard, strong and impact resistant. They have good machining characteristics which permit its fabrication into gears, handles, bushings and furniture.

The following two precesses are commonly used for producing laminated products.

1. *High pressure process.* In this process, the resinoid material (*e.g.,* phenol formaldehyde) is first dissolved by a solvent (*e.g.,* alcohol) to convert it into a liquid varnish. The rolls of paper or fabric are passed through this liquid bath for impregnation, as shown in Fig. 23.8. It is then squeezed through a set of rollers and then dried in drying ovens at a temperature much lower (*i.e.,* 150°C) than the curing temperature of the resin, to evaporate the solvent. In order to facilitate lamination, the sheets are cut into convenient sizes and piled one over another (depending upon the desired thickness of the final sheet) between the polished metal plates at the top and bottom. These are then bonded together by a hydraulic press at about 180°C and under high pressure from 8 MPa to 14 MPa. Thus the laminated sheets are produced.

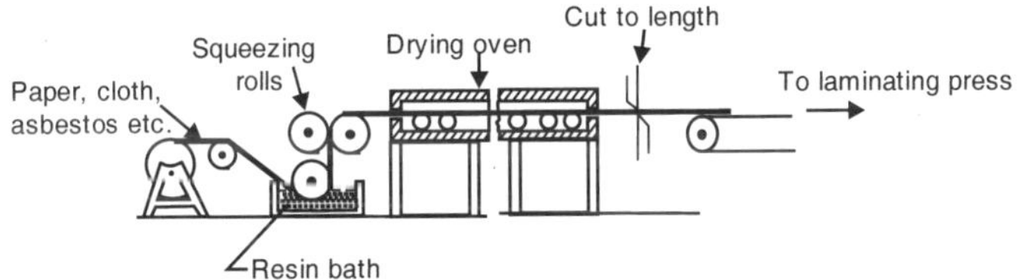

Fig. 23.8. Process for preparing sheets for lamination.

The tubes are made by rolling the impregnated material around a steel mandrel of a desired diameter and then placed in a tube mould for heating till they are cured.

The rods are produced by rolling and heating the impregnated material inside the cylindrical moulds, followed by grinding to size.

The paper base laminates are used in electrical products. The fabric base materials, which are stronger and harder, are better for stressed parts. The gears made of convass base are quiet in operation.

2. *Low pressure process.* It is also known as reinforced plastic moulding. In this process, a pressure upto 3 MPa is applied. The reinforcing materials used are glass fibres, asbestos, nylon, cotton, mat etc. The commonly used thermosetting resins are phenolics, furane, silicones, epoxies and polyesters. The pigments are added to obtain different colour shades. The fillers like clay (usually china clay), wood flour, silica sand, calcium carbonate and aluminium silicates are used to obtain better physical properties, better surface finish, reduced mould shrinkage and to lower cost.

Some of the processes used for low pressure laminates are as follows :

(*a*) *Contact moulding.* It is the simplest of all the processes. In this process, the layers of reinforcing material are coated with catalysed resin and placed in layers on the prepared form. The number of layers depends upon the strength required in the finished product. In order to speed up the production of parts, both the resin and the reinforcing material are sprayed by means of guns, instead of being coated by hand. The main advantages of this process are :

* The thermosetting resin mostly used for laminated plastics are phenolics, silicones, epoxies and melamines.

(*i*) It requires minimum equipment,

(*ii*) The mould cost is low,

(*iii*) There is no size limit of the product, and

(*iv*) The changes in design can be readily made.

(*b*) *Pressure bag moulding.* In this process, the lay-up of resin impregnated material is placed in the mould. It is covered by a bag made of rubber. A pressure plate is secured to the top of the mould. The air or steam pressure upto to 0.35 MPa is applied between the bag and the plate to inflate the bag against the lay up. The advantages of this process are :

(*i*) The cylindrical shapes can be easily made,

(*ii*) It provides dense and voidless mouldings,

(*iii*) The undercuts are possible,

(*iv*) The cores and inserts can be used, and

(*v*) It retains the advantages of contact moulding.

(*c*) *Vacuum bag moulding.* In this process, the lay-up of resin impregnated material is placed over the form mould. It is covered by a bag made of cellophane or polyvinyl film and secured to the mould. The vacuum is drawn through the ports provided in the mould. This results atmospheric pressure to act on the bag and forces the lay-up to conform to the mould. The advantage of this process are :

(*i*) It gives better surface finish,

(*ii*) There is less air entrapement and less voids, and

(*iii*) It retains the advantages of contact moulding.

(*d*) *Autoclave forming.* It is a modification of either the pressure method or vacuum bag method. The main difference is that after the lay-up has been made and covered with a plastic film, the entire assembly is placed in a steam or hot air autoclave at 0.35 to 0.7 MPa. The additional pressure permits higher glass content and improved air removal. The advantages of this process are:

(*i*) It provides a dense and void free mouldings,

(*ii*) The cores and inserts can be used,

(*iii*) The undercuts are possible, and

(*iv*) It retains the advantages of contact moulding.

(*e*) *Matched-die moulding.* It is a high speed process and is used for large production. The parts made by matched-die moulding have greater accuracy, high strength and good surface finish. The dies are usually made of steel for small parts and cast iron for larger parts. They may also be made from reinforced plastics or wood, for small production.

23·15 Machining of Plastics

The machining is essential when accuracy greater than can be obtained by moulding methods, is required or when features such as tapped holes which can not be obtained by moulding. Most plastic materials can be machined by using metal-working tools and machines. Since plastic materials have a low thermal conductivity and a high coefficient of expansion, therefore, the heat produced in cutting must be kept at an absolute minimum. In order to minimise this heat, it is advisable to use cutting fluid and to grind all tools with larger clearance angles than required for metal cutting. In machining plastic materials, light cuts and high cutting speeds with low feed rates should be used. Some of the operations used for plastics are as follows :

1. *Sawing.* In sawing plastic materials, coarse-tooth hacksaw blade should be used to prevent clogging the teeth. The brittle plastic materials such as acrylics are cut with fine-tooth saw to avoid splintering the edges.

2. *Turning.* The high speed steel cutting tools in a centre lathe can be used to turn plastic materials. A back rake angle of the tool should be about 20° for turning all thermoplastics. A 0° rake angle can be used on the more brittle plastics, while a 15° rake angle used on softer plastics aids the flow of material over the tool face. The cutting edges must be kept sharp. Since high and heavy cuts generate too much heat, and also cause the plastic to deform under tool forces, therefore, a cutting speed of about 150 m/min with feed rates between 0.1 and 0.25 mm/revolution should be used.

3. *Drilling.* In drilling plastics, standard high speed drills are used but they must be cleared frequently to remove swarf. The drills with 20° helix angle reduce the effect of swarf clogging in the flutes and a point angle of about 90° give better finish on break-through with the softer plastics. All the drills should have a clearance angle of 15° to 20°. A cutting speed of about 40 m/min with a feed rate of 0.1 mm per revolution should be used with drilling plastic materials.

4. *Reaming.* The helical flute reamers are used. The reamer should be sharp, otherwise the material tends to be pushed away rather than cut.

5. *Tapping.* The holes drilled or moulded in plastics can be tapped with high-speed steel ground-thread taps. In tapping softer plastic materials, there is a tendency for the material to be pushed away rather than cut. In order to prevent this, special taps about 0.05 mm to 0.13 mm oversize are used.

6. *Threading.* The threading can be carried out with a single-point tool in a centre lathe with the same angles as for turning. The high speed steel dies can also be used, but care must be taken to ensure that the thread is being cut and the material is not merely being pushed aside. It may be noted that the threads cut directly in plastics materials will not withstand high loads and will wear out if the screw fasteners are removed and replaced several times. When high strength and reliability are required, then the threaded inserts are used. These can be pushed into a predrilled hole.

23·16 Joining of Plastics

Following are the various methods of joining plastics :

1. *Hot gas welding.* In this method, the surfaces to be joined are clamped together. The hot air, from a special electrically heated torch or gun, at a temperature of about 300°C is directed at the surfaces and the filler rod (usually of the same material) is pressed into the joint as the area becomes tacky. The downward pressure of the filler rod makes the weld which fuses and solidifies as the welding proceeds, as shown in Fig. 23.9. It may be noted that more than one weld run may be necessary, depending upon the thickness of the material to be joined. The filler rods are generally circular in section and for small work, it is about 3 mm diameter.

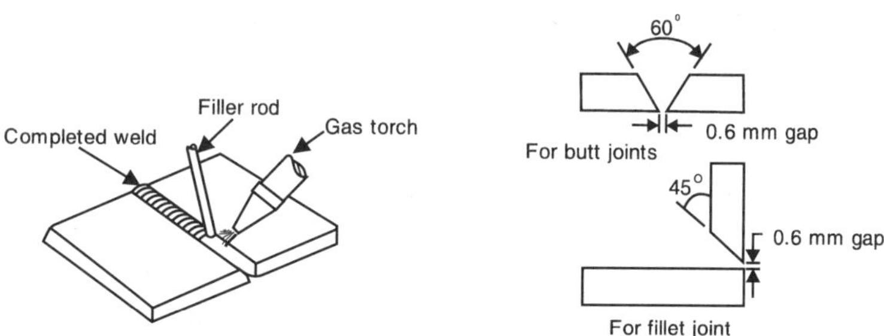

Fig. 23.9. Hot gas welding.

The surfaces to the welded have to be prepared to accept the filler rod. The edges are chamfered to an included angle of 60° for butt joint and for fillet joints, an angle of 45° is used. The surfaces to be welded must be clean and free from grease.

The hot gas welding method is used to weld PVC, polypropylene and polythene sheets to fabricate tanks, vessels, pipes for all types of fluids and ducting.

2. *Hot sealing.* In this method, a heated metal strip or bar at a temperature between 180°C and 230°C, is used. The plastic film or sheets are brought in contact with the heated strip until they are sufficiently soft. At this time, the plastic film or sheets are removed from the heat and held firmly until the melted plastic cools and makes a firm joint.

This method is used on nylon and with polyethylene sheet such as in the manufacture of polyethylene bags.

3. *Solvent welding.* In this method, the solvent is applied to the plastic surfaces to be joined which dissolve the edges and produce tacky surfaces. They are then brought together, moved slightly to obtain mixing and held in contact in correct position to dry and harden. The drying time ranges from a few minutes to hours.

The main uses of solvents are with polystyrene used in the manufacture of toys, display signs, ornaments and models.

The main disadvantage of solvent welding is the risk of some of the solvent reaching surfaces other than those being joined and leaving a mark. Care must be taken when using solvents as many are flammable and give off toxic vapours.

4. *Friction welding.* This method is limited to round plastic parts which can be rotated while in contact to generate heat. The frictional heat melts the interfaces and when the motion is stopped, the parts fuse together. A speed of 3 to 4.5 m/s and a pressure ranging from 0.14 to 1.4 MPa must be matched to the material to produce the right amount of heat.

5. *Ultrasonic welding.* In this method, high frequency sound waves (about 20 kHz) are used. It works best on those thermoplastic materials which have relatively low melting points and modulus of elasticity greater than 1380 MN/m^2. In this method, a sonic tip is placed at the centre or the part which transmits sound waves to the joint and converted into frictional heat. The operation takes 3 to 4 seconds.

QUESTIONS

1. What are the different types of plastics and their properties ?

2. What are the general properties of plastics that account for their use as engineering material ?

3. Describe the various types of thermosetting plastics.

4. Discuss the function of the following in plastics :

(*a*) Fillers; (*b*) Stabilisers,

(*c*) Plasticisers, and (*d*) Accelators.

5. Describe with neat sketches the following plastic processing methods, stating their advantages and application :

(*a*) Compression moulding, and (*b*) Injection moulding.

6. What are the advantages of injection moulding over the compression moulding process of plastics ?

7. What are laminated plastics ? Give their important properties and uses.

8. Explain the difference between high pressure and low pressure plastic laminating process. Describe any one of them.

9. Write short notes on machining and joining of plastics.

OBJECTIVE TYPE QUESTIONS

1. The thermosetting plastics are those plastics which
(*a*) are formed into shape under heat and pressure and results in a permanently hard product
(*b*) do not become hard with the application of heat and pressure and no chemical change occurs
(*c*) are flexible and can withstand considerable wear under suitable conditions
(*d*) are used as a friction lining for clutches and brakes

2. Which of the following are thermosetting resins ?
(*a*) Amino resins (*b*) Phenolic resins
(*c*) Cellulose derivatives (*d*) both (*a*) and (*b*)
(*e*) both (*b*) and (*c*)

3. Which of the following are thermoplastic resins ?
(*a*) Silicon resions (*b*) Polymides
(*c*) Cellulose derivatives (*d*) both (*a*) and (*b*)
(*e*) both (*b*) and (*c*)

4. Polyvinylchloride (PVC) is a ...... material.
(*a*) thermoplastic (*b*) thermosetting

5. The bottles from thermoplastic material are made by
(*a*) compression moulding (*b*) injection moulding
(*c*) blow moulding (*d*) extrusion moulding

6. The long plastic rods, tubes and sheets are made by
(*a*) compression moulding (*b*) injection moulding
(*c*) blow moulding (*d*) extrusion moulding

7. The process of producing film or thin sheets by squeezing a thermoplastic material between the revolving cylinders, is known as
(*a*) laminating (*b*) calendering
(*c*) contact moulding (*d*) pressure bag moulding

8. The process of producing plastic products in the moulds without the application of pressure, is known as
(*a*) casting (*b*) moulding (*c*) calendering (*d*) laminating

9. For turning all thermoplastics, the back rake angle of tool should be about
(*a*) 10° (*b*) 20° (*c*) 30° (*d*) 40°

10. Which of the following statement is wrong ?
(*a*) The apparent density of a powdered metal is the mass per unit volume of loose or unpacked powder.
(*b*) The catalysts are used to accelerate the chemical reaction during the process of polymerisation of plastics.
(*c*) The plasticisers are added to plastics to improve the flowability of the plastic in the mould.
(*d*) none of the above

ANSWERS

1. (*a*)	2. (*d*)	3. (*e*)	4. (*a*)	5. (*c*)
6. (*d*)	7. (*b*)	8. (*a*)	9. (*b*)	10. (*c*)

24

Metallic and Non-metallic Coatings

1. Introduction. 2. Cleaning of Metals. 3. Chemical Methods of Cleaning. 4. Mechanical Methods of Cleaning. 5. Ultrasonic Cleaning. 6. Metallic Coatings. 7. Hot Dipping. 8. Electroplating. 9. Metal Spraying. 10. Metal Cladding. 11. Cementation. 12. Anodizing. 13. Non-metallic Coatings. 14. Paints. 15. Varnishes. 16. Lacquers. 17. Enamels. 18. Plastic Coatings. 19. Chemical Coatings. 20. Vitreous Enamel Coatings. 21. Ceramic Coatings.

24·1 Introduction

When metals are exposed to the environment containing liquid and gases etc., the surface of the metal starts deteriorating. This type of deterioration or destruction may be due to the direct chemical attack or electro-chemical attack, and is known as *corrosion*. There are many examples of corrosion in day-to-day life, but the two most familiar examples are the rusting of iron and the formation of green film on the surface of copper. The rusting of iron takes place, when the iron is exposed to atmospheric conditions. During this exposure, a layer of reddish scale and powder of oxide is formed and the iron becomes weak. The formation of green film on the surface of copper takes place when it is exposed to moist air containing carbon dioxide.

Thus, it is very necessary to prevent or reduce the rate of corrosion. For this, a protective coating is applied on the base metal. The protective coatings may be broadly classified into the following two categories :

1. Metallic coating, and 2. Non-metallic (organic or inorganic) coating.

These coatings are also done for better appearance and decorative purposes.

Notes : 1. The type of corrosion which occurs due to the chemical reaction between the surface of the material and the gases or vapours, is called *dry corrosion*. The gases which take part in corrosive attack are oxygen, nitrogen, sulphur dioxide, hydrogen sulphide, halogens and vapours at high temperatures.

2. The type of corrosion that takes places as result of reaction (called electrochemical reaction) of metal with water or aqueous solution of salts, acids or bases, is known as *wet corrosion*. These aqueous solutions are capable of conducting electric current and they are called electrolytes.

24·2 Cleaning of Metals

Before metallic or non-metallic coating is applied, the first step is to clean the surface of the base metal, in order to provide good adhesion and finish. There are many types of contaminants such as dirt, oil, grease, oxide, scale etc. which ultimately affect the life of the product. The selection of a cleaning process mainly depends upon the type of contaminants, degree of cleanliness required and the composition of the metal. The cleaning may be done by chemical methods or by mechanical methods as discussed in the following pages.

24·3 Chemical Methods of Cleaning

The cleaning of metals by chemical methods may be done by alkalines, acids and solvents. Some of the chemical cleaning methods are described below :

1. *Alkaline cleaning*. The alkaline cleaning is most efficient and economical in removing oil and grease by saponification or emulsification or both. The alkaline cleaners are prepared from cleaning agents such as caustic soda or sodium metasilicate which are added to some type of soap to

aid in emulsification. The common methods of applying alkaline cleaners include electro cleaning, spray cleaning, dip-tank cleaning and steam cleaning. After cleaning, the parts must be thoroughly rinsed with water, otherwise, any trace of alkali, left on the surface of metal, may corrode it.

The alkaline cleaners are used on all metals except zinc, lead, tin, brass and aluminium as these metals are chemically active and may be attacked by alkaline cleaners.

2. *Acid pickling.* The acid pickling is the most common method of removing oxide scale or flux residue from the surface of metals. Before this process is done, alkaline cleaning of part is done to remove dirt, greases and oils in order to obtain an even removal of oxides during the pickling process. In this method, either diluted sulphuric acid, hydrochloric acid, phosphoric acid or muriatic acid is sprayed on the part or the part is dipped in a tank, agitated and then washed and rinsed thoroughly. It may be noted that sulphuric acid and the muriatic acid solutions can be used while hot, but the hydrochloric acid solution is used at room temperature.

The acid pickling is used for removing oil and grease. In some cases, it is used for removing light rust.

Notes : (*a*) To decrease the action of acid upon the metal surface particularly aluminium, inhibitors such as ether, liquid, glycol or detergents are added to the pickling solution.

(*b*) Generally, after the pickling process, parts are dipped in weak alkaline solution to neutralize the acid, left on the surface of the parts, and then rinsed.

3. *Solvent cleaning.* The solvent cleaning is primarily used to remove oil, grease and dirt on metals such as zinc, lead and aluminium. It may be performed by either emulsified solvent or chlorinated solvent. In *emulsified solvent cleaning process*, an organic solvent is mixed with a hydrocarbon-soluble emulsifying agent (*i.e.*, soap and kerosene mixture with a small amount of water). The cleaning is done by dipping the part in the solution and then rinsed. To remove heavy soil or dirt, the part is stirred in a tank or sprayed with hot solution, and then rinsed. The parts which needs to be electroplated afterwards must be treated with alkaline cleaner to remove any organic matter left on the surface .

The *chlorinated solvent cleaning* (also known as vapour degreasing) is an effective and widely used method of removing heavy oils and greases. This method provides a high degree of surface cleanliness. In this method, either a solution of trichloroethylene (boiling point 85°C), or perchloroethylene (boiling point 120°C) is heated in a tank to its boiling point and the vapours are produced. The parts placed in a wire basket are suspended in these vapours. This causes the vapours to condense on the metal surfaces and removes all the oil and grease. The disadvantage of this method is that the vapours do not remove the solid dirt and the organic matter, hence vapour degreasing is followed by alkaline cleaning.

24·4 Mechanical Methods of Cleaning

The cleaning of metals by mechanical methods may be performed as discussed below :

1. *Abrasive blast cleaning.* The abrasive blast cleaning is widely used for removing all classes of scale and rust from forgings, castings, weldments and heat-treated parts. Cleaning is done either in dry or wet abrasive blasting machines. The abrasive blast cleaning equipment consists of cleaning cabinet, an abrasive reserviour and an air nozzle. The abrasive particles such as sand, steel grit or shot are drawn through the hose to join the high velocity air stream at the nozzle and strike against the surface of the metal to be cleaned. The used abrasives slides down the centrally sloped bottom of the cabinet and return to the reservoir.

In case of wet abrasive blasting machines, the wet abrasive slurry is fed into the blasting gun by a centrifugal pump, which forces it to exit through the nozzle along with compressed air. Silica, soft silicon dioxide, quartz or aluminium oxide are used as wet abrasives.

Note : Sometimes, cleaning is performed by means of a high velocity air blast, which is directed by hand.

2. *Tumbling.* The tumbling in dry abrasive is used for removing rust and scale from small metal parts of simple shape. In this method, the workpieces are placed in a drum along with abrasive materials. The abrasive materials can be granite chips, aluminium oxide pellets or sand. In operation, the drum is rotated and the movement of the workpieces and the abrasive materials against each other produces by friction a fine cutting action, which removes rust and scale from the product.

Tumbling is the least expensive cleaning method. However, parts with complex shape, deep recess and other irregularities cannot be descaled uniformly by tumbling. It may require several hours of tumbling to complete the cleaning, but if slugs of various sizes and shapes are mixed, good results can be achieved. Slug reaches all corners and cavities and clean it.

It may be noted that by this process cleaning is not possible. The entire surface of the part is subjected to cleaning during this process.

3. *Barrel rolling.* Barrel rolling and tumbling are principally the same processes with the only difference that in former process barrel is loaded about 40 to 60 percent capacity, while in tumbling it is nearly full.

The abrasives used in this are slag, cinders, sharp sand, granite chips, silicon carbide or aluminium oxide along with water, dilute acid or soap solution. Sometimes, scrap punching or mineral matter is also added for wet rolling. For dry rolling, hard wood dust or leather scrap is added to keep workpieces separate from each other.

In operation, as the barrel rotates, the parts roll over and falls to the bottom of the barrel. This motion cuts down the surface of parts.

4. *Power brushing.* Power brushing is used to remove light rust, heat treatment scale, weld flux, machining burns and other unwanted contaminants. One of the major advantage of power brush is that it can be used in manual, semi-automatic or integrated methods.

In manual brushing, the part is fed against a rotating wheel brush by hand whereas in automatic machines, parts move past a series of rotating brushes, removing very little metal. When the brushes are used with fine abrasive compounds, they can produce fine finish on metal surfaces.

The selection of the brush is based upon the surface to be cleaned. Wire wheel brushes made of steel, brass and stainless steel are used to remove contaminants from nickel alloys, brass alloys, copper, bronze, brass, steel and stainless steel parts.

5. *Buffing and polishing.* The basic purpose of buffing and polishing operation is to obtain highly reflective and polished surfaces. However, buffing produces a relatively more reflective, smooth, uniform surface with a high brilliant lustre as compared to polishing. The buffing wheels are made from very soft materials like felt, muslin or linen and are made in various forms like radial, parallel, concentric, square etc. The abrasive used for buffing is a very fine powder of crushed flint, quartz, silicon carbide, aluminium oxide and iron oxide (red-rouge). To reduce abrading or roughing action, the abrasive is blended with some lubricant like grease, wax, glue, oil or tallon, to give a flowing action.

The polishing process involves the use of dry abrasive in oil or some other lubricant. Generally, the polishing wheels are made from canvas, muslin, felt and leather. As a result of polishing operation, the surface layer of the metal is locally plastically deformed, due to which the surface inequalities are removed, scratches are finished and compressive residual stresses are induced in the metal. Consequently the fatigue resistance of the parts is increased in addition to obtaining a highly reflective surface.

In both buffing and polishing operations, workpiece is held in fixtures, on a rotating circular work table. The table moves past a series of adjustable polishing and buffing wheels. The wheels are adjusted, according to the shape of the workpiece, to polish all its surfaces. Sometimes the polishing and buffing wheels are mounted on floor polishing lathes for ordinary finish.

24·5 Ultrasonic Cleaning

The ultrasonic cleaning is done to remove insoluble particles left on the surface of work from grinding, polishing, lapping, honing, buffing and drawing, at some inaccessible cavities such as indentations, slots, blind holes, internal threads, bores and some precision parts such as carburetors, fuel injector parts, hydraulic needle valves etc. This method is an expensive one because of the high cost of equipment and higher maintenance cost, so its use is restricted to applications in which other methods have proved inadequate or unsuitable.

In this method, penetrating power of the solvent or aqueous cleaner is accelerated by high frequency sound wave energy. For this, an ultrasonic generator which produces high frequency electrical energy (generally 20 kHz, but in certain extreme cases as high as 100 kHz), a transducer which changes the electrical energy into high frequency sound waves and a tank containing liquid cleaner are used. The principal of ultrasonic cleaning is shown in Fig. 24.1

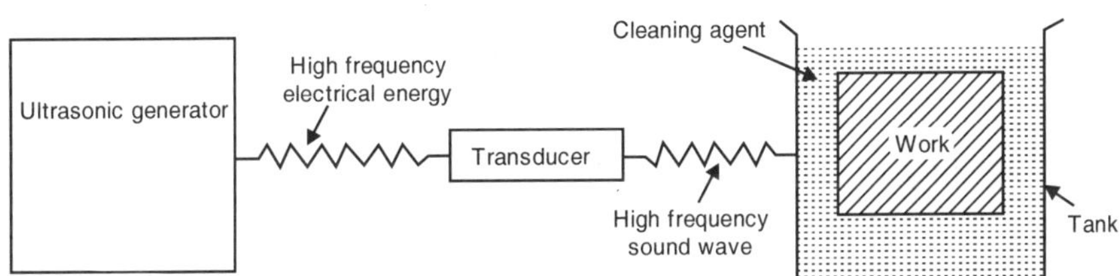

Fig. 24.1. Principle of ultrasonic cleaning.

The parts to be cleaned are placed in the tank and the ultrasonic energy alternately compresses and expands the solution and produces several concurrent effects. The dirt and grease are ripped apart by ultrasonic scrubbing action, which are emulsified. The cleaning agents may be detergent, acids or solvents like cyclohexane and trichloroethylene.

24·6 Metallic Coatings

These are the coatings of metals, which are applied on the base metal to prevent corrosion as well as for decorative purposes. The metallic coatings may be obtained by the following methods :

1. Hot dipping, 2. Electroplating, 3. Metal spraying, 4. Metal cladding, and 5. Cementation.

The methods are discussed, in detail, in the following pages.

24·7 Hot Dipping

It is mainly used to apply coatings of low melting metals such as zinc, tin, cadmium, aluminium or lead on iron, steel and copper which have relatively higher melting temperatures. In this method, the article is cleaned and dipped in a bath of molten metal. It is then taken out from the bath and finished properly. The most common processes of hot dipping are galvanising and tin plating as discussed below :

1. *Galvanising.* It is a process of providing thin layer of zinc coating (highly corrosion resistant) on iron or low carbon steel by dipping them in a bath of molten zinc. It improves the resistance against corrosion due to atmosphere and water. Since zinc is anodic to iron or steel (which serves as cathodic), therefore the zinc coating (also called anodic coating) is obtained by electrochemical reaction between the zinc and the iron or steel. The process is carried out as follows :

First of all, the work is cleaned by pickling with dilute sulphuric acid, to remove any scale, rust and impurities. After washing and drying, it is dipped in a bath of molten zinc maintained at a temperature of about 450°C. The surface of the bath is kept covered with a molten flux layer usually of zinc ammonium chloride to prevent oxide formation. The duration of immersion in a zinc bath depends upon the size of the work to be coated. It is then taken out from the bath and passed through a pair of rollers to remove any excess of zinc. Finally, the work is cooled and dried.

The galvanising process is widely used for sheetings for roofs and walls of building, structural parts, pipes, containers, fencing materials, nails, wires etc.

Note. Since zinc gets dissolved in dilute acids to form highly toxic compound, therefore, the galvanised utensils should not be used for preparing and storing foodstuffs especially acidic ones.

2. *Tinplating or Tinning*. It is a process of providing a thin layer of tin coating on the iron or steel sheets. In this process, the metal sheet is first treated with dilute sulphuric acid (pickling) to remove any oxide film. It is then fluxed by passing through a bath of zinc chloride. The flux helps the molten tin to adhere to the metal sheet. Now the sheet is dipped into a tank of molten tin which is maintained at a temperature of approximately 320ºC. This process gives a coating thickness of about 0.003 mm. Finally, the sheet is passed through a series of rollers to remove any excess of tin and produce a thin layer of uniform thickness on the sheet.

The tin plating or tinning is generally done on metal sheets used for manufacturing containers for storing foodstuffs, ghee, oils, kerosene and packing food materials. The tinned-copper sheets are employed for making cooking utensils and refrigeration equipments.

24·8 Electroplating

It is the process of depositing a very thin layer of metal coating, on the base metal, by passing a direct current through an electrolyte solution containing some salt of the coating metal. Now-a-days, the electroplating (sometimes known as electrodepositing) is one of the best methods for the commercial production of a metallic coating.

In this process, the article to be electroplated is first immersed in hot alkaline solution to remove oil and grease. It is then treated with dilute hydrochloric acid or sulphuric acid to remove scale and oxides. It may be noted that hydrochloric acid is used for nickel and copper plating, while sulphuric acid is used for chromium plating. The cleaned article is made to act as cathode whereas the coating metal as an anode. The cathode and anode are dipped in a solution containing some salt of the coating material *i.e.*, electrolyte, as shown in Fig 24.2. When direct current is passed, the coating metal ions migrate to the cathode and get deposited there. Thus a thin layer of coating metal is obtained on the article, which is made as cathode. The thickness of coating can be 0.005 mm or more. The commonly used metals, which are used for electroplating, are copper, nickel, chromium, silver, gold, cadmium and tungsten.

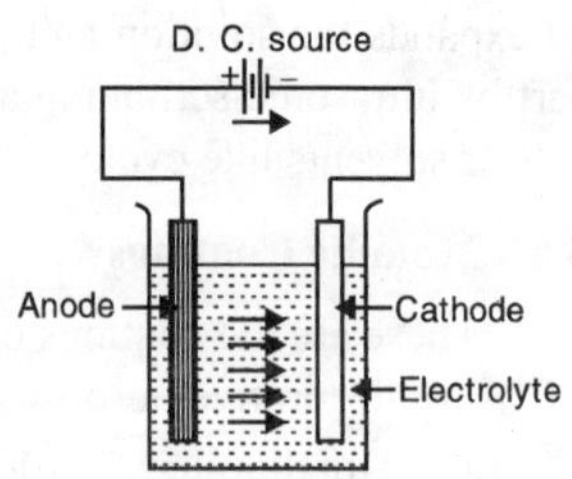

Fig. 24.2. Electroplating.

The electroplating is extensively used for decorative purposes and for protecting iron and steel articles from corrosion.

24·9 Metal Spraying

It is a process of providing a thin coating by depositing an atomized metal on the surface of the base metal. In this process, the coating metal in the molten state is sprayed on the prepared surface of the base metal. Following are the two common methods of metal spraying :

1. *Wire gun method*. In this method, the coating metal in the wire form is sprayed from a gun torch, as shown in Fig. 24.3.

The wire is drawn through the gun and nozzle by a pair of rollers and melted by an oxy-acetylene flame. A blast of compressed air is used to atomize the molten metal and deposit it uniformly on the prepared surface. Though any metal that can be drawn into wire, can be used in this gun but the mostly used are alloys such as steel, bronze, aluminium and nickel.

2. *Powder metal method.* This method is also known as thermospray method. In this method, the coating metal in the finely divided powdered form is fed from a container through a rubber hose to the spray gun. The powdered metal gets melted almost instantly as it passes through the oxy-acetylene or hydrogen gas flame. In this, no compressed air is required as there is sufficient force of the gas flame to carry the atomized metal at high speed to the surface being sprayed. This method is, however, limited to low melting metals like zinc, lead, tin etc.

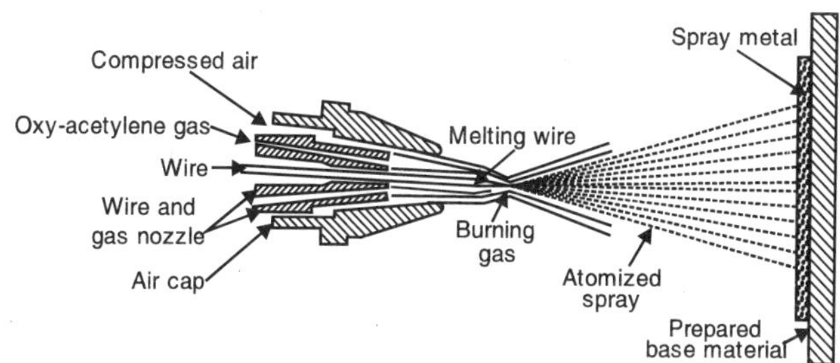

Fig. 24.3. Metal spraying.

The metal spraying method has the following advantages over other coating methods :

1. The coating can be applied to large and irregularly shaped objects.
2. The coating can be applied to fabricated structure and there is no possibility of damage to the coating.
3. The coating can be applied even on to non-metallic objects made of glass, wood, plastic etc.

24·10 Metal Cladding

It is a process of providing a comparatively thicker layer of coating on the metal surface. The main object of cladding is to produce a corrosion resistant surface. All the corrosion resisting metals (like nickel, copper, lead, silver etc.) and alloys (like stainless steel, nickel alloys, copper alloys, lead alloys etc.) can be used as cladding material. The base metal on which cladding is done are mild steel, aluminium, copper, nickel and their alloys.

The cladding is, usually, done by hot rolling. In this method, thin sheets of the coating metal and the base metal are arranged in the form of a sandwitch. These are then passed through rollers under heat and pressure. This method is widely used in the aircraft industry of alclad sheeting, in which a plate of duralumin is sandwitched between the two layers of pure aluminium. It is also used to make bi-metal strips for temperature controlled devices.

24·11 Cementation

It is a process of providing thin layer of powdered metal coating by heating the base metal. The temperature to which the base metal is heated, is always more than the melting temperature of the coating metal. The most common processes of cementation are as follows :

1. *Sheradising.* It is a process of providing thin layer of zinc on steel parts. In this process, the articles to be coated are first cleaned and then packed with zinc dust in a drum. The drum is then closed tightly and rotated slowly for two to three hours. The drum is kept at a temperature of about 350°C either by gas heating or electricity. At the end of the process, the surface of the articles are covered with a thin layer of zinc.

The sheradising improves the resistance against corrosion. It is used especially for small steel articles such as bolts, screws, nuts, washers, etc.

2. *Chromizing*. It is a process of providing thin layer of chromium on steel parts. This process is carried out by heating chromium powder (55%) and alumina (45%) with the steel parts at about 1300 – 1400ºC for 3 to 4 hours. The alumina is used to prevent coalescence of chromium particles.

The chromizing greatly improves the resistance against oxidation. It is extensively applied for turbine blades.

3. *Calorizing*. It is process of providing thin layer of aluminium on steel parts. In this process, the steel parts are first cleaned by sand blasting and then heated in a tightly packed drum with a mixture of aluminium powder and aluminium oxide together with a small quantity of ammonium chloride as a flux. The process is carried out in a reducing atmosphere of hydrogen at 900º for 24 hrs. Upon heating, the aluminium diffuses deeper into the metal and the coating becomes thicker but it may be noted that because of diffusion the average aluminium content is lower.

The calorizing improves the resistance to oxidation at high temperatures. It is very suitable for treating steel parts for furnaces, oil refineries, driers and kilns etc.

Note. Calorised coatings are quite resistant to attack by sulphurous gases. Such materials are used in oil refining industries. Other applications include furnace parts, carburising and heat treatment boxes and pyrometric equipment where oxidation occurs at high rate.

24·12 Anodizing

It is a process by which an oxide film is deposited (created) on a metal surface, for providing corrosion resistant and decorative finish. Generally, aluminium and its allots are anodized since aluminium has a natural tendency to get oxidised when exposed to atmosphere but anodizing can also be done on metals like lead, magnesium, copper, zinc, bronze etc.

An electrolytic circuit is used in the process, in which the workpiece is made an anode. The electrolyte which is generally aqueous solution of sulphuric acid, oxalic acid and chromic acid which give up oxygen on electrolysis. Lead or graphite is made as cathode. When a direct current (D.C.) is passed through the electrolyte, oxygen is liberated at the surface of the anode, which impinges on the workpiece, forming a porous oxide film, giving it cellular structure appearance. It may be noted that to close the pores, the oxide film is further subjected to a process called *sealing*, in order to make it permanent and durable. This can be done by treating the oxide film with boiling water-steam bath (85ºC to 100ºC) or dilute nickel acetate solution. The coating produced by this method have greyish green colour. Other colours can also be obtained by using a suitable dye material.

The thickness and density of the oxide layer depends on the anodizing conditions and on the conditions of the metal being anodized. The higher the proportion of acid in the solution, the thicker will be the anodized film but the film be porous and soft. The film thickness also increases with current density, the anodizing time (upto a certain limiting value), and with the lowering of the temperature of the solution. Also, an alternating current produces a thin anodizing layer whereas a direct current produce a coating, having deeper penetration.

The anodized metal can be directly used for drawing and forming operations, without destroying the coating or its protective qualities, as the coating is an integral part of the metal. Air craft materials, gift wares, home appliances, hardware, aluminium door frames, channels, etc. are anodized by this process.

24·13 Non-metallic Coatings

These are coatings of non-metals, which are applied on the base metal. Following are the various types of non-metallic coatings :

1. Paints, 2. Varnishes, 3. Lacquers, 4. Enamels, 5. Plastic coatings, 6. Chemical coatings, 7. Vitreous enamel coatings, and 8. Ceramic coatings.

These are discussed, in detail, in the following pages.

24·14 Paints

The paints are fluids designed to cover the materials (like steel plates, resin etc.) to be coated with coating film which protects them and improves their appearance. In general, following are the purposes of applying a paint :

1. *Protection of the object.* The paints are rust proof, corrosion proof and have good resistance against oil, weather and shock.

2. *Appearance of the object.* The paints have good colour, high gloss, smoothness and cubic effect.

3. *Others.* The paints have good sound proof properties.

The paint commonly used for engineering purposes is an oil paint. It is a fluid paste prepared by dissolving a base into a vehicle along with a colouring pigment. The bases used in oil paints are white lead, zinc white, red lead etc. and the vehicles are linseed oil, poppy oil, nut oil, tung oil etc. The colouring pigments include black, blue, brown etc. The other ingredients of an oil paint are solvent or thinner, drier and inert filler. The solvent or thinner (turpentine oil, petroleum spirit etc.) is used to modify the consistency of the paint. The drier (litharge, lead acetate, manganese sulphate etc.) enables the paint to dry quickly. The inert filler (powdered chalk, charcoal, silica etc.) are used to make the paint economical and of desired quality.

Before applying a coat of paint, the primer (lead oxide) has to be applied first to the surface, in order to secure proper cohesion of the paint. When a thin coating is applied on any surface, the thinner evaporates from the painted surface and the drying oil gets oxidised, thus forming a dry thin pigmented film.

A large number of ready made paints are available in the market under various trade marks. These paints are usually too thick to be applied in a thin uniform layer. They are, therefore, made thin by adding a suitable quantity of oil while using them The paint may be applied either by a brush, spray gun or dipping. The spraying with a spray gun is rapid and gives a smooth painted surface.

24·15 Varnishes

The varnish is a homogeneous mixture of natural or synthetic resin in a particular solvent. The commonly used resins are copal, amber, lac or shellac, dammer etc. and the solvents are linseed oil, turpentine oil, methylated spirit or alcohol. A drier (litharge) is also added to help in quick drying of varnish.

The varnish is used as a protective and decorative coating of suitable surfaces. When it dries, it leaves a hard, transparent, glossy, lustrous and durable film. The following are the main two types of varnishes :

1. Oil varnishes, and 2. Spirit varnishes.

The *oil varnish* is a homgeneous solution of one or more resins in a drying oil (linseed oil) and a volatile solvent (turpentine oil) and drier. This type of varnish takes longer period for drying, but the film produced is hard, quite lustrous and durable.

The *spirit varnish* is obtained by dissolving the resin (lac or shallac) in a volatile solvent (methylated spirit). This type of varnish dries quite rapidly, but the film produced is brittle and thus has the tendancy to crack or peel off. Such varnishes are used, usually, for polishing wooden surfaces.

24·16 Lacquers

The lacquer is a solution made by dissolving nitrocellulose, resin and plasticizer in a solvent with or without the colouring pigments. In a lacquer, nitrocellulose provides toughness and resistance to abrasion ; the resins (such as alkyd, copal, dammer, ester, gum etc.) increase adhesion and hardness and the plasticizer (castor oil) is added to improve elasticity and plasticity. The solvent is usually a mixture of ketone, alcohol and a hydrocarbon.

Since the lacquer has quick drying property due to the evaporation of solvent, therefore it is widely used for giving a finish to automobile bodies. It is also used for preparing artificial leather and for painting of interior wood work, metal surfaces, furniture etc. for decoration purposes.

24·17 Enamels

The enamel is a pigmented varnish which dry into a very lustrous and glossy finish, having very good flow. In other words, it is an intimate dispersion of pigment in a varnish. The properties of an enamel depends largely upon the varnish's vehicle and resin. The following two types of enamels are available, depending upon the method of drying used :

1. Air-drying enamel, and 2. Baking enamel.

When enamel dries at room temperature by oxidation or polymerization or by both, it is then known as air-drying enamel.

When enamel dries at elevated temperature either in the presence or absence of oxygen, it is then known as baking enamel. It provides a hard and more abrasive resistant finish than air drying enamel.

It may be noted that the drying process mainly depends upon the purpose for which the product is finally used. The enamels are available in all colours and can be applied easily. They have the ability to resist corrosive atmospheres and attack of most chemical agents. These are mostly used in metal processing industry.

24·18 Plastic Coatings

The plastic coatings have good resistance to corrosion, good adhesion ability, high resistance to abrasion and attractive appearance. The plastic coated metals are used in chemical plant equipments, tank, pipe lines, valve bodies, water treatment plants etc. Following plastics are generally used for coating purposes :

1. Plasticized polyvinyl chloride (P.V.C) ; 2. Nylon ; 3. Penton ; 4. Cellulose acetate butyrate (C.A.B) ; 5. Polyethylene (Polythene) ; 6. Polytetra fluro ethylene (P.T.F.E.) ; 7. Polytrifluro monochloro ethylene (P.T.F.C.E.) ; 8. Polyester/Polymethane ; 9. Polypropylene ; and 10. Acrylics.

The various methods of coating the plastics on metals are discussed below :

1. *Fine powder dipping*. In this method, the hot parts are dipped into a gas agitated bed of fine powder. The part is further heated to fuse the coated particles so that a smooth film is formed on the part. By this method, polyethylene, P.V.C., nylon or C.A.B. are coated on various metals.

2. *Liquid plastisol method* (*Dip coating*). In this method, the hot parts are dipped in a liquid plastisol tank. The plastisol will form a gelled coating on the part. The coating thickness upto 12 mm can be achieved by this method. Generally electrical parts, tool handles, automotive bumpers, plate and dish racks are coated by this process.

3. *Electrostatic spraying*. The electrostatic spray method consists of a powder reservoir or hopper, powder feed mechanism, a spraying gun and a generator. The powder is fed to the gun by means of an air pump. The spray gun is applied with high voltage source so that the particles impelled at high velocity gets strongly charged. The part which is to be coated is earthed, therefore charged particles are attracted toward the part thus forming a layer on the part.

4. *Vacuum coating*. In this method, vacuum is used to attract the powder on the surface of a preheated part. No air is entrapped in coating as the process is done in absence of air. Generally nylon and penton are coated by this method.

24·19 Chemical Coatings

In this process, the metal to be coated is cleaned and dipped in a chemical solution in the same way as the plastic coating. The chemical solution is, generally, prepared by dissolving a metal having higher resistance against corrosion in some suitable chemical. The chemical solution thus prepared reacts with the metal surface and produces an adherent coating of the metal compound. The following two types of the chemical coatings are important :

1. *Chromatic coatings*. In this process, the metal to be coated is dipped in the solution of sodium dichromate and concentrated nitric acid. The chromatic coated surface provides adequate resistance to corrosion. It is generally provided for protection of zinc, aluminium, magnesium parts etc.

2. *Phosphate coatings*. In this process, the metal to be coated is dipped in a hot solution of manganese dihydrogen phosphate at 90°C for about 45 minutes. The phosphate coated surface provides adequate resistance to corrosion. It is generally provided for the protection of iron, steel, zinc, aluminium, tin parts etc.

The phosphate coating has an excellent property *i.e.*, it has the ability to retain paint and provide lubrication when necessary. It is usually done before painting the surface to give an organic finish and to provide corrosion resistance to the surfaces of refrigerators, washing machines, automobile parts, outdoor electrical fittings etc.

24·20 Vitreous Enamel Coatings

The vitreous enamels are glass-like materials having wide range of composition. The constituents of enamel may be borax, alumina, iron oxide, cobalt oxide, soda titanium, salt petre, quartz, tin oxide, felspar, fluospar, magnesia and other substances which are required to give colour or other physical properties. All these constituents are smelted together at 1200°C and are grounded in the form of powder with water to form a fine slurry. The slurry is then applied on cleaned objects by dipping or spraying. It is then dried and heated in a furnace to a temperature around 850°C.

The vitreous enamel coatings have good corrosion and chemical (except alkalis and hydrofluoric acid) resistance. They may be applied to both ferrous ad non-ferrous surfaces.

24·21 Ceramic Coatings

The ceramic coating is similar in character to vitreous enamels, but it has high refractoriness. In this process, the coating material (high refractory oxides such as aluminium oxide, chromium oxide, cobalt oxide or beryllium oxide) is applied to the metal by dipping or spraying.

The ceramic coating protects the metal against oxidation and corrosion at high temperatures. It is used in the making of nozzles, vanes and blades of gas turbines, air-crafts, rocket propulsion systems, thermo-couple tubes, cutting tools etc.

QUESTIONS

1. What do you understand by corrosion ? Name the different methods used to prevent it.

2. Discuss the various methods used for cleaning of metals.

3. What method is used for

(*a*) removing oil and grease from the metal surfaces, and

(*b*) removing oxides, dirt and scale from the metal surfaces.

4. Explain the following metallic coatings :

(*a*) Galvanizing, and (*b*) Electroplating.

5. What are the main ingredients of an oil paint ? Explain their functions.

6. Write short notes on :

(*a*) Varnishes ; (*b*) Lacquers, and (*c*) Enamels.

7. Why plastic and chemical coatings are applied on metals ?

8. What is the difference between vitreous enamel coating and ceramic coating ?

OBJECTIVE TYPE QUESTIONS

1. The process of providing a thin layer of zinc coating on iron and steel is called
(*a*) electroplating (*b*) galvanising (*c*) metal cladding (*d*) cementation

2. The process of providing a thin layer of phosphate on steel is called
(*a*) perkerizing (*b*) terne coating (*c*) calorising (*d*) anodizing

3. In order to clean the surface of robust parts, the method preferred is
(*a*) tumbling (*b*) abrasive blasting
(*c*) chemical cleaning (*d*) none of these

4. In electroplating
(*a*) the base metal acts as cathode and the coating metal as anode
(*b*) the cathode and anode are dipped in a solution containing some salt of the coating material, *i.e.*, electrolyte
(*c*) a direct current is passed through the electrolyte
(*d*) all of the above

5. Which of the following statement is correct ?
(*a*) The process of providing a thin layer of tin on iron or steel sheets is known as electroplating.
(*b*) Galvanised utensils are used for preparing food.
(*c*) The process of providing a comparatively thicker layer of coating on metal surface is called metal cladding.
(*e*) all of the above
(*f*) none of the above

6. Calorizing is a process of providing thin layer of
(*a*) aluminium (*b*) chromium (*c*) zinc
(*d*) both (*a*) and (*b*) (*e*) both (*b*) and (*c*)

7. In an oil point, turpentine oil is used as a
(*a*) base (*b*) vehicle (*c*) pigment (*d*) solvent

8. The constituent which provides a body to an oil paint is called
(*a*) base (*b*) vehicle (*c*) pigment (*d*) solvent

9. The solution made by dissolving nitrocellulose, resin and plasticizer in a solvent, is called
(*a*) paint (*b*) enamel (*c*) lacquer (*d*) varnish

10. Which of the following protective coating is applied to protect the metal against oxidation and corrosion at high temperatures ?
(*a*) paint (*b*) ceramic coating
(*c*) vitreous enamel coating (*d*) enamel

ANSWERS

1. (*b*)	2. (*a*)	3. (*a*)	4. (*d*)	5. (*c*)
6. (*a*)	7. (*d*)	8. (*a*)	9. (*c*)	10. (*b*)

25

Pipes and Pipe Fittings

1. Introduction. 2. Classification of Pipes. 3. Indian Standard Specification for Pipes. 4. Pipe Fittings. 5. Taps and Valves. 6. Symbols for Pipe Fittings and Valves. 7. Pipe Vice. 8. Pipe Wrenches. 9. Pipe Cutting. 10. Pipe Threads and Thread Cutting. 11. Bending of Pipes. 12. Pipe Joints.

25·1 Introduction

The pipes are hollow circular cylinders and are used for transporting various fluids such as water, steam, gases, oils and other chemicals from one place to another. The pipes are usually connected to vessels from which they transport the fluid. The extensive use of the pipes is found in water supply and sewage disposal systems, oil refineries, food processing plants, chemical plants and power stations. The pipes are also used as structural elements such as columns and hand rails.

The pipes are made of large number of materials such as cast iron, wrought iron or malleable iron, steel, brass, copper, aluminium, lead, concrete, plastic, rubber etc. The choice of the material depends upon the purpose for which the pipe is used.

25·2 Classification of Pipes

The pipes, according to the material and their use, are classified as follows :

1. *Cast iron pipes.* The cast iron pipes are used only for low pressure applications because of its low resistance to shocks. They are best suited for water, gas and sewage systems. The cast iron pipes are made by the following two methods :

(*a*) Sand casting, and (*b*) Spun (centrifugal) process.

In sand casting process, the molten metal is poured into sand lined moulds, which are set in either horizontal or vertical positions. The horizontal positioned pipes are 100 per cent stronger in tension and 50 percent stronger in rupture than vertical positioned pipes.

In spun process, pipes may be cast in either sand mould or in a water cooled mould. The spun process produces finer grained, denser structured and lighter pipes with more uniform thickness.

The C.I. pipes for water, gas and sewages are available in various sizes from 80 mm to 1500 mm nominal bore and lengths upto 4 metres. The cast iron pipes are generally used for pressures upto 1 MPa but can be used even upto 2·5 MPa. The cast iron pipes used for draining rain water are available in sizes from 50 mm to 150 mm nominal diameters and of overall length 1.8 metres.

2. *Wrought iron or Malleable iron pipes.* The wrought iron or malleable iron pipes are used for carrying water and gas at low pressures for domestic purposes. The pipes are galvanised by dipping them in a molten zinc. Such pipes are also known as *galvanised iron pipes.* Galvanizing prevents corrosion of pipes. Wrought iron pipes are manufactured by rolling wrought iron flat plates to the required diameter and finally the joint is welded and finished.

These pipes are available in various sizes from 6 mm to 150 mm nominal bore and length from 4 to 7 metres.

3. *Steel pipes.* The steel pipes are used for carrying water, steam, gas, sewage and air at a high pressure. As compared to cast iron and wrought iron pipes, the steel pipes have greater strength, less liable to breakage and longer lengths upto 12 metres. The steel pipes may be manufactured either seamless or by welding process. The seamless pipes are made by piercing an ingot while red hot and

after putting it on a mandrel, rolling it through dies to give the required size. The welded pipes are made out of a rolling steel plates which are rolled into a cylindrical shape and finally the joint is welded and finished.

The steel pipes are available in various sizes from 50 mm to 1800 mm nominal bore. They may be obtained as black pipes or galvanised pipes.

4. *Copper and Brass pipes.* The copper and brass pipes are generally used for hot water systems, refrigeration systems and general plumbing work. Copper and brass pipes are corrosion resistant and are suitable for processing plants where scale and oxidation of steel pipes are harmful.

These pipes are either solid-drawn or made from sheet with the seam (longitudinal joint) welded or brazed. The copper and brass pipes can be easily bent or coiled to the desired shape. These pipes are available in straight lengths upto 4 metres.

It may be noted that the copper pipes of less than 50 mm diameter are called tubes. They are made as hard temper and as soft tubing. The hard temper tubing is much stiffer than soft tubing and should be used where rigidity is desired. The soft tubing can be easily bent and therefore it is generally used where frequent bending is required. The hard temper copper tubing is available in lengths upto 6 metres and soft tubing in coils of 15 metres.

5. *Aluminium pipes.* The aluminium pipes are used in aircraft and marine applications, because of its light weight. They are also used for furniture and decorative work.

6. *Lead pipes.* The lead pipes are used for sanitary purposes. These pipes are made by expanding molten lead through a die over a mandrel. The lead pipes can be easily bent to the desired shape and the joints are made by soldering process. The lead pipes are available with nominal internal diameter ranging from 10 mm to 125 mm and thickness from 2 to 10 mm.

7. *Concrete pipes.* The reinforced cement concrete pipes are widely used for water mains, sewers, culverts and in irrigation. These pipes are made in three types, *i.e.,* light duty, medium duty and heavy duty. The light duty pipes are made with nominal internal diameter ranging from 80 mm to 2200 mm and thickness from 25 to 110 mm. The medium duty pipes have 80 mm to 2600 mm nominal internal diameter and thickness from 25 to 215 mm. The heavy duty pipes have 80 mm to 1200 mm nominal internal diameter and thickness from 25 mm to 65 mm.

8. *Plastic pipes.* The plastic pipes are also called P.V.C. pipes. These pipes are cheap and light in weight. They are extensively used for sanitary purpose and electric wire conduits. These pipes are resistant to most inorganic acids, alkalies and salts as well as to many organic chemicals and do not corrode. Plastic pipes are widely used in chemical industry.

9. *Rubber pipes.* The rubber pipes are flexible and are available in coils. These pipes are used in household work for water, oil and gas.

25·3 Indian Standard Specification for Pipes

According to Indian standards [IS : 1239 (Part I)-1990], the pipe size is designated by the nominal bore of the pipe. For example, a 15 mm pipe means that the diameter of bore is 15 mm. For each nominal bore, there are three types of pipes, *i.e.,* light, medium and heavy depending upon the wall thickness of the pipe.

25·4 Pipe Fittings

The pipe fittings are used to join adjacent lengths of pipes. They are also frequently used to provide changes in direction, to provide branch connections at different angles or to effect a change in size. They are made of cast iron, malleable iron, cast or forged steel, non-ferrous alloys and other materials for special applications. The ferrous fittings are made for threaded, welded or flanged joints. The non-ferrous fittings are made for threaded, solder, flared or flanged joints.

The various screwed fittings commonly used with malleable iron or steel pipes are as follows :

1. *Tee.* It is used to connect a branch pipe at right angles to the main pipe line. The tee may be of equal size or of reducing size.

2. *Elbow.* It is used to connect pipes at right angles. The elbow may be of equal size or of reducing size.

3. *Cross.* It is used to connect four pipes at right angles to each other.

4. *Socket.* It is used to connect two pipes of the same diameter.

5. *Reducer or reducing socket.* It is used to connect two pipes of different diameters.

6. *Nipple or coupler.* It is a small length pipe having outside threads at the ends. It is used to get extra length of the pipe, by screwing it with the two pipes having internal threads.

7. *Union.* It is used to connect two pipes and can be disconnected easily when desired.

8. *Plug.* It has external threads and is used with a tee or cross to close the opening of a pipe which is left for further use.

9. *Cap.* It has internal threads and is used to close the threaded end of the pipe.

All the above mentioned fittings are shown in Fig. 25.1.

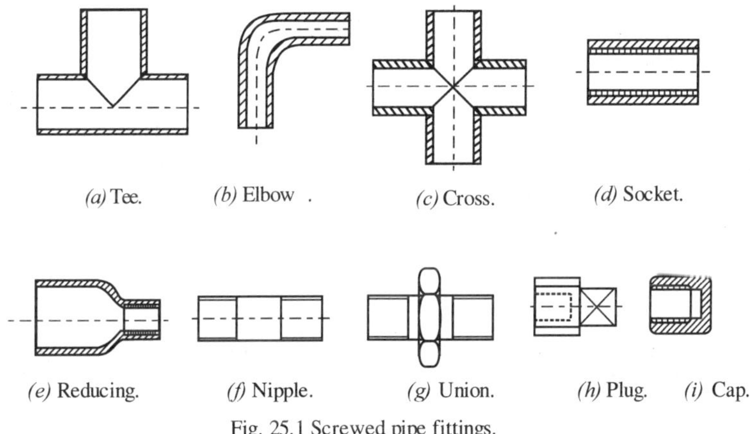

(a) Tee. *(b)* Elbow. *(c)* Cross. *(d)* Socket.

(e) Reducing. *(f)* Nipple. *(g)* Union. *(h)* Plug. *(i)* Cap.

Fig. 25.1 Screwed pipe fittings.

The flanged fittings used with cast iron pipes are shown in Fig. 25.2.

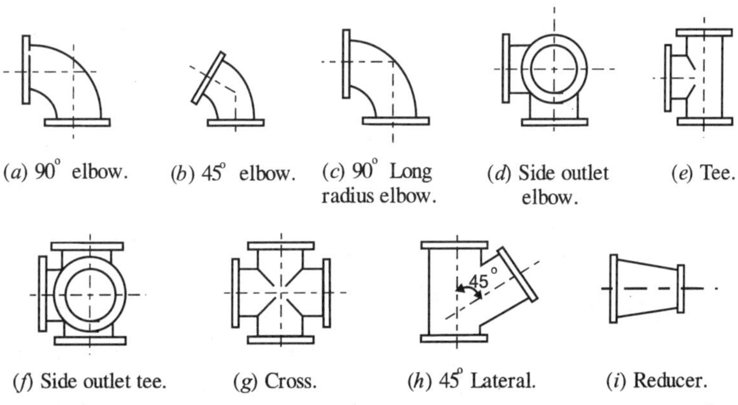

(*a*) 90° elbow. (*b*) 45° elbow. (*c*) 90° Long radius elbow. (*d*) Side outlet elbow. (*e*) Tee.

(*f*) Side outlet tee. (*g*) Cross. (*h*) 45° Lateral. (*i*) Reducer.

Fig. 25.2. Flanged pipe fittings.

25·5 Taps and Valves

The *taps* are generally used for domestic water supply purposes. The basic function of tap is to stop and regulate the flow of water. The taps are usually made out of brass and cast iron. In domestic taps, washers are often worn out and requires periodic replacement.

The *valves* are used to stop or regulate the flow of fluids in a pipe line. These can be operated by hand, fluid pressure or by some mechanism. The valves which are operated by fluid pressure are known as non-return valves as they allow the fluid to flow in one direction only. Small valves are made of brass, gun metal and bronze whereas large valves are made of cast iron or cast steel. The common types of valves are float valves, gate valves, globe valves and check valves. These valves are discussed below.

1. *Float valve.* These valves are used to prevent overflow of water from the tank. A PVC or metallic hollow ball is used as a float. When the water level in tank is at minimum, the float is at lowest position and allows water to flow in the tank. As water level in the tank rises, the float also rises with it and becomes buoyant when the maxinum permissible water level in the tank is reached. The upward force of water to the float is transferred to the lever (which is connected to the float), which, in turn, moves the plunger to close inlet opening thereby disconnecting the water supply.

2. *Gate valves.* These valves have a stem which rises up or the gate moves along the stem for a straight line flow, with the rotation of hand wheel. The leakage is prevented by gland and packing.

3. *Globe valves.* These valves are used to control the flow of steam from the boiler to the engine. The lift of the valve is perpendicular to the flow of steam and is operated by a hand wheel. The size of valve is specified by the size of the passage through which the steam flows or by the size of the valve seat. These valves are also known as steam stop valves.

4. *Check valves.* These types of valves are used as non return valves as described earlier. The lift of the valve is in perpendicular direction to the valve seat.

25·6 Symbols for Pipe Fittings and Valves

In order to prepare a working drawing of a piping system, the following symbols for the pipe fittings and valves are used :

Name of fitting	Symbol		Name of fitting	Symbol	
	Screwed	Flanged		Screwed	Flanged
Elbow 90°			Lateral		
Elbow 45°			Union		
			Sleeve		
Elbow Turned up			Cap		
Elbow turned down					
Reducing elbow			Reducer		
Tee			Gate valve		

Name of fitting	Symbol: Screwed	Symbol: Flanged	Name of fitting	Symbol: Screwed	Symbol: Flanged
Tee-outlet up			Check valve		
Tee-outlet down			Globe valve		
Cross			Stop cock		

The application of these symbols is shown in Fig. 25.3.

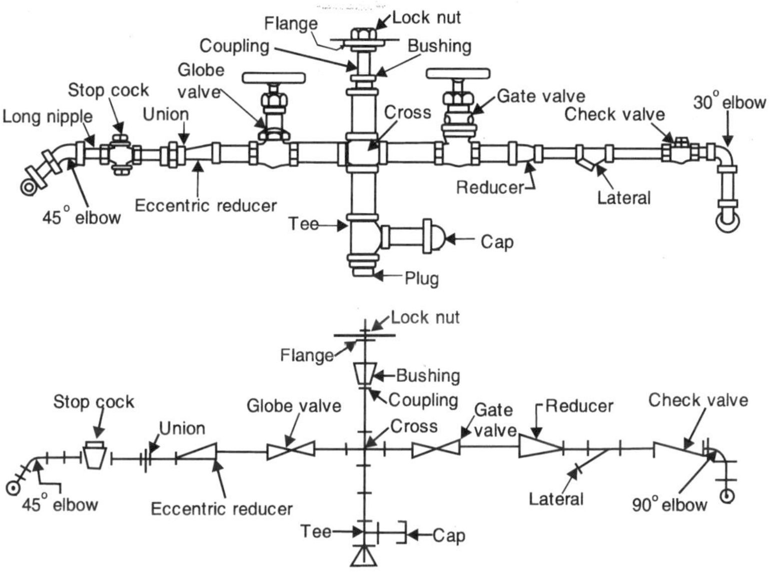

Fig. 25.3. Application of symbols of pipe fitting and valves.

25·7 Pipe Vice

The pipe vice, as shown in Fig. 25.4, is used for holding pipes for various purposes such as cutting, threading etc. The body (also known as column) of the vice is integral with base which has holes for fixing. The threads in the body for the screwed spindle are properly and accurately cut. The screwed spindle has square or trapezoidal threads. The pipe vice consists of two serrated jaws, one of which is fixed and the other is movable. The jaws are "V" shaped and it grips the work at four points. The handle should be straight and slide freely in the hole provided at the end of the screwed spindle.

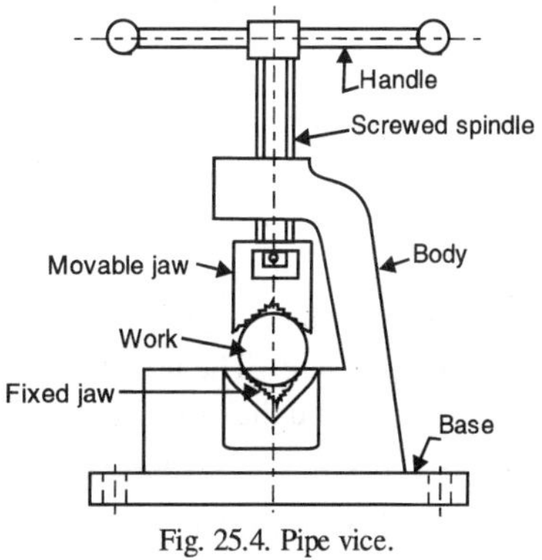

Fig. 25.4. Pipe vice.

25·8 Pipe Wrenches

Pipe wrenches are used to turn pipes, round bars etc. or for assembling and disassembling them from fittings. Pipe wrenches are made in several designs but most commonly used are adjustable type pipe wrenches, which have movable part and a main body.

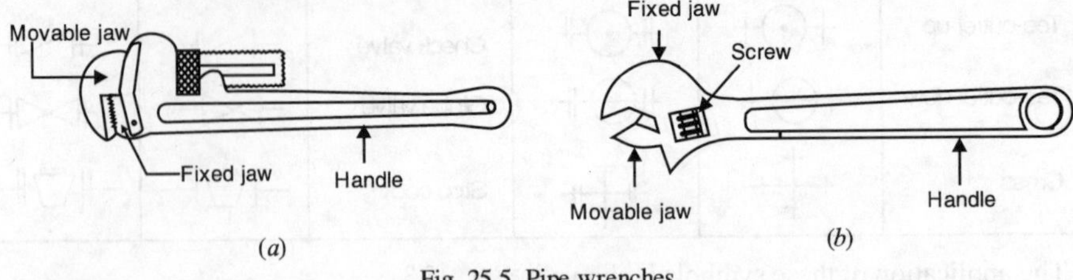

Fig. 25.5. Pipe wrenches.

The adjustable pipe wrench as shown in Fig. 25.5(*a*) has a definite range through which its jaws can be adjusted. The opening of jaws, depends upon the movable part of the wrench. Also, the jaws of the wrench have serrations which enables positive grip over pipes and tubes. Another type of adjustable pipe wrench is shown in Fig. 25.5(*b*) and is suitable for smaller and thinner objects.

25·9 Pipe Cutting

The pipes are available in standard lengths in markets. In order to meet the requirements, it is sometimes desired to cut or extend the standard lengths to suit our requirements. Pipe can be cut by two tools :

1. Hacksaw, and 2. Pipe cutter.

Now we will discuss the method of pipe cutting by the above mentioned tools briefly.

1. *Hacksaw.* For cutting pipes with hacksaw, the pipe is held firmly in vice and the required length is cut with the help of a hacksaw. It may be noted that while cutting, the hacksaw cuts the pipe perpendicular to the axis of the pipe as shown in Fig. 25.6.

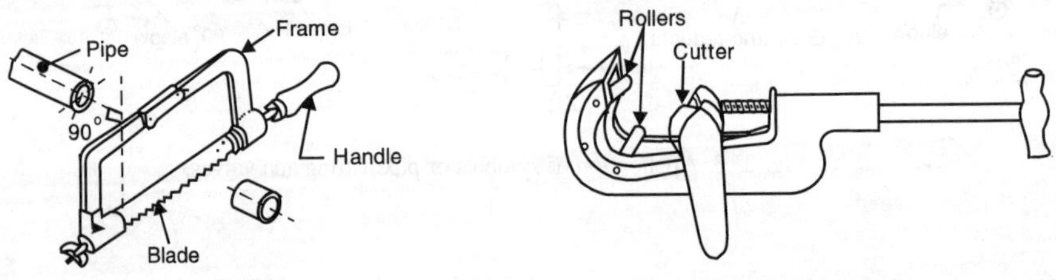

Fig. 25.6. Hacksaw. Fig. 25.7. Pipe cutter.

2. *Pipe cutter.* A standard pipe cutter is a special tool to cut pipes and tubes. It consists of forged hook shaped frame which carries three cutting wheels which are very hard and have sharp cutting edges along their periphery. Out of the three cutting wheels, two are permanently fixed on the body on pin bearings, and the third one which is mounted on the projected part of the handle can be adjusted at a desired distance from the fixed cutting wheels to accommodate different sizes of pipes.

For cutting pipe with this tool, the pipe is held firmly in vice and then the tool is rotated around the pipe with pressure. The pressure can be constantly increased by the operator while he revolves the pipe cutter around the pipe. Due to the continuous action of the cutter wheel on the pipe, the pipe is cut along a circle. A standard pipe cutter is shown in Fig. 25.7.

25·10 Pipe Threads and Thread Cutting

A pipe is threaded at its ends for connecting pipe fittings. The pipe threads are similar to

British Standard Whitworth (B.S.W.) threads but have finer pitches. In external pipe threading, threads are specified by the bore of the pipe.

The threads are cut externally on the pipes with a die and a stock, as already discussed in Chapter on Bench Work and Fitting.

25·11 Bending of Pipes

Bending of pipe is done by means of a pipe bending equipment. Small diameter (*i.e.*, upto 40 mm) pipes can be easily bent by this equipment. The pipe bending equipment, as shown in Fig. 25.8, consists of a base plate which is either fixed on a table or supported on four cast iron legs. The operating lever is stopped at the extreme left position by stopper pin 'X'. The pipe is inserted between guide roller of the operating handle and the main roller at this position. The lever is then rotated to the right, thereby bending the pipe to the desirable amount. The lower part of the pipe is supported by the guide pin 'Y'.

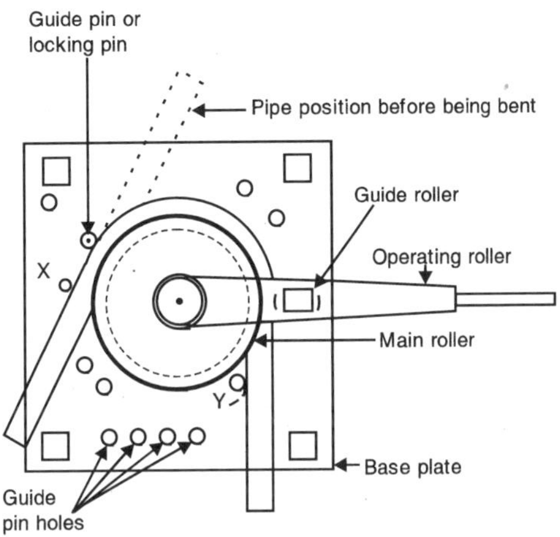

Fig. 25.8. Bending of pipe.

Notes : 1. It is not possible to make sharp bends by this equipment.

2. The curvature of the bend is limited to the radius of the main roller only.

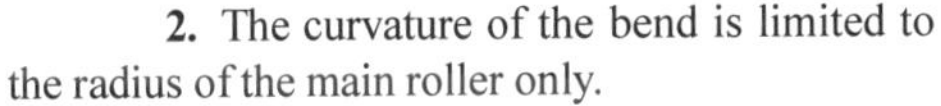
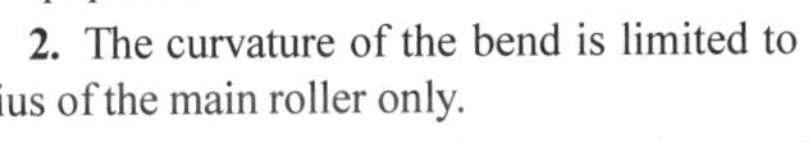

3. The large diameter pipes can be bend using hydraulic pipe bending machines.

25·12 Pipe Joints

We have already discussed that the length of pipes available are limited. Therefore, various lengths of pipes have to be joined to suit any particular installation. There are various forms of pipe joints used in common practice, but the most common of them are discussed below :

1. *Socket or coupler joint.* The most common method of joining pipes is by means of a socket or a coupler as shown in Fig. 25.9. A socket is a small piece of pipe threaded inside. It is screwed on half way on the threaded end of one pipe and the other pipe is then screwed in the remaining half of socket. In order to prevent leakage, jute or hemp is wound around the threads at the end of each pipe. This type of joint is mostly used for pipes carrying water at low pressure and where the overall smallness of size is most essential.

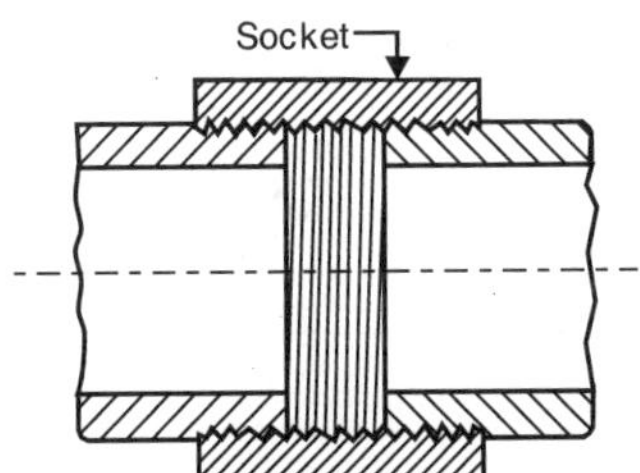

Fig. 25.9. Socket or coupler joint.

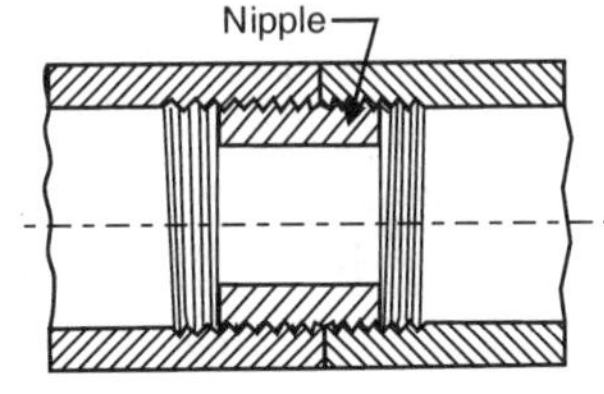

Fig. 25.10. Nipple joint.

2. *Nipple joint.* In this type of joint, a nipple which is a small piece of pipe threaded outside

is screwed in the internally threaded end of each pipe, as shown in Fig. 25.10. The disadvantage of this joint is that it reduces the area of flow.

3. *Union joint.* In order to disengage pipes joined by a socket, it is necessary to unscrew pipe from one end. This is sometimes inconvenient when pipes are long.

The union joint, as shown in Fig. 25.11, provides the facility of disengaging the pipes by simply unscrewing a coupler nut.

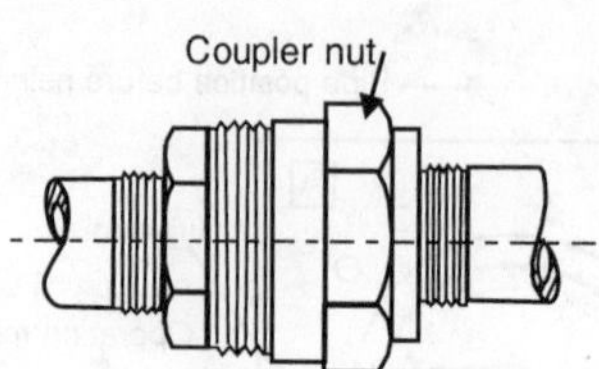

Fig. 25.11. Union joint.

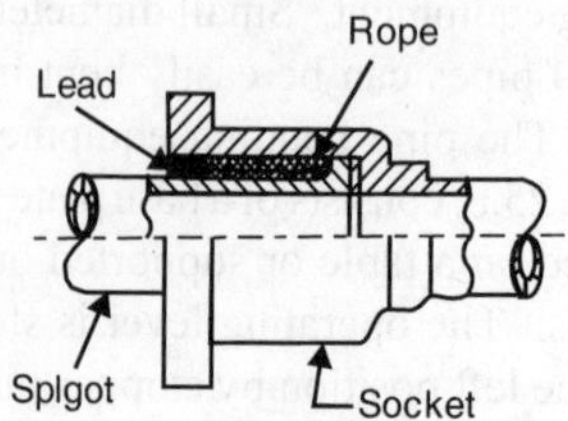

Fig. 25.12. Spigot and socket joint.

4. *Spigot and socket joint.* A spigot and socket joint, as shown in Fig. 25.12, is chiefly used for pipes which are buried in the earth. Such pipe lines are laid straight as far as possible. One of the important features of this joint is its flexibility as it adopts itself to small changes in level due to settlement of earth which takes place due to climate and other conditions.

In this type of joint, the spigot end of one pipe fits into the socket end of the other pipe. The remaining space between the two is filled with a jute rope and a ring of lead. When the lead solidifies it is caulked-in tightly.

5. *Expansion joint.* The pipes carrying steam at high pressure are usually joined by means of expansion joint. This joint is used in steam pipes to take up expansion and contraction of pipe line due to change of temperature.

In order to allow for change in length, steam pipes are not rigidly clamped but supported on rollers. The rollers may be arranged on wall bracket, hangers or floor stands. The expansion bends, as shown in Fig. 25.13(*a*) and (*b*) are useful in a long pipe line. These pipe bends will spring in either direction and readily accommodate themselves to small movements of the actual pipe ends to which they are attached.

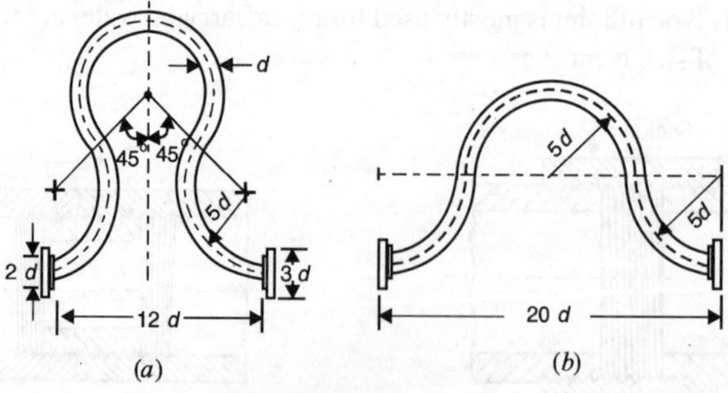

Fig. 25.13. Expansion bends.

The copper corrugated expansion joint, as shown in Fig. 25.14(*a*), is used on short lines and is satisfactory for limited service. An expansion joint as shown in Fig. 25.14(*b*) (also known as gland and stuffing box arrangement), is the most satisfactory when the pipes are well supported and cannot sag.

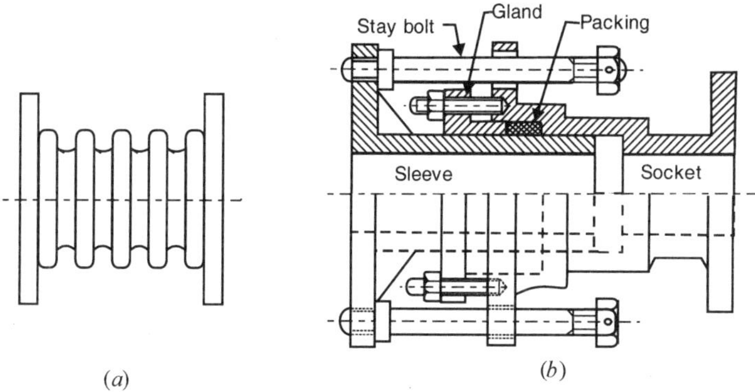

Fig. 25.14. Expansion joints.

6. *Flanged joint.* It is one of the most widely used pipe joint . A flanged joint may be made with flanges cast integral with the pipes or loose flanges welded or screwed. Fig. 25.15 shows two cast iron pipes with integral flanges at their ends. The flanges are connected by means of bolts. The flanges have been standardised for pressures upto 2 N/mm^2. The flange faces are machined to ensure correct alignment of the pipes. The joint may be made leakproof by placing a gasket of soft material, rubber or convass between the flanges. The flanges are made thicker than the pipe walls, for strength. The pipes may be strengthened for high pressure duty by increasing the thickness of pipe for a short length from the flange, as shown in Fig. 25.16.

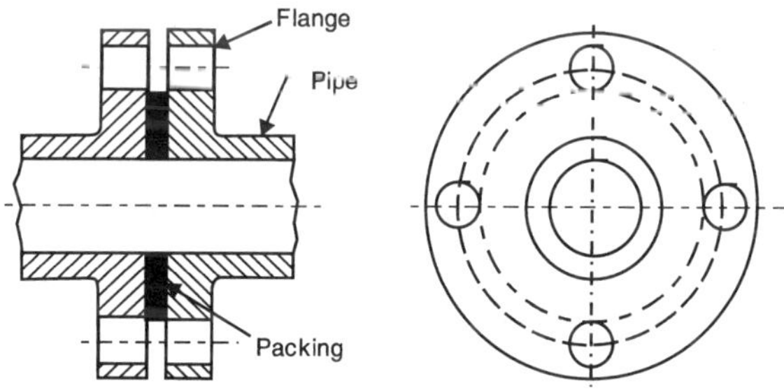

Fig. 25.15. Flanged joint.

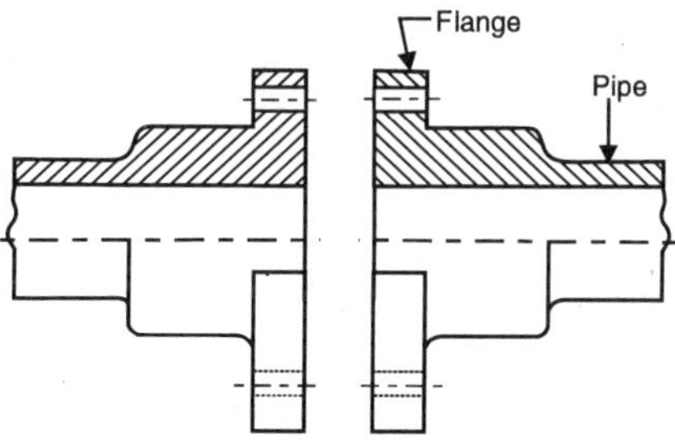

Fig. 25.16. Flanged joint.

For even high pressures and for large diameters, the flanges are further strengthened by ribs or stiffeners as shown in Fig. 25.17(*a*). The ribs are placed between the bolt holes.

For larger size pipes, separate loose flanges screwed on the pipes as shown in Fig. 25.17(*b*) are used instead of integral flanges.

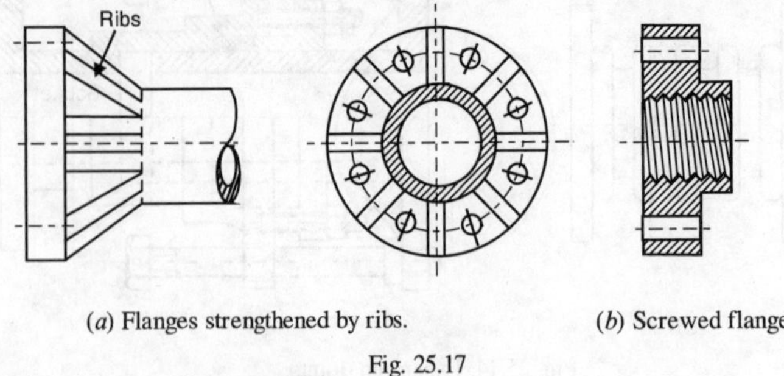

(*a*) Flanges strengthened by ribs. (*b*) Screwed flange

Fig. 25.17

7. *Hydraulic pipe joint.* This type of joint has oval flanges and are fastened by means of two bolts, as shown in Fig. 25.18. The oval flanges are usually used for small pipes upto 175 mm diameter. The flanges are generally cast integral with the pipe ends. Such joints are used to carry fluid pressure varying from 5 to 14 N/mm^2. Such a high pressure is found in hydraulic applications like riveting, pressing, lifts etc. The hydraulic machines used in these installations are pumps, accumulators, intensifiers etc.

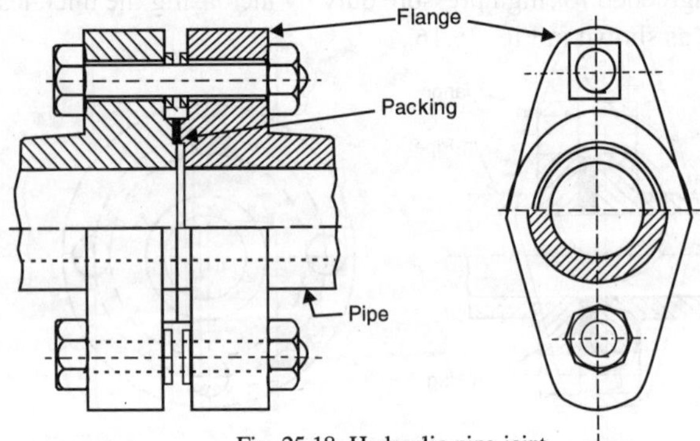

Fig. 25.18. Hydraulic pipe joint.

8. *Wiped joint.* In the wiped joint, as shown in Fig. 25.19, one end of the pipe is opened slightly wider while the end of other pipe is finished to a sharp edge. The two ends of the pipes are placed against each other. The outer surface of the pipes at the joint are polished, for a length approximately equal to the diameter of the pipes. The solder is melted and poured over the joint and rapidly wiped with a piece of cloth smeared with grease. The solder sticks to the polished surfaces of the pipes, thus joining them together.

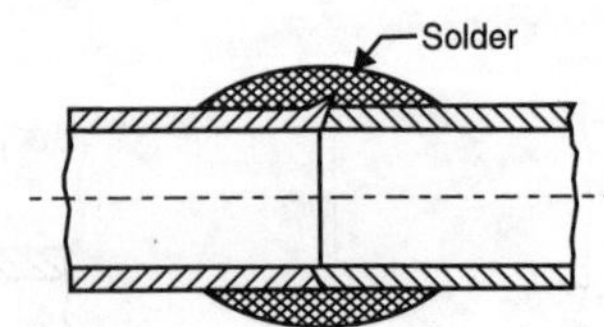

Fig. 25.19. Wiped joint.

The wiped joint is commonly used for joining lead pipes.

QUESTIONS

1. What are the different types of pipes ? Give their uses.

2. How a pipe is specified according to Indian standards ?

3. Explain, with neat sketches, the common types of fittings used with galvanised iron pipes.

4. Explain, with neat sketches, the flanged types of fittings used with cast iron pipes.

5. What is the function of a valve in a pipe line ? Describe different types of valves.

6. Give the symbols for the following screwed and flanged pipe fittings :

(*a*) Elbow 90°; (*b*) Reducing elbow ; (*c*) Tee ; (*d*) Cross ; (*e*) Union ; (*f*) Sleeve ; and (*g*) Reducer.

7. Write a short note on pipe threads and thread cutting.

8. Explain the following types of pipe joints :

(*a*) Socket joint ; (*b*) Nipple joint ; (*c*) Spigot and socket joint ; (*d*) Flanged joint ; and (*e*) Wiped joint.

OBJECTIVE TYPE QUESTIONS

1. The pipes commonly used for supplying water for household purposes are
(*a*) copper pipes (*b*) aluminium pipes
(*c*) galvanised iron pipes (*d*) plastic pipes

2. The pipes commonly used for draining rain water are
(*a*) cast iron pipes (*b*) steel pipes
(*c*) rubber pipes (*d*) brass pipes

3. Pipes used in water supply generally run under
(*a*) pressure (*b*) gravity
(*c*) both pressure and gravity (*d*) none of these

4. According to Indian standards, pipes are specified by their
(*a*) outside diameter (*b*) nominal bore
(*c*) wall thickness (*d*) material

5. Lead pipes are used for
(*a*) house hold work (*b*) rain water drainage
(*c*) decorative purposes (*d*) sanitary purposes

6. A pipe fitting used to branch off pipe at right angles from the supply line is called a
(*a*) tee (*b*) bend (*c*) cross (*d*) union

7. It is required to connect four pipes at right angles to each other. The fitting used should be a
(*a*) tee (*b*) bend (*c*) cross (*d*) union

8. Which of the following joint is commonly used for pipes carrying water at low pressure ?
(*a*) union joint (*b*) spigot and socket joint
(*c*) coupler joint (*d*) nipple joint

9. The pipes which are burried in the earth should be joined with
(*a*) union joint (*b*) spigot and socket joint
(*c*) coupler joint (*d*) nipple joint

10. The wiped joint is commonly used for joining
(*a*) lead pipes (*b*) plastic pipes
(*c*) Steel pipes (*d*) rubber pipes

ANSWERS

1. (*c*)	2. (*a*)	3. (*b*)	4. (*b*)	5. (*d*)
6. (*a*)	7. (*c*)	8. (*c*)	9. (*b*)	10. (*a*)

26

Machine Tools (Introduction)

1. Introduction. 2. Lathe. 3. Working Principle of a Lathe. 4. Types of Lathe. 5. Principal Parts of a Lathe. 6. Size and Specifications of a Lathe. 7. Lathe Accessories. 8. Lathe Tools. 9. Lathe Operations. 10. Terms used in Lathe Machining. 11. Shaper. 12. Principle Parts of a Shaper. 13. Working Principle and Operation of a Shaper. 14. Classifications of Shapers. 15. Specifications of a Shaper. 16. Quick Return Mechanism of a Shaper (Crank and Slotted Lever Type). 17. Planer. 18. Types of Planers. 19. Principal Parts of a Planer. 20. Size and Specifications of a Planer. 21. Planer Tools. 22. Work-holding Devices. 23. Difference Between a Planer and a Shaper. 24. Drilling Machine. 25. Working Principle of a Drilling Machine. 26. Size of a Drilling Machine. 27. Specification of a Drilling Machine. 28. Drilling Operations. 29. Drilling Speed and Feed. 30. Machining and Operation Time for Drilling Machine.

26·1 Introduction

A machine which performs the material removal operation with tools, to produce desired shape and size of the workpiece, is known as a *machine tool*. The purpose of machine tools is to save time, cost of production and to get better output which can not be obtained with hand tools.

The various types of machine tools commonly used in workshops are lathes, shapers, planers, drilling machines, grinding machines and milling machines. In this chapter, we shall only discuss lathes, shapers, planers and drilling machines in brief.

26·2 Lathe

The lathe is one of the most important machines in any workshop. Its main objective is to remove material from outside by rotating the work against a *cutting tool. Though a lathe is used to produce cylindrical work, yet it may also be used for many other purposes such as drilling , threading, grinding, milling etc.

26·3 Working Principle of a Lathe

The working principle of a lathe is shown in Fig. 26.1. In a lathe, the workpiece is held in a chuck or between centres and rotated about its axis at a uniform speed. The cutting tool held in the tool post is fed into the workpiece for a desired depth and in the desired direction (*i.e.,* in the linear, transverse or lateral direction). Since there exists a relative motion between the workpiece and the cutting tool, therefore the material is removed in the form of chips and the desired shape in obtained.

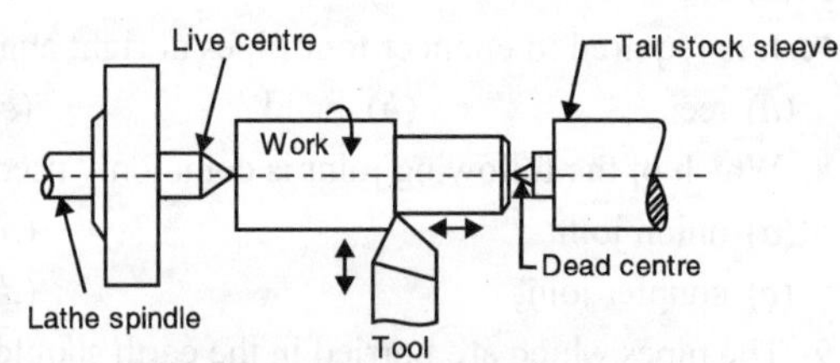

Fig. 26.1. Working principle of a lathe.

26·4 Types of Lathe

Though the fundamental principle of operation of all lathes is same and perform the same function, yet they are classified according to the design, type of drive, arrangement of gears, and purpose. The following are important types of lathes.

* The cutting tool material should be harder than the work material.

1. *Speed lathe.* It is one of the simplest type of all lathes. It is driven by power and consists of a bed, a headstock, a tail stock and an adjustable slide for supporting the tool. Since the tool is fed into the work by hand and cuts are very small, therefore this type of lathe is driven at high speeds usually from 1200 to 3600 r.p.m. The work may be held between centres or attached to a face plate on the headstock. The speed lathe is used mainly for wood working, centering, metal spinning, polishing etc.

2. *Engine or centre lathe.* It is a general purpose lathe and is widely used in workshops. It differs from a speed lathe that it has additional mechanism for controlling the spindle speed and for supporting and controlling the feed of the fixed cutting tool. The cutting tool may be fed both in cross and longitudinal direction with reference to the lathe axis with the help of a carriage.

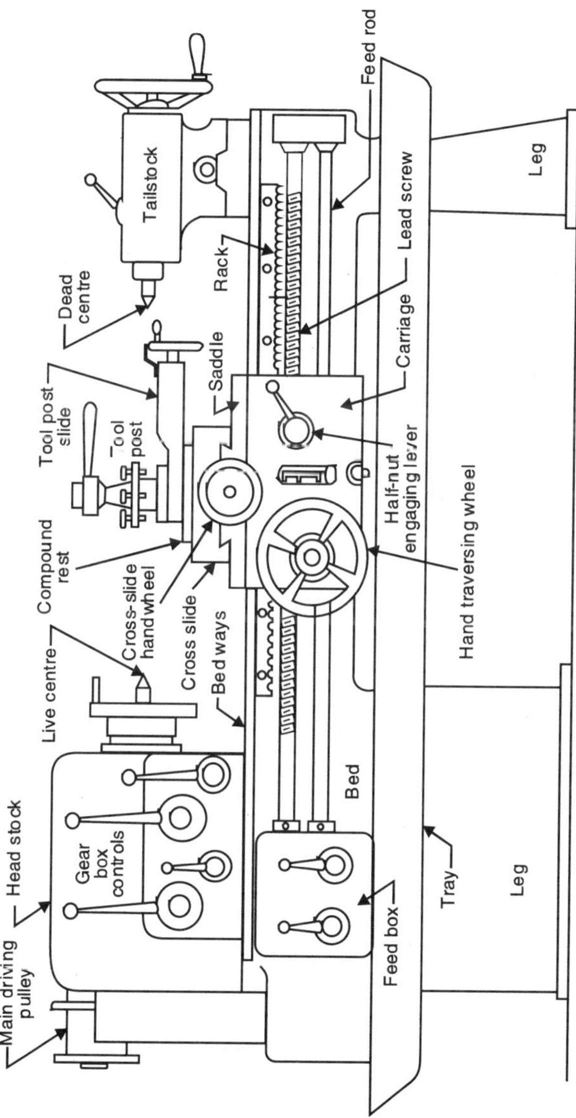

Fig. 26.2. Geared-head centre lathe.

The engine lathe, depending upon the design of headstock for receiving power, may be classified as belt driven lathe, motor driven lathe and geared head lathe.

3. *Bench lathe.* The bench lathe is so small that it can be mounted on a bench. All the types of operation can be performed on this lathe that may be done on an ordinary speed or engine lathe. This is used for small work usually requiring considerable accuracy such as in the production of gauges, punches and beds for press tools.

4. *Tool room lathe.* The tool room lathe is similar to an engine lathe and is equipped with all the accessories needed for accurate tool work. It has an individually driven geared headstock with a wide range of spindle speeds. Since this lathe is used for precision work on tools, gauges, dies, jigs and other small parts, therefore greater skill is needed to operate the lathe.

5. *Capstan and turret lathe.* The capstan and turret lathes are the modification of engine lathes and is particularly used for mass production of identical parts in the minimum time. These lathes are semi-automatic and are fitted with multi tool holding devices, called capstan and turret heads. The advantage of capstan and turret lathe is that several different types of operations can be performed on a workpiece without resetting of work or tools.

6. *Automatic lathes.* The automatic lathes are so designed that the tools are automatically fed to the work and withdrawn after all the operations are complete to finish the work. Since the entire operation is automatic, these lathes require little attention of the operator. These lathes are used for mass production of identical parts.

7. *Special purpose lathes.* The work which can not be conveniently accommodated or machined on a standard lathe, the special purpose lathes are used. The *gap bed lathe* which has a removable section in the bed in front of the headstock to provide a space or gap, is used to swing extra large diameter jobs. The *crankshaft lathe* is especially used for turning crankshafts. The *wheel lathe* which is of large design is especially used for finishing the journals and for turning the locomotive driving wheels. The *axle lathe* is used for turning car axles. The *manufacturing lathe,* which is equipped with front and rear tool blocks and other features, is used for turning duplicate parts rapidly.

26·5 Principal Parts of a Lathe

The various parts of a geared head centre lathe are shown in Fig. 26.2. The principal parts are as follows :

1. *Bed.* The bed is the base or foundation of the lathe. It is a heavy, rigid casting made in one piece. It holds or supports all other parts of the lathe. The top of the bed is planed to form guide ways for the carriage and tail stock. The guide ways are of two types, *i.e.,* wide flat guide ways as shown in Fig. 26.3(*a*) or inverted Vee guide ways. Generally, the combination of both the flat and inverted Vee guide ways is used, as shown in Fig. 26.3(*b*).

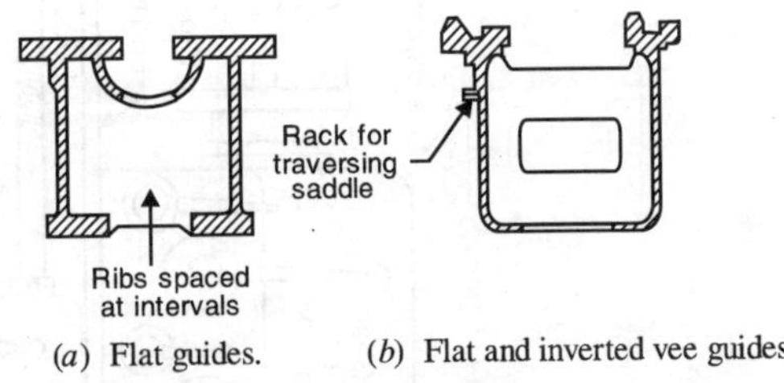

(*a*) Flat guides. (*b*) Flat and inverted vee guides

Fig. 26.3. Bed.

2. *Headstock.* The headstock is permanently fastened on the inner ways at the left hand end of the bed. It serves to support the spindle and driving arrangements. All lathes receive their power through the headstock, which may be equipped with a step-cone pulleys or a geared head drive. The modern lathes are provided with all-geared type headstock to get large variation of spindle speeds. Fig. 26.4 shows a 9-speed all-geared headstock.

The headstock spindle is made hollow, as shown in Fig. 26.5, in order to allow the long bar or work holding devices to pass through. The front end of the hole is tapered for holding centres and other tools having a standard morse tapered shank. A tapered sleeve fits into the tapered spindle hole. The spindle nose may of threaded design or flanged type, but the former type is commonly used on lathes. The threaded nose enables the chucks, driving plates and face plates to be screwed.

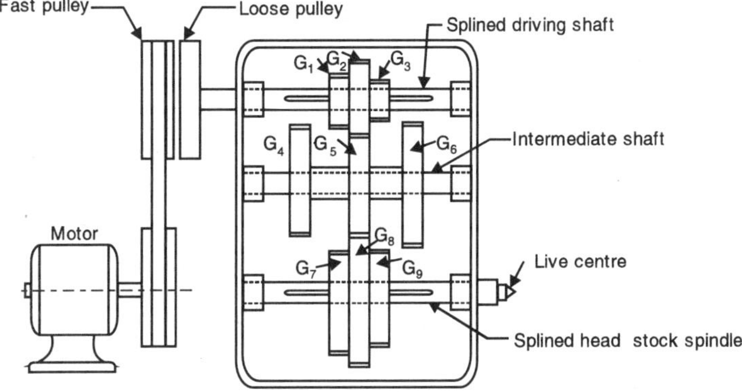

Fig. 26.4. 9-speed all-geared headstock.

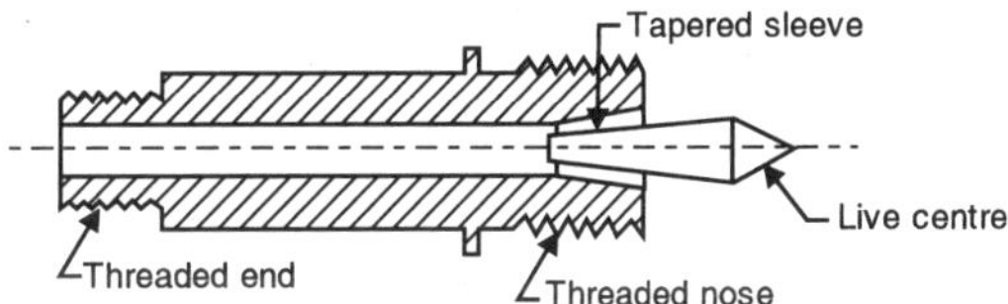

Fig. 26.5. Headstock spindle.

3. *Tailstock.* The tailstock is the counterpart of the headstock and is situated at the right hand end of the bed. It is used for supporting the work when turning on centres or when a long component is to be held in a chuck. It is also used for holding and feeding the tools such as drills, reamers, taps etc.

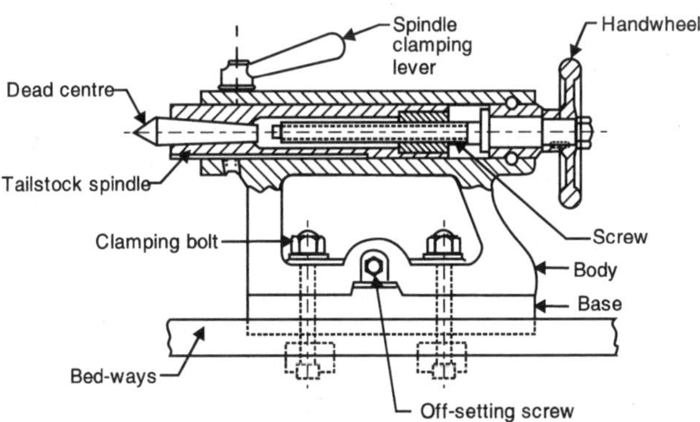

Fig. 26.6. Tailstock.

The tailstock, as shown in Fig. 26.6, consists of two parts *i.e.,* the base and the body. The two parts are held together by clamping bolts. The body of the tailstock is a casting and placed on its base which is shaped to fit the innerways of the bed. The tailstock body is adjustable and can be moved towards or away from the operator. The tailstock spindle is a hollow shaft and is mounted in the tailstock body. The left hand end of the hole is tapered to hold the dead centre and other tools having the same taper as that of spindle hole. The spindle can be firmly clamped in position by means of a clamping lever.

4. *Carriage.* The carriage controls and supports the cutting tool. Since the carriage moves on the ways of the bed, therefore its motion is parallel to the rotational axis of the spindle. This motion is commonly used for producing cylindrical surfaces. The carriage, as shown in Fig. 26.7, has the following five major parts :

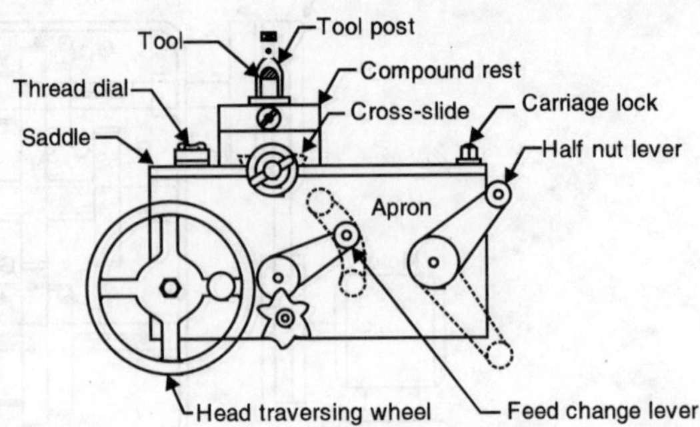

Fig. 26.7. Carriage.

(*a*) *Saddle.* It is an H-shaped casting that fits over the bed and slides along the ways between the headstock and tailstock.

(*b*) *Cross-slide.* It is mounted on the saddle. It supports the compound rest. The cross-slide also has a dovetail that fits over the saddle dovetail. The transverse movement of the cross-slide is obtained by turning the cross-slide hand wheel.

(*c*) *Compound rest.* It is mounted on top of the cross-slide. It can be swivelled to any desired angle in a horizontal plane and its base is graduated in degrees to indicate its angular setting.

(*d*) *Tool post.* It is mounted above the compound rest. A T-slot is machined in the compound rest to accomodate the tool post. It serves to rigidly clamp the cutting tool or tool holder, in a desired position.

(*e*) *Apron.* It is fastened to the saddle and hangs over the front of the bed. It contains gears, clutches and levers for operating the carriage by hand and power feeds. The apron hand wheel can be turned to move the carriage longitudinally back and forth by hand.

5. *Feed mechanism.* The feed mechanism of a lathe is employed for imparting various feeds to the cutting tool. The feed *i.e.,* the movement of the tool relative to the work may be longitudinal, cross and angular, as shown in Fig. 26.8.

Fig. 26. 8. Longitudinal, cross and angular feeds.

The feed is said to be longitudinal when the tool moves parallel to the work. It is obtained by the movement of the carriage which can be operated by hand or power.

The feed is said to be cross when the tool moves perpendicular to the work. It is obtained with the help of a cross-slide which can be operated by hand or power.

The feed is said to be angular when the tool moves at an angle to the work. It is obtained by swivelling the compound slide which can be operated by hand only.

The feed mechanism consists of feed reverse lever, tumbler reversing mechanism, change gears, feed gear box, quick change gear box, lead screw, feed rod, apron mechanism and half nut mechanism.

26·6 Size and Specifications of a Lathe

The size of a lathe is determined by

1. The height of the centres above the top of the bed. This is shown by H in Fig. 26.9.
2. The swing or the maximum diameter of the work that can be rotated over the ways of the bed. This is equal to twice the height of the centres. This is shown by D in Fig. 26.9.
3. The maximum length of the work that can be accomodated between the lathe centres. This is shown by L in Fig. 26.9.
4. The maximum diameter of the work that can be rotated over the lathe saddle. This is shown by D_1 in Fig. 26.9.
5. The maximum diameter of the work that can be accomodated in the gap for a gap bed lathe. This is shown by D_2 in Fig. 26.9.

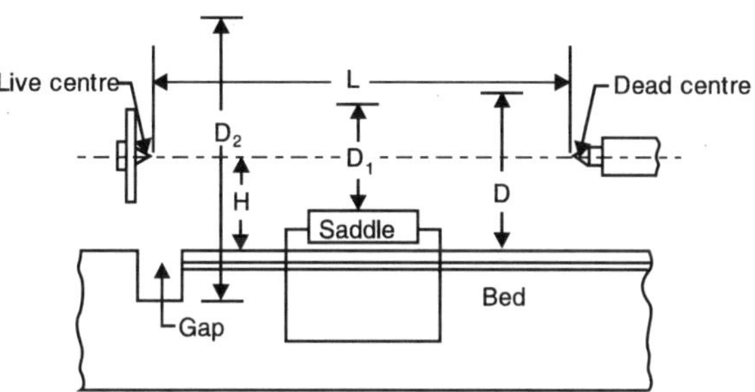

Fig. 26.9. Lathe specifications.

In addition to the above, the following specifications are necessary to provide while ordering a lathe :

1. The length, width and depth of bed.
2. The depth and width of the gap, if it is a gap lathe.
3. The swing over gap.
4. The number and range of spindle speeds.
5. The number of feeds.
6. The lead screw diameter.
7. The number and range of metric threads that can be cut.
8. The tailstock spindle travel.
9. The tailstock spindle set over.
10. The back gear-ratio.
11. The power rating of electric motor.

26·7 Lathe Accessories

The lathe accessories are used for holding and supporting the work or for holding the cutting tool. The various lathe accessories are discussed as follows :

1. *Centres.* There are two types of centres *i.e.,* live centre and dead centre. A centre which fits into the headstock spindle and revolves with the work is called *live centre* while the centre which is used in a tailstock spindle and does not revolve is called *dead centre.* These centres have a standard Morse taper shank at one end and a 60° point at the other end. The various types of centres, used for different works, are shown in Fig. 26.10.

2. *Chucks.* It is an important device used for holding and rotating the workpiece in a lathe.

The workpieces which are too short to be held between centres, are clamped in a chuck. It is attached to the lathe spindle by means of two bolts with the back plate screwed on to the spindle nose. There are many types of the chuck, but the following two are commonly used.

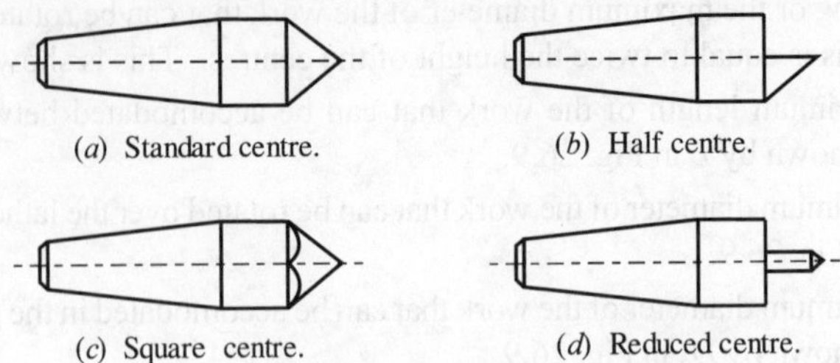

(*a*) Standard centre. (*b*) Half centre.

(*c*) Square centre. (*d*) Reduced centre.

Fig. 26.10. Centres.

(*a*) *Three jaw universal chuck.* The three jaw universal chuck, as shown in Fig. 26.11(*a*) is also called self centering chuck or scroll chuck. Thus chuck is used for holding round and hexagonal work.

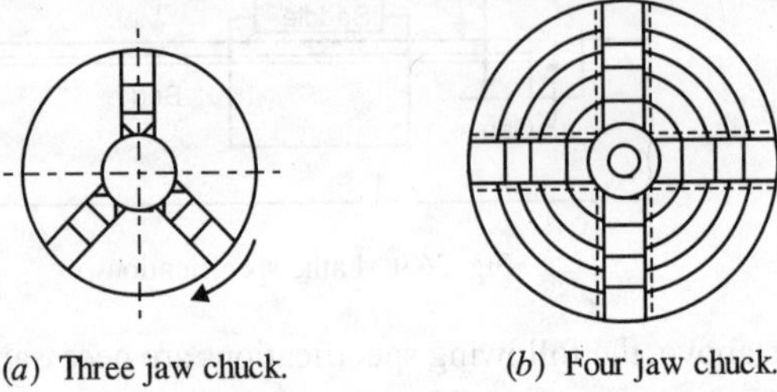

(*a*) Three jaw chuck. (*b*) Four jaw chuck.

Fig.26.11. Chucks.

(*b*) *Four jaw independent chuck.* The four jaw independent chuck, as shown in Fig. 26.11(*b*), has four reversible jaws, each of which may be independently adjusted to accomodate the work it supports. This type of chuck can hold square, round and irregular shape of work in either a concentric or eccentric position.

Note : The other type of the chucks are combination chucks, magnetic chuck, collect chuck, drill chuck, and air or hydraulic chuck.

3. *Lathe dog or carrier.* The work placed on a mandrel or held between centres is rotated positively by clamping the dog or carrier to the end of the work. This is engaged with a pin attached to the drive plate or face plate. The lathe dog or carrier may be of straight type or bent type as shown in Fig. 26.12(*a*) and (*b*) respectively.

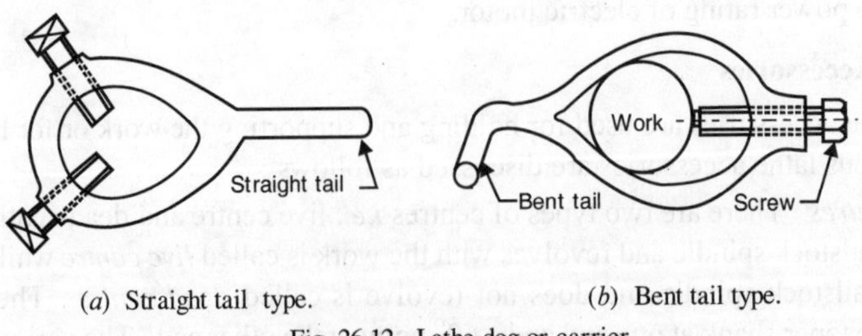

(*a*) Straight tail type. (*b*) Bent tail type.

Fig. 26.12. Lathe dog or carrier.

4. *Drive plate.* The drive plate, as shown in Fig. 26.13, is a circular plate which is bored out and threaded so that it can be attached to the spindle nose. It also carries a hole for the pin which is

used only when the work is held in a lathe dog having straight tail. When bent-tail dog is used, this pin is taken out and the bent portion of the tail is inserted into the hole.

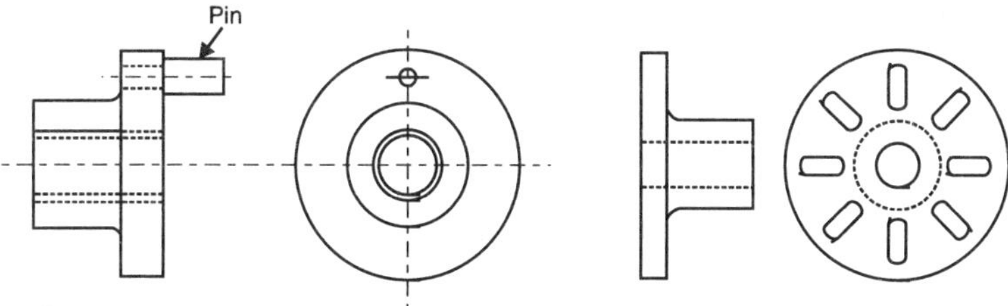

Fig. 26. 13. Drive plate.

Fig. 26. 14. Face plate.

5. *Face plate.* The face plate, as shown in Fig. 26.14, is similar to drive plate except that it is larger in diameter. It contains more open slots or T-slots so that bolts may be used to clamp the workpiece to the face of the plate. The face plate is used for holding workpieces which cannot be conveniently held in a chuck.

6. *Angle plate.* An angle plate is simply a cast iron plate with two faces planed at right angles to each other and having slots in various positions for the clamping bolts. It is always used with the face plate for holding such parts which can not be clamped against the vertical surface of the face plate.

7. *Mandrels.* The lathe mandrel is a cylindrical bar with centre hole at each end. It is used to hold hollow workpieces to machine their external surface. The work revolves with the mandrel which is mounted between the centres of the lathe. The various types of mandrels used for different classes of work are shown in Fig. 26.15.

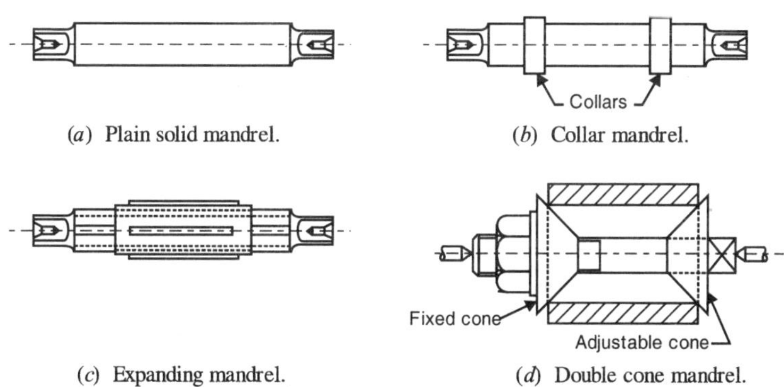

(*a*) Plain solid mandrel. (*b*) Collar mandrel.

(*c*) Expanding mandrel. (*d*) Double cone mandrel.

Fig. 26.15. Types of mandrels.

8. *Rests.* When a long slender workpieces are to be turned between the lathe centres, it must be supported at some intermediate point in order to prevent springing or bending of the workpiece. This auxiliary support is known as *rest*.

The following two types of rests are commonly used in an engine lathe.

(*a*) Fixed steady, and (*b*) Travelling steady or follower rest.

The *fixed steady rest* is shown in Fig. 26.16(*a*). Though its main function is to provide a support for long thin workpiece that is turned between centres, yet it can be used for supporting one end of a long workpiece for facing, drilling and boring operations. It may be noted that when steady rest is used, the carriage can not be fed to the full length of the work.

The *travelling steady* or *follower rest,* as shown in Fig. 26.16(*b*), is used to support the work which is close to the cutting tool. It is bolted to the cross-slide on the saddle and travels with the tool as it feeds along the work.

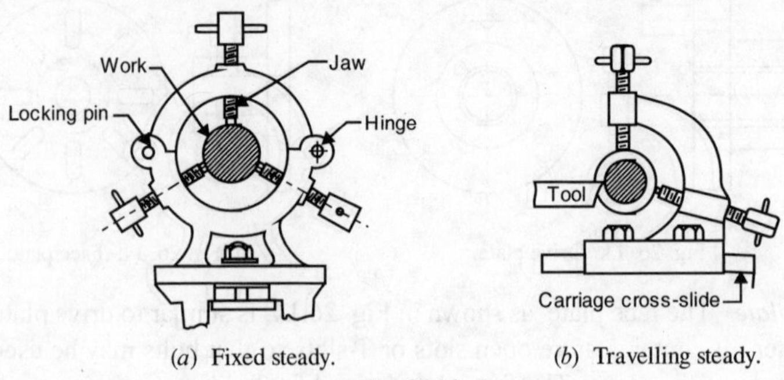

(*a*) Fixed steady. (*b*) Travelling steady.

Fig. 26.16. Rests.

26·8 Lathe Tools

The tool used in a lathe, for general purpose work, is a *single point tool, but for special operations multi-point tools may be used. The material used for lathe tools should have hardness, toughness, heat resistance and low wear. The commonly used materials are high carbon steel, high speed steel, cemented carbides, diamond tips and ceramics.

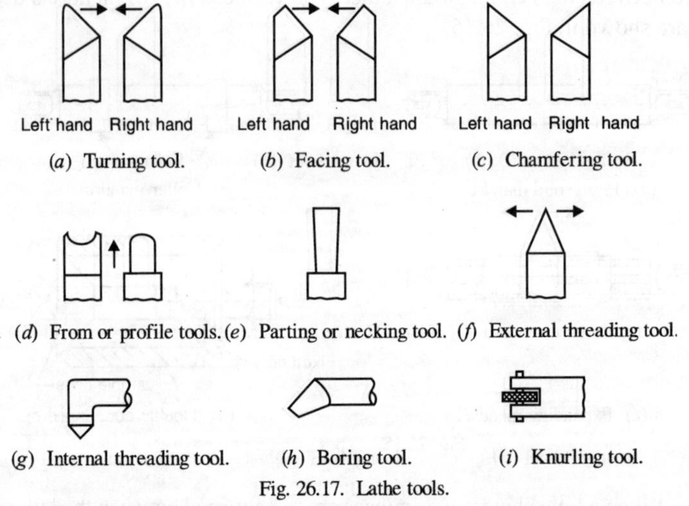

(*a*) Turning tool. (*b*) Facing tool. (*c*) Chamfering tool.

(*d*) From or profile tools. (*e*) Parting or necking tool. (*f*) External threading tool.

(*g*) Internal threading tool. (*h*) Boring tool. (*i*) Knurling tool.

Fig. 26.17. Lathe tools.

The lathe tools, depending upon the nature of operation done by the tool, are classified as follows :

1. Turning tool (left hand or right hand),
2. Facing tool (left hand or right hand),
3. Chamfering tool (left hand or right hand),
4. Form or profile tool,

---

* A tool which has primarily one cutting edge is known as a single-point cutting tool, such as lathe tools, shaper tools, planer tools, boring tools etc.

5. Parting or necking tool,
6. External threading tool,
7. Internal threading tool,
8. Boring tool, and
9. Knurling tool.

These tools are shown in Fig. 26.17.

26·9 Lathe Operations

The most common operations which can be carried out on a lathe are facing, plain turning, step turning, taper turning, drilling, reaming, boring, undercutting, threading, and knurling. These operations are discussed as follows :

1. *Facing.* This operation is almost essential for all works. In this operation, as shown in Fig. 26.18, the workpiece is held in the chuck and the facing tool is fed from the centre of the workpiece towards the outer surface or from the outer surface to the centre, with the help of a cross-slide.

2. *Plain turning.* It is an operation of removing excess amount of material from the surface of the cylindrical workpiece. In this operation, as shown in Fig. 26.19, the work is held either in the chuck or between centres and the longitudinal feed is given to the tool either by hand or power.

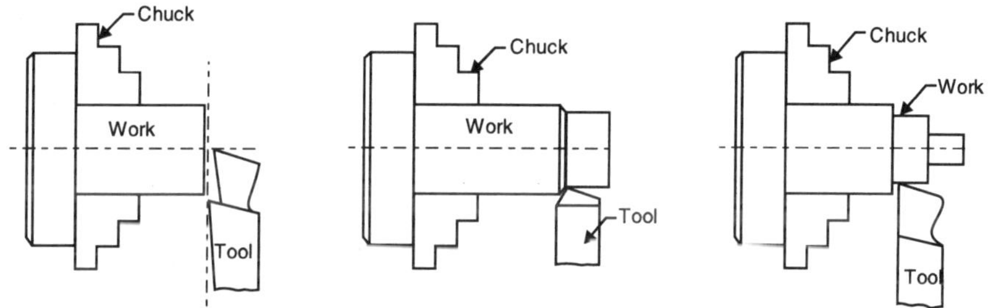

Fig. 26.18. Facing. Fig. 26.19. Plain turning. Fig. 26.20. Step turning.

3. *Step turning.* It is an operation of producing various steps of different diameters in the workpiece, as shown in Fig. 26.20. This operation is carried out in the similar way as plain turning.

4. *Taper turning.* It is an operation of producing an external conical surface on a workpiece. A small taper may be produced with the help of a forming tool or chamfering tool, but the larger tapers are produced by swivelling the compound rest, as shown in Fig. 26.21, at the required angle or by offsetting the tailstock or by taper turning attachment.

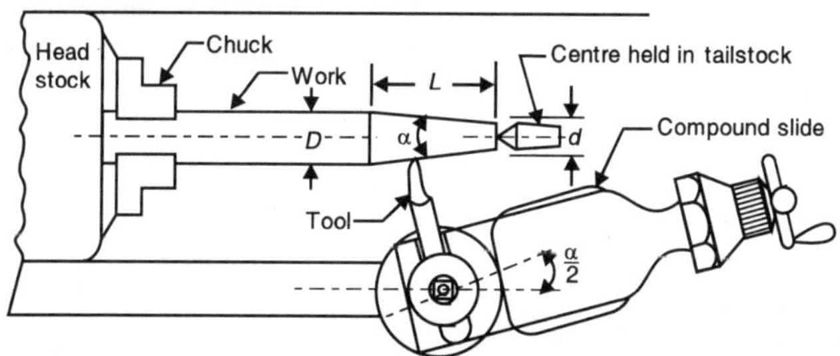

Fig. 26.21. Taper turning.

Note. It D is the larger diameter, d is the smaller diameter and L is the length of taper, as shown in Fig. 26.21, then

$$\text{Taper} = \frac{D-d}{L}$$

5. *Drilling.* It is an operation of making a hole in a workpiece with the help of a drill. In this operation, as shown in Fig. 26.22, the workpiece is held in a chuck and the drill is held in the tailstock. The drill is fed manually, into the rotating workpiece, by rotating the tailstock hand wheel.

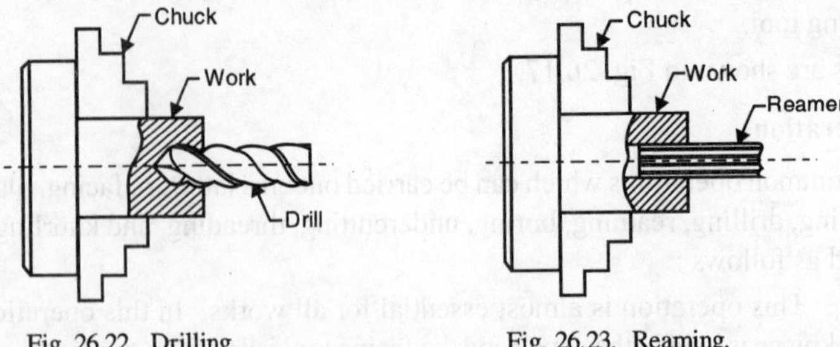

Fig. 26.22. Drilling. Fig. 26.23. Reaming.

6. *Reaming.* It is an operation of finishing the previously drilled hole. In this operation, as shown in Fig. 26.23, a reamer is held in the tailstock and it is fed into the hole in the similar way as for drilling.

7. *Boring.* It is an operation of enlarging of a hole already made in a workpiece. In this operation, as shown in Fig. 26.24, a boring tool or a bit mounted on a rigid bar is held in the tool post and fed into the work by hand or power in the similar way as for turning.

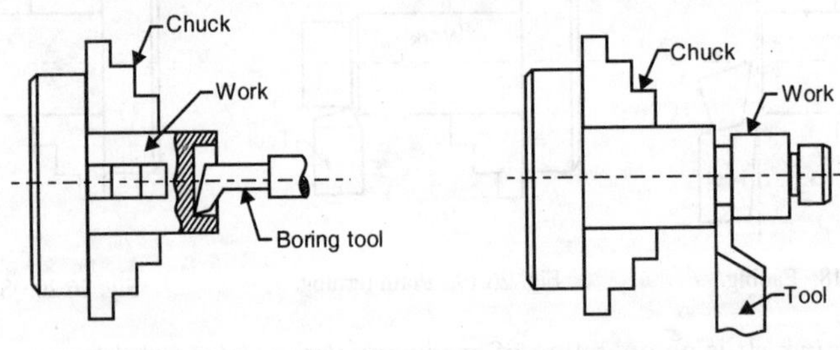

Fig. 26.24. Boring. Fig. 26.25. Undercutting.

8. *Undercutting or Grooving.* It is an operation of reducing the diameter of a workpiece over a very narrow surface. In this operation, as shown in Fig. 26.25, a tool of appropriate shape is fed into the revolving work upto the desired depth at right angles to the centre line of the workpiece.

9. *Threading.* It is an operation of cutting helical grooves on the external cylindrical surface of workpiece. In this operation, as shown in Fig 26.26, the work is held in a chuck or between centres and the threading tool is fed longitudinally to the revolving work. The longitudinal feed is equal to the pitch of the thread to be cut.

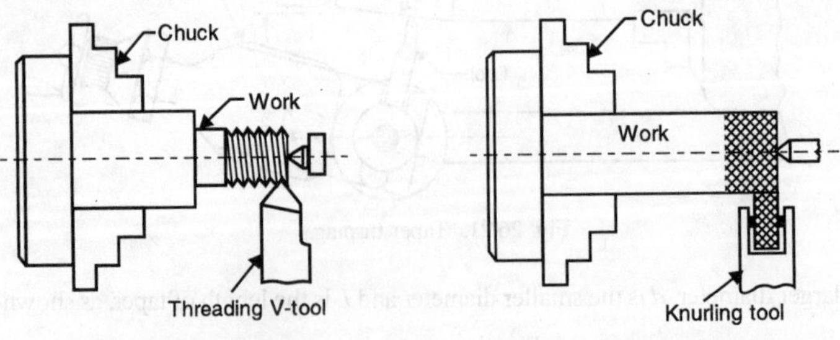

Fig. 26.26. Threading. Fig. 26.27. Knurling.

10. *Knurling.* It is an operation of providing knurled surface on the workpiece. In this operation, as shown in Fig. 26.27, a knurled tool is moved longitudinally to a revolving workpiece surface. The projections on the knurled tool reproduce depressions on the work surface.

26·10 Terms Used in Lathe Machining

The following terms are commonly used while machining a workpice on lathe.

1. *Cutting speed.* It is defined as the speed at which the metal is removed by the tool from the workpiece. In other words, it is the peripheral speed of the work past the cutting tool. It is usually expressed in metres per minute.

2. *Feed.* It is defined as the distance which the tool advances for each revolution of the work. It is usually expressed in millimetres.

3. *Depth of cut.* It is defined as the depth of penetration of the tool into the workpiece during machining. In other words, it is the perpendicular distance measured from the machined surface to the unmachined surface of the workpiece. It is usually expressed in millimetres.

26·11 Shaper

The shaper (also called shaping machine) is a reciprocating type of machine tool used for producing small flat surfaces with the help of a single point tool reciprocating over the stationary workpiece. The flat surface may be horizontal, inclined or vertical. The reciprocating motion of the tool is obtained either by the crank and slotted lever quick return motion mechanism or Whitworth quick return motion mechanism.

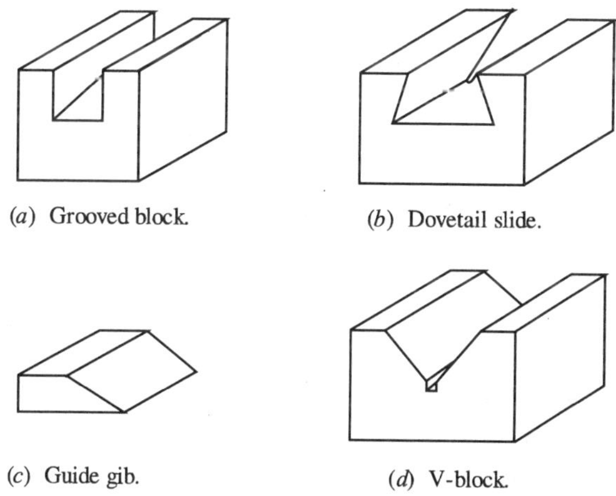

(*a*) Grooved block. (*b*) Dovetail slide.

(*c*) Guide gib. (*d*) V-block.

Fig. 26.28. Parts produced on a shaper.

Some examples of the parts produced on a shaper are shown in Fig. 26.28.

26·12 Principal Parts of a Shaper

The principal parts of a shaper, as shown in Fig. 26.29, are as follows :

1. *Base.* It is a heavy structure of cast iron which supports other parts of a shaper.

2. *Column.* It is a box-like structure made up of cast iron and mounted upon the base. It contains the driving mechanism and is provided with two machined guide ways on the top of it on which the ram reciprocates.

3. *Ram.* It is a reciprocating member which reciprocates on the guideways provided above the column. It carries a tool-slide on its head and a mechanism for adjusting the stroke length.

4. *Tool head.* It is attached to the front portion of the ram with the help of a nut and a bolt. It is used to hold the tool rigidly. It also provides the vertical and angular movements to the tool for cutting.

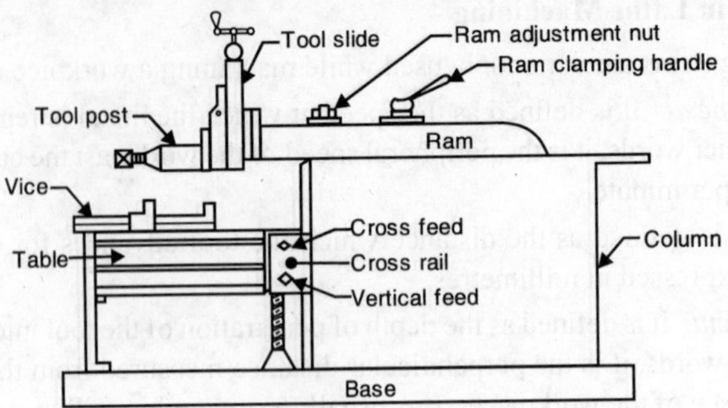

Fig. 26.29. Principal parts of a shaper.

5. *Cross-rail.* It is attached to the front vertical portion of the column. It is used for the following two purposes :

(*a*) It helps in elevating the table over the column in the upward direction, and

(*b*) The table can be moved in a direction perpendicular to the axis of the ram over this cross-rail.

6. *Table.* It is used for holding the work piece. It can be adjusted horizontally and vertically with the help of spindles.

26·13 Working Principle and Operation of a Shaper

We have already discussed that in a shaper, a single point cutting tool reciprocates over the stationary workpiece. The tool is held in the tool post of the reciprocating ram and performs the cutting operation during its *forward stroke. It may be noted that during the **backward stroke of the ram, the tool does not remove material from the workpiece. Both these strokes (*i.e.*, forward and backward strokes) form one working cycle of the shaper.

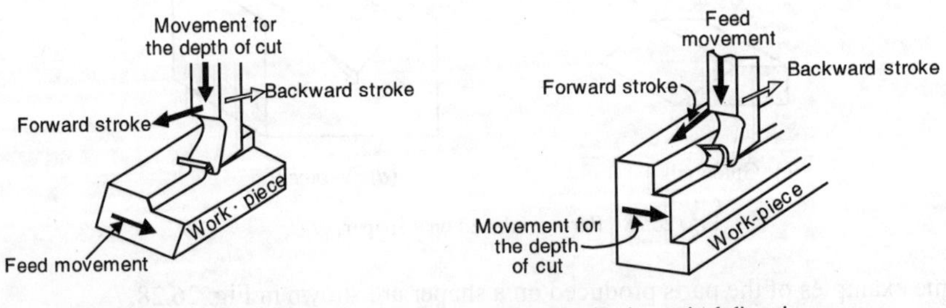

(*a*) Shaping in horizontal direction. (*b*) Shaping in vertical direction.

Fig. 26.30. Working principle and operation of a shaper.

For shaping in horizontal direction, as shown in Fig. 26.30(*a*), the clamped workpiece is fed against the reciprocating tool after every cutting cycle. The depth of cut is adjusted by moving the tool downward towards the workpiece. For shaping in vertical direction, as shown in Fig. 26.30(*b*),

* The forward stroke is also called working stroke or cutting stroke.

** The backward stroke is also called return or idle stroke.

the tool is fed vertically towards the workpiece after every cutting cycle. The depth of cut is adjusted by moving the workpiece sideways.

26·14 Classifications of Shapers

The shapers are classified as follows :

1. *According to the ram driving mechanism.* According to the ram driving mechanism, the shapers are classified as follows :

(*a*) *Crank shaper.* In a crank shaper, a crank and a slotted lever quick return motion mechanism is used to give reciprocating motion to the ram. The crank arm is adjustable and is arranged inside the body of a bull gear (also called crank gear).

(*b*) *Geared shaper.* In a geared shaper, the ram carries a rack below it, which is driven by a spur gear. This type of shaper is not widely used.

(*c*) *Hydraulic shaper.* In a hydraulic shaper, a hydraulic system is used to drive the ram. This type of shaper is more efficient than the crank and geared shaper.

2. *According to position and travel of ram.* According to the position and travel of ram, the shapers are classified as follows :

(*a*) *Horizontal shaper.* In a horizontal shaper, the ram moves or reciprocates in a horizontal direction. This type of shaper is mainly used for producing flat surfaces.

(*b*) *Vertical shaper.* In a vertical shaper, the ram reciprocates vertically in the downward as well as in upward direction. This type of shaper is very convenient for machining internal surfaces, keyways, slots or grooves.

3. *According to the direction of cutting stroke.* According to the direction of cutting stroke, the shapers are classified as follows :

(*a*) *Push-cut shaper.* In a push-cut shaper, the ram pushes the tool across the work during cutting operation. In other words, forward stroke is the cutting stroke and the backward stroke is an idle stroke. This is the most general type of shaper used in common practice.

(*b*) *Draw-cut shaper.* In a draw-cut shaper, the ram draws or pulls the tool across the work during cutting operation. In other words, the backward stroke is the cutting stroke and forward stroke is an idle stroke.

4. *According to the design of the table.* According to the design of the table, the shapers are classified as :

(*a*) *Standard or plain shaper.* In a standard or plain shaper, the table has only two movements *i.e.,* horizontal and vertical, to give the feed. It can not be swivelled or tilted.

(*b*) *Universal shaper.* In a universal shaper, in addition to the above two movements, the table can be swivelled about a horizontal axis parallel to the ram and the upper portion of the table can be tilted about the other horizontal axis perpendicular to the first axis. This type of shaper is mostly used in tool room work.

26·15 Specifications of a Shaper

The shaper is specified as follows :

1. Maximum length of stroke is millimetres,
2. Size of the table, *i.e.,* length, width and depth of the table,
3. Maximum vertical and horizontal travel of the table,
4. Maximum number of strokes per minute,
5. Power of the drive motor,
6. Type of quick return mechanism,
7. Floor space required, and
8. Weight.

26·16 Quick Return Mechanism of a Shaper (Crank and Slotted Lever Type)

The crank and slotted lever quick return motion mechanism for a shaper is shown in Fig. 26.31. An electric motor drives the pinion at a uniform angular speed which, in turn, drives a bull gear. The crank *CB* is contained in the bull gear and may be adjusted in length by a screw mechanism. The crank rotates at a uniform angular speed with the bull gear about the fixed centre *C*. A sliding block is attached to the crank-pin at *B* which is fitted within the slotted lever. This sliding block slides in the guideways of the slotted lever, the bottom end of which is pivoted at *A* to the base of the shaper while the upper end *P* is forked and connected to the ram by a short link *PR*.

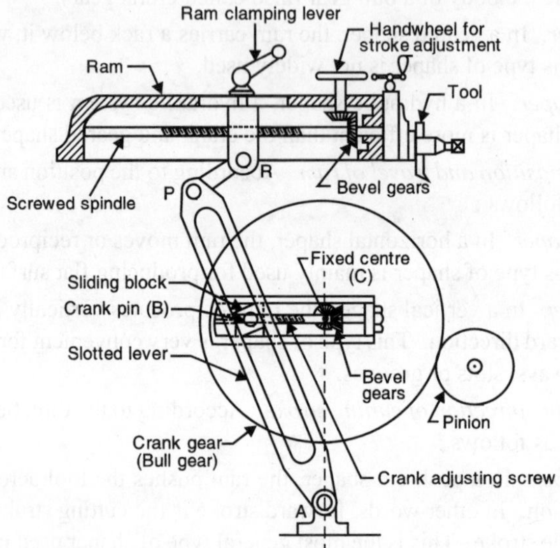

Fig. 26.31. Crank and slotted lever quick return motion mechanism

When the bull gear rotates, the crank and the sliding block attached to the crank pin also rotates and at the same time slides up and down in the slotted lever causing it to oscillate about the pivoted point *A*. The short link *PR* transmits motion from *AP* to ram which carries the tool and reciprocate along the line of stroke R_1R_2 as shown in Fig. 26.32. Thus in a shaper, the rotary motion of the crank or bull gear is converted into reciprocating motion of the ram. The line of stroke (*i.e.*, R_1R_2) is perpendicular to the line *AQ* produced.

The principle of quick return motion can be easily understood from the configuration diagram of the mechanism as shown in Fig. 26.32. In the extreme

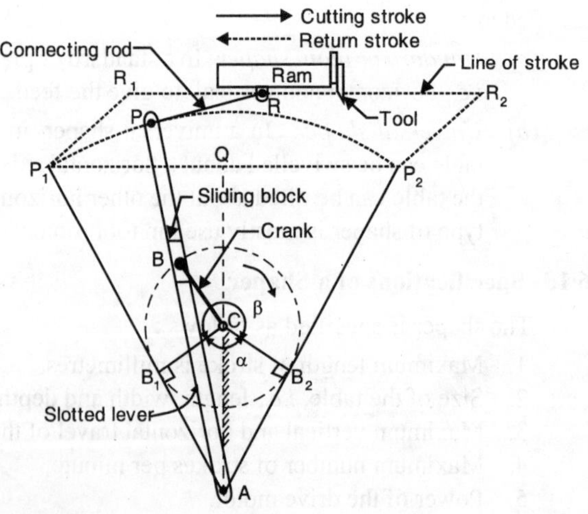

Fig. 26.32. Principle of quick return motion.

positions, the slotted lever AP occupies the positions AP_1 and AP_2 which are tangential to the crank-pin circle and the cutting tool is at the end of the stroke. The forward or cutting stroke occurs when the crank rotates from the position CB_1 to CB_2 (or through an angle β) in the clockwise direction. The backward or return stroke occurs when the crank rotates from the position CB_2 to CB_1 (or through an angle α) in the clockwise direction. Since the crank rotates at a uniform speed, therefore,

$$\frac{\text{Time of cutting stroke}}{\text{Time of return stroke}} = \frac{\beta}{\alpha} = \frac{\beta}{360^\circ - \beta} \text{ or } \frac{360^\circ - \alpha}{\alpha}$$

Since the tool travels a distance of R_1R_2 during cutting and return stroke, therefore travel of the tool or length of stroke,

$$R_1R_2 = P_1P_2 = 2P_1Q = 2AP \times \frac{CB}{AC}$$

Note : From Fig. 26.32, we see that the angle β made by the forward or cutting stroke is greater than the angle α described by the backward or return stroke. Since the crank rotates with uniform angular speed, therefore the return stroke is completed within shorter time. Thus, it is a quick return motion mechanism.

26·17 Planer

The planer (also called planing machine) is a machine tool used to produce plane and flat surface by a single point cutting tool. It is similar to a shaper but its size is very large and is adopted for producing flat surfaces of much larger work than a shaper. The fundamental difference between a planer and shaper is that in a planer, the tool remains stationary and the work reciprocates whereas in the shaper, the work remains stationary and the tool reciprocates.

26·18 Types of Planers

The various types of planers commonly used are

1. Standard or double housing planer,
2. Open side planer,
3. Pit planer,
4. Edge or plate planer, and
5. Divided table planer.

Of all the above types of planers, the standard or double housing planer is most widely used in workshops. In this type of planer, as shown in Fig. 26.33, there are two housings (or columns), one on each side of the table. The housings are connected at the top by a cast iron support. The table reciprocates on accurate guide ways provided on a long heavy bed. The two tool heads which carry the tool are mounted on the cross rails. These tools may be fed either by hand or by power in crosswise or vertical direction.

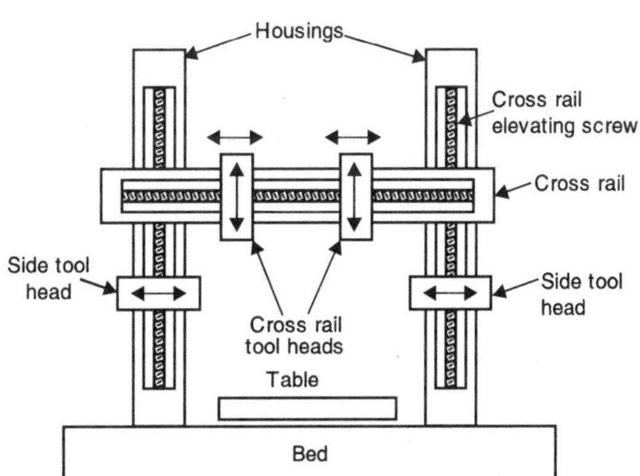

Fig. 26.33. Standards or double housing planer.

In large planers, there are four tool heads, two on the cross rails and one each on the vertical face of the housings. They can also be moved either in a vertical or horizontal direction to apply the feed. All the four tool heads may be controlled and operated individually. On large work, all the tool heads are often operated simultaneously.

26·19 Principal Parts of a Planer

The principal parts of a double housing planer, as shown in Fig. 26.33, are as follows :

1. *Bed.* The bed of a planer is a box like casting having cross ribs. It provides the foundation for the machine and supports the housing and all other moving parts. The upper part of the bed is provided with precision Vee type guide ways on which the table slides.

2. *Table.* The planer table is a heavy rectangular precision machined casting. It reciprocates along the ways of the bed and supports the work. The top face of the table is provided with T-slots and holes uniformly spaced for fixing the work and work holding devices over the table with the help of T-bolts, clamps etc. The table may be driven by rack and gear, by rack and double helical gear or by hydraulic system.

3. *Housing.* The housings are rigid column-like castings placed on each side of the bed. The front face of the housings is accurately machined to provide precision ways. The housings enclose the cross-rail elevating screws, vertical and cross-feed screws for the tool heads, counter balancing weights for the cross rail etc.

4. *Cross-rail.* The cross-rail is mounted on the precision machined ways of the two housings. It may be raised or lowered on the housings to accomodate work of different heights on the table and to allow for the adjustment of the tools.

5. *Tool heads.* The tool heads are mounted on the cross-rail or housings by means of a saddle which slides along the rail or housing ways. The saddle may be made to move transversely on the cross-rail to give crossfeed. Two tool heads are provided on the cross-rail so that the two tools may remove the material from the work simultaneously. In addition to these two tool heads, there are other two tools heads which are mounted on each housing.

26·20 Size and Specifications of a Planer

The planer size is generally given by the following :

1. Width of the table in mm, 2. Maximum distance of the table to the cross rail in metres, and 3. Maximum stroke of the table in mm.

For example, a standard or double housing planer of size 750 mm × 350 mm × 2.5 m means that a planer can machine a work of width 750 mm, height 350 mm and length 2.5 m. The double housing planers are also available in size upto 3000 mm × 3000 mm × 18.25 m.

In addition to the above planer size, the following particulars are also given to specify the planer :

1. Length of bed, 2. Length of the table, 3. Type of drive, 4. Power of the motor, and 5. Number of speed and feed etc.

26·21 Planer Tools

The planer tools are mostly made of high speed steel. The cemented carbide tipped tools are also used on planers for production work. The planer tools are similar to those used on shaper, but, as the depth of cut are heavier and cutting strokes are large, the planer tools are large in size. A planer tool many be classified as roughing or finishing and right hand or left hand type, Fig. 26.34 shows the various types of tools used on planers.

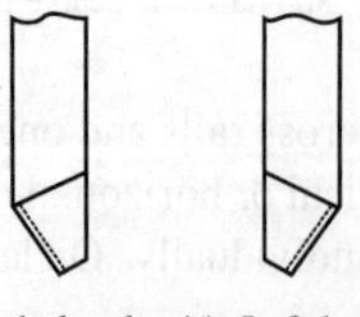

(*i*) Right hand (*ii*) Left hand
(*a*) Straight roughing tools.

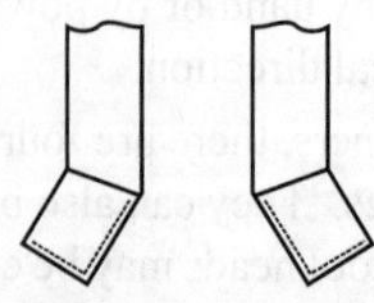

(*i*) Right hand (*ii*) Left hand
(*b*) Bent roughing tools.

Fig. 26.34. Planer tools.

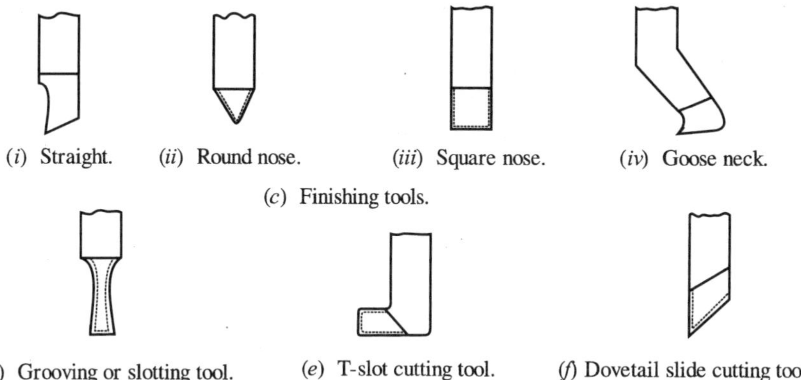

(*i*) Straight. (*ii*) Round nose. (*iii*) Square nose. (*iv*) Goose neck.

(*c*) Finishing tools.

(*d*) Grooving or slotting tool. (*e*) T-slot cutting tool. (*f*) Dovetail slide cutting tool.

Fig. 26.34. Planer tools.

26·22 Work-holding Devices

Since the cutting forces in planing a work are quite heavy, therefore it is essential that the work is connected rigidly to the table so that it does not shift its position during planing. The various work holding devices are vices, step blocks, stops, stop pins, jacks, V-blocks, T-bolts and clamps etc. Some of these devices are shown in Fig. 26.35.

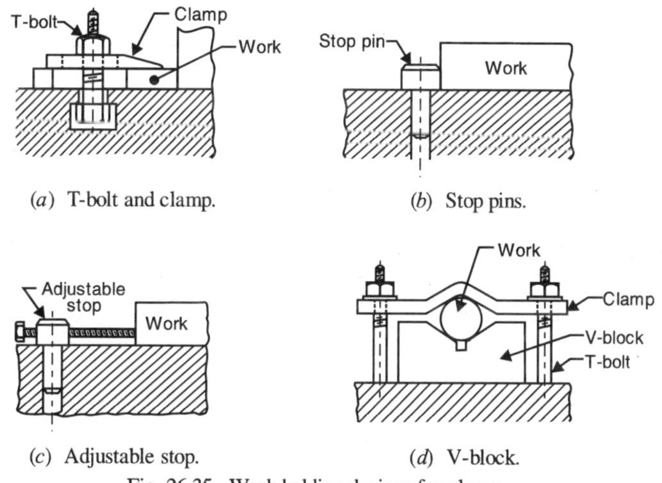

(*a*) T-bolt and clamp. (*b*) Stop pins.

(*c*) Adjustable stop. (*d*) V-block.

Fig. 26.35. Work holding devices for planer.

26·23 Difference Between a Planer and a Shaper

Following are the main differences between a planer and a shaper :

Planer	*Shaper*
1. It is a heavy duty machine tool.	1. It is a light duty machine tool.
2. It requires more floor area.	2. It requires less floor area.
3. It is adopted for large works.	3. It is used for small works.
4. More than one cutting tool can be used at a time.	4. Only one cutting tool is used at a time.
5. Tool is fixed and work moves.	5. Work is fixed and the tool moves.

Planer	Shaper
6. Planer table is either driven by gears or by hydraulic means.	6. It is normally driven by crank and slotted lever quick return motion mechanism.
7. Heavier feeds are applied.	7. Lighter feeds are applied.
8. It can take deep cuts.	8. It cannot take deep cuts.
9. Work setting requires much of skill and takes longer time.	9. Work may be clamped easily and quickly.
10. Tool used are of larger size.	10. Tool used are of smaller size.

26·24 Drilling Machine

The basic purpose of a drilling machine* is to drill cylindrical holes in workpieces (metallic and non-metallic materials). The holes are cut out of the material with a cutting tool, which is known as drill. The drill is fixed in a rotating spindle and can be fed towards the workpiece which may be fixed to the table or to the base of the machine. The speed of the spindle and the feed can be adjusted according to the workpiece. Some examples of drilling work are shown in Fig. 26.36.

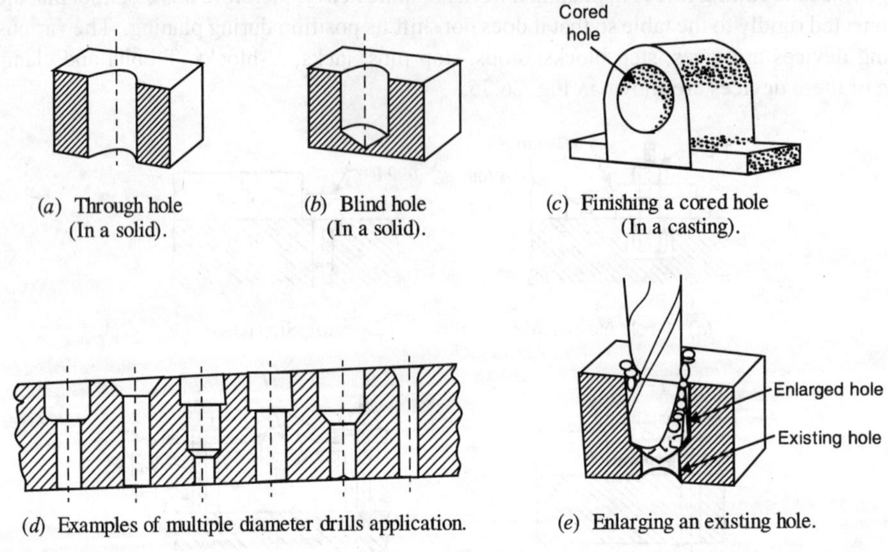

(*a*) Through hole (In a solid).

(*b*) Blind hole (In a solid).

(*c*) Finishing a cored hole (In a casting).

(*d*) Examples of multiple diameter drills application.

(*e*) Enlarging an existing hole.

Fig. 26.36. Examples of drilling work.

26·25 Working Principle of a Drilling Machine

We have already discussed that a metal cutting operation that produces cylindrical holes or enlarges existing holes with an end cutting tool is called as drilling. The end cutting tool used for drilling holes in the workpiece is called as twist drill* and is provided with two cutting edges. In order to cut off chips, two motions are given to the drill simultaneously as shown in Fig. 26.37.

1. *Rotary motion*. The rotary motion is called as main motion or cutting motion. In some cases, this motion is given to workpieces also (*e.g.,* while drilling on the lathe). The cutting motion is generally measured in m/min. The highest cutting speed is on the periphery of the drill and it decreases towards the centre of the drill.

---

* For details refer to Chapter on Bench Work and Fitting.

** For details refer to Chapter on Bench Work and Fitting.

2. *Liner motion towards the fixed workpiece.* The motion is called as feed and it controls the thickness of the chip. In small bench type drilling machines, instead of giving feed to the drill, the workpiece is moved towards the rotating drill by raising the table. The feed is generally measured in mm/rev. As the drill is provided with two cutting edges, the thickness of the chip is half the feed.

By the simultaneous double action of cutting (or main motion) and the feed, each cutting edge of the drill describes a spiral and thereby produces a constant flow of chips.

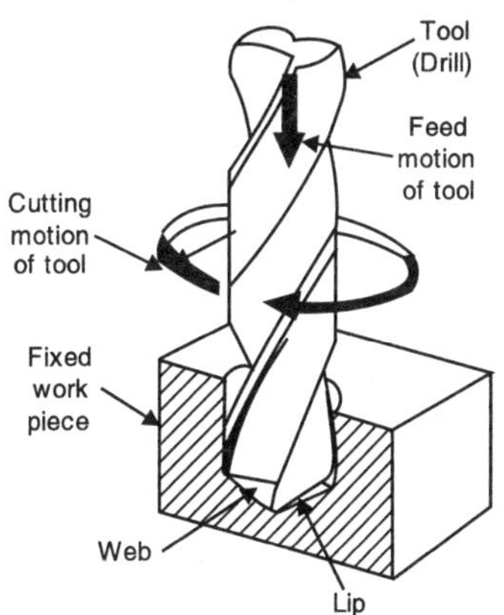

Fig. 26.37. Motions of a twist drill

26·26 Size of a Drilling Machine

Portable drilling machines are designated by the maximum diameter of drill they can hold. *Sensitive and upright drilling machines* are specified by the diameter of the largest workpiece that can be accomodated which is twice the clearance between the centre line of the drill and the machine pillar or frame. The size of *radial drilling machine* is designated by the length of the radial arm and the column diameter.

26·27 Specification of a Drilling Machine

A heavy duty drilling machine is specified by following parameters.

1. Drilling capacity,
2. Taper in spindle (Morse no.),
3. Distance between spindle and column (maximum and minimum) in case of radial drilling machine,
4. Transverse of spindle,
5. Minimum distance between spindle and table,
6. Minimum distance between spindle and base plate,
7. Working surface of table (*i.e.,* diameter),
8. Range of spindle speeds,
9. Range of power feed per revolution,
10. Motor speed, and
11. Motor power.

26·28 Drilling Operations

The most common operations which can be carried on a drilling machine are drilling, boring, reaming, tapping, counter boring, spot facing and countersinking. These operations are discussed as follows :

1. *Drilling.* It is an operation of producing a circular hole in a workpiece by forcing a drill against it.

2. *Boring.* It is an operation of enlarging a hole that has already been drilled by a single point tool, so as to make it true to required size.

3. *Reaming.* It is an operation of slightly enlarging a machined hole to proper size with a smooth finish. The reamer is an accurate tool and is not designed to remove much metal. The reaming allowance is usually 0.2 mm only. The reaming tool is shown in Fig. 26.38(*a*).

4. *Tapping.* It is an operation of producing internal threads in a hole by means of a tool called tap. A tap is shown in Fig. 26.38 (*b*).

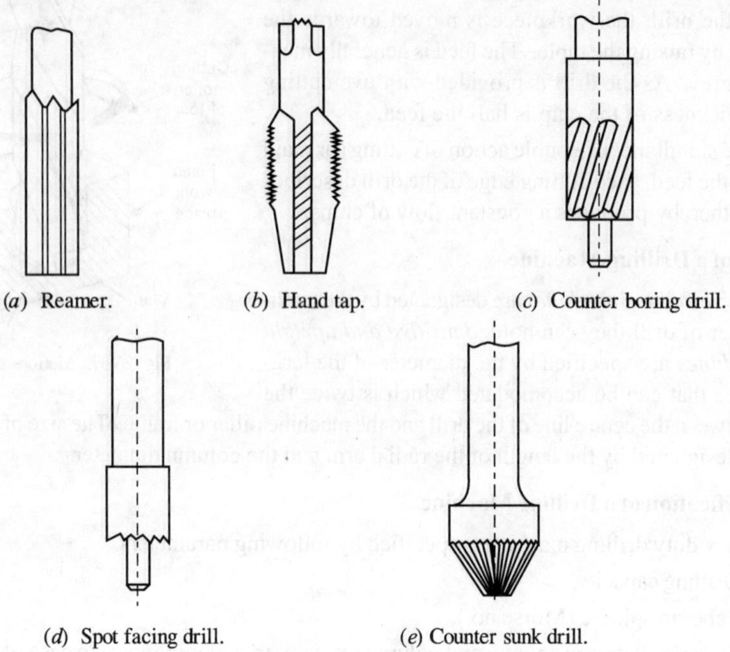

(*a*) Reamer. (*b*) Hand tap. (*c*) Counter boring drill.

(*d*) Spot facing drill. (*e*) Counter sunk drill.

Fig. 26.38. Drilling operations.

5. *Counter boring.* It is an operation of enlarging the mouth of a drilled hole to set bolt heads and nuts below the surface so that they may not project out from the surface level. It is done by a counter boring tool as shown in Fig. 26.38 (*c*) which is similar to an end mill or a single blade ground to give two cutting edges and is guided by a pilot pin which fits into the drilled hole.

6. *Spot facing.* It is an operation to finish off a small portion of rough surface around a drilled hole to provide smooth seat for bolt head. Spot facing tool is a single blade, as shown in Fig. 26.38 (*d*) (which is little larger than the bolt head diameter) ground to give two cutting edges and is guided by a stem which runs in a pilot hole.

7. *Counter-sinking.* It is an operation to bevel the top of a drilled hole for making a conical seat for a flat head screw. Ordinary flat drill ground to correct angle is used for countersinking holes.. A counter sunk drill is shown in Fig. 26.38 (*e*).

26·29 Drilling Speed and Feed

The *cutting speed* in drilling is the peripheral speed of the drill relative to the work. Mathematically,

$$\text{Cutting speed } (S) = \frac{\pi D N}{1000} \text{ m / min}$$

where D = Diameter of drill in mm, and

N = Work speed in r.p.m.

The cutting speed depends upon the material being cut, the tool material, capacity of machine, quality of surface required and the coolant used.

Feed of a drill is the distance it moves into the work with each revolution of the spindle. It is generally measured in mm/revolution of the spindle.

26·30 Machining and Operation Time for Drilling Machine

Machining time (t_m) is the operation time of the machine *i.e.*, the time during which the cutting edge of the drill cuts the chips off. It is generally measured in minutes. Mathematically,

$$t_m = \frac{l + 0.3\,d}{\text{Feed} \times N}$$

where l = Depth of hole to be drilled in mm,

d = Diameter of drill in mm, and

N = Revolutions of drill per min.

Example 26.1. *How long will it take to drill a hole of 10 mm diameter in a mild steel plate of 50 mm thickness when the cutting speed of drill is 30 m/min and feed is 0.15 mm/rev. ?*

Solution. Given : d = 10 mm ; l = 50 mm ; S = 30 m/min ; Feed = 0.15 mm/ rev

We know that R.P.M. of drill,

$$N = \frac{1000S}{\pi\,d} = \frac{1000 \times 30}{\pi \times 10} = 955 \text{ r.p.m.}$$

∴ Drilling time for the operation,

$$t_m = \frac{l + 0.3d}{\text{Feed} \times N} = \frac{50 + 0.3 \times 10}{0.15 \times 955} = \frac{53}{143.25} = 0.37 \text{ min } \textbf{Ans.}$$

QUESTIONS

1. Define lathe as a machine tool. Give a brief description of types of lathe.
2. What is working principle of a lathe ?
3. Describe, with suitable sketches, the functions of the following lathe parts :
   (*a*) Lathe bed; (*b*) Head stock;
   (*c*) Tail stock; and (*d*) Carriage.
4. How a centre lathe is specified ?
5. Write short notes on the following lathe accessories :
   (*a*) Centres ; (*b*) Chucks ; (*c*) Face plate ; (*d*) Mandrels ; and (*e*) Rests.
6. Describe, with the help of suitable sketches, the various types of tools used in a lathe.
7. Name the various operations that can be performed on a centre lathe. Describe, briefly, the following operations :
   (*a*) Facing ; (*b*) Plain turning; (*c*) Taper turning; (*d*) Drilling, and
   (*e*) Threading.
8. Define cutting speed, feed and depth of cut in case of lathe machining.
9. Describe, with a neat sketch, the principal parts of a shaper.
10. Discuss the working principle and operations of a shaper.
11. Explain the crank and slotted lever quick return motion mechanism for a shaper.
12. What is the fundamental difference between a planer and a shaper ?
13. How do you specify a shaper ?
14. How is a drilling machine specified ?
15. Describe the working principle of a drilling machine.
16. Differentiate between drilling, boring and reaming operations.
17. Differentiate between single point and multi-point cutting tools.

OBJECTIVE TYPE QUESTIONS

1. The lathe bed is made of
(*a*) mild steel (*b*) alloy steel
(*c*) pig iron (*d*) chilled cast iron

2. The angle between the lathe centres is
(*a*) 30° (*b*) 45°
(*c*) 60° (*d*) 90°

3. The lathe centres are provided with standard taper known as
(*a*) Morse taper (*b*) Seller's taper
(*c*) Chapman taper (*d*) Brown and Sharpe taper

4. The chuck used for setting up of heavy and irregular shaped work should be
(*a*) three jaw universal chuck (*b*) four jaw independent chuck
(*c*) magnetic chuck (*d*) drill chuck

5. A mandrel is used to hold
(*a*) an eccentric work (*b*) a heavy work
(*c*) a thin work (*d*) none of these

6. An operation of reducing the diameter of a workpiece over a very narrow surface, is known as
(*a*) counter-boring (*b*) grooving
(*c*) knurling (*d*) facing

7. An operation of producing an external conical surface on a workpiece is known as
(*a*) plain turning (*b*) step turning
(*c*) taper turning (*d*) facing

8. For turning small taper on long workpieces, the suitable method is
(*a*) by a form tool (*b*) by swivelling the compound rest
(*c*) by offsetting the tail stock (*d*) by a taper turning attachment

9. In a shaper
(*a*) tool is stationary and the work reciprocates
(*b*) work is stationary and the tool reciprocates
(*c*) tool and work both reciprocates
(*d*) none of the above

10. In a shaper, the metal is removed during
(*a*) forward stroke (*b*) return stroke
(*c*) both the forward and return strokes (*d*) neither the forward stroke nor return stroke

11. In a planer
(*a*) tool is stationary and the work reciprocates
(*b*) work is stationary and the tool reciprocates
(*c*) tool and work both reciprocates
(*d*) none of the above

12. Drilling is an example of
(*a*) simple cutting (*b*) orthogonal cutting
(*c*) oblique cutting (*d*) none of these

13. Which of the following operations can be performed by a drilling machine ?

(*a*) Spot facing (*b*) Reaming

(*c*) Tapping (*d*) all of these

14. Portable drilling machine is specified by

(*a*) the maximum diameter of the drill which can be held

(*b*) the length of arm and column diameter

(*c*) the diameter of the largest workpiece that can be drilled

(*d*) the drilling area, the size and number of holes a machine can drill

15. Feed of drill is

(*a*) equal to half the radius of the drill

(*b*) the distance moved by the drill into the workpiece on every revolution of the drill

(*c*) the distance moved by the drill per second

(*d*) equal to one-fourth the radius of the drill

ANSWERS

1. (*d*)	2. (*c*)	3. (*a*)	4. (*b*)	5. (*d*)
6. (*b*)	7. (*c*)	8. (*a*)	9. (*b*)	10. (*a*)
11. (*a*)	12. (*c*)	13. (*d*)	14. (*a*)	15. (*b*)

Index